Digital Speech Transmission

Digital Speech Transmission

Enhancement, Coding and Error Concealment

Peter Vary

Institute of Communication Systems and Data Processing
RWTH Aachen University, Germany

Rainer Martin

Institute of Communication Acoustics
Ruhr-Universität Bochum, Germany

JOHN WILEY & SONS, LTD

Telephone (+44) 1243 779777

Email (for orders and customer service enquiries): cs-books@wiley.co.uk
Visit our Home Page on www.wiley.com

Other Wiley Editorial Offices

John Wiley & Sons Inc., 111 River Street, Hoboken, NJ 07030, USA

Jossey-Bass, 989 Market Street, San Francisco, CA 94103-1741, USA

Wiley-VCH Verlag GmbH, Boschstr. 12, D-69469 Weinheim, Germany

John Wiley & Sons Australia Ltd, 42 McDougall Street, Milton, Queensland 4064, Australia

John Wiley & Sons (Asia) Pte Ltd, 2 Clementi Loop #02-01, Jin Xing Distripark, Singapore 129809

John Wiley & Sons Canada Ltd, 22 Worcester Road, Etobicoke, Ontario, Canada M9W 1L1

Wiley also publishes its books in a variety of electronic formats. Some content that appears in print may not be available in electronic books.

Library of Congress Cataloging-in-Publication Data

Vary, Peter.
Digital speech transmission : enhancement, coding, and error concealment/
Peter Vary and Rainer Martin.
p. cm.
Includes bibliographical references and index.
ISBN 0-471-56018-9 (cloth : alk. paper)
1. Speech processing systems. 2. Signal processing—Digital techniques.
3. Error-correcting codes (Information theory) I. Martin, Rainer. II.
Title.
TK7882.S65.V37 2005
621.39′9—dc22

2005034991

British Library Cataloguing in Publication Data

A catalogue record for this book is available from the British Library

ISBN-13 978-0-471-56018-9
ISBN-10 0-471-56018-X

Typeset by the authors using LATEX software
Printed and bound in Great Britain by Antony Rowe Ltd, Chippenham, Wiltshire
This book is printed on acid-free paper responsibly manufactured from sustainable forestry in which at least two trees are planted for each one used for paper production.

Contents

Preface

The digital processing, storage, and transmission of speech signals have gained great practical importance. The main application areas are digital mobile radio, acoustic human–machine communication, and digital hearing aids. In fact, these applications are the driving force behind many scientific and technological developments in this field. A specific characteristic of these application areas is that theory and practice are closely linked; there is a seamless transition from theory to system simulations using general-purpose computers and to system realizations with programmable processors.

This book has been written for electrical engineers, information technology engineers, as well as for engineering students. It summarizes recent developments in the broad field of digital speech transmission and is based to a large extent on joint research of the authors. This book is used in courses at RWTH Aachen University and Ruhr-Universität Bochum. Portions of this volume are translated and revised from the German edition of *Digitale Sprachsignalverarbeitung*, by P. Vary, U. Heute, and W. Hess, with kind permission of Teubner Verlag. The reader will find supplementary information, publications, programs, and audio samples on the following web sites:

`http://www.ind.rwth-aachen.de`
`http://www.rub.de/ika`

The scope of the individual subjects treated in the book often exceeds that of the lectures; recent research results, standards, problems of realization, and applications have been included, as well as many suggestions for further reading. To gain maximum benefit from the text, the reader should be familiar with the fundamentals of digital signal processing and statistical signal and system description. A summary of spectral analysis, digital filter banks, as well as stochastic signals and estimation is provided.

The authors are grateful to all members of staff and students who contributed to the book through research results, discussions, and editorial work. We thank Dr.-Ing. Tim Fingscheidt, Dr.-Ing. Peter Jax, and Dr.-Ing. Marc Adrat for contributions to Chapters 9 and 10. Horst Krott prepared most of the diagrams. Dipl.-Ing. David Bauer and Dipl.-Ing. Laurent Schmalen helped us with LaTeX editing, and Diplom-Anglistin Heike Hagena and Christina Storms, MA, supported us in translating the texts. These contributions are gratefully acknowledged.

We would like to express our thanks for the friendly assistance of the editors of John Wiley & Sons.

Finally, we would especially like to thank Dr.-Ing. Christiane Antweiler, whose tireless efforts resulted in completion, consistent layout, and content coordination of the entire manuscript.

Aachen and Bochum, January, 2006

Peter Vary and Rainer Martin

1

Introduction

Language is the most essential means of human communication. It is used in two modes: as spoken language (*speech communication*) and as written language (*textual communication*). In our modern information society both modes are greatly enhanced by technical systems and devices. E-mail, short-messaging, and the world-wide web have revolutionized textual communication while

- digital mobile radio systems,
- acoustic human–machine communication, and
- digital hearing aids

have significantly expanded the possibilities and convenience of speech communication.

Digital processing of speech signals for the purpose of transmission (or storage) is a branch of information technology and an engineering science which draws on various other disciplines such as physiology, phonetics, linguistics, acoustics, and psychoacoustics. It is this multidisciplinary aspect which makes digital speech processing a challenging as well as rewarding task.

The goal of this book is a comprehensive discussion of fundamental issues, standards, and recent trends in speech communication technology. Speech communi-

---

Digital Speech Transmission: Enhancement, Coding and Error Concealment
Peter Vary and Rainer Martin

cation technology helps to mitigate a number of physical constraints and technological limitations, most notably

- bandwidth limitations of the telephone channel,
- shortage of radio frequencies,
- acoustic background noise,
- interfering acoustic echo signals from loudspeaker(s), and
- (residual) transmission errors caused by the radio channel.

The enormous advances in signal processing technology have contributed to the success of speech signal processing. At present, integrated digital signal processors allow economic real-time implementations of complex algorithms which require several thousand operations per speech sample. For this reason advanced speech signal processing functions can be implemented in cellular phones as illustrated in Fig. 1.1.

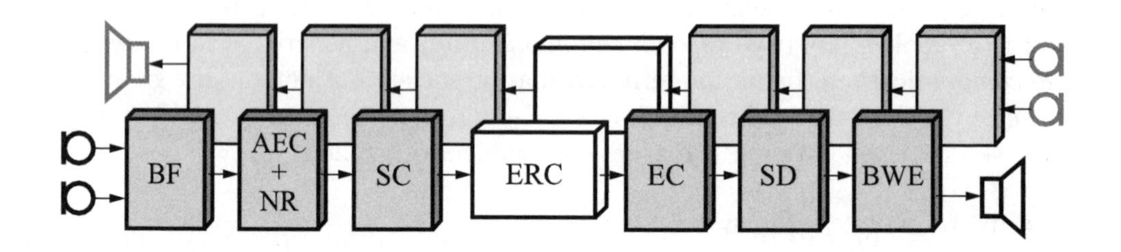

Figure 1.1: Speech signal processing in a handsfree mobile terminal
BF : Beamforming
AEC : Acoustic Echo Cancellation
NR : Noise Reduction
SC : Speech Coding
ERC : Equivalent Radio Channel
EC : Error Concealment
SD : Speech Decoding
BWE: Bandwidth Extension

The handsfree terminal in Fig. 1.1 facilitates communication via a microphone and a loudspeaker. Handsfree telephone facilities are installed in motor vehicles in order to enhance road safety and to increase convenience in general.

At the near end of the transmission system, three different pre-processing steps are taken to improve communication in the presence of ambient noise and loudspeaker signals. In the first step, two or more microphones are used to enhance the near-end speech signal by **beamforming** (BF). Specific characteristics of the interference, such as the spatial distribution of the sound sources and the statistics of the spatial sound field, are exploited.

Acoustic echoes occur when the far-end signal leaks from the loudspeaker of the handsfree set into the microphone(s) via the acoustic path. As a consequence, the far-end speaker will hear his or her own voice delayed by twice the signal propagation time of the telephone network. Therefore, in a second step, the acoustic echo must be compensated by an adaptive digital filter, the **acoustic echo canceller** (AEC).

The third module of the pre-processing chain is **noise reduction** (NR). Single channel noise reduction systems are most effective for short-term stationary noise. They are based on optimal filters and estimation techniques.

Speech coding (SC), **error concealment** (EC), and **speech decoding** (SD) facilitate the efficient use of the mobile radio channel. Speech coding algorithms for mobile communications with typical bit rates between 4 and 13 bit/s are explicitly based upon a model of speech production and exploit properties of the hearing mechanism.

At the receiving side of the transmission system, the speech quality is ensured by means of error correction (channel decoding), which is not within the scope of this book. In Fig. 1.1 the (inner) channel coding/decoding as well as modulation/demodulation and transmission over the physical radio channel are modeled as an **equivalent radio channel** (ERC). In spite of channel coding, quite frequently residual errors remain. The negative auditive effects of these errors can be reduced by **error concealment** (EC) techniques. In many cases, these effects can be reduced by exploiting both residual source redundancy and information about the instantaneous quality of the transmission channel.

Finally, the decoded signal might be subjected to artificial **bandwidth extension** (BWE) which expands narrowband (0.3 – 3.4 kHz) to wideband (0.05 – 7.0 kHz) speech. With the introduction of true wideband speech coding into telephone networks this step will be of significant importance as, for a long transition period, narrowband and wideband speech terminals will coexist.

Some of these processing functions find applications in multimedia terminals and digital hearing aids.

The book is organized as follows. The first part (Chapters 2–5) deals with the *fundamentals* of speech processing: models of speech production and hearing, spectral transformations, filter banks, and stochastic processes.

The second part (Chapters 6–8) covers the issue of *speech coding*. Quantization, differential waveform encoding, linear prediction, and especially the concepts of code excited linear prediction (CELP) are discussed. Finally, some of the most relevant speech codec standards are presented. Recent developments such as the *Adaptive Multi-Rate* (AMR, narrowband and wideband) codec for GSM and UMTS or variable rate coding for Internet telephony are addressed.

The third part of the book (Chapters 9–13) is concerned with the measures of speech enhancement of Fig. 1.1: error concealment, single channel noise reduction, acoustic echo cancellation, multi-channel noise reduction and beamforming, and, finally, artificial bandwidth extension.

2

Models of Speech Production and Hearing

Modern digital speech communication is largely based on knowledge of speech production, hearing, and perception. In this chapter, we will discuss some fundamental aspects in so far as they are of importance for optimizing speech processing algorithms such as speech coding, speech enhancement, or feature extraction for automatic speech recognition.

In particular, we will study the mechanism of speech production and the typical characteristics of speech signals. The digital speech production model will be derived from acoustical and physical considerations. The resulting *all-pole model of the vocal tract* is the key element of most of the current speech coding algorithms and standards. The parameters of the all-pole model can also be used for speech recognition, which, however, will not be treated here.

Furthermore, we will analyze how the human auditory system works, and we will focus on perceptual fundamentals which can be exploited to improve the quality and the cost effectiveness of speech processing algorithms to be discussed in later chapters. With respect to perception, the main aspects to be considered in digital speech transmission are the *masking effect* and the spectral resolution of the auditory system.

Digital Speech Transmission: Enhancement, Coding and Error Concealment
Peter Vary and Rainer Martin

As a detailed discussion of the acoustic theory of speech production, phonetics, psychoacoustics, and perception is beyond the scope of this book, the reader is referred to the literature (e.g., [Fant 1970], [Flanagan 1972], [Rabiner, Schafer 1978], [Picket 1980], [Zwicker, Fastl 1999]).

2.1 Organs of Speech Production

The production of speech sounds involves the manipulation of an airstream. The acoustic representation of speech is a sound pressure wave originating from the physiological speech production system. A simplified schematic of the human speech organs is given in Fig. 2.1.

The main components and their functions are:

- lungs: the energy generator,
- trachea: for energy transport,
- larynx with vocal cords: the signal generator, and
- vocal tract: (pharynx, oral and nasal cavities) the acoustic filter.

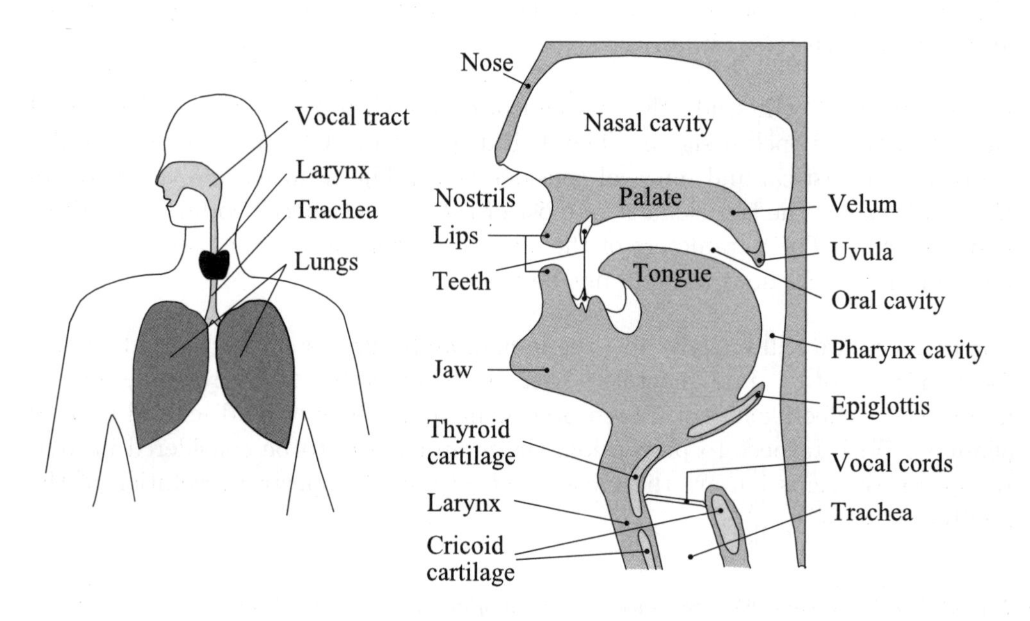

Figure 2.1: Organs of speech production

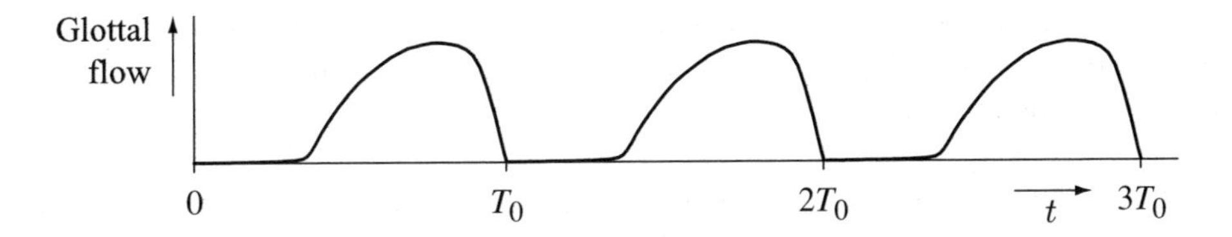

Figure 2.2: Glottis airflow during voiced sounds

By contraction, the *lungs* produce an airflow which is modulated by the *larynx*, processed by the *vocal tract*, and radiated via the lips and the nostrils. The *larynx* provides several biological and sound production functions. In the context of speech production, its purpose is to control the stream of air that enters the vocal tract via the *vocal cords*.

Speech sounds are produced by means of various mechanisms. *Voiced sounds* are produced if the airflow is interrupted periodically by the movements (vibration) of the vocal cords (see Fig. 2.2). This self-sustained oscillation, i.e., the repeated opening and closing of the vocal cords, can be explained by the so-called *Bernoulli effect* as in fluid dynamics: as airflow velocity increases, local pressure decreases. At the beginning of each cycle the area between the vocal cords, which is called the *glottis*, is almost closed by means of appropriate tension of the vocal cords. Then an increased air pressure builds up below the glottis, forcing the vocal cords to open. As the vocal cords diverge, the velocity of the air through the glottis increases suddenly, which causes a drop of the local pressure. Thus, the vocal cords are closed again and the next cycle can start if the airflow from the lungs and the tension of the vocal cords are sustained. Due to the abrupt periodic interruptions of the airflow, as schematically illustrated in Fig. 2.2, the resulting excitation (pressure wave) of the vocal tract has a fundamental frequency of $f_0 = 1/T_0$ and has a large number of harmonics. These are spectrally shaped according to the frequency response of the acoustic vocal tract. The length T_0 of the cycle is called *pitch period*.

Unvoiced sounds are generated by a constriction at the open glottis or along the vocal tract causing a non-periodic turbulent air flow.

Plosive sounds are caused by building up the air pressure behind a complete constriction somewhere in the vocal tract, followed by a sudden opening. The released flow may create a voiced or an unvoiced sound or even a mixture of both, depending on the actual constellation of the articulators.

The *vocal tract* can be subdivided into three sections: the pharynx, the oral cavity, and the nasal cavity. As the entrance to the nasal cavity can be closed by the velum, a distinction is often made in the literature between the nasal tract (from velum to nostrils) and the vocal tract (from trachea to lips, including the pharynx cavity).

In what follows we will define the *vocal tract* as a variable acoustic resonator including the nasal cavity with the velum either open or closed, depending on the specific sound to be produced. From the engineering point of view, the resonance frequencies are varied by changing the size and the shape of the vocal tract using different constellations and movements of the *articulators*, i.e., tongue, teeth, lips, velum, lower jaw, etc. A human can produce quite different sounds based on different vocal tract constellations and different excitations.

2.2 Characteristics of Speech Signals

Most languages can be described as a set of elementary linguistic units which are called *phonemes*. A phoneme is defined as the smallest unit which differentiates between two words in one language. The acoustic representation associated with a phoneme is called a *phone*. American English, for instance, consists of about 42 phonemes, which are subdivided into four classes:

Vowels are voiced and belong to the speech sounds with the largest energy. They exhibit a quasi-periodic time structure caused by oscillation of the vocal cords. The duration varies from 40 to 400 ms. Vowels can be distinguished by the time-varying resonance characteristics of the vocal tract. The resonance frequencies are also called *formant frequencies*. Examples: /a/ as in "father", /i/ as in "eve".

Diphthongs involve a gliding transition of the articulators from one vowel to another vowel. Examples: /oU/ as in "boat", /ju/ as in "you".

Approximants are a group of voiced phonemes for which the airstream escapes through a relatively narrow aperture in the vocal tract without friction. They can thus be regarded as intermediate between vowels and consonants [Gimson, Cruttenden 1994]. Examples: /w/ in "wet", /r/ in "ran".

Consonants are produced with stronger constriction of the vocal tract than vowels. All kinds of excitation can be observed. Consonants are subdivided in *nasals*, *stops*, *fricatives*, *aspirates*, and *affricatives*. Examples of these five subclasses: /m/ as in "more", /t/ as in "tea", /f/ as in "free", /h/ as in "hold", and /tʃ/ as in "chase".

Each of these classes may be further divided into subclasses which are related to the interaction of the articulators within the vocal tract. The phonemes can further be classified as either *continuant* (excitation of a more or less non-time-varying vocal tract) or *non-continuant* (rapid vocal tract changes). The class of continuant sounds consists of vowels and fricatives (voiced and unvoiced). The non-continuant sounds are represented by diphthongs, semivowels, stops, and affricates.

For the purpose of speech signal processing, the articulatory and phonetic aspects are not as important as the typical characteristics of the signal waveforms, namely, the categories

- voiced,
- unvoiced,
- mixed voiced/unvoiced,
- plosive, and
- silence.

Voiced sounds are characterized by their fundamental frequency, i.e., the frequency of vibration of the vocal cords, and by the specific pattern of amplitudes of the spectral harmonics.

In the speech signal processing literature the fundamental frequency is usually called *pitch* and the respective period is called *pitch period*. It should be noted that in the field of psychoacoustics the term pitch is used differently, i.e., for the perceived fundamental frequency of a sound, whether or not that frequency is actually present in the waveform (e.g., [Deller Jr. et al. 2000]). The pitch of male voices usually ranges from 50 to 250 Hz, while for female voices it is in the interval 120–500 Hz.

Unvoiced sounds are determined mainly by their characteristic spectral envelopes. Voiced and unvoiced excitation do not exclude each other. They may occur simultaneously, e.g., in fricative sounds.

The distinctive feature of *plosive sounds* is the dynamically transient change of the vocal tract. Immediately before the transition, no sound is radiated from the lips for a short period because of a total constriction in the vocal tract. There might be a small amount of low-frequency components radiated through the throat. Then the sudden change with release of the constriction produces a plosive burst.

Some typical speech waveforms are shown in Fig. 2.3.

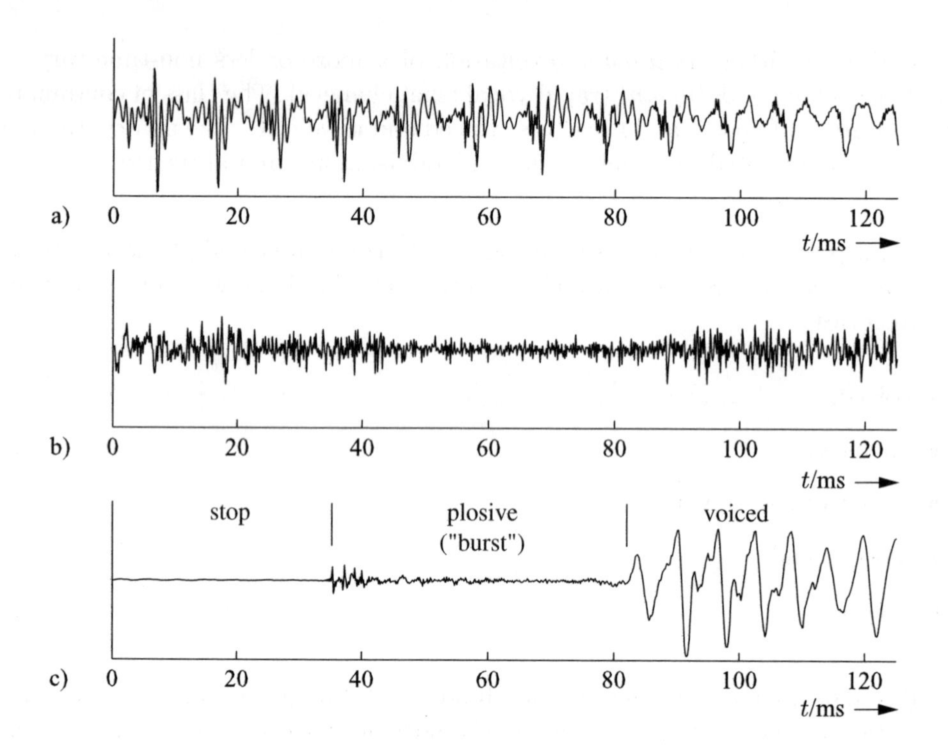

Figure 2.3: Characteristic waveforms of speech signals (each line = 100 ms) [Vary et al. 1998]
a) Voiced (vowel with transition to voiced consonant)
b) Unvoiced (fricative)
c) Transition stop–plosive–vowel

2.3 Model of Speech Production

The purpose of developing a model of speech production is not to obtain an accurate description of the real anatomy and physiology of the human speech system. We would rather like to achieve a simplifying mathematical representation for reproducing the essential characteristics of speech signals.

In analogy to the human speech production system as discussed in Section 2.1, it seems reasonable to design a parametric two-stage model consisting of an *excitation source* and a *vocal tract filter*, see also [Rabiner, Schafer 1978], [Parsons 1986], [Quatieri 2001], [Deller Jr. et al. 2000]. The final digital *source-filter model* as illustrated in Fig. 2.4 will be derived below.

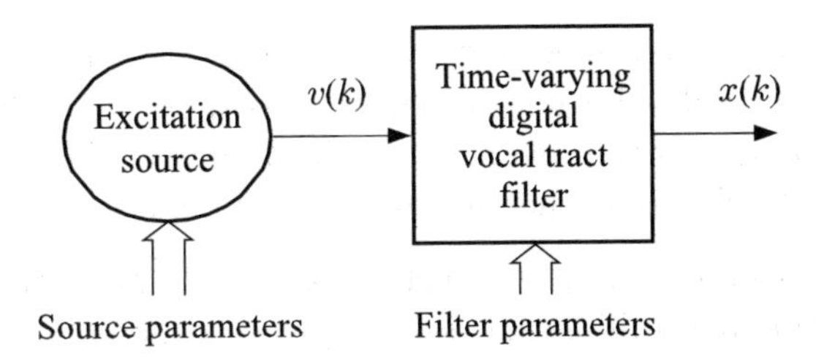

Figure 2.4: Digital source–filter model

The model consists of two components:

- the *excitation source* featuring mainly the influence of the lungs and the vocal cords (voiced, unvoiced, mixed) and
- the *time-varying digital vocal tract filter* approximating the behavior of the vocal tract (spectral envelope and dynamic transitions).

In the first and simple model, the *excitation generator* only has to deliver either white noise or a periodic sequence of *pitch-pulses* for synthesizing unvoiced and voiced sounds respectively, whereas the vocal tract is modeled as a time-varying digital filter.

2.3.1 Acoustic Tube Model of the Vocal Tract

The digital source–filter model of Fig. 2.4, especially the vocal tract filter, will be derived from the physics of sound propagation inside an acoustic tube in this section. To estimate the complexity in terms of the necessary filter degree, we start with the extremely simplifying physical model of Fig. 2.5. The pharynx and

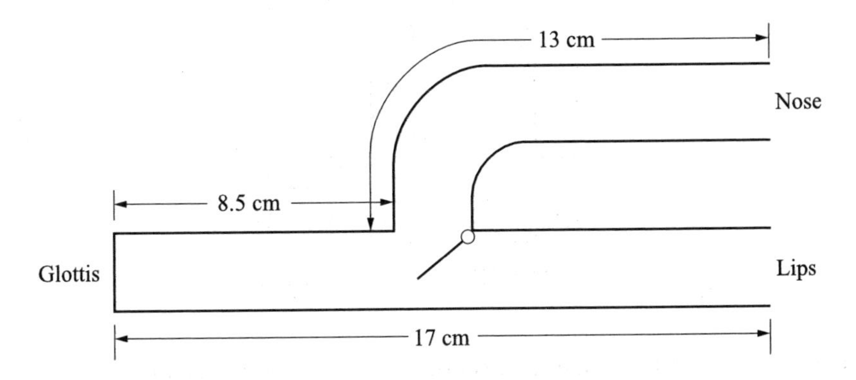

Figure 2.5: Simplified physical model of the vocal tract

oral cavities are represented by a lossless tube with constant cross-section and the nasal cavity by a second tube which can be closed by the velum. The length of $L = 17\,\text{cm}$ corresponds to the average length of the vocal tract of a male adult. The tube is (almost) closed at the glottis side and open at the lips.

In the case of a non-nasal sound the velum is closed. Then the wavelength λ_i of each resonance frequency fulfills the standing wave condition

$$(2i-1)\cdot\frac{\lambda_i}{4} = L\,; \quad i = 1, 2, 3, \ldots\,. \tag{2.1}$$

For $L = 17\,\text{cm}$ we compute the resonance frequencies

$$f_i = \frac{c}{\lambda_i} = (2i-1)\frac{c}{4L} \in \{500, 1500, 2500, 3500, \ldots\}\,\text{Hz} \tag{2.2}$$

where the sound velocity is given by $c = 340\,\text{m/s}$.

Taking (2.2) into account, as well as the fact that conventional telephone speech has a frequency limitation of about 4000 Hz, we would have to consider only four resonance frequencies within the frequency band of telephone speech. Thus, the overall filter degree for synthesizing telephone speech is roughly only $n = 8$ as each resonance frequency corresponds to a pole-pair or second-order filter section. As a rule of thumb we can state the need for "one resonance per kHz".

In a second step, we improve our acoustic tube model as shown in Fig. 2.6. For simplicity, the nasal cavity is not considered (velum is closed). The cylindrical tube

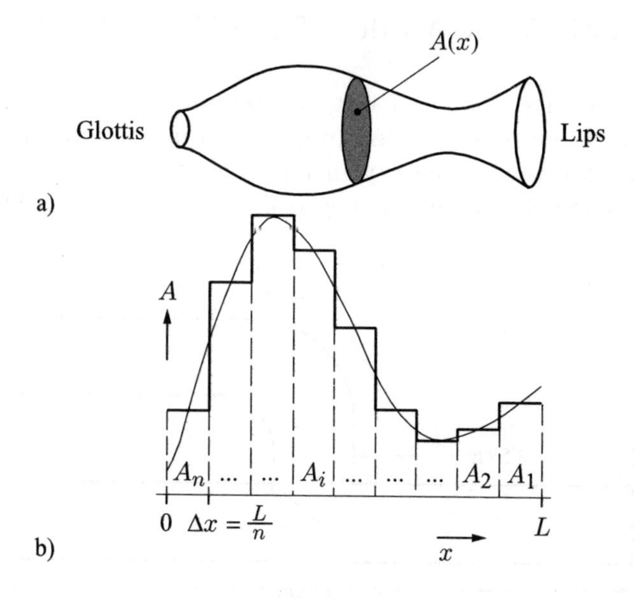

Figure 2.6: a) Tube model with continuous area function $A(x)$
b) Stepwise approximation of $A(x)$

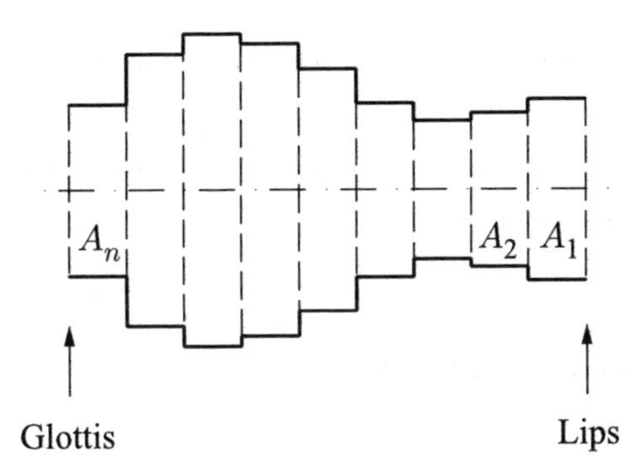

Figure 2.7: Representation of the vocal tract by a concatenation of uniform lossless acoustic tube segments each of length $\Delta x = L/n$

of Fig. 2.5 is replaced by a tube with continuous area function $A(x)$. This area function can be approximated by a piecewise constant contour which corresponds to a concatenation of n short cylindrical tubes each of length $\Delta x = L/n$ but of (possibly) different constant area A_i, $i = 1, 2, \ldots, n$, as shown in Fig. 2.7.

The generation of sound is related to the vibration and perturbation of air particles. When describing sound propagation through the concatenated tube segments, we have to deal with the *particle velocity* $u(x,t)$ (unit m/s) and the *pressure*, strictly speaking the pressure fluctuations $p(x,t)$ (unit $\mathrm{N/m^2}$) about the ambient or average (atmospheric) pressure. There is a vast amount of literature on relations between the velocity and the pressure (e.g., [Fant 1970]) inside a tube segment, taking different degrees of simplifying assumptions into account.

It is not within the scope of this book to discuss this theory as such, as we are interested primarily in developing the discrete-time filter structure which is widely used for digital speech synthesis and speech coding. It will be outlined here that this filter can be derived under simplifying assumptions from the acoustic vocal tract model of Fig. 2.7.

If we make the usual assumptions of *lossless tube segments*, fixed area A in both time and space, no friction, and plane wave propagation, the sound velocity and the pressure are related by the following partial differential equations, e.g., [O'Shaughnessy 2000]:

$$-\frac{\partial p}{\partial x} = \rho \cdot \frac{\partial u}{\partial t} \tag{2.3-a}$$

$$-\frac{\partial u}{\partial x} = \frac{1}{\rho c^2} \cdot \frac{\partial p}{\partial t} \,. \tag{2.3-b}$$

These *plane wave equations* are based on elementary physical laws relating the mass, energy, and velocity of an infinitesimal small volume of air particles.

The equations (2.3-a) and (2.3-b) are the starting point for the derivation of the discrete-time vocal tract filter.

First, we introduce the *volume velocity* v as the product of the particle velocity u and the cross-sectional area A

$$v = u \cdot A \tag{2.4}$$

and modify (2.3-a) and (2.3-b) as follows:

$$-\frac{\partial p}{\partial x} = \frac{\rho}{A} \cdot \frac{\partial v}{\partial t} \tag{2.5-a}$$

$$-\frac{\partial v}{\partial x} = \frac{A}{\rho c^2} \cdot \frac{\partial p}{\partial t} \,. \tag{2.5-b}$$

Next, the combination of both equations (2.5-a) and (2.5-b) leads to differential equations of second order for the pressure p and the volume velocity v

$$\frac{\partial^2 p}{\partial x^2} = \frac{1}{c^2} \cdot \frac{\partial^2 p}{\partial t^2} \tag{2.6-a}$$

$$\frac{\partial^2 v}{\partial x^2} = \frac{1}{c^2} \cdot \frac{\partial^2 v}{\partial t^2} \,. \tag{2.6-b}$$

Pressure p and volume velocity v are dependent on time t and space x. The differential equations (2.6-a) and (2.6-b) can be solved by combining a forward and backward traveling wave f and b, respectively, both moving at velocity c. The general solution for the volume velocity v and the pressure p is then given by

$$v(x,\, t) = f\left(t - \frac{x}{c}\right) - b\left(t + \frac{x}{c}\right) \tag{2.7-a}$$

$$p(x,\, t) = \left[f\left(t - \frac{x}{c}\right) + b\left(t + \frac{x}{c}\right)\right] \cdot Z \quad \text{with} \quad Z = \frac{\rho \cdot c}{A} \,, \tag{2.7-b}$$

where the quantity Z represents an *acoustic impedance*.

This solution can be applied to each of the tube segments which can be identified by the index i, where boundary conditions exist at the transitions between adjacent segments. In each of these segments we use a local space coordinate x, ranging from 0 to Δx as shown in Fig. 2.8, and consider the junction of segment i with segment $i-1$. Pressure and volume velocity must be continuous both in time and in space. At the junction, the following constraints have to be fulfilled:

$$v_i(x = \Delta x,\, t) = v_{i-1}(x = 0,\, t) \tag{2.8-a}$$

$$p_i(x = \Delta x,\, t) = p_{i-1}(x = 0,\, t) \,. \tag{2.8-b}$$

Using the general solution (2.7-a) and (2.7-b) with the boundary conditions (2.8-a), (2.8-b) and introducing the notation $\tau = \Delta x / c$ for the propagation time through

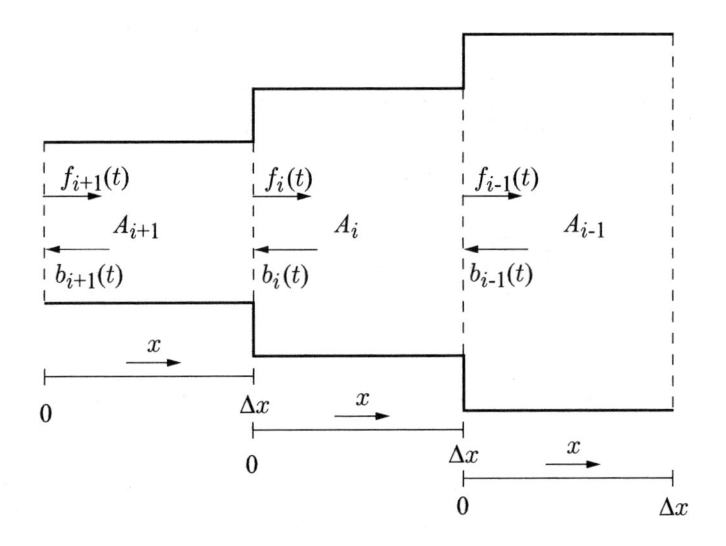

Figure 2.8: Wave equations and junctions of tube segments

the tube segment, we get the boundary conditions in terms of the forward and backward traveling waves

$$f_i(t-\tau) - b_i(t+\tau) = f_{i-1}(t) - b_{i-1}(t) \tag{2.9-a}$$

$$Z_i \cdot \left[f_i(t-\tau) + b_i(t+\tau)\right] = Z_{i-1} \cdot \left[f_{i-1}(t) + b_{i-1}(t)\right]. \tag{2.9-b}$$

Note that we have eliminated the space coordinate as we are now considering the forward and backward traveling waves at the junction only. With respect to segment i, we are interested in the input–output dependencies, i.e., the two functional relations

$$f_{i-1} = F_1(f_i,\, b_{i-1}) \tag{2.10-a}$$

$$b_i = F_2(f_i,\, b_{i-1}). \tag{2.10-b}$$

After some elementary algebraic operations we obtain

$$f_{i-1}(t) = (1 + r_i) \cdot f_i(t-\tau) + r_i \cdot b_{i-1}(t) \tag{2.11-a}$$

$$b_i(t) = -r_i \cdot f_i(t-2\tau) + (1 - r_i) \cdot b_{i-1}(t-\tau), \tag{2.11-b}$$

where

$$r_i = \frac{Z_i - Z_{i-1}}{Z_i + Z_{i-1}} = \frac{A_{i-1} - A_i}{A_{i-1} + A_i}, \tag{2.12}$$

which is called the *reflection coefficient.*[1] From $A_i > 0$ it readily follows that $-1 \leq r_i \leq +1$.

[1]Note, in the literature the reflection coefficient is partly defined with the opposite sign.

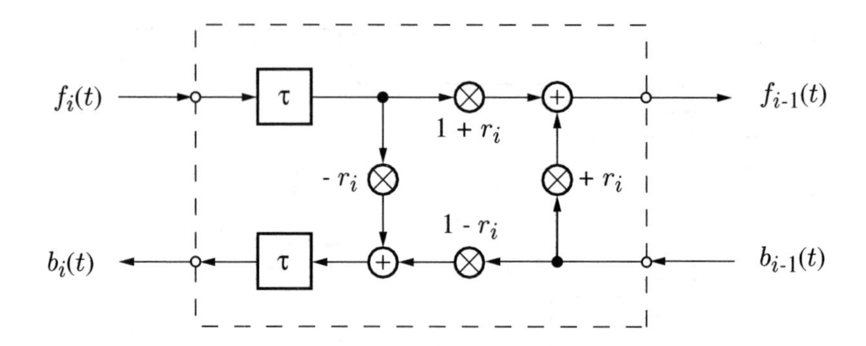

Figure 2.9: Illustration of the solution (2.11-a) and (2.11-b)

The solutions (2.11-a), (2.11-b) are called the *Kelly–Lochbaum* equations [Kelly, Lochbaum 1962]. Kelly and Lochbaum used this structure, which is illustrated in Fig. 2.9, for computer generation of synthetic speech in 1962.

The forward traveling wave $f_{i-1}(t)$, *leaving* segment i on the right and *entering* segment $i - 1$, consists of two components:

- a portion of the backward traveling wave $b_{i-1}(t)$ which is partly reflected at the junction with weight $+r_i$ and
- a portion of the forward traveling wave $f_i(t)$ which is delayed by τ due to the segment length Δx and partly propagated into segment $i-1$ with weight $1 + r_i$.

The backward traveling wave $b_i(t)$, *leaving* segment i on the left and *entering* segment $i + 1$, consists also of two components:

- a portion of the backward traveling wave $b_{i-1}(t)$ which is partly propagated at the junction with weight $1 - r_i$ and delayed by τ and
- a portion of the forward traveling wave $f_i(t)$ which is partly reflected at the junction $i/(i - 1)$ with weight $-r_i$ and delayed by 2τ due to traveling back and forth through segment i.

Note that the reflection coefficient r_i can take positive and negative values depending on the relative sizes of the areas A_i and A_{i-1}.

We illustrate the complete solution in Fig. 2.10. Special considerations are needed at the terminating glottis and at the lips. We can model the free space beyond the lips as an additional tube with index 0, with an infinite area A_0, and with infinite length. Thus, the first reflection coefficient becomes

$$r_1 = \frac{A_0 - A_1}{A_0 + A_1} = +1\,. \tag{2.13}$$

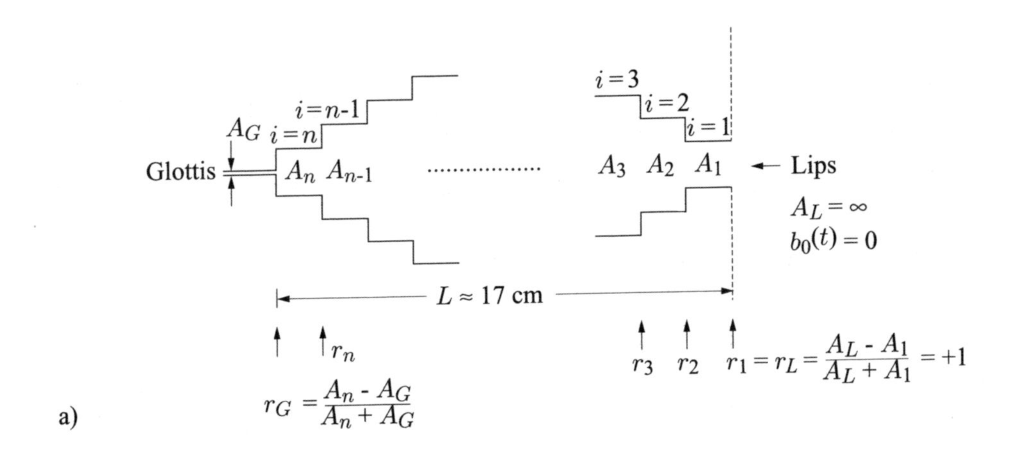

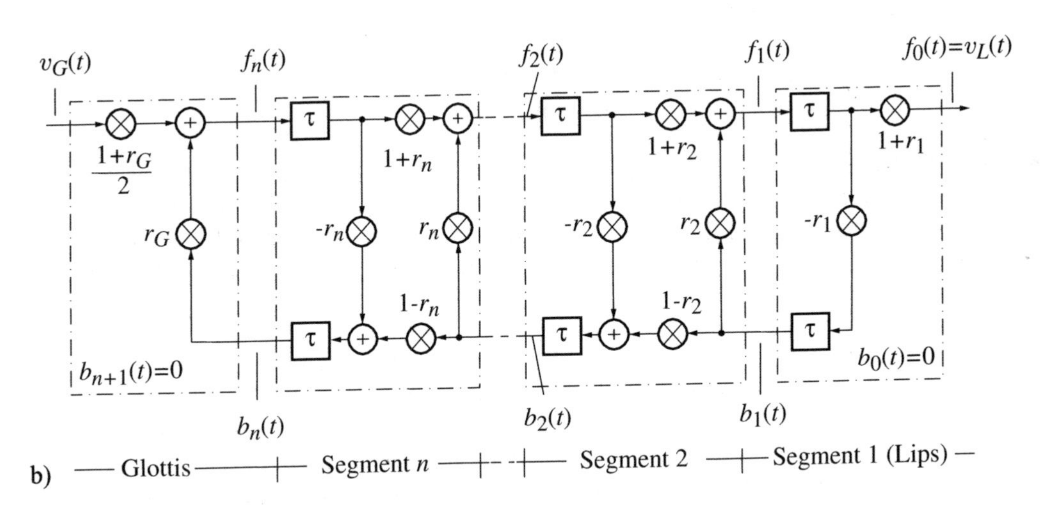

Figure 2.10: Solution of the wave equations for the complete concatenated set of tube segments
a) Acoustical tube model
b) Blockdiagram

A backward traveling wave does not exist, i.e., $b_0(t) = 0$. Taking into consideration the solution of Eq. (2.7-a), we get for the segment $i = 0$ (free space beyond the lips, segment with infinite length, no backward traveling wave $b_0(t)$, infinite area $A_0 = \infty$) at the local space coordinate $x = 0$

$$v_L(t) \doteq v_0(0,\, t) = f_0(t)\,. \tag{2.14}$$

At the glottis, the left side of segment $i = n$ is almost closed. The backward traveling wave $b_n(t)$ is partly reflected according to the effective reflection coefficient r_G. The active excitation $v_G(t)$ might be modeled by a vibrating wall or piston, driven solely by the airflow from the lungs, i.e., $b_{n+1} = 0$. Following once more the for-

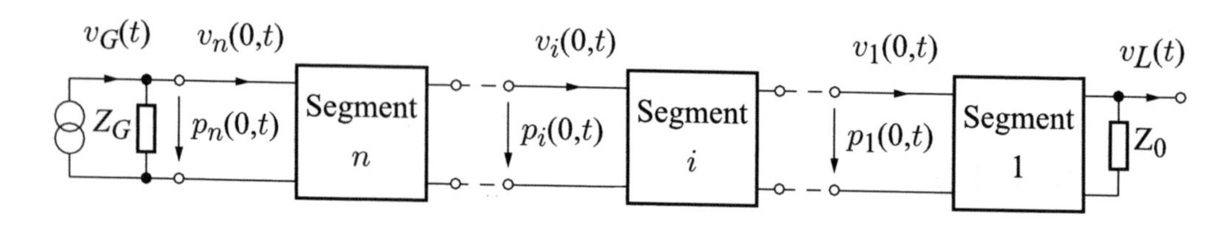

Figure 2.11: Concatenation of electrical line segments as analogy to the concatenation of acoustic tube segments

mal solution of Eq. (2.7-a), the volume velocity of the virtual segment $n+1$ is dependent on the forward traveling wave only, i.e., strictly speaking,

$$v_G(t) \doteq f_{n+1}(t-\tau) = v_{n+1}(\Delta x,\, t)\,. \tag{2.15}$$

In (2.15) the delay parameter τ is irrelevant; it is introduced just for formal reasons to model a virtual segment $n+1$ of length Δx.

The volume velocity v_G is weighted by the factor $(1+r_G)/2$. The division of $(1+r_G)$ by 2 may be considered as a matter of normalization as the concatenated tube is a passive device. This normalization by 0.5 can be derived by considering the acoustic impedances of the excitation generator and the acoustic load of the concatenated tube.

There is an analogy between the concatenation of acoustic tube segments and the serial connection of homogeneous electrical line segments (e.g., [Rabiner, Schafer 1978]). Basically, the mathematical description by the differential equations (2.6-a), (2.6-b), (2.7-a), and (2.7-b) is the same, where the pressure $p_i(t)$ corresponds to the voltage and the volume velocity $v_i(t)$ has the meaning of a current.

A detailed discussion of this analogy is beyond the scope of this book. Nevertheless, the block diagram of Fig. 2.11 can be used to outline the derivation of the normalization by 0.5.

The excitation source is a current source with "current" $v_G(t)$ and impedance Z_G. Thus the "input current" $v_n(0,t)$ to the "line segment $i=n$" is given by

$$v_n(0,\, t) = v_G(t) \;-\; \frac{p_n(0,\, t)}{Z_G}\,. \tag{2.16}$$

Using this relation as the boundary constraint and the solution (2.7-a) and (2.7-b) for segment n at the space variable $x=0$, i.e.,

$$v_n(0,\, t) = f_n(t) \;-\; b_n(t) \tag{2.17-a}$$

$$p_n(0,\, t) = \big[f_n(t) \;+\; b_n(t)\big]\cdot Z_n \quad \text{with} \quad Z = \frac{\rho\cdot c}{A}\,, \tag{2.17-b}$$

we derive from (2.16) and (2.17-a) with a few algebraic steps the equation

$$f_n(t) = v_G(t) \cdot \frac{Z_G}{Z_n + Z_G} + b_n(t) \cdot \frac{Z_G - Z_n}{Z_n + Z_G} \tag{2.18-a}$$

$$= v_G(t) \cdot \frac{1 + r_G}{2} + b_n(t) \cdot r_G \tag{2.18-b}$$

with

$$r_G = \frac{Z_G - Z_n}{Z_n + Z_G} . \tag{2.18-c}$$

This analogy between the acoustical system and the electrical transmission line can also be used for further refinements of the model, in particular in modeling the glottal excitation and the load effects from lip radiation. For more detailed studies of this analogy the reader is referred to the literature (e.g., [Rabiner, Schafer 1978]).

2.3.2 Digital All-Pole Model of the Vocal Tract

So far we have derived a solution for the wave equations of the propagation of sound in the concatenated lossless tube segments. We obtained a functional relation between the volume velocity $v_L(t)$ at the lips and the volume velocity $v_G(t)$ at the glottis as described by the block diagram of Fig. 2.10. From this representation an equivalent discrete-time or digital filter model can easily be derived.

The structure in Fig. 2.10 may be interpreted as an analog network consisting of delay elements, multipliers, and adders, which in the case of constant coefficients r_G, r_L, and r_i, $i = 1, 2, \ldots, n$, can be interpreted as a linear time-invariant system (LTI system). Any LTI system may be characterized by its impulse response or its corresponding frequency response. Obviously, the impulse response of the system in Fig. 2.10 is discrete in time, as the internal signals $f_i(t)$ and $b_i(t)$, $i = 1, 2, \ldots, n$, are delayed by multiples of the basic one-way delay τ caused by each of the tube segments.

If we apply a Dirac impulse $\delta(t)$ as stimulus signal to the system input, i.e., $v_G(t) = \delta(t)$, we observe a response $v_L(t)$ at the output which is the impulse response $h_0(t)$. Although time t is still a continuous variable, it is obvious that the output takes non-zero values at multiples of the basic delay time τ only. The first non-zero value occurs at time $t = n \cdot \tau$ after the propagation time through the n tube segments. Due to the feedback structure, the output signal is a sequence of weighted Dirac impulses with a minimum distance between the pulses of 2τ. More specifically, the system response can be written as

$$v_L(t) = h_0(t) = \sum_{\kappa=0}^{\infty} \delta(t - n\tau - 2\kappa\tau) \cdot h(\kappa) , \tag{2.19}$$

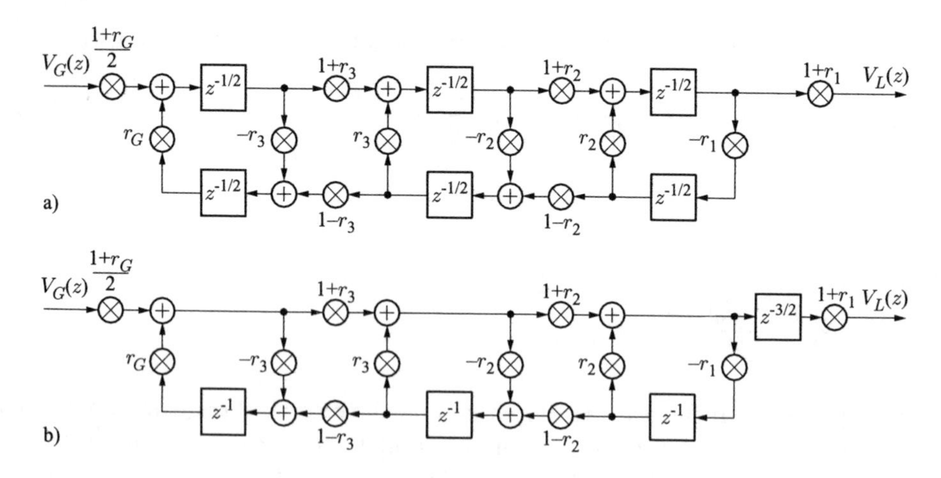

Figure 2.12: Discrete-time models of the vocal tract (example $n = 3$)
a) Structure derived from the wave equations; $z^{-1/2} \equiv$ delay by τ
b) Modified structure with delay elements $T = 2\tau \equiv z^{-1}$

where the values $h(\kappa)$ denote the weights and the first weight is given by

$$h(0) = \frac{1 + r_G}{2} \cdot \prod_{i=1}^{n} (1 + r_i) . \tag{2.20}$$

Thus, the impulse response of the tube model is already discrete in time, and the signal-flow graph of the equivalent digital filter may be considered as a one-to-one translation as shown in Fig. 2.12-a. Note that in Fig. 2.12 we have introduced the z-transforms $V_G(z)$ and $V_L(z)$ of the sequences $v_G(\kappa \cdot T)$ and $v_L(\kappa \cdot T)$, with $T = 2 \cdot \tau$ respectively. Therefore, a delay by τ is denoted in Fig. 2.12-a by $z^{-1/2}$. The time discretization τ of the block diagram is intimately connected with the spatial discretization Δx of the tube model. However, the effective time discretization (sampling interval) of the output signal v_L is $T = 2\tau$. Thus, the digital filter would have to be operated at twice the sampling rate, i.e.,

$$\frac{1}{\tau} = \frac{2}{T} = 2 \cdot f_s . \tag{2.21}$$

By systematically shifting the delay elements $z^{-1/2}$ across the nodes, the equivalent signal-flow graph of Fig. 2.12-b can be derived. This structure contains only internal delays of $T = 2\tau$ corresponding to z^{-1} and one external delay at the output of $n/2 \cdot \tau = 3/2 \cdot \tau$ corresponding to $z^{-3/2}$. If the number n of segments is not even, the delay by $n/2 \cdot \tau$ would imply an interpolation by half a sampling interval. In any case the output delay will not have any influence on the quality of the synthesized speech. Therefore, the delay by $n/2 \cdot \tau$ can be omitted in practical implementations and the digital filter can operate at the sampling rate f_s.

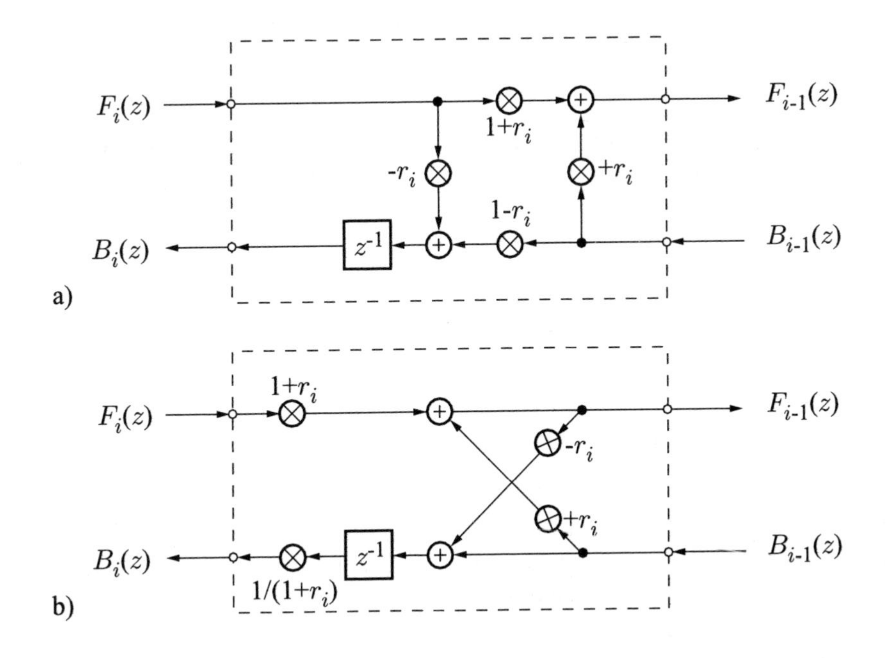

Figure 2.13: Equivalent filter sections
a) Ladder structure
b) Lattice structure

The structure of Fig. 2.12-b without the output delay according to $z^{-n/2}$ is used for speech synthesis and model-based speech coding. This special structure is called the *ladder structure*.

It is known from network theory that there are digital filter structures having the same transfer function but different signal-flow graphs. They may differ, for instance, with respect to the number of arithmetic operations needed to calculate one output sample or with respect to numerical problems in the case of finite precision arithmetic (fixed point arithmetic).

Two equivalent structures are shown in Fig. 2.13. The first structure is the *ladder structure* we have derived above. The equivalence can readily be checked by the analysis of the block diagram of the so-called *lattice structure* of Fig. 2.13-b:

$$F_{i-1}(z) = (1 + r_i) \cdot F_i(z) \; + \; r_i \cdot B_{i-1}(z) \tag{2.22-a}$$

$$B_i(z) = \frac{1}{1 + r_i} \cdot \left[-(1 + r_i) \cdot r_i \cdot F_i(z) \; + \; (1 - r_i^2) \cdot B_{i-1}(z)\right] \cdot z^{-1} \tag{2.22-b}$$

$$= \left[-r_i \cdot F_i(z) \; + \; (1 - r_i) \cdot B_{i-1}(z)\right] \cdot z^{-1} \, . \tag{2.22-c}$$

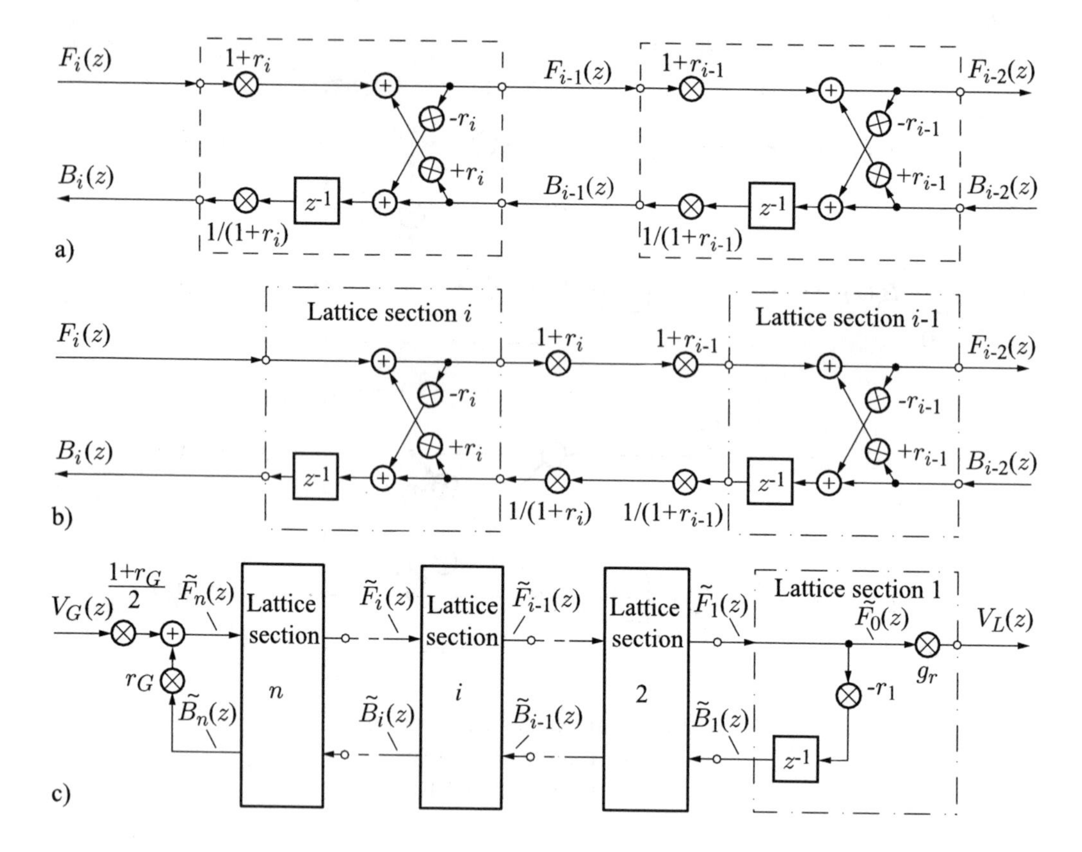

Figure 2.14: Vocal tract model as a cascade of lattice filter sections
a) Lattice sections with four multipliers
b) Lattice section i with shifted scaling multipliers
c) Lattice sections with two multipliers and gain scaling by g_r at the output

The *lattice structure* also requires four multiplications and four additions per section. However, the pre- and post-scaling factors $1+r_i$ and $1/(1+r_i)$ can be shifted and merged along the cascade of filter sections (cf. Fig. 2.14-a,b) in such a way that they can be combined to a single scaling or gain factor

$$g_r = \prod_{i=1}^{n} (1 + r_i) \tag{2.23}$$

at the filter output cascade as shown in Fig. 2.14-c.

Now we need only two multiplications and additions per section and one scaling multiplication by g_r, i.e., a total number of $2n + 1$ operations per output sample $v_L(\kappa \cdot T)$, where we use a basic multiply-and-add operation (MAC operation) as available in many signal processors.

In comparison to the ladder structure, we have reduced the computational complexity by a factor of almost 2. The ladder and the lattice structure are stable as long as the reflection coefficients are limited to the range

$$-1 < r_i < +1 . \tag{2.24}$$

This stability can be guaranteed due to the underlying physical tube model of the vocal tract.

The input and output signals of the 2-multiplier-lattice elements of Fig. 2.14-c are different from those of the original lattice structure of Fig. 2.14-a by the corresponding scaling factors. Therefore, the z-transforms of the intermediate signals of the lattice sections are denoted by $\tilde{F}_i(z)$ and $\tilde{B}_i(z)$.

The lattice structure of Fig. 2.14 can be translated into the direct-form structure. The equivalent direct-form coefficients can be derived from a vector product representation of the input–output relation of the 2-multiplier-lattice elements of Fig. 2.14-b,c. For reasons of simplicity we use the notation $\tilde{F}_i$ instead of $\tilde{F}_i(z)$, etc.

$$\tilde{F}_{i-1} = \tilde{F}_i + r_i \cdot \tilde{B}_{i-1} \tag{2.25-a}$$

$$\tilde{B}_i = -r_i \cdot \tilde{F}_{i-1} \cdot z^{-1} + \tilde{B}_{i-1} \cdot z^{-1} . \tag{2.25-b}$$

From (2.25-a) and (2.25-b) we derive the following representation

$$\begin{pmatrix} \tilde{F}_i \\ \tilde{B}_i \end{pmatrix} = \begin{pmatrix} +1 & -r_i \\ -r_i \cdot z^{-1} & +z^{-1} \end{pmatrix} \cdot \begin{pmatrix} \tilde{F}_{i-1} \\ \tilde{B}_{i-1} \end{pmatrix} . \tag{2.26}$$

Furthermore, we have constraints at the lips

$$V_L = \tilde{F}_0 \cdot g_r \tag{2.27-a}$$

$$\tilde{B}_0 = 0 \tag{2.27-b}$$

$$\begin{pmatrix} \tilde{F}_1 \\ \tilde{B}_1 \end{pmatrix} = \begin{pmatrix} +1 & -r_1 \\ -r_1 \cdot z^{-1} & +z^{-1} \end{pmatrix} \cdot \begin{pmatrix} \tilde{F}_0 \\ 0 \end{pmatrix} \tag{2.27-c}$$

$$= \begin{pmatrix} +1 & -r_1 \\ -r_1 \cdot z^{-1} & +z^{-1} \end{pmatrix} \cdot \begin{pmatrix} 1 \\ 0 \end{pmatrix} \cdot \frac{V_L}{g} \tag{2.27-d}$$

and at the glottis

$$\tilde{F}_n = V_G \cdot \frac{1 + r_G}{2} + r_G \cdot \tilde{B}_n , \tag{2.28-a}$$

i.e.,

$$V_G = \frac{2}{1 + r_G} \cdot [\tilde{F}_n - r_G \cdot \tilde{B}_n] \tag{2.28-b}$$

$$= \frac{2}{1 + r_G} \cdot \begin{pmatrix} 1 & -r_G \end{pmatrix} \cdot \begin{pmatrix} \tilde{F}_n \\ \tilde{B}_n \end{pmatrix} , \tag{2.28-c}$$

where $\begin{pmatrix} 1 & -r_G \end{pmatrix}$ denotes a row vector with two elements. From (2.26), (2.27-d), and (2.28-c) we finally derive the following expression

$$\begin{aligned} V_G(z) &= \frac{2}{1+r_G} \cdot \begin{pmatrix} 1 & -r_G \end{pmatrix} \prod_{i=1}^{n} \begin{pmatrix} +1 & -r_i \\ -r_i \cdot z^{-1} & +z^{-1} \end{pmatrix} \begin{pmatrix} 1 \\ 0 \end{pmatrix} \cdot \frac{V_L(z)}{g_r} \\ &= \frac{2}{1+r_G} \cdot \frac{1}{g_r} \cdot D(z) \cdot V_L(z) \end{aligned} \tag{2.29}$$

where $D(z)$ can be evaluated as

$$D(z) = 1 - \sum_{i=1}^{n} c_i \cdot z^{-i} = 1 - C(z). \tag{2.30}$$

By inversion of (2.29) we find with (2.23) the frequency response $H(z)$ and its dependency on z

$$H(z) = \frac{V_L(z)}{V_G(z)} = \frac{1+r_G}{2} \cdot \frac{g_r}{D(z)} \tag{2.31-a}$$

$$= \frac{1+r_G}{2} \cdot \frac{g_r}{1 - C(z)} \tag{2.31-b}$$

$$= \frac{1+r_G}{2} \cdot \frac{\prod_{i=1}^{n}(1 + r_i)}{1 - \sum_{i=1}^{n} c_i \cdot z^{-i}}. \tag{2.31-c}$$

The direct-form vocal tract transfer function $H(z)$ has n poles and no non-trivial zeros ($z = 0$). This type of filter is called an *all-pole filter*. The corresponding direct-form implementation is illustrated in Fig. 2.15.

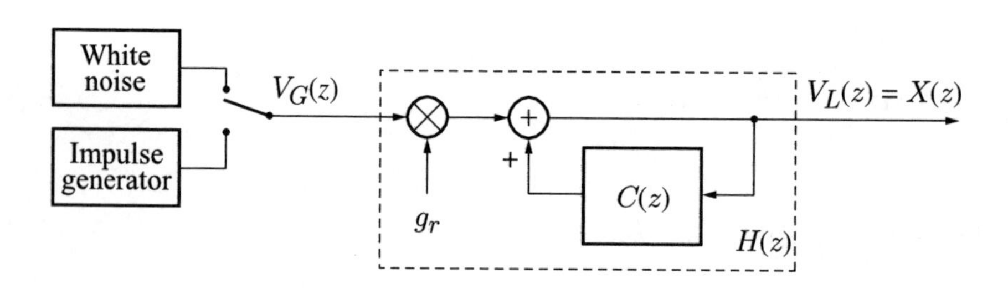

Figure 2.15: All-pole vocal tract model

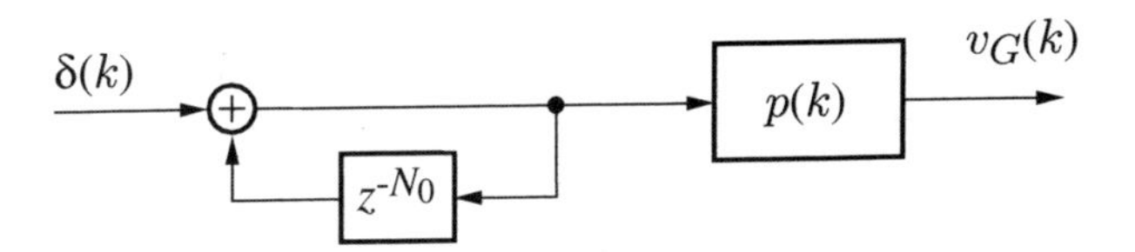

Figure 2.16: Glottis model for voiced excitation

For producing unvoiced sounds, a noise-like excitation signal $v_G(k)$ is required. A sequence of pitch pulses according to Fig. 2.2 is needed for producing purely voiced sounds.

The periodic voiced excitation signal $v_G(k)$ can be generated as illustrated in Fig. 2.16 where

$$\delta(k) = \begin{cases} 1 & k = 0 \\ 0 & k \neq 0 \end{cases} \tag{2.32}$$

denotes the unit impulse. The output signal is given by

$$v_G(k) = \sum_{\lambda} p(k - \lambda \cdot N_0) \tag{2.33}$$

with the glottis impulse response $p(k)$ and the pitch period N_0. The impulse response $p(k)$ determines the shape of a single pitch cycle. In practical implementations of speech codecs, usually the glottis impulse response is merged with the impulse response of the vocal tract filter. Furthermore, it should be noted that the model of Fig. 2.15 is to a certain degree too simple as, e.g., some speech sounds require a mixed voiced/unvoiced excitation. This is taken into consideration in the different speech coding algorithms by replacing the source switch in Fig. 2.15 by a weighted superposition of the noisy and periodic excitation.

2.4 Anatomy of Hearing

The peripheral hearing organ is divided into three sections (e.g., [Hudde 2005]):

- outer ear,
- middle ear, and
- inner ear,

as illustrated in Fig. 2.17.

The *outer ear* consists of the pinna, the outer ear canal, and the ear drum. The pinna protects the opening and contributes to the directivity of hearing in combination with the head and the shoulders. The outer ear canal is a nearly uniform

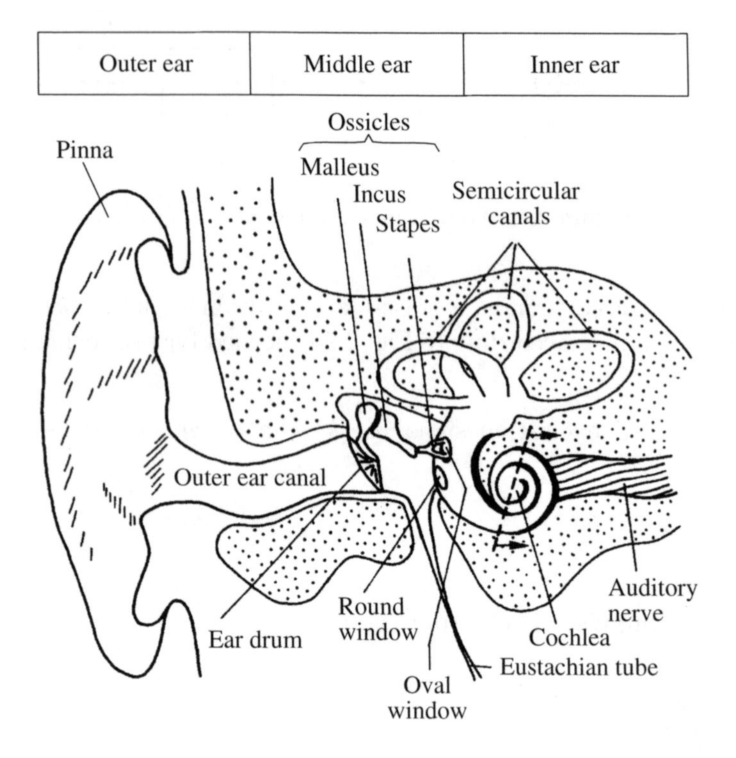

Figure 2.17: Schematic drawing of the ear ([Zwicker, Fastl 1999])

tube, closed at the inner end, with a length up to 3 cm and a diameter of about 0.7 cm. This tube transmits the sound to the ear drum. Within the frequency range of speech it has a single resonance frequency between 3 kHz and 4 kHz. This is the reason for the increased sensitivity of the ear in this frequency range. The ear drum is a stiff, conical membrane, which vibrates because of the forces of the oscillating air particles.

The *middle ear* is an air-filled cavity which is connected to the outer ear by the ear drum and to the inner ear by the round window and the oval window. It contains three tiny bones, the ossicles, which provide the acoustic coupling between the ear drum and the oval window. A mechanical impedance transformation by a factor of about 15 from the airborne sound to the fluids of the inner ear is performed according to the area ratio of ear drum and the oval window. Consequently, oscillations of the air particles by small forces and large displacements are transformed into large forces and small displacements.

Additionally, the middle ear is connected to the upper throat via the Eustachian tube, which is opened briefly during swallowing, to equalize the air pressure in the middle ear to that of the environment. This is necessary to adjust the resting point of the ear drum and the working point of the middle ear ossicles.

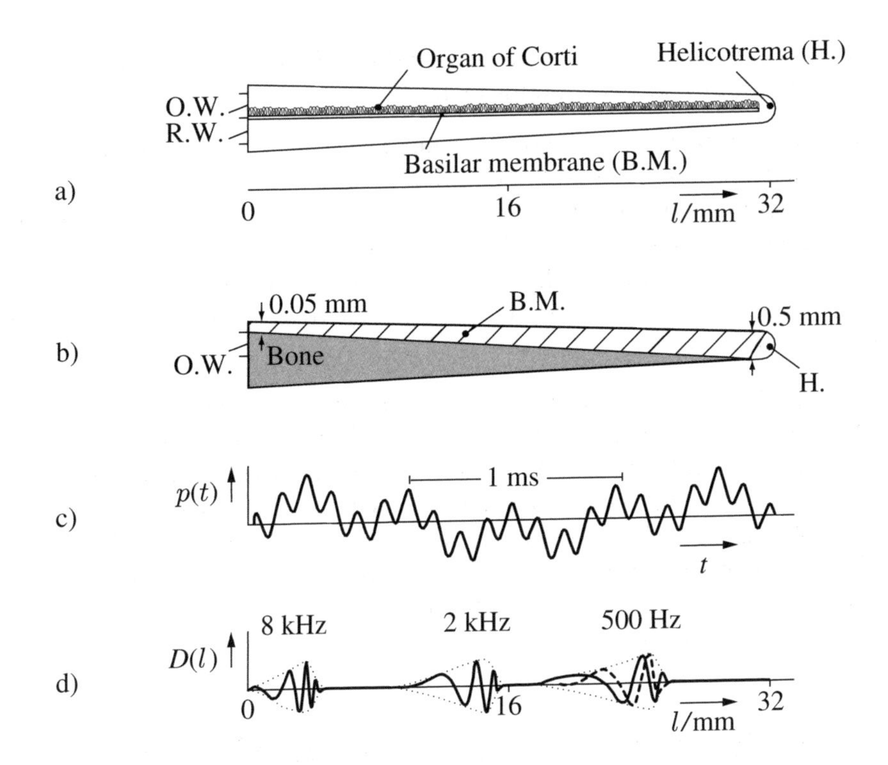

Figure 2.18: Frequency-to-place transformation along the basilar membrane (adapted from [Zwicker, Fastl 1999]); O.W. = Oval Window; R.W. = Round Window
a) Top-down view into the unwound cochlea
b) Side view
c) Tripple-tone audio signal (sound pressure, 500 Hz, 2000 Hz, 8000 Hz)
d) Displacement $D(l)$ by excitation (traveling waves and envelopes)

The *inner ear* consists of the organs for the sense of equilibrium and the sense of orientation with the semicircular canals (which are not involved in the hearing process) and the *cochlea* with the round and the oval window. The cochlea contains the *basilar membrane* with the *organ of Corti*, which converts mechanical vibrations into impulses in the auditory nerve. The cochlea is formed like a snail-shell and consists of 2.5 turns. Figure 2.18 illustrates the schematic view of the "untwisted" cochlea, which has a length of about 32 mm. The movements of the air particles are transmitted by the *ossicles* through the oval window (O.W.) to the incompressible fluids which drive the basilar membrane (B.M.). The uncoiled cochlea is separated by the basilar membrane into an upper and a lower chamber (*scala vestibuli* and *scala tympani*, Fig. 2.18-a) which are connected at the far end by an opening which is called the *helicotrema*.

The organ of Corti, which sits above the basilar membrane, senses vibrations with about 3600 inner hair cells and about 26000 outer hair cells and passes the information on to the auditory nerve and to the brain via neural synapses.

Through the helicotrema and the round window (R.W.), pressure compensation of the traveling fluidic waves is performed.

The basilar membrane is about 0.05 mm wide at the oval window and about 0.5 mm wide at the helicotrema (see Fig. 2.18-b). The basilar membrane performs a transformation of sound frequency to place by means of a traveling wave mechanism. High frequencies stimulate the membrane and thus the hair cells of the organ of Corti near the oval window, while the resonances for low frequencies are near to the helicotrema. This transformation corresponds to a spectral analysis using a (non-uniform) filter bank.

A schematic drawing of the frequency-to-place transformation is shown in Fig. 2.18-c,d for an audio signal (sound pressure $p(t)$), consisting of three sinusoidal tones of 500, 2000, and 8000 Hz. The three signal components cause vibrations in separate regions of the basilar membrane in such a way that the envelope of the dynamical displacement $D(l)$ reaches local maxima at certain locations. The envelope of the displacements is quite steep towards the helicotrema, while in the direction of the oval window a flat descent can be observed. This behavior determines the characteristics of the masking effect to be discussed in Section 2.5.3.

2.5 Psychoacoustic Properties of the Auditory Organ

2.5.1 Hearing and Loudness

Speech is carried by small temporal variations of the sound pressure $p(t)$. The physical unit of $p(t)$ is the pascal (Pa), where the relevant range for hearing covers more than seven decades from the threshold of hearing at 10^{-5} Pa to the threshold of pain at 10^2 Pa. The normalized *sound pressure* p/p_0 or equivalently the normalized *sound intensity* I/I_0 are measured on a logarithmic scale

$$L = 20 \cdot \log_{10}\left(\frac{p}{p_0}\right) = 10 \cdot \log_{10}\left(\frac{I}{I_0}\right). \tag{2.34}$$

The level L is given in decibels (dB), with the reference sound pressure $p_0 = 20\,\mu\text{Pa}$ and the reference sound intensity $I_0 = 10^{-12}\,\text{W/m}^2$.

The audible ranges of frequency and sound pressure level can be visualized in the so-called *hearing area* as shown in Fig. 2.19. The abscissa represents the frequency f on a logarithmic scale, the ordinate the sound pressure level L.

The threshold in quiet is dependent on frequency. The highest sensitivity between 3 and 4 kHz is due to the resonance frequency of the outer ear canal as mentioned above.

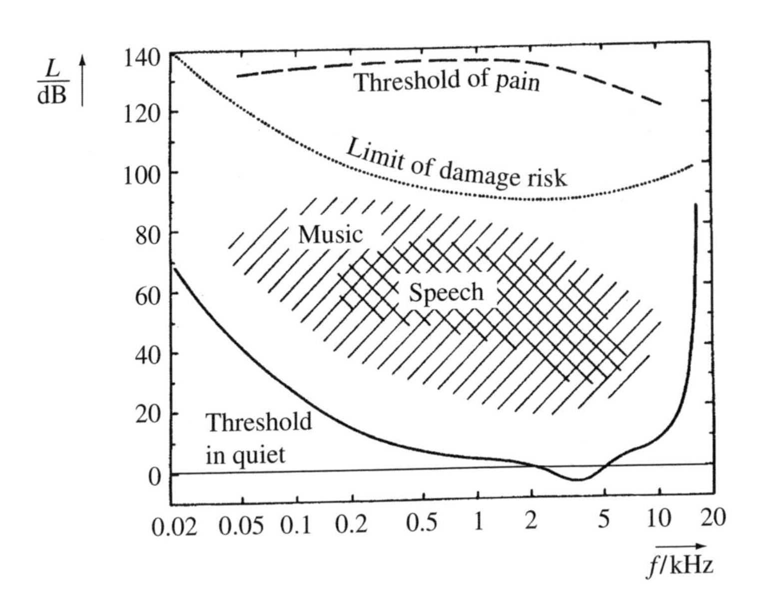

Figure 2.19: The hearing area ([Zwicker, Fastl 1999])

The dotted line indicates the limit of damage risk for an "average person", which also strongly depends on frequency and takes the smallest values in the range between 1 and 5 kHz. The areas of speech and music sounds are indicated by different hatched patterns.

In general, the perceived *loudness level* L_L is a function of both frequency f and sound pressure level L as shown in Fig. 2.20.

Contours of constant subjective loudness can be found by auditive comparison of a sinusoidal test tone at different frequencies and amplitudes with a sinusoidal reference at 1 kHz. At the reference frequency, the loudness level is identical to the sound pressure level. The loudness level L_L of a test tone is the same as the loudness level of a reference 1 kHz tone with sound pressure level $L = L_L$. The loudness level L_L is given in the pseudo-unit *phon*.

In psychoacoustics the perceived loudness, which is defined in the unit *sone*, is often evaluated by auditive tests. The relation between loudness N in sones and loudness level L_L in phons can be approximated by

$$N \approx 2^{(L_L - 40)/10} . \tag{2.35}$$

A perceived loudness of $N = 1$ sone corresponds to a loudness level of $L_L = 40$ phon.

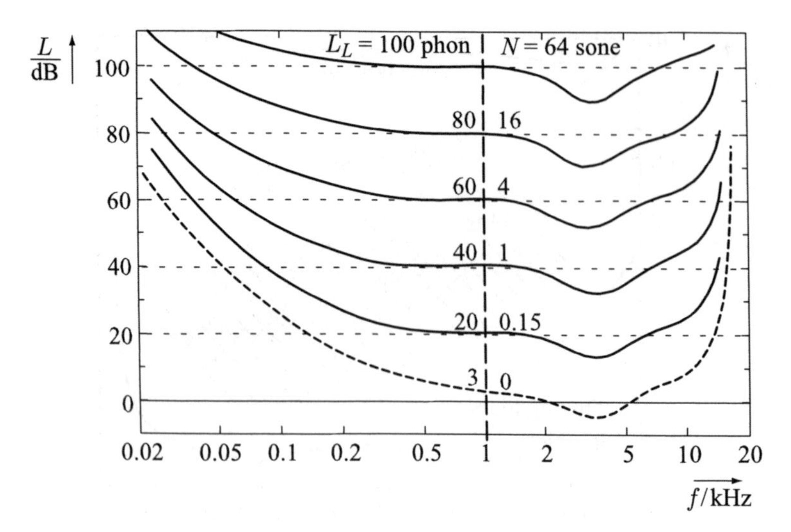

Figure 2.20: Contours of equal loudness level L_L for sinusoidal tones ([Zwicker, Fastl 1999])

2.5.2 Spectral Resolution

The spectral resolution of the ear is related to the frequency-to-place transformation on the basilar membrane. The frequency resolution can be analyzed by specific auditive tests.

The just noticeable variations in frequency can be measured as follows. Sinusoidal frequency modulation with a modulation frequency of about 4 Hz and different values of the modulation index and thus different frequency deviations Δf is applied to a "carrier sine wave" of frequency f_0. The test persons are asked to detect the difference between the unmodulated carrier signal with frequency f_0 and the modulated tone. As the frequency-modulated signal has two main spectral components at $f_0 \pm \Delta f$, the effective variation in frequency is $2 \cdot \Delta f$.

At frequencies $f_0 < 500$ Hz, a constant, just noticeable difference of $2 \cdot \Delta f = 3.6$ Hz can be measured. Above 500 Hz this value increases proportionally according to

$$2 \cdot \Delta f \approx 0.007 \cdot f_0 \,. \tag{2.36}$$

We can distinguish very small changes in frequency of about 0.7% in this range. In the frequency range up to 16 kHz, about 640 frequency steps can be distinguished.

Another important aspect is the effective spectral resolution, which is responsible for a certain loudness perception. By using test signals either consisting of several sinusoidal tones closely spaced in frequency or consisting of bandpass noise with adjustable bandwidth, it can be shown that the ear integrates the excitation over a certain frequency interval. By means of listening experiments, 24 intervals can be identified in the audible frequency range up to 16 kHz which are called the *critical bands*, e.g., [Zwicker, Fastl 1999]. In technical terms they can be described as a filter bank with non-uniform spectral resolution. There is a strong correlation between the critical bands and the excitation patterns of the basilar membrane. If the basilar membrane of length $d = 32\,\text{mm}$ is divided into 24 uniform intervals of length $d_B = d/24 = 4/3\,\text{mm}$, each segment is equivalent to a critical band. The critical bands do not correspond to 24 discrete filters, but they quantify the effective frequency-dependent frequency resolution of loudness perception. The uniform subdivision of the basilar membrane is widely known as the *Bark scale* in honor of the famous physicist H. G. Barkhausen. Table 2.1 lists the Bark number b (*critical-band rate*), the center frequency f_c, and the bandwidth Δf_b of each critical band. Within the frequency band $100\,\text{Hz} \leq f \leq 3400\,\text{Hz}$ of voice telephony, there are 16 critical bands.

The Bark scale in Table 2.1 is approximately related to linear frequency f according to the following analytical expressions:

$$\frac{b}{\text{Bark}} = 13 \arctan\left(0.76 \frac{f}{\text{kHz}}\right) + \arctan\left(\frac{f}{7.5\,\text{kHz}}\right) \tag{2.37}$$

and

$$\frac{\Delta f_c}{\text{Bark}} = 25 + 75 \cdot \left[1 + 1.4 \left(\frac{f}{\text{kHz}}\right)^2\right]^{0.69} . \tag{2.38}$$

The critical-band feature and the masking effect resulting from it can be exploited to improve the performance and to reduce the complexity of speech signal processing algorithms such as speech recognition, speech coding, or speech enhancement.

Table 2.1: Critical bands (adapted from [Zwicker, Fastl 1999])
(b = critical-band rate; f_c = center frequency; Δf_c = bandwidth)

b/Bark	f_c/Hz	Δf_c
0.5	50	100
1.5	150	100
2.5	250	100
3.5	350	100
4.5	450	110
5.5	570	120
6.5	700	140
7.5	840	150
8.5	1000	160
9.5	1170	190
10.5	1370	210
11.5	1600	240

b/Bark	f_c/Hz	Δf_c
12.5	1850	280
13.5	2150	320
14.5	2500	380
15.5	2900	450
16.5	3400	550
17.5	4000	700
18.5	4800	900
19.5	5800	1100
20.5	7000	1300
21.5	8500	1800
22.5	10500	2500
23.5	13500	3500

2.5.3 Masking

Masking occurs in many everyday situations, where a dominant sound renders a weaker sound inaudible. One example is music at "disco level" which might completely mask the ringing of a mobile phone. In that situation the music signal is called the *masker* and the ringing tone is called the *test tone*.

The masking effect can easily be demonstrated by using a narrow bandpass noise or a sinusoidal tone with fixed frequency $f_M = f_c$ as masker and a sinus with variable frequency f_T or a fixed narrowband noise as test tone. Narrowband means that in the masking experiment the bandwidth of the noise (being either masker or test signal) does not exceed the critical bandwidth at this frequency.

One example is given in Fig. 2.21 for a narrowband noise as masker centered at $f_M = 1\,\text{kHz}$, having a bandwidth of $\Delta f_M = 160\,\text{Hz}$ and a fixed sound pressure level L_M. For any frequency f_T of the test tone, the sinusoidal tone is masked by the noise as long as its level L_T is below the masking threshold. The masking curves are not symmetric with frequency f_T. They are steeper towards lower frequencies than towards higher frequencies. This follows, as discussed in Sec. 2.4, from the shape of the traveling wave envelope on the basilar membrane (see also Fig. 2.18). The dips in the curves at masker levels L_M of 80 dB and 100 dB are caused by non-linear effects of the hearing system, if the masker is a sinusoidal signal.

As mentioned earlier, noise can also be masked by a sinus signal. Furthermore, the masking threshold can be evaluated for complex test sounds such as speech

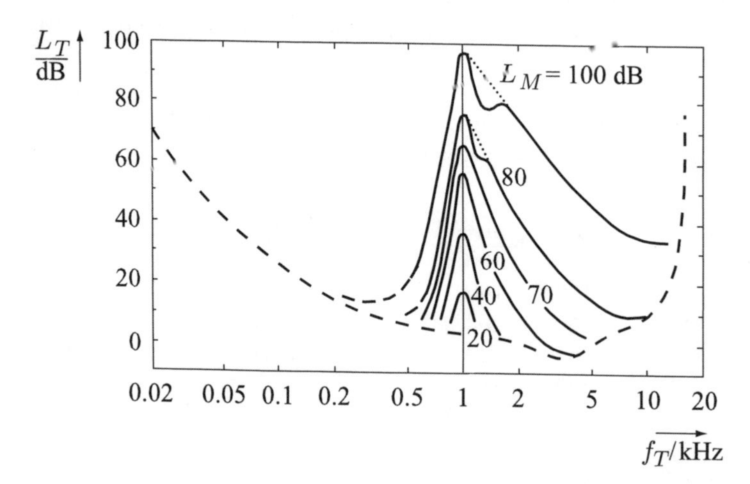

Figure 2.21: Level L_T of a sinusoidal test tone of frequency f_T masked by
—— a sinusoidal masker with $f_M = 1$ kHz
.... narrowband noise with level L_M and with critical bandwidth, centered at $f_M = f_c = 1$ kHz ([Zwicker, Fastl 1999])

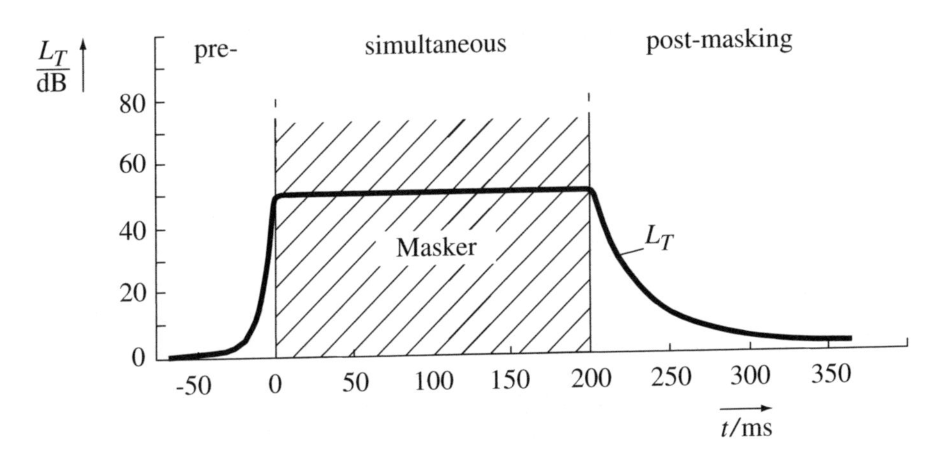

Figure 2.22: Pre- and post-masking: necessary level L_T for the audibility of a sinusoidal burst (test tone) masked by wideband noise (adapted from [Zwicker, Fastl 1999])

signals. This is widely used in speech and audio coding to mask quantization noise, i.e., to improve at a given bit rate the subjective quality of the decoded signal.

The masking effect discussed so far is called masking in the frequency domain or *simultaneous masking* as the masker and the test signal are present at the same time.

Apart from this, *non-simultaneous masking* can be observed in the time domain. The test sound is switched on before the masker occurs or after the masker is switched off. The onset of the test tone may become inaudible or masked. This is called *pre-masking* and *post-masking*. The effect of pre-masking is not very strong but post-masking is quite pronounced. The latter can be explained by the fact that the basilar membrane and the organ of Corti need some time to recover the threshold in quiet after the masker has been switched off. The principle of masking with wideband noise as masker and sinusoidal bursts as test tone is illustrated in Fig. 2.22.

Pre- and post-masking are exploited in speech and audio processing, such as frequency-domain-based coding, to hide pre- and post-echoes originating from non-perfect reconstruction.

Bibliography

Deller Jr., J. R.; Proakis, J. G.; Hansen, J. H. L. (2000). *Discrete-time Processing of Speech Signals*, 2nd edn, IEEE Press, New York.

Fant, G. (1970). *Acoustic Theory of Speech Production*, 2nd edn, Mouton, The Hague.

Flanagan, J. L. (1972). *Speech Analysis, Synthesis, and Perception*, 2nd edn, Springer Verlag, Berlin.

Gimson, A. C.; Cruttenden, A. (1994). *Gimson's Pronunciation of English*, 5th revised edn, Arnold, London.

Hudde, H. (2005). A Functional View on the Peripheral Human Hearing Organ, *in* J. Blauert (ed.), *Communication Acoustics*, Springer Verlag, Berlin.

Kelly, J. L.; Lochbaum, C. C. (1962). Speech Synthesis, *Proceedings of Fourth International Congress on Acoustics*, Copenhagen, Denmark, pp. 1–4.

O'Shaughnessy, D. (2000). *Speech Communications*, IEEE Press, New York.

Parsons, T. W. (1986). *Voice and Speech Processing*, McGraw-Hill, New York.

Picket, J. M. (1980). *The Sounds of Speech Communication*, Pro-Ed, Inc., Austin, Texas.

Quatieri, T. F. (2001). *Discrete-Time Speech Signal Processing*, Prentice Hall, Englewood Cliffs, New Jersey.

Rabiner, L. R.; Schafer, R. W. (1978). *Digital Processing of Speech Signals*, Prentice Hall, Englewood Cliffs, New Jersey.

Vary, P.; Heute, U.; Hess, W. (1998). *Digitale Sprachsignalverarbeitung*, B. G. Teubner, Stuttgart (in German).

Zwicker, E.; Fastl, H. (1999). *Psychoacoustics*, 2nd edn, Springer Verlag, Berlin.

3

Spectral Transformations

Spectral transformations are key to many speech processing algorithms. The purpose of the spectral transformation is to represent a signal in a domain where certain signal properties are better accessible or a specific processing task is more efficiently accomplished. In this chapter, we will summarize the definitions and properties of the Fourier transform (FT) for continuous and discrete time signals as well as the discrete Fourier transform (DFT), its fast realizations, and the z-transform. No extensive derivations will be given. For more elaborate expositions of this material the reader is referred to the many excellent textbooks on digital signal processing. The chapter concludes with an introduction into the real and the complex cepstrum and applications.

3.1 Fourier Transform of Continuous Signals

The *Fourier transform* (FT) provides an analysis of signals in terms of its spectral components. It relates a continuous (time) domain signal $x_a(t)$ of, in general, infinite support to its Fourier domain representation $X_a(j\omega)$

$$X_a(j\omega) = \int_{-\infty}^{\infty} x_a(t) e^{-j\omega t} \mathrm{d}t \tag{3.1}$$

Digital Speech Transmission: Enhancement, Coding and Error Concealment
Peter Vary and Rainer Martin

with $\omega = 2\pi f$ denoting the radian frequency. The inverse transformation is given by

$$x_a(t) = \frac{1}{2\pi} \int_{-\infty}^{\infty} X_a(j\omega) e^{j\omega t} \mathrm{d}\omega \, . \tag{3.2}$$

The FT converges in the mean square if

$$\int_{-\infty}^{\infty} |x_a(t)|^2 \mathrm{d}t < \infty \, . \tag{3.3}$$

This condition includes all signals of finite duration. When $x(t)$ is absolutely summable and other conditions, known as the Dirichlet conditions, e.g., [Oppenheim et al. 1996], are met, the inverse FT reconstructs a signal which is identical to the original signal except for a finite number of discontinuities. $X_a(j\omega)$ is, in general, a continuous and non-periodic function of frequency. The FT has a number of well-known properties, which are summarized in Table 3.1.

Table 3.1: Properties of the FT

Property	Time domain	Frequency domain
Definition	$x(t) = \frac{1}{2\pi} \int_{-\infty}^{\infty} X(j\omega)\, e^{j\omega t} \mathrm{d}\omega$	$X(j\omega) = \int_{-\infty}^{\infty} x(t)\, e^{-j\omega t} \mathrm{d}t$
Linearity	$a\, x_1(t) + b\, x_2(t)$	$a\, X_1(j\omega) + b\, X_2(j\omega)$
Conjugation	$x^*(t)$	$X^*(-j\omega)$
Symmetry	$x(t)$ is a real-valued signal	$X(j\omega) = X^*(-j\omega)$
Even part	$x_e(t) = 0.5\,(x(t) + x(-t))$	$\mathrm{Re}\{X(j\omega)\}$
Odd part	$x_o(t) = 0.5\,(x(t) - x(-t))$	$j\,\mathrm{Im}\{X(j\omega)\}$
Convolution	$x_1(t) * x_2(t)$	$X_1(j\omega) \cdot X_2(j\omega)$
Time shift	$x(t - t_0)$	$e^{-j\omega t_0} X(j\omega)$
Modulation	$x(t)\, e^{j\omega_M t}$	$X(j(\omega - \omega_M))$
Scaling	$x(at),\ \ a \in \mathbb{R},\, a \neq 0$	$\frac{1}{\lvert a \rvert} X\left(j\frac{\omega}{a}\right)$
Parseval's theorem	$\int_{-\infty}^{\infty} x(t) y^*(t) \mathrm{d}t = \frac{1}{2\pi} \int_{-\infty}^{\infty} X(j\omega) Y^*(j\omega) \mathrm{d}\omega$	

3.2 Fourier Transform of Discrete Signals

The *Fourier transform of discrete signals* (FTDS)[1] is derived using the representation of sampled signals as pulse trains

$$x_s(t) = \sum_{k=-\infty}^{\infty} x_a(kT)\, \delta_a(t - kT) \tag{3.4}$$

and the shifting property of the Dirac impulse[2]

$$\int_{-\infty}^{\infty} \delta_a(t - t_0)\, f(t)\, \mathrm{d}t = f(t_0)\,. \tag{3.5}$$

For a sampling period of $T = \frac{1}{f_s}$ we obtain from the continuous time FT (3.1)

$$\begin{aligned} X_s(j\omega) &= \int_{-\infty}^{\infty} x_s(t)\, e^{-j\omega t}\, \mathrm{d}t = \sum_{k=-\infty}^{\infty} x_a(kT) \int_{-\infty}^{\infty} \delta_a(t - kT)\, e^{-j\omega t}\, \mathrm{d}t \\ &= \sum_{k=-\infty}^{\infty} x_a(kT)\, e^{-j\omega kT}\,. \end{aligned}$$

$X_s(j\omega)$ is a continuous and periodic function of the radian frequency ω and hence also of frequency f. To see this we note that the complex phasor

$$e^{-j\omega kT} = e^{-j2\pi fkT} = \cos(2\pi fkT) - j\sin(2\pi fkT)$$

is periodic in f with period $f = \frac{1}{T} = f_s$. Therefore, we have

$$X_s\left(j\left(\omega + \frac{2\pi\ell}{T}\right)\right) = X_s(j\omega) \text{ for any } \ell \in \mathbb{Z}.$$

To facilitate the treatment of sampled signals in the Fourier domain we normalize the frequency variable f on the sampling rate f_s and introduce the *normalized radian frequency* $\Omega = \omega T = 2\pi fT = 2\pi\frac{f}{f_s}$. We then obtain the FTDS and the inverse transform with $x(k) = x_a(kT)$ as

$$X(e^{j\Omega}) = \sum_{k=-\infty}^{\infty} x(k)\, e^{-j\Omega k} \quad \text{and} \quad x(k) = \frac{1}{2\pi}\int_{-\pi}^{\pi} X(e^{j\Omega})\, e^{j\Omega k}\, \mathrm{d}\Omega\,. \tag{3.6}$$

---

[1] The FTDS is also known as the *discrete time Fourier transform* (DTFT), e.g. [Oppenheim et al. 1996]. We prefer the more general terminology.

[2] We will denote the continuous time Dirac impulse by $\delta_a(t)$ and use $\delta(k)$ for the discrete unit impulse (3.8).

Note that the inverse transform is evaluated over one period of the spectrum only. $X(e^{j\Omega})$ is a complex quantity and may be written in terms of its real and imaginary parts

$$X(e^{j\Omega}) = \mathrm{Re}\{X(e^{j\Omega})\} + j\,\mathrm{Im}\{X(e^{j\Omega})\} = X_R(e^{j\Omega}) + jX_I(e^{j\Omega})$$

or in terms of its magnitude and phase

$$X(e^{j\Omega}) = |X(e^{j\Omega})|\, e^{j\phi(\Omega)}\,.$$

$|X(e^{j\Omega})|$ is called the amplitude spectrum and $\phi(\Omega)$ is called the phase spectrum. The principal value of the phase is denoted by $\arg\{X(e^{j\Omega})\} \in \{-\pi, \pi\}$. Frequently, we will also use the logarithm of the amplitude spectrum $20\log_{10}(|X(e^{j\Omega})|)$. The properties of the FTDS are summarized in Table 3.2.

As an example we compute the FTDS of a discrete rectangular pulse

$$x(k) = \sum_{\ell=-(N-1)/2}^{(N-1)/2} \delta\,(k-\ell) \tag{3.7}$$

Table 3.2: Properties of the FTDS

Property	Time domain	Frequency domain
Definition	$x(k) = \frac{1}{2\pi}\int_{-\pi}^{\pi} X(e^{j\Omega})\, e^{j\Omega k} \mathrm{d}\Omega$	$X(e^{j\Omega}) = \sum_{k=-\infty}^{\infty} x(k)\, e^{-j\Omega k}$
Linearity	$a\,x_1(k) + b\,x_2(k)$	$a\,X_1(e^{j\Omega}) + b\,X_2(e^{j\Omega})$
Conjugation	$x^*(k)$	$X^*(e^{-j\Omega})$
Symmetry	$x(k)$ is a real-valued signal	$X(e^{j\Omega}) = X^*(e^{-j\Omega})$
Even part	$x_e(k) = 0.5\,(x(k) + x(-k))$	$\mathrm{Re}\{X(e^{j\Omega})\}$
Odd part	$x_o(k) = 0.5\,(x(k) - x(-k))$	$j\,\mathrm{Im}\{X(e^{j\Omega})\}$
Convolution	$x_1(k) * x_2(k)$	$X_1(e^{j\Omega}) \cdot X_2(e^{j\Omega})$
Time shift	$x(k-k_0)$	$e^{-j\Omega k_0} X(e^{j\Omega})$
Modulation	$x(k)\, e^{j\Omega_M k}$	$X(e^{j(\Omega-\Omega_M)})$
Parseval's theorem	$\sum_{k=-\infty}^{\infty} x(k)y^*(k) = \frac{1}{2\pi}\int_{-\pi}^{\pi} X(e^{j\Omega})Y^*(e^{j\Omega})\mathrm{d}\Omega$	

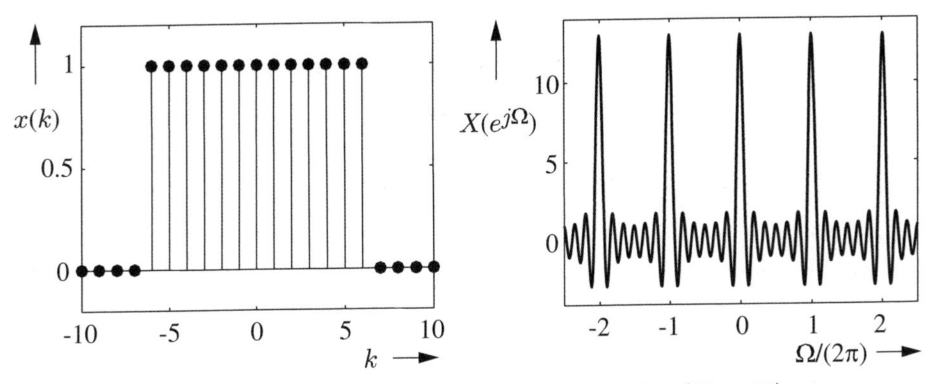

Figure 3.1: FT of a discrete rectangular pulse ($N = 13$)

where N is odd and $\delta(k)$ is the discrete unit impulse sequence, i.e.,

$$\delta(k) = \begin{cases} 1 & k = 0 \\ 0 & k \neq 0 \,. \end{cases} \tag{3.8}$$

With

$$X(e^{j\Omega}) = \sum_{k=-(N-1)/2}^{(N-1)/2} e^{-j\Omega k} = e^{j\Omega(N-1)/2} \sum_{k=0}^{N-1} e^{-j\Omega k} = \frac{e^{j\Omega N/2} - e^{-j\Omega N/2}}{e^{j\Omega/2} - e^{-j\Omega/2}}$$

we obtain

$$X(e^{j\Omega}) = \frac{\sin(\Omega N/2)}{\sin(\Omega/2)} \,. \tag{3.9}$$

For $N = 13$, the sequence $x(k)$ and the FTDS $X(e^{j\Omega})$ are plotted in Fig. 3.1. For the above example, $X(e^{j\Omega})$ is a real-valued function.

3.3 Linear Shift Invariant Systems

Systems map one or more input signals onto one or more output signals. In the case of a single input and a single output we have

$$y(k) = T\{x(k)\} \tag{3.10}$$

where $T\{\cdot\}$ defines the mapping from any input sequence $x(k)$ to the corresponding output sequence $y(k)$.

A system is *linear* if and only if for any two signals $x_1(k)$ and $x_2(k)$ and any $a, b \in \mathbb{C}$ we have

$$T\{a\,x_1(k) + b\,x_2(k)\} = a\,T\{x_1(k)\} + b\,T\{x_2(k)\} \,. \tag{3.11}$$

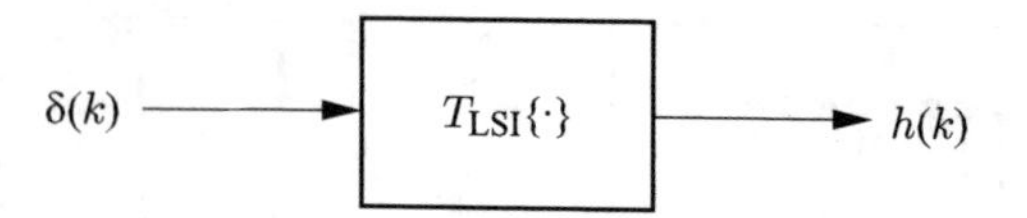

Figure 3.2: LSI system

A system is *memoryless* when $y(k = k_0)$ depends only on $x(k = k_0)$ for any k_0.

A system is *shift invariant* if for any k_0

$$y(k - k_0) = T\{x(k - k_0)\}\,. \tag{3.12}$$

A system which is linear and shift invariant is called a *linear shift invariant* (LSI) system.

An LSI system is defined by a mapping $T_{\mathrm{LSI}}\{\cdot\}$ as shown in Fig. 3.2. It may also be characterized by its impulse response $h(k) = T_{\mathrm{LSI}}\{\delta(k)\}$ if we assume that the system is at rest before any signal is applied. The latter condition implies that initially all system state variables are zero.

Using the linearity and the shift invariance, we obtain

$$\begin{aligned} y(k) &= T_{\mathrm{LSI}}\{x(k)\} \\ &= T_{\mathrm{LSI}}\left\{\sum_{l=-\infty}^{\infty} x(l)\,\delta(k-l)\right\} = \sum_{l=-\infty}^{\infty} x(l)\,T_{\mathrm{LSI}}\{\delta(k-l)\} \\ &= \sum_{l=-\infty}^{\infty} x(l)\,h(k-l) = x(k) * h(k)\,. \end{aligned}$$

This relation is known as *convolution* and denoted by $*$:

$$\begin{aligned} y(k) = x(k) * h(k) &= \sum_{l=-\infty}^{\infty} x(l)\,h(k-l) \\ &= \sum_{l=-\infty}^{\infty} h(l)\,x(k-l) = h(k) * x(k)\,. \end{aligned} \tag{3.13}$$

An especially useful class of systems is specified via a *difference equation* with constant coefficients a_μ and b_ν

$$\begin{aligned} y(k) &= b_0 x(k) + b_1 x(k-1) + \cdots + b_N x(k-N) \\ &\quad + a_1 y(k-1) + a_2 y(k-2) + \cdots + a_M y(k-M) \end{aligned} \tag{3.14}$$

where input and output samples with index $l \leq k$ contribute to the current output sample $y(k)$. This equation may be cast in a more compact form

$$\sum_{\mu=0}^{M} \widetilde{a}_{\mu}\, y(k-\mu) = \sum_{\nu=0}^{N} b_{\nu}\, x(k-\nu) \quad \widetilde{a}_0 = 1 \text{ and } \widetilde{a}_{\mu} = -a_{\mu} \quad \forall\, \mu > 0\,. \tag{3.15}$$

The maximum of N and M is called the order of the system. Additionally, we must specify M initial conditions to obtain a unique solution. In most practical cases, however, we assume that the system is causal and that it is at rest before an input signal is applied.

3.3.1 Frequency Response of LSI Systems

The complex exponential $x(k) = e^{j(\Omega_0 k + \phi_0)}$ is an eigenfunction of an LSI system, i.e.,

$$y(k) = \sum_{\ell=-\infty}^{\infty} h(\ell)\, e^{j(\Omega_0(k-\ell)+\phi_0)} = e^{j(\Omega_0 k+\phi_0)} \sum_{\ell=-\infty}^{\infty} h(\ell)\, e^{-j\Omega_0 \ell} \tag{3.16}$$

$$= e^{j(\Omega_0 k+\phi_0)}\, H(e^{j\Omega_0}) \tag{3.17}$$

where the *frequency response*

$$H(e^{j\Omega}) = \sum_{k=-\infty}^{\infty} h(k)\, e^{-j\Omega k}$$

is the FT of the discrete impulse response $h(k)$ at $\Omega = \Omega_0$ and

$$H(e^{j\Omega}) = \mathrm{Re}\{H(e^{j\Omega})\} + j\,\mathrm{Im}\{H(e^{j\Omega})\} = |H(e^{j\Omega})|\, e^{j\phi(\Omega)}\,.$$

3.4 The z-transform

The FTDS does not converge for signals such as the unit step $u(k)$, which is defined as

$$u(k) = \begin{cases} 1 & k \geq 0 \\ 0 & k < 0\,. \end{cases} \tag{3.18}$$

Therefore, we extend the FTDS into the complex plane by substituting

$$j\Omega = j\omega T \rightarrow sT \quad \text{with} \quad s = \alpha + j\omega\,.$$

To avoid the periodic repetition of the resulting spectral function along the imaginary axis, it is common practice to map the left half plane into the unit circle and to map periodic repetitions onto each other. This is achieved by introducing the complex variable $z = e^{sT} = e^{\alpha T} e^{j\omega T} = re^{j\Omega}$ with $r = e^{\alpha T}$. The *two-sided z-transform* of a discrete sequence $x(k)$ is then given by

$$\mathcal{Z}\{x(k)\} = X(z) = \sum_{k=-\infty}^{\infty} x(k) z^{-k} . \tag{3.19}$$

Alternatively, we could use the Laplace transform [Oppenheim et al. 1996] of a sampled signal $x_s(t)$ as in (3.4) and obtain

$$\int_{-\infty}^{\infty} x_s(t)\, e^{-st} \mathrm{d}t = \sum_{k=-\infty}^{\infty} x_a(kT) \int_{-\infty}^{\infty} \delta_a(t-kT)\, e^{-st} \mathrm{d}t = \sum_{k=-\infty}^{\infty} x(k)\, e^{-sTk}$$

which, with the above mapping $z = e^{sT}$, results in the definition (3.19) of the z-transform.

The z-transform converges for a given z when

$$|X(z)| = \left| \sum_{k=-\infty}^{\infty} x(k)\, z^{-k} \right|$$

attains a finite value, i.e.,

$$\left| \sum_{k=-\infty}^{\infty} x(k)\, z^{-k} \right| < \infty .$$

Since

$$X(z) = \sum_{k=-\infty}^{\infty} x(k)\, z^{-k} = \sum_{k=-\infty}^{\infty} x(k)\, e^{-\alpha Tk}\, e^{-j\omega Tk} = \sum_{k=-\infty}^{\infty} x(k)\, r^{-k}\, e^{-j\omega Tk}$$

the z-transform converges when

$$\sum_{k=-\infty}^{\infty} |x(k)\, r^{-k}| < \infty . \tag{3.20}$$

The region of the complex plane for which the above condition holds is known as the region of convergence (ROC). The ROC depends only on $|z|$ since within the ROC we have

$$\sum_{k=-\infty}^{\infty} \left|x(k)\, z^{-k}\right| = \sum_{k=-\infty}^{\infty} |x(k)| \left|z^{-k}\right| = \sum_{k=-\infty}^{\infty} |x(k)|\, |z|^{-k} < \infty . \tag{3.21}$$

Table 3.3: Properties of the z-transform

Properties	Time domain	z-domain	ROC
Definition	$x(k) = \oint_C \frac{X(z)}{2\pi j} z^{k-1} \mathrm{d}z$	$X(z) = \sum_{k=-\infty}^{\infty} x(k)\, z^{-k}$	R_x
Linearity	$a\, x_1(k) + b\, x_2(k)$	$a\, X_1(z) + b\, X_2(z)$	Contains at least $R_{x_1} \cap R_{x_2}$; it may be larger, when poles and zeros cancel
Conjugation	$x^*(k)$	$X^*(z^*)$	R_x
Time reversal	$x(-k)$	$X(z^{-1})$	$1/R_x$
Convolution	$x_1(k) * x_2(k)$	$X_1(z) \cdot X_2(z)$	Contains at least $R_{x_1} \cap R_{x_2}$; it may be larger, when poles and zeros cancel
Time shifting	$x(k + k_0)$	$z^{k_0} X(z)$	R_x (it may not contain $z = 0$ or $z = \infty$)

Note that the z-transform of a sequence with finite support with $|x(k)| < \infty$ does converge in the whole complex plane except for $z = 0$. Other properties of the z-transform are summarized in Table 3.3. The last column of Table 3.3 specifies the ROC. For example, when the ROC of $X_1(z)$ is R_{x_1} and the ROC of $X_2(z)$ is R_{x_2} then the ROC of $aX_1(z)+bX_2(z)$ is at least equal to the intersection of both ROC's. Since poles of the individual z-transforms may cancel, it can be larger than the intersection of R_{x_1} and R_{x_2}. Finally, we state the initial value theorem: when $x(k)$ is causal we have $x(0) = \lim_{z \to \infty} X(z)$. The proofs for this and the theorems in Table 3.3 can be found, for instance, in [Oppenheim et al. 1999].

3.4.1 Relation to FT

In general, the z-transform is equivalent to the FT of $x(k)\, r^{-k}$. Moreover, for $z = e^{j\Omega}$ we obtain

$$X(e^{j\Omega}) = \sum_{k=-\infty}^{\infty} x(k)\, e^{-j\Omega k},$$

which is the FT of a discrete signal $x(k)$, provided that the unit circle is within the ROC.

The z-transform may be interpreted as a Laurent series expansion of $X(z)$. $X(z)$ is an analytic (holomorphic, regular) function within the ROC. Therefore, the z-transform and all its derivatives are continuous functions in the ROC. More details on complex analysis can be found, for instance, in [Churchill, Brown 1990].

We may also define a *one-sided z-transform*

$$\mathcal{Z}_1 \{x(k)\} = \sum_{k=0}^{\infty} x(k)\, z^{-k}$$

which is identical to the two-sided transform for causal signals.

3.4.2 Properties of the ROC

For a z-transform which is a rational function of z we summarize the following properties of the ROC [Oppenheim, Schafer 1975]:

- The ROC is a ring or a disk centered around the origin. It cannot contain any poles and must be a connected region.
- If $x(k)$ is a right-sided sequence, the ROC extends outward from the outermost pole to $|z| < \infty$.
- If $x(k)$ is a left-sided sequence, the ROC extends from the innermost pole to $z = 0$.
- The FT of $x(k)$ converges absolutely if and only if the ROC includes the unit circle.

3.4.3 Inverse z-transform

The inverse z-transform is based on the Cauchy integral theorem

$$\frac{1}{2\pi j} \oint_C z^{k-1} \mathrm{d}z = \begin{cases} 1 & k = 0 \\ 0 & k \neq 0 \end{cases} \tag{3.22}$$

and may be written as

$$x(k) = \frac{1}{2\pi j} \oint_C X(z)\, z^{k-1} \mathrm{d}z\,. \tag{3.23}$$

To prove the inverse transform we note that

$$x(k) = \frac{1}{2\pi j} \oint_C X(z) z^{k-1} \mathrm{d}z = \frac{1}{2\pi j} \oint_C \sum_{m=-\infty}^{\infty} x(m)\, z^{-m}\, z^{k-1} \mathrm{d}z$$

$$= \sum_{m=-\infty}^{\infty} x(m) \frac{1}{2\pi j} \oint_C z^{k-m-1} \mathrm{d}z = x(k)\,.$$

In general, the evaluation of the above integral is cumbersome. For rational z-transforms methods based on

- long division,
- partial fraction expansions, and
- table look-up

are much more convenient. A description of these methods can be found, for instance, in [Proakis, Manolakis 1992], [Oppenheim et al. 1999]. A selection of transform pairs is given in Table 3.4.

Table 3.4: Selected z-transform pairs

Sequence	z-transform	ROC
$\delta(k)$	1	all z
$\delta(k-k_0),\ k_0 > 0$	z^{-k_0}	all z except 0
$\delta(k-k_0),\ k_0 < 0$	z^{-k_0}	all z except ∞
$u(k)$	$\frac{1}{1-z^{-1}}$	$\lvert z\rvert > 1$
$a^k u(k)$	$\frac{1}{1-az^{-1}}$	$\lvert z\rvert > \lvert a\rvert$
$\cos(\Omega_0 k)u(k)$	$\frac{1-\cos(\Omega_0)z^{-1}}{1-2\cos(\Omega_0)z^{-1}+z^{-2}}$	$\lvert z\rvert > 1$
$\sin(\Omega_0 k)u(k)$	$\frac{\sin(\Omega_0)z^{-1}}{1-2\cos(\Omega_0)z^{-1}+z^{-2}}$	$\lvert z\rvert > 1$
$r^k \cos(\Omega_0 k)u(k)$	$\frac{1-r\cos(\Omega_0)z^{-1}}{1-2r\cos(\Omega_0)z^{-1}+r^2z^{-2}}$	$\lvert z\rvert > r$
$r^k \sin(\Omega_0 k)u(k)$	$\frac{r\sin(\Omega_0)z^{-1}}{1-2r\cos(\Omega_0)z^{-1}+r^2z^{-2}}$	$\lvert z\rvert > r$

3.4.4 z-transform Analysis of LSI Systems

When states are initially zero, the z-transform of the output signal $y(k)$ of an LSI system $T_{\mathrm{LSI}}\{\cdot\}$ is given by

$$Y(z) = H(z)X(z) \tag{3.24}$$

where $X(z)$ and $H(z)$ are the z-transforms of the input signal $x(k)$ and the impulse response $h(k)$, respectively. This a direct consequence of the convolution theorem as stated in Table 3.3. We call $H(z)$ the system response (or transfer function) of the system T_{LSI} (see also Fig. 3.3).

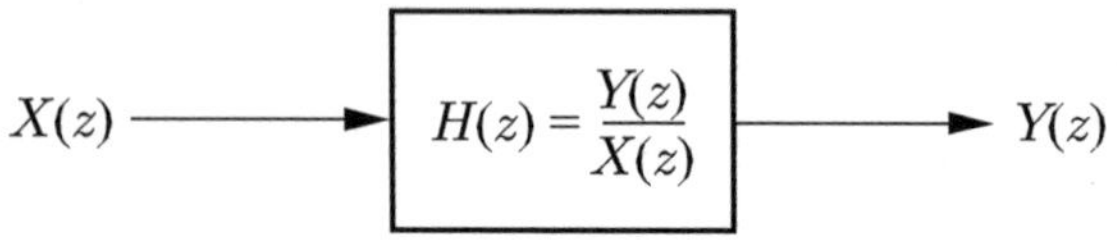

$$X(z) \longrightarrow \boxed{H(z) = \frac{Y(z)}{X(z)}} \longrightarrow Y(z)$$

Figure 3.3: Input–output relation of LSI systems in the z-domain

An especially important class of LSI systems is characterized by a rational system response $H(z)$ with constant coefficients

$$H(z) = \frac{B(z)}{\widetilde{A}(z)} = \frac{\sum\limits_{\nu=0}^{N} b_\nu\, z^{-\nu}}{\sum\limits_{\mu=0}^{M} \widetilde{a}_\mu\, z^{-\mu}} = \frac{\sum\limits_{\nu=0}^{N} b_\nu\, z^{-\nu}}{1 - \sum\limits_{\mu=1}^{M} a_\mu\, z^{-\mu}} = b_0 \frac{\prod\limits_{\nu=1}^{N} (1 - z_{0,\nu}\, z^{-1})}{\prod\limits_{\mu=1}^{M} (1 - z_{\infty,\mu}\, z^{-1})}$$

where we have set $\widetilde{a}_0 = 1$ and $\widetilde{a}_k = -a_k$. We distinguish the following cases:

- General recursive (zeros and poles):

$$H(z) = \frac{B(z)}{1 - A(z)} = \frac{\sum\limits_{\nu=0}^{N} b_\nu\, z^{-\nu}}{1 - \sum\limits_{\mu=1}^{M} a_\mu\, z^{-\mu}} = \frac{z^M}{z^N} \frac{\sum\limits_{\nu=0}^{N} b_\nu\, z^{N-\nu}}{z^M - \sum\limits_{\mu=1}^{M-1} a_\mu\, z^{M-\mu}} \; .$$

- Non-recursive (all zeros):

$$H(z) = B(z) = \sum_{\nu=0}^{N} b_\nu\, z^{-\nu} = \frac{\sum\limits_{\nu=0}^{N} b_\nu\, z^{N-\nu}}{z^N} \; .$$

- Purely recursive (all poles):

$$H(z) = \frac{b_0}{1 - A(z)} = \frac{b_0}{1 - \sum\limits_{\mu=1}^{M} a_\mu\, z^{-\mu}} = \frac{b_0\, z^M}{z^M - \sum\limits_{\mu=0}^{M-1} a_\mu\, z^{M-\mu}} \; .$$

Note that the frequently used terminology in the brackets does not account for zeros and poles at $z = 0$. Furthermore:

- $H(z)$ does not uniquely specify the impulse response of the system. We must also specify an ROC.
- If the system is stable, the impulse response must be absolutely summable. Therefore, the ROC must include the unit circle.
- If the system is causal, the ROC must extend from the outermost pole outward.
- For a stable and causal system, we must require that all poles lie within the unit circle.
- For a stable and invertible system, we must require that all poles and all zeros are within the unit circle. Systems which meet this condition are also called *minimum-phase* systems.

3.5 The Discrete Fourier Transform

The FTDS is not directly suited for numerical computations since it requires in principle an input signal of infinite support and delivers a continuous spectral function $X(e^{j\Omega})$.

By contrast, the *discrete Fourier transform* (DFT) computes a finite set of discrete Fourier coefficients X_μ from a finite number of signal samples. The DFT coefficients represent the spectrum of the input signal at equally spaced points on the frequency axis. However, when the support of the signal is larger than the transformation length, the DFT coefficients are not identical to the FT of the complete signal at these frequencies. Nevertheless, the DFT is an indispensable tool for numerical harmonic analysis of signals of finite or infinite support.

The coefficients X_μ of the DFT are computed via the finite sum

$$X_\mu = \sum_{k=0}^{M-1} x(k)\, e^{-j\frac{2\pi\mu k}{M}}, \quad \mu = 0, \ldots, M-1. \tag{3.25}$$

The M signal samples are recovered by the inverse relationship

$$x(k) = \frac{1}{M} \sum_{\mu=0}^{M-1} X_\mu\, e^{j\frac{2\pi\mu k}{M}}, \quad k = 0, \ldots, M-1. \tag{3.26}$$

The DFT coefficients X_μ are periodical, i.e., $X_{\mu+\lambda M} = X_\mu$ for $\lambda \in \mathbb{Z}$. The same is true for the signal $x(k)$ reconstructed from M complex DFT coefficients, since, using (3.26), we have $x(k + \lambda M) = x(k)$. Therefore, we may extend the M samples of $x(k)$, $k = 0, \ldots, M-1$, into an M-periodic signal $x_{\widetilde{M}}(k)$

where $x_{\widetilde{M}}(k) = x([k]_{\mathrm{mod}\, M})$. We then have

$$x_{\widetilde{M}}(k) = \frac{1}{M} \sum_{\mu=0}^{M-1} X_\mu e^{j\frac{2\pi\mu}{M}k} \tag{3.27}$$

for any k.

The coefficients of the DFT are spaced by $\Delta\Omega = \frac{2\pi}{M}$ on the normalized frequency axis Ω. When the signal samples $x(k)$ are generated by means of sampling a continuous signal $x_a(t)$ with sampling period $T = \frac{1}{f_s}$, the coefficients of the DFT are spaced by $\Delta f = \frac{f_s}{M}$, i.e., $f_\mu = \frac{\mu f_s}{M}$.

When the signal $x(k)$ is real valued, the DFT coefficients have a number of symmetry properties which can be exploited to reduce the complexity of frequency domain algorithms. These and other properties of the DFT are summarized in Table 3.5. For a real input sequence the symmetry properties of the DFT are

Table 3.5: Properties of the DFT

Property	Time domain	Frequency domain
Definition	$x(k) = \frac{1}{M} \sum_{\mu=0}^{M-1} X_\mu \, e^{j2\pi\frac{\mu k}{M}}$ $k = 0 \ldots M-1$	$X_\mu = \sum_{k=0}^{M-1} x(k) \, e^{-j2\pi\frac{\mu k}{M}}$ $\mu = 0 \ldots M-1$
Linearity	$a\,x(k) + b\,y(k)$	$a\,X_\mu + b\,Y_\mu$
Symmetry	$x(k)$ is real valued	$X_\mu = X^*_{[-\mu]_{\mathrm{mod}\, M}}$
Convolution	$\sum_{\ell=0}^{M-1} x(\ell)\, y([k-\ell]_{\mathrm{mod}\, M})$	$X_\mu Y_\mu$
Multiplication	$x(k)\, y(k)$	$\frac{1}{M} \sum_{\ell=0}^{M-1} X_\ell \, Y_{[\mu-\ell]_{\mathrm{mod}\, M}}$
Delay	$x([k+k_0]_{\mathrm{mod}\, M})$	$e^{+j2\pi\frac{\mu k_0}{M}} X_\mu$
Modulation	$x(k)\, e^{-j2\pi\frac{k\mu_0}{M}}$	$X_{[\mu+\mu_0]_{\mathrm{mod}\, M}}$
Parseval's theorem	$\sum_{k=0}^{M-1} x(k) y^*(k) = \frac{1}{M} \sum_{\mu=0}^{M-1} X_\mu Y^*_\mu$	

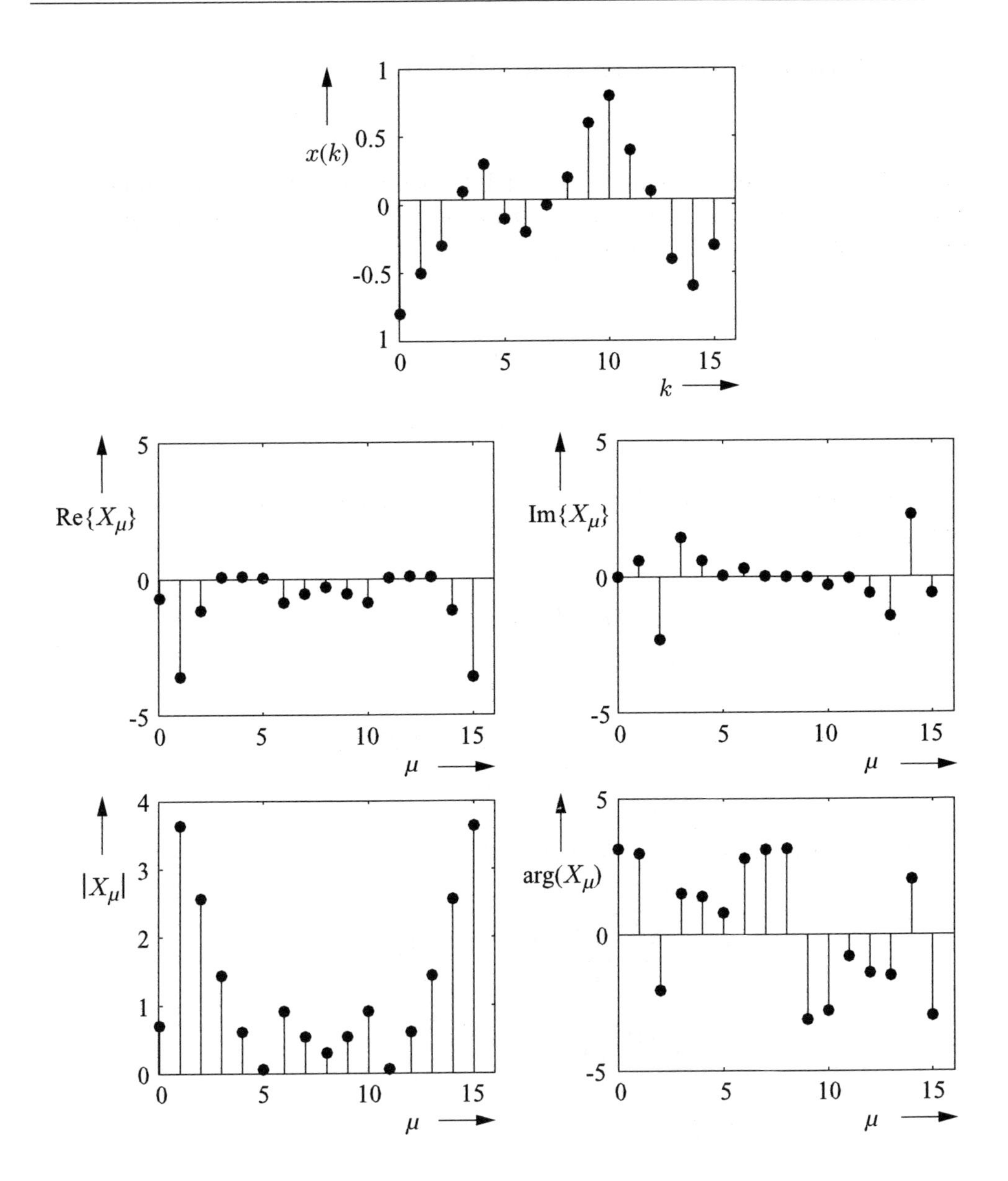

Figure 3.4: Real part, imaginary part, magnitude, and phase of DFT coefficients of a real-valued sequence $x(k)$

illustrated in Fig. 3.4. While the real part and the magnitude are even symmetric, the imaginary part and the phase are odd symmetric.

In what follows, we will discuss some of the properties in more detail.

3.5.1 Linear and Cyclic Convolution

The multiplication of two sequences of DFT coefficients, $X_\mu Y_\mu$, corresponds to a *cyclic convolution* of the corresponding time domain sequences $x(k)$ and $y(k)$. The cyclic convolution may be written as

$$\begin{aligned}\sum_{\ell=0}^{M-1} x(\ell)\, y([k-\ell]_{\mathrm{mod}\, M}) &= \sum_{\ell=0}^{M-1} x(\ell)\, y_{\widetilde{M}}(k-\ell)\\ &= x(k)\, R_M(k) * y_{\widetilde{M}}(k)\end{aligned}$$

where $*$ denotes the aperiodic (linear) convolution and $R_M(k)$ denotes a rectangular window

$$R_M(k) = \begin{cases} 1 & 0 \le k \le M-1 \\ 0 & \text{otherwise}. \end{cases} \tag{3.28}$$

To see this, we write the inverse DFT (IDFT) of $X_\mu Y_\mu$ as

$$\begin{aligned}\mathrm{IDFT}\,\{X_\mu Y_\mu\} &= \frac{1}{M}\sum_{\mu=0}^{M-1} X_\mu Y_\mu\, e^{j\frac{2\pi k}{M}\mu}\\ &= \frac{1}{M}\sum_{\mu=0}^{M-1}\sum_{\ell=0}^{M-1} x(\ell)\, e^{-j\frac{2\pi \ell}{M}\mu}\, Y_\mu\, e^{j\frac{2\pi k}{M}\mu}\\ &= \sum_{\ell=0}^{M-1} x(\ell)\frac{1}{M}\sum_{\mu=0}^{M-1} Y_\mu\, e^{j\frac{2\pi(k-\ell)}{M}\mu}\\ &= \sum_{\ell=0}^{M-1} x(\ell)\, y_{\widetilde{M}}(k-\ell)\,.\end{aligned}$$

The cyclic convolution is therefore equivalent to an aperiodic convolution of M samples of one sequence with the M-periodic extension of the other sequence.

The cyclic convolution of two sequences is illustrated in Fig. 3.5 and Fig. 3.6, where the result in Fig. 3.6 corresponds to a linear convolution. To obtain the linear convolution of two sequences of lengths N and L, the DFT length must be larger than or equal to $N + L - 1$. In the example this is achieved for $M \geq 15$. Significant cyclic effects are observed for $M < 15$ as seen in Fig. 3.5.

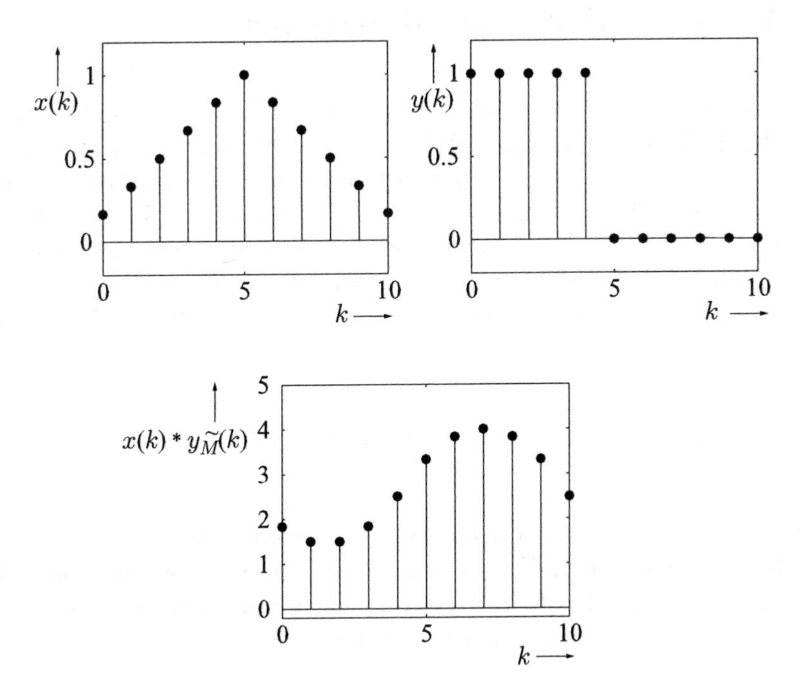

Figure 3.5: Cyclic convolution of two sequences $x(k)$ and $y(k)$ with $M = 11$

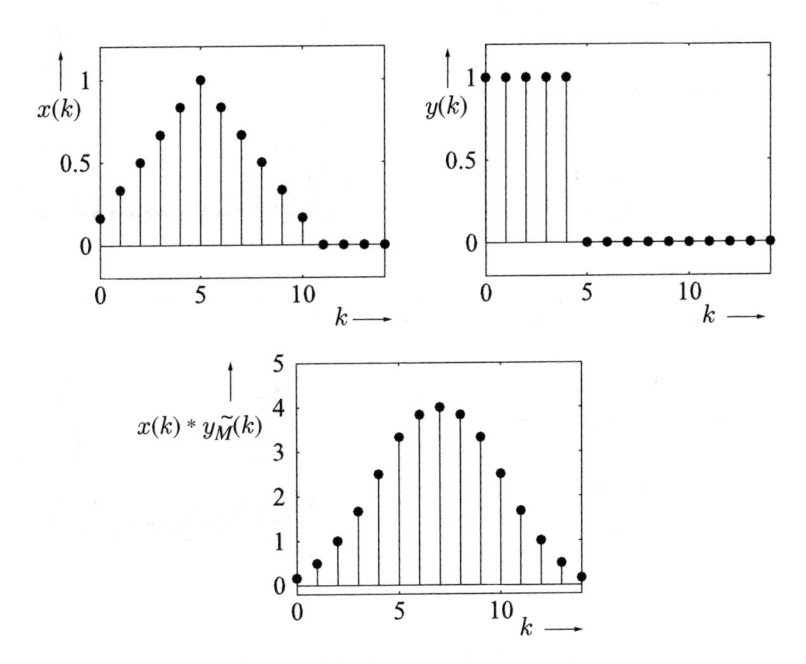

Figure 3.6: Cyclic convolution of two sequences $x(k)$ and $y(k)$ with $M = 15$. The DFT length is chosen such that a linear convolution is obtained.

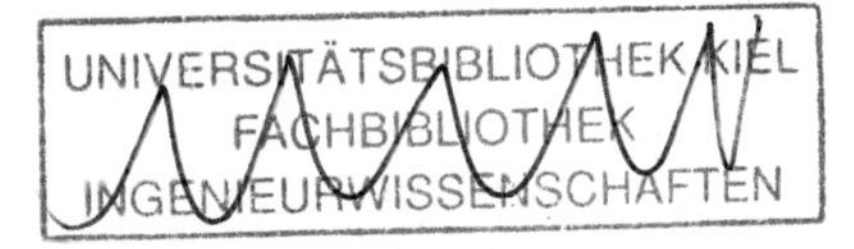

3.5.2 The DFT of Windowed Sequences

Quite frequently, the DFT is applied to signals which extend beyond the length M of the DFT. In this case, the DFT uses only M samples of the signal. The application of a window of length M to the time domain signal corresponds to a convolution of the FTDS of the complete signal with the FTDS of the window function in the spectral domain. After applying a window $w(k)$ to sequence $x(k)$, the DFT coefficients X_μ are equal to the spectrum of the windowed sequence $w(k)x(k)$ at the discrete frequencies $\Omega_\mu = \frac{2\pi\mu}{M}$. For $\mu = 0, \ldots, M-1$, we may write this as

$$X_\mu = \left[X(e^{j\Omega}) * W(e^{j\Omega})\right]_{\Omega=\Omega_\mu} = \sum_{k=0}^{M-1} w(k)\, x(k)\, e^{-j\frac{2\pi\mu k}{M}} . \tag{3.29}$$

We find that the DFT spectrum is a sampled version of the spectrum $X(e^{j\Omega}) * W(e^{j\Omega})$ where $W(e^{j\Omega})$ is the FTDS of the window function $w(k)$. The spread of the spectrum due to this convolution in the frequency domain is also known as *spectral leakage* and illustrated in Fig. 3.7.

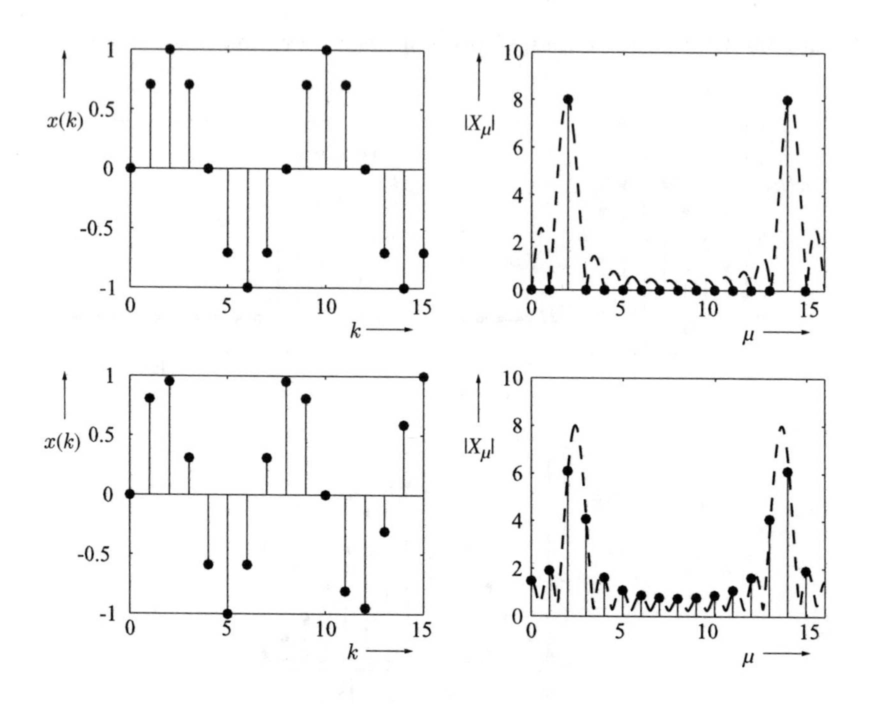

Figure 3.7: DFT and FTDS (dashed) of a finite segment of a sinusoid
Top: Integer multiple of period is equal to the DFT length M
Bottom: Integer multiple of period is not equal to the DFT length M

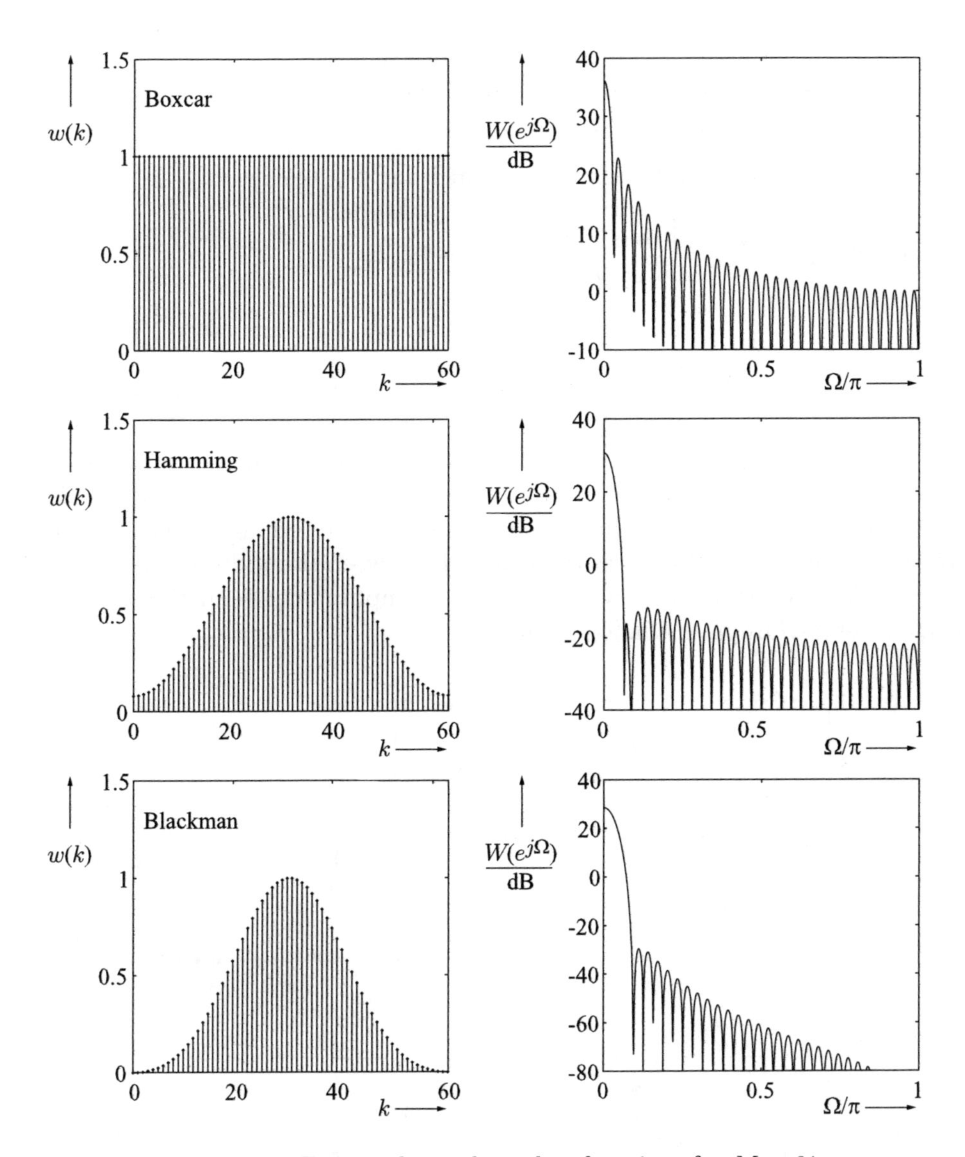

Figure 3.8: Frequently used window functions for $M = 61$

Spectral leakage may be reduced

- by increasing the DFT length M or
- by using a tapered window $w(k)$.

Compared to the rectangular window, tapered windows possess a wider main lobe in their frequency response (Fig. 3.8) and thus lead to reduced frequency resolution. Some of the frequently used window functions may be written as

$$w(k) = a - (1-a)\cos\left(k\frac{2\pi}{M-1}\right), \quad k = 0 \ldots M-1, \tag{3.30}$$

with $a = 1$ for the rectangular (boxcar) window,
$a = 0.54$ for the Hamming window, and
$a = 0.5$ for the Hann window.

The frequency response of these window functions is given by

$$W(e^{j\Omega}) = e^{-j\frac{M-1}{2}\Omega}\left[a\frac{\sin\left(\frac{M}{2}\Omega\right)}{\sin\left(\frac{1}{2}\Omega\right)} - \frac{1-a}{2}\left(\frac{\sin\left(\frac{M}{2}(\Omega - \frac{2\pi}{M-1})\right)}{\sin\left(\frac{1}{2}(\Omega - \frac{2\pi}{M-1})\right)} + \frac{\sin\left(\frac{M}{2}(\Omega + \frac{2\pi}{M-1})\right)}{\sin\left(\frac{1}{2}(\Omega + \frac{2\pi}{M-1})\right)}\right)\right].$$

Other well-known windows are the Blackman, the Tukey, and the Kaiser window, e.g., [Oppenheim et al. 1999]. The Blackman window is quite similar to the Hann and Hamming windows, but it has one additional cosine term to further reduce the ripple ratio. Figure 3.8 illustrates the trade-off between the width of the main lobe and the side-lobe attenuation while Fig. 3.9 exemplifies the reduction of spectral leakage when a Hamming window is used.

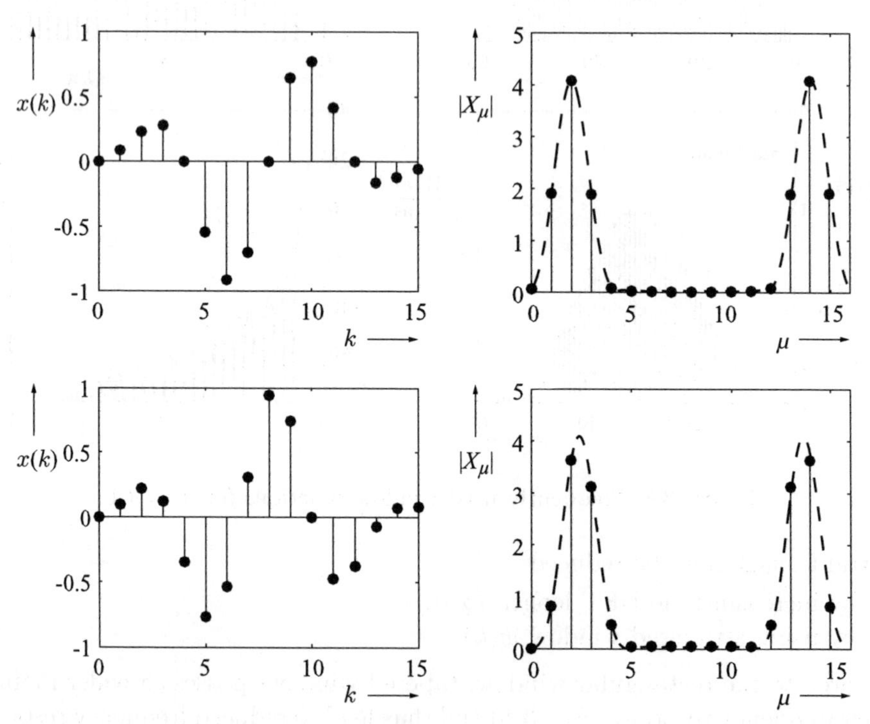

Figure 3.9: DFT and FTDS (dashed) of Hamming windowed sinusoids
Top: Integer multiple of period is equal to DFT length
Bottom: Integer multiple of period is not equal to DFT length
In both cases, spectral leakage is significantly reduced.

3.5.3 Spectral Resolution and Zero Padding

The DFT length M is identical to the number of discrete bins in the frequency domain. These bins are spaced on the normalized frequency axis according to

$$\Delta\Omega = \frac{2\pi}{M}\,. \tag{3.31}$$

However, the spacing between frequency bins must not be confused with the spectral resolution of the DFT. Spectral resolution may be defined via the capability of the DFT to separate closely spaced sinusoids as illustrated in Fig. 3.10. In general, the spectral resolution depends on the number of samples of the transformed sequence $x(k)$ and the window function $w(k)$. Spectral resolution of the DFT is best for the boxcar window. Tapered windows reduce the spectral resolution but also the spectral leakage. The spectral resolution of the DFT may be increased by increasing the length M of the data window, i.e., using more signal samples to compute the DFT coefficients.

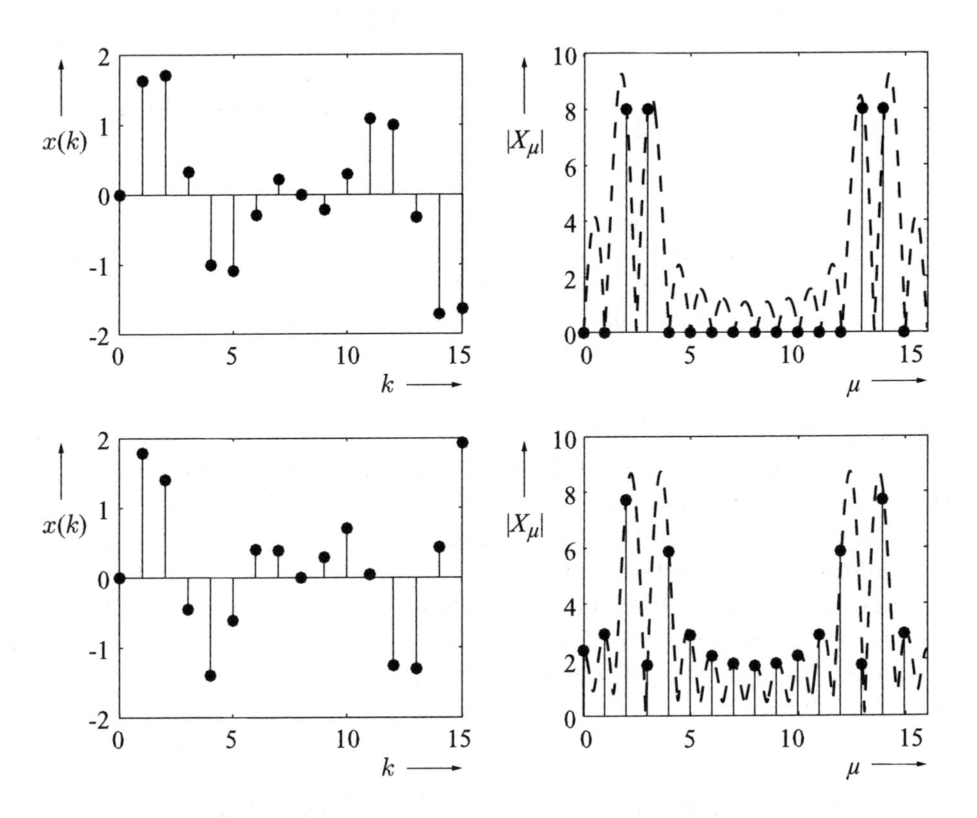

Figure 3.10: Resolution of two closely spaced sinusoids at frequencies Ω_1 and Ω_2
Top: $\Omega_1 = 2\pi/8$, $\Omega_2 = 5\pi/16$ Bottom: $\Omega_1 = 2.2\pi/8$, $\Omega_2 = 2.2\pi/8 + 2\pi/16$

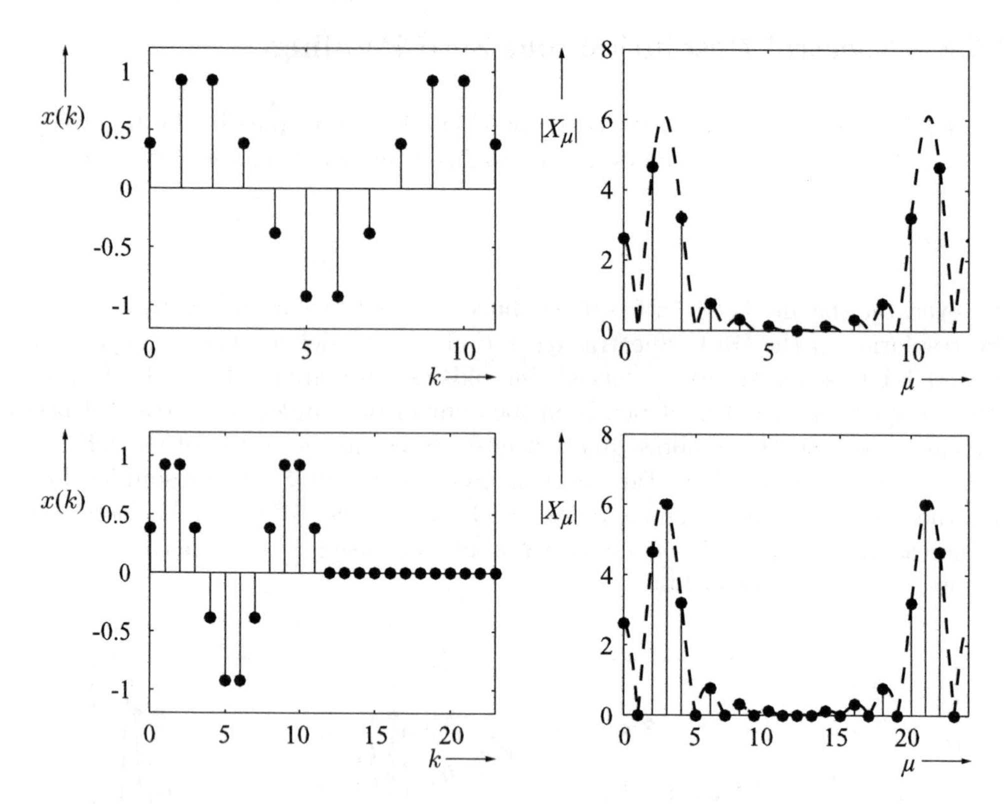

Figure 3.11: Zero padding a sequence $x(k)$
Top: Sequence (left) and DFT coefficients (right) without zero padding
Bottom: Sequence (left) and DFT coefficients (right) after zero padding
The dashed line indicates the FTDS spectrum of the sequence, which is the same for both cases.

The technique known as *zero padding* increases the number of frequency bins in the DFT domain. It docs not increase the spectral resolution in the sense that closely spaced sinusoids can be better separated. Obviously, appending zeros to a segment of a signal, where the signal is in general much longer than the DFT length, does not add information about this signal. This is illustrated in Fig. 3.11. However, for finite length signals, the DFT and zero padding allow us to compute the FTDS without error for any number of frequency bins. This is useful, for example, when the frequency response of a *finite impulse response* (FIR) filter must be computed with high resolution.

3.5.4 Fast Computation of the DFT: The FFT

Fast Fourier transform (FFT) algorithms were used by C. F. Gauß in the 19th century, temporarily forgotten, and rediscovered [Cooley, Tukey 1965] when digital computers emerged [Cooley 1992]. FFT algorithms provide for an exact computation of the DFT, however, at a significantly reduced complexity.

In fact, the FFT is not a single algorithm but a family of different algorithms which rely on the principle of "divide and conquer" and symmetry considerations. The basic idea of the FFT algorithm is to divide the overall DFT computation into smaller subtasks, compute these subtasks, and then recombine the results. FFT algorithms can be classified with respect to

- their design procedure (*decimation-in-time, decimation-in-frequency*)
- their radix (2, 3, 4, ...)
- memory requirements (in-place vs. not in-place)
- addressing schemes (bit-reversal vs. linear).

It should be pointed out that FFT algorithms do not exist only for DFT lengths $M = 2^p$ (powers of two) although these constitute the most widespread form of the FFT. Efficient algorithms are available for other DFT lengths $M \neq 2^p$ as well, e.g., [Oppenheim, Schafer 1975]. In what follows we demonstrate the basic idea for a radix-2 decimation-in-time algorithm [Cochran et al. 1967].

3.5.5 Radix-2 Decimation-in-Time FFT

The *decimation-in-time* algorithm splits the sequence of spectral coefficients

$$X_\mu = \sum_{k=0}^{M-1} x(k)\, e^{-j\frac{2\pi\mu k}{M}}, \quad \mu = 0, \dots, M-1,$$

where $M = 2^p$, into an even-indexed and odd-indexed subsequence

$$X_\mu = \sum_{k=0}^{M/2-1} x(2k)\, e^{-j\frac{4\pi\mu k}{M}} + \sum_{k=0}^{M/2-1} x(2k+1)\, e^{-j\frac{2\pi\mu(2k+1)}{M}}$$

which may be rewritten as two DFTs of length $M/2$

$$X_\mu = \sum_{k=0}^{M/2-1} x(2k)\, e^{-j\frac{2\pi\mu k}{M/2}} + e^{-j\frac{2\pi\mu}{M}} \sum_{k=0}^{M/2-1} x(2k+1)\, e^{-j\frac{2\pi\mu k}{M/2}}.$$

While the computational effort of the DFT is M^2 complex multiplications and additions, the effort is now reduced to $2\left(\frac{M}{2}\right)^2 + M = \frac{M^2}{2} + M$ multiplications and additions.

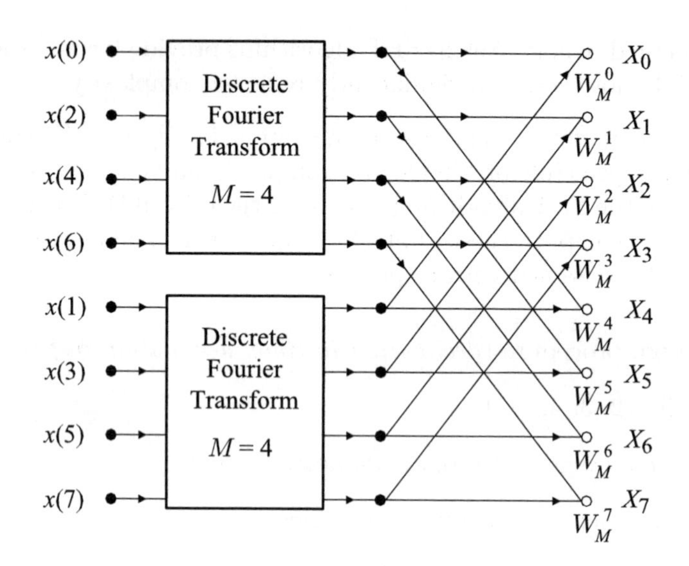

Figure 3.12: Signal-flow graph after the first decimation step ($M = 8$)

To prepare for the next decomposition we set $M_2 = M/2$, $x_{2e}(k) = x(2k)$, $x_{2o}(k) = x(2k+1)$, and define a complex phasor $W_M = e^{-j\frac{2\pi}{M}}$. We then obtain

$$X_\mu = \sum_{k=0}^{M_2-1} x_{2e}(k)\, e^{-j\frac{2\pi\mu k}{M_2}} + W_M^\mu \sum_{k=0}^{M_2-1} x_{2o}(k)\, e^{-j\frac{2\pi\mu k}{M_2}}\,, \tag{3.32}$$

a signal-flow graph of which is shown in Fig. 3.12. Two DFTs of length $M_2 = M/2$ are combined by means of the "twiddle factors" W_M^μ, $\mu = 0 \ldots M-1$. Since these DFTs of length M_2 are M_2-periodic, we compute them for $\mu = 0 \ldots M_2 - 1$ only and reuse the results for $\mu = M_2 \ldots M-1$.

In the next decomposition step we split each of the two sequences $x_{2e}(k)$ and $x_{2o}(k)$ into two subsequences of length $M/4$. After this decomposition, which is shown in Fig. 3.13, the computational effort is reduced to

$$2\left(2\left(\frac{M}{4}\right)^2 + \frac{M}{2}\right) + M = \frac{M^2}{4} + 2M$$

complex multiplications and additions. After $p-1$ decimation steps we have $2^{(p-1)}$ sequences of length 2. Hence, the computational effort is

$$\frac{M^2}{2^p} + (p-1)M = M + (p-1)M = pM = p\,2^p = M\log_2(M) \tag{3.33}$$

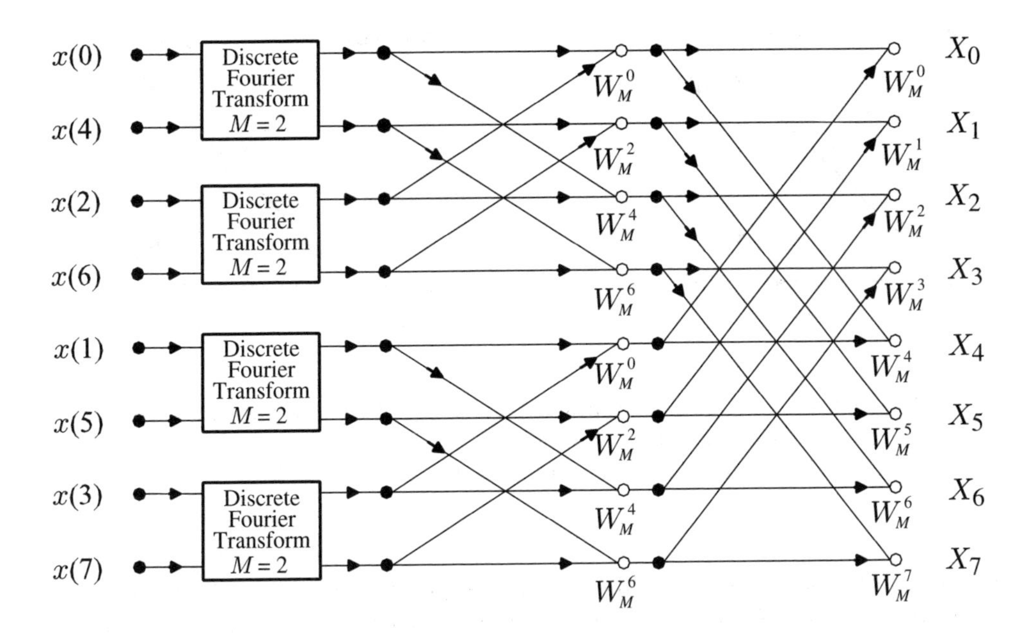

Figure 3.13: Signal-flow graph after the second decimation step ($M = 8$)

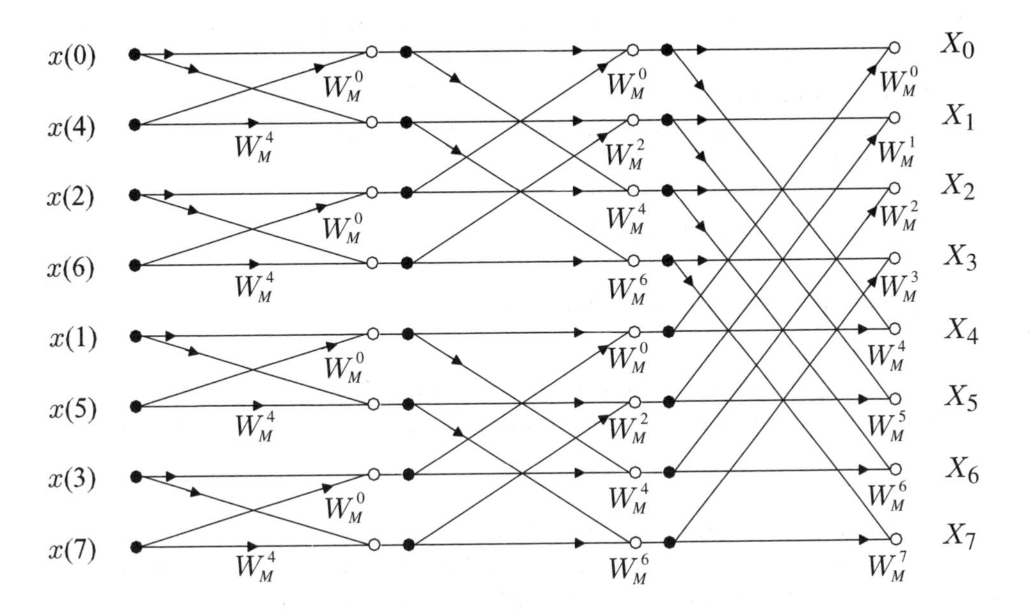

Figure 3.14: Signal-flow graph after the third decimation step ($M = 8$)

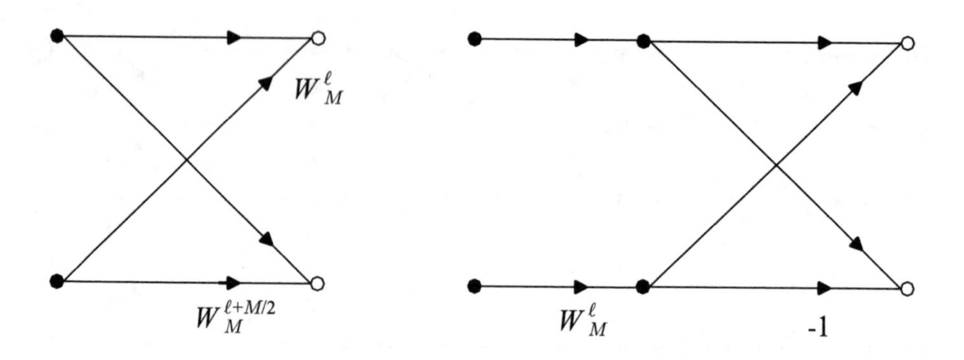

Figure 3.15: The basic computational element of the radix-2 FFT:
Left: The butterfly
Right: An efficient implementation

complex multiplications and additions. The signal-flow graph of the final result is shown in Fig. 3.14. For example, when the DFT length is $M = 1024$, we obtain a complexity reduction of $\frac{M \log_2(M)}{M^2} \approx 0.01$.

The regular structure which we find at all stages of the algorithm is called a "butterfly". The left plot and the right plot in Fig. 3.15 show the butterfly as it was used in the preceding development and in a form which requires only one complex multiplication, respectively. The latter makes use of the relation $W_M^{\ell+M/2} = -W_M^\ell$. Without further optimizations, the computational complexity of the FFT algorithm is therefore

$$\begin{aligned} &M \log_2(M) \quad \text{complex additions and} \\ &\frac{M}{2} \log_2(M) \quad \text{complex multiplications,} \end{aligned} \tag{3.34}$$

where one complex addition requires two real-valued additions, and one complex multiplication needs two real-valued additions and four real-valued multiplications.

As a result of the repeated decimations, the input sequence $x(k)$ is not in its natural order. However, the elements of the scrambled input sequence can be addressed using the bit-reversal addressing mode which is supported on most digital signal processors (DSPs). Bit-reversal addressing reads the address bits of each input element in reverse order. For example, the address of the second and the seventh element of the sequence 0 0 1 and 1 1 0, respectively, are read in reverse order as 1 0 0 and 0 1 1. The latter addresses correspond to the actual position of these elements in the scrambled sequence.

3.6 Fast Convolution

The use of the convolution theorem of the DFT presents an attractive alternative to the direct computation of the convolution sum. In conjunction with the FFT this method is called *fast convolution.*

The linear convolution of two complex-valued sequences of length N requires two FFTs and one inverse FFT (IFFT) of length $M = 2N - 1$ and M complex multiplications. If the computational effort for one FFT or IFFT is $KM \log_2(M)$, where K depends on the algorithm and the computer hardware, we then have

$$3KM \log_2(M) + M$$

complex operations for a fast convolution. This compares favourably to N^2 operations for the direct computation of the convolution sum when N is large. For example, for $N = 1024$ and $K = 1$ we need 69594 complex operations for the fast convolution compared to 1048576 complex operations for the direct computation.

3.6.1 Fast Convolution of Long Sequences

Another, even more interesting case is the convolution of a very long causal sequence $x(k)$, e.g., a speech signal, with a relatively short impulse response $h(k)$ of an FIR filter with N non-zero taps. This convolution may be performed in segments using a DFT of length $M > N$. In this case, the impulse response of the filter is transformed only once, leading to additional savings.

There are two basic methods (and numerous variations thereof) for the fast convolution of long sequences which are known as the *overlap-add* and *overlap-save* techniques.

The overlap-add method uses non-overlapping segments of the input sequence and adds partial results to reconstruct the output sequence. In contrast, the overlap-save method uses overlapping segements of the input sequence and reconstructs the output sequence without overlapping the results of the partial convolutions. In what follows we will briefly illustrate these two techniques.

3.6.2 Fast Convolution by Overlap-Add

The overlap-add technique segments the input signal $x(k)$ into non-overlapping shorter segments $x_1(k)$, $x_2(k)$, ... of length $N_x < M$, with

$$x_\ell(k) = \begin{cases} x\big(k + (\ell - 1)N_x\big) & 0 \le k < N_x \\ 0 & \text{elsewhere}. \end{cases} \tag{3.35}$$

Each of these segments is convolved with the impulse response $h(k)$. The results of these convolutions, $y_1(k)$, $y_2(k)$, ..., are then overlap-added to yield the final result

$$\begin{aligned} y(k) &= \sum_{\nu=0}^{N-1} h(\nu)\, x(k-\nu) \\ &= \sum_{\nu=0}^{N-1} h(\nu) \sum_{\ell} x_\ell\big(k-(\ell-1)N_x-\nu\big) \\ &= \sum_{\ell} \underbrace{\sum_{\nu=0}^{N-1} h(\nu)\, x_\ell\big(k-(\ell-1)N_x-\nu\big)}_{=y_\ell\left(k-(\ell-1)N_x\right)} . \end{aligned}$$

With appropriate zero padding the convolutions of the shorter segments with the impulse response $h(k)$ may be performed in the frequency domain. The overlap-add method requires that the convolution is linear. Therefore, we must choose $N + N_x - 1 \leq M$. This is illustrated in Fig. 3.16, where the top part of the graph shows the linear convolution of the full sequence while the lower parts depicts the partial convolutions of zero-padded shorter segments.

3.6.3 Fast Convolution by Overlap-Save

For the overlap-save procedure we use segments $x_1(k)$, $x_2(k)$, ... of length $N_x = M > N$ which contain $N-1$ ("saved") samples of the previous segment and $M - N + 1$ new samples,

$$x_\ell(k) = \begin{cases} x\big(k+(\ell-1)M-\ell(N-1)\big) & 0 \leq k < M \\ 0 & \text{elsewhere} . \end{cases} \tag{3.36}$$

After padding the impulse response with $M - N$ zeros the cyclic convolution of these segments with impulse response $h(k)$ yields $N-1$ invalid samples and $M-N+1$ valid samples. The latter are concatenated with the valid samples of the previously computed segment. In this way, all of the output signal is constructed. The overlap-save procedure is illustrated in Fig. 3.17. The invalid samples of the output sequences $y_1(k)$, $y_2(k)$, ... are marked x and are discarded. Only the valid samples (marked •) are concatenated.

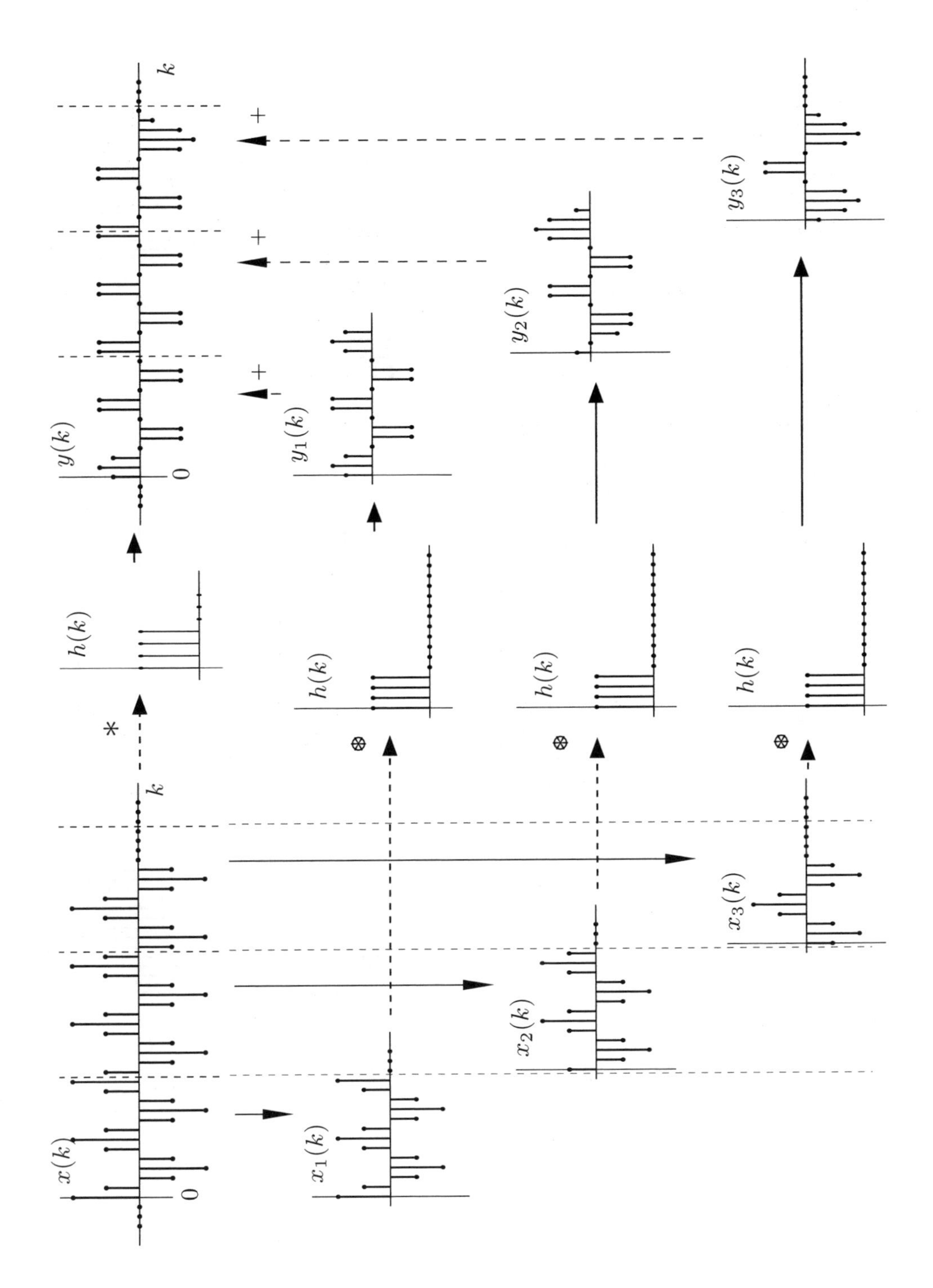

Figure 3.16: Illustration of the overlap-add technique. $N_x = 13$, $N = 4$, and $M = 16$. $\circledast$ denotes a fast convolution.

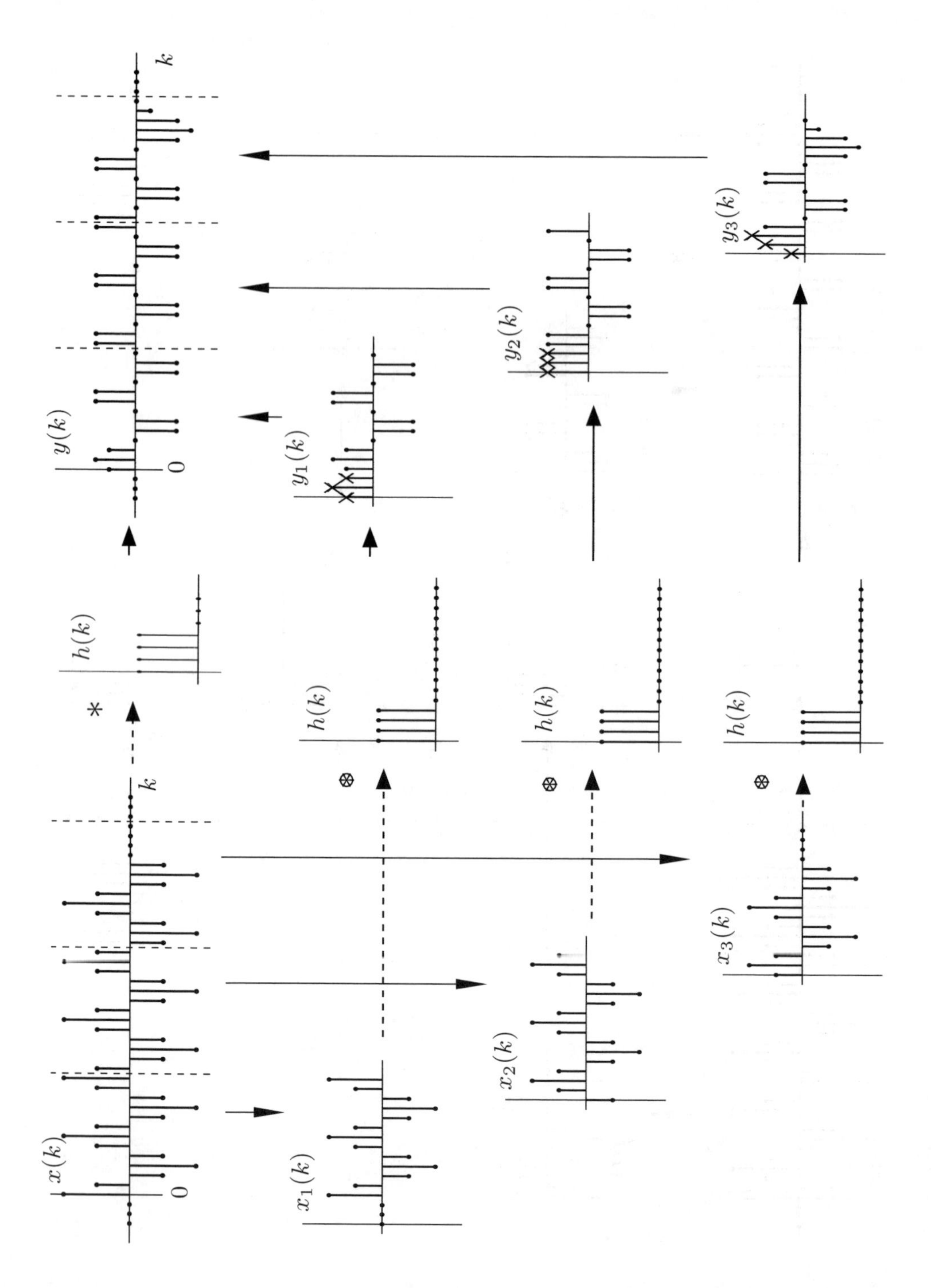

Figure 3.17: Illustration of the overlap-save technique. $N_x = M = 16$ and $N = 4$. Samples marked x are discarded. ⊛ denotes a fast convolution.

3.7 Cepstral Analysis

The convolution of two discrete time sequences corresponds to a multiplication of their spectra. It turns out that the logarithm of the spectrum and its inverse transform are also very useful as the logarithm of the product corresponds to a sum of the logarithms. The inverse transform of the logarithm of the spectrum is called the *cepstrum*. The processing related to the computation of the cepstrum is a special case of the more general concept of homomorphic processing [Oppenheim, Schafer 1975]. Cepstral analysis is also useful in the context of stochastic signals, especially when the power spectrum of these signals is obtained from an autoregressive model. It has been used in speech coding for the quantization of the spectral envelope [Hagen 1994], and in speech recognition for the computation of spectral features [Davis, Mermelstein 1980], [Jankowski et al. 1995]. In this section we briefly summarize some properties and applications of the cepstrum.

3.7.1 Complex Cepstrum

The *complex cepstrum* $x_{cc}(k)$ of a sequence $x(k)$ with $X(z) = \mathcal{Z}\{x(k)\}$ is defined as the sequence whose z-transform yields the logarithm of $X(z)$,

$$\sum_{k=-\infty}^{\infty} x_{cc}(k) z^{-k} = \ln(X(z)) = X_{cc}(z) \tag{3.37}$$

where we use the natural complex logarithm $\ln(X(z)) = \log_e(X(z))$. The logarithm of the z-transform may be written in terms of the magnitude and the phase of $X(z)$ as

$$\ln(X(z)) = \ln(|X(z)|\, e^{j\phi(z)}) = \ln(|X(z)|) + j\phi(z)\,. \tag{3.38}$$

We further assume that $X_{cc}(z) = \ln(X(z))$ is a valid z-transform and thus converges in some region of the complex z-plane. For practical reasons we also require that $x(k)$ and $x_{cc}(k)$ are real and stable sequences. Then, both $X(z)$ and $X_{cc}(z)$ converge on the unit circle and $\ln(|X(e^{j\Omega})|)$ and $\phi(\Omega)$ are even and odd functions of Ω, respectively. In this case we may obtain $x_{cc}(k)$ from an inverse FT

$$x_{cc}(k) = \frac{1}{2\pi} \int_{-\pi}^{\pi} \left[\ln(|X(e^{j\Omega})|) + j\phi(\Omega)\right] e^{j\Omega k}\, d\Omega\,. \tag{3.39}$$

Since singularities of $\ln(X(z))$ are found at the poles and the zeros of $X(z)$, a stable and causal sequence $x_{cc}(k)$ is only obtained when both poles and zeros are within the unit circle. As a consequence, $x(k)$ is a minimum-phase sequence if and only if its complex cepstrum is causal [Oppenheim et al. 1999].

The definition of the complex logarithm in (3.37) is not at all trivial as $X_{cc}(z)$ must be an analytic and hence continuous function in the ROC. Since the imaginary part of $X_{cc}(z)$ is identical to the phase of $X(z)$, we cannot substitute the principal value $-\pi < \arg\{X(z)\} \leq \pi$ for the phase in general. The complex logarithm must therefore be defined such that it is invertible and analytic in the ROC [Oppenheim, Schafer 1975].

It can be shown that the complex cepstrum satisfies

$$x(k) = \sum_{\ell=-\infty}^{\infty} \left(\frac{\ell}{k}\right) x_{cc}(\ell)\, x(k-\ell)\,, \quad k \neq 0\,. \tag{3.40}$$

When $x(k)$ is a minimum-phase sequence with $x(k) = 0$ for $k < 0$ and $X(z)$ is a rational function, we have $x_{cc}(k) = 0$ for $k < 0$. For a minimum-phase sequence, the complex cepstrum may be computed recursively

$$x_{cc}(k) = \begin{cases} 0 & k < 0 \\ \ln(x(0)) & k = 0 \\ \dfrac{x(k)}{x(0)} - \displaystyle\sum_{\ell=0}^{k-1} \frac{\ell}{k}\, x_{cc}(\ell)\, \frac{x(k-\ell)}{x(0)} & k > 0\,. \end{cases} \tag{3.41}$$

3.7.2 Real Cepstrum

Because of the difficulties in the definition and evaluation of the complex logarithm, the *real cepstrum*, or *cepstrum* for short, is preferred in many applications. Also, for minimum-phase signals the real cepstrum captures all information about the signal since the phase is determined by the magnitude spectrum [Oppenheim, Schafer 1975]. The real cepstrum is defined as

$$c_x(k) = \frac{1}{2\pi} \int_{-\pi}^{\pi} \ln(|X(e^{j\Omega})|)\, e^{j\Omega k}\, \mathrm{d}\Omega \tag{3.42}$$

where we assume that all singularities of $\ln(|X(z)|)$ are within the unit circle. This implies that $x(k)$ as well as $x_{cc}(k)$ are causal, minimum-phase sequences. It therefore suffices to evaluate the real part of $X_{cc}(z)$ which is $\ln(|X(z)|)$. Note that

$$c_x(0) = \frac{1}{2\pi} \int_{-\pi}^{\pi} \ln\left(|X(e^{j\Omega})|\right) \mathrm{d}\Omega\,, \tag{3.43}$$

which, for an all-pole signal model $X(e^{j\Omega}) = \sigma/(1 - A(e^{j\Omega}))$, evaluates to $c_x(0) = \ln(\sigma)$ [Markel, Gray 1976]. For the all-pole model we might transform the coefficients a_k into cepstral coefficients $c_x(k)$ and vice versa using the recursive relation

$$c_x(k) = a_k + \sum_{\ell=1}^{k-1} \frac{k-\ell}{k} c_x(k-\ell)\, a_\ell = a_k + \sum_{\ell=1}^{k-1} \frac{\ell}{k} c_x(\ell)\, a_{k-\ell}, \quad k \geq 1 \qquad (3.44)$$

with $c_x(0) = \ln(\sigma)$ and $c_x(k) = c_x(-k)$ for $k < 0$. This recursion can be used in both directions. By definition, the summations in (3.44) yield zero for $k = 1$.

Since $\ln(|X(e^{j\Omega})|)$ is a real and even function of Ω, $c_x(k)$ is also real and even and may be obtained as the even part of $x_{cc}(k)$,

$$c_x(k) = \frac{x_{cc}(k) + x_{cc}^*(-k)}{2}, \qquad (3.45)$$

since $\ln\left(\left|X(e^{j\Omega})\right|^2\right) = \ln\left(X(e^{j\Omega})\right) + \ln\left(X^*(e^{j\Omega})\right)$. Furthermore, $x_{cc}(k) = 0$ for $k < 0$ and

$$x_{cc}(k) = c_x(k)\,(2\,u(k) - \delta(k)) \qquad (3.46)$$

for $k \geq 0$ where $u(k)$ and $\delta(k)$ are the unit step and the unit impulse sequences, respectively.

3.7.3 Applications of the Cepstrum

3.7.3.1 Construction of Minimum-Phase Sequences

The above relations may be used to construct a minimum-phase sequence from a given non-minimum-phase sequence $x(k)$. Any rational z-transform $X(z)$ of a stable sequence $x(k)$ may be decomposed into

$$X(z) = X_{\min}(z) X_{\mathrm{AP}}(z)$$

where $X_{\min}(z)$ and $X_{\mathrm{AP}}(z)$ correspond to a minimum-phase and an all-pass sequence, respectively. In general, the ROC of $X(z)$ is an annular region in the z-plane which contains the unit circle. The corresponding complex cepstrum is a two-sided sequence. With

$$\ln(X(z)) = \ln(X_{\min}(z)) + \ln(X_{\mathrm{AP}}(z))$$

we obtain the minimum-phase sequence by extracting the causal part of the complex cepstrum. An example is shown in Fig. 3.18.

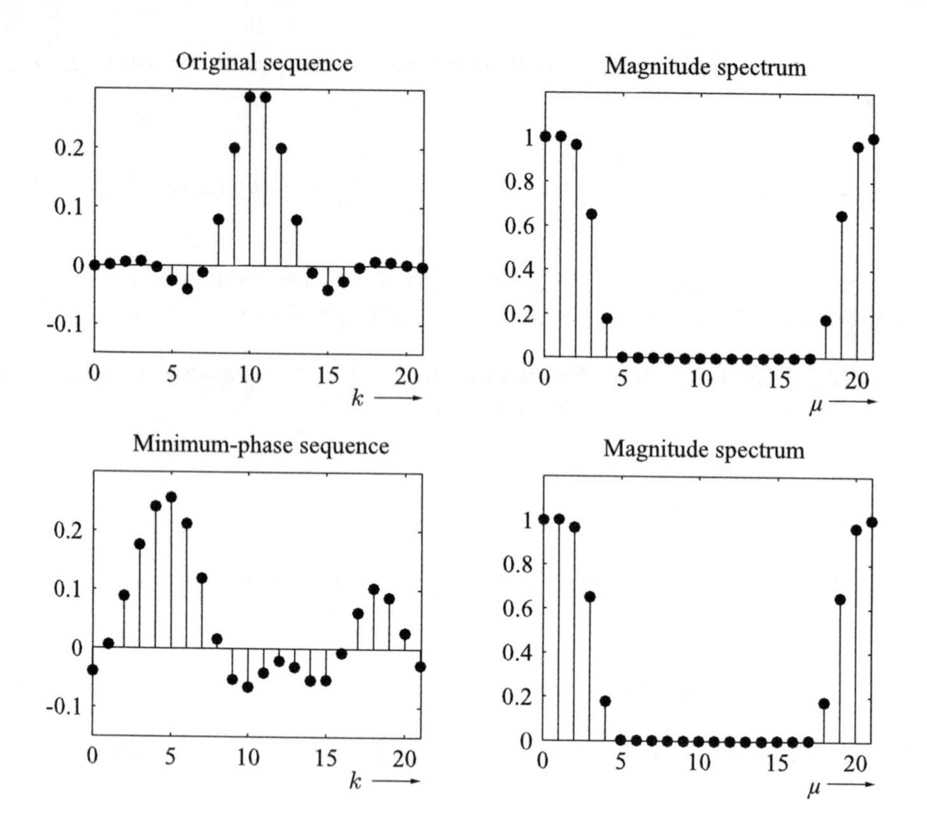

Figure 3.18: Linear-phase sequence (top) and corresponding minimum-phase sequence (bottom). Both sequences have the same magnitude spectrum.

3.7.3.2 Deconvolution by Cepstral Mean Subtraction

When a speech signal is recorded in a reverberant acoustic environment, it may be written as the convolution of the original, unreverberated speech signal $s(k)$ with the impulse response $h(k)$ of this environment,

$$x(k) = h(k) * s(k) \quad \Leftrightarrow \quad X(e^{j\Omega}) = H(e^{j\Omega})\, S(e^{j\Omega})\,.$$

A similar situation occurs when a speech signal is recorded via a telephone line. In this case, network echoes contribute to what is called *convolutive noise* in the speech recognition community. The use of the real cepstrum obviously leads to

$$\log\left(|X(e^{j\Omega})|\right) = \log\left(|H(e^{j\Omega})|\right) + \log\left(|S(e^{j\Omega})|\right) \Leftrightarrow c_x(k) = c_h(k) + c_s(k)$$

or, using the DFT, to the short-term relation

$$\log\left(|X_\mu(\lambda)|\right) \approx \log\left(|H_\mu(\lambda)|\right) + \log\left(|S_\mu(\lambda)|\right) \tag{3.47}$$

provided that circular convolutive effects are negligibly small. μ denotes the frequency bin and λ the frame index, respectively. The deconvolution might be now achieved by subtracting $\log(|H_\mu(\lambda)|)$ from $\log(|X_\mu(\lambda)|)$. When the impulse response of the reverberant system is constant over time, an estimate of $\log(|H_\mu(\lambda)|)$ is given by the expectation of $\log(|X_\mu(\lambda)|) - \log(|S_\mu(\lambda)|)$. In a practical realization of this method, we approximate the expectation of $\log(|X_\mu(\lambda)|)$ by a time average and compute $\mathrm{E}\{\log(|S_\mu(\lambda)|)\}$ from undisturbed speech data. We then obtain

$$\log\widehat{(|S_\mu(\lambda)|)} = \log(|X_\mu(\lambda)|) - \left(\frac{1}{N}\sum_{i=\lambda-N+1}^{\lambda} \log(|X_\mu(i)|) - \mathrm{E}\{\log(|S_\mu(\lambda)|)\}\right).$$

Note that the cepstral coefficients of zero mean Gaussian signals are not zero mean [Stockham et al. 1975], [Ephraim, Rahim 1999]. This technique and its variations [Stockham et al. 1975], [Acero, Huang 1995], [Rahim et al. 1996] are frequently used in speech recognition. They are less successful in combating acoustic echoes, as the acoustic echo path is typically of high order and not stationary.

3.7.3.3 Computation of the Spectral Distortion Measure

When we are given two magnitude squared spectra $|H(e^{j\Omega})|^2$ and $|\widehat{H}(e^{j\Omega})|^2$, where the latter may be an approximation of the former, the *total log spectral distance* (SD) of the two spectra is defined as

$$\mathrm{SD} = \sqrt{\frac{1}{2\pi}\int_{-\pi}^{\pi}\left[20\log_{10}|H(e^{j\Omega})| - 20\log_{10}|\widehat{H}(e^{j\Omega})|\right]^2 d\Omega}\,. \qquad (3.48)$$

This distance (or distortion) measure is used, for instance, for the quantization of the spectral envelopes of speech signals [Hagen 1994]. When $H(e^{j\Omega})$ and $\widehat{H}(e^{j\Omega})$ represent minimum-phase signals with

$$\int_{-\pi}^{\pi} \ln|H(e^{j\Omega})| d\Omega = \int_{-\pi}^{\pi} \ln|\widehat{H}(e^{j\Omega})| d\Omega = 0\,, \qquad (3.49)$$

the distortion measure may be computed using cepstral coefficients [Markel, Gray 1976]:

$$\begin{aligned}\mathrm{SD} &= \sqrt{\frac{1}{2\pi}\int_{-\pi}^{\pi}\left[20\log_{10}|H(e^{j\Omega})| - 20\log_{10}|\widehat{H}(e^{j\Omega})|\right]^2 d\Omega} \\ &= 20\log_{10}(e)\sqrt{2}\sqrt{\sum_{k=1}^{\infty}\left(c_h(k) - \widehat{c}_h(k)\right)^2}\,. \end{aligned} \qquad (3.50)$$

To prove this relation we use $\log_{10}(x) = \log_e(x)\log_{10}(e)$ and note that

$$\mathrm{SD} = \sqrt{\frac{1}{2\pi}\int_{-\pi}^{\pi}\left[20\log_{10}|H(e^{j\Omega})| - 20\log_{10}|\widehat{H}(e^{j\Omega})|\right]^2 \mathrm{d}\Omega}$$

$$= 20\log_{10}(e)\sqrt{\frac{1}{2\pi}\int_{-\pi}^{\pi}\left[\sum_{k=-\infty}^{\infty} c_h(k)\,e^{-jk\Omega} - \sum_{k=-\infty}^{\infty} \widehat{c}_h(k)\,e^{-jk\Omega}\right]^2 \mathrm{d}\Omega}\,.$$

Parseval's theorem (see Table 3.2)

$$\frac{1}{2\pi}\int_{-\pi}^{\pi}\left(\sum_{k=-\infty}^{\infty}(c_h(k)-\widehat{c}_h(k))\,e^{-jk\Omega}\right)\left(\sum_{k=-\infty}^{\infty}(c_h(k)-\widehat{c}_h(k))\,e^{-jk\Omega}\right)^* \mathrm{d}\Omega$$

$$= \sum_{k=-\infty}^{\infty}(c_h(k)-\widehat{c}_h(k))\,(c_h(k)-\widehat{c}_h(k))^*$$

yields

$$\mathrm{SD} = 20\log_{10}(e)\sqrt{\frac{1}{2\pi}\int_{-\pi}^{\pi}\left|\sum_{k=-\infty}^{\infty}(c_h(k)-\widehat{c}_h(k))\,e^{-jk\Omega}\right|^2 \mathrm{d}\Omega}$$

$$= 20\log_{10}(e)\sqrt{\sum_{k=-\infty}^{\infty}(c_h(k)-\widehat{c}_h(k))^2}\,. \tag{3.51}$$

The relation $c_h(0) = 0$, $c_h(-k) = c_h(k)$, finally leads to

$$\mathrm{SD} = 20\log_{10}(e)\sqrt{2}\,\sqrt{\sum_{k=1}^{\infty}(c_h(k)-\widehat{c}_h(k))^2}\,. \tag{3.52}$$

This spectral distortion measure [Quackenbush et al. 1988] is widely used to assess the performance of vector quantizers in speech coding [Kleijn, Paliwal 1995] as well as the quality of speech enhancement algorithms, e.g., [Gustafsson et al. 2002], [Cohen 2004].

Bibliography

Acero, A.; Huang, X. (1995). Augmented Cepstral Normalization for Robust Speech Recognition, *IEEE International Workshop on Automatic Speech Recognition and Understanding*, pp. 147–148.

Churchill, R. V.; Brown, J. W. (1990). *Introduction to Complex Variables and Applications*, 5th edn, MacGraw-Hill, New York.

Cochran, W. T.; Cooley, J. W.; Favin, D. L.; Helms, H. D.; Kaenel, R. A.; Lang, W. W.; Maling, G. C.; Nelson, D. E.; Rader, C. M.; Welch, P. D. (1967). What is the Fast Fourier Transform?, *IEEE Transactions on Audio and Electroacoustics*, vol. AU-15, June, pp. 45–55.

Cohen, I. (2004). Speech Enhancement Using a Noncausal A Priori SNR Estimator, *Signal Processing Letters*, vol. 11, no. 9, pp. 725–728.

Cooley, J. (1992). How the FFT Gained Acceptance, *IEEE Signal Processing Magazine*, vol. 9, pp. 10–13.

Cooley, J.; Tukey, J. (1965). An Algorithm for the Machine Calculation of Complex Fourier Series, *Mathematics of Computation*, vol. 19, pp. 297.

Davis, S.; Mermelstein, P. (1980). Comparison of Parametric Representations for Monosyllabic Word Recognition in Continuously Spoken Sentences, *IEEE Transactions on Acoustics, Speech and Signal Processing*, vol. 28, no. 4, pp. 357–366.

Ephraim, Y.; Rahim, M. (1999). On Second Order Statistics and Linear Estimation of Cepstral Coefficients, *IEEE Transactions on Speech and Audio Processing*, vol. 7, no. 2, pp. 162–176.

Gustafsson, S.; Martin, R.; Jax, P.; Vary, P. (2002). A Psychoacoustic Approach to Combined Acoustic Echo Cancellation and Noise Reduction, *IEEE Transactions on Speech and Audio Processing*, vol. 10, no. 5, July, pp. 245–256.

Hagen, R. (1994). Spectral Quantization of Cepstral Coefficients, *Proceedings of the IEEE International Conference on Acoustics, Speech, and Signal Processing (ICASSP)*, Adelaide, pp. 509–512.

Jankowski, C.; Vo, H.-D.; Lippmann, R. (1995). A Comparison of Signal Processing Front Ends for Automatic Word Recognition, *IEEE Transactions on Speech and Audio Processing*, vol. 3, no. 4, pp. 286–293.

Kleijn, W. B.; Paliwal, K. K. (eds.) (1995). *Speech Coding and Synthesis*, Elsevier, Amsterdam.

Markel, J. D.; Gray, A. H. (1976). *Linear Prediction of Speech*, Springer Verlag, Berlin, Heidelberg, New York.

Oppenheim, A.; Schafer, R. (1975). *Digital Signal Processing*, Prentice Hall, Englewood Cliffs, New Jersey.

Oppenheim, A.; Schafer, R.; Buck, J. (1999). *Discrete-Time Signal Processing*, 2nd edn, Prentice Hall, Englewood Cliffs, New Jersey.

Oppenheim, A.; Willsky, A.; Nawab, H. (1996). *Signals and Systems*, Prentice Hall, Upper Saddle River, New Jersey.

Proakis, J. G.; Manolakis, D. G. (1992). *Digital Signal Processing: Principles, Algorithms and Applications*, 2nd edn, Macmillan, New York.

Quackenbush, S. R.; Barnwell III, T. P.; Clements, M. A. (1988). *Objective Measures of Speech Quality*, Prentice Hall, Englewood Cliffs, New Jersey.

Rahim, M.; Juang, B.-H.; Chou, W.; Buhrke, E. (1996). Signal Condition Techniques for Robust Speech Recognition, *Signal Processing Letters*, vol. 3, no. 4, pp. 107–109.

Stockham, T.; Cannon, T.; Ingebretsen, R. (1975). Blind Deconvolution Through Digital Signal Processing, *Proceedings of the IEEE*, vol. 63, no. 4, pp. 678–692.

4

Filter Banks for Spectral Analysis and Synthesis

4.1 Spectral Analysis Using Narrowband Filters

In contrast to the transform-based methods of the previous chapter we will now discuss spectral analysis using narrowband filters. Two equivalent measurement procedures are illustrated in Fig. 4.1. We are interested in the temporal evolution of a signal $x(k)$ at a certain frequency Ω_μ. The first approach (see Fig. 4.1-a) consists of three steps: narrow bandpass filtering, spectral shifting by Ω_μ, and sampling rate decimation by r.

The intermediate bandpass signal $v_\mu(k)$, which is centered at Ω_μ, is shifted to the baseband by complex demodulation, i.e., by multiplication with $e^{-j\Omega_\mu k}$. The resulting complex-valued baseband signal $\overline{x}_\mu(k)$ is called the *subband signal*. This signal gives information about the signal $x(k)$ at frequency Ω_μ and time instant k. The resolution in time and frequency is determined by the impulse response $h_\mu^{BP}(k)$ and the frequency response $H_\mu^{BP}(e^{j\Omega})$ of the bandpass, respectively.

The same result can be obtained by applying first modulation and then feeding the modulated signal to an equivalent lowpass filter with frequency response H as shown in Fig. 4.1-b. The intermediate output signals $\overline{x}_\mu(k)$ of both systems are

Digital Speech Transmission: Enhancement, Coding and Error Concealment
Peter Vary and Rainer Martin

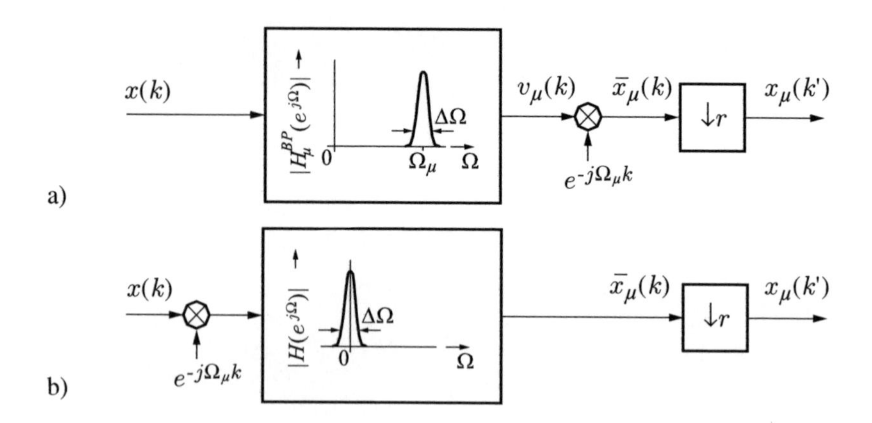

Figure 4.1: Spectral analysis by
a) Bandpass filtering and subsequent complex demodulation
b) Complex demodulation and subsequent lowpass filtering

identical if the frequency responses of the bandpass (BP) and the lowpass filters satisfy the following relation:

$$H_\mu^{BP}(e^{j\Omega}) = H\left(e^{j(\Omega-\Omega_\mu)}\right) . \tag{4.1-a}$$

According to the modulation theorem of the Fourier transform the corresponding time domain impulse responses are related by

$$h_\mu^{BP}(k) = h(k)\, e^{j\Omega_\mu k} . \tag{4.1-b}$$

The signal $\overline{x}_\mu(k)$ in Fig. 4.1-a can be formulated explicitly as

$$\overline{x}_\mu(k) = e^{-j\Omega_\mu k} \sum_{\kappa=-\infty}^{\infty} x(\kappa)\, h_\mu^{BP}(k-\kappa) \tag{4.2-a}$$

$$= e^{-j\Omega_\mu k} \sum_{\kappa=-\infty}^{\infty} x(\kappa)\, h(k-\kappa)\, e^{j\Omega_\mu (k-\kappa)} \tag{4.2-b}$$

$$= \sum_{\kappa=-\infty}^{\infty} x(\kappa)\, e^{-j\Omega_\mu \kappa}\, h(k-\kappa) \tag{4.2-c}$$

where (4.2-c) describes the lowpass approach of Fig. 4.1-b. If the complex samples $\overline{x}_\mu(k)$ are calculated at frequencies

$$\Omega_\mu = \Delta\Omega \cdot \mu, \quad \mu = 0, 1, 2, \ldots, M-1$$

and if the whole frequency range $0 \leq \Omega \leq 2\pi$ is covered, then each set of M samples $\overline{x}_\mu(k)$, $\mu = 0, 1, 2, \ldots, M-1$, at any time instant k is called the *short-term spectrum.*

The complex samples $\overline{x}_\mu(k)$ may be represented either by their real and imaginary parts or by (short-term) magnitude and phase

$$\overline{x}_\mu(k) = \overline{x}_{Re,\mu}(k) + j \cdot \overline{x}_{Im,\mu}(k) \tag{4.3-a}$$

$$= |\overline{x}_\mu(k)| \cdot e^{j\varphi_\mu(k)} . \tag{4.3-b}$$

If the filter is causal and stable, the impulse response $h(k)$ is zero for $k < 0$, and $|h(k)|$ will decay with time k. The time dependency of the short-term spectrum is due to the position k of the time-reversed impulse response $h(k - \kappa)$ which acts as a sliding window. As shown examplarily in Fig. 4.2, the impulse response $h(k - \kappa)$ weights the most recent signal samples $x(\kappa)$ up to the observation time instant k.

Because of the decay of $|h(k)|$, the older parts of the signal have less influence than the most recent samples. Furthermore, the sampling rate of the subband signals

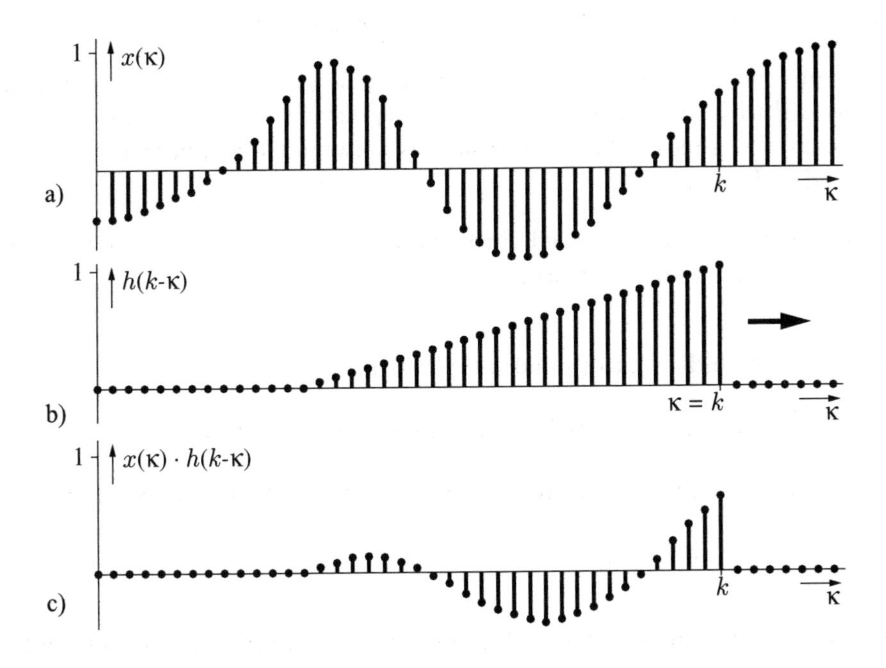

Figure 4.2: a) Signal $x(\kappa)$
b) Impulse response $h(k - \kappa)$
c) Weighted signal history $x(\kappa) \cdot h(k - \kappa)$

$\overline{x}_\mu(k)$ can be decimated due to the narrow bandwidth $\Delta\Omega$ of the frequency selective filter H. By decimating the sampling rate by r, we obtain the output sequence

$$x_\mu(k') = \overline{x}_\mu(k = k' \cdot r) . \tag{4.4}$$

The bandwidth of the signals $\overline{x}_\mu(k)$ is limited to the passband width $\Delta\Omega$ of the filter. Usually $\Delta\Omega$ is markedly smaller than π. If the stopband attenuation of the selective filter is sufficiently high, the output signal $\overline{x}_\mu(k)$ can be calculated at the reduced rate

$$f_s' = f_s \, \frac{\Delta\Omega}{2\pi} . \tag{4.5}$$

Due to the computational complexity of any further processing of $\overline{x}_\mu(k)$, the sampling rate should be reduced – according to the sampling theorem – as much as possible by an integer factor $r > 1$ with

$$r \leq r_{\max} = \frac{2\pi}{\Delta\Omega} ; \quad r \in \mathbb{N} . \tag{4.6}$$

In the extreme case with $r = r_{\max}$, which is called *critical decimation*, aliasing can only be avoided by using ideal lowpass or bandpass filters. With non-ideal filters the decimation factor r has to be chosen somewhat smaller.

For ease of analytical description a special version $\tilde{x}_\mu(k)$ of the decimated sequence is introduced with the original sampling rate f_s but with $r-1$ zeros filled in between the decimated samples $x_\mu(k')$. The relations between these different sequences are illustrated in Fig. 4.3.

The following notations are used throughout this chapter for decimated and upsampled versions of any sequence:

- k, κ: time indices at original sampling rate f_s
- k', κ': time indices at reduced sampling rate $f_s' = f_s/r$
- decimated sequence at sampling rate f_s'/r *without* intermediate zero samples

$$x_\mu(k') = \overline{x}_\mu(k = k' \cdot r) \tag{4.7}$$

- upsampled sequence at sampling rate f_s *with* intermediate zero samples

$$\tilde{x}_\mu(k) = \begin{cases} \overline{x}_\mu(k) & \text{if } k = k' \cdot r, \quad k' = 0, \pm 1, \pm 2, \ldots \\ 0 & \text{else} . \end{cases} \tag{4.8}$$

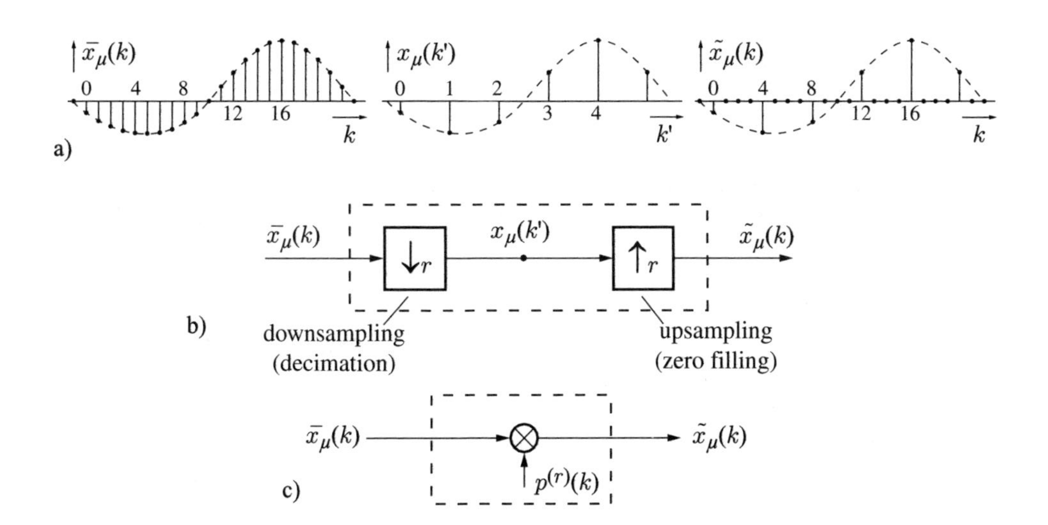

Figure 4.3: Decimation and zero filling for $r = 4$ with $p^{(r)}(k) = \sum_{k'=-\infty}^{\infty} \delta(k - k'r)$
a) Signals
b) Block diagram
c) Equivalent model

The process of sampling rate reduction by r with subsequent upsampling by r (zero filling) can be described analytically by multiplying $\overline{x}_\mu(k)$ with a periodic pulse sequence $p^{(r)}(k)$ (Fig. 4.3)

$$\tilde{x}_\mu(k) = \overline{x}_\mu(k) \cdot p^{(r)}(k) \tag{4.9}$$

$$p^{(r)}(k) = \sum_{k'=-\infty}^{\infty} \delta(k - k'r)\,. \tag{4.10}$$

The sequence $p^{(r)}(k)$ can be expressed as

$$p^{(r)}(k) \;=\; \frac{1}{r}\sum_{i=0}^{r-1} e^{jki\frac{2\pi}{r}}\,, \tag{4.11}$$

as the complex exponential is periodic in i. According to the modulation theorem, we get

$$\tilde{X}_\mu(e^{j\Omega}) \;=\; \frac{1}{r}\sum_{i=0}^{r-1} \overline{X}_\mu\left(e^{j(\Omega - \frac{2\pi}{r} i)}\right)\,. \tag{4.12}$$

An example with $r = 4$ is illustrated in Fig. 4.4.

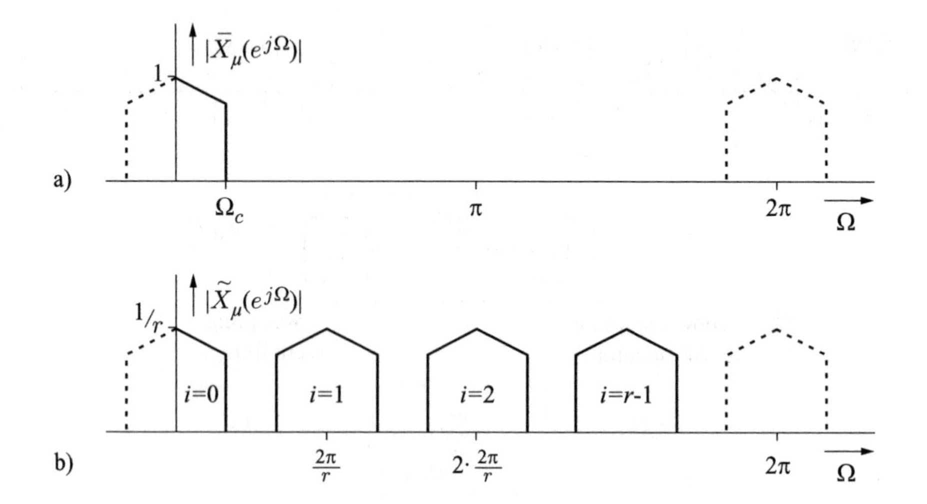

Figure 4.4: Spectra before and after decimation (with zero filling) for $r = 4$

In (4.8) the decimation grid is aligned to the origin of the time axis, i.e., with $k = 0$. In some cases it will be necessary that upsampled sequences have a constant time shift by k_0 so that $k = 0$ is not a decimation instant. In any case upsampled sequences with or without time shift but with intermediate zero samples will be marked by " ~ ".

Note that a displacement of the decimation grid by k_0 samples, which will be needed in the context of *polyphase network filter banks* (see Section 4.2), can be described as follows:

$$p^{(r)}(k - k_0) = \frac{1}{r} \sum_{i=0}^{r-1} e^{j\frac{2\pi}{r}(k-k_0)i} \tag{4.13-a}$$

$$= \begin{cases} 1 & \text{if } k = k' \cdot r + k_0, \quad k' = 0, \pm 1, \pm 2, \dots \\ 0 & \text{else} . \end{cases} \tag{4.13-b}$$

As we have kept the zero samples in the time domain, the spectrum is periodic with $2\pi/r$. If these zeros are deleted, we have to consider the normalized frequency $\Omega' = r \cdot \Omega$ (period 2π) corresponding to the reduced sampling rate $f_s' = f_s/r$.

4.1.1 Short-Term Spectral Analyzer

The behavior of the system of Fig. 4.1 can be described in the frequency domain by considering the Fourier transforms of the signals $x(k)$, $\overline{x}_\mu(k)$, and $\tilde{x}_\mu(k)$. For reasons of simplicity, we assume an ideal bandpass with center frequency Ω_μ. The

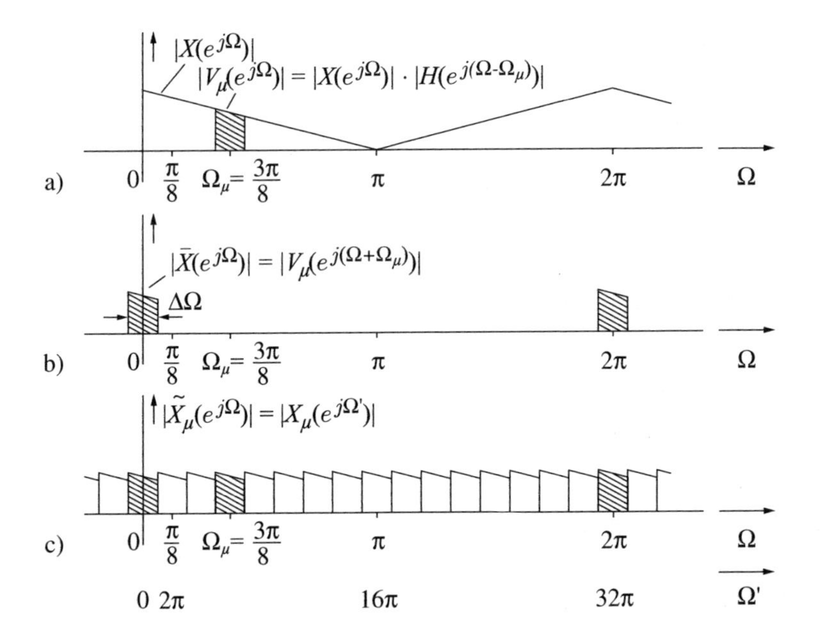

Figure 4.5: Spectral relations of the short-term spectrum analyzer of Fig. 4.1-a [Vary, Wackersreuther 1983]
a) Selection of a narrow band by bandpass filtering
b) Frequency shift of the filter output by complex demodulation with $e^{-j\Omega_\mu k}$
c) Periodic spectrum after critical decimation by $r = M = 16$.
The normalized frequency axis $f_s^{'} = \frac{f_s}{r} = \frac{f_s}{16}$ is denoted by $\Omega^{'} = r\Omega$.

spectral relations are explained exemplarily by Fig. 4.5. In this example a passband width of

$$\Delta\Omega = \frac{2\pi}{16}$$

and a center frequency of

$$\Omega_\mu = \Omega_3 = \frac{2\pi}{16} \cdot 3$$

are considered; Ω_μ is an integer multiple of $2\pi/r$. Then the frequency shift of the bandpass signal $v_\mu(k)$ by Ω_μ is not explicitly necessary, as the decimation process of (4.12) implicitly produces the required component in the baseband. This effect can easily be explained in the time domain, if we take into consideration that the complex exponential with frequency

$$\Omega_\mu = \frac{2\pi}{M}\,\mu$$

is periodic with length M/μ, where M is an integer multiple of r according to

$$M = m \cdot r\,; \qquad m \in \mathbb{N}.$$

Then the decimation process delivers the samples (see also Fig. 4.1-a)

$$x_\mu(k') = \overline{x}_\mu(k = k'r) = v_\mu(k'r)\; e^{-j\frac{2\pi}{m\cdot r}\mu\cdot k'\cdot r} \tag{4.14-a}$$

$$= v_\mu(k'r)\; e^{-j\frac{2\pi}{m}\mu\cdot k'}\,. \tag{4.14-b}$$

Two cases are of special interest:

a) Critical Decimation

$$r \;=\; M\,, \qquad \text{i.e., } m = 1\,.$$

Equation (4.14-b) results in

$$x_\mu(k') \;=\; v_\mu(k'r)\,.$$

In this case we can omit the demodulation process in Fig. 4.1-a, i.e., the multiplication by $e^{-j\Omega_\mu k}$, and apply the decimation process immediately to $v_\mu(k)$. This procedure is called *integer-band sampling.*

b) Half-Critical Decimation

Due to non-ideal filter characteristics a decimation factor of

$$r \;=\; M/2\,, \qquad \text{i.e., } m = 2\,,$$

is often chosen and (4.14-b) results in

$$x_\mu(k') = v_\mu(k'r)\; e^{-j\pi\mu\cdot k'} \;=\; v_\mu(k'r)\cdot(-1)^{\mu\cdot k'}\,. \tag{4.15}$$

For channels with even-frequency index μ we get the same result as before. Explicit demodulation (frequency shift) of $v_\mu(k)$ is not necessary, the decimation process can be applied to $v_\mu(k)$. However, if the channel index μ is not even, the decimated samples $v_\mu(k'r)$ have to be multiplied by $(-1)^{k'}$ to obtain the same result as in the case with explicit demodulation. At the reduced sampling rate $f_s' = f_s/r$

this corresponds to a frequency shift by π, and the lowpass frequency spectra of channels having odd indices are mirrored. This should be taken into consideration if we process the decimated samples $v_\mu(k'r)$ without the correction (4.15).

So far we have discussed the measurement of a single component $x_\mu(k')$ (fixed μ) of the short-term spectrum. If we are interested in the complete spectrum we need M parallel filters with different center frequencies $\Omega_\mu = 2\pi\mu/M$, $\mu = 0, 1, \ldots, M-1$, to achieve a uniform spectral resolution with

$$\Delta\Omega = \frac{2\pi}{M}.$$

According to (4.1), the impulse responses of these bandpass filters are modulated versions of the lowpass impulse response $h(k)$, which is called the *prototype impulse response*. The block diagram of the complete short-term spectrum analyzer (*analysis filter bank*) with M bandpass impulse responses

$$h_\mu^{BP}(k) = h(k)\, e^{j\frac{2\pi}{M}\mu k}, \quad \mu = 0, 1, \ldots, M-1 \tag{4.16}$$

is shown in Fig. 4.6. In what follows, this block diagram will serve as a *reference model*, although this implementation is obviously suboptimal with respect to the computational effort:

- most of the filter output samples $v_\mu(k)$ are discarded by decimation
- the impulse responses h_μ^{BP} and $h_{M-\mu}^{BP}$ are complex conjugates in pairs.

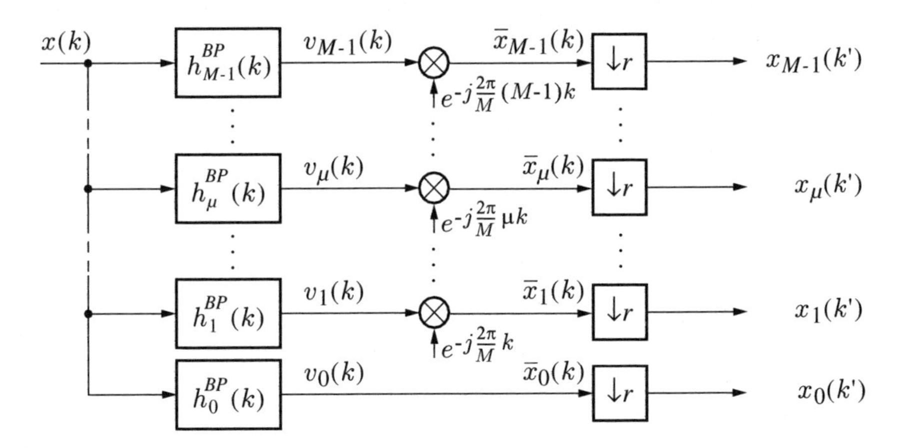

Figure 4.6: Short-term spectral analyzer with M parallel channels and decimation by r

If the bandpasses are FIR (*finite impulse response*) filters, the computational effort can significantly be reduced by calculating output samples only at the decimated rate. Furthermore, if the input signal $x(k)$ is real-valued, the output samples $x_\mu(k')$ and $x_{M-\mu}(k')$ are complex conjugates of each other. We only need to calculate the samples for $\mu = 0, 1, \ldots, M/2$. We will see later on that the computational complexity can be further reduced by using the FFT algorithm in combination with any FIR or IIR (*infinite impulse response*) prototype impulse response $h(k)$ (see Section 4.2).

Nevertheless, this reference system is well suited for studying the filter design issues.

4.1.2 Prototype Filter Design for the Analysis Filter Bank

First of all we have to specify the desired spectral resolution, i.e., the number M of channels, the passband width $\Delta\Omega$, and the stopband attenuation. We can use any filter design method such as [Dehner 1979], [Parks et al. 1979], [Parks, Burrus 1987], [Schüssler 1994], as available in MATLAB® to approximate the desired frequency response $H(z = e^{j\Omega})$ according to an error criterion (e.g., "Minimum Mean Square", "Min-Max"-, "Equi-Ripple"- or "Tschebyscheff"-behavior).

However, the overall frequency response of the filter bank has to be taken into account as a second design criterion.

A reasonable design constraint is that the overall magnitude response should be flat, or that in case of a linear-phase prototype filter with impulse response $h(k)$ the overall impulse response $h_A(k)$ of the analysis filter bank should be a mere delay of k_0 samples:

$$h_A(k) = \sum_{\mu=0}^{M-1} h_\mu^{BP}(k) \stackrel{!}{=} \delta(k - k_0) . \tag{4.17-a}$$

By inserting (4.16) we get

$$h_A(k) = \sum_{\mu=0}^{M-1} h_\mu^{BP}(k) \tag{4.17-b}$$

$$= \sum_{\mu=0}^{M-1} h(k)\, e^{j\frac{2\pi}{M}\mu k} \tag{4.17-c}$$

$$h_A(k) = h(k) \cdot \sum_{\mu=0}^{M-1} e^{j\frac{2\pi}{M}\mu k} \tag{4.17-d}$$

$$= h(k) \cdot M \cdot p^{(M)}(k) \tag{4.17-e}$$

with

$$p^{(M)}(k) = \frac{1}{M} \sum_{\mu=0}^{M-1} e^{j\frac{2\pi}{M}\mu k}$$

$$= \begin{cases} 1 & \text{if } k = \lambda M, \ \lambda = 0, \pm 1, \pm 2, \ldots \\ 0 & \text{else} . \end{cases} \tag{4.17-f}$$

As the bandpass responses are modulated versions of the prototype impulse response $h(k)$, the effective overall impulse response $h_A(k)$ has non-zero samples only at $k = \lambda M$. Therefore, an ideal overall response can be obtained if the prototype lowpass filter with $k_0 = \lambda_0 \cdot M$ satisfies the condition

$$h(\lambda M) = \begin{cases} 1/M & \lambda = \lambda_0 \\ 0 & \lambda \neq \lambda_0 . \end{cases} \tag{4.18}$$

These filters are also known as M-th band filters. The prototype impulse response has equidistant zeros and a non-zero sample at time instant $k_0 = \lambda_0 \cdot M$. The samples $h(k)$ with $k \neq \lambda \cdot M$ have no direct influence on the effective overall frequency response. They can be designed to optimize the frequency selectivity of the filter bank.

The design criterion (4.18) can easily be met if we use the "modified Fourier approximation" method, e.g., [Mitra 1998]. The prototype impulse response of odd length L, e.g., $L = 4 \cdot M + 1$, is obtained by multiplying the non-causal impulse response $h_{LP}(k)$ of the ideal lowpass filter with cutoff frequency $\Omega_c = 2\pi/M$ by any window $w(k)$ of finite length L centered symmetrically around $k = 0$. Finally, the impulse response is made causal by delaying the product $h_{LP}(k) \cdot w(k)$ by $k_0 = (L-1)/2$ samples:

$$h(k) = h_{LP}(k - k_0) \cdot w(k - k_0) .$$

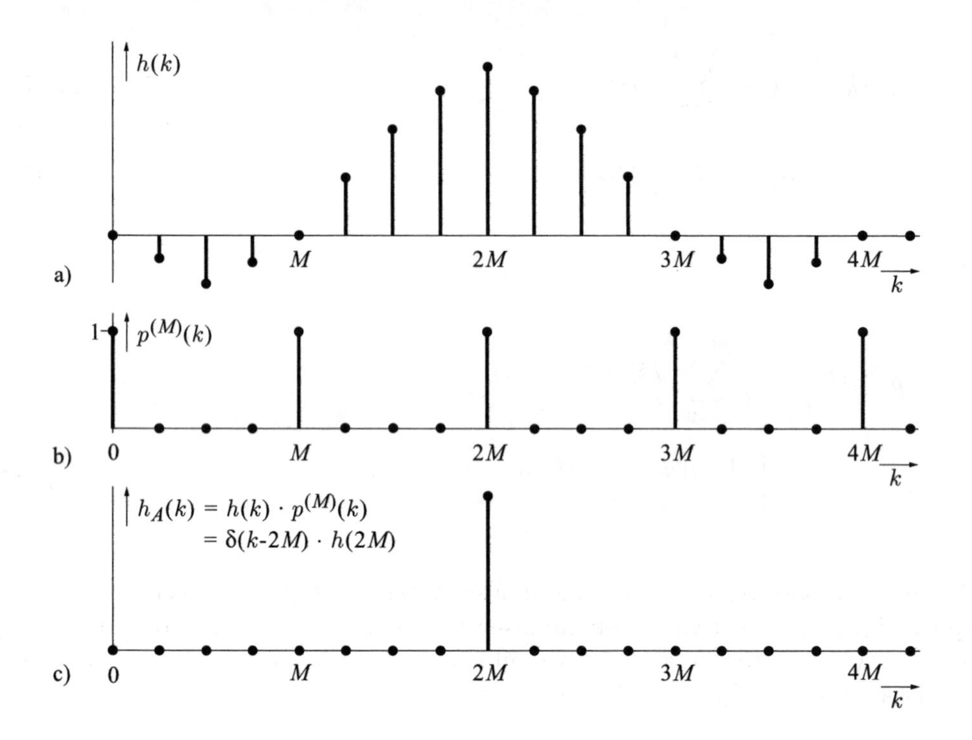

Figure 4.7: Design criterion for a perfect overall frequency response $h_A(k)$
a) Prototype impulse response $h(k) = h_{LP}(k - k_0) \cdot w(k - k_0)$: $L = 4 \cdot M + 1,\ M = 4$, Hamming window $w(k)$
b) Decimation function $p^{(M)}(k)$
c) Effective overall impulse response $h_A(k)$

A design example is illustrated in Fig. 4.7.

If the impulse response $h(k)$ of the prototype lowpass filter of order $4M$ (length $L = 4M + 1$) follows (4.18), the effective overall impulse response $h_A(k)$ corresponds to a pure delay of $k_0 = 2M$ samples.

4.1.3 Short-Term Spectral Synthesizer

In some applications, such as noise suppression, transform, or subband coding, we need to recover a time domain signal $y(k)$ from the samples $x_\mu(k')$ of the short-term spectrum. For any frequency index μ, the samples $x_\mu(k')$ are to be considered

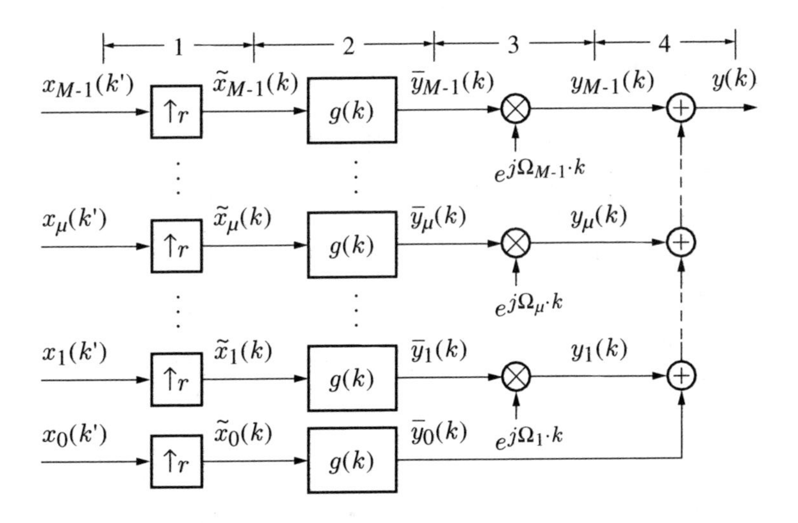

Figure 4.8: Short-term spectral synthesizer with M parallel channels

as decimated subband signals associated with center frequency Ω_μ. If we assume a decimation factor $r = M/m$, the subband signals $x_\mu(k^{'})$ can be represented as lowpass signals at the reduced sampling rate $f_s^{'} = f_s/r$. As indicated in Fig. 4.8, the process of reconstruction or resynthesis consists of the following steps:

1. Upsampling from $f_s^{'}$ to $f_s = f_s^{'} \cdot r$ by filling in $r-1$ zeros in between adjacent samples of $x_\mu(k^{'})$.

2. Interpolation by applying a filter with impulse response $g(k)$.

3. Frequency shift of the interpolated lowpass signal $\overline{y}_\mu(k)$ by Ω_μ (modulation).

4. Superposition of the interpolated bandpass signals $y_\mu(k)$.

The interpolation is achieved by applying the samples $\tilde{x}_\mu(k)$ to a lowpass filter with (two-sided) passband width $\Delta\Omega$ and impulse response $g(k)$. The spectral relations are described by Fig. 4.9 (in case of an ideal lowpass $g(k)$).

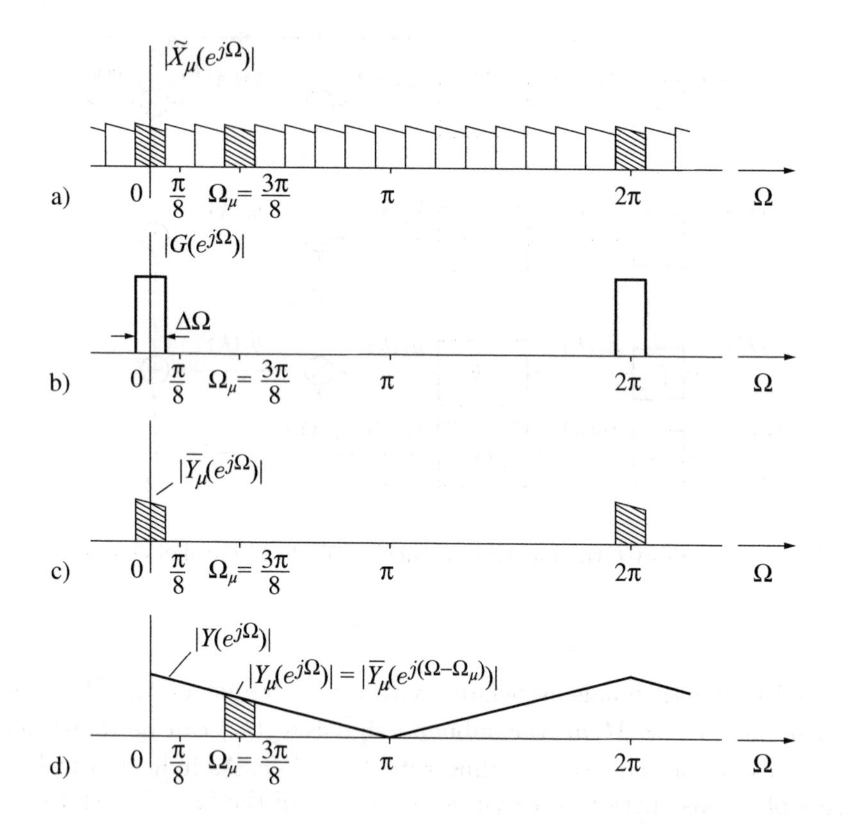

Figure 4.9: Spectral relations of short-term spectral synthesis
a) Magnitude spectrum of subband signal $\tilde{x}_\mu(k)$
b) Magnitude frequency response of the interpolator $g(k)$
c) Magnitude spectrum of the interpolated subband signal $\overline{y}_\mu(k)$
d) $y(k)$: superposition of the frequency-shifted signals $y_\mu(k)$

4.1.4 Short-Term Spectral Analysis and Synthesis

If the subband signals are not modified at all, the signal $x(k)$ can be reconstructed perfectly at the output. As some algorithmic delay caused by filter operations cannot be avoided, it should be a delayed version of the input signal

$$y(k) \stackrel{!}{=} x(k - k_0) .$$

The overall model of the spectral analysis–synthesis system without any modification of the subband signals is shown in Fig. 4.10-a. The more detailed sub-channel

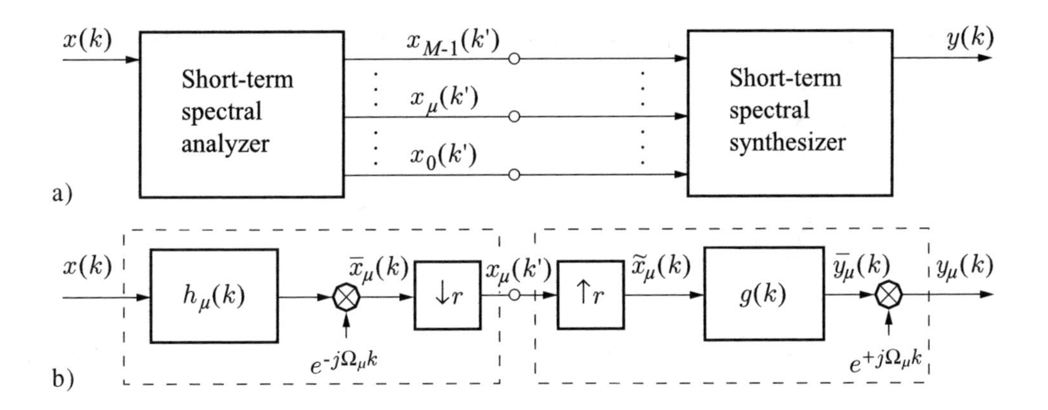

Figure 4.10: Reference model of the analysis-synthesis system
a) Overall system
b) Sub-channel model

model for the contribution $y_\mu(k)$ to the reconstructed output signal $y(k)$ is illustrated in Fig. 4.10-b. With the spectral representation of the intermediate signal $\overline{x}_\mu(k)$

$$\overline{X}_\mu(e^{j\Omega}) = X(e^{j(\Omega+\frac{2\pi}{M}\mu)}) \cdot H_\mu^{BP}(e^{j(\Omega+\frac{2\pi}{M}\mu)}) \tag{4.19-a}$$

$$= X(e^{j(\Omega+\frac{2\pi}{M}\mu)}) \cdot H(e^{j\Omega}) \tag{4.19-b}$$

we get the baseband contribution $\overline{Y}_\mu(e^{j\Omega})$ of channel μ under the assumptions that

- the decimation factor r is chosen according to the bandwidth $\Delta\Omega$ of the prototype lowpass filter
- the stopband attenuation of $h(k)$ is sufficiently high so that any aliasing due to the decimation process can be neglected
- the interpolation filter $g(k)$ is designed properly, i.e., the spectral repetitions due to the zero filling process are sufficiently suppressed

as follows (see also, e.g., [Vaidyanathan 1993]):

$$\overline{Y}_\mu(e^{j\Omega}) = \frac{1}{r} \cdot \tilde{X}_\mu(e^{j\Omega}) \cdot G(e^{j\Omega}) \tag{4.20-a}$$

$$= \frac{1}{r} \cdot X(e^{j(\Omega+\frac{2\pi}{M}\mu)}) \cdot H(e^{j\Omega}) \cdot G(e^{j\Omega}) \tag{4.20-b}$$

and finally the spectrum of the output signal (cf. Fig. 4.9 and Fig. 4.10)

$$Y(e^{j\Omega}) = \sum_{\mu=0}^{M-1} Y_\mu(e^{j\Omega}) \tag{4.21-a}$$

$$= \sum_{\mu=0}^{M-1} \overline{Y}_\mu(e^{j(\Omega - \frac{2\pi}{M}\mu)}) \tag{4.21-b}$$

$$= \frac{1}{r} \cdot \sum_{\mu=0}^{M-1} X(e^{j\Omega}) \cdot H(e^{j(\Omega - \frac{2\pi}{M}\mu)}) \cdot G(e^{j(\Omega - \frac{2\pi}{M}\mu)}) \tag{4.21-c}$$

$$= X(e^{j\Omega}) \cdot \frac{1}{r} \cdot \sum_{\mu=0}^{M-1} H(e^{j(\Omega - \frac{2\pi}{M}\mu)}) \cdot G(e^{j(\Omega - \frac{2\pi}{M}\mu)}) \tag{4.21-d}$$

$$= X(e^{j\Omega}) \cdot H_{AS}(e^{j\Omega}) \,. \tag{4.21-e}$$

The frequency response $H_{AS}(e^{j\Omega})$ denotes the effective overall response of the short-term analysis–synthesis system.

4.1.5 Prototype Filter Design for the Analysis–Synthesis Filter Bank

From (4.21-d) a criterion for perfect reconstruction (neglecting aliasing and interpolation errors but allowing a delay of k_0 samples) can be derived,

$$H_{AS}(e^{j\Omega}) = \frac{1}{r} \cdot \sum_{\mu=0}^{M-1} H(e^{j(\Omega - \frac{2\pi}{M}\mu)}) \cdot G(e^{j(\Omega - \frac{2\pi}{M}\mu)}) \tag{4.22-a}$$

$$\stackrel{!}{=} e^{-jk_0 \cdot \Omega} \,, \tag{4.22-b}$$

which corresponds to

$$h_{AS} \stackrel{!}{=} \delta(k - k_0) \tag{4.23}$$

in the time domain.

Equation (4.22-a) can be interpreted as the superposition of M frequency responses

$$Q_\mu(e^{j\Omega}) = H(e^{j(\Omega - \frac{2\pi}{M}\mu)}) \cdot G(e^{j(\Omega - \frac{2\pi}{M}\mu)}) \stackrel{!}{=} Q(e^{j(\Omega - \frac{2\pi}{M}\mu)}) \,, \tag{4.24}$$

which are weighted by the constant factor $1/r$. The corresponding time domain responses

$$q_\mu(k) = q(k) \cdot e^{j\frac{2\pi}{M}\mu k} \tag{4.25}$$

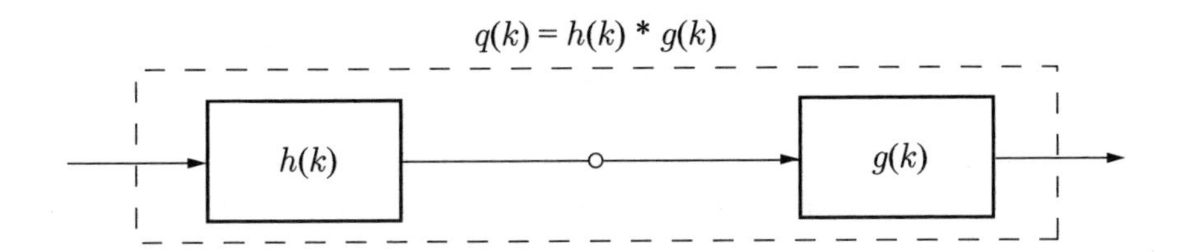

Figure 4.11: Effective prototype impulse response of the analysis–synthesis system

are modulated versions of the resulting overall prototype impulse response $q(k)$, which is the convolution of $h(k)$ and $g(k)$ as indicated in Fig. 4.11. Finally, the overall impulse response $h_{AS}(k)$ of the analysis–synthesis system should be a delay of k_0 samples. In analogy to (4.17-a) we get

$$h_{AS}(k) = \frac{1}{r} \cdot \sum_{\mu=0}^{M-1} q_\mu(k) \tag{4.26-a}$$

$$= q(k) \cdot \frac{1}{r} \cdot \sum_{\mu=0}^{M-1} e^{j\frac{2\pi}{M}\mu k} \tag{4.26-b}$$

$$= q(k) \cdot \frac{1}{r} \cdot M \cdot p^{(M)}(k) \tag{4.26-c}$$

$$\stackrel{!}{=} \delta(k - k_0)\,. \tag{4.26-d}$$

According to (4.26-c), the effective overall impulse response $h_{AS}(k)$ has non-zero samples only at times $k = \lambda M$. Therefore, an ideal overall response can be obtained if $q(k)$ with $k_0 = \lambda_0 \cdot M$ satisfies the following condition:

$$q(\lambda M) = \begin{cases} r/M & \lambda = \lambda_0 \\ 0 & \lambda \neq \lambda_0\,. \end{cases} \tag{4.27}$$

The convolutional product $q(k) = h(k) * g(k)$ should have equidistant zeros and a non-zero sample at time instant k_0, i.e., $q(k)$ should be the impulse response of an M-th band filter. The samples $q(k)$ with $k \neq \lambda \cdot M$ have no direct influence on the effective overall frequency response. They can be chosen to optimize the frequency selectivity $\Delta\Omega$ of the filter bank.

The impulse response $q(k)$ can easily be designed, e.g., by the "modified Fourier approximation" method as described in Section 4.1.1. However, the decomposition of $q(k)$ into $h(k)$ and $g(k)$ is not trivial, especially if identical linear-phase prototypes $h(k) = g(k)$ are used. In this case the product transfer function $Q(z)$ should have double zeros only. Different exact and approximate solutions have been proposed in [Wackersreuther 1985], [Wackersreuther 1986], [Wackersreuther 1987], [Nguyen 1994], [Kliewer 1996]; see also [Vaidyanathan 1993].

An alternative method is to split $Q(z)$ into a minimum-phase response $h(k)$ and a maximum-phase component $g(k)$ (or vice versa) by cepstral domain techniques as proposed in [Boite, Leich 1981].

4.1.6 Filter Bank Interpretation of the DFT

The discrete Fourier transform (DFT) may be interpreted as a special case of the short-term spectral analyzer of Fig. 4.1. We consider the calculation of the DFT of a block of M samples from an infinite sequence $x(k)$. M samples are extracted from $x(k)$ by applying a sliding window function $w(\lambda)$, $\lambda = 0, 1, \ldots, M-1$, of finite length so that the most recent sample $x(k)$ is weighted by $w(M-1)$ and the oldest sample $x(k-M+1)$ by $w(0)$.

The input sequence to the DFT at time k is defined as

$$x(k-M+1),\ x(k-M+2),\ \ldots,\ x(k-M+1+\lambda),\ \ldots,\ x(k)$$

and the μ-th DFT coefficient $X_\mu(k)$ at time k is given by

$$X_\mu(k) = \sum_{\lambda=0}^{M-1} x(k-M+1+\lambda)\ w(\lambda)\ e^{-j\frac{2\pi}{M}\mu\lambda}\,. \tag{4.28-a}$$

For the sake of compatibility with the reference configuration of Fig. 4.6, we select a window function which is the time-reversed version of a prototype lowpass impulse response $h(k)$ of length M:

$$w(\lambda) = h(M-1-\lambda)\,; \quad \lambda = 0, 1, \ldots, M-1 \tag{4.28-b}$$

and get

$$X_\mu(k) = \sum_{\lambda=0}^{M-1} x(k-M+1+\lambda)\ h(M-1-\lambda)\ e^{-j\frac{2\pi}{M}\mu\lambda}\,. \tag{4.28-c}$$

With the substitution

$$k-M+1+\lambda = \kappa, \quad \text{i.e.,} \quad M-1-\lambda = k-\kappa \quad \text{and} \quad \lambda = \kappa - k + M - 1$$

we obtain

$$X_\mu(k) = \sum_{\kappa=k-M+1}^{k} x(\kappa)\; h(k-\kappa)\; e^{-j\frac{2\pi}{M}\mu(\kappa-k+M-1)} \tag{4.28-d}$$

$$= e^{j\frac{2\pi}{M}\mu(k+1)} \cdot \sum_{\kappa=-\infty}^{\infty} x(\kappa)\; h(k-\kappa)\; e^{-j\frac{2\pi}{M}\mu\kappa} \tag{4.28-e}$$

$$= e^{j\frac{2\pi}{M}\mu(k+1)} \cdot \overline{x}_\mu(k)\,. \tag{4.28-f}$$

In (4.28-d) we can replace the explicit summation limits by $\pm\infty$ as the window $h(k)$ implicitly performs the necessary limitation. Thus, the final expression for $X_\mu(k)$ is – except for the modulation factor (whose magnitude is 1) – identical to the intermediate signal $\overline{x}_\mu(k)$ of the bandpass or lowpass short-time analyzer of Fig. 4.1 for

$$\Omega_\mu \;=\; \frac{2\pi}{M}\cdot\mu \qquad \text{and} \qquad h_{LP}(k) \;=\; h(k)$$

(see also (4.2-a) and (4.2-c)). Thus, the result

$$X_\mu(k) \;=\; \text{DFT}\Big\{x(\kappa)\; h(k-\kappa)\Big\}\,; \qquad \mu \;=\; 0,1,\ldots,M-1$$

of the DFT of any windowed signal segment may be interpreted in terms of modulation, filtering, and decimation. At time instant k we obtain a set of (modulated) short-term spectral values

$$X_\mu(k) \;=\; e^{j\frac{2\pi}{M}\mu(k+1)} \cdot \overline{x}_\mu(k)\,, \tag{4.29}$$

where k determines the position of the sliding window (see also Fig. 4.2). If the position of the window is shifted between successive DFT calculations by $r \;=\; M$, successive DFT calculations do not overlap in time. The "DFT-observation instances" are

$$k \;=\; k^{'}\cdot M-1\,, \qquad k^{'} \;\in\; \mathbb{N}\,.$$

In this case, which corresponds to *critical decimation*, the modulation factor vanishes

$$X_\mu(k \;=\; k^{'}\cdot M-1) \;=\; \overline{x}_\mu(k \;=\; k^{'}\cdot M-1) \;=\; x_\mu(k^{'}) \tag{4.30}$$

and the DFT delivers – with respect to magnitude and phase – exactly the same samples as the short-term analyzers of Fig. 4.1.

Note that in (4.30) the decimation grid has a time shift of $\Delta k = -1$, i.e., $k = -1$ and $k = k^{'}\cdot M - 1$ are on the decimation grid, but neither $k = 0$ nor $k = k^{'}\cdot M$.

If the DFT calculation overlaps, e.g., by half a block length (*half-critical decimation*, $r = M/2$), the output samples are calculated at the time instances

$$k = k' \cdot M/2 - 1, \qquad k' \in \mathbb{Z}.$$

Then the modulation factor has to be taken into consideration as follows:

$$\begin{aligned} X_\mu(k = k' \cdot M/2 - 1) &= e^{j\pi\mu k'} \cdot \overline{x}_\mu(k = k' \cdot M/2 - 1) \\ &= (-1)^{\mu k'} \cdot \overline{x}_\mu(k = k' \cdot M/2 - 1), \end{aligned}$$

if we are interested in the samples $x_\mu(k')$ according to Fig. 4.1. In both cases, $r = M$ and $r = M/2$, the samples are produced by the DFT on a decimation grid $k' \cdot r - 1$, which is displaced from the origin ($k = 0$) by one sample interval. A block diagram of the windowed DFT is given in Fig. 4.12. The samples $x(k)$ are fed into a delay chain (delay operator z^{-1}) and the delayed samples $x(k - \lambda)$ are multiplied by the window coefficients $w(M - 1 - \lambda) = h(\lambda)$. It is obvious that the delay chain has to be operated at the original sampling rate f_s, whereas the window multiplications and the DFT calculations can be carried out at the reduced sampling rate $f_s' = f_s/r$ if decimation by r is applied to the output.

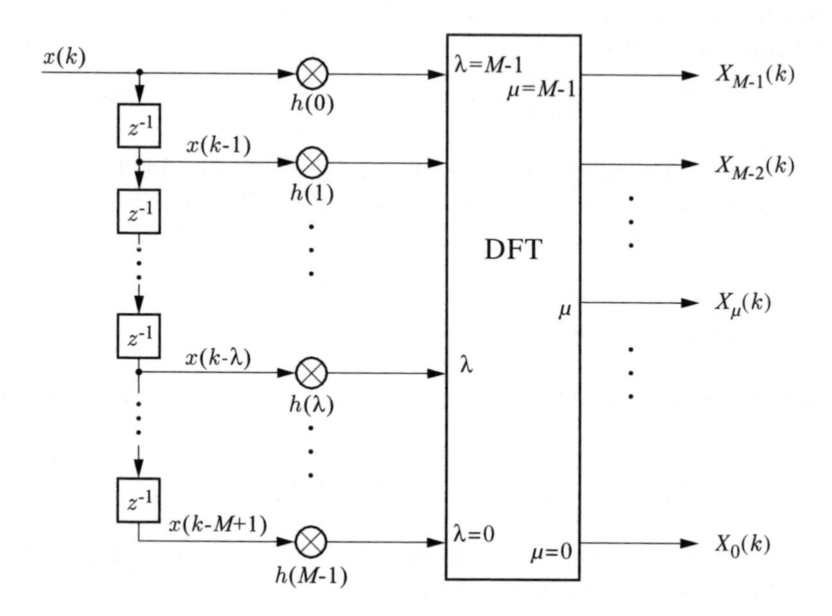

Figure 4.12: Implementation of the sliding window DFT

4.2 Polyphase Network Filter Banks

The *polyphase network filter bank* (PPN filter bank) is a very efficient implementation of the short-term spectral analyzer and the short-term spectral synthesizer of Sections 4.1.1 and 4.1.3, respectively. The computational complexity of the two systems given in Fig. 4.6 and Fig. 4.8 can be reduced significantly.

The key points are:

1. The output samples of the analyzer (Fig. 4.6) are calculated only at the reduced sampling rate $f_s^{'} = f_s/r$.
2. At each (decimated) time instant $k^{'}$ the complete set of output samples $x_\mu(k^{'})$, $\mu = 0, 1 \ldots, M-1$, can be obtained by a single FFT of length M.
3. The output of the synthesizer (Fig. 4.8) can be calculated using one FFT of length M per r samples, which performs interpolation and spectral translation in combination with the impulse response $g(k)$.

If FIR prototype impulse responses $h(k)$ and $g(k)$ are used, there are two equivalent implementations:

A: overlapping of windowed segments

B: polyphase filtering.

Concept B is the more general one, as it can be applied to FIR as well as to IIR prototype impulse responses. The term polyphase network (PPN) is widely used in the literature (e.g., [Bellanger et al. 1976], [Vary 1979], [Vary, Heute 1980], [Vary, Heute 1981], [Vaidyanathan 1990], [Vaidyanathan 1993]) because different phase characteristics of decimated subsequences and partial impulse responses can be identified, which play an important role.

4.2.1 PPN Analysis Filter Bank

In this section the two equivalent approaches A and B will be derived from the reference short-term spectral analyzer of Fig. 4.1 using an FIR prototype impulse response $h_{LP}(k) = h(k)$ of length L $(k = 0, 1, \ldots, L-1)$, where L may be larger than M, the number of channels. For reasons of simplicity, we define

$$L = N \cdot M \; ; \quad N \in \mathbb{N} \tag{4.31}$$

and append zero samples if the length of $h(k)$ is not an integer multiple of M.

Approach A: Overlapping of Windowed Segments

The lowpass signal $\overline{x}_\mu(k)$ of the μ-th channel before sampling rate decimation can be formulated equivalently either by bandpass filtering and complex post-modulation (Fig. 4.1-a, Eq. (4.2-a)) or by complex pre-modulation and lowpass filtering (Fig. 4.1-b, Eq. (4.2-c)) as

$$\overline{x}_\mu(k) = e^{-j\frac{2\pi}{M}\mu k} \cdot \sum_{\kappa=-\infty}^{\infty} x(k-\kappa) \cdot e^{j\frac{2\pi}{M}\mu\kappa} \cdot h(\kappa) \tag{4.32-a}$$

$$= \sum_{\kappa=-\infty}^{\infty} x(\kappa) \cdot e^{-j\frac{2\pi}{M}\mu\kappa} \cdot h(k-\kappa)\,. \tag{4.32-b}$$

Due to the length $L = N \cdot M$ of the FIR impulse response $h(k)$, the summation index κ in (4.32-a) is limited to the range

$$\kappa = 0, 1, \ldots, N \cdot M - 1\,.$$

The key to the derivation of the overlap structure A consists in the index substitution

$$\kappa = \nu \cdot M + \lambda\,; \quad \lambda = 0, 1, \ldots, M-1$$
$$\nu = 0, 1, \ldots, N-1\,.$$

Thus, (4.32-a) can be rearranged as follows:

$$\overline{x}_\mu(k) = e^{-j\frac{2\pi}{M}\mu k} \cdot \sum_{\nu=0}^{N-1}\sum_{\lambda=0}^{M-1} x(k-\nu M-\lambda) \cdot h(\nu M+\lambda) \cdot e^{j\frac{2\pi}{M}\mu\lambda} \tag{4.32-c}$$

$$= e^{-j\frac{2\pi}{M}\mu k} \cdot \sum_{\lambda=0}^{M-1} \underbrace{\left(\sum_{\nu=0}^{N-1} x(k-\nu M-\lambda) \cdot h(\nu M+\lambda)\right)}_{u_\lambda(k)} e^{j\frac{2\pi}{M}\mu\lambda} \tag{4.32-d}$$

$$= e^{-j\frac{2\pi}{M}\mu k} \cdot \sum_{\lambda=0}^{M-1} u_\lambda(k) \cdot e^{j\frac{2\pi}{M}\mu\lambda} \tag{4.32-e}$$

$$= e^{-j\frac{2\pi}{M}\mu k} \cdot \left[\sum_{\lambda=0}^{M-1} u_\lambda(k) \cdot e^{-j\frac{2\pi}{M}\mu\lambda}\right]^* \tag{4.32-f}$$

$$= e^{-j\frac{2\pi}{M}\mu k} \cdot [\mathrm{DFT}\{u_\lambda(k)\}]^* \tag{4.32-g}$$

$$= e^{-j\frac{2\pi}{M}\mu k} \cdot [U_\mu(k)]^* \tag{4.32-h}$$

$$= W_M^{\mu k} \cdot [U_\mu(k)]^* \tag{4.32-i}$$

where $[..]^*$ denotes the complex conjugate operation,

$$W_M^{\mu k} = e^{-j\frac{2\pi}{M}\mu k} \tag{4.33}$$

are the complex post-modulation terms, and

$$u_\lambda(k) = \sum_{\nu=0}^{N-1} x(k - \nu M - \lambda) \cdot h(\nu M + \lambda) \; ; \quad \lambda = 0, 1, \ldots, M-1 \tag{4.34}$$

is the real-valued input sequence to the DFT.

The resulting expression (4.32-g) looks like the DFT of the intermediate sequence $u_\lambda(k)$, $\lambda = 0, \ldots, M-1$, of length M. As the sequence $u_\lambda(k)$ is real-valued, we may use either the DFT or the inverse DFT. In the latter case a scaling factor of $1/M$ has to be taken into account. Here we prefer the DFT for the analysis part of the filter bank.

In conclusion, the complete set of samples $\overline{x}_\mu(k)$, $\mu = 0, 1, \ldots, M-1$, can be calculated very efficiently for any FIR prototype impulse response $h(\kappa)$ of length L by applying the FFT algorithm to the sequence $u_\lambda(k)$, $\lambda = 0, 1, \ldots, M-1$, at the fixed but arbitrary time instant k. If we are interested not only in the magnitude samples $|\overline{X}_\mu(k)|$, but also in the magnitude and phase or real and imaginary parts of $\overline{X}_\mu(k)$, respectively, then the post-modulation by $W_M^{\mu k}$ has to be carried out according to (4.32-i).

In a pre-processing step the intermediate sequence u_λ according to (4.34) has to be determined as illustrated in Fig. 4.13 by overlapping N windowed segments, each

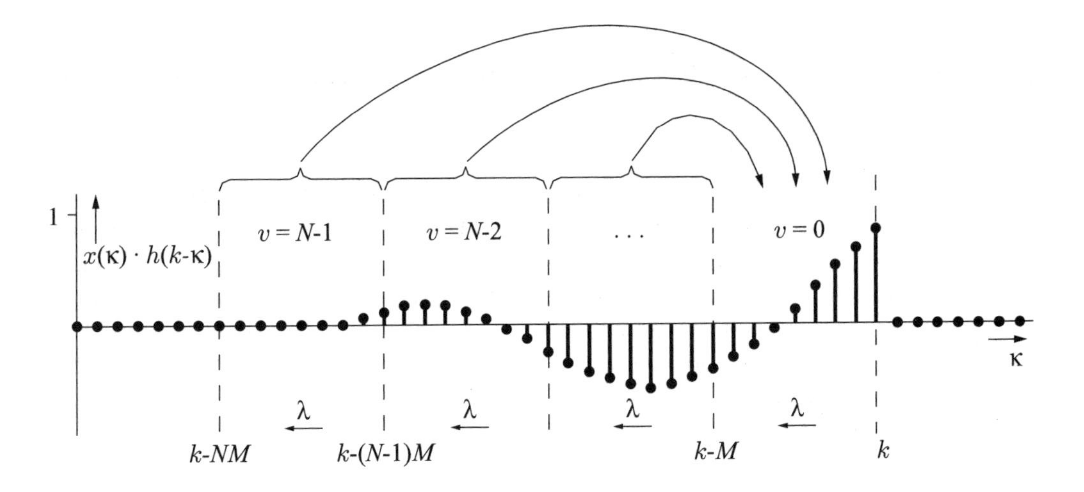

Figure 4.13: Overlapping of windowed segments; example $M = 8, N = 4$

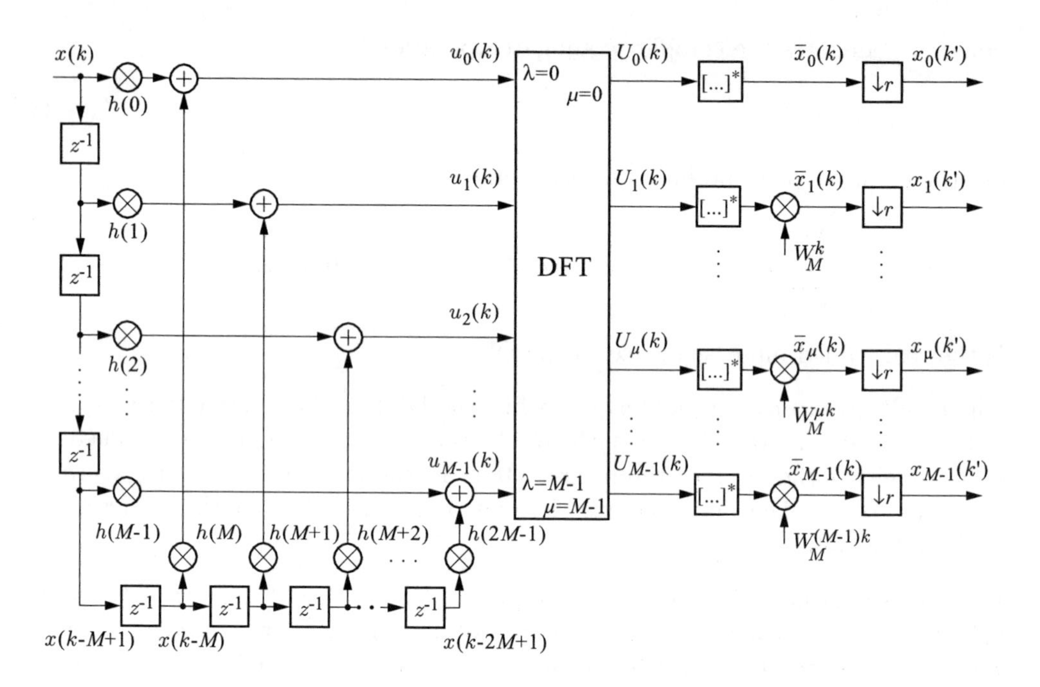

Figure 4.14: Polyphase network (PPN) filter bank according to approach A; example $L = N \cdot M = 2M$; $[...]^*$ denotes complex conjugation

of length M ($\nu = 0, 1, \ldots, N-1$). The complete pre-processing step requires $M \cdot N$ multiplications and additions which are not more than the number of operations needed to calculate one output sample of a single lowpass filter of length $L = N \cdot M$.

The corresponding block diagram is shown in Fig. 4.14 for the special case $L = N \cdot M = 2M$.

So far we have not considered the fact that the narrowband signals $\overline{x}_\mu(k)$ may be represented at a reduced sampling rate. In the algorithm of Fig. 4.14 the pre-processing step and the DFT are carried out at each time instant k and the sample rate is decimated afterwards by r. However, the output samples can be calculated immediately at the reduced sampling rate, i.e., the pre-processing step, the DFT, the "$[...]^*$-complex conjugation", and the post-modulation has to be calculated only at the decimated time instances, e.g., at

$$k \;=\; k' \cdot r\,, \tag{4.35}$$

while the delay chain at the input has to be operated at the original sampling rate.

The two decimation cases of special interest are critical decimation ($r = M$) and half-critical decimation ($r = \frac{M}{2}$), with the post-modulation factors

$$W_M^{\mu \cdot k' \cdot r} = e^{-j\frac{2\pi}{M}\mu k' r} = \begin{cases} 1 & r = M \\ (-1)^{\mu k'} & r = M/2 \,. \end{cases} \tag{4.36}$$

Note that in Fig. 4.14 the order of the input index λ is reversed in comparison to Fig. 4.12. This difference is due to the fact that the decimation grid with $k = k' \cdot r$ is not displaced with respect to the origin and the simple post-modulation factors of (4.36) are desired here. If we considered the expression (4.32-i) for the decimated instances $k = k' \cdot r - 1$ a post-modulation term

$$e^{-j\frac{2\pi}{M}\mu(k' r-1)} = W_M^{\mu k' \cdot r} \cdot e^{+j\frac{2\pi}{M}\mu}$$

would result. The combined effect of the second factor $e^{+j\frac{2\pi}{M}\mu}$ with the "[...]*-complex conjugation" is equivalent to a cyclic shift and an inversion of the order of the DFT input sequence $u_\lambda(k)$. If this is taken into consideration, it can easily be shown that for $L = M$ the two block diagrams of Fig. 4.14 and Fig. 4.12 are equivalent as given by (4.29).

In any case the output samples $x_\mu(k')$ are exactly the same as those of the reference structure of Fig. 4.1.

Approach B: Polyphase Filtering

The second approach is an alternative interpretation of (4.34) in terms of convolution instead of overlapping weighted segments. This allows us to use an FIR or even an IIR prototype lowpass filter. For an FIR prototype filter the actual difference lies in the organization of the two nested loops (ν and λ) within the pre-processing step.

The alternative implementation B consists of reorganizing the block diagram of Fig. 4.14 and the key equation (4.32) by taking the decimation process into account.

We introduce M subsequences

$$\tilde{x}_\lambda(k) = \begin{cases} x(k) & k = i \cdot M + \lambda \\ 0 & \text{else} \end{cases} \qquad \lambda = 0, 1, \ldots, M-1 \tag{4.37}$$

decimated and upsampled by M and furthermore M partial impulse responses

$$\tilde{h}_\lambda(k) = \begin{cases} h(k) & k = i \cdot r - \lambda \\ 0 & \text{else} \end{cases} \qquad \lambda = 0, 1, \ldots, M-1\,, \tag{4.38}$$

decimated and upsampled by r which are defined on decimation grids with different displacements in time by $\pm\lambda$. There are M partial impulse responses; however, for $r = M/2$ only $M/2$ responses are different ($\tilde{h}_{M/2+i} = \tilde{h}_i$, $i = 0, 1, \ldots, M/2$).

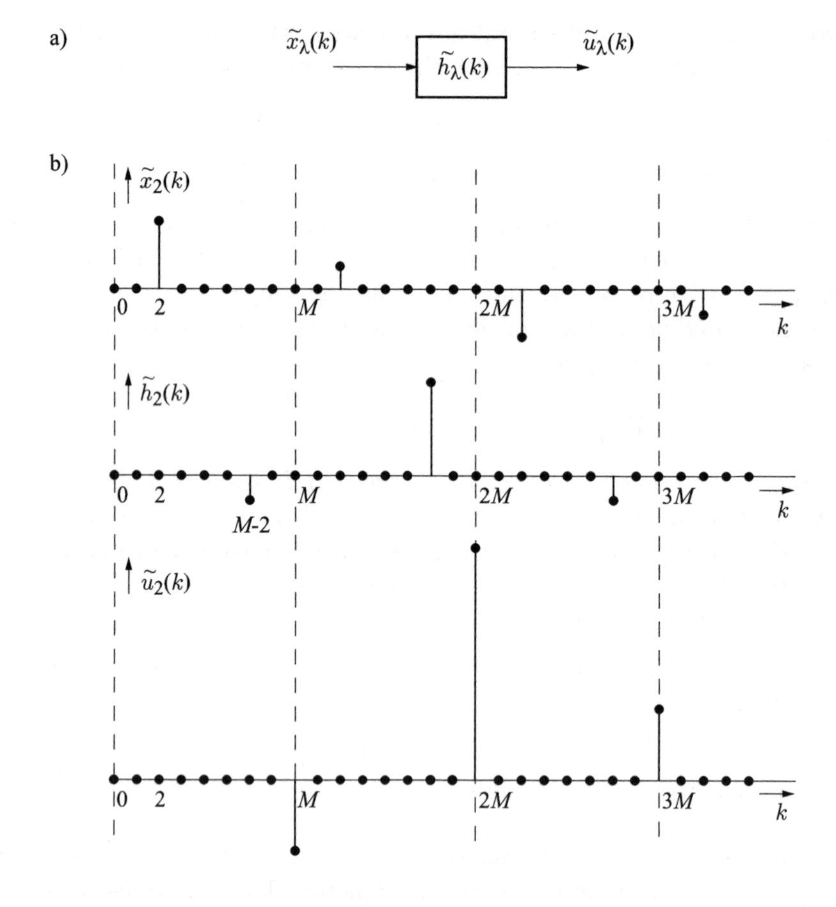

Figure 4.15: Polyphase filtering: convolution of subsequences with partial impulse responses
a) Block diagram
b) Example $M = 8,\ \lambda = 2,\ r = M$

Each intermediate sequence $\tilde{u}_\lambda(k)$ can be interpreted for any fixed index λ as a time sequence, i.e., as convolution of the subsequence $\tilde{x}_\lambda(k)$ with the partial impulse response $\tilde{h}_\lambda(k)$

$$\tilde{u}_\lambda(k) = \tilde{x}_\lambda(k) * \tilde{h}_\lambda(k) \begin{cases} \neq 0 & k = k^{'} \cdot r \\ = 0 & \text{else} . \end{cases} \tag{4.39}$$

For the special case $r = M$, Fig. 4.15 illustrates exemplarily that the intermediate signals $\tilde{u}_\lambda(k)$ take non-zero values only at the decimation instances

$$k = k^{'} \cdot M \ ; \quad k^{'} \in \mathbb{Z} . \tag{4.40}$$

Assuming a causal signal $x(k)$ and a causal impulse response $h(k)$, the first non-zero value of $\tilde{x}_\lambda(k)$ is at $k = \lambda$ and the first non-zero sample of $\tilde{h}_\lambda(k)$ is at

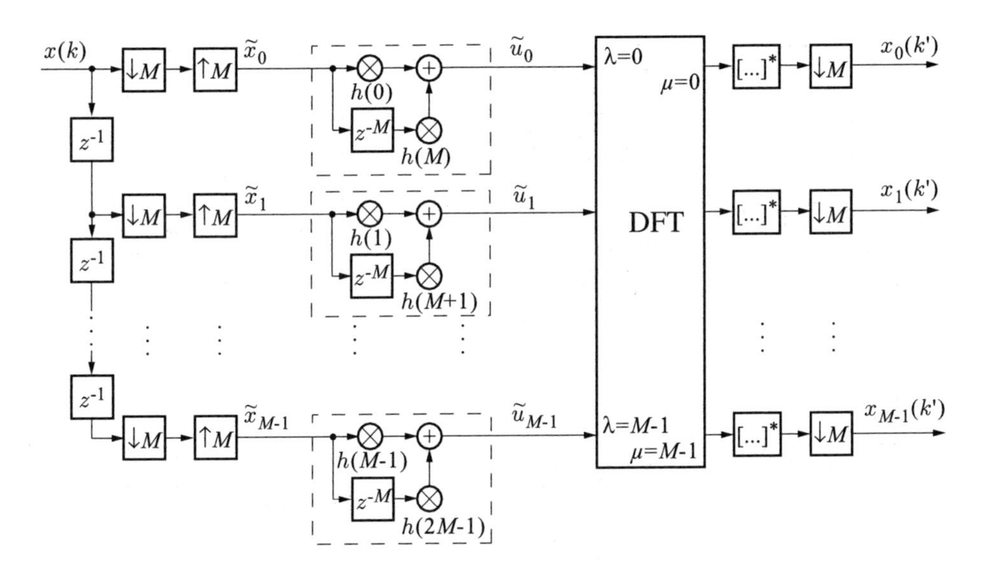

Figure 4.16: Basic polyphase network (PPN) filter bank according to approach B; example $L = N \cdot M = 2M,\ r = M$

$k = M - \lambda$ ($\lambda = 1, \ldots, M-1$) or $k = 0$ ($\lambda = 0$). As a result all partial convolutions produce non-zero output signals $\tilde{u}_\lambda(k)$ on the same decimation time grid

$$k = k' \cdot M . \tag{4.41}$$

Thus, the partial filters perform different phase shifts, this is why the term *polyphase filtering* was introduced ([Bellanger et al. 1976]).

For the special case $r = M$ and $L = 2M$, the block diagram of approach B is shown in Fig. 4.16. In comparison to Fig. 4.14 only the pre-processing has been reorganized, taking the decimation by $r = M$ as well as the post-modulation terms $W_M^{\mu \cdot k' \cdot M} = 1$ (see also (4.36)) into account.

Because of (4.37) and (4.38) the zero samples of $\tilde{h}_\lambda(k)$, $\tilde{x}_\lambda(k)$, and $\tilde{u}_\lambda(k)$ need not be processed. This is illustrated in Fig. 4.17. The decimation by $r = M$ takes place at the taps of the delay line with unit delays $T = 1/f_s$, whereas the partial filters and the DFT run at the reduced sample rate $f_s' = f_s/M$.

In the general case with decimation by r, there are $r - 1$ intermediate zero coefficients between two non-zero coefficients of $\tilde{h}_\lambda(k)$ and $M - 1$ intermediate zeros between the samples of the subsequences $\tilde{x}_\lambda(k)$. As the input samples to the partial filters, running at the reduced sample rate, are needed at the time instances $k = k' \cdot r$, we first have to decimate the delayed versions of the input signal by $r = M$ and then to upsample them by $m = M/r$.

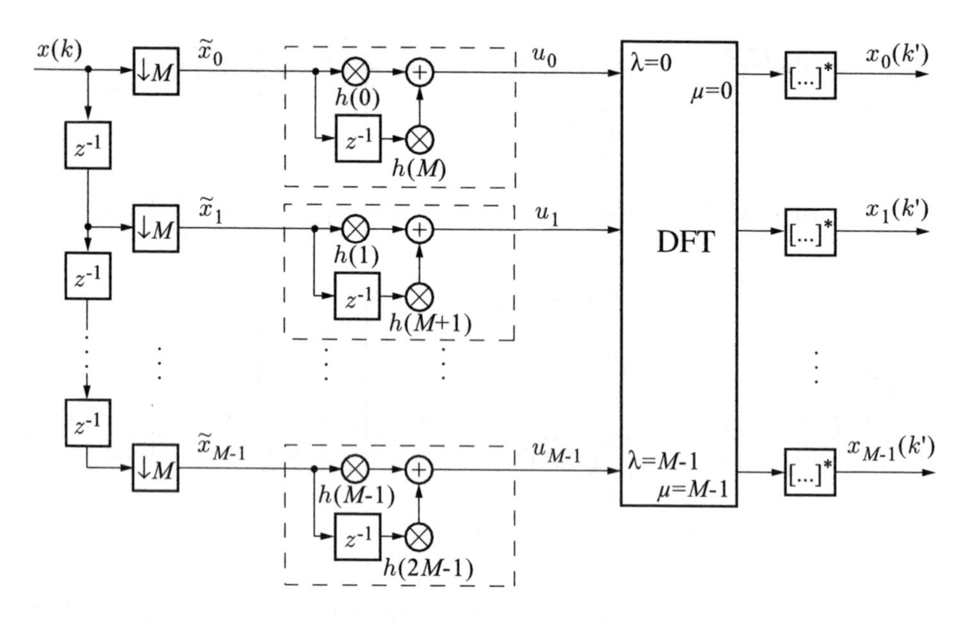

Figure 4.17: Efficient implementation of the polyphase network (PPN) filter bank with partial filters running at the reduced sampling rate $f_s^{'} = f_s/M$; example $L = N \cdot M = 2M, \ r = M$

The solution for the general case producing output samples $x_\mu(k^{'})$ at the reduced sampling rate $f_s^{'} = f_s/r$, with $r = \frac{M}{m}$ and any prototype lowpass filter with impulse response $h(k)$, is shown in Fig. 4.18. The decimated partial impulse responses are defined by

$$h_\lambda(k^{'}) = h(k^{'} \cdot r - \lambda) \ ; \qquad \lambda = 0, 1, \ldots, M-1 \, . \tag{4.42}$$

Usually half-critical decimation $r = M/2$ is chosen to avoid spectral aliasing in the subband signals $x_\mu(k^{'})$. The decimated partial filters $h_\lambda(k^{'})$ and the DFT have to be computed at the reduced sampling rate f_s/r only. For $r = M/2$ the signals $\tilde{x}_\lambda(k^{'} \cdot r)$ are obtained by decimation of the delayed versions of $x(k)$ by M and subsequent upsampling by 2 to achieve the reduced sampling rate. Therefore, we have $M/r - 1 = 1$ zero sample in between two decimated samples.

The PPN approach of Fig. 4.18 may be interpreted as a generalization of the windowed DFT (Fig. 4.12) so that the window multiplication in Fig. 4.12 is replaced by a convolution (polyphase filter) with the decimated partial impulse responses $h_\lambda(k^{'})$.

If the prototype lowpass impulse response is of length $L = M$, both systems are identical according to (4.29).

The main advantage of the PPN concept is that the spectral selectivity, i.e., the frequency response $H(e^{j\Omega})$, can be designed independently of the number M of

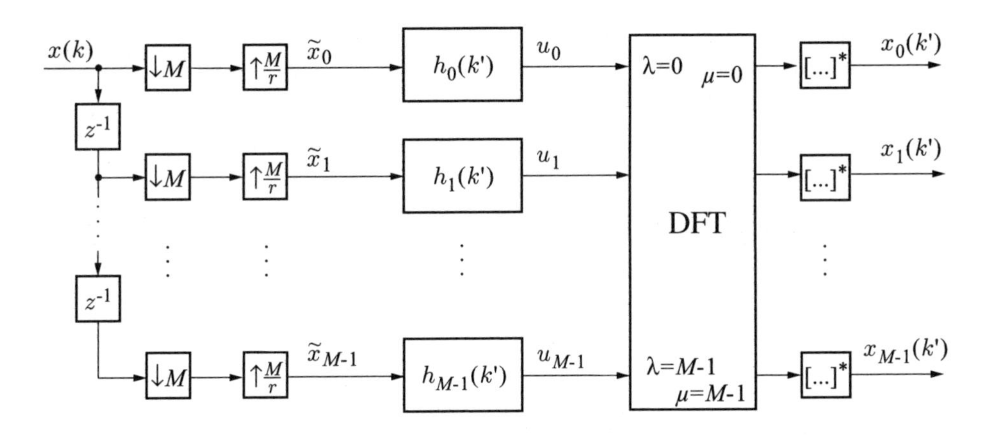

Figure 4.18: General solution: polyphase network (PPN) filter bank with partial filters $h_\lambda(k^{'})$ running at the reduced sampling rate f_s/r; $r = \frac{M}{m}$

channels and that even IIR prototype filters are possible (e.g., [Vary 1979]). The discussion of the latter possibility is beyond the scope of this book.

Finally, it should be noted that instead of the DFT the *discrete cosine transform* (DCT) or some generalized versions (GDFT, GDCT) (e.g., [Crochiere, Rabiner 1983], [Vary et al. 1998]) can be used.

With the *generalized* DCT the individual bandpass impulse responses can be formulated as

$$h_\mu^{BP}(k) = h(k)\ \cos\left(\frac{\pi}{M}(\mu + \mu_0)(k + k_0)\right); \tag{4.43}$$

$$\mu = 0, 1, \ldots M-1; \quad \mu_0, k_0 \in \{0, 1/2\}\ .$$

4.2.2 PPN Synthesis Filter Bank

The short-term spectral synthesizer of Fig. 4.8 can be implemented very efficiently using the inverse discrete Fourier transform (IDFT) and a polyphase network (PPN) as shown in Fig. 4.19. The PPN consists of partial impulse responses $g_\lambda(k^{'})$, which are obtained by subsampling the impulse response $g(k)$ of the interpolation filter of the reference structure given in Fig. 4.8. The impulse response $g(k)$ can also be regarded as the prototype impulse response of the synthesis filter bank.

The structure of this efficient PPN filter bank can be derived straightforwardly from the reference synthesizer of Fig. 4.8. The output signal $y(k)$ is obtained as

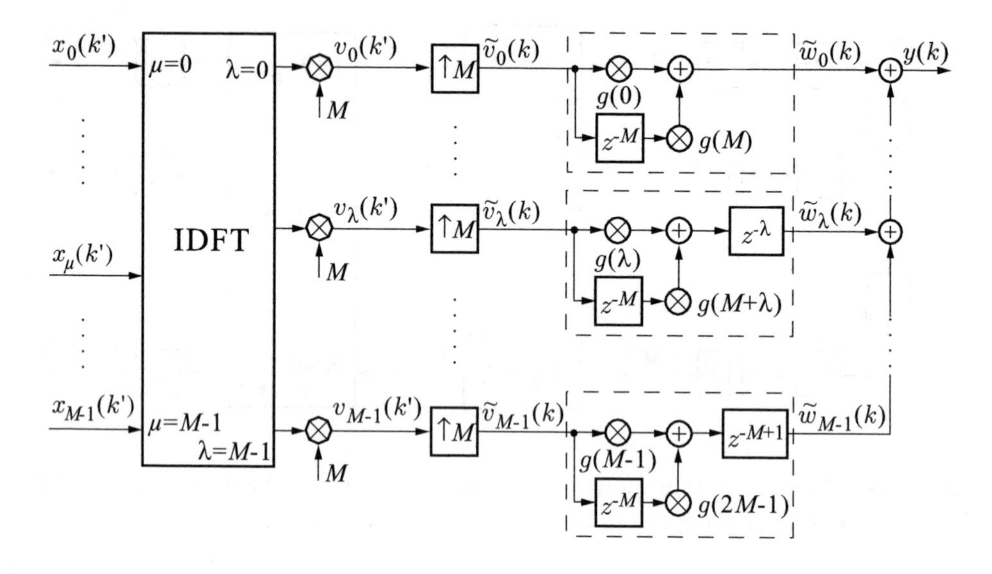

Figure 4.19: Basic structure of the polyphase synthesis filter bank:
example $r = M,\ L = N \cdot M = 2M$, convolution of upsampled subsequences $\tilde{v}_\lambda(k) \neq 0$ only for $k = k' \cdot M$ with partial impulse responses of length $N = 2$

the superposition of the interpolated and modulated subband signals:

$$y(k) = \sum_{\mu=0}^{M-1} y_\mu(k) \tag{4.44-a}$$

$$= \sum_{\mu=0}^{M-1} \overline{y}_\mu(k) \cdot e^{j\frac{2\pi}{M}\mu k} \tag{4.44-b}$$

$$= \sum_{\mu=0}^{M-1} \sum_{\kappa=0}^{k} \tilde{x}_\mu(\kappa) \cdot g(k-\kappa) \cdot e^{j\frac{2\pi}{M}\mu k}, \tag{4.44-c}$$

with the causal FIR $g(k)$ of length L, where L might be larger than M.

The key points in the derivation of the structure of the PPN synthesis filter bank of Fig. 4.19 from the reference structure of Fig. 4.8 are

1. substitution $k = i \cdot M + \lambda; \quad \lambda \in \{0, 1, \ldots, M-1\}; \quad i \in \mathbb{Z}$
2. exchange of the order of the two summations in (4.44-c).

Due to the periodicity of the complex exponential function with

$$e^{+j\frac{2\pi}{M}\mu(i \cdot M+\lambda)} = e^{+j\frac{2\pi}{M}\mu\lambda}$$

we obtain

$$y(i \cdot M + \lambda) = \sum_{\kappa=0}^{i \cdot M+\lambda} \underbrace{\left(\sum_{\mu=0}^{M-1} \tilde{x}_\mu(\kappa) \cdot e^{+j\frac{2\pi}{M}\mu\lambda} \right)}_{\tilde{v}_\lambda(\kappa)} \cdot g(i \cdot M + \lambda - \kappa). \qquad (4.44\text{-d})$$

For each fixed but arbitrary value of κ, the expression inside the brackets is the IDFT of the complex samples $\tilde{x}_\mu(\kappa)$, $\mu = 0, 1, \ldots, M-1$. If the IDFT routine includes a scaling factor of $1/M$, this can be corrected by subsequent multiplication with M according to

$$\begin{aligned} \tilde{v}_\lambda(\kappa) &= \sum_{\mu=0}^{M-1} \tilde{x}_\mu(\kappa) \cdot e^{+j\frac{2\pi}{M}\mu\lambda} \\ &= M \cdot \text{IDFT}\{\tilde{x}_\mu(\kappa)\}\,; \quad \kappa = \text{fixed}\,. \end{aligned}$$

It should be noted that $\tilde{x}_\mu(\kappa)$ is an upsampled sequence with (in the general case) $r-1$ intermediate zero samples between each pair of the decimated samples according to (4.8). The sequence $\tilde{v}_\lambda(\kappa)$ has the same temporal structure

$$\tilde{v}_\lambda(\kappa) \begin{cases} \neq 0 & \kappa = \kappa' \cdot r \\ = 0 & \text{else}\,. \end{cases} \qquad (4.44\text{-e})$$

Finally, we get for each fixed index $\lambda = 0, 1, \ldots, M-1$ and variable time or frame index i

$$\begin{aligned} y(i \cdot M + \lambda) &= \sum_{\kappa=0}^{i \cdot M+\lambda} \tilde{v}_\lambda(\kappa) \cdot g(i \cdot M + \lambda - \kappa) & (4.44\text{-f}) \\ &= \tilde{v}_\lambda(\kappa) * \tilde{g}_\lambda(\kappa)\,. & (4.44\text{-g}) \end{aligned}$$

At the time instances $k = i \cdot M + \lambda$ the output sample $y(k)$ is determined solely by the sequence $\tilde{v}_\lambda(\kappa)$ filtered with the partial impulse response

$$\tilde{g}_\lambda(\kappa) = \begin{cases} g(\kappa) & \kappa = \kappa' \cdot r + \lambda \\ 0 & \text{else} \end{cases} \qquad \lambda = 0, 1, \ldots, M-1, \qquad (4.45)$$

i.e., a decimated and upsampled version of the interpolator impulse response $g(k)$.

Therefore, we only need to deal with the decimated sequences $x_\mu(\kappa')$ and $v_\lambda(\kappa')$ and to carry out the IDFT every r-th sampling interval.

The corresponding block diagram is shown for the special case $r = M$ and $L = 2M$ in Fig. 4.19. For the sake of compatibility with (4.44-f), the input sequences $\tilde{v}_\lambda(k)$ to the partial filters are obtained here from the sequences $v_\lambda(k')$ by upsampling.

However, as $\tilde{v}_\lambda(k)$ as well as the partial impulse $\tilde{g}_\lambda(k)$ responses have $r-1$ zero samples between each pair of non-zero samples, only every r-th output sample of each of the partial filters can take non-zero values according to

$$\tilde{w}_\lambda(\kappa)\begin{cases} \neq 0 & \kappa=\kappa^{'}\cdot r+\lambda \\ =0 & \text{else}\ . \end{cases} \tag{4.46}$$

The superposition of the filter output samples $\tilde{w}_\lambda(k)$ of Fig. 4.19 can be performed by the superposition

$$\begin{aligned} y(k) &= \sum_{\lambda=0}^{M-1} \tilde{w}_\lambda(k) \quad ; \quad \text{any } k \\ &= \tilde{w}_\lambda(k) \quad ; \quad k=i\cdot M+\lambda\ . \end{aligned} \tag{4.47}$$

Furthermore, the polyphase filters can also be run at the decimated sampling rate and upsampling of $v_\lambda(k^{'})$ is not required. This leads to the final and efficient solution as given in Fig. 4.20, with

$$g_\lambda(k^{'}) = M\cdot g(k^{'}\cdot r+\lambda) \tag{4.48}$$

$$G_\lambda(z) = M\cdot\sum_{k^{'}=0}^{\infty} g_\lambda(k^{'})\cdot z^{-k^{'}}\ . \tag{4.49}$$

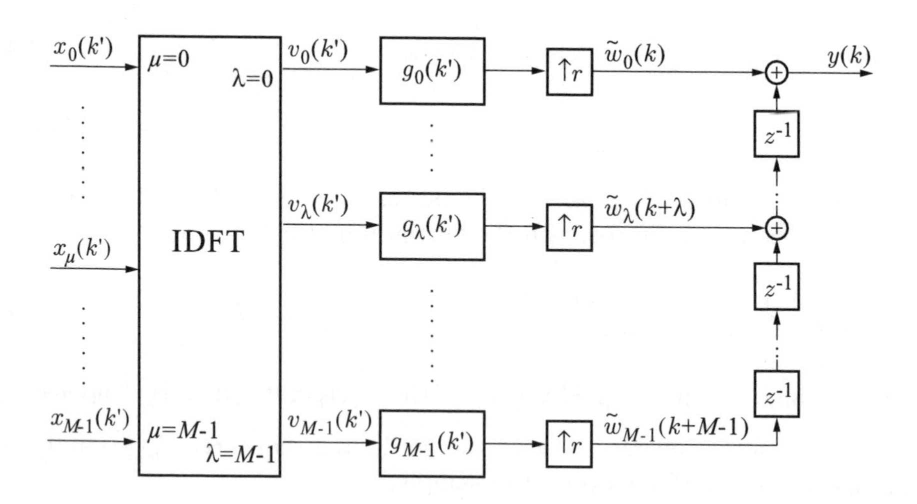

Figure 4.20: General solution of the efficient polyphase synthesis filter bank: DFT and partial filters are running at the reduced sampling rate f_s/r

Note that in comparison to Fig. 4.19 the scale factor M is now applied by scaling the partial filters for reasons of simplicity. The delays $z^{-\lambda}$ of the individual filter branches are provided by the output delay chain.

The chain of delay elements and adders at the output performs the superposition of the samples $\tilde{w}_\lambda(k)$ according to (4.47).

The conventional IDFT synthesis turns out to be a special case of the structure of Fig. 4.20 if the interpolator has rectangular impulse response

$$g(k) = \begin{cases} 1 & 0 \leq k < L-1 \\ 0 & \text{else} . \end{cases} \tag{4.50}$$

Note that in contrast to the conventional analysis and synthesis by DFT and IDFT, the polyphase concept allows us with little additional complexity to improve significantly the spectral selectivity and the interpolation task. In the analysis stage, each of the M window multiplications actually has to be replaced by a short convolution with only $N = L/M$ coefficients $h_\lambda(k')$. The typical parameter selection is $N = 2, \ldots, 4$. In the synthesis stage we need L/r multiplications for each of the partial filters $g_\lambda(k')$. The design criteria and procedures for the prototype lowpass $h(k)$ and the interpolator $g(k)$ as described in Sections 4.1.1 and 4.1.4 apply to the PPN analysis and PPN synthesis filter bank without any modification.

4.3 Quadrature Mirror Filter Banks

The objective of the QMF approach is the spectral decomposition of the signal $x(k)$ into $M = 2^K$ *real-valued* subband signals with maximum sampling rate decimation as well as the reconstruction (synthesis) of the signal from the subband signals. The analysis and the synthesis filter banks are based on special half-band filters which are called *quadrature mirror filters* (QMFs) (e.g., [Esteban, Galand 1977], [Vaidyanathan 1993]). These filters are used in a tree structure with decimation/interpolation by $r = 2$ in each stage of the tree.

4.3.1 Analysis–Synthesis Filter Bank

We first consider the special case with $M = 2$, which is the basic block of the tree structure.

The input signal is split by a lowpass filter with impulse response $h_{LP}(k)$ and a complementary highpass filter with impulse response $h_{HP}(k)$ into a lowpass signal $\overline{x}_0(k)$ and a highpass signal $\overline{x}_1(k)$. As the lowpass and highpass filters are half-band

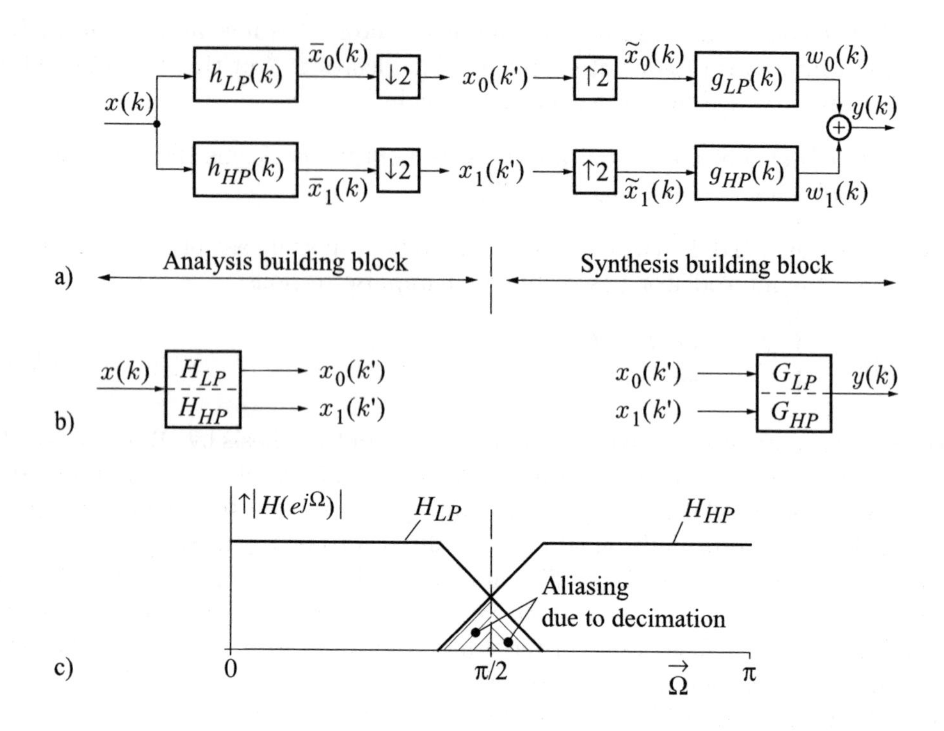

Figure 4.21: QMF analysis–synthesis bank
a) Block diagram of basic building block $M = 2$
b) Simplified block diagram
c) Schematic of frequency responses

filters, the sample rate can be decimated by a factor $r = 2$ (see Fig. 4.21):

$$x_0(k^{'}) = \overline{x}_0(2 \cdot k^{'}) \quad ; \quad x_1(k^{'}) = \overline{x}_1(2 \cdot k^{'} + \rho) \quad ; \quad \rho \in \{0, 1\} . \tag{4.51}$$

It is assumed that in the lowpass channel the even-numbered samples are selected and in the highpass channel either the even samples ($\rho = 0$) or the odd samples ($\rho = 1$) .

In the synthesis building block the subband signals are upsampled by a factor 2 (insertion of zero samples) and interpolated by lowpass and highpass filters with impulse responses $g_{LP}(k)$ and $g_{HP}(k)$. The interpolated signals $w_0(k)$ and $w_1(k)$ are added. As the filters are non-ideal, aliasing cannot be avoided (see Fig. 4.21-c). However, the disturbing aliasing components can be eliminated within the synthesis process as shown below.

With the basic blocks of Fig. 4.21-b a tree-structured analysis–synthesis filter bank can be constructed as illustrated in Fig. 4.22 for $M = 2^K = 8$. The input signal

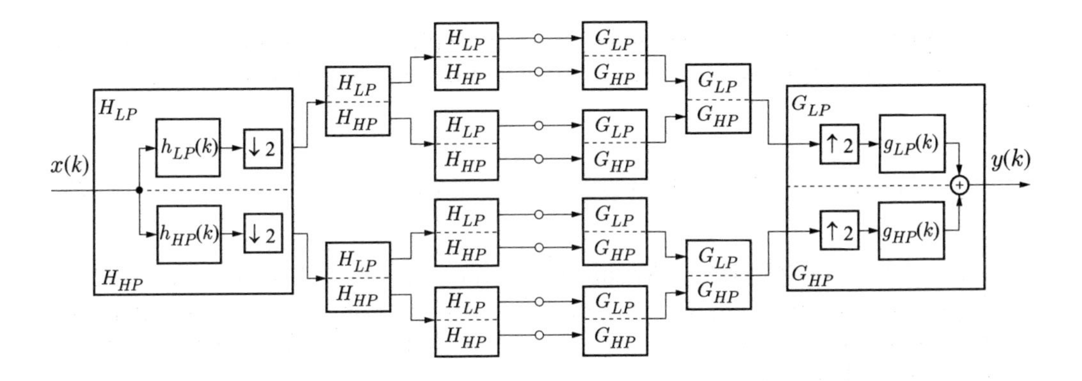

Figure 4.22: Tree-structured QMF analysis–synthesis bank, $N = 2^K = 8$ channels

$x(k)$ is decomposed into $M = 8$ subband signals by $K = 3$ stages. In each stage the sampling rate is decimated by $r = 2$.

4.3.2 Compensation of Aliasing and Signal Reconstruction

In what follows the analysis–synthesis blocks of Fig. 4.21-a will be discussed in detail.

We assume a lowpass filter with impulse response $h_{LP}(k) = h(k)$ and transfer function $H_{LP}(z) = H(z)$. Furthermore, the highpass filter is obtained from the lowpass filter by modulation

$$h_{HP}(k) = e^{\pm j\pi k} \cdot h_{LP}(k) = (-1)^k \cdot h(k) \tag{4.52-a}$$

$$H_{HP}(e^{j\Omega}) = H(e^{+j(\Omega \mp \pi)}) \tag{4.52-b}$$

$$H_{HP}(z) = H(-z)\,. \tag{4.52-c}$$

The two subband signals $\overline{x}_0(k)$ and $\overline{x}_1(k)$ are described in the z-domain according to Fig. 4.21-a as

$$\overline{X}_0(z) = X(z) \cdot H(z) \tag{4.53}$$

$$\overline{X}_1(z) = X(z) \cdot H(-z)\,. \tag{4.54}$$

The sampling rate decimation with subsequent upsampling by $r = 2$ can be formulated analytically by multiplication of $\overline{x}_0(k)$ and $\overline{x}_1(k)$ with a sampling function in the time domain:

$$\tilde{x}_0(k) = \overline{x}_0(k) \cdot p^{(2)}(k - \rho) \quad ; \quad \rho \in \{0, 1\} \tag{4.55-a}$$

$$\tilde{x}_1(k) = \overline{x}_1(k) \cdot p^{(2)}(k - \rho)\,. \tag{4.55-b}$$

The decimation in the two channels can be in phase ($\rho = 0$ or $\rho = 1$ in both channels) or out of phase ($\rho = 0$ in channel 0 and $\rho = 1$ in channel 1, or vice versa) with

$$p^{(2)}(k-\rho) = \frac{1}{2}\left[1 + (-1)^{\rho} \cdot (-1)^{k}\right] \; ; \quad \rho \in \{0, 1\} \, . \tag{4.56}$$

Because $(-1)^k = e^{+j\pi k}$ the z-domain representation of (4.55-a) and (4.55-b) can easily be obtained by using the modulation theorem. For reasons of simplicity we assume that the lowpass signal $\overline{x}_0(k)$ is decimated by $p^{(2)}(k)$ and the highpass signal $\overline{x}_1(k)$ either by applying $p^{(2)}(k)$ or $p^{(2)}(k-1)$:

$$\tilde{X}_0(z) = \frac{1}{2}\Big[\overline{X}_0(z) + \overline{X}_0(-z)\Big] \tag{4.57-a}$$

$$\tilde{X}_1(z) = \frac{1}{2}\Big[\overline{X}_1(z) + (-1)^{\rho} \cdot \overline{X}_1(-z)\Big] \, . \tag{4.57-b}$$

The schematics of the frequency characteristics are given in Fig. 4.23. Note that after decimation the frequency axis has been normalized to $\Omega' = r \cdot \Omega = 2 \cdot \Omega$ and that the highpass signal occurs in a mirrored version in the baseband $0 \le \Omega' \le \pi$.

Within the synthesis building block the two signals $\tilde{x}_0(k)$ and $\tilde{x}_1(k)$ are interpolated using the filters with the z-transforms $G_{LP}(z)$ and $G_{HP}(z)$ to obtain the intermediate signals $w_0(k)$ and $w_1(k)$ at the original sampling rate.

The reconstructed output signal $y(k)$ is finally given in the z-domain by

$$Y(z) = W_0(z) + W_1(z) \tag{4.58-a}$$

$$= G_{LP}(z) \cdot \tilde{X}_0(z) + G_{HP}(z) \cdot \tilde{X}_1(z) \tag{4.58-b}$$

$$= G_{LP}(z) \cdot \frac{1}{2}\Big[X(z)H(z) + X(-z)H(-z)\Big]$$

$$+ \; G_{HP}(z) \cdot \frac{1}{2}\Big[X(z)H(-z) + (-1)^{\rho} \cdot X(-z)H(z)\Big] \tag{4.58-c}$$

$$= \frac{1}{2}X(z) \cdot \Big[G_{LP}(z)H(z) + G_{HP}(z)H(-z)\Big]$$

$$+ \; \frac{1}{2}X(-z) \cdot \Big[G_{LP}(z)H(-z) + (-1)^{\rho} \cdot G_{HP}(z)H(z)\Big] \, . \tag{4.58-d}$$

The first part of (4.58-d) constitutes the desired signal and the second part the disturbing aliasing component $\big(X(-z) \hat{=} X(e^{j(\Omega-\pi)})\big)$. The aliasing component can be compensated as follows:

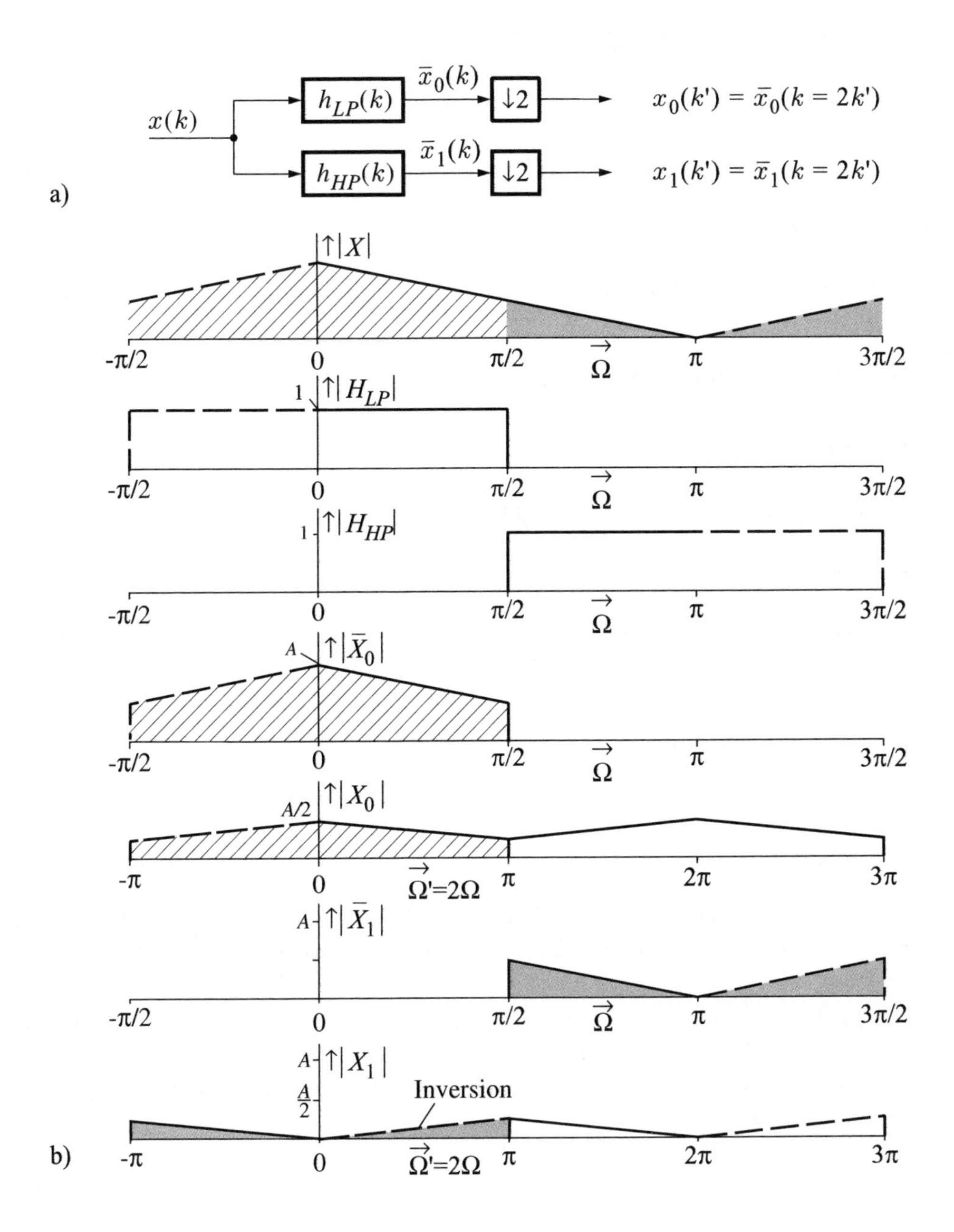

Figure 4.23: Spectral relations in the QMF analysis block
a) Block diagram
b) Schematics of spectra
(Ω' = normalized frequency after decimation)

a) Aliasing compensation for $\rho = 0$

The requirement

$$G_{LP}(z) \cdot H(-z) + G_{HP}(z) \cdot H(z) \stackrel{!}{=} 0 \tag{4.59}$$

can be fulfilled if

$$G_{LP}(z) = H(z) \text{ and } G_{HP}(z) = -H(-z) = -G_{LP}(-z)\,. \tag{4.60-a}$$

b) Aliasing compensation for $\rho = 1$

The requirement

$$G_{LP}(z) \cdot H(-z) - G_{HP}(z) \cdot H(z) \stackrel{!}{=} 0$$

can be met if

$$G_{LP}(z) = H(z) \text{ and } G_{HP}(z) = H(-z)\,. \tag{4.60-b}$$

The general solution for a) and b) is

$$G_{LP}(z) = H(z) \tag{4.60-c}$$
$$G_{HP}(z) = -(-1)^{\rho} \cdot H(-z)\,.$$

Hence, only the lowpass filter with transfer function $H(z)$ has to be designed. Thus, the effective transfer function of the analysis–synthesis system is given by

$$H_{AS}(z) = \frac{1}{2}\left[H^2(z) - (-1)^{\rho} \cdot H^2(-z)\right]\,. \tag{4.60-d}$$

Perfect reconstruction of the signal is achieved if the overall frequency response is a delay by k_0 samples, i.e.,

$$H_{AS}(z) \stackrel{!}{=} z^{-k_0} \tag{4.61}$$

according to $y(k) = x(k - k_0)$.

The desired behavior (4.61) can be approximated with an accuracy which is sufficient for practical applications by appropriate design of a linear-phase FIR $h(k)$ of length $L = n + 1$ with

$$h(k) = h(n-k)\,; \quad k = 0, 1, \ldots, n\,. \tag{4.62}$$

In the case of $\rho = 0$, it can be shown that an impulse response of even length and in the case of $\rho = 1$ a filter with odd length are needed to avoid $|H_{AS}(e^{j\frac{\pi}{2}})| = 0$.

The desired compensation of the aliasing components does not necessarily require the selection $G_{LP}(z) = H(z)$. A more general condition can be derived from the second part of (4.58-d) with

$$\frac{G_{LP}(z)}{G_{HP}(z)} \stackrel{!}{=} -(-1)^{\rho} \frac{H(z)}{H(-z)} . \tag{4.63}$$

This condition provides new degrees of freedom for the optimization. Various solutions can be found in the literature (e.g., [Jain, Crochiere 1983], [Jain, Crochiere 1984], [Johnston 1980], [Wackersreuther 1987], [Smith, Barnwell 1986]).

4.3.3 Efficient Implementation

The QMF analysis and synthesis filter bank with $M = 2$ can be considered as a special case of the PPN filter bank with DFT length $M = 2$. Therefore, the computationally efficient approaches according to Fig. 4.17 and Fig. 4.20 can be applied here.

For reasons of simplicity we will consider only the case $\rho = 0$ and the filter design according to (4.60-a) with even length $L = n + 1$. Taking into account the sampling rate decimation and the fact that the relation between the lowpass response $h_{LP}(k) = h(k)$ and the highpass response is given in the time domain by (see (4.52-a))

$$h_{HP}(k) = (-1)^k \cdot h(k) , \tag{4.64}$$

we get (see also Fig. 4.21)

$$\begin{aligned}
\overline{x}_0(k = 2k') &= \sum_{\kappa=0}^{n} h(\kappa) x(2k' - \kappa) \\
&= \sum_{\kappa'=0}^{\frac{n-1}{2}} h(2\kappa') x(2k' - 2\kappa') + \sum_{\kappa'=0}^{\frac{n-1}{2}} h(2\kappa' + 1) x(2k' - 2\kappa' - 1) \\
&= a(k') + b(k')
\end{aligned} \tag{4.65}$$

$$\begin{aligned}
\overline{x}_1(k = 2k') &= \sum_{i=0}^{n} h(\kappa) \cdot (-1)^{\kappa} \cdot x(2k' - \kappa) \\
&= a(k') - b(k') .
\end{aligned} \tag{4.66}$$

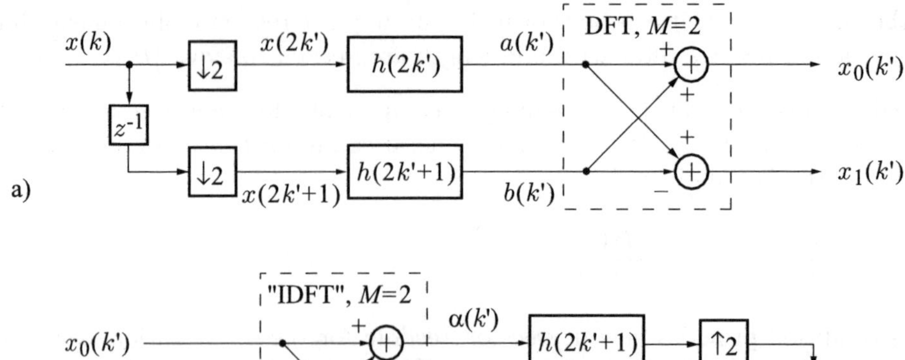

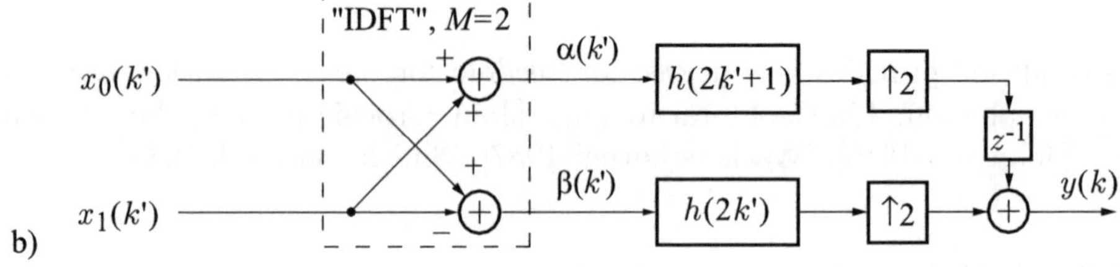

Figure 4.24: Efficient implementation of QMF bank
a) Analysis filter bank
b) Synthesis filter bank

Both decimated samples, $\overline{x}_0(k = 2k^{'})$ and $\overline{x}_1(k = 2k^{'})$, can be calculated from the quantities $a(k^{'})$ and $b(k^{'})$. $a(k^{'})$ is the result of the convolution of the even samples of $x(k)$ with the even samples of $h(k)$, and $b(k^{'})$ results from the convolution of the respective odd samples. If the original impulse response is decomposed into its two polyphase components, the overall computational complexity for calculating $\overline{x}_0(2k^{'})$ and $\overline{x}_1(2k^{'})$ is only slightly higher than the complexity of a single convolution of length $n+1$. The corresponding block diagram for the QMF analysis bank is illustrated in Fig. 4.24-a.

The derivation of the efficient structure of the synthesis block is slightly more complicated. Again, we consider the case $\rho = 0$ and the filter selection

$$g_{LP}(k) = h(k) \quad ; \quad g_{HP}(k) = -(-1)^k \cdot h(k) \tag{4.67}$$

with even length $L = n+1$. Furthermore, we have to take into consideration that the interpolator input signals $\tilde{x}_0(k)$ and $\tilde{x}_1(k)$ in Fig. 4.21-a have non-zero values only at even time instances, i.e.,

$$\tilde{x}_\mu(k) = \begin{cases} x_\mu(k^{'}) & k = 2k^{'}\,; \qquad \mu \in \{0,1\} \\ 0 & k = 2k^{'}+1\,. \end{cases} \tag{4.68}$$

We get

$$y(k) = \sum_{\kappa=0}^{n} h(\kappa) \cdot \tilde{x}_0(k-\kappa) \; + \; \sum_{\kappa=0}^{n} -(-1)^{\kappa} \cdot h(\kappa) \cdot \tilde{x}_1(k-\kappa) \tag{4.69}$$

$$= \sum_{\kappa'=0}^{\frac{n-1}{2}} h(2\kappa') \cdot \left[\tilde{x}_0(k-2\kappa') \; - \; \tilde{x}_1(k-2\kappa')\right]$$

$$+ \sum_{\kappa'=0}^{\frac{n-1}{2}} h(2\kappa'+1) \cdot \left[\tilde{x}_0(k-2\kappa'-1) + \tilde{x}_1(k-2\kappa'-1)\right] \tag{4.70}$$

$$= \begin{cases} \sum_{\kappa'=0}^{\frac{n-1}{2}} h(2\kappa') \cdot \beta\left(k' - \kappa'\right) & k = 2k' \\ \sum_{\kappa'=0}^{\frac{n-1}{2}} h(2\kappa'+1) \cdot \alpha\left(k' - \kappa'\right) & k = 2k'+1. \end{cases}$$

Due to (4.68), the first summation in (4.70) contributes to the even time instances $k = 2k'$ and the second summation to the odd instances $k = 2k' + 1$ only.

The block diagram of the efficient structure is given in Fig. 4.24-b. The multiplexing, i.e., interlacing of the even and the odd samples, can be described by upsampling by factor the $r = 2$ in combination with the delay and sum operation.

An example application is the QMF bank of the G.722 wideband speech codec (see Appendix A) with $M = 2$ channels and a prototype lowpass filter with $n = 24$. The frequency responses of the lowpass and highpass filters, and the overall frequency response of the analysis–synthesis filter bank, are given in Fig. 4.25.

The QMF concept can easily be modified to achieve non-uniform frequency resolution by leaving out some of the filters of the tree structure. For a QMF bank with $M = 8$, the structure of the analysis filter bank and a schematic of the frequency response are illustrated in Fig. 4.26. It should be noted that this filter bank structure is closely related to the wavelet transform, e.g., [Burrus et al. 1998], [Vetterli, Kovačević 1995], [Vaidyanathan 1993].

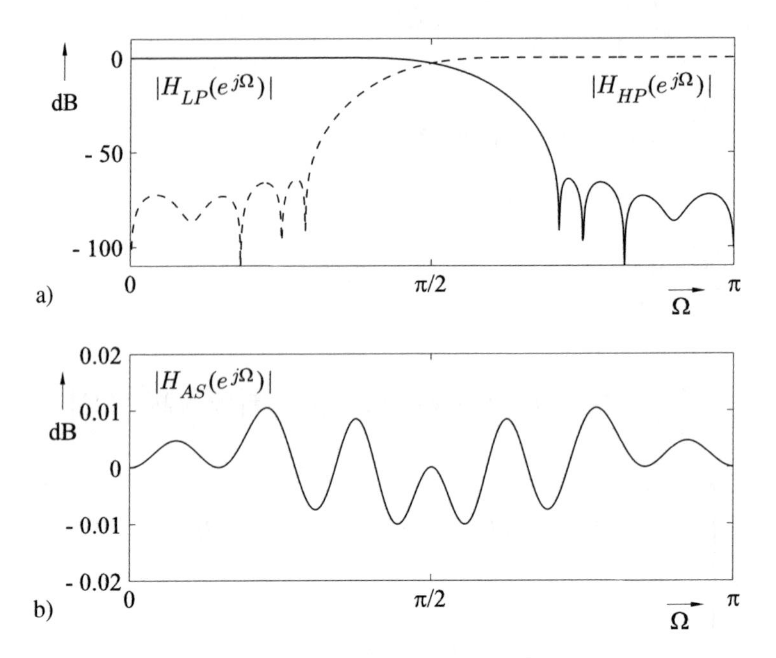

Figure 4.25: Example: QMF bank of the G.722 wideband speech codec with $M = 2$ channels
a) Magnitude response of the half-band filters
b) Resulting overall magnitude response

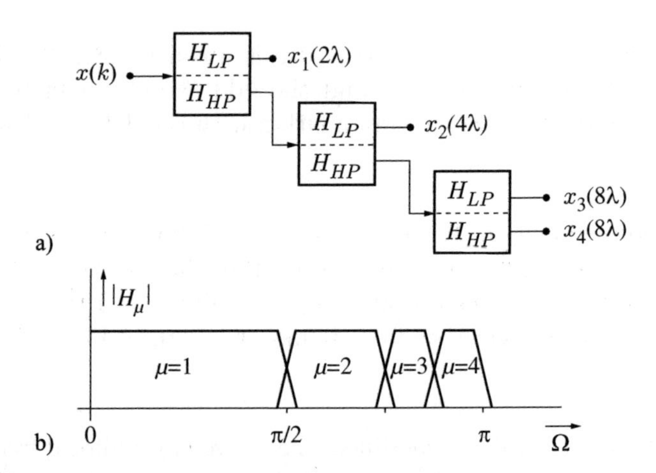

Figure 4.26: QMF tree structure for non-uniform frequency resolution
a) Block diagram
b) Schematic of the effective frequency resolution

Bibliography

Bellanger, M.; Bonnerot, G.; Coudreuse, M. G. (1976). Digital Filtering by Polyphase Network: Application to Sampling-rate Alteration and Filter Banks, *IEEE Transactions on Acoustics, Speech and Signal Processing*, vol. 24, pp. 109–114.

Boite, R.; Leich, H. (1981). A New Procedure for the Design of High-order Minimum-phase FIR Digital or CCD Filters, *IEEE Transactions on Signal Processing*, vol. 3, pp. 101–108.

Burrus, C. S.; Gopinath, R. A.; Guo, H. (1998). *Introduction to Wavelets and Wavelet Transforms: A Primer*, Prentice Hall, Upper Saddle River, New Jersey.

Crochiere, R. E.; Rabiner, L. R. (1983). *Multirate Digital Signal Processing*, Prentice Hall, Englewood Cliffs, New Jersey.

Dehner, G. (1979). Program for the Design of Recursive Digital Filters, *Programs for Digital Signal Processing*, IEEE Press, New York.

Esteban, D.; Galand, C. (1977). Application of Quadrature Mirror Filters to Split-band Voice Coding Schemes, *Proceedings of the IEEE International Conference on Acoustics, Speech, and Signal Processing (ICASSP)*, pp. 191–195.

Jain, V. K.; Crochiere, R. E. (1983). A Novel Approach to the Design of Analysis-synthesis Filter Banks, *Proceedings of the IEEE International Conference on Acoustics, Speech, and Signal Processing (ICASSP)*, pp. 228–231.

Jain, V. K.; Crochiere, R. E. (1984). Quadrature-mirror Filter Design in the Time Domain, *IEEE Transactions on Acoustics, Speech and Signal Processing*, vol. 32, pp. 353–361.

Johnston, J. D. (1980). A Filter Family Designed for Use in Quadrature-mirror Filter Banks, *Proceedings of the IEEE International Conference on Acoustics, Speech, and Signal Processing (ICASSP)*, Denver, USA, pp. 291–294.

Kliewer, J. (1996). Simplified Design of Linear-phase Prototype Filters for Modulated Filter Banks, *Signal Processing VIII: Theories and Applications*, G. Ramponi, G. L. Sicuranza, S. Carrato, S. Marsi (eds.), Elsevier, Amsterdam, pp. 1191–1194.

Mitra, K. M. (1998). *Digital Signal Processing - A Computer Based Approach*, McGraw-Hill, New York.

Nguyen, T. (1994). Near-perfect-reconstruction Pseudo-QMF Banks, *IEEE Transactions on Signal Processing*, vol. 42, pp. 64–75.

Parks, T. W.; Burrus, C. S. (1987). *Digital Filter Design*, John Wiley & Sons, Ltd, Chichester.

Parks, T. W.; McClellan, J. H.; Rabiner, L. R. (1979). FIR Linear-Phase Filter-Design Program, *Programs for Digital Signal Processing*, IEEE Press, New York.

Schüssler, H. W. (1994). *Digitale Signalverarbeitung I*, Springer Verlag, Berlin (in German).

Smith, M. J. T.; Barnwell, T. P. (1986). Exact Reconstruction Techniques for Tree-structured Subband Coders, *IEEE Transactions on Acoustics, Speech and Signal Processing*, vol. 34, pp. 434–441.

Vaidyanathan, P. P. (1990). Multirate Digital Filters, Filter Banks, Polyphase Networks, and Applications: A Tutorial, *Proceedings of the IEEE*, vol. 78, pp. 56–93.

Vaidyanathan, P. P. (1993). *Multirate Systems and Filterbanks*, Prentice Hall, Englewood Cliffs, New Jersey.

Vary, P. (1979). On the Design of Digital Filter Banks Based on a Modified Principle of Polyphase, *International Journal of Electronics and Communications (AEÜ, Archiv für Elektronik und Übertragungstechnik)*, vol. 33, pp. 293–300.

Vary, P.; Heute, U. (1980). A Short-time Spectrum Analyzer with Polyphase-network DFT, *Signal Processing*, vol. 2, pp. 55–65.

Vary, P.; Heute, U. (1981). A Digital Filter Bank with Polyphase Network and FFT Hardware: Measurements and Applications, *Signal Processing*, vol. 3, pp. 307–319.

Vary, P.; Heute, U.; Hess, W. (1998). *Digitale Sprachsignalverarbeitung*, B. G. Teubner, Stuttgart (in German).

Vary, P.; Wackersreuther, G. (1983). A Unified Approach to Digital Polyphase Filter Banks, *International Journal of Electronics and Communications (AEÜ, Archiv für Elektronik und Übertragungstechnik)*, vol. 37, pp. 29–34.

Vetterli, M.; Kovačević, J. (1995). *Wavelets and Subband Coding*, Prentice Hall, Engelwood Cliffs, New Jersey.

Wackersreuther, G. (1985). On the Design of Filters for Ideal QMF and Polyphase Filter Banks, *International Journal of Electronics and Communications (AEÜ, Archiv für Elektronik und Übertragungstechnik)*, vol. 39, pp. 123–130.

Wackersreuther, G. (1986). Some New Aspects of Filters for Filter Banks, *IEEE Transactions on Acoustics, Speech and Signal Processing*, vol. 34, pp. 1182–1200.

Wackersreuther, G. (1987). *Ein Beitrag zum Entwurf digitaler Filterbänke*, PhD thesis. Ausgewählte Arbeiten über Nachrichtensysteme, vol. 64, H. W. Schüßler (ed.), Universität Erlangen (in German).

5

Stochastic Signals and Estimation

In this chapter, we will review the basic concepts and tools which are required to deal with stochastic signals such as speech signals. Among these are random variables and stochastic processes as well as power spectra and fundamentals of estimation theory. The objective is to provide a compilation of useful concepts and theorems. More extensive discussions of these subjects can be found in many excellent textbooks, for instance, [Papoulis, Unnikrishna Pillai 2001] and [Melsa, Cohn 1978].

5.1 Basic Concepts

5.1.1 Random Events and Probability

Modern theory of probability [Kolmogorov 1933] defines the probability $P(A_i)$ of an event A_i on the basis of set-theoretic concepts and axioms, not on the basis of *observed* random phenomena. It thus facilitates the treatment of random processes as it provides a clear conceptual separation between *observed* random phenomena and theoretical *models* of such phenomena.

Digital Speech Transmission: Enhancement, Coding and Error Concealment
Peter Vary and Rainer Martin

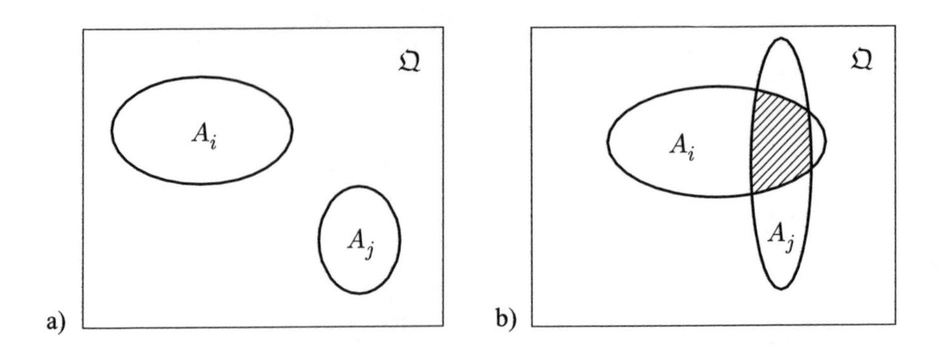

Figure 5.1: Two random events A_i and A_j in the probability space $\mathfrak{Q}$
a) Events are mutually exclusive
b) Events are not mutually exclusive

Given a set $\mathfrak{Q}$ of elementary events $\{\xi_1, \xi_2, \ldots\}$ and a set $\mathfrak{F}$ of subsets $\{A_1, A_2, \ldots\}$ of $\mathfrak{Q}$, we call the subsets A_i in $\mathfrak{F}$ random events. We assume that $\mathfrak{F}$ includes $\mathfrak{Q}$, the empty set ϕ, any union $\cup A_i$ of subsets, and the complement $\overline{A}_i$ of any subset A_i.[1] The complement $\overline{A}_i$ of a subset A_i is defined as the set $\mathfrak{Q}$ without the elementary events in A_i. We may then assign a probability measure $P(A_i)$ to a random event A_i in such a way that

- the probability is a non-negative real number: $P(A_i) \geq 0$;
- the probability of the certain event is one: $P(\mathfrak{Q}) = 1$;
- the probability that either A_i or A_j occurs is $P(A_i \cup A_j) = P(A_i) + P(A_j)$, provided that the two events A_i and A_j are mutually exclusive.

The last condition is illustrated in Fig. 5.1-a. When two random events A_i and A_j are disjoint, the probability of A_i or A_j is the sum of the individual probabilities. If these events are not mutually exclusive (Fig. 5.1-b), the sum of the probabilities of the individual events is not equal to the probability of the event that A_i or A_j occurs. The triple $(\mathfrak{Q}, \mathfrak{F}, P)$ is called a probability space.

The joint probability of two events A_i and A_j is the probability that A_i and A_j occur and is denoted by $P(A_i, A_j)$. With respect to the set representation, the joint probability is a measure assigned to the intersection of events A_i and A_j in a space which contains all possible joint events (product space).

In the general case of N events A_i, $i = 1 \ldots N$, we denote the probability of the simultaneous occurrence of these events as $P(A_1, A_2, \ldots, A_N)$.

[1]These properties define a σ-algebra [Kolmogorov 1933].

5.1.2 Conditional Probabilities

Conditional probabilities capture the notion of *a priori* information. The conditional probability of an event B, given an event A, with $P(A) \neq 0$, is defined as the joint probability $P(B, A)$ normalized on the probability of the given event A

$$P(B \mid A) = \frac{P(B, A)}{P(A)} . \tag{5.1}$$

When N mutually exclusive events $A_1, A_2, \ldots, A_N$ partition the set of elementary events Ω in such a way that $P(A_1) + P(A_2) + \cdots + P(A_N) = 1$, we may write the *total probability* of an arbitrary event B as

$$P(B) = P(B \mid A_1)P(A_1) + P(B \mid A_2)P(A_2) + \cdots + P(B \mid A_N)P(A_N) .$$

Furthermore, for any of these events A_i we may write

$$\begin{aligned} P(A_i \mid B) &= \frac{P(B, A_i)}{P(B)} \\ &= \frac{P(B \mid A_i)P(A_i)}{P(B)} \\ &= \frac{P(B|A_i)P(A_i)}{P(B|A_1)P(A_1) + P(B|A_2)P(A_2) + \cdots + P(B|A_N)P(A_N)} . \end{aligned}$$

This is also known as *Bayes' theorem.*

5.1.3 Random Variables

A random variable x maps elementary events ξ_i onto real numbers and, thus, is the basic vehicle for the mathematical analysis of random phenomena. In our context, random variables are used to represent samples of signals, parameters, and other quantities. A random variable may be continuous or discrete valued. While the former can attain any range of values on the real line, the latter is confined to countable and possibly finite sets of numbers.

Random variables may be grouped into vectors. Vectors of random variables may have two elements (bivariate) or a larger number of elements (multivariate). We may also define complex-valued random variables by mapping the real and the imaginary parts of a complex variable onto the elements of a bivariate vector of real-valued random variables.

5.1.4 Probability Distributions and Probability Density Functions

The *(cumulative) distribution function* P_x of a random variable x is defined as the probability that x is smaller than or equal to a threshold value u,

$$P_x(u) = P(x \leq u) . \tag{5.2}$$

We obtain the whole distribution function by allowing u to range from $-\infty$ to ∞. Obviously, P_x is non-negative and non-decreasing with $P_x(-\infty) = 0$. $P_x(\infty)$ corresponds to the certain event, therefore $P_x(\infty) = 1$.

The *probability density function* (PDF) is defined as the derivative (when it exists) of the distribution function with respect to the threshold u

$$p_x(u) = \frac{\mathrm{d}P_x(u)}{\mathrm{d}u} . \tag{5.3}$$

A PDF always satisfies

$$\int_{-\infty}^{\infty} p_x(u)\,\mathrm{d}u = 1 . \tag{5.4}$$

Using Dirac impulses, a PDF may be defined for probability distribution functions $P_x(u)$ with discontinuities.

The joint PDF of N random variables x_1, x_2, ..., x_N is defined as

$$p_{x_1 \cdots x_N}(u_1, \ldots, u_N) = \frac{\mathrm{d}^N P_{x_1 \cdots x_N}(u_1, \ldots, u_N)}{\mathrm{d}u_1 \ldots \mathrm{d}u_N} \tag{5.5}$$

where $P_{x_1 \cdots x_N}(u_1, \ldots, u_N)$ is given by

$$P_{x_1 \cdots x_N}(u_1, \ldots, u_N) = P(x_1 \leq u_1, \ldots, x_N \leq u_N) . \tag{5.6}$$

Given a joint PDF, we compute marginal densities by integrating over one or more variables. For example, given $p_{x_1 \cdots x_N}(u_1, \ldots, u_N)$ we compute the marginal density $p_{x_1 \cdots x_{N-1}}(u_1, \ldots, u_{N-1})$ as

$$p_{x_1 \cdots x_{N-1}}(u_1, \ldots, u_{N-1}) = \int_{-\infty}^{\infty} p_{x_1 \cdots x_N}(u_1, \ldots, u_N)\,\mathrm{d}u_N . \tag{5.7}$$

5.1.5 Conditional PDFs

We define the conditional probability density function $p_{x|A}(u \mid A)$ of a random variable x given an event A via the relation

$$p_{x|A}(u \mid A)\, P(A) = p_{xA}(u, A) \tag{5.8}$$

where $p_{xA}(u, A)$ is defined as

$$p_{xA}(u, A) = \frac{\mathrm{d}P_{xA}(u, A)}{\mathrm{d}u} = \frac{\mathrm{d}P(x \leq u, A)}{\mathrm{d}u}\,. \tag{5.9}$$

The event A may comprise a random variable y which is equal to a value v. For a discrete random variable we may restate the above relation in terms of densities using the simplified notation

$$p_{x|y}(u \mid v)\, P(y = v) = p_{xy}(u, v)\,. \tag{5.10}$$

This is called the *mixed form* of the Bayes' theorem, which also holds for continuous random variables, as

$$p_{x|y}(u \mid v)\, p_y(v) = p_{xy}(u, v)\,. \tag{5.11}$$

Furthermore, we write the density version of Bayes' theorem as

$$p_{x|y}(u \mid v)\, p_y(v) = p_{xy}(u, v) = p_{y|x}(v \mid u)\, p_x(u)\,, \tag{5.12}$$

or with (5.7) as

$$p_{x|y}(u \mid v) \int_{-\infty}^{\infty} p_{x,y}(u, v)\,\mathrm{d}u = p_{x|y}(u \mid v) \int_{-\infty}^{\infty} p_{y|x}(v \mid u)\, p_x(u)\,\mathrm{d}u\,. \tag{5.13}$$

Any joint probability density may be factored into conditional densities as

$$\begin{aligned} p(x_1, \ldots, x_N) &= p(x_1|x_2, \ldots, x_N)\, p(x_2, \ldots, x_N) \\ &= p(x_1|x_2, \ldots, x_N)\, p(x_2|x_3, \ldots, x_N)\, p(x_3, \ldots, x_N) \\ &= p(x_1|x_2, \ldots, x_N)\, p(x_2|x_3, \ldots, x_N) \cdots p(x_{N-1} \mid x_N)\, p(x_N)\,. \end{aligned}$$

For three random variables, x_1, x_2, and x_3, the above *chain rule* simplifies to

$$p(x_1, x_2, x_3) = p(x_1 \mid x_2, x_3)\, p(x_2 \mid x_3)\, p(x_3)\,. \tag{5.14}$$

5.2 Expectations and Moments

For continuous random variables, the *mean* of a random variable x is given by its *expected value* $\mathrm{E}\{x\}$,

$$\mu_x = \mathrm{E}\{x\} = \int_{-\infty}^{\infty} u\, p_x(u)\, \mathrm{d}u\,, \tag{5.15}$$

and for a discrete random variable x which assumes one of M values u_i, $i = 1 \dots M$, by

$$\mu_x = \sum_{i=1}^{M} u_i\, p_x(u_i)\,. \tag{5.16}$$

The expectation is easily evaluated if the PDF of the random variable is known. In the remainder of this chapter we will be mostly concerned with continuous random variables. All of these results can easily be adapted to discrete random variables by exchanging the integration for a summation.

More generally, we may write the expectation of any function $f(x)$ of a random variable x as

$$\mathrm{E}\{f(x)\} = \int_{-\infty}^{\infty} f(u)\, p_x(u)\, \mathrm{d}u\,. \tag{5.17-a}$$

Thus, despite the transformation of x into $y = f(x)$, we may still use the probability density function $p_x(u)$ to compute the expected value $\mathrm{E}\{f(x)\}$. More specifically, with

$$f(x) = (x - x_0)^m$$

we obtain the m-th *central moment* with respect to x_0. For $x_0 = 0$ and $m = 2$ we have the *power* of x

$$\mathrm{E}\{x^2\} = \int_{-\infty}^{\infty} u^2\, p_x(u)\, \mathrm{d}u\,, \tag{5.17-b}$$

and for $x_0 = \mu_x$ and $m = 2$ the *variance*

$$\sigma_x^2 = \mathrm{E}\{(x - \mu_x)^2\} = \mathrm{E}\{x^2\} - \mu_x^2. \tag{5.17-c}$$

5.2.1 Conditional Expectations and Moments

The expected value of a random variable x conditioned by an event $y = v$ is defined as

$$\mathrm{E}_{x|y}\{x \mid y\} = \int_{-\infty}^{\infty} u \, p_{x|y}(u \mid v) \, \mathrm{d}u \,, \tag{5.18}$$

and, when no confusion is possible, also written as $\mathrm{E}\{x \mid y\}$. Using the relation

$$\mathrm{E}_{f(x)|y}\{f(x) \mid y\} = \int_{-\infty}^{\infty} f(u) \, p_{x|y}(u \mid v) \, \mathrm{d}u \tag{5.19}$$

and $f(x) = (x - x_0)^m$ we may compute any *conditional* central moment.

5.2.2 Examples

In what follows, we will discuss some of the frequently used PDFs and their moments. For an extensive treatment of these and other PDFs we refer the reader to [Papoulis, Unnikrishna Pillai 2001], [Johnson et al. 1994], [Kotz et al. 2000].

5.2.2.1 The Uniform Distribution

When a random variable x is uniformly distributed in the range $[X_1, X_2]$, its probability density is given by

$$p_x(u) = \begin{cases} \dfrac{1}{X_2 - X_1} & u \in [X_1, X_2] \\ 0 & u \notin [X_1, X_2] \,. \end{cases} \tag{5.20}$$

Figure 5.2 illustrates this density. For the uniform density we find the mean, the power, and the variance as follows:

$$\text{mean:} \qquad \mu_x = \frac{1}{2}[X_1 + X_2] \,, \tag{5.21}$$

$$\text{power:} \qquad \mathrm{E}\{x^2\} = \frac{1}{3}\left[X_1^2 + X_1 X_2 + X_2^2\right] , \tag{5.22}$$

$$\text{variance:} \qquad \sigma_x^2 = \frac{1}{12}(X_2 - X_1)^2 \,. \tag{5.23}$$

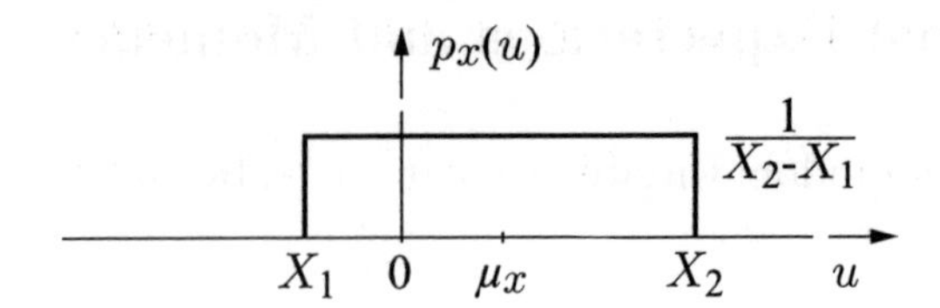

Figure 5.2: Uniform probability density

5.2.2.2 The Gaussian Density

The *Gaussian density* is defined as

$$p_x(u) = \frac{1}{\sqrt{2\pi}\sigma_x} \exp\left(-\frac{(u-\mu_x)^2}{2\sigma_x^2}\right) \tag{5.24}$$

and parameterized by its mean μ_x and variance $\sigma_x^2 > 0$. The Gaussian density is plotted for three different variances in Fig. 5.3-a. The power of a Gaussian distributed random variable is given by $\mathrm{E}\left\{x^2\right\} = \sigma_x^2 + \mu_x^2$.

By definition, a complex Gaussian random variable is a pair of real-valued random variables which are jointly Gaussian distributed. This definition includes the case of two independent real-valued Gaussian random variables which represent the real and the imaginary parts of the complex variable.

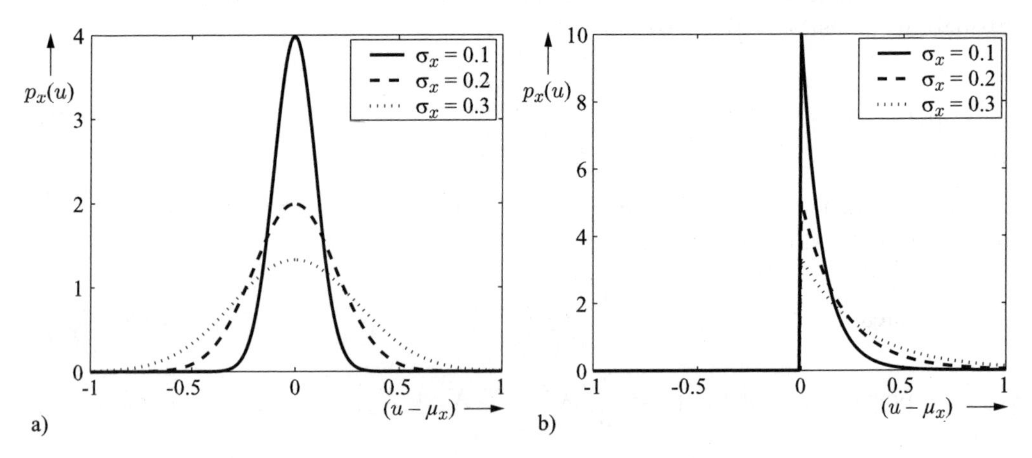

Figure 5.3: a) Gaussian probability density functions
b) Exponential probability density functions

5.2.2.3 The Exponential Density

For $\sigma_x > 0$ the *(one-sided) exponential density* (see Fig. 5.3-b) is given by

$$p_x(u) = \begin{cases} \frac{1}{\sigma_x} \exp\left(-\frac{u}{\sigma_x}\right) & u \geq 0 \\ 0 & u < 0 \end{cases} \tag{5.25}$$

where σ_x^2 is again the variance of the density function. The mean of an exponentially distributed random variable is given by σ_x, and hence its power is $2\sigma_x^2$.

5.2.2.4 The Laplace Density

For $\sigma_x > 0$, the *two-sided exponential density* (also known as the *Laplace density*) is defined as

$$p_x(u) \;=\; \frac{1}{\sqrt{2}\sigma_x} \exp\left(-\sqrt{2}\,\frac{|u-\mu_x|}{\sigma_x}\right) \tag{5.26}$$

where μ_x is the mean and σ_x^2 is the variance. The Laplace density function is plotted in Fig. 5.4-a.

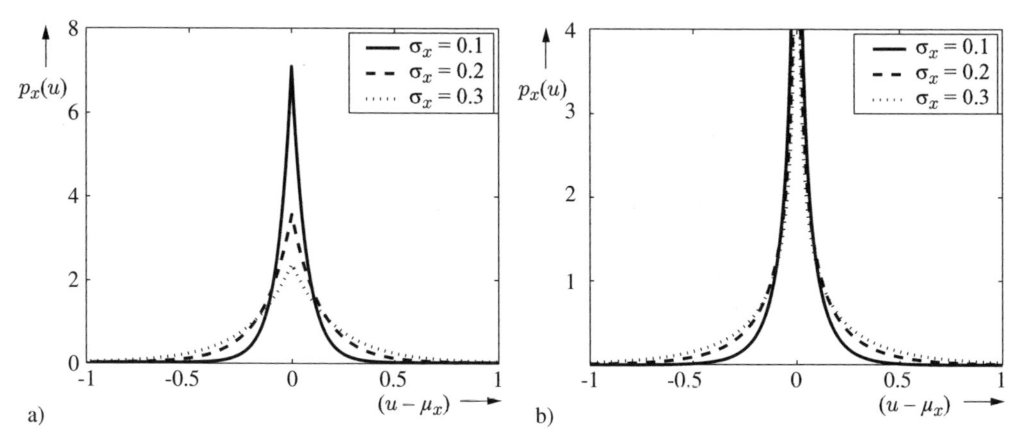

Figure 5.4: a) Laplace probability density functions
b) Gamma probability density functions

5.2.2.5 The Gamma Density

For $\sigma_x > 0$, the *gamma density*[2] (see Fig. 5.4-b) is given by

$$p_x(u) = \frac{\sqrt[4]{3}}{2\sqrt{2\pi\,\sigma_x}} \frac{1}{\sqrt{|u-\mu_x|}} \exp\left(-\frac{\sqrt{3}}{2}\frac{|u-\mu_x|}{\sigma_x}\right), \tag{5.27}$$

where μ_x is the mean and σ_x^2 is the variance.

5.2.3 Transformation of a Random Variable

Frequently, we will consider the PDF of a random variable $y = f(x)$ which is a deterministic function of another random variable x. To specify the PDF $p_y(u)$, we must first solve $u = f(x)$ for the real roots $x^{\langle k\rangle}$, $k = 1 \ldots N(u)$, where $N(u)$ depends on the value u. When for a specific value of u no real roots exist, we set $N(u) = 0$. The PDF of y is then given by [Papoulis, Unnikrishna Pillai 2001]

$$p_y(u) = \begin{cases} \dfrac{p_x(x^{\langle 1\rangle})}{|f'(x^{\langle 1\rangle})|} + \cdots + \dfrac{p_x(x^{\langle N(u)\rangle})}{|f'(x^{\langle N(u)\rangle})|} & N(u) > 0 \\ 0 & N(u) = 0 \end{cases} \tag{5.28}$$

where $f'(x)$ is the derivative of $f(x)$ with respect to x.

For any invertible function $u = f(x)$, we only have one single root $x = f^{-1}(u)$ and

$$\frac{1}{f'(x)_{|x=f^{-1}(u)}} = \frac{\mathrm{d}f^{-1}(u)}{\mathrm{d}u}. \tag{5.29}$$

Hence,

$$p_y(u) = \left|\frac{\mathrm{d}f^{-1}(u)}{\mathrm{d}u}\right| p_x\left(f^{-1}(u)\right). \tag{5.30}$$

As an example, we compute the PDF of the square of a random variable, i.e., $y = f(x) = x^2$ with $f'(x) = 2x$. Since for $u < 0$ the equation $u = x^2$ has no real roots we have $p_y(u) = 0$ for $u < 0$. For $u > 0$ we find two real roots $x^{\langle 1\rangle} = \sqrt{u}$ and $x^{\langle 2\rangle} = -\sqrt{u}$. Therefore, we may write

$$p_y(u) = \begin{cases} \dfrac{p_x(\sqrt{u})}{2\sqrt{u}} + \dfrac{p_x(-\sqrt{u})}{2\sqrt{u}} & u > 0 \\ 0 & u \le 0\,. \end{cases} \tag{5.31}$$

[2] Our definition is a special case of the more general gamma density function as defined in [Johnson et al. 1994]

Thus, if x is a Gaussian random variable with

$$p_x(u) = \frac{1}{\sqrt{2\pi}\sigma_x} \exp\left(-\frac{u^2}{2\sigma_x^2}\right) \tag{5.32}$$

we find the density of $y = x^2$ as

$$p_y(u) = \begin{cases} \frac{1}{\sqrt{2\pi u}\sigma_x} \exp\left(-\frac{u}{2\sigma_x^2}\right) & u > 0 \\ 0 & u \leq 0 . \end{cases} \tag{5.33}$$

5.2.4 Relative Frequencies and Histograms

In contrast to the abstract concept of probabilities and PDFs, the relative frequencies of events and the histogram are closely linked to experiments and observed random phenomena. When we consider an experiment with L possible outcomes $\{A_1, A_2, \ldots, A_L\}$ and repeat this experiment N times, the relative frequency $\frac{N_i}{N}$ may serve as an estimate of the probability of event A_i.

A histogram is a quantized representation of the absolute or relative frequencies of events. Frequently, the events under consideration consist of a random variable x within a given range of values. The histogram divides the range of the variable into discrete bins and displays the combined frequencies of values which fall into one of these bins.

Figure 5.5 depicts two histograms of computer-generated Gaussian noise. While the histogram on the left hand side displays absolute frequencies, the histogram on the left hand side uses relative frequencies.

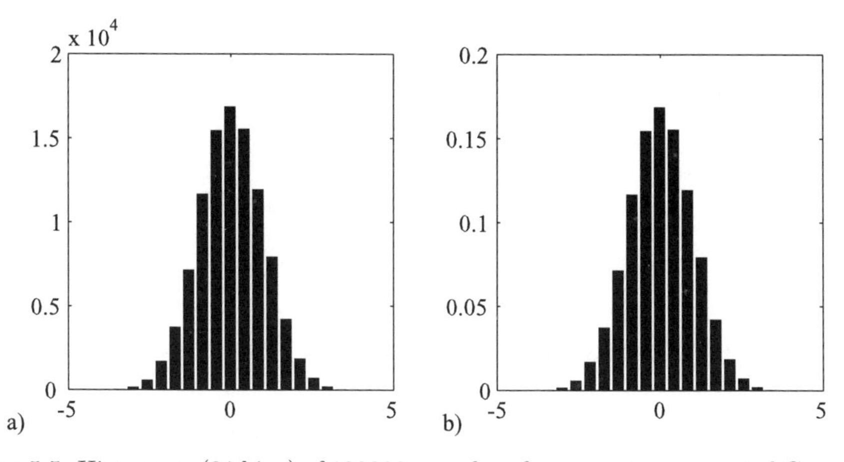

Figure 5.5: Histogram (21 bins) of 100000 samples of a computer-generated Gaussian-distributed random process with zero mean and unit variance
a) Absolute frequencies
b) Relative frequencies

The random variable x and the histogram provide for a numerical representation of events, which makes them accessible for mathematical analysis such as the computation of average values etc. We may also fit a PDF to a histogram of relative frequencies. This may then serve as a parametric model for the observed frequencies of the random phenomenon.

5.3 Bivariate Statistics

As a special case of the multivariate PDF in (5.5) we define *the joint probability density function* of two random variables x and y as the derivative of the bivariate probability distribution function $P_{xy}(u, v)$

$$p_{xy}(u, v) = \frac{\mathrm{d}^2 P_{xy}(u, v)}{\mathrm{d}u \, \mathrm{d}v} \tag{5.34}$$

where P_{xy} is given by

$$P_{xy}(u, v) = P(x \leq u, y \leq v), \tag{5.35}$$

and u and v are threshold values for x and y, respectively.

5.3.1 Marginal Densities

We obtain the monovariate density of either random variable by integrating $p_{xy}(u, v)$ over the other random variable,

$$p_x(u) = \int_{-\infty}^{\infty} p_{xy}(u, v) \, \mathrm{d}v, \tag{5.36}$$

$$p_y(v) = \int_{-\infty}^{\infty} p_{xy}(u, v) \, \mathrm{d}u. \tag{5.37}$$

5.3.2 Expectations and Moments

The *cross-correlation* of two random variables x and y is given by

$$\varphi_{xy} = \mathrm{E}\{x \cdot y\} = \int_{-\infty}^{\infty} \int_{-\infty}^{\infty} u \, v \, p_{xy}(u, v) \, \mathrm{d}u \, \mathrm{d}v \tag{5.38}$$

and the *cross-covariance* by

$$\psi_{xy} = \mathrm{E}\left\{(x-\mu_x)(y-\mu_y)\right\} = \varphi_{xy} - \mu_x\mu_y \,. \tag{5.39}$$

The normalized cross-covariance r_{xy} is given by

$$r_{xy} = \frac{\psi_{xy}}{\sigma_x\sigma_y} \tag{5.40}$$

with $|r_{xy}| \leq 1$.

5.3.3 Uncorrelatedness and Statistical Independence

Two random variables x and y are referred to as uncorrelated if and only if their cross-covariance is zero, i.e.,

$$\mathrm{E}\left\{(x-\mu_x)(y-\mu_y)\right\} \;=\; 0 \,. \tag{5.41}$$

Two random variables x and y are statistically independent if and only if their joint probability density $p_{xy}(u,v)$ factors into the marginal densities $p_x(u)$ and $p_y(v)$,

$$p_{xy}(u,v) = p_x(u)\,p_y(v) \,. \tag{5.42}$$

Two statistically independent random variables x and y are always uncorrelated since

$$\mathrm{E}\left\{(x-\mu_x)(y-\mu_y)\right\} \;=\; \int_{-\infty}^{\infty}\int_{-\infty}^{\infty}(u-\mu_x)(v-\mu_y)\,p_x(u)\,p_y(v)\,\mathrm{d}u\,\mathrm{d}v \tag{5.43}$$

$$= \int_{-\infty}^{\infty}(u-\mu_x)\,p_x(u)\,\mathrm{d}u\int_{-\infty}^{\infty}(v-\mu_y)\,p_y(v)\,\mathrm{d}v = 0 \,. \tag{5.44}$$

The converse is in general not true. A notable exception, however, is the bivariate (or multivariate) Gaussian density. Whenever jointly Gaussian random variables are uncorrelated, they are also statistically independent.

When two random variables x and y are statistically independent, the PDF of their sum $z = x + y$ is the convolution of the individual PDFs,

$$p_z(w) = \int_{-\infty}^{\infty} p_x(w-u)\,p_y(u)\,\mathrm{d}u \,. \tag{5.45}$$

As an example, we compute the density function of the sum $z = x_1^2 + x_2^2$ of two squared zero mean Gaussian random variables x_1 and x_2 which have the same variance σ^2 and are statistically independent. From (5.33) we recall that the PDF of $y = x^2$ is given by

$$p_y(u) = \begin{cases} \frac{1}{\sqrt{2\pi u}\sigma} \exp\left(-\frac{u}{2\sigma^2}\right) & u > 0 \\ 0 & u \leq 0\,. \end{cases} \tag{5.46}$$

Assuming independence, we compute the density function of z by convolving the density (5.46) with itself. The resulting density is the *exponential density*, a special case of the more general χ^2-*density* [Johnson et al. 1994]:

$$p_z(u) = \begin{cases} \frac{1}{2\sigma^2} \exp\left(-\frac{u}{2\sigma^2}\right) & u \geq 0 \\ 0 & u < 0\,. \end{cases} \tag{5.47}$$

Note that the above density also arises for the squared magnitude of a complex zero mean Gaussian random variable when the real and imaginary parts are independent and of the same variance σ^2.

5.3.4 Examples of Bivariate PDFs

Two widely used bivariate density functions are the uniform and the Gaussian density.

5.3.4.1 The Bivariate Uniform Density

Figure 5.6 depicts the joint density of two statistically independent, uniformly distributed random variables x and y. If x and y are uniform in $[X_1, X_2]$ and $[Y_1, Y_2]$ respectively, their joint probability density is given by

$$p_{xy}(u, v) = \begin{cases} \frac{1}{(X_2 - X_1)(Y_2 - Y_1)} & u \in [X_1, X_2] \text{ and } v \in [Y_1, Y_2] \\ 0 & \text{elsewhere}\,. \end{cases} \tag{5.48}$$

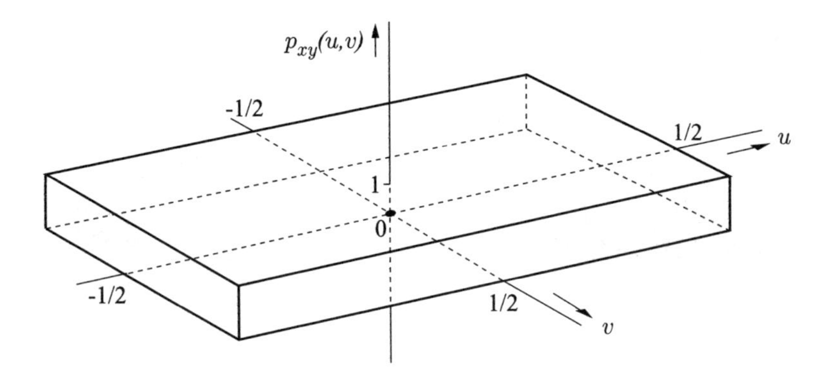

Figure 5.6: Bivariate uniform probability density of two statistically independent random variables x and y with $X_1 = Y_1 = -0.5$ and $X_2 = Y_2 = 0.5$

5.3.4.2 The Bivariate Gaussian Density

The Gaussian density of two random variables x and y may be parameterized by the means μ_x and μ_y, the variances σ_x^2 and σ_y^2, and the normalized cross-covariance r_{xy}. It is given by

$$p_{xy}(u, v) = \frac{1}{2\pi\sigma_x\sigma_y\sqrt{1 - r_{xy}^2}} \cdot \exp\left(-\frac{(u-\mu_x)^2}{2(1-r_{xy}^2)\sigma_x^2} - \frac{(v-\mu_y)^2}{2(1-r_{xy}^2)\sigma_y^2} + r_{xy}\frac{(u-\mu_x)(v-\mu_y)}{(1-r_{xy}^2)\sigma_x\sigma_y}\right) \tag{5.49}$$

and depicted in Fig. 5.7 for various values of σ_x, σ_y, and r_{xy}.

If x and y are uncorrelated, i.e., $r_{xy} = 0$, the bivariate density factors into two monovariable Gaussian densities. Therefore, two uncorrelated Gaussian random variables are also statistically independent. Furthermore, the marginal densities of a bivariate Gaussian density are Gaussian densities.

5.3.5 Functions of Two Random Variables

Given two functions $y_1 = f_1(x_1, x_2)$ and $y_2 = f_2(x_1, x_2)$ of two random variables x_1 and x_2, we find the joint density function of y_1 and y_2 in terms of the joint density of x_1 and x_2 as [Papoulis, Unnikrishna Pillai 2001]

$$p_{y_1y_2}(u, v) = \frac{p_{x_1x_2}(x_1^{\langle 1\rangle}, x_2^{\langle 1\rangle})}{|\mathfrak{J}(x_1^{\langle 1\rangle}, x_2^{\langle 1\rangle})|} + \cdots + \frac{p_{x_1x_2}(x_1^{\langle N\rangle}, x_2^{\langle N\rangle})}{|\mathfrak{J}(x_1^{\langle N\rangle}, x_2^{\langle N\rangle})|} \tag{5.50}$$

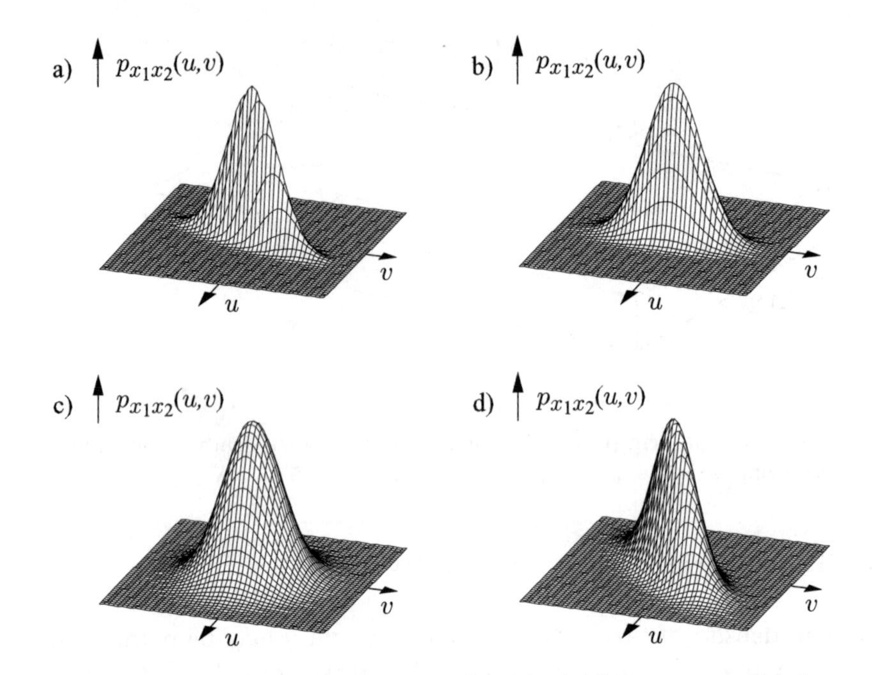

Figure 5.7: Bivariate Gaussian densities with $\mu_x = \mu_y = 0$ and
a) $\sigma_x = 0.1,\ \sigma_y = 0.25,\ r_{xy} = -0.75$
b) $\sigma_x = 0.1,\ \sigma_y = 0.25,\ r_{xy} = 0$
c) $\sigma_x = \sigma_y = 0.25,\ r_{xy} = 0$
d) $\sigma_x = \sigma_y = 0.25,\ r_{xy} = -0.75$

where $x_1^{\langle k\rangle}$ and $x_2^{\langle k\rangle}$, $k = 1 \dots N$, are the real roots of the simultaneous equations $u = f_1(x_1, x_2)$ and $v = f_2(x_1, x_2)$. $|\mathfrak{J}(x_1^{\langle k\rangle}, x_2^{\langle k\rangle})|$ is the determinant of the Jacobian matrix

$$\mathfrak{J}(x_1^{\langle k\rangle}, x_2^{\langle k\rangle}) = \begin{pmatrix} \dfrac{\partial y_1}{\partial x_1}_{|x_1 = x_1^{\langle k\rangle}} & \dfrac{\partial y_1}{\partial x_2}_{|x_2 = x_2^{\langle k\rangle}} \\ \dfrac{\partial y_2}{\partial x_1}_{|x_1 = x_1^{\langle k\rangle}} & \dfrac{\partial y_2}{\partial x_2}_{|x_2 = x_2^{\langle k\rangle}} \end{pmatrix}. \tag{5.51}$$

When there is no solution to the simultaneous equations, we obtain $p_{y_1y_2}(u, v) = 0$.

In the special case of a linear transform with $u = ax_1 + bx_2$, $v = cx_1 + dx_2$, and $|ad - bc| \neq 0$ we have single roots $x_1 = Au + Bv$ and $x_2 = Cu + Dv$ and obtain

$$p_{y_1y_2}(u, v) = \frac{1}{|ad - bc|} p_{x_1x_2}(Au + Bv, Cu + Dv)\,. \tag{5.52}$$

As an example, we consider the additive signal model $y = x + n$ where x and n are statistically independent. We are interested in the computation of $p_{xy}(u, v)$ and of $p_{y|x}(v \mid u)$. We define $x_1 = x$ and $x_2 = n$ and have $u = x_1 = x$, $v = x_1 + x_2 = x + n$,

$a = c = d = 1$, and $b = 0$. The inversion of these equations yields $x_1 = u$ and $x_2 = v - u$. Since $|ad - bc| = 1$, we find

$$p_{xy}(u, v) = p_{xn}(u, v - u) = p_x(u)\, p_n(v - u) \quad \text{and} \quad p_{y|x}(v \mid u) = p_n\,(v - u)\,.$$

5.4 Probability and Information

5.4.1 Entropy

The *entropy* $H(x)$ of a discrete random variable x with $x = u_i$, $i = 1 \dots M$, describes the average uncertainty that one of M values is attained. In the case of a discrete random variable it is defined as

$$H(x) = -\sum_{i=1}^{M} p_x(u_i) \log(p_x(u_i)) \tag{5.53}$$

and measured in "bits" if the logarithm is computed with respect to base 2, and in "nats" if the logarithm is computed with respect to base e. As uncertainty is equivalent to information, $H(x)$ is also a measure of information. The uncertainty of a random variable x with fixed amplitude limits is maximal if the random variable is uniformly distributed within these limits.

For a continuous random variable with a non-zero probability density on the interval $[a, b]$ uncertainty is measured in terms of its *differential entropy*

$$H(x) = -\int_a^b p_x(u) \log(p_x(u))\, du\,. \tag{5.54}$$

5.4.2 Kullback–Leibler Divergence

The *Kullback–Leibler divergence* $D(p_x \,||\, p_y)$ is a measure of similarity of two discrete PDFs $p_x(u)$ and $p_y(u)$,

$$D(p_x \,||\, p_y) = \sum_{i=1}^{M} p_x(u_i) \log\left(\frac{p_x(u_i)}{p_y(u_i)}\right)\,. \tag{5.55}$$

Clearly, the Kullback–Leibler divergence is not symmetric, i.e.,

$$D(p_x \,||\, p_y) \neq D(p_y \,||\, p_x)\,.$$

The Kullback–Leibler divergence is always non-negative, and zero if and only if $p_x = p_y$. Frequently, the symmetric *Kullback–Leibler distance*

$$D_S(p_x \mid\mid p_y) = \frac{1}{2}\left(D(p_x \mid\mid p_y) + D(p_y \mid\mid p_x)\right) \tag{5.56}$$

is used.

5.4.3 Mutual Information

The *mutual information* $\mathfrak{I}(x, y)$ measures the information about a random variable x conveyed by another random variable y. It gives an indication of the dependence of two random variables and may be expressed as the Kullback–Leibler divergence between the joint probability density and the marginal densities of the two random variables, i.e.,

$$\mathfrak{I}(x, y) = D(p_{xy} \mid\mid p_x \cdot p_y) = \sum_{i=1}^{M}\sum_{\ell=1}^{N} p_{xy}(u_i, v_\ell) \log\left(\frac{p_{xy}(u_i, v_\ell)}{p_x(u_i)p_y(v_\ell)}\right). \tag{5.57}$$

The mutual information measure is symmetric in p_x and p_y and non-negative. It is equal to zero if the two random variables are statistically independent.

5.5 Multivariate Statistics

For the treatment of multivariate statistics it is convenient to introduce vector notation. We now consider vectors of random variables

$$\mathbf{x} = \begin{pmatrix} x_1 \\ x_2 \\ \vdots \\ x_N \end{pmatrix} \tag{5.58}$$

and define the first- and second-order statistics in terms of the vector components. The *mean vector* is given by

$$\boldsymbol{\mu}_x = \mathrm{E}\{\mathbf{x}\} = \begin{pmatrix} \mathrm{E}\{x_1\} \\ \mathrm{E}\{x_2\} \\ \vdots \\ \mathrm{E}\{x_N\} \end{pmatrix} = \begin{pmatrix} \mu_{x_1} \\ \mu_{x_2} \\ \vdots \\ \mu_{x_N} \end{pmatrix} \tag{5.59}$$

and the *covariance matrix* by

$$\mathbf{C}_{xx} = \mathrm{E}\left\{(\mathbf{x}-\boldsymbol{\mu}_x)(\mathbf{x}-\boldsymbol{\mu}_x)^T\right\} \tag{5.60}$$

$$= \begin{pmatrix} \psi_{x_1x_1} & \psi_{x_1x_2} & \cdots & \psi_{x_1x_N} \\ \psi_{x_2x_1} & \psi_{x_2x_2} & \cdots & \psi_{x_2x_N} \\ \vdots & \vdots & \ddots & \vdots \\ \psi_{x_Nx_1} & \psi_{x_Nx_2} & \cdots & \psi_{x_Nx_N} \end{pmatrix}. \tag{5.61}$$

The *correlation matrix* $\mathbf{R}_{xx}$ is then computed as

$$\mathbf{R}_{xx} = \mathrm{E}\left\{\mathbf{x}\mathbf{x}^T\right\} = \mathbf{C}_{xx} + \boldsymbol{\mu}_x\boldsymbol{\mu}_x^T. \tag{5.62}$$

5.5.1 Multivariate Gaussian Distribution

The multivariate Gaussian probability density of a vector-valued random variable $\mathbf{x}$ with M components is defined by

$$\mathfrak{N}(\mathbf{x}, \boldsymbol{\mu}_x, \mathbf{C}_{xx}) = \frac{1}{\sqrt{(2\pi)^M \mid \mathbf{C}_{xx} \mid}} \exp\left(-\frac{1}{2}(\mathbf{x}-\mu_x)^T \mathbf{C}_{xx}^{-1}(\mathbf{x}-\mu_x)\right) \tag{5.63}$$

where μ_x and $\mathbf{C}_{xx}$ denote the mean vector and the covariance matrix respectively. $\mid \mathbf{C}_{xx} \mid$ is the determinant of the covariance matrix.

5.5.2 χ^2-distribution

When N independent and identically distributed zero mean Gaussian random variables with variance σ^2 are added, the resulting random variable

$$\chi^2 = x_1^2 + x_2^2 + \cdots + x_N^2 \tag{5.64}$$

is χ^2-distributed with N degrees of freedom. The χ^2*-density* is given by

$$p_{\chi^2}(u) = \begin{cases} \dfrac{u^{N/2-1}\exp\left(-\dfrac{u}{2\sigma^2}\right)}{(\sqrt{2\sigma^2})^N\Gamma(N/2)} & u \geq 0 \\ 0 & u < 0 \end{cases} \tag{5.65}$$

where $\Gamma(\cdot)$ is the complete gamma function. With [Gradshteyn, Ryzhik 2000, Theorem 3.381.4]

$$\int_0^\infty x^n e^{-ax}\, \mathrm{d}x = \frac{\Gamma(n+1)}{a^{n+1}}, \quad \mathrm{Re}\{a\} > 0,\ \mathrm{Re}\{n\} > 0 \tag{5.66}$$

we find the mean of a χ^2-distributed random variable as $\mu_{\chi^2} = N\sigma^2$ and the variance as $\sigma^2_{\chi^2} = 2N\sigma^4$.

The normalized sum of squares

$$\widetilde{\chi}^2 = \frac{1}{N}(x_1^2 + x_2^2 + \cdots + x_N^2) \tag{5.67}$$

is also χ^2-distributed with

$$p_{\widetilde{\chi}^2}(u) = \begin{cases} \dfrac{u^{N/2-1}\exp\left(-\dfrac{uN}{2\sigma^2}\right)}{(\sqrt{2\sigma^2/N})^N\Gamma(N/2)} & u \geq 0 \\ 0 & u < 0 \end{cases} \tag{5.68}$$

with mean σ^2 and variance $\frac{2}{N}\sigma^4$. The same density arises when we divide the sum of N squared independent Gaussians, each of which has the variance $\sigma^2/2$, by $\frac{N}{2}$. This is the case, for example, when we average $K = N/2$ independent magnitude, squared, complex Gaussian random variables and when the mean of the squared magnitude of each complex Gaussian is equal to σ^2.

5.6 Stochastic Processes

An indexed sequence of random variables $\ldots, x(k-1), x(k), x(k+1), \ldots, k \in \mathbb{Z}$, is called a stochastic process. For any index k, a random experiment determines the value of $x(k)$. However, the outcome of an experiment at $k = k_1$ may also depend on the outcomes for $k \neq k_1$. Then, the variables are statistically dependent. In our context, stochastic processes are used to model sampled stochastic signals such as speech signals. An observed speech sample is then interpreted as a specific instantiation of the underlying random process. Stochastic processes are characterized in terms of their monovariate distribution function or their moments for each k, as well as their multivariate statistical properties. In general, all of these quantities are functions of the sampling time k.

5.6.1 Stationary Processes

A stochastic process is called *strict sense stationary* if all its statistical properties (such as moments) are invariant with respect to a variation of k.

A stochastic process is *wide sense stationary* if its first- and second-order moments are invariant to a variation of the independent variable k, i.e.,

$$\mathrm{E}\{x(k)\} = \mu_x \quad \forall\, k \quad \text{and} \tag{5.69}$$

$$\mathrm{E}\{x(k)x(k+\lambda)\} = \varphi_{xx}(\lambda) \quad \forall\, k\,. \tag{5.70}$$

Clearly, speech signals are neither strict nor wide sense stationary. However, within sufficiently short observation intervals, the first- and second-order statistical properties of speech signals show only little variations. Speech signals can therefore be considered to be *short-time wide sense stationary.*

5.6.2 Auto-correlation and Auto-covariance Functions

The *auto-correlation function* quantifies the amount of correlation between the variables of a stochastic process. It is defined as

$$\varphi_{xx}(k_1, k_2) = \mathrm{E}\{x(k_1)\, x(k_2)\} = \int_{-\infty}^{\infty}\int_{-\infty}^{\infty} u\, v\, p_{x(k_1)x(k_2)}(u, v)\, \mathrm{d}u\, \mathrm{d}v \tag{5.71}$$

and is related to the *auto-covariance function*

$$\psi_{xx}(k_1, k_2) = \mathrm{E}\{(x(k_1) - \mu_x(k_1))\, (x(k_2) - \mu_x(k_2))\} \tag{5.72}$$

by

$$\varphi_{xx}(k_1, k_2) = \psi_{xx}(k_1, k_2) + \mu_x(k_1)\mu_x(k_2)\,. \tag{5.73}$$

For a wide sense stationary process the auto-correlation function depends only on the difference $\lambda = k_2 - k_1$ of indices k_1 and k_2 and not on their absolute values. Thus, for a wide sense stationary process we may define the auto-correlation function for any k_1 as

$$\begin{aligned}\varphi_{xx}(\lambda) &= \mathrm{E}\{x(k_1)\, x(k_1 + \lambda)\} \\ &= \int_{-\infty}^{\infty}\int_{-\infty}^{\infty} u\, v\, p_{x(k_1)x(k_1+\lambda)}(u, v)\, \mathrm{d}u\, \mathrm{d}v \;=\; \psi_{xx}(\lambda) + \mu_x^2\,,\end{aligned} \tag{5.74}$$

with

$$\begin{aligned}\psi_{xx}(\lambda) &= \mathrm{E}\{(x(k_1) - \mu_x)\, (x(k_1 + \lambda) - \mu_x)\} \\ &= \varphi_{xx}(\lambda) - \mu_x^2\,.\end{aligned} \tag{5.75}$$

The auto-correlation function, as well as the auto-covariance function, are symmetric, i.e., $\varphi_{xx}(\lambda) = \varphi_{xx}(-\lambda)$. The auto-correlation function attains its maximum and its maximum absolute value for $\lambda = 0$, thus $|\varphi_{xx}(\lambda)| \leq \varphi_{xx}(0)$.

For $\lambda = 0$ we have

$$\varphi_{xx}(0) \;=\; \psi_{xx}(0) + \mu_x^2 \;=\; \sigma_x^2 + \mu_x^2\,. \tag{5.76}$$

In general, we find the moments of a function g of $x(k_1)$ and $x(k_1 + \lambda)$ as

$$\mathrm{E}\{g(x(k_1), x(k_1+\lambda))\} = \int_{-\infty}^{\infty}\int_{-\infty}^{\infty} g(u, v)\, p_{x(k_1)x(k_1+\lambda)}(u, v)\, \mathrm{d}u\, \mathrm{d}v\,. \tag{5.77}$$

5.6.3 Cross-correlation and Cross-covariance Functions

In analogy to the above definitions, the *cross-correlation* and the *cross-covariance functions* may be used to characterize the second-order statistics of two (different) stochastic processes $x(k)$ and $y(k)$ and may be written as

$$\begin{aligned}
\varphi_{xy}(k_1, k_2) &= \mathrm{E}\{x(k_1)\, y(k_2)\} \\
&= \int_{-\infty}^{\infty}\int_{-\infty}^{\infty} u\, v\, p_{x(k_1)y(k_2)}(u, v)\, \mathrm{d}u\, \mathrm{d}v \\
&= \psi_{xy}(k_1, k_2) + \mu_x(k_1)\mu_y(k_2)\,.
\end{aligned} \tag{5.78}$$

As before, we may simplify our notation for stationary processes $x(k)$ and $y(k)$

$$\varphi_{xy}(\lambda) = \mathrm{E}\{x(k_1)y(k_1+\lambda)\} \tag{5.79}$$

and

$$\begin{aligned}
\psi_{xy}(\lambda) &= \mathrm{E}\{(x(k_1) - \mu_x)(y(k_1+\lambda) - \mu_y)\} \\
&= \varphi_{xy}(\lambda) - \mu_x\mu_y\,.
\end{aligned} \tag{5.80}$$

5.6.4 Multivariate Stochastic Processes

Similarly to the definition of vectors of random variables, we may define vector-valued (multivariate) stochastic processes as

$$\mathbf{x}(k) = \begin{pmatrix} x_1(k) \\ x_2(k) \\ \vdots \\ x_N(k) \end{pmatrix}. \tag{5.81}$$

The first- and second-order statistics are then also dependent on index k. For example, the mean is given by

$$\boldsymbol{\mu}_x(k) = \mathrm{E}\{\mathbf{x}(k)\} = \begin{pmatrix} \mathrm{E}\{x_1(k)\} \\ \mathrm{E}\{x_2(k)\} \\ \vdots \\ \mathrm{E}\{x_N(k)\} \end{pmatrix} \tag{5.82}$$

and the correlation matrix $\mathbf{R}_{xx}(k_1, k_2)$ of the process $\mathbf{x}(k)$ at indices k_1 and k_2 by

$$\mathbf{R}_{xx}(k_1, k_2) = \mathrm{E}\left\{\mathbf{x}(k_1)\,\mathbf{x}^T(k_2)\right\} \tag{5.83}$$

$$= \begin{pmatrix} \mathrm{E}\{x_1(k_1)x_1(k_2)\} & \mathrm{E}\{x_1(k_1)x_2(k_2)\} & \cdots & \mathrm{E}\{x_1(k_1)x_N(k_2)\} \\ \mathrm{E}\{x_2(k_1)x_1(k_2)\} & \mathrm{E}\{x_2(k_1)x_2(k_2)\} & \cdots & \mathrm{E}\{x_2(k_1)x_N(k_2)\} \\ \vdots & \vdots & \ddots & \vdots \\ \mathrm{E}\{x_N(k_1)x_1(k_2)\} & \mathrm{E}\{x_N(k_1)x_2(k_2)\} & \cdots & \mathrm{E}\{x_N(k_1)x_N(k_2)\} \end{pmatrix}$$

which is in general not a symmetric matrix. In analogy to the monovariate case, we may also define a covariance matrix, i.e.,

$$\mathbf{C}_{xx}(k_1, k_2) \;=\; \mathrm{E}\left\{(\mathbf{x}(k_1) - \boldsymbol{\mu}_x(k_1))\,(\mathbf{x}(k_2) - \boldsymbol{\mu}_x(k_2))^T\right\}\,. \tag{5.84}$$

When $\mathbf{x}(k)$ is a stationary process, the above quantities will not depend on the absolute values of the time indices k_1 and k_2 but on their difference $\lambda = k_2 - k_1$ only.

An interesting special case arises when the elements of the random vector $\mathbf{x}(k)$ are successive samples of one and the same monovariate process $x(k)$,

$$\mathbf{x}(k) = \begin{pmatrix} x(k) \\ x(k-1) \\ \vdots \\ x(k-N+1) \end{pmatrix}. \tag{5.85}$$

In this case the correlation matrix is given by

$$\mathbf{R}_{xx}(k_1, k_2) = \mathrm{E}\left\{\mathbf{x}(k_1)\,\mathbf{x}^T(k_2)\right\} \tag{5.86}$$

$$= \begin{pmatrix} \varphi_{xx}(k_1, k_2) & \varphi_{xx}(k_1, k_2-1) & \cdots & \varphi_{xx}(k_1, k_2-N+1) \\ \varphi_{xx}(k_1-1, k_2) & \cdots & \cdots & \varphi_{xx}(k_1-1, k_2-N+1) \\ \vdots & \vdots & \ddots & \vdots \\ \varphi_{xx}(k_1-N+1, k_2) & \cdots & \cdots & \varphi_{xx}(k_1-N+1, k_2-N+1) \end{pmatrix}$$

which is a symmetric matrix for $k_1 = k_2 = k$, i.e., $\mathbf{R}_{xx}(k,k) = \mathbf{R}_{xx}^T(k,k)$.

When the monovariate process $x(k)$ is wide sense stationary the elements of the correlation matrix are independent of the absolute time indices and depend only on the index difference $\lambda = k_2 - k_1$. Then,

$$\mathbf{R}_{xx}(\lambda) = \mathrm{E}\left\{\mathbf{x}(k_1)\,\mathbf{x}^T(k_1+\lambda)\right\} \tag{5.87}$$
$$= \begin{pmatrix} \varphi_{xx}(\lambda) & \varphi_{xx}(\lambda-1) & \cdots & \varphi_{xx}(\lambda-N+1) \\ \varphi_{xx}(\lambda+1) & \varphi_{xx}(\lambda) & \cdots & \varphi_{xx}(\lambda-N+2) \\ \vdots & \vdots & \ddots & \vdots \\ \varphi_{xx}(\lambda+N-1) & \varphi_{xx}(\lambda+N-2) & \cdots & \varphi_{xx}(\lambda) \end{pmatrix}.$$

All matrix elements a_{ij} which have the same difference $i-j$ of their row and columns indices are identical. A matrix of this structure is called a *Toeplitz* matrix. For $k_1 = k_2 = k$, i.e., $\lambda = 0$, we obtain a correlation matrix

$$\mathbf{R}_{xx} = \mathrm{E}\left\{\mathbf{x}(k)\,\mathbf{x}^T(k)\right\} \tag{5.88}$$
$$= \begin{pmatrix} \varphi_{xx}(0) & \varphi_{xx}(-1) & \cdots & \varphi_{xx}(-N+1) \\ \varphi_{xx}(1) & \varphi_{xx}(0) & \cdots & \varphi_{xx}(-N+2) \\ \vdots & \vdots & \ddots & \vdots \\ \varphi_{xx}(N-1) & \varphi_{xx}(N-2) & \cdots & \varphi_{xx}(0) \end{pmatrix}$$

which is a symmetric Toeplitz matrix. This matrix is completely specified by its first row or column. For a discrete time stochastic process, the correlation matrix is always non-negative definite and almost always positive definite [Haykin 1996].

In analogy to the above definitions, the cross-correlation matrix of two vector-valued processes $\mathbf{x}(k)$ and $\mathbf{y}(k)$ is given by

$$\mathbf{R}_{xy}(k_1, k_2) = \mathrm{E}\left\{\mathbf{x}(k_1)\,\mathbf{y}(k_2)^T\right\} \tag{5.89}$$

with the above special cases defined accordingly.

5.7 Estimation of Statistical Quantities by Time Averages

5.7.1 Ergodic Processes

Quite frequently we cannot observe more than a single instantiation of a stochastic process. Then, it is not possible to estimate its statistics by averaging over an ensemble of observations. However, if the process is stationary, we might replace ensemble averages by time averages. Whenever the statistics of a stationary random process may be obtained with probability one from time averages over a single observation, the random process is called (strict sense) *ergodic*. This definition implies that specific instances of the random process may not be suited to obtain time averages. These instances, however, occur with probability zero.

Ergodicity is an indispensable prerequisite for many practical applications of statistical signal processing, yet in general it is difficult to prove.

5.7.2 Short-Time Stationary Processes

Strictly speaking, the estimation of statistical quantities via time averaging is only admissible when the signal is stationary and ergodic. As speech signals and many noise signals are not stationary and hence not ergodic, we must confine time averages to short segments of the signal where stationarity is not grossly violated. To apply time averaging we must require that the signal is at least *short-time stationary*.

We define the *short-time mean* of M successive samples of a stochastic process $x(k)$ as

$$\overline{x}(k) = \frac{1}{M} \sum_{\kappa=k-M+1}^{k} x(\kappa) \tag{5.90}$$

and the *short-time variance*

$$\widehat{\sigma}^2(k) = \frac{1}{M-1} \sum_{\kappa=k-M+1}^{k} (x(\kappa) - \overline{x}(k))^2 \,. \tag{5.91}$$

When $x(k)$ is a wide sense stationary and uncorrelated random process, i.e. $\mathrm{E}\left\{(x(k_i) - \mu_{x(k_i)})(x(k_j) - \mu_{x(k_j)})\right\} = 0$ for all $i \neq j$, it can be shown that these estimates are unbiased, i.e.,

$$\mathrm{E}\left\{\overline{x}(k)\right\} = \mathrm{E}\left\{x(k)\right\} \quad \text{and} \quad \mathrm{E}\left\{\widehat{\sigma}^2(k)\right\} = \mathrm{E}\left\{(x(k) - \mu_x)^2\right\} \,.$$

It is not at all trivial to identify short-time stationary segments. On the one hand one is tempted to make the segment of assumed stationarity as long as possible in order to reduce the error variance of the estimate. On the other hand the intrinsic non-stationary nature of speech and many noise signals does not allow the use of long averaging segments without introducing a significant bias. Thus, we have to find a balance between bias and error variance.

Similar averaging procedures may be employed to estimate the cross- and auto-correlation functions on a short-time basis, i.e.,

$$\widehat{\varphi}_{xy}(\lambda, k) = \frac{1}{M} \sum_{\kappa=k-M+1}^{k} x(\kappa)\, y(\kappa + \lambda)\,, \tag{5.92}$$

and

$$\widehat{\varphi}_{xx}(\lambda, k) = \frac{1}{M} \sum_{\kappa=k-M+1}^{k} x(\kappa)\, x(\kappa + \lambda)\,. \tag{5.93}$$

The latter estimate is not symmetric in λ. A symmetric estimate might be obtained by first extracting a signal segment of length M

$$\widetilde{x}(u) = \begin{cases} x(u) & \text{if } k - M + 1 \leq u \leq k \\ 0 & \text{else} \end{cases} \tag{5.94}$$

and then computing the sum of products

$$\begin{aligned} \widehat{\varphi}_{xx}(\lambda, k) &= \frac{1}{M - |\lambda|} \sum_{\kappa=k-M+1}^{k} \widetilde{x}(\kappa)\, \widetilde{x}(\kappa + \lambda) \\ &= \frac{1}{M - |\lambda|} \sum_{\kappa=0}^{M-1} \widetilde{x}(k + \kappa - M + 1)\, \widetilde{x}(k + \kappa + \lambda - M + 1) \end{aligned} \tag{5.95}$$

for $0 \leq \lambda < M$ on the assumption that the signal $\widetilde{x}(k)$ is zero outside this segment. It can be shown that this estimate is unbiased. For a stationary random process and $M \to \infty$ it approaches $\widehat{\varphi}_{xx}(\lambda)$ with probability one.

5.8 Power Spectral Densities

We define the *auto-power spectral density* of a wide sense stationary stochastic process $x(k)$ as the Fourier transform of its auto-correlation function,

$$\Phi_{xx}(e^{j\Omega}) = \sum_{\lambda=-\infty}^{\infty} \varphi_{xx}(\lambda)\, e^{-j\Omega\lambda}\,, \tag{5.96}$$

and the *cross-power spectral density* of two processes $x(k)$ and $y(k)$ as the Fourier transform of their cross-correlation function

$$\Phi_{xy}(e^{j\Omega}) = \sum_{\lambda=-\infty}^{\infty} \varphi_{xy}(\lambda)\, e^{-j\Omega\lambda} \tag{5.97}$$

whenever these transforms exist. The auto-correlation function may be obtained from an inverse Fourier transform of the power spectral density,

$$\varphi_{xx}(\lambda) = \frac{1}{2\pi} \int_{-\pi}^{\pi} \Phi_{xx}(e^{j\Omega})\, e^{j\Omega\lambda}\, d\Omega\,, \tag{5.98}$$

and the signal power from

$$\varphi_{xx}(0) = \frac{1}{2\pi} \int_{-\pi}^{\pi} \Phi_{xx}(e^{j\Omega})\, d\Omega\,. \tag{5.99}$$

5.8.1 White Noise

For an uncorrelated and stationary noise signal $n(k)$, we obtain $\mathrm{E}\{n(k)n(k+\lambda)\} = \delta(\lambda)\, \varphi_{nn}(0)$. Thus, we may write

$$\Phi_{nn}(e^{j\Omega}) = \sum_{\lambda=-\infty}^{\infty} \delta(\lambda)\, \varphi_{nn}(0)\, e^{-j\Omega\lambda} = \varphi_{nn}(0)\,. \tag{5.100}$$

As the power spectral density is constant over the full range of frequencies, uncorrelated noise is also called *white noise.*

5.9 Estimation of the Power Spectral Density

An estimate of the power spectrum may be obtained from the estimate $\widehat{\varphi}_{xx}(\lambda, k)$ of the auto-correlation function as in (5.95) and a subsequent Fourier transform [Oppenheim et al. 1999]. Although this auto-correlation estimate is unbiased, its Fourier transform is not an unbiased estimate of the power spectral density. This is a result of the finite segment length involved in the computation of the auto-correlation.

5.9.1 The Periodogram

An estimate of the power spectral density of a wide sense stationary random process may be obtained by computing the Fourier transform of a finite signal segment $x(k)$,

$$X(e^{j\Omega}, k) = \sum_{\ell=0}^{M-1} x(k+\ell-M+1)\, e^{-j\Omega\ell} \tag{5.101}$$

where we now include the dependency on the time index k in our notation. The magnitude squared Fourier transform normalized on the transform length is called a *periodogram* and denoted by

$$I(e^{j\Omega}, k) = \frac{1}{M} |X(e^{j\Omega}, k)|^2 . \tag{5.102}$$

The periodogram $I(e^{j\Omega}, k)$ is identical to the normalized magnitude squared DFT coefficients of this signal segment at discrete equispaced frequencies $\Omega_\mu = 2\pi\mu/M$. For a real-valued signal $x(k)$, the periodogram may be written as

$$\begin{aligned} I(e^{j\Omega}, k) &= \frac{1}{M} \left| \sum_{\kappa=0}^{M-1} x(k+\kappa-M+1)\, e^{-j\Omega\kappa} \right|^2 \qquad (5.103) \\ &= \frac{1}{M} \sum_{\kappa=-M+1}^{M-1} \sum_{\ell=0}^{M-1} \widetilde{x}(k+\ell-M+1)\, \widetilde{x}(k+\ell+\kappa-M+1)\, e^{-j\Omega\kappa} \end{aligned}$$

where $\widetilde{x}(u)$ is defined as in (5.94).

The inner sum in (5.103) is recognized as an estimate of the auto-correlation function (5.95). Thus, the periodogram corresponds to the Fourier transform of the windowed estimated auto-correlation function,

$$I(e^{j\Omega}, k) = \sum_{\lambda=-M+1}^{M-1} \left(\frac{M-|\lambda|}{M} \right) \widehat{\varphi}_{xx}(\lambda, k) e^{-j\Omega\lambda} . \tag{5.104}$$

Using a tapered analysis window $w(k)$ on the signal segment, we obtain a *modified periodogram* which is defined as

$$I_M(e^{j\Omega}, k) = \frac{1}{\sum_{\ell=0}^{M-1} w^2(\ell)} \left| \sum_{\ell=k-M+1}^{k} w(\ell) x(\ell) e^{-j\Omega\ell} \right|^2 . \tag{5.105}$$

In conclusion we note that

- because of the finite limits of summation the periodogram is a biased estimator of the power spectrum,
- the periodogram is asymptotically ($M \to \infty$) unbiased, and,
- since its variance does not approach zero for $M \to \infty$ the periodogram is not *consistent* [Oppenheim et al. 1999]. In fact, the variance of the periodogram does not depend on the transform length. In the computation of the periodogram, the additional data associated with a larger transform length increases the frequency resolution but does not reduce the estimation error.

5.9.2 Smoothed Periodograms

To reduce the variance of the power spectrum estimate, some form of smoothing is required. A variance reduction might be obtained by smoothing successive (modified) periodograms $I_M(e^{j\Omega}, k)$ over time or by smoothing a single periodogram over frequency. We briefly outline two methods of time domain smoothing.

Non-recursive Smoothing in Time The *non-recursively smoothed periodogram*, defined as

$$P^{(MA)}(e^{j\Omega}, k) = \frac{1}{K_{MA}} \sum_{\kappa=0}^{K_{MA}-1} I_M(e^{j\Omega}, k + (\kappa - K_{MA} + 1)\, r)\,, \qquad (5.106)$$

uses a sliding signal frame of length M and a frame advance of r. It can be shown that for uncorrelated signal frames its variance approaches zero for $K_{MA} \to \infty$. For a fixed total number of signal samples, the variance of this estimate may be reduced by overlapping successive signal segments [Welch 1967]. For example, for half-overlapping signal segments the variance is reduced by a factor of 11/18 [Welch 1967].

Recursive Smoothing in Time To compute the Welch periodogram, K_{MA} periodograms must be stored. The *recursively smoothed periodogram*,

$$P^{(AR)}(e^{j\Omega}, k + r) = \alpha\, P^{(AR)}(e^{j\Omega}, k) + (1 - \alpha)\, I_M(e^{j\Omega}, k + r)\,, \qquad (5.107)$$

is more memory efficient. It can be interpreted as an infinite sum of periodograms weighted over time by an exponential window. This corresponds to a first-order low-pass filter and for stability reasons we have $0 < \alpha < 1$. For $\alpha \approx 0$, little smoothing is applied and hence the variance of the estimate is close to the variance of the periodogram. For $\alpha \approx 1$ and stationary signals, the variance of the estimate is small. The above procedure may also be used for smoothing short-time stationary periodograms and for tracking the mean value of the periodograms over larger periods of time. Then, we must strike a balance between smoothing for variance reduction and tracking of non-stationary signal characteristics.

5.10 Statistical Properties of Speech Signals

The statistical properties of speech signals have been thoroughly investigated and are well documented, e.g., [Jayant, Noll 1984], [Brehm, Stammler 1987]. They are of interest, for instance, in the design of optimal quantizers. In general, speech signals can be modeled as non-stationary stochastic processes. The power, the

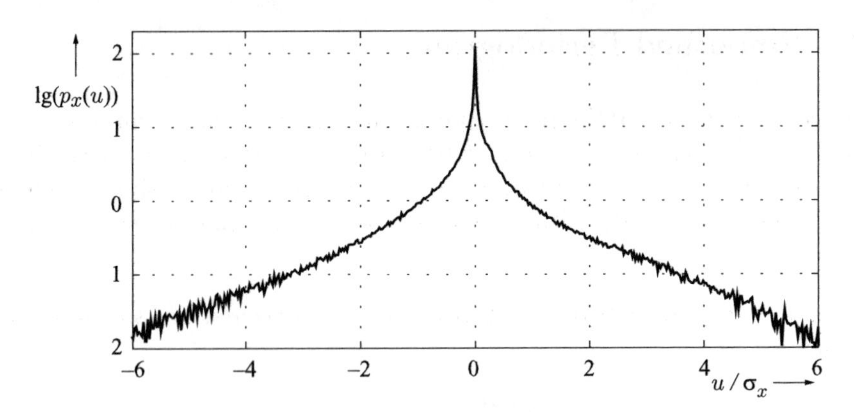

Figure 5.8: Logarithmic plot of relative histogram of normalized speech amplitudes ($n = 489872$ speech samples, 1024 histogram bins)

correlation properties, and the higher-order statistics vary from one speech sound to the next. For most practical purposes statistical parameters may be estimated on short, quasi-stationary segments of the speech signal.

It is also instructive to consider the long-term histogram of speech signals as shown in Fig. 5.8. Small amplitudes are much more frequent than large amplitudes and the tails of the density do not decay as fast as for a Gaussian signal. Thus, the PDF of speech signals in the time domain is well modeled by *supergaussian* distributions such as the Laplace, gamma, or K_0 PDFs [Brehm, Stammler 1987].

5.11 Statistical Properties of DFT Coefficients

Frequently, we analyze and process speech signals in the spectral domain using short-term spectral analysis and the DFT. It is therefore of interest to investigate the statistical properties of DFT coefficients. We assume that a signal $x(k)$ is transformed into the frequency domain by applying a window $w(k)$ to a frame of M consecutive samples of $x(k)$ and by computing the DFT of size M on the windowed data. This sliding window DFT analysis results in a set of frequency domain signals $X_\mu(k)$ (see Chapter 4) which can be written as

$$X_\mu(k) = \sum_{\kappa=0}^{M-1} w(\kappa)\, x(k - M + 1 + \kappa)\, e^{-j\frac{2\pi\mu\kappa}{M}} \tag{5.108}$$

where μ is the frequency bin index, $\mu \in \{0, 1, \ldots, M-1\}$. The index μ is related to the normalized center frequency Ω_μ of each frequency bin by $\Omega_\mu = 2\pi\mu/M$. Thus, the complex Fourier coefficients $X_\mu(k)$ constitute a random process.

Using the above definitions, individual variables may be split into their real and imaginary parts

$$X_\mu(k) = \mathrm{Re}\{X_\mu(k)\} + j\,\mathrm{Im}\{X_\mu(k)\} \tag{5.109}$$

or into their magnitude and phase

$$X_\mu(k) = R_\mu(k)\, e^{j\theta_\mu(k)} . \tag{5.110}$$

To simplify notations, we will drop the dependency of all the above quantities on the time index k whenever this is possible.

5.11.1 Asymptotic Statistical Properties

For the discussion of asymptotic properties of the DFT coefficients we assume

- that the transform length M approaches infinity, $M \to \infty$, and
- that the transform length M is much larger than the span of correlation of signal $x(k)$.

The latter condition excludes, for instance, periodic signals from our discussion. If the input signal is sufficiently random we may conclude from the *central limit theorem* that for $\mu \notin \{0, M/2\}$, the real and imaginary parts of the DFT coefficients X_μ can be modeled as mutually independent, zero-mean Gaussian random variables [Brillinger 1981] with variance $0.5\,\sigma^2_{X_\mu} = 0.5\,\mathrm{E}\left\{|X_\mu|^2\right\}$, i.e.,

$$\begin{aligned} p_{\mathrm{Re}\{X_\mu\}}(u) &= \frac{1}{\sqrt{\pi}\sigma_{X_\mu}} \exp\left(-\frac{u^2}{\sigma^2_{X,\mu}}\right) , \\ p_{\mathrm{Im}\{X_\mu\}}(v) &= \frac{1}{\sqrt{\pi}\sigma_{X_\mu}} \exp\left(-\frac{v^2}{\sigma^2_{X,\mu}}\right) . \end{aligned} \tag{5.111}$$

For a real-valued input signal $x(k)$ and $\mu \in \{0, M/2\}$ the imaginary part of X_μ is zero and the real part is also Gaussian distributed with variance $\sigma^2_{X,\mu} = \mathrm{E}\left\{|X_\mu|^2\right\}$.

For $\mu \notin \{0, M/2\}$ the joint distribution of the real and imaginary parts is given by

$$p_{\mathrm{Re}\{X_\mu\},\,\mathrm{Im}\{X_\mu\}}(u, v) = \frac{1}{\pi\sigma^2_{X,\mu}} \exp\left(-\frac{u^2 + v^2}{\sigma^2_{X,\mu}}\right) \tag{5.112}$$

or with $z = u + jv$ by

$$p_{X_\mu}(z) = \frac{1}{\pi\sigma^2_{X,\mu}} \exp\left(-\frac{|z|^2}{\sigma^2_{X,\mu}}\right) . \tag{5.113}$$

Conversion to polar coordinates, $X_\mu = R_\mu e^{j\theta_\mu}$, yields a *Rayleigh density* for the magnitude $|X_\mu| = R_\mu$,

$$p_{R_\mu}(u) = \begin{cases} \dfrac{2u}{\sigma^2_{X,\mu}} \exp\left(-\dfrac{u^2}{\sigma^2_{X,\mu}}\right) & u \geq 0 \\ 0 & u < 0 \end{cases} \tag{5.114}$$

and a uniform distribution for the principal value of the phase θ_μ, $0 \leq \theta_\mu \leq 2\pi$,

$$p_{\theta_\mu}(u) = \begin{cases} \dfrac{1}{2\pi} & 0 \leq u \leq 2\pi \\ 0 & \text{elsewhere}\,. \end{cases} \tag{5.115}$$

Since for the Gaussian model the magnitude and the phase are statistically independent, the joint density is the product of the component densities [Papoulis, Unnikrishna Pillai 2001],

$$\begin{aligned} p_{R_\mu,\theta_\mu}(u,v) &= p_{R_\mu}(u)\, p_{\theta_\mu}(v) \\ &= \begin{cases} \dfrac{u}{\pi\sigma^2_{X,\mu}} \exp\left(-\dfrac{u^2}{\sigma^2_{X,\mu}}\right) & u \geq 0 \text{ and } 0 \leq v \leq 2\pi \\ 0 & \text{elsewhere}\,. \end{cases} \end{aligned} \tag{5.116}$$

Furthermore, each magnitude squared frequency bin $|X_\mu|^2 = R^2_\mu$ is an exponentially distributed random variable with PDF

$$p_{R^2_\mu}(u) = \begin{cases} \dfrac{1}{\sigma^2_{X,\mu}} \exp\left(-\dfrac{u}{\sigma^2_{X,\mu}}\right) & u \geq 0 \\ 0 & u < 0\,. \end{cases} \tag{5.117}$$

5.11.2 Signal-plus-Noise Model

In applications such as speech enhancement we consider an observed signal $x(k) = s(k) + n(k)$ which is a sum of a desired signal $s(k)$ and a noise signal $n(k)$ where $n(k)$ is statistically independent of the desired signal $s(k)$. Obviously, this also leads to an additive noise model in the Fourier or in the DFT domain, $X_\mu(k) = S_\mu(k) + N_\mu(k)$. We now compute the conditional density for the observed DFT coefficients X_μ given the desired coefficients $S_\mu = A_\mu e^{j\alpha_\mu} = A_\mu\left(\cos(\alpha_\mu) + j\sin(\alpha_\mu)\right)$ on the Gaussian assumption. Since the desired signal and the noise are additive and statistically independent, the conditional densities for the real and imaginary parts are given by

$$\begin{aligned} p_{\mathrm{Re}\{X_\mu\}|\mathrm{Re}\{S_\mu\}}(u \mid \mathrm{Re}\{S_\mu\}) &= \frac{1}{\sigma_{N_\mu}\sqrt{\pi}} \exp\left(-\frac{(u - A_\mu \cos(\alpha_\mu))^2}{\sigma_{N,\mu}^2}\right) \\ p_{\mathrm{Im}\{X_\mu\}|\mathrm{Im}\{S_\mu\}}(v \mid \mathrm{Im}\{S_\mu\}) &= \frac{1}{\sigma_{N_\mu}\sqrt{\pi}} \exp\left(-\frac{(v - A_\mu \sin(\alpha_\mu))^2}{\sigma_{N,\mu}^2}\right) \end{aligned} \quad (5.118)$$

where we conditioned on the real and the imaginary parts of the desired coefficients. With $z = u + jv$ the conditional joint density is given by

$$\begin{aligned} &p_{\mathrm{Re}\{X_\mu\},\mathrm{Im}\{X_\mu\}|S_\mu}(u, v \mid S_\mu) \\ &\quad = \frac{1}{\pi\sigma_{N,\mu}^2} \exp\left(-\frac{|z - A_\mu \exp(j\alpha_\mu)|^2}{\sigma_{N,\mu}^2}\right) \\ &\quad = \frac{1}{\pi\sigma_{N,\mu}^2} \exp\left(-\frac{|z|^2 + A_\mu^2 - 2A_\mu \mathrm{Re}\{\exp(-j\alpha_\mu)z\}}{\sigma_{N,\mu}^2}\right). \end{aligned} \quad (5.119)$$

Since a rotation in the complex plane does not change the magnitude

$$|z|^2 = |z\exp(-j\alpha_\mu)|^2 = \mathrm{Re}\{z\exp(-j\alpha_\mu)\}^2 + \mathrm{Im}\{z\exp(-j\alpha_\mu)\}^2, \quad (5.120)$$

the conditional joint density can be also written as

$$\begin{aligned} &p_{\mathrm{Re}\{X_\mu\},\mathrm{Im}\{X_\mu\}|S_\mu}(u, v \mid S_\mu) \\ &= \frac{1}{\pi\sigma_{N,\mu}^2} \exp\left(-\frac{(\mathrm{Re}\{\exp(-j\alpha_\mu)z\} - A_\mu)^2 + \mathrm{Im}\{\exp(-j\alpha_\mu)z\}^2}{\sigma_{N,\mu}^2}\right) \end{aligned} \quad (5.121)$$

which leads to a *Rician* PDF for the conditional magnitude [Papoulis, Unnikrishna Pillai 2001], [McAulay, Malpass 1980]

$$p_{R_\mu|S_\mu}(u \mid S_\mu) = \begin{cases} \dfrac{2u}{\sigma_{N,\mu}^2} \exp\left(-\dfrac{u^2 + A_\mu^2}{\sigma_{N,\mu}^2}\right) I_0\left(\dfrac{2A_\mu u}{\sigma_{N,\mu}^2}\right) & u \geq 0 \\ 0 & u < 0 \end{cases} \quad (5.122)$$

where $I_0(\cdot)$ denotes the modified Bessel function of the first kind. When no speech is present, the magnitude obeys a Rayleigh distribution as before.

5.11.3 Statistical Properties of DFT Coefficients for Finite Frame Lengths

We now investigate the probability distribution of speech coefficients in the DFT domain. Unlike in the previous section, we consider a short transform length M as it is used in mobile communications and other applications. There, the asymptotic assumptions are not well fulfilled, especially for voiced speech sounds which exhibit a high degree of correlation [Martin 2002], [Martin, Breithaupt 2003], [Martin 2005b].

We suggest that also in the short-term Fourier domain (frame size < 100ms) the Laplace and gamma densities are much better models for the PDF of the real and imaginary parts of the DFT coefficients than the commonly used Gaussian density. In this section, we will briefly review these densities and provide examples of experimental data.

Let $\text{Re}\{S_\mu\} = S_R$ and $\text{Im}\{S_\mu\} = S_I$ denote the real and the imaginary part of a clean speech DFT coefficient respectively. To enhance the readability we will drop both the frame index k and the frequency index μ and consider a single DFT coefficient at a given time instant. $\sigma_s^2/2$ denotes the variance of the real and imaginary parts of the DFT coefficient. Then, the Laplacian and the gamma densities (real and imaginary parts) are given by

$$
\begin{aligned}
p_{S_R}(u) &= \frac{1}{\sigma_s}\exp\left(-\frac{2|u|}{\sigma_s}\right) \\
p_{S_I}(v) &= \frac{1}{\sigma_s}\exp\left(-\frac{2|v|}{\sigma_s}\right)
\end{aligned}
\tag{5.123}
$$

and

$$
\begin{aligned}
p_{S_R}(u) &= \frac{\sqrt[4]{3}}{2\sqrt{\pi\sigma_s}\sqrt[4]{2}}|u|^{-\frac{1}{2}}\exp\left(-\frac{\sqrt{3}|u|}{\sqrt{2}\sigma_s}\right) \\
p_{S_I}(v) &= \frac{\sqrt[4]{3}}{2\sqrt{\pi\sigma_s}\sqrt[4]{2}}|v|^{-\frac{1}{2}}\exp\left(-\frac{\sqrt{3}|v|}{\sqrt{2}\sigma_s}\right),
\end{aligned}
\tag{5.124}
$$

respectively. The gamma density diverges when the argument approaches zero but provides otherwise a good fit to the observed data.

Figures 5.9 and 5.10 plot the histogram of the real part of the DFT coefficients ($M = 256$, $f_s = 8000\,\text{Hz}$) of clean speech averaged over three male and three female speakers. Since speech has a time-varying power, the coefficients represented in the histogram are selected such as to range in a narrow power interval. For the depicted histogram, coefficients are selected in a 2 dB wide interval across all frequency bins except for the lowest and the highest bins. Thus, the histogram and

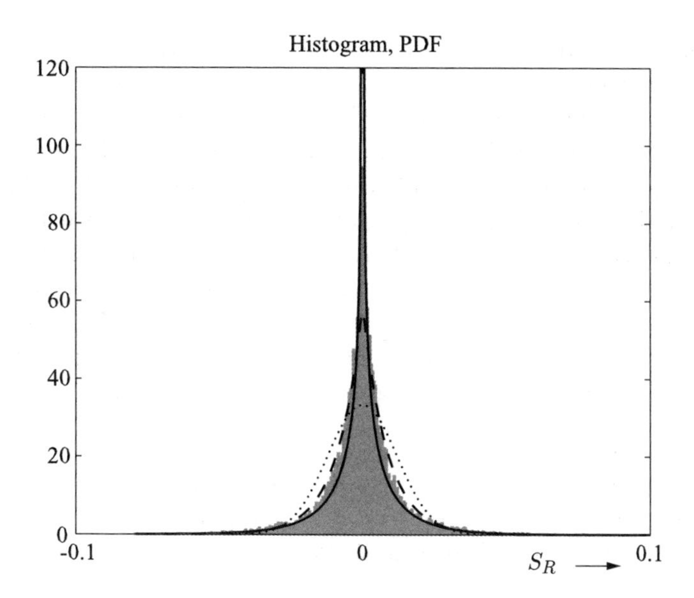

Figure 5.9: Gaussian (dotted), Laplace (dashed), and gamma (solid) density fitted to a histogram of the real part of clean speech DFT coefficients (shaded, $M = 256$, $f_s = 8000\,\text{Hz}$) [Martin 2002]; © 2002 IEEE

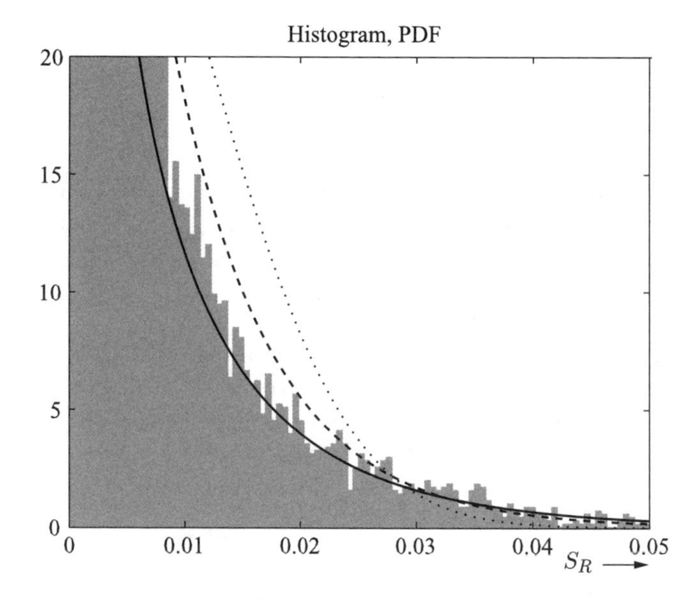

Figure 5.10: Gaussian (dotted), Laplace (dashed), and gamma (solid) density fitted to a histogram of the real part of clean speech DFT coefficients (shaded, $M = 256$, $f_s = 8000\,\text{Hz}$) [Martin 2002]; © 2002 IEEE

the corresponding model densities are in fact conditioned on the measured SNR of the spectral coefficients. The full histogram in Fig. 5.9 as well as the enlarged section in Fig. 5.10 show that indeed the Laplace and gamma densities provide a much better fit to the DFT data than the Gaussian distribution. This result is also reflected by the estimated Kullback-Leibler divergence $D(p_x \,||\, p_y)$ [Kullback 1997] between the histogram data $p_H(x)$ and one of the above PDF candidates $p_{S_R}(u)$. We find that the Kullback-Leibler distance is about 3 times smaller for the gamma density and about 6 times smaller for the Laplace density than for the Gaussian density [Martin 2005b]. An even better match of the measured PDF can be obtained with a linear combination of the Laplacian and the gamma model, where the weights 0.7 and 0.3 are assigned to the Laplacian and the gamma model respectively, or with a generalized gamma density [Shin et al. 2005].

Similar approximations may be developed for the magnitude $A = |S|$ of the DFT coefficients. A good fit to the observed data is obtained for the generalized gamma density [Lotter, Vary 2003], [Dat et al. 2005]

$$p_A(u) = \frac{\gamma^{\nu+1}}{\Gamma(\nu+1)} \frac{u^\nu}{\sigma_S^{\nu+1}} \exp\left\{-\gamma \frac{u}{\sigma_S}\right\} \tag{5.125}$$

where for $\mathrm{E}\left\{A^2\right\} = \sigma_S^2$ the constraint $\gamma = \sqrt{(\nu+1)(\nu+2)}$ applies. The parameters ν, μ determine the shape of the PDF and thus allow to adapt the underlying PDF of the conditional estimator to the real distribution.

5.12 Optimal Estimation

Quite often we are faced with the task of estimating the value of a random variable x when observations of another random vector $\mathbf{y} = (y_1, \dots, y_N)^T$ are given. The resulting estimator $\widehat{x} = f(\mathbf{y})$ maps the random observed variables $\mathbf{y}$ onto the estimated variable $\widehat{x}$ which is then in turn random.

In the context of speech processing, various optimization criteria are used [Vary 2004]. The *maximum likelihood* (ML) estimator selects a value in the range of x such that the conditional joint probability density of the observed variables is maximized, i.e.,

$$\widehat{x} = \arg\max_x p_{\mathbf{y}|x}(\mathbf{y} \mid x) \,. \tag{5.126}$$

The *maximum a posteriori* (MAP) estimator is defined by

$$\widehat{x} = \arg\max_x p_{x|\mathbf{y}}(x \mid \mathbf{y}) = \arg\max_x \frac{p_{\mathbf{y}|x}(\mathbf{y} \mid x)\, p_x(x)}{p_{\mathbf{y}}(\mathbf{y})} \tag{5.127}$$

where now the *a priori* distribution $p_x(x)$ of the unknown variable x is employed. When the *a priori* density $p_x(x)$ is uniformly distributed, the MAP estimator results in the same estimate as the ML estimator.

In a more general setting we might strive to minimize the statistical expectation

$$\mathrm{E}_{x\mathbf{y}}\left\{C(x,\widehat{x})\right\} = \int_{-\infty}^{\infty} \cdots \int_{-\infty}^{\infty} C(u, f(\mathbf{y}))p_{x\mathbf{y}}(u, y_1, \ldots, y_N)\, \mathrm{d}u\, \mathrm{d}y_1 \ldots \mathrm{d}y_N \tag{5.128}$$

of a *cost function* $C(x,\widehat{x})$ over the probability spaces of x and $\mathbf{y}$. The most prominent of these estimators is the *minimum mean square error* (MMSE) estimator where $C(x,\widehat{x}) = (x-\widehat{x})^2$. MMSE solutions to the estimation problem will be discussed in more detail below.

5.12.1 MMSE Estimation

We show that the optimal solution $\widehat{x}$ in the MMSE sense is given by the expectation of x conditioned on the vector of observations $\mathbf{y} = (y_1, \ldots, y_N)^T$, i.e.,

$$\widehat{x} = \mathrm{E}_x\left\{x \mid \mathbf{y}\right\} = \int_{-\infty}^{\infty} u\, p_{x|\mathbf{y}}(u \mid y_1, \ldots, y_N)\, \mathrm{d}u\,. \tag{5.129}$$

The optimal solution $\widehat{x} = f(\mathbf{y})$ is a function of the joint statistics of the observed variables and x. The computation of the mean square error $\mathrm{E}_{x\mathbf{y}}\left\{(x-\widehat{x})^2\right\}$ requires averaging over the probability space of x as well as of $\mathbf{y}$. Thus, we might expand the mean square error into

$$\mathrm{E}_{x\mathbf{y}}\left\{(x-\widehat{x})^2\right\} = \int_{-\infty}^{\infty} \cdots \int_{-\infty}^{\infty} (u-\widehat{x})^2 p_{x\mathbf{y}}(u, y_1, \ldots, y_N)\, \mathrm{d}u\, \mathrm{d}y_1 \ldots \mathrm{d}y_N \tag{5.130}$$

$$= \int_{-\infty}^{\infty} \cdots \int_{-\infty}^{\infty} (u-\widehat{x})^2 p_{x|\mathbf{y}}(u \mid y_1, \ldots, y_N) p_{\mathbf{y}}(y_1, \ldots, y_N)\, \mathrm{d}u\, \mathrm{d}y_1 \ldots \mathrm{d}y_N\,.$$

Since PDFs are non-negative, it is sufficient to minimize the inner integral

$$\int_{-\infty}^{\infty} (u-\widehat{x})^2 p_{x|\mathbf{y}}(u \mid y_1, \ldots, y_N)\, \mathrm{d}u \tag{5.131}$$

for any given vector $\mathbf{y}$ of observations. Setting the first derivative with respect to $\widehat{x}$ to zero then yields the desired result

$$\widehat{x} = \int_{-\infty}^{\infty} u\, p_{x|\mathbf{y}}(u \mid y_1, \dots, y_N)\, du = \mathrm{E}_x\{x \mid \mathbf{y}\}\,. \tag{5.132}$$

In general, $\widehat{x}$ is a non-linear function of the observed values $\mathbf{y}$.

5.12.2 Optimal Linear Estimator

We simplify the estimation procedure by constraining the estimate $\widehat{x}$ to be a linear combination of the observed data, i.e.,

$$\widehat{x} = \mathbf{h}^T \mathbf{y} \tag{5.133}$$

where $\mathbf{h}^T = (h_1, \dots, h_N)$ is a vector of constant weights. The expansion of $\mathrm{E}\left\{(x-\widehat{x})^2\right\}$ leads to

$$\mathrm{E}\left\{(x-\widehat{x})^2\right\} = \mathrm{E}\left\{x^2\right\} - 2\mathrm{E}\left\{x\,\mathbf{y}^T\right\}\mathbf{h} + \mathbf{h}^T\mathrm{E}\left\{\mathbf{y}\mathbf{y}^T\right\}\mathbf{h} \tag{5.134}$$

and the minimization of the mean square error for an invertible autocorrelation matrix $\mathbf{R}_{yy}$ to

$$\mathbf{h} = \mathbf{R}_{yy}^{-1}\,\mathbf{r}_{xy} \tag{5.135}$$

where $\mathbf{r}_{xy}$ is defined as

$$\mathbf{r}_{xy} = \left(\mathrm{E}\{x\,y_1\}, \dots, \mathrm{E}\{x\,y_N\}\right)^T. \tag{5.136}$$

Thus, for a given vector of observations $\mathbf{y}$, we compute the estimated value as

$$\widehat{x} = \mathbf{y}^T\mathbf{R}_{yy}^{-1}\,\mathbf{r}_{xy}\,. \tag{5.137}$$

In contrast to the general non-linear solution, the probability densities of the signal and the noise are not involved in the computation of the linearly constrained solution. The weight vector $\mathbf{h} = \mathbf{R}_{yy}^{-1}\,\mathbf{r}_{xy}$ is a function of second-order statistics but not a function of the observed vector $\mathbf{y}$ itself.

The linear estimator may be extended to the more general non-homogeneous case

$$\widehat{x} = \mathbf{h}^T\mathbf{y} + a\,. \tag{5.138}$$

We obtain

$$\mathbf{h} = \mathbf{C}_{yy}^{-1}\mathbf{c}_{xy} = (\mathbf{R}_{yy}^{-1} - \boldsymbol{\mu}_y\boldsymbol{\mu}_y^T)^{-1}(\mathbf{r}_{xy} - \boldsymbol{\mu}_y\mu_x) \tag{5.139}$$

and

$$a = \mu_x - \boldsymbol{\mu}_y^T\mathbf{C}_{yy}^{-1}\mathbf{c}_{xy} \,. \tag{5.140}$$

The estimate is therefore given by

$$\widehat{x} = (\mathbf{y}^T - \boldsymbol{\mu}_y)\mathbf{C}_{yy}^{-1}\mathbf{c}_{xy} + \mu_x \,. \tag{5.141}$$

If we assume an additive signal plus noise model, $y = x + n$, where the signal x and the noise n are mutually uncorrelated signals with $\mathrm{E}\{x\} = \mu_x$, $\mathrm{E}\{n\} = 0$, and $\mathrm{E}\{x\,n\} = 0$, we may further simplify the above result. It is especially instructive to interpret the solution in the case of a single observation y. Then, we have

$$\widehat{x} = (y - \mu_x)\frac{\sigma_x^2}{\sigma_x^2 + \sigma_n^2} + \mu_x = (y - \mu_x)\frac{\xi}{1+\xi} + \mu_x \tag{5.142}$$

where $\xi = \sigma_x^2/\sigma_n^2$ is called the *a priori* signal-to-noise ratio (SNR). For $\xi \gg 1$ we have $\widehat{x} \approx y$. The estimate is approximately equal to the observed variable. For a low *a priori* SNR, $\xi \gtrapprox 0$, we have $\widehat{x} \approx \mu_x$. In this case, the best estimate is the unconditional mean of the desired variable x.

5.12.3 The Gaussian Case

We will show that for the additive noise model, a scalar observation y, and jointly Gaussian signals the non-linear MMSE estimator is identical to the linearly constrained estimator. Since x and n are uncorrelated and have zero mean, the PDF of $y = x + n$ is given by

$$p_y(v) = \frac{1}{\sqrt{2\pi(\sigma_x^2 + \sigma_n^2)}} \exp\left(-\frac{v^2}{2(\sigma_x^2 + \sigma_n^2)}\right)$$

because the convolution of two Gaussians is a Gaussian. The conditional density $p_{y|x}(v \mid u) = p_n(v-u)$ and the density of the undisturbed signal x may be written as

$$p_n(v - u) = \frac{1}{\sqrt{2\pi}\sigma_n} \exp\left(-\frac{(v-u)^2}{2\sigma_n^2}\right) \tag{5.143}$$

and

$$p_x(u) = \frac{1}{\sqrt{2\pi}\sigma_x} \exp\left(-\frac{u^2}{2\sigma_x^2}\right) \tag{5.144}$$

respectively. The non-linear MMSE estimate is therefore given by

$$\widehat{x} = \int_{-\infty}^{\infty} u\, p_{x|y}(u \mid v)\, \mathrm{d}u \tag{5.145}$$

$$= \frac{1}{p_y(v)} \int_{-\infty}^{\infty} u\, p_{y|x}(v \mid u)\, p_x(u)\, \mathrm{d}u \tag{5.146}$$

$$= K \int_{-\infty}^{\infty} u \exp\left(-u^2 \frac{\sigma_x^2 + \sigma_n^2}{2\sigma_x^2 \sigma_n^2} + \frac{vu}{\sigma_n^2}\right) \mathrm{d}u \tag{5.147}$$

with

$$K = \frac{1}{p_y(v)} \frac{\exp\left(-v^2/(2\sigma_n^2)\right)}{2\pi\sqrt{\sigma_n^2 \sigma_x^2}} .$$

With [Gradshteyn, Ryzhik 2000, Theorem 3.462]

$$\int_{-\infty}^{\infty} x \exp\left(-px^2 + 2qx\right) \mathrm{d}x = \frac{q}{p}\sqrt{\frac{\pi}{p}} \exp\left(\frac{q^2}{p}\right), \qquad \mathrm{Re}\{p\} > 0 \tag{5.148}$$

we obtain the estimator in terms of the observed random variable y as

$$\widehat{x} = \frac{\sigma_x^2}{\sigma_x^2 + \sigma_n^2}\, y = \frac{\xi}{1+\xi}\, y \tag{5.149}$$

with ξ as defined above. This result can be extended to multiple observations and multivariate estimates as well.

5.12.4 Joint Detection and Estimation

In a practical application, the desired signal x is not always present in the observed noisy signal y. The optimal estimator must be adapted to this uncertainty about the presence of the desired signal and must deliver an optimal estimate regardless of whether the signal is present or not. In a slightly more general setting, we assume that there are two versions x_0 and x_1 of the desired signal, which are present in the observed signal with *a priori* probabilities $P(H^{(0)})$ and $P(H^{(1)}) = 1 - P(H^{(0)})$ respectively, where we denote the presence of x_0 and the presence of x_1 as the two hypotheses

$$H^{(0)}: \quad x = x_0 \quad \text{and} \tag{5.150}$$

$$H^{(1)}: \quad x = x_1 . \tag{5.151}$$

The two versions of the signal x are treated as random variables with possibly different probability density functions. With these assumptions, the PDF of x may be written as

$$p_x(u) = p_{x|H^{(0)}}(u \mid H^{(0)})P(H^{(0)}) + p_{x|H^{(1)}}(u \mid H^{(1)})P(H^{(1)}) . \tag{5.152}$$

Again, we use a quadratic cost function $C(x, \widehat{x}) = (\widehat{x} - x)^2$ where $\widehat{x}$ is in general a function of the observed variable $y = x + n$. We minimize the total (Bayes) cost

$$\mathfrak{J} = \int_{-\infty}^{\infty}\int_{-\infty}^{\infty} C(u, \widehat{x}(v)) p_{xy}(u, v) \, du \, dv \tag{5.153}$$

$$= \int_{-\infty}^{\infty}\int_{-\infty}^{\infty} (\widehat{x}(v) - u)^2 \Big(p_{xy|H^{(0)}}(u, v \mid H^{(0)})P(H^{(0)}) + p_{xy|H^{(1)}}(u, v \mid H^{(1)})P(H^{(1)})\Big) \, du \, dv . \tag{5.154}$$

Setting the first derivative of the inner integral with respect to $\widehat{x}$ to zero, we obtain

$$\int_{-\infty}^{\infty} (\widehat{x}(v) - u)\Big(p_{xy|H^{(0)}}(u, v \mid H^{(0)})P(H^{(0)}) + p_{xy|H^{(1)}}(u, v \mid H^{(1)})P(H^{(1)})\Big) \, du = 0$$

and substituting

$$\begin{aligned} p_{xy|H^{(0)}}(u, v \mid H^{(0)})P(H^{(0)}) &= p_{y|H^{(0)}}(v \mid H^{(0)}) p_{x|y,H^{(0)}}(u \mid v, H^{(0)})P(H^{(0)}) \\ p_{xy|H^{(1)}}(u, v \mid H^{(1)})P(H^{(1)}) &= p_{y|H^{(1)}}(v \mid H^{(1)}) p_{x|y,H^{(1)}}(u \mid v, H^{(1)})P(H^{(1)}) \end{aligned}$$

yields

$$\begin{aligned} \widehat{x}(v) \Big[p_{y|H^{(0)}}(v \mid H^{(0)})P(H^{(0)}) + p_{y|H^{(1)}}(v \mid H^{(1)})P(H^{(1)})\Big] \\ = p_{y|H^{(0)}}(v \mid H^{(0)})P(H^{(0)}) \int_{-\infty}^{\infty} u p_{x|y,H^{(0)}}(u \mid v, H^{(0)}) \, du \\ + p_{y|H^{(1)}}(v \mid H^{(1)})P(H^{(1)}) \int_{-\infty}^{\infty} u p_{x|y,H^{(1)}}(u \mid v, H^{(1)}) \, du . \end{aligned} \tag{5.155}$$

We introduce the generalized likelihood ratio

$$\Lambda(v) = \frac{p_{y|H^{(1)}}(v \mid H^{(1)})}{p_{y|H^{(0)}}(v \mid H^{(0)})} \frac{P(H^{(1)})}{P(H^{(0)})} \tag{5.156}$$

and obtain the solution [Middleton, Esposito 1968]

$$\widehat{x}(v) = \mathrm{E}_x\left\{x \mid v, H^{(0)}\right\} \frac{1}{1+\Lambda(v)} + \mathrm{E}_x\left\{x \mid v, H^{(1)}\right\} \frac{\Lambda(v)}{1+\Lambda(v)}. \tag{5.157}$$

The joint MMSE detection and estimation problem leads to a linear combination of the MMSE estimators for the two hypotheses $H^{(0)}$ and $H^{(1)}$. The weights of the two estimators are in the range $[0, 1]$ and are determined as a function of the generalized likelihood ratio $\Lambda(v)$. Since the likelihood ratio is in general a continuous function of the observed data v, (5.157) comprises a *soft decision* weighting of the two conditional estimators.

Bibliography

Brehm, H.; Stammler, W. (1987). Description and Generation of Spherically Invariant Speech-Model Signals, *Signal Processing*, vol. 12, pp. 119–141.

Brillinger, D. R. (1981). *Time Series: Data Analysis and Theory*, Holden-Day, New York.

Dat, T. H.; Takeda, K.; Itakura, F. (2005). Generalized Gamma Modeling of Speech and its Online Estimation for Speech Enhancement, *Proceedings of the IEEE International Conference on Acoustics, Speech, and Signal Processing (ICASSP)*, vol. IV, pp. 181–184.

Gradshteyn, I. S.; Ryzhik, I. M. (2000). *Table of Integrals, Series, and Products*, 6th edn, Academic Press, San Diego, California.

Haykin, S. (1996). *Adaptive Filter Theory*, 3rd edn, Prentice Hall, Englewood Cliffs, New Jersey.

Jayant, N. S.; Noll, P. (1984). *Digital Coding of Waveforms*, Prentice Hall, Englewood Cliffs, New Jersey.

Johnson, N. L.; Kotz, S.; Balakrishnan, N. (1994). *Continuous Univariate Distributions*, John Wiley & Sons, Ltd., Chichester.

Kolmogorov, A. (1933). *Grundbegriffe der Wahrscheinlichkeitsrechnung*, Ergebnisse der Mathematik und ihrer Grenzgebiete, Springer Verlag, Berlin (in German).

Kotz, S.; Balakrishnan, N.; Johnson, N. L. (2000). *Continuous Multivariate Distributions*, 2nd edn, vol. 1, John Wiley & Sons, Ltd., Chichester.

Kullback, S. (1997). *Information Theory and Statistics*, Dover, New York. Republication of the 1968 edition.

Lotter, T.; Vary, P. (2003). Noise Reduction by Maximum A Posteriori Spectral Amplitude Estimation with Supergaussian Speech Modeling, *Proceedings of the International Workshop on Acoustic Echo and Noise Control (IWAENC)*.

Martin, R. (2002). Speech Enhancement Using MMSE Short Time Spectral Estimation with Gamma Distributed Speech Priors, *Proceedings of the IEEE International Conference on Acoustics, Speech, and Signal Processing (ICASSP)*, Orlando, Florida, USA, pp. 253–256.

Martin, R. (2005). Speech Enhancement based on Minimum Mean Square Error Estimation and Supergaussian Priors, *IEEE Transactions on Speech and Audio Processing*, vol. 13, no. 5, pp. 845–856.

Martin, R.; Breithaupt, C. (2003). Speech Enhancement in the DFT Domain Using Laplacian Speech Priors, *Proceedings of the International Workshop on Acoustic Echo and Noise Control (IWAENC)*.

McAulay, R. J.; Malpass, M. L. (1980). Speech Enhancement Using a Soft-Decision Noise Suppression Filter, *IEEE Transactions on Acoustics, Speech and Signal Processing*, vol. 28, no. 2, December, pp. 137–145.

Melsa, J. L.; Cohn, D. L. (1978). *Decision and Estimation Theory*, McGraw-Hill, New York.

Middleton, D.; Esposito, R. (1968). Simultaneous Optimum Detection and Estimation of Signals in Noise, *IEEE Transactions on Information Theory*, vol. 14, no. 3, pp. 434–444.

Oppenheim, A.; Schafer, R. W.; Buck, J. R. (1999). *Discrete-Time Signal Processing*, 2nd edn, Prentice Hall, Englewood Cliffs, New Jersey.

Papoulis, A.; Unnikrishna Pillai, S. (2001). *Probability, Random Variables, and Stochastic Processes*, 4th edn, McGraw-Hill, New York.

Shin, J. W.; Chang, J.; Kim, N. S. (2005). Statistical Modeling of Speech Signals Based on Generalized Gamma Distribution, *IEEE Signal Processing Letters*, vol. 12, no. 3, pp. 258–261.

Vary, P. (2004). Advanced Signal Processing in Speech Communication, *Proceedings of the European Signal Processing Conference (EUSIPCO)*, Vienna, Austria, pp. 1449–1456.

Welch, P. D. (1967). The Use of Fast Fourier Transform for the Estimation of Power Spectra: A Method Based on Time Averaging over Short, Modified Periodograms, *IEEE Transactions on Audio and Electroacoustics*, vol. 15, no. 2, pp. 70–73.

6

Linear Prediction

This chapter is concerned with the estimation of the spectral envelope of speech signals and its parametric representation. By far the most successful technique, known as *linear predictive* analysis, is based on autoregressive (AR) modeling. Linear prediction (LP) enables us to estimate the coefficients of AR filters and is thus closely related to the model of speech production. It is in fact a key component of all speech compression algorithms, e.g. [Jayant, Noll 1984], [Vary et al. 1998]. AR modeling, in conjunction with linear prediction, not only is used successfully in speech coding, but has also found numerous applications in spectral analysis, in speech recognition, and in the enhancement of noisy signals. The application of LP techniques in speech coding is quite natural, as (simplified) models of the vocal tract correspond to AR filters. The underlying algorithmic task of LP modeling is to solve a set of linear equations. Fast and efficient algorithms, such as the Levinson–Durbin algorithm [Durbin 1960], [Makhoul 1975], are available and will be explained in detail. Speech signals are stationary only for relatively short periods of 20 to 400 ms. To account for the variations of the vocal tract, the prediction filter must be adapted on a short-term basis. We discuss block-oriented and sequential methods.

Digital Speech Transmission: Enhancement, Coding and Error Concealment
Peter Vary and Rainer Martin

6.1 Vocal Tract Models and Short-Term Prediction

At the outset of our considerations we recall that speech may be modeled as the output of a linear, time-varying filter excited either by periodic pulses or by random noise, as shown in Fig. 6.1.

The filter with impulse response $h(k)$, which is assumed time invariant at first, is excited with the signal

$$v(k) = g \cdot u(k)\,, \tag{6.1}$$

where the gain factor g controls the amplitude, and thus the power of the excitation signal $v(k)$.

To synthesize *voiced* segments, a periodic impulse sequence

$$u(k) = \sum_{i=-\infty}^{+\infty} \delta(k - iN_0) \tag{6.2}$$

with the period N_0 is used; to synthesize *unvoiced* segments, a white noise signal $u(k)$ with variance $\sigma_u^2 = 1$ is applied.

In general, the relation between the excitation signal $v(k)$ and the output signal $x(k)$ is described in the time domain by the difference equation

$$x(k) = \sum_{i=0}^{m'} b_i\, v(k-i) - \sum_{i=1}^{m} c_i\, x(k-i)\,. \tag{6.3}$$

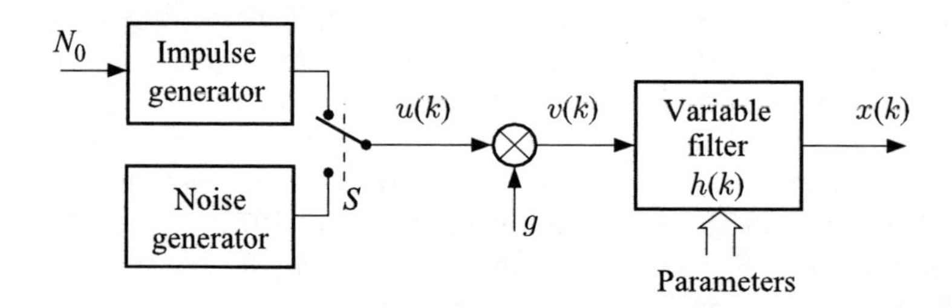

Figure 6.1: Discrete time model of speech production

N_0	:	pitch period
S	:	voiced/unvoiced decision
g	:	gain
$h(k)$	:	impulse response
$x(k)$	:	speech signal
$v(k)$	:	excitation signal

The output signal $x(k)$ is generated by a linear combination of the instantaneous excitation sample $v(k)$, m' delayed samples $v(k-i)$ $(i = 1, 2, \ldots, m')$, and m past output values $x(k-i)$ $(i = 1, 2, \ldots, m)$. In the z-domain, this leads with $c_0 = 1$ to the transfer function

$$\frac{X(z)}{V(z)} = H(z) = \frac{\sum_{i=0}^{m'} b_i \; z^{-i}}{\sum_{i=0}^{m} c_i \; z^{-i}} = \frac{z^m \sum_{i=0}^{m'} b_{m'-i} \; z^i}{z^{m'} \sum_{i=0}^{m} c_{m-i} \; z^i} . \tag{6.4}$$

This function has the form of a general, recursive digital filter with infinite impulse response (IIR). Depending on the choice of the coefficients, the following signal models can be distinguished:

a) All-Zero Model

For $c_0 = 1$ and $c_i \equiv 0$ $(i = 1, 2, \ldots, m)$, the filter is purely non-recursive, i.e., in the time domain we obtain

$$x(k) = \sum_{i=0}^{m'} b_i \; v(k-i) \tag{6.5-a}$$

and in the frequency domain

$$H(z) = \frac{\sum_{i=0}^{m'} b_{m'-i} \; z^i}{z^{m'}} = \frac{1}{z^{m'}} \cdot \prod_{i=1}^{m'} (z - z_{0i}) . \tag{6.5-b}$$

The transfer function has an m'-th order pole at $z = 0$ and it is determined by its zeros z_{0i} alone. This model is known as the *all-zero* model or *moving-average* model (MA model).

b) All-Pole Model

With $b_0 \neq 0$ and $b_i \equiv 0$ $(i = 1, 2, \ldots, m')$ a recursive filter with

$$x(k) = b_0 \; v(k) - \sum_{i=1}^{m} c_i \; x(k-i) \tag{6.6-a}$$

and

$$H(z) = b_0 \frac{z^m}{\sum_{i=0}^{m} c_{m-i} \; z^i} = b_0 \frac{z^m}{\prod_{i=1}^{m} (z - z_{\infty i})} \tag{6.6-b}$$

results.

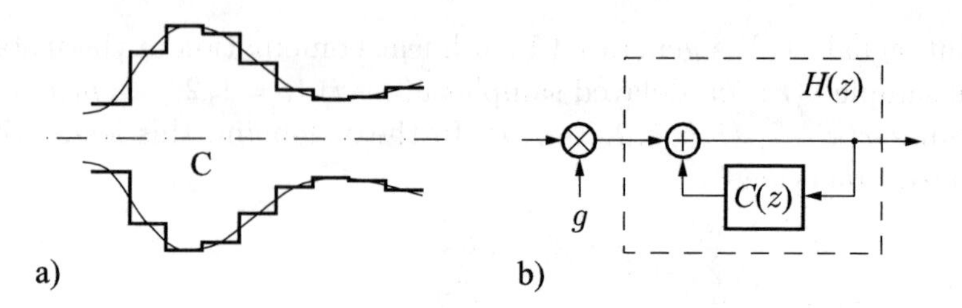

Figure 6.2: All-pole filter as model of the lossless tube
a) Acoustical tube model
b) Corresponding digital AR filter

Except for the m-th order zero at $z = 0$, this filter has only poles $z_{\infty i}$. Therefore, it is called an *all-pole* model. In signal theory, (6.6-a) is associated with an *autoregressive process* or *AR process*. This process corresponds to the model of speech production introduced in Section 2.3, neglecting the nasal tract and the glottal and labial filters. The corresponding discrete-time filter is the equivalent model of the lossless acoustical tube as illustrated in Fig. 6.2.

In this case, the effective transfer function of (6.4) for $b_0 = 1$ is

$$H(z) = \frac{1}{1 - C(z)} \tag{6.7-a}$$

with

$$C(z) = -\sum_{i=1}^{m} c_i \cdot z^{-1} . \tag{6.7-b}$$

c) Pole-Zero Model

The general case is described by (6.3) and (6.4) and represents a mixed *pole–zero* model. In statistics, this is called an *autoregressive moving-average* model (ARMA model).

This would be the adequate choice, if we include the pharynx (C), the nasal cavity (D), and the mouth cavity (E) separately in our model, as indicated in Fig. 6.3-a. As the radiations of the nostrils and the lips superimpose each other, we have a parallel connection of the tubes (D) and (E) and a serial connection of the two with tube (C).

The three tubes can be modeled by three individual AR filters with the transfer functions

$$H_C(z) = \frac{1}{1 - C(z)} , \quad H_D(z) = \frac{1}{1 - D(z)} , \quad H_E(z) = \frac{1}{1 - E(z)} , \tag{6.8}$$

where $D(z)$ and $E(z)$ are defined analog to (6.7-b).

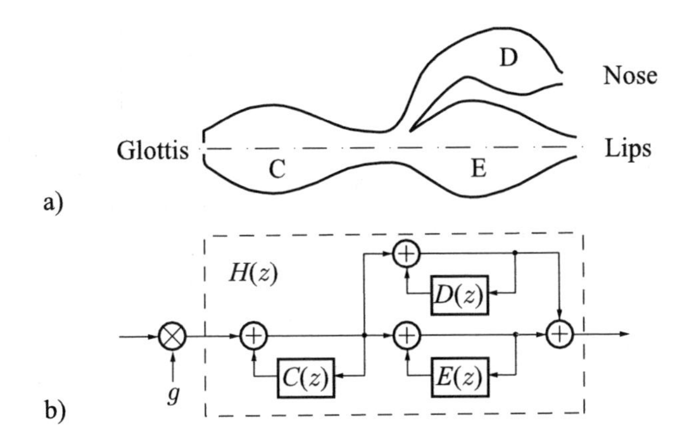

Figure 6.3: Pole–zero filter as model of connected lossless tube sections
a) Acoustical tube model
b) Corresponding digital ARMA filter

Thus, we obtain the overall transfer function

$$H(z) = \frac{1}{1 - C(z)} \cdot \left(\frac{1}{1 - D(z)} + \frac{1}{1 - E(z)} \right) \tag{6.9-a}$$

$$= \frac{2 - D(z) - E(z)}{(1 - C(z))\,(1 - D(z))\,(1 - E(z))} \tag{6.9-b}$$

$$= \frac{\prod_{i=1}^{m'} (z - z_{0i})}{\prod_{i=1}^{m} (z - z_{\infty i})} \tag{6.9-c}$$

having poles and zeros.

However, in connection with coding and synthesis of speech signals, the *all-pole* model is most frequently used in practice. This would mean that the nasal cavity is neglected and sections (C) and (E) of Fig. 6.3-a can jointly be modeled by one single all-pole model of corresponding degree.

The predominance of the all-pole model in practical applications will be further justified by the following considerations (e.g., [Deller Jr. et al. 2000]). First, the general pole–zero filter according to (6.4), which is assumed to be causal and stable, will be examined. In the z-plane, the poles are located inside the unit circle, while the zeros may also lie outside the unit circle. Such a situation is illustrated in Fig. 6.4-a for two pole–zero pairs.

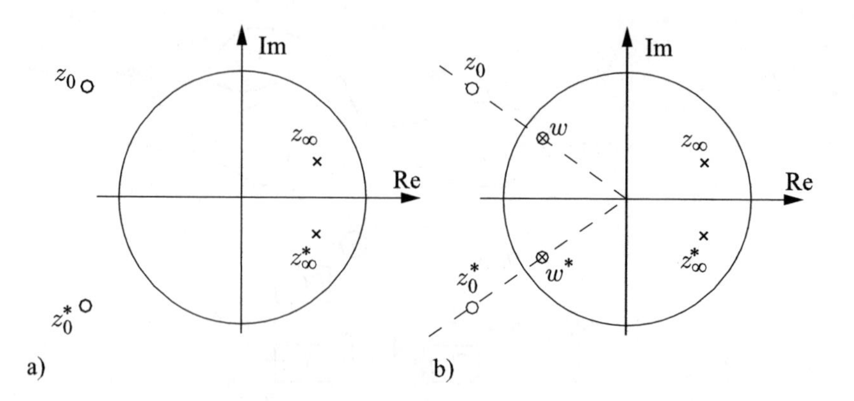

Figure 6.4: Pole–zero locations
a) Original filter $H(z)$
b) Decomposition into minimum-phase and all-pass filter
o = location of a zero
× = location of a pole

The transfer function $H(z)$ can now be split into a *minimum-phase* system $H_{\text{min}}(z)$ and an all-pass transfer function $H_{\text{Ap}}(z)$ according to

$$H(z) = H_{\text{min}}(z) \cdot H_{\text{Ap}}(z) . \tag{6.10}$$

In order to do so, the zeros z_0 and z_0^* outside the unit circle are first reflected into the circle to the positions $1/z_0^*$ and $1/z_0$. Then, they are compensated by poles at the same (reflected) position, as depicted in Fig. 6.4-b. At $z = w$ and $z = w^*$, we now have a zero, as well as a pole. The two locations of the zeros and the poles at

$$w = \frac{1}{z_0^*} \qquad \text{and} \qquad w^* = \frac{1}{z_0} \tag{6.11}$$

are assigned to a minimum-phase system having zeros inside the unit circle

$$H_{\text{min}}(z) = \frac{(z-w)(z-w^*)}{(z-z_\infty)(z-z_\infty^*)} . \tag{6.12}$$

The two zeros outside the unit circle and the poles at w and w^* form an all-pass system

$$H_{\text{Ap}}(z) = \frac{(z-z_0)(z-z_0^*)}{(z-w)(z-w^*)} , \tag{6.13}$$

with a constant magnitude

$$\left|H_{\text{Ap}}(z = e^{j\Omega})\right| = |z_0|^2 . \tag{6.14}$$

For the synthesis of speech it is sufficient to realize only the minimum-phase system, since speech perception is relatively insensitive to phase changes caused by all-pass filtering. Therefore, a minimum-phase pole–zero filter can be used to model the speech synthesis process according to Fig. 6.1. This leads to three important consequences:

1. Since the poles and zeros are located within the unit circle, a stable inverse filter exists with

$$H_{\text{min}}^{-1}(z) \;=\; \frac{1}{H_{\text{min}}(z)} \,. \tag{6.15}$$

   The vocal tract model filter $1/H_{\text{min}}(z)$ can thus be inverted to recover an excitation signal.

2. Every minimum-phase pole–zero filter can be described exactly by an all-pole filter of infinite order, which in turn can be approximated by an m-th order filter (e.g., [Marple Jr. 1987]). This justifies the use of an all-pole filter for the synthesis of speech in practical applications.

3. The coefficients of an all-pole filter can be derived from the speech samples $x(k)$ by solving a set of linear equations. Fast algorithms are available to fulfill this task (see Section 6.3).

In accordance with these considerations, the model of speech production is usually based on an all-pole filter. The coefficients of this filter can be determined, as will be shown below, using *linear prediction* techniques. The prediction implies the above-mentioned inverse filtering of the speech signal $x(k)$, so that, apart from the filter parameters, an excitation signal $v(k)$ according to the model in Fig. 6.1 can be obtained for speech synthesis or coding.

The all-pole model, which is defined by difference equation (6.6-a), shows that successive samples $x(k)$ are correlated. With given coefficients c_i, each sample $x(k)$ is determined by the preceding samples $x(k-1),\ x(k-2), \ldots, x(k-m)$, and $v(k)$ which is also called *innovation*. For reasons of simplicity, we assume that $b_0 = g = 1$. Therefore, it must be possible to estimate or predict the present sample $x(k)$ despite the contribution of the innovation $v(k)$ by a weighted linear combination of previous samples:

$$\hat{x}(k) \;=\; \sum_{i=1}^{n} a_i \; x(k-i) \,. \tag{6.16}$$

This operation is called *linear prediction* (LP). The predicted signal $\hat{x}(k)$ is a function of the unknown coefficients a_i and the previous samples $x(k-i),\ i = 1, 2, \ldots, n$. The difference

$$d(k) \;\doteq\; x(k) - \hat{x}(k) \tag{6.17}$$

is termed *prediction error* signal.

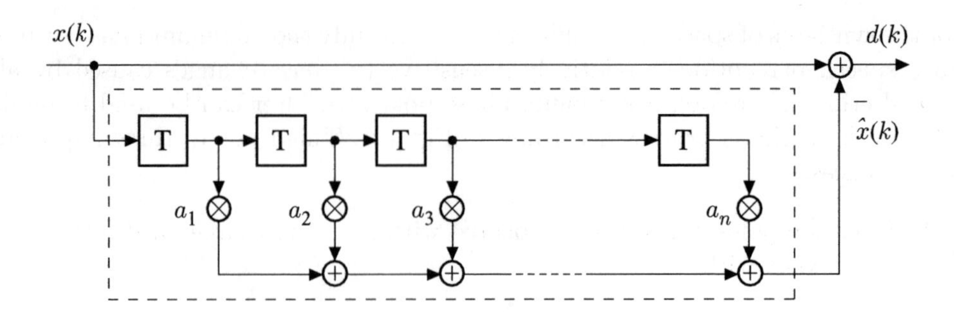

Figure 6.5: Linear prediction with a non-recursive filter of order n

The prediction task and the generation of the error signal can be described, as shown in Fig. 6.5, by non-recursive filtering of signal $x(k)$. The overall filter with input $x(k)$ and output $d(k)$ is called the *LP-analysis* filter. As the number n of predictor coefficients $a_i, i = 1, 2, \ldots, n$ is relative small, the memory of the predictor is quite short according to

$$n \cdot T = n \cdot \frac{1}{f_s} . \tag{6.18}$$

A typical figure is $n = 8$, $f_s = 8\,\text{kHz}$ and $n \cdot T = 1\,\text{ms}$. Therefore, the predictor of Fig. 6.5 is usually called the *short-term predictor.*

Referring to (6.6-a) and (6.16) for $n = m$ we obtain

$$d(k) = x(k) \; - \; \hat{x}(k) \tag{6.19-a}$$

$$= v(k) \; - \; \sum_{i=1}^{m} c_i \; x(k-i) \; - \; \sum_{i=1}^{m} a_i \; x(k-i) \tag{6.19-b}$$

$$= v(k) \; - \; \sum_{i=1}^{m} \Big(c_i \; + \; a_i\Big) \; x(k-i) . \tag{6.19-c}$$

If we could adjust the predictor coefficients a_i in such a way that

$$a_i \;\; = \;\; - \, c_i \, , \tag{6.20-a}$$

we obtain

$$d(k) \;\; = \;\; v(k) . \tag{6.20-b}$$

In this case, the LP-analysis filter would compensate the vocal tract filter. Therefore, the coefficients a_i implicitly describe the (instantaneous) spectral envelope of the speech signal. In the following section, it will be shown how to calculate optimal coefficients a_i by performing a system identification according to (6.20-a).

6.2 Optimal Prediction Coefficients for Stationary Signals

In this section it will be shown that prediction coefficients a_i, which minimize the mean square error

$$\mathrm{E}\{d^2(k)\} = \mathrm{E}\{(x(k) - \hat{x}(k))^2\}, \tag{6.21}$$

also obey (6.20-a). Furthermore, an algorithm for calculating the coefficients a_i from the speech signal $x(k)$ is derived.

For the sake of simplicity, it is first presumed that the unknown coefficients c_i of the model filter, or the impulse response $h(k)$ of the vocal tract respectively, are time invariant and that the order $m = n$ is known. Furthermore, a real-valued, stationary, uncorrelated, and zero-mean excitation signal $v(k)$ (white noise, unvoiced stationary speech) is assumed. Then, the case of periodic pulse excitation $v(k)$ (voiced stationary speech) will be considered as well.

6.2.1 Optimum Prediction

Within the optimization procedure, the auto-correlation functions of the sequences $v(k)$ and $x(k)$ and their cross-correlation are of interest:

$$\varphi_{vv}(\lambda) = \mathrm{E}\{v(k)\ v(k\pm\lambda)\} = \begin{cases} \sigma_v^2 & \text{for } \lambda = 0 \\ 0 & \text{for } \lambda = \pm 1, \pm 2, \ldots \end{cases} \tag{6.22}$$

$$\varphi_{xx}(\lambda) = \mathrm{E}\{x(k)\ x(k\pm\lambda)\} = \sigma_v^2 \sum_{i=0}^{\infty} h(i)\ h(i\pm\lambda) \tag{6.23}$$

$$\varphi_{vx}(\lambda) = \mathrm{E}\{v(k)\ x(k+\lambda)\} = \mathrm{E}\left\{\sum_{i=0}^{\infty} h(i)\ v(k+\lambda-i)\ v(k)\right\} \tag{6.24-a}$$

$$= \sum_{i=0}^{\infty} h(i)\ \varphi_{vv}(\lambda-i) = h(\lambda)\ \sigma_v^2\,. \tag{6.24-b}$$

The predictor coefficients a_i, $i = 1, 2, \ldots, n$ are chosen to minimize the mean square prediction error according to (6.21).

Taking (6.16) and (6.17) into account, the partial derivative of (6.21) with respect to the coefficient a_λ is set to 0

$$\frac{\partial \mathrm{E}\{d^2(k)\}}{\partial a_\lambda} = \mathrm{E}\left\{2\,d(k)\;\frac{\partial d(k)}{\partial a_\lambda}\right\} \tag{6.25-a}$$

$$= \mathrm{E}\left\{-2\,d(k)\;x(k-\lambda)\right\} \stackrel{!}{=} 0\;; \quad \lambda = 1,2,\ldots,n \tag{6.25-b}$$

and the second derivation results in

$$\frac{\partial^2 \mathrm{E}\{d^2(k)\}}{\partial a_\lambda^2} = \mathrm{E}\left\{2\,x^2(k-\lambda)\right\} \geq 0\,, \tag{6.25-c}$$

proving that the solution is a minimum. Equation (6.25-b) has a single solution

$$a_i \;=\; -\,c_i\,, \tag{6.26}$$

as with (6.20-b), (6.24-b), (6.25-b), and the assumed causality ($h(-\lambda) = 0$; $\lambda = 1,2,\ldots,n$) the following holds

$$\mathrm{E}\{-2\,d(k)\;x(k-\lambda)\} = -2\,\mathrm{E}\{v(k)\;x(k-\lambda)\} \tag{6.27-a}$$

$$= -2\,\varphi_{vx}(-\lambda) \;=\; 0\,. \tag{6.27-b}$$

The prediction process compensates the model filter, as was shown with (6.20-b). The complete filter of Fig. 6.5 with the input signal $x(k)$ and the output signal $d(k)$ can be interpreted as the inverse of the vocal tract model.

If, in case of voiced segments, the excitation signal is not a white noise but a periodic pulse train

$$v(k) = g \cdot \sum_{i=-\infty}^{\infty} \delta\,(k - i \cdot N_0) \tag{6.28}$$

with $N_0 > n$ then (6.22) is valid for $\lambda = 0, \pm 1, \pm 2, \ldots, \pm(N_0 - 1)$ with

$$\sigma_v^2 = \frac{g^2}{N_0}\,. \tag{6.29}$$

Then (6.27-b) is fulfilled for the interesting range of $\lambda = 1, 2, \ldots, n < N_0$ too.

With the above assumptions, the power minimization of the prediction error implies in the unvoiced as well as in the voiced case a system identification process, as the coefficients a_i of the optimal non-recursive predictor match the coefficients c_i of the recursive model filter.

Equation (6.25-b) can be transformed into an explicit rule to calculate the predictor coefficients a_i. For this purpose, $d(k)$ and $x(k)$ according to (6.17) and (6.16) are inserted into (6.25-b)

$$-2\,\mathrm{E}\{d(k)\ x(k-\lambda)\} = -2\,\mathrm{E}\left\{\left(x(k) - \sum_{i=1}^{n} a_i\ x(k-i)\right)\ x(k-\lambda)\right\} \tag{6.30-a}$$

$$= -2\,\varphi_{xx}(\lambda) + 2\sum_{i=1}^{n} a_i\ \varphi_{xx}(\lambda - i)\ \stackrel{!}{=}\ 0\,. \tag{6.30-b}$$

For $\lambda = 1, 2, \ldots, n$, this results in the so-called *normal equations* in vector matrix notation

$$\begin{pmatrix} \varphi_{xx}(1) \\ \varphi_{xx}(2) \\ \vdots \\ \varphi_{xx}(n) \end{pmatrix} = \begin{pmatrix} \varphi_{xx}(0) & \varphi_{xx}(-1) & \ldots & \varphi_{xx}(1-n) \\ \varphi_{xx}(1) & \varphi_{xx}(0) & \ldots & \varphi_{xx}(2-n) \\ \vdots & \vdots & \ddots & \vdots \\ \varphi_{xx}(n-1) & \varphi_{xx}(n-2) & \ldots & \varphi_{xx}(0) \end{pmatrix} \begin{pmatrix} a_1 \\ a_2 \\ \vdots \\ a_n \end{pmatrix} \tag{6.30-c}$$

or

$$\boldsymbol{\varphi}_{xx} = \mathbf{R}_{xx}\ \mathbf{a}\,, \tag{6.30-d}$$

respectively, where $\boldsymbol{\varphi}_{xx}$ denotes the correlation vector and $\mathbf{R}_{xx}$ the correlation matrix. $\mathbf{R}_{xx}$ is a real-valued, positive-definite Toeplitz matrix.

The solution of the normal equations provides the optimal coefficient vector

$$\mathbf{a}_{\text{opt}} = \mathbf{R}_{xx}^{-1}\ \boldsymbol{\varphi}_{xx}\,. \tag{6.31}$$

The power of the prediction error in the case of optimal predictor coefficients can be calculated explicitly:

$$\sigma_d^2 = \mathrm{E}\{(x(k) - \hat{x}(k))^2\} \tag{6.32-a}$$

$$= \mathrm{E}\{x^2(k) - 2\,x(k)\,\hat{x}(k) + \hat{x}^2(k)\}\,. \tag{6.32-b}$$

In vector notation

$$\hat{x}(k) = \mathbf{a}^T\ \mathbf{x}(k-1)$$

with $\mathbf{a} \doteq (a_1, a_2, \ldots, a_n)^T$ and $\mathbf{x}(k-1) \doteq (x(k-1), x(k-2), \ldots, x(k-n))^T$, we get

$$\sigma_d^2 = \sigma_x^2 - 2\,\mathbf{a}^T\boldsymbol{\varphi}_{xx} + \mathbf{a}^T\mathbf{R}_{xx}\,\mathbf{a}\,. \tag{6.32-c}$$

Inserting (6.30-d) results in

$$\sigma_d^2 = \sigma_x^2 - \mathbf{a}^T\ \boldsymbol{\varphi}_{xx} \tag{6.32-d}$$

$$= \sigma_x^2 - \sum_{\lambda=1}^{n} a_\lambda\ \varphi_{xx}(\lambda)\,. \tag{6.32-e}$$

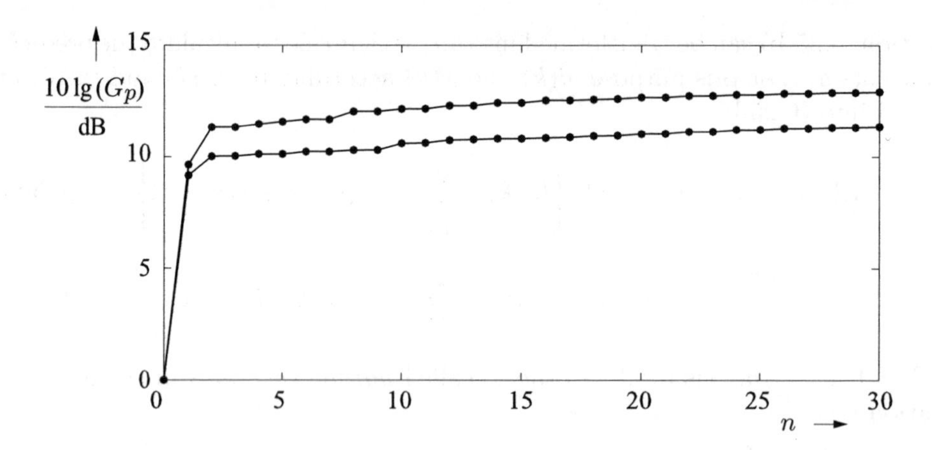

Figure 6.6: Logarithmic prediction gain $10\lg\left(\sigma_x^2/\sigma_d^2\right)$ of a time-invariant prediction filter for two different speakers (signal length 30 s, $f_s = 8\,\text{kHz}$)

Alternatively, (6.32-d) can be reformulated with (6.31) into

$$\sigma_d^2 = \sigma_x^2 - \boldsymbol{\varphi}_{xx}^T \, \mathbf{R}_{xx}^{-1} \, \boldsymbol{\varphi}_{xx} \,, \tag{6.32-f}$$

where the optimal coefficients a_i are not explicitly needed to calculate σ_d^2.

The power ratio

$$G_p = \frac{\sigma_x^2}{\sigma_d^2} \tag{6.33}$$

is called the *prediction gain*. This gain is a measure for the bit rate reduction that can be achieved through predictive coding techniques (cf. Section 8.3.3). Figure 6.6 depicts the prediction gain for two speech signals with a duration of 30 seconds each as a function of the predictor order n. For each predictor order $n = 1, 2, \ldots, 30$, a set of coefficients which is optimal for the respective complete signal was calculated according to (6.31).

It becomes apparent that for predictor orders $n \geq 2$ the prediction gain increases rather slowly. The achievable prediction gain depends to a certain extent on the speaker.

6.2.2 Spectral Flatness Measure

The achievable prediction gain is related to the *spectral flatness* of the signal $x(k)$. No prediction gain can be achieved for white noise which has a constant power spectral density

$$\Phi_{xx}(e^{j\Omega}) = \text{const.} = \sigma_x^2 \,. \tag{6.34}$$

This power spectral density is completely flat; the signal is not correlated. With increasing correlation the prediction gain increases while the spectral flatness decreases.

A first measure of spectral flatness is the average of the logarithmic normalized power spectral density, e.g., [Markel, Gray 1976], [Sluijter 2005]

$$\mathrm{sf}_x = \frac{1}{2\pi} \int_{-\pi}^{\pi} \ln \left[\frac{\Phi_{xx}\left(e^{j\Omega}\right)}{\sigma_x^2} \right] \, d\Omega \tag{6.35}$$

with

$$\sigma_x^2 = \frac{1}{2\pi} \int_{-\pi}^{\pi} \Phi_{xx}\left(e^{j\Omega}\right) \, d\Omega \,. \tag{6.36}$$

The measure (6.35) is bounded according to

$$-\infty < \mathrm{sf}_x \leq 0$$

where the maximum value $\mathrm{sf}_x = 0$ applies in case of white noise.

A more convenient spectral flatness measure γ_x^2 with

$$0 \leq \gamma_x^2 \leq 1 \tag{6.37}$$

is obtained from (6.35), e.g., [Makhoul, Wolf 1972], [Markel, Gray 1976] by

$$\gamma_x^2 = \exp\left(\mathrm{sf}_x\right) \tag{6.38}$$

$$= \exp\left(\frac{1}{2\pi} \int_{-\pi}^{\pi} \ln\left(\Phi_{xx}(e^{j\Omega})\right) \, d\Omega \; - \; \ln\left(\sigma_x^2\right) \right) . \tag{6.39}$$

Taking (6.36) into account, the widely used form of the spectral flatness measure results:

$$\gamma_x^2 = \frac{\exp\left(\frac{1}{2\pi} \int_{-\pi}^{\pi} \ln\left(\Phi_{xx}(e^{j\Omega})\right) \, d\Omega \right)}{\frac{1}{2\pi} \int_{-\pi}^{\pi} \Phi_{xx}(e^{j\Omega}) \, d\Omega} \,. \tag{6.40}$$

In the case of $\Phi_{xx}(e^{j\Omega}) = \text{const.} = \sigma_x^2$ we obtain $\gamma_x^2 = 1$.

The expression (6.40) is the ratio of the geometric and the arithmetic mean of the power spectral density $\Phi_{xx}(e^{j\Omega})$. This becomes more obvious when the integrals in (6.40) are approximated by summations and when the power spectral density is replaced by the periodogram, i.e., the squared magnitude of the discrete Fourier transform of a segment of $x(k)$ of length M:

$$\gamma_x^2 \simeq \frac{\exp\left(\frac{1}{M}\sum\limits_{\mu=0}^{M-1} \ln|X(e^{j\Omega_\mu})|^2\right)}{\frac{1}{M}\sum\limits_{\mu=0}^{M-1} |X(e^{j\Omega_\mu})|^2} \tag{6.41-a}$$

$$= \frac{\left[\prod\limits_{\mu=0}^{M-1} |X(e^{j\Omega_\mu})|^2\right]^{\frac{1}{M}}}{\frac{1}{M}\sum\limits_{\mu=0}^{M-1} |X(e^{j\Omega_\mu})|^2}, \quad \Omega_\mu = \frac{2\pi}{M}\mu,\ \mu = 0, 1, \ldots, M-1\,. \tag{6.41-b}$$

The spectral flatness measure is related to the achievable prediction gain as shown below.

We consider the LP-analysis filter with input $x(k)$ and output $d(k)$ as shown in Fig. 6.7. If $x(k)$ is a random process with power spectral density $\Phi_{xx}(e^{j\Omega})$, the following relations are valid:

$$\Phi_{dd}\left(e^{j\Omega}\right) = |A_0\left(e^{j\Omega}\right)|^2 \cdot \Phi_{xx}\left(e^{j\Omega}\right) \tag{6.42}$$

$$\frac{1}{2\pi}\int\limits_{-\pi}^{\pi} \ln\left(\Phi_{dd}\left(e^{j\Omega}\right)\right)\, d\Omega = \frac{1}{2\pi}\int\limits_{-\pi}^{\pi} \ln|A_0\left(e^{j\Omega}\right)|^2\, d\Omega \tag{6.43}$$

$$+\ \frac{1}{2\pi}\int\limits_{-\pi}^{\pi} \ln\left(\Phi_{xx}\left(e^{j\Omega}\right)\right)\, d\Omega\,. \tag{6.44}$$

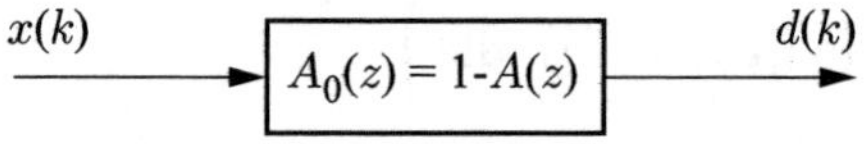

x(k) → [A0(z) = 1-A(z)] → d(k)

Figure 6.7: LP-analysis filter with $A_0(z) = 1 - A(z) = 1 - \sum\limits_{i=1}^{n} a_i \cdot z^{-i} = \frac{1}{z^n}\prod\limits_{i=1}^{n}(z - z_i)$

It can be shown that the average value of the logarithm of $|A_0(z)|$ is zero if $A_0(z)$ has all its zeros within the unit circle, e.g., [Markel, Gray 1976], [Itakura, Saito 1970]

$$\frac{1}{2\pi}\int_{-\pi}^{\pi} \ln |A_0\left(e^{j\Omega}\right)|^2 \, \mathrm{d}\Omega = 0 . \tag{6.45}$$

Therefore, the ratio of the spectral flatness measures of $x(k)$ and $d(k)$ can be written as

$$\frac{\gamma_x^2}{\gamma_d^2} = \frac{\exp\left(\frac{1}{2\pi}\int_{-\pi}^{\pi} \ln\left(\Phi_{xx}\left(e^{j\Omega}\right)\right) \, \mathrm{d}\Omega\right)}{\sigma_x^2} \cdot \frac{\sigma_d^2}{\exp\left(\frac{1}{2\pi}\int_{-\pi}^{\pi} \ln\left(\Phi_{dd}\left(e^{j\Omega}\right)\right) \, \mathrm{d}\Omega\right)}$$

$$= \frac{\sigma_d^2}{\sigma_x^2} = \frac{1}{G_p} .$$

For perfect prediction (requiring infinite predictor degree) the residual signal $d(k)$ has a perfectly flat spectrum with $\gamma_d^2 = 1$. Thus the following relation is valid

$$\lim_{n \to \infty} \left\{ \frac{1}{G_p} \right\} = \gamma_x^2 . \tag{6.46}$$

The spectral flatness measure γ_x^2 is identical to the inverse of the theoretically maximum prediction gain.

6.3 Predictor Adaptation

Speech signals can be considered as stationary only for relatively short time intervals between 20 and 400 ms; thus, the coefficients of the model filter change quickly. Therefore, it is advisable to optimize the predictor coefficients frequently. Here, we distinguish between block-oriented and sequential methods.

6.3.1 Block-Oriented Adaptation

The optimization of the coefficients a_i is performed for short signal segments (also called blocks or frames) which consist of N samples each. Considering the vocal tract variations, the block length is usually set to $T_B = 10 - 30$ ms. With a sampling frequency of $f_s = 8$ kHz, this corresponds to a block size of $N = 80 - 240$ signal samples.

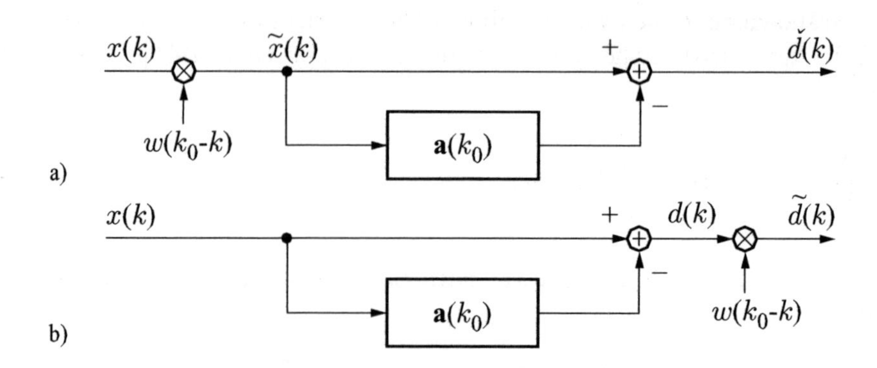

Figure 6.8: Block-oriented adaptation
a) Auto-correlation method (stationary method)
b) Covariance method (non-stationary method)

Basically, the solution (6.31) for stationary signals can be used for the block-oriented adaptation. At time instant k_0 the coefficient vector

$$\mathbf{a}(k_0) = (a_1(k_0), a_2(k_0), \ldots, a_n(k_0))^T$$

is determined. As $x(k)$ is not limited to N samples, we can, for the optimization, limit either the input signal $x(k)$ or the prediction error signal $d(k)$ to a finite time interval. These two cases are often referred to in the literature as the *auto-correlation method* and *covariance method* [Makhoul 1975]. The terms *stationary approach* and *non-stationary approach* or *auto-* and *cross-correlation method*, however, are more appropriate.

The distinction between the two methods is illustrated in Fig. 6.8. For the basic version of the *auto-correlation method* (Fig. 6.8-a) a frame $\tilde{x}(k)$ which includes the last N samples up to the time instant k_0 is extracted from the sequence $x(k)$ using a window $w(k)$ of length N

$$\tilde{x}(k) = x(k)\, w(k_0 - k)\,, \tag{6.47}$$

while for the *covariance method* (Fig. 6.8-b) the finite segment is extracted from the prediction error signal

$$\tilde{d}(k) = d(k)\, w(k_0 - k)\,. \tag{6.48}$$

These two different definitions of the time-limited segment lead to the two above-mentioned alternative approaches, which will be discussed in the next two sections.

6.3.1.1 Auto-correlation Method

The signal $\tilde{x}(k)$ attains non-zero values only within the time interval

$$k_1 = k_0 - N + 1 \leq k \leq k_0 . \tag{6.49}$$

Thus, $\tilde{x}(k)$ has finite energy. Due to the finite order n of the non-recursive predictor, the prediction error signal $\check{d}(k)$ is limited to the finite interval

$$k_1 = k_0 - N + 1 \leq k \leq k_0 + n = k_2 \tag{6.50}$$

and also has finite energy. Therefore, the optimization procedure can be simplified by minimizing the energy of the signal $\check{d}(k)$

$$\sum_{k=k_1}^{k_2} \check{d}^2(k) \stackrel{!}{=} \min . \tag{6.51}$$

In analogy to the derivation for stationary signals, the solution (6.31) is valid by replacing the auto-correlation function

$$\varphi_{xx}(\lambda) = \mathrm{E}\{x(k)\, x(k+\lambda)\}$$

by the short-term auto-correlation which may be defined for signals with finite energy as

$$r_\lambda = \sum_{k=k_1}^{k_0} \tilde{x}(k)\, \tilde{x}(k+\lambda) \tag{6.52-a}$$

$$= \sum_{k=k_1}^{k_0-\lambda} x(k)\, x(k+\lambda) \qquad \forall\, \lambda = 0, 1, \ldots, n , \tag{6.52-b}$$

Due to the symmetry

$$r_\lambda = r_{-\lambda} \tag{6.53}$$

the normal equations (6.30-c) can be rewritten as

$$\begin{pmatrix} r_1 \\ r_2 \\ r_3 \\ r_4 \\ \vdots \\ r_n \end{pmatrix} = \begin{pmatrix} r_0 & r_1 & r_2 & r_3 & \cdots & r_{n-1} \\ r_1 & r_0 & r_1 & r_2 & \cdots & r_{n-2} \\ r_2 & r_1 & r_0 & r_1 & \cdots & r_{n-3} \\ r_3 & r_2 & r_1 & r_0 & \cdots & r_{n-4} \\ \vdots & \vdots & \vdots & \vdots & \ddots & \vdots \\ r_{n-1} & r_{n-2} & r_{n-3} & r_{n-4} & \cdots & r_0 \end{pmatrix} \begin{pmatrix} a_1 \\ a_2 \\ a_3 \\ a_4 \\ \vdots \\ a_n \end{pmatrix} \tag{6.54-a}$$

or according to (6.30-d)

$$\mathbf{r} = \mathbf{R}\ \mathbf{a} \tag{6.54-b}$$

respectively, where φ_{xx} is replaced by $\mathbf{r}$ and $\mathbf{R}_{xx}$ by $\mathbf{R}$. Like the correlation matrix $\mathbf{R}_{xx}$, the short-term correlation matrix $\mathbf{R}$ has a symmetric Toeplitz structure. This facilitates the use of very efficient algorithms such as the Levinson–Durbin algorithm (see Section 6.3.1.3) for the solution of the set of equations.

Because of the formal analogy with the solution (6.31) for stationary signals, the term *stationary approach* is used. For the calculation of the short-term autocorrelation values r_λ, often non-rectangular windows $w(k)$ are used, e.g., the Hamming window or asymmetric windows, e.g., [Barnwell III 1981], which gives more weight to the most recent samples than to the older ones.

6.3.1.2 Covariance Method

For the so-called *covariance method*, the energy of the error signal $\tilde{d}(k)$ in Fig. 6.8-b is minimized over the finite interval of length N

$$\sum_{k=k_1}^{k_0} \tilde{d}^2(k) \;=\; \sum_{k=k_1}^{k_0} d^2(k) \;\stackrel{!}{=}\; \min\,. \tag{6.55}$$

Compared to (6.51), the upper limit of the sum changes. Note that, in addition to this, the signal $x(k)$ is now applied to the filter which is not restricted to the time interval $k_1 \leq k \leq k_0$.

With

$$d(k) \;=\; x(k) \;-\sum_{i=1}^{n} a_i\; x(k-i)\,, \tag{6.56}$$

the partial derivative of (6.55) with respect to a specific coefficient a_λ yields

$$\frac{\partial}{\partial a_\lambda} \sum_{k=k_1}^{k_0} d^2(k) = 2 \sum_{k=k_1}^{k_0} d(k)\; \frac{\partial d(k)}{\partial a_\lambda}; \qquad \lambda = 1, 2, \ldots, n \tag{6.57-a}$$

$$= -2 \sum_{k=k_1}^{k_0} d(k)\; x(k-\lambda) \tag{6.57-b}$$

$$= -2 \sum_{k=k_1}^{k_0} \left(x(k) - \sum_{i=1}^{n} a_i\; x(k-i) \right) x(k-\lambda) \stackrel{!}{=} 0\,. \tag{6.57-c}$$

For $\lambda = 1, 2, \ldots, n$ this results in

$$\sum_{k=k_1}^{k_0} x(k)\, x(k-\lambda) \;=\; \sum_{i=1}^{n} a_i \sum_{k=k_1}^{k_0} x(k-i)\, x(k-\lambda)\,. \tag{6.58}$$

With the abbreviation

$$\hat{r}_{i,\lambda} \;=\; \sum_{k=k_1}^{k_0} x(k-i)\, x(k-\lambda) \tag{6.59}$$

and the symmetry

$$\hat{r}_{i,\lambda} \;=\; \hat{r}_{\lambda,i}\,, \tag{6.60}$$

we obtain the set of equations

$$\begin{pmatrix} \hat{r}_{0,1} \\ \hat{r}_{0,2} \\ \hat{r}_{0,3} \\ \vdots \\ \hat{r}_{0,n} \end{pmatrix} = \begin{pmatrix} \hat{r}_{1,1} & \hat{r}_{1,2} & \hat{r}_{1,3} & \cdots & \hat{r}_{1,n} \\ \hat{r}_{1,2} & \hat{r}_{2,2} & \hat{r}_{2,3} & \cdots & \hat{r}_{2,n} \\ \hat{r}_{1,3} & \hat{r}_{2,3} & \hat{r}_{3,3} & \cdots & \hat{r}_{3,n} \\ \vdots & \vdots & \vdots & \ddots & \vdots \\ \hat{r}_{1,n} & \hat{r}_{2,n} & \hat{r}_{3,n} & \cdots & \hat{r}_{n,n} \end{pmatrix} \begin{pmatrix} a_1 \\ a_2 \\ a_3 \\ \vdots \\ a_n \end{pmatrix}, \tag{6.61-a}$$

or, in compact notation,

$$\hat{\mathbf{r}} = \hat{\mathbf{R}}\,\mathbf{a}\,. \tag{6.61-b}$$

In contrast to the stationary approach (6.54-a), the calculation of the short-term correlation values $\hat{r}_{i,\lambda}$ is not shift invariant

$$\hat{r}_{i,\lambda} \;\neq\; \hat{r}_{i+i_0,\lambda+i_0}\,. \tag{6.62}$$

Therefore, this method of calculating the predictor coefficients is also called the *non-stationary approach.*

While for the auto-correlation method the number of product terms $x(i)\, x(i+\lambda)$ used in (6.52-b) decreases for increasing λ, the computation of $\hat{r}_{\lambda,i}$ is always based on N product terms. Consequently, in addition to the N samples from $k = k_1$ to $k = k_0$, n preceding values $x(k)$ are needed for this approach. The covariance method thus provides more precise estimates of the short-term correlation than the auto-correlation method. The resulting matrix $\hat{\mathbf{R}}$ is still symmetric, but the Toeplitz form of $\mathbf{R}$ is lost. Thus, the matrix inversion requires more complex methods, such as the Cholesky decomposition. It should be noted that stability of the resulting synthesis filter cannot be guaranteed. However simple stabilizing methods exist, such as adding a small constant in the diagonal of $\hat{\mathbf{R}}$.

Various recursion algorithms are available to solve the set of equations for both the auto-correlation method and the covariance method. Here, the Levinson–Durbin algorithm for the autocorrelation method will be developed as one possible solution. Comparable results are provided by similar algorithms, such as those of Schur, Burg, Le Roux, and Gueguen ([Kay 1988], [Marple Jr. 1987]).

6.3.1.3 Levinson–Durbin Algorithm

In this section we will derive the *Levinson–Durbin* algorithm for the auto-correlation method [Levinson 1947], [Durbin 1960]. This algorithm starts from a known solution $\mathbf{a}^{(p-1)}$ of the set of equations (6.54-a) for the predictor order $(p-1)$ leading to the solution $\mathbf{a}^{(p)}$ for the order p. Beginning with the (trivial) solution for $p=0$ we can find the solution for the actually desired predictor order n iteratively with low computational effort. As an example, we will now examine the step from $p-1=2$ to $p=3$.

In order to simplify the notation, we replace the predictor coefficients by

$$\alpha_i^{(p)} \doteq -a_i^{(p)} \; ; \quad i \in \{1, 2, \dots, p\} . \tag{6.63}$$

For $p=2$, the normal equations

$$\begin{pmatrix} r_1 \\ r_2 \end{pmatrix} + \begin{pmatrix} r_0 & r_1 \\ r_1 & r_0 \end{pmatrix} \cdot \begin{pmatrix} \alpha_1^{(2)} \\ \alpha_2^{(2)} \end{pmatrix} = \begin{pmatrix} 0 \\ 0 \end{pmatrix} \tag{6.64-a}$$

can be written equivalently as

$$\begin{pmatrix} r_1 & r_0 & r_1 \\ r_2 & r_1 & r_0 \end{pmatrix} \cdot \begin{pmatrix} 1 \\ \alpha_1^{(2)} \\ \alpha_2^{(2)} \end{pmatrix} = \begin{pmatrix} 0 \\ 0 \end{pmatrix} . \tag{6.64-b}$$

In analogy to (6.32-e), the following expression holds for the short-term energy of the prediction error:

$$\mathrm{E}^{(2)} = r_0 + \sum_{i=1}^{2} \alpha_i^{(2)} r_i \; . \tag{6.65}$$

Thus, (6.64-b) can be extended to

$$\begin{pmatrix} r_0 & r_1 & r_2 \\ r_1 & r_0 & r_1 \\ r_2 & r_1 & r_0 \end{pmatrix} \cdot \begin{pmatrix} 1 \\ \alpha_1^{(2)} \\ \alpha_2^{(2)} \end{pmatrix} = \begin{pmatrix} \mathrm{E}^{(2)} \\ 0 \\ 0 \end{pmatrix} . \tag{6.66-a}$$

or in vector matrix notation

$$\mathbf{R}^{(3)} \cdot \boldsymbol{\alpha}^{(2)} = \mathbf{e}^{(2)} . \tag{6.66-b}$$

Because of the symmetry property of the correlation matrix of dimension $(p+1) \times (p+1)$, the following representation is valid, too:

$$\begin{pmatrix} r_0 & r_1 & r_2 \\ r_1 & r_0 & r_1 \\ r_2 & r_1 & r_0 \end{pmatrix} \cdot \begin{pmatrix} \alpha_2^{(2)} \\ \alpha_1^{(2)} \\ 1 \end{pmatrix} = \begin{pmatrix} 0 \\ 0 \\ \mathrm{E}^{(2)} \end{pmatrix} . \tag{6.67}$$

For $p = 3$, we now try the following solution approach:

$$\boldsymbol{\alpha}^{(3)} \doteq \begin{pmatrix} 1 \\ \alpha_1^{(3)} \\ \alpha_2^{(3)} \\ \alpha_3^{(3)} \end{pmatrix} = \begin{pmatrix} 1 \\ \alpha_1^{(2)} \\ \alpha_2^{(2)} \\ 0 \end{pmatrix} + k_3 \begin{pmatrix} 0 \\ \alpha_2^{(2)} \\ \alpha_1^{(2)} \\ 1 \end{pmatrix} \tag{6.68}$$

with the unknown constant k_3. Obviously, the coefficients $\alpha_1^{(3)}$, $\alpha_2^{(3)}$, and $\alpha_3^{(3)}$ are only a function of the respective coefficients for $p = 2$ and k_3.

Extending (6.66-b) from $p = 2$ to 3 results in

$$\mathbf{R}^{(4)} \, \boldsymbol{\alpha}^{(3)} = \mathbf{e}^{(3)} \tag{6.69-a}$$

and with (6.68) we get

$$\begin{pmatrix} r_0 & r_1 & r_2 & r_3 \\ r_1 & r_0 & r_1 & r_2 \\ r_2 & r_1 & r_0 & r_1 \\ r_3 & r_2 & r_1 & r_0 \end{pmatrix} \cdot \left[\begin{pmatrix} 1 \\ \alpha_1^{(2)} \\ \alpha_2^{(2)} \\ 0 \end{pmatrix} + k_3 \begin{pmatrix} 0 \\ \alpha_2^{(2)} \\ \alpha_1^{(2)} \\ 1 \end{pmatrix} \right]$$

$$= \begin{pmatrix} \mathrm{E}^{(2)} \\ 0 \\ 0 \\ q \end{pmatrix} + k_3 \begin{pmatrix} q \\ 0 \\ 0 \\ \mathrm{E}^{(2)} \end{pmatrix} \stackrel{!}{=} \begin{pmatrix} \mathrm{E}^{(3)} \\ 0 \\ 0 \\ 0 \end{pmatrix} \tag{6.69-b}$$

with the known quantity $q = r_3 + r_2 \, \alpha_1^{(2)} + r_1 \, \alpha_2^{(2)}$.

For the determination of k_3 and $\mathrm{E}^{(3)}$ we exploit (6.69-b) and find the conditions

$$\mathrm{E}^{(2)} + k_3 \, q = \mathrm{E}^{(3)}$$

$$q + k_3 \, \mathrm{E}^{(2)} = 0$$

or

$$k_3 = -\frac{q}{\mathrm{E}^{(2)}} \tag{6.70-a}$$

$$\mathrm{E}^{(3)} = \mathrm{E}^{(2)} \left(1 + k_3 \frac{q}{\mathrm{E}^{(2)}}\right)$$

$$= \mathrm{E}^{(2)} \left(1 - {k_3}^2\right) \tag{6.70-b}$$

respectively. Using k_3 as expressed in (6.70-a), $\boldsymbol{\alpha}^{(3)}$ can be calculated by (6.68).

In the next step, the solution for $p = 4$ can be evaluated. The coefficient k_4 and the energy of the prediction error $\mathrm{E}^{(4)}$ are determined in analogy to (6.68)–(6.70-b).

The complete algorithm is summarized in general form. Starting from the solution for $p = 0$, i.e., no prediction, the solution for $p{=}n$ is computed in n steps.

1. Computation of $n{+}1$ values r_i of the short-term auto-correlation.

2. $p = 0$, i.e., no prediction or

$$d(k) = x(k)$$

$$\mathrm{E}^{(0)} = r_0$$

$$\alpha_0^{(0)} \doteq 1\,.$$

3. For $p \geq 1$, computation of

(a)
$$q = \sum_{i=0}^{p-1} \alpha_i^{(p-1)}\, r_{p-i}$$
$$k_p = -\frac{q}{\mathrm{E}^{(p-1)}}$$

(b)
$$\alpha_0^{(p)} = 1$$
$$\alpha_i^{(p)} = \alpha_i^{(p-1)} + k_p\, \alpha_{p-i}^{(p-1)} \qquad \forall\; 1 \leq i \leq p-1$$
$$\alpha_p^{(p)} = k_p$$

(c)
$$\mathrm{E}^{(p)} = \mathrm{E}^{(p-1)} \left(1 - {k_p}^2\right)$$

(d)
$$p = p + 1\,.$$

4. Repetition of step 3, if $p \leq n$.

5. $a_i = -\alpha_i^{(n)} \qquad \forall\; 1 \leq i \leq n\,.$

The parameters k_p are called reflection coefficients. Besides an index decrement by one ($k_{i-1} = r_i$), they are identical to those reflection coefficients which characterize the tube model of the vocal tract developed in Chapter 2. The first reflection coefficient of the tube model corresponds to $k_0 = 1$.

Reflection coefficients have a number of interesting properties:

- They are limited by one in magnitude, i.e., $|k_p| \leq 1$. This is a consequence of using proper auto-correlation sequences as input to the Levinson–Durbin algorithm. Therefore, it can be concluded that the short-term energy $\mathrm{E}^{(p)}$ is reduced from iteration to iteration, i.e.,

$$\mathrm{E}^{(p)} = \mathrm{E}^{(p-1)} \, (1 - {k_p}^2) \leq \mathrm{E}^{(p-1)} \, . \tag{6.71}$$

- The transfer function

$$1 - A(z) = 1 - \sum_{i=1}^{n} a_i \; z^{-i}$$

  of the LP-analysis filter has all of its roots inside the unit circle, if and only if $|k_p| < 1 \; \forall \; p$ [Hayes 1996]. The reflection coefficients are therefore key for checking the stability of the inverse LP filter and for the development of inherently stable filter structures. In fact, the well-known Schur–Cohn stability test [Haykin 1996] employs the Levinson–Durbin recursion.

- Given the reflection coefficients k_p ($p = 1 \ldots n$), the AR model coefficients $a_i = -\alpha_i^{(n)}$ ($i = 1 \ldots n, \; a_0 = 1$) can be calculated by the Levinson–Durbin algorithm.

- Given the AR model coefficients $a_i = -\alpha_i^{(n)}$, the reflections coefficients k_p can be computed by the following recursion:

  1. Initialization with $\alpha_i^{(n)} = -a_i$ for $1 \leq i \leq n$.
  2. For $p = n, n-1, \ldots, 1$, computation of

$$\text{(a)} \qquad k_p \;=\; \alpha_p^{(p)}$$

$$\text{(b)} \qquad \alpha_i^{(p-1)} \;=\; \frac{\alpha_i^{(p)} - k_p \, \alpha_{p-i}^{(p)}}{1 - k_p^2} \, ; \qquad 1 \leq i \leq p-1 \, .$$

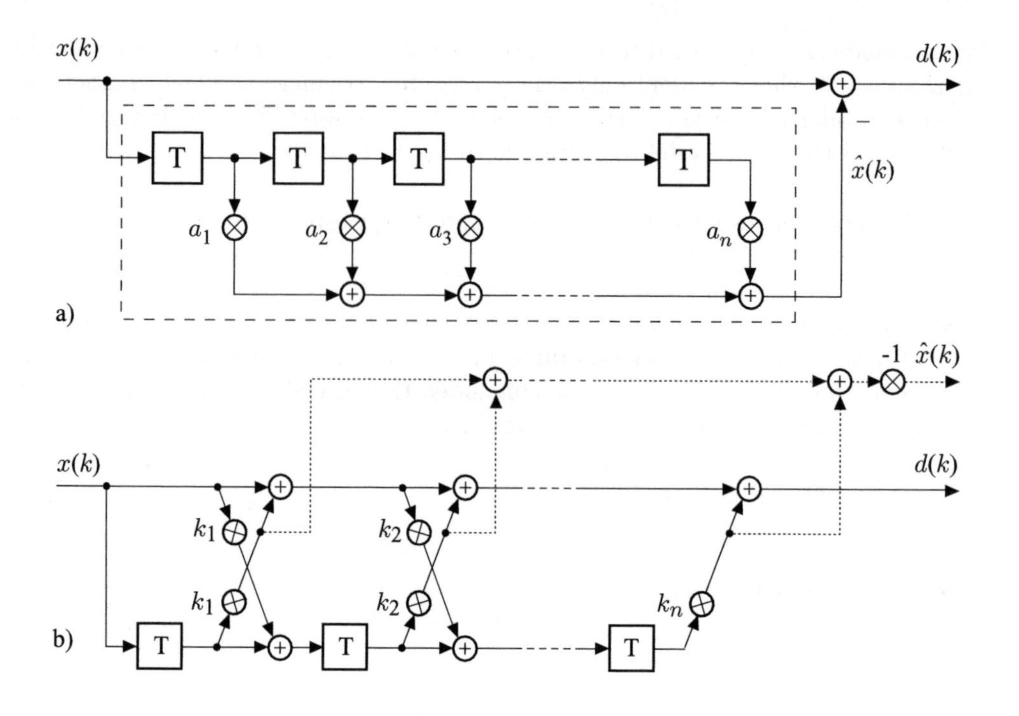

Figure 6.9: Block diagram of the linear predictor
a) Direct structure
b) Lattice structure

Since the Levinson–Durbin algorithm provides the predictor coefficients, as well as the reflection coefficients, the predictor can be realized alternatively in direct form (Fig. 6.9-a) or e.g., in the so-called lattice structure (Fig. 6.9-b).

The effect of the prediction is illustrated for one example in Fig. 6.10. It depicts the speech signal $x(k)$ and the prediction error signal $d(k)$ in the time and frequency domain, as well as the corresponding magnitude responses of the LP-analysis and synthesis filter. In this example, the $n = 8$ predictor coefficients were calculated every 20 ms using the auto-correlation method and a rectangular window of length $N = 160$. The LP-analysis filter performs a reduction of the dynamic range (Fig. 6.10-b) in the time domain and spectral flattening (Fig. 6.10-d, *whitening effect*) in the frequency domain. The corresponding frequency responses of the analysis and synthesis filter are shown in Fig. 6.10-e and 6.10-f. Obviously, the LP-synthesis filter describes the spectral envelope of $x(k)$, i.e., the frequency response of the vocal tract filter. The LP-analysis filter produces an error signal $d(k)$, which still has a quasi-periodic structure. Furthermore the spectral envelope of the error signal is almost flat (see Fig. 6.10-d).

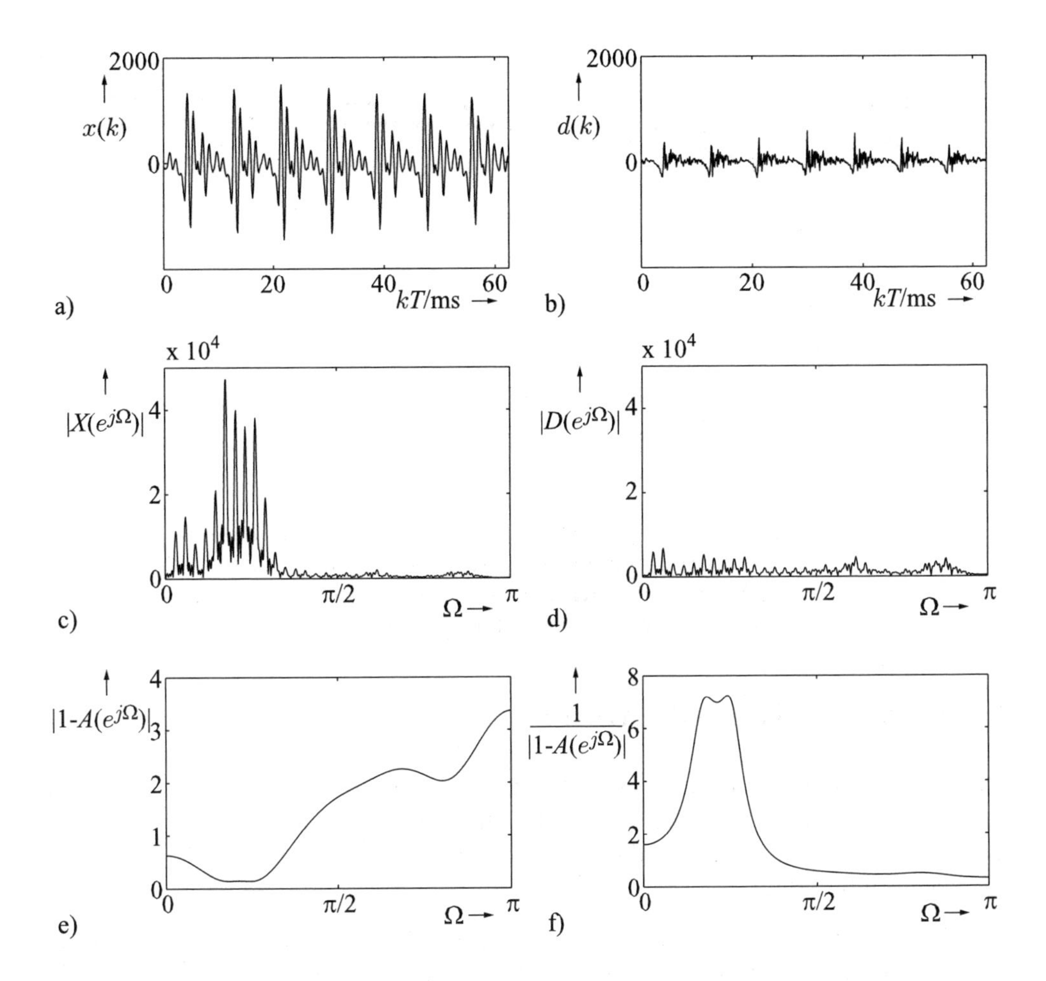

Figure 6.10: Example of the effect of linear prediction with block adaptation
a) Speech signal $x(k)$
b) Prediction error signal $d(k)$
c) Short-term spectral analysis of the speech signal
d) Short-term spectral analysis of the prediction error signal
e) Magnitude response of the LP-analysis filter
f) Magnitude response of the LP-synthesis filter

Figure 6.11 shows the achievable prediction gain as a function of the predictor order n for two sample speech signals. The predictors were adapted by the autocorrelation method. In comparison to the results obtained for a time-invariant predictor as in Fig. 6.6, a distinctly higher prediction gain results due to the block adaptation. Furthermore, it can be observed that with an adaptive adjustment the prediction gain saturates at a filter order of $n = 8$–10. An additional increase of the prediction order provides no appreciable further gain. This confirms the vocal tract model of Section 2.3.

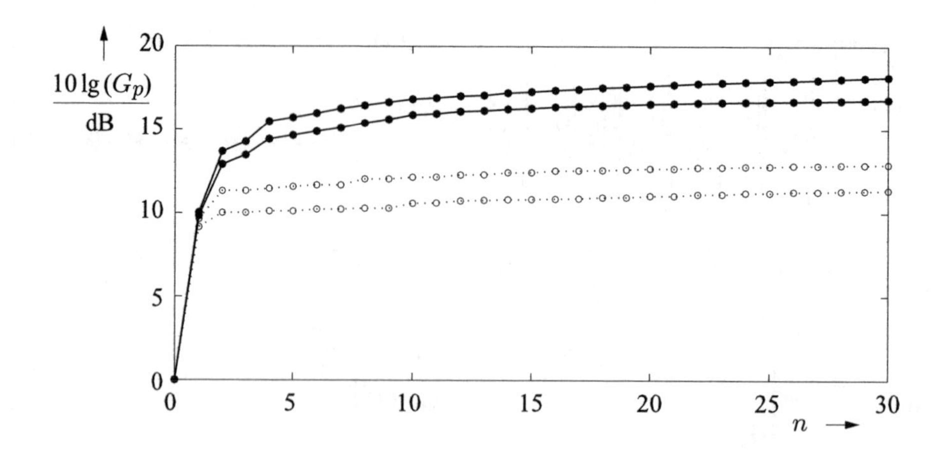

Figure 6.11: Logarithmic prediction gain for block adaptation with auto-correlation based predictors for two speakers (solid) and for a time-invariant prediction filter (dashed, Fig. 6.6); (signal length 30 s, $N = 160$, $f_s = 8$ kHz)

6.3.2 Sequential Adaptation

With the block-oriented adaptation the predictor coefficients a_i $(i = 1, 2, \dots, n)$ are recalculated for blocks of N samples. In this section an alternative method, the *least-mean-square* (LMS) algorithm, will be derived, in which the coefficients are sequentially adapted in each sample interval.

In a first step we consider a predictor with a single fixed coefficient a. The power of the prediction error signal $d(k) = x(k) - a\, x(k-1)$ can be expressed as

$$\sigma_d^2 = \sigma_x^2 - 2\,a\,\varphi_{xx}(1) + a^2\,\sigma_x^2$$

according to (6.32-c). The power σ_d^2 is a second-order function of the coefficient a, which is depicted in Fig. 6.12.

With (6.31), the minimum mean square error is reached in point C for

$$a_{\text{opt}} = \frac{\varphi_{xx}(1)}{\varphi_{xx}(0)} = \frac{\varphi_{xx}(1)}{\sigma_x^2}.$$

Starting from points A or B, the minimum (point C) can be approached iteratively by taking the gradient

$$\nabla = \frac{\partial \sigma_d^2}{\partial a} = -2\,\varphi_{xx}(1) + 2\,a\,\varphi_{xx}(0) .$$

into consideration. After inserting a_{opt} in the above equation, we obtain

$$\nabla = 2\,\varphi_{xx}(0) \cdot (a - a_{\text{opt}}) .$$

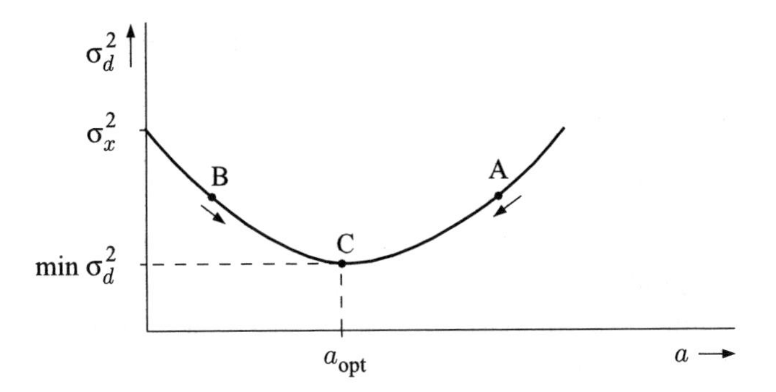

Figure 6.12: Error performance surface of the adaptive filter

The gradient is proportional to the difference of the instantaneous, i.e., time-variant coefficient a from the optimum a_{opt}. To reduce the mean square error σ_d^2, the instantaneous coefficient $a(k)$ must thus be corrected in the direction of the negative gradient according to

$$a(k+1) \;=\; a(k) - \vartheta \cdot \nabla\,. \tag{6.72}$$

Here, the constant ϑ denotes the stepsize, which controls the size of the incremental correction.

This procedure, which is known in the literature as the *steepest descent algorithm*, can be generalized and applied to the n-th order prediction as follows. With the signal vector

$$\mathbf{x}(k-1) = \Big(x(k-1), x(k-2), \ldots, x(k-n)\Big)^T \tag{6.73}$$

and an arbitrary but fixed coefficient vector

$$\mathbf{a}(k) = \mathbf{a} \;=\; \Big(a_1, a_2, \ldots, a_n\Big)^T\,, \tag{6.74}$$

the prediction $\hat{x}(k)$ can be described as

$$\hat{x}(k) = \sum_{i=1}^{n} x(k-i)a_i \tag{6.75-a}$$

$$= \mathbf{a}^T \mathbf{x}(k-1)\,, \tag{6.75-b}$$

and the following expression results for the power of the prediction error:

$$\sigma_d^2 \;=\; \mathrm{E}\Big\{\big(x(k) - \mathbf{a}^T\,\mathbf{x}(k-1)\big)^2\Big\}\,. \tag{6.76}$$

The gradient with respect to the coefficient vector **a** becomes

$$\nabla = -2\,\mathrm{E}\left\{\left(x(k) - \mathbf{a}^T\,\mathbf{x}(k-1)\right)\mathbf{x}(k-1)\right\} \tag{6.77-a}$$

$$= -2\,\boldsymbol{\varphi}_{xx} + 2\,\mathbf{R}_{xx}\,\mathbf{a}\,. \tag{6.77-b}$$

The gradient indicates the direction of the steepest ascent of the mean square error. For an iterative minimization of the mean square error, the instantaneous coefficient vector $\mathbf{a}(k)$ must consequently be corrected in the direction of the negative gradient. In analogy to (6.72) this results in

$$\mathbf{a}(k+1) \;=\; \mathbf{a}(k) + 2\,\vartheta\Big(\boldsymbol{\varphi}_{xx} - \mathbf{R}_{xx}\,\mathbf{a}(k)\Big)\,. \tag{6.78}$$

The steepest descent algorithm requires knowledge of the auto-correlation values $\varphi_{xx}(\lambda)$ for $\lambda = 0, 1, \ldots, n$, in the form of the correlation vector $\boldsymbol{\varphi}_{xx}$ and the auto-correlation matrix $\mathbf{R}_{xx}$.

One member of the family of *stochastic gradient algorithms* is the so-called *least-mean-square* (LMS) algorithm. This algorithm is particularly interesting for practical applications, as the auto-correlation values are not explicitly required.

For the LMS algorithm, a simple estimator for the mean square error σ_d^2 is used, i.e., the *instantaneous squared error*

$$\hat{\sigma}_d^2(k) = d^2(k) \tag{6.79-a}$$

$$= \left(x(k) - \mathbf{a}^T(k)\,\mathbf{x}(k-1)\right)^2. \tag{6.79-b}$$

In analogy to (6.77-b), the *instantaneous gradient* results in

$$\hat{\nabla} = -2\,\underbrace{\left(x(k) - \mathbf{a}^T(k)\,\mathbf{x}(k-1)\right)}_{d(k)}\,\mathbf{x}(k-1) \tag{6.80-a}$$

$$= -2\,d(k)\,\mathbf{x}(k-1)\,, \tag{6.80-b}$$

where instead of $\mathbf{R}_{xx}$ and $\boldsymbol{\varphi}_{xx}$ the prediction error $d(k)$ and the state variables $\mathbf{x}(k-1)$ are needed (Fig. 6.13).

Consequently, the LMS algorithm for the coefficient vector reads

$$\mathbf{a}(k+1) \;=\; \mathbf{a}(k) + 2\,\vartheta\,d(k)\;\mathbf{x}(k-1) \tag{6.81-a}$$

or for the individual coefficient

$$a_i(k+1) \;=\; a_i(k) + 2\;\vartheta\,d(k)\;x(k-i) \qquad \forall\;\; 1 \le i \le n \tag{6.81-b}$$

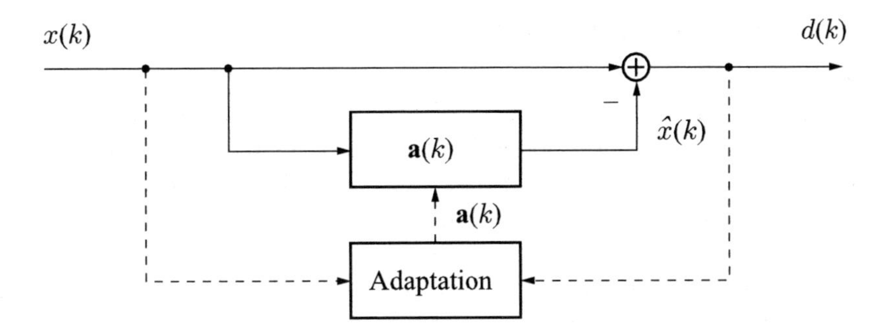

Figure 6.13: Predictor with sequential adaptation

respectively, with the effective stepsize parameter 2ϑ.

For stability reasons the stepsize parameter must be limited to the range

$$0 < \vartheta < \frac{1}{\|\mathbf{x}(k-1)\|^2}$$

(e.g., [Haykin 1996]).

Assuming a stationary AR process $x(k)$, the coefficient vector $\mathbf{a}$ converges with a sufficiently small stepsize towards the optimal solution according to (6.31)

$$\mathbf{a} \rightarrow \mathbf{a}_{\text{opt}} = \mathbf{R}_{xx}^{-1}\, \boldsymbol{\varphi}_{xx} \,.$$

Due to its low complexity the LMS algorithm is of great practical significance (see also Section 8.3.4 and Chapter 13). A time-variant stepsize is often used to improve the convergence behavior.

Numerous further adaptation algorithms are discussed in the literature, which differ regarding their convergence characteristics and their complexity. Here, the *recursive least-square* (RLS) algorithm is mentioned as an example. It can be derived from the steepest descent algorithm as well. In this case, instead of the autocorrelation values $\varphi_{xx}(\lambda)$, estimated values $\hat{\varphi}_{xx}(\lambda)$ are used, which are determined by recursive computation with exponential windowing. This method is characterized by a high convergence speed. The improvement in performance, however, is achieved at the expense of a large increase in computational complexity, i.e., the number of operations per iteration grows with the square of the filter order n. In comparison, for the LMS algorithm, there is only a linear increase in complexity (e.g., Chapter 13, [Haykin 1996]).

6.4 Long-Term Prediction

As shown in Section 6.3 by example of Fig. 6.10, we can extract the spectral envelope of $x(k)$ by short-term prediction using $n = 8$–10 coefficients. According to the speech production model (Fig. 6.1) the resulting LP-synthesis filter represents the vocal tract, and the prediction error signal $d(k)$ the excitation. Thus, the remaining quasi-periodic spectral fine structure as in Fig. 6.10-d is determined by the excitation generator of the speech production model. This periodic structure is associated with the long-term correlation of the speech signal $x(k)$ or of the prediction error signal $d(k)$ respectively, as illustrated in Fig. 6.14.

The short-term prediction discussed in Section 6.3 exploits the short-term correlation $\varphi_{xx}(\kappa)$ ($\kappa = 0 \ldots n$, with $n = 8$–10). Obviously, the prediction error signal $d(k)$ (Fig. 6.10-b) still exhibits long-term correlation, which is due to the pitch period $T_0 = 1/f_0$ of voiced segments. With fundamental frequencies in the range of $50\,\text{Hz} \leq f_0 \leq 250\,\text{Hz}$ and a sampling rate of $f_s = 8\,\text{kHz}$ the periods T_0 have lengths of $N_0 = 32$–160 samples. As T_0 is large in comparison to $n \cdot T$ (memory of the short-term predictor), the prediction over the time span T_0 is called *long-term prediction* (LTP).

The high correlation of subsequent signal periods can be used for a further improvement of the prediction gain by estimating the most recent signal period from the preceding one. If the instantaneous period length N_0 is known, the long-term prediction error signal can be calculated as follows:

$$d'(k) = d(k) - b \cdot d(k - N_0) = d(k) - \hat{d}(k) \tag{6.82}$$

with a weighting factor b.

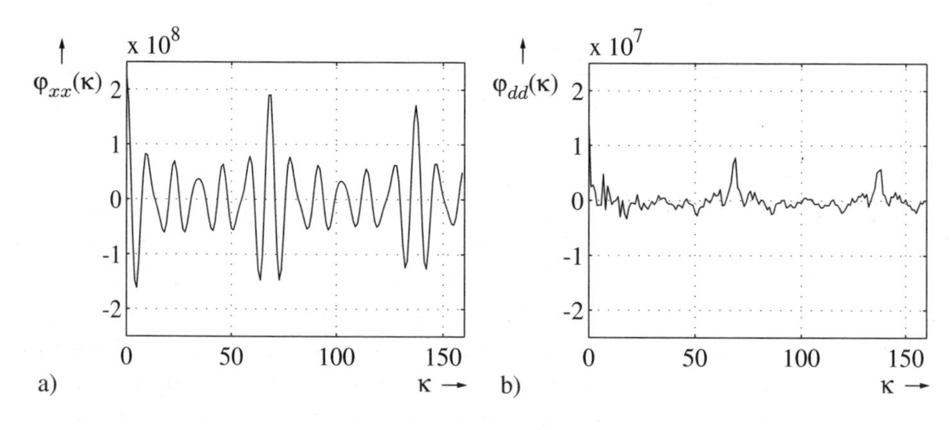

Figure 6.14: Auto-correlation function of the
a) Speech signal $x(k)$
b) Error signal $d(k)$
for the example signals depicted in Fig. 6.10

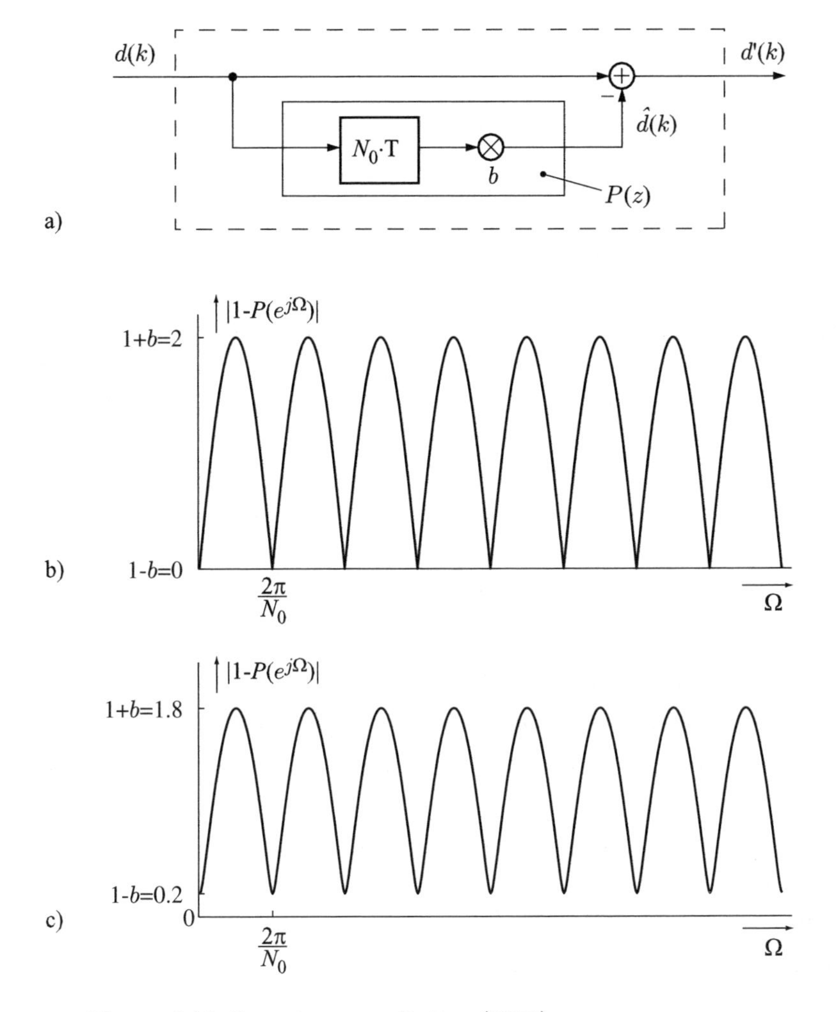

Figure 6.15: Long-term prediction (LTP)
a) Block diagram of the LTP-analysis filter
b) Magnitude response $|1 - P(e^{j\Omega})|$ for $b = 1$
c) Magnitude response $|1 - P(e^{j\Omega})|$ for $b = 0.8$

Figure 6.15-a depicts the block diagram of the corresponding LTP-analysis filter. The frequency response can be computed as follows:

$$1 - P(e^{j\Omega}) = 1 - b \cdot e^{-jN_0\Omega} \tag{6.83-a}$$

$$= \sqrt{(1 - b \cdot \cos(N_0\Omega))^2 + b^2 \cdot \sin^2(N_0\Omega)} \cdot e^{-j\varphi_P(\Omega)} \tag{6.83-b}$$

with

$$\varphi_P = -\arctan\left(\frac{b \cdot \sin(N_0\Omega)}{1 - b \cdot \cos(N_0\Omega)}\right). \tag{6.83-c}$$

The magnitude response is of particular interest. In the special case $b = 1$, with

$$|1 - P(e^{j\Omega})| = 2 \cdot \left|\sin\left(\frac{N_0 \Omega}{2}\right)\right| \tag{6.83-d}$$

the plot outlined in Fig. 6.15-b results in equidistant zeros at

$$\Omega_i = \frac{2\,\pi}{N_0}\, i\,, \qquad i \in \mathbb{Z}\,. \tag{6.84}$$

Accordingly, the LTP-analysis filter is a comb filter with equally spaced notches at Ω_i. As the length of the instantaneous pitch period $T_0 = 1/f_0$ is, in general, not an exact integer multiple of the sampling interval T, the notches are not necessarily exactly at the normalized fundamental frequency $\Omega_0 = 2\pi/N_0$ and its harmonics Ω_i.

The two parameters N_0 and b are chosen in such a way as to minimize the energy

$$\sum_k d'^2(k) = \sum_k \Big(d(k) - b \cdot d(k - N_0)\Big)^2 \tag{6.85}$$

of the error signal $d'(k)$ for short blocks.

In analogy to short-term prediction, the summation limits in (6.85) can be chosen according to either the auto-correlation method or the covariance method (see Section 6.3.1).

In what follows, the covariance method is applied. The energy of the error signal $d'(k)$ is minimized over an interval of length L. Most frequently, the length of the interval is $L \cdot T = 5\,\text{ms}$ or $L = 40$ for $f_s = 8\,\text{kHz}$ respectively.

For each fixed but arbitrary value N_0, the optimal coefficient b can be determined by minimizing the energy of the error signal

$$\frac{\partial}{\partial b} \sum_{k=k_0-L+1}^{k_0} d'^2(k) - \sum_{k=k_0-L+1}^{k_0} -2d(k-N_0)\Big(d(k) - b \cdot d\,(k - N_0)\Big) \stackrel{!}{=} 0\,. \tag{6.86-a}$$

With the abbreviation $k_1 = k_0 - L + 1$ we finally get

$$b = \frac{\sum\limits_{k=k_1}^{k_0} d(k)\; d(k - N_0)}{\sum\limits_{k=k_1}^{k_0} d^2(k - N_0)} = \frac{\mathrm{R}(N_0)}{\mathrm{S}(N_0)}\,, \tag{6.86-b}$$

where $\mathrm{R}(N_0)$ is the short-term auto-correlation function for $\lambda = N_0$ according to the covariance method (cf. Section 6.3.1.2) and $\mathrm{S}(N_0)$ is the energy of the current frame of the error signal $d(k)$.

Inserting the optimal coefficient b into (6.85), the resulting error energy can be computed as a function of the parameter N_0

$$\sum_{k=k_0-L+1}^{k_0} d'^2(k) = \mathrm{S}(0) - \frac{\mathrm{R}^2(N_0)}{\mathrm{S}(N_0)} . \tag{6.87}$$

This expression can be utilized to determine the optimal delay N_0. In (6.87), only the second term depends on N_0. Thus, in a first step this term is maximized through variation of N_0 in the relevant range of, for example, $32 \leq N_0 < 160$. In this range N_0 can take, e.g., $2^7 = 128$ different values, which allows coding with 7 bits only. Subsequently, the weighting coefficient b can be determined for the delay N_0 according to (6.86-b). This coefficient can be quantized quite coarsely (see also Appendix A).

A signal example is depicted in Fig. 6.16, showing the input signal $x(k)$, the first error signal $d(k)$ after short-term prediction, and the second error signal $d'(k)$ after long-term prediction. By the use of the second predictor, a further significant dynamic reduction is achieved. Chapter 8 will reveal how this additional prediction gain can be exploited for bit rate reductions in the sense of model-based and psychoacoustically motivated source coding. The effect of the two-stage prediction is illustrated once more in the frequency domain by Fig. 6.17.

Essentially, the formant structure is removed by means of block adaptive short-term prediction or through filtering with the transfer function $1 - A(z)$, leaving a spectrally flattened version of the input signal with an almost periodic respectively harmonic structure over a wide range of frequencies. The subsequent processing with the LTP-analysis filter with the transfer function $1 - P(z)$ causes a further power reduction and an almost complete elimination of the harmonic structure, finally resulting in a spectrally flat noise like a residual signal. In this example, the short-term predictor of the order $n = 8$ was adapted every 20 ms ($N = 160$ samples), while the parameters of the long-term predictor were computed every 5 ms ($L = 40$ samples).

Recall that the delay N_0 represents in principle the pitch period, which is approximated by an integer multiple of the sampling interval. For this reason, general approaches for pitch detection may be used [Hess 1983].

In fact, however, the minimization according to (6.87) is based on a criterion which does not aim at approximating the true pitch period, but at minimizing the energy of the prediction error.

As a consequence, the lowest error energy might in some circumstances be achieved with a delay N_0, which does not correspond to the true, but to, for instance, half of the actual pitch period (or to twice the fundamental frequency).

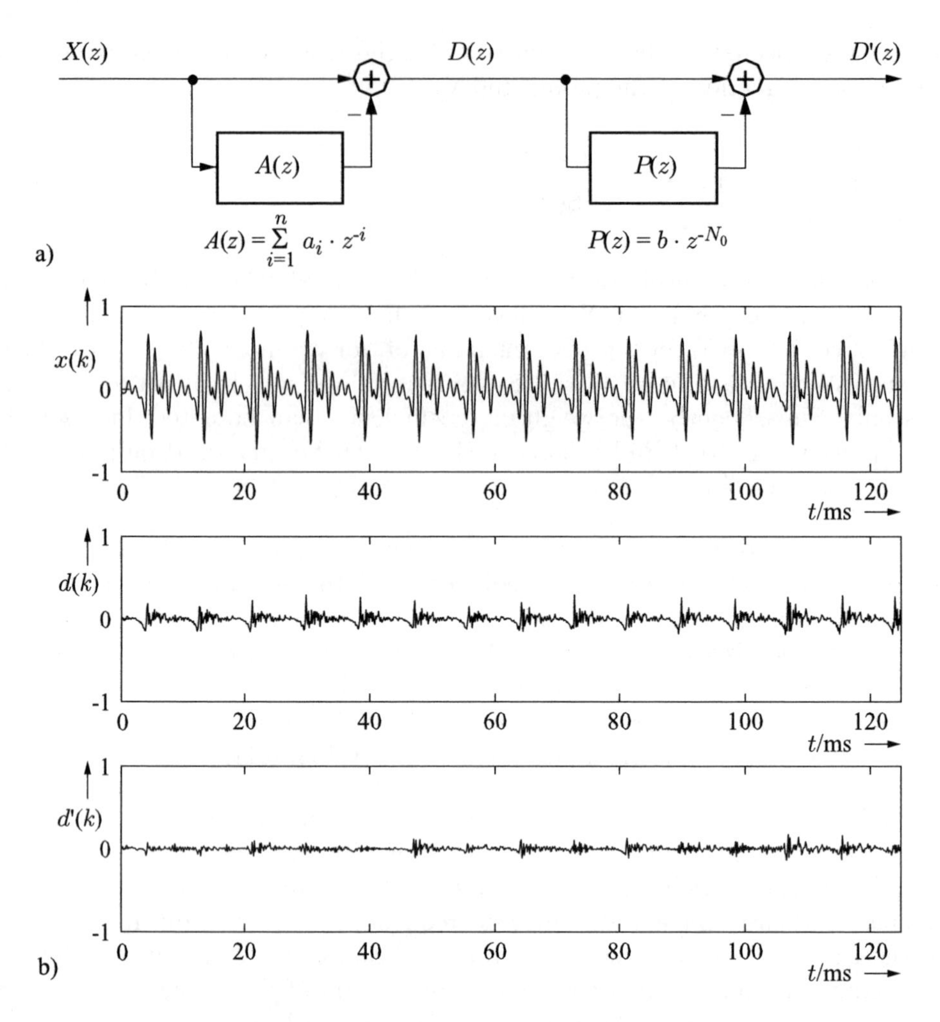

Figure 6.16: Example of short-term and long-term prediction
a) Block diagram
b) Time signals

Just as for the short-term prediction, the prediction gain can be improved by increasing the filter order. Often a long-term predictor with three coefficients according to

$$P(z) = b_{-1} \cdot z^{-(N_0-1)} + b_0 \cdot z^{-N_0} + b_{+1} \cdot z^{-(N_0+1)} \tag{6.88}$$

is used. With its interpolating effect this predictor generally provides an improved prediction, as the true pitch period in most cases is not an integer multiple of the sampling interval or the fundamental frequency f_0 is not an integer fraction of the sampling frequency, respectively. In this case, however, three coefficients must be transmitted, which requires a correspondingly higher bit rate.

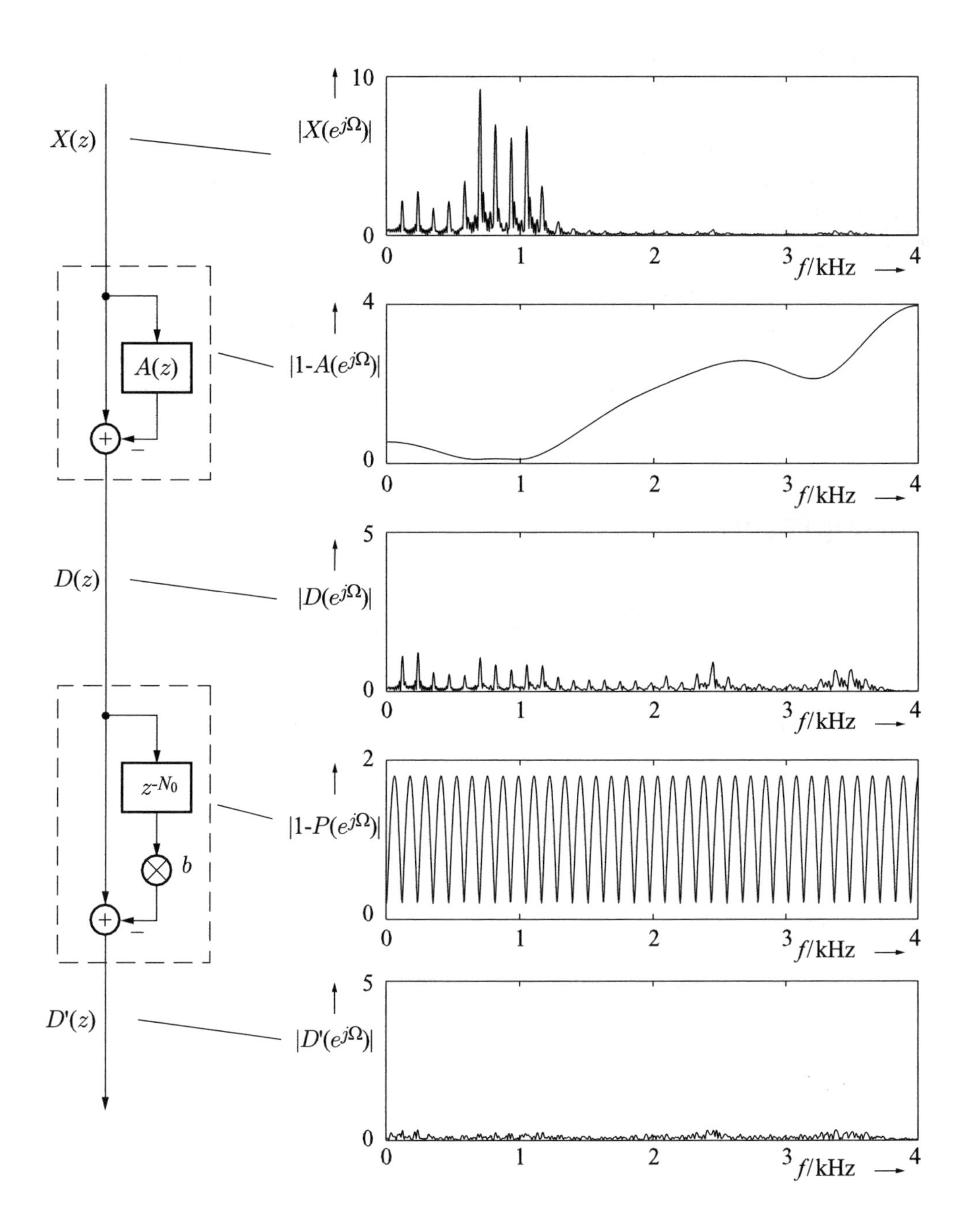

Figure 6.17: Example: spectral impact of short-term and long-term prediction for the vowel "a"

A similar improvement of the prediction gain can also be achieved with a predictor with one coefficient as in Fig. 6.15, if the sampling frequency of the first residual signal $d(k)$ is increased by interpolation. In this way, the time resolution is improved accordingly. In the literature, this technique is called *high-resolution LTP analysis* (see [Marques et al. 1990] and [Marques et al. 1989]). Interpolating by a factor of 4, the word length for the representation of the parameter N_0 grows only by 2 bits. Regarding the bit rate, the high-resolution LTP analysis must therefore be preferred to the multiple-tap long-term prediction according to (6.88).

In analogy to the short-term predictor structures, which will be discussed in Section 8.3, the long-term prediction can alternatively be implemented as a forward predictor (*open-loop*) or backward predictor (*closed-loop*). When quantizing the error signal, these two alternatives differ regarding the quantization error on the receiver side (cf. Section 8.3.3).

Bibliography

Barnwell III, T. P. (1981). Recursive Windowing for Generating Autocorrelation Coefficients for LPC Analysis, *IEEE Transactions on Acoustics, Speech and Signal Processing*, vol. 29, no. 5, October, pp. 1062–1066.

Deller Jr., J. R.; Proakis, J. G.; Hansen, J. H. L. (2000). *Discrete-time Processing of Speech Signals*, 2nd edn, IEEE Press, New York.

Durbin, J. (1960). The Fitting of Time-series Models, *Revue de l'Institut International de Statistique*, vol. 28, no. 3, pp. 233–243.

Hayes, M. H. (1996). *Advanced Digital Signal Processing*, John Wiley & Sons, Ltd., Chichester.

Haykin, S. (1996). *Adaptive Filter Theory*, Prentice Hall, Englewood Cliffs, New Jersey.

Hess, W. J. (1983). *Pitch Determination of Speech Signals*, Springer Verlag, Berlin.

Itakura, F.; Saito, S. (1970). A Statistical Method for Estimation of Speech Spectral Density and Formant Frequencies, *Electronics and Communications in Japan*, vol. 52-A, pp. 36–43.

Jayant, N. S.; Noll, P. (1984). *Digital Coding of Waveforms*, Prentice Hall, Englewood Cliffs, New Jersey.

Kay, S. (1988). *Modern Spectral Estimation*, Prentice Hall, Englewood Cliffs, New Jersey.

Levinson, N. (1947). The Wiener RMS (Root Mean Square) Error Criterion in Filter Design and Prediction, *Journal of Mathematical Physics*, vol. 25, no. 4, pp. 261–278.

Makhoul, J. (1975). Linear Prediction: A Tutorial Review, *IEEE Proceedings*, vol. 63, April, pp. 561–580.

Makhoul, J.; Wolf, J. (1972). Linear Prediction and the Spectral Analysis of Speech, *Technical Report 2304*, Bolt, Beranek, and Newman Inc., Boston, Massachusetts.

Markel, J. D.; Gray, A. H. (1976). *Linear Prediction of Speech*, Springer Verlag, Berlin, Heidelberg, New York.

Marple Jr., S. L. (1987). *Digital Spectral Analysis: With Applications*, Prentice Hall, Englewood Cliffs, New Jersey.

Marques, J. S.; Trancoso, I. M.; Tribolet, J. M.; Almeida, L. B. (1990). Improved Pitch Prediction with Fractional Delays in CELP Coding, *Proceedings of the IEEE International Conference on Acoustics, Speech, and Signal Processing (ICASSP)*, Albuquerque, New Mexico, USA, pp. 665–668.

Marques, J. S.; Tribolet, J. M.; Trancoso, I. M.; Almeida, L. B. (1989). Pitch Prediction with Fractional Delays in CELP Coding, *Proc. of European Conference on Speech Technology*, Paris, France, pp. 509–512.

Sluijter, R. J. (2005). *The Development of Speech Coding and the First Standard Coder for Public Mobile Telephony*, PhD thesis, Technical University Eindhoven.

Vary, P.; Heute, U.; Hess, W. (1998). *Digitale Sprachsignalverarbeitung*, B. G. Teubner, Stuttgart (in German).

7

Quantization

7.1 Analog Samples and Digital Representation

In the context of quantization we first of all have to deal with a sequence of samples $s(k) = s_a(t = kT)$ of the analog signal $s_a(t)$ which is a function of the continuous time variable t. The samples $s_a(kT)$ at the discrete time instants $kT = k/f_s$ are continuous quantities, where T is the *sampling period*, i.e., the time between successive samples and $f_s = 1/T$ is the *sampling frequency*. By quantization we convert each sample $s_a(kT)$ into a quantized version $s(k)$ which can take only one out of $K_0 = 2^{w_0}$ different discrete values.

The sequence of samples $s(k)$ is now discrete with respect to time and amplitude.

The complete process of analog-to-digital conversion (A/D conversion) which is shown in Fig. 7.1-a consists of the three steps:

- lowpass filtering according to the sampling theorem with cutoff frequency $f_c \leq f_s$,
- sampling at frequency f_s,
- quantization with word length w_0, i.e., $K_0 = 2^{w_0}$ different quantization levels s_i, with $i = 0, \ldots, K_0 - 1$.

Digital Speech Transmission: Enhancement, Coding and Error Concealment
Peter Vary and Rainer Martin

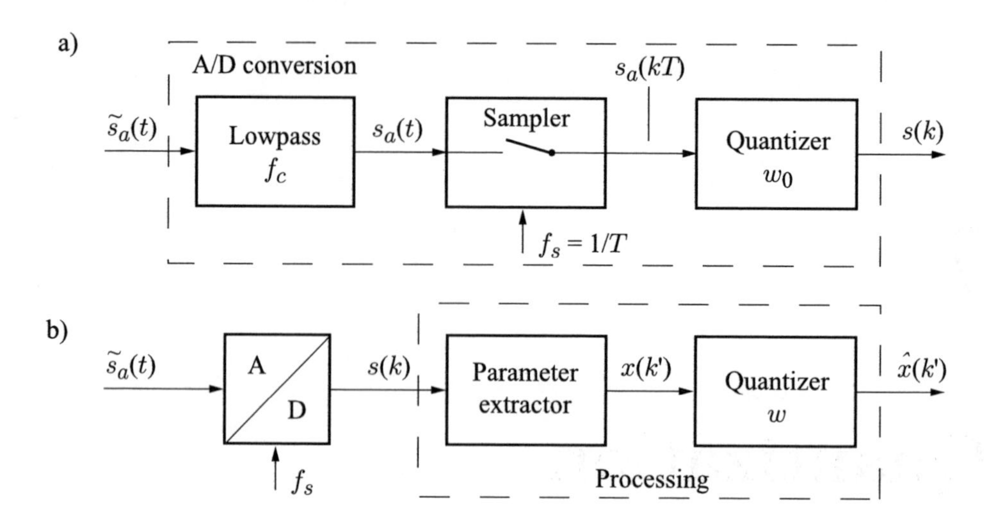

Figure 7.1: A/D conversion and parameter extraction
a) A/D conversion of an analog signal:
lowpass filtering, sampling at f_s, and
quantization with $K_0 = 2^{w_0}$ levels
b) Parameter extraction:
block processing at time instances k' and
quantization with $K = 2^w$ levels

Quantization is not only needed for A/D conversion, but also whenever any parameter $x(k')$ which has been extracted from the signal samples $s(k)$ has to be represented at a reduced word length $w \leq w_0$, e.g., for the purpose of source encoding (compression). It is assumed that the parameter $x(k')$ is calculated by processing blocks of samples $s(k)$, where k' is denoting the block time index (see Fig. 7.1-b).

Various quantization techniques are available whose basic principles will be discussed in this chapter. To simplify the representation, no distinction will be made between signal samples $s(k)$ and parameters $x(k')$. We will consider either values x which are applied to a scalar quantizer with $K = 2^w$ quantizer reproduction levels or vectors $\mathbf{x} = (x_1, x_2, \ldots, x_L)^T$ consisting of L values $x_\lambda (\lambda = 1, 2, \ldots, L)$ which are applied to a vector quantizer with $K = 2^w$ quantizer reproduction vectors. The quantity x might represent, for instance, a signal sample which is provided by an A/D converter with $w_0 = 13$ bit resolution and must be represented with a word length of $w = 8$ bits. The parameter x might be, for example, a predictor coefficient, which has been calculated with a digital signal processor using 16 bit fixed point arithmetic. For the purpose of transmission, this parameter has to be represented, e.g, with a word length of $w = 5$.

7.2 Uniform Quantization

First, we consider a symmetric quantizer which maps the input range

$$x_{\min} \leq x \leq +x_{\max} \qquad \text{with } x_{\min} = -x_{\max}\,, \tag{7.1}$$

to the output range

$$\hat{x}_{\min} \leq \hat{x} \leq +\hat{x}_{\max} \qquad \text{with } \hat{x}_{\min} = -\hat{x}_{\max} \tag{7.2}$$

where the quantized value is denoted by $\hat{x}$, it can take one out of $K = 2^w$ quantizer reproduction levels (in short *quantization levels*) $\hat{x}_i$. The amplitude range is subdivided into $K = 2^w$ uniform intervals of width

$$\Delta x = \frac{2\,x_{\max}}{K} = \frac{x_{\max}}{2^{w-1}}\,, \tag{7.3}$$

where Δx is called the *quantizer stepsize*. The quantization operation can be described, as depicted in Fig. 7.2-a, by a staircase function

$$\hat{x} = f(x) \in \left\{ \hat{x}_i = \pm(2i-1) \cdot \frac{\Delta x}{2} \right\} \tag{7.4}$$

representing the *quantizer characteristic*. The quantized value $\hat{x}$ differs from x by the *quantization error* or *quantization noise*

$$e = \hat{x} - x = f(x) - x\,, \tag{7.5}$$

according to

$$\hat{x} = e + x\,. \tag{7.6}$$

The quantization operation can be modeled by the quantization noise $e(k)$ which is added to $x(k)$ (see Fig. 7.2-b). In Fig. 7.2-c, $x(k)$, $\hat{x}(k)$, and $e(k)$ are plotted for a sinusoidal signal. In this example, the sampling rate is much higher than the frequency of the sinusoidal signal which results in a smooth shape of $x(k)$. Due to the coarse quantization with $K = 8$ levels and the relatively high sampling rate, each step of the staircase-shaped signal $\hat{x}(k)$ consists of several identical quantizer reproduction levels.

The quantizer of Fig. 7.2 is not able to represent $x = 0$ exactly, as the smallest magnitude of $\hat{x}$ is $\frac{\Delta x}{2}$. With a slight modification of $f(x)$ the value $\hat{x} = 0$ can be allowed. However, in this case the symmetry is lost if the number K of quantization stepsizes is even.

Three different quantizer characteristics are depicted in Fig. 7.3 for $w = 4$ or $K = 16$, respectively. Figure 7.3-a represents the symmetric case of a *midrise* quantizer with eight quantization levels in the positive and negative range. In contrast to this, the *midtread* quantizer of Fig. 7.3-b allows the accurate representation of $x = 0$, but takes seven levels in the positive and eight levels in the negative range. The magnitude truncation characteristic of Fig. 7.3-c is symmetric and allows the accurate representation of $x = 0$.

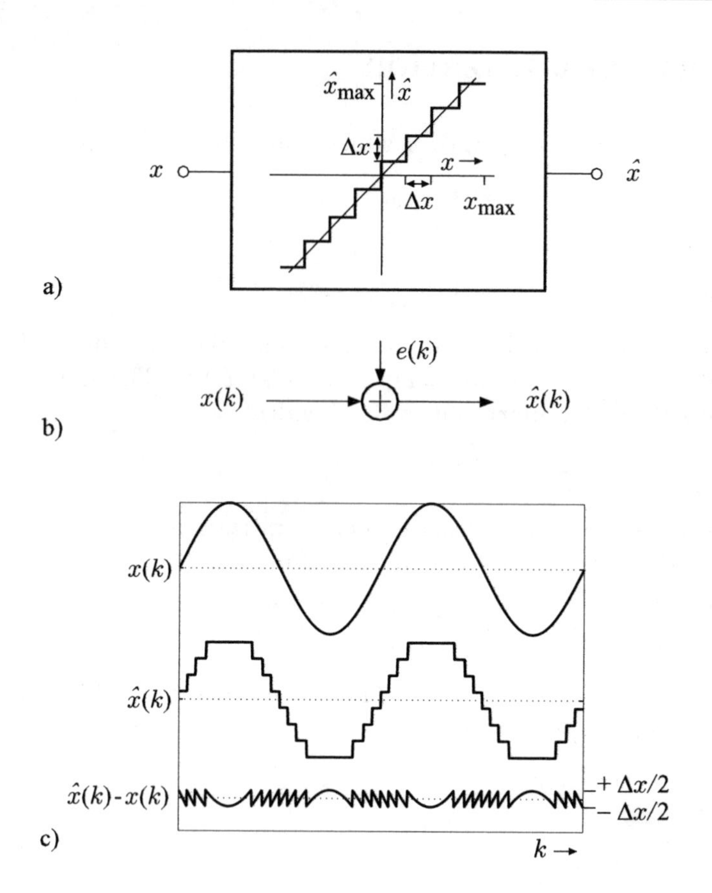

Figure 7.2: Description of a uniform quantization operation
a) Quantizer characteristic for $w = 3$, i.e., $K = 8$
b) Quantizer model with additive quantization noise
c) Example: sinusoidal signal

For the graphs in Fig. 7.3-a and 7.3-b the quantization is performed by a rounding operation, whereas the quantizer characteristic in Fig. 7.3-c corresponds to magnitude truncation. The parameters of these quantization operations are compiled in Fig. 7.3.

For both characteristics of Fig. 7.3-a and 7.3-b the quantization error is limited to

$$|\,e\,| \leq \frac{\Delta x}{2}\,. \tag{7.7}$$

In contrast, for the characteristic in Fig. 7.3-c, the maximum quantization error is twice as large, i.e.,

$$\max|e| = \Delta x\,. \tag{7.8}$$

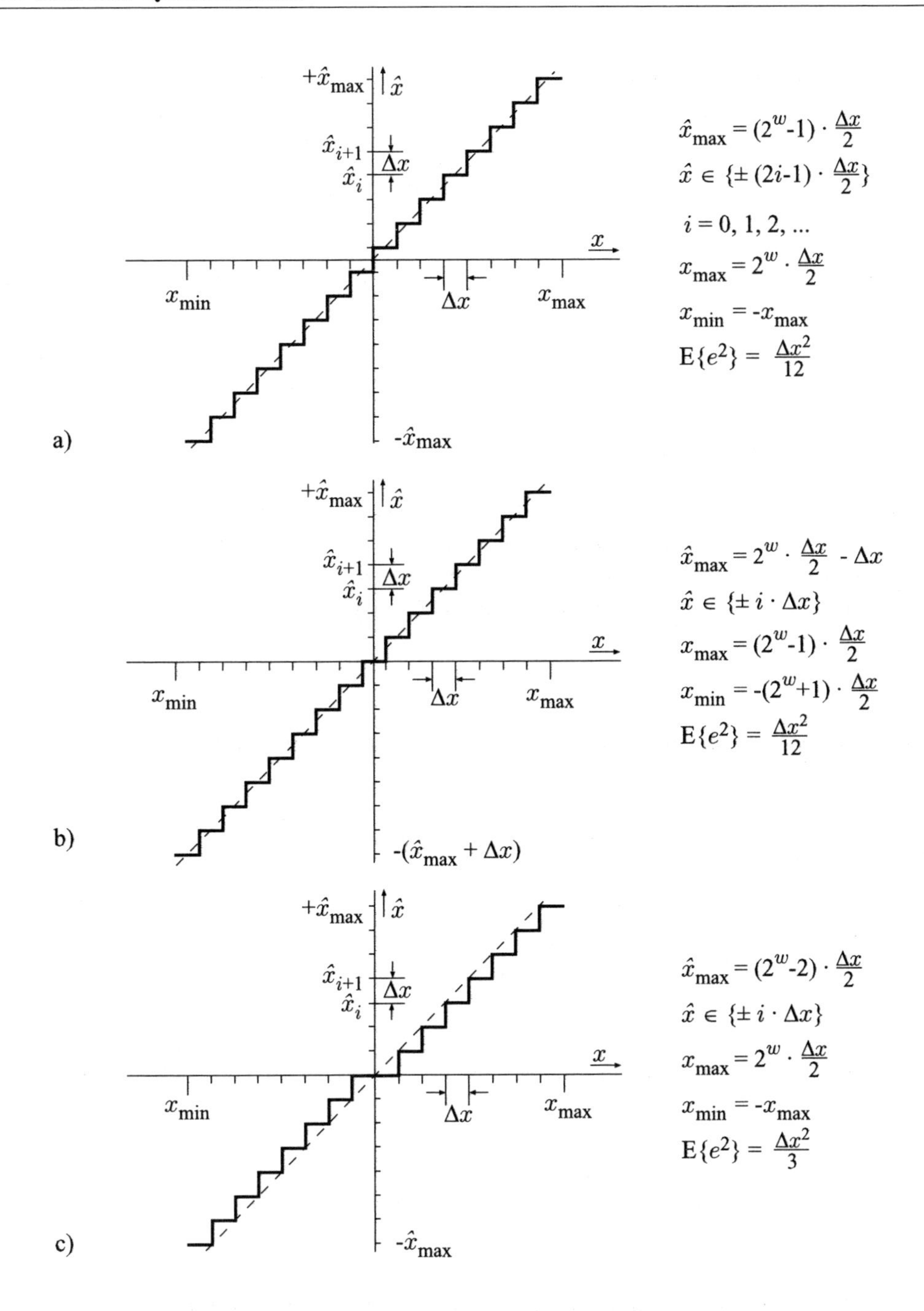

Figure 7.3: Uniform quantizer characteristics with $w = 4$
a) Uniform *midrise* quantizer: $\hat{x}_i = (2i-1)\frac{\Delta x}{2}$; $i = 0, \pm 1, \pm 2, \ldots$
b) Uniform *midtread* quantizer: $\hat{x}_i = i\,\Delta x$; $i = 0, \pm 1, \pm 2, \ldots$
c) Uniform quantizer with magnitude truncation

The quantization characteristics depicted in Fig. 7.3 can be represented in the range $x_{\min} \leq x \leq x_{\max}$ analytically as follows:

$$\hat{x} = f_a(x) = \text{sign}(x) \cdot \Delta x \cdot \left[\text{int}\left(\frac{|x|}{\Delta x}\right) + 0.5 \right], \tag{7.9-a}$$

$$\hat{x} = f_b(x) = \text{sign}(x) \cdot \Delta x \cdot \text{int}\left(\frac{|x|}{\Delta x} + 0.5\right), \tag{7.9-b}$$

$$\hat{x} = f_c(x) = \text{sign}(x) \cdot \Delta x \cdot \text{int}\left(\frac{|x|}{\Delta x}\right), \tag{7.9-c}$$

where $\text{int}(x)$ the integer part of x.

Although there is a deterministic relation between the actual sample $x(k)$ and the resulting quantization error $e(k)$, the quantization error is usually modeled as a statistical quantity. We assume that the mean value of the sequence $x(k)$ is zero and that $x(k)$ has a symmetric probability density function

$$p_x(+u) = p_x(-u). \tag{7.10}$$

We are interested in quantifying the performance of the quantizer in terms of the resulting signal-to-noise ratio.

With the probability density function $p_x(u)$ of signal $x(k)$, we obtain the power S of the signal as

$$S = \mathrm{E}\{x^2\} = \int_{-\infty}^{+\infty} u^2 \, p_x(u) \, du \tag{7.11}$$

and due to the quantization characteristic $\hat{x} = f(x)$ for the quantization noise power N

$$N = \mathrm{E}\{e^2(x)\} = \int_{-\infty}^{+\infty} \Big(f(u) - u \Big)^2 p_x(u) \, du. \tag{7.12}$$

We assume a symmetric quantization characteristic as shown, for example, in Fig. 7.3-a. The overload amplitude for this type of quantizer characteristic is

$$\pm\, x_{\max} = \pm \left(\hat{x}_{\max} + \frac{\Delta x}{2} \right). \tag{7.13}$$

Exploiting the symmetry, the following equation is valid for the quantization noise power:

$$N = 2 \int_{0}^{x_{\max}} \left(f(u) - u\right)^2 p_x(u)\,\mathrm{d}u \; + \; 2 \int_{x_{\max}}^{\infty} \left(\hat{x}_{\max} - u\right)^2 p_x(u)\,\mathrm{d}u \tag{7.14-a}$$

$$= N_Q \; + \; N_O\,. \tag{7.14-b}$$

As $x(k)$ is not necessarily limited to the dynamic range $\pm x_{\max}$, the noise power can be divided into two components N_Q and N_O, which are caused by quantization (Q) or by overload and clipping (O), respectively. In the following, we will assume that no overload errors occur or that this effect can be neglected, i.e., $N_O = 0$ applies.

In the positive range the given uniform midrise quantizer has $\frac{K}{2} = 2^{w-1}$ quantization levels

$$\hat{x}_i \;=\; f(x) \;=\; i \cdot \Delta x \;-\; \frac{\Delta x}{2}\,, \qquad i \;=\; 1, 2, \ldots, \frac{K}{2}\,, \tag{7.15}$$

which are assigned to the i-th interval

$$(i-1) \cdot \Delta x \;\leq\; x \;<\; i \cdot \Delta x \tag{7.16-a}$$

or,

$$\hat{x}_i \;-\; \frac{\Delta x}{2} \;\leq\; x \;<\; \hat{x}_i \;+\; \frac{\Delta x}{2}\,. \tag{7.16-b}$$

The contributions of the individual quantization intervals to the total quantization noise power N_Q according to (7.14-a) result from integration over each interval. We obtain

$$N_Q \;=\; 2 \sum_{i=1}^{\frac{K}{2}} \int_{(i-1)\cdot\Delta x}^{i\cdot\Delta x} \left(\hat{x}_i \;-\; u\right)^2 p_x(u)\,\mathrm{d}u\,. \tag{7.17}$$

With the substitution $z = u - \hat{x}_i$ this expression can be simplified to

$$N_Q \;=\; 2 \sum_{i=1}^{\frac{K}{2}} \int_{-\frac{\Delta x}{2}}^{+\frac{\Delta x}{2}} z^2 \; p_x(z + \hat{x}_i)\,\mathrm{d}z\,. \tag{7.18}$$

The quantization noise power depends on the probability density function $p_x(u)$ of the sequence $x(k)$. For the special case of uniform distribution it can be easily computed. With

$$\max|x| = x_{\max} \tag{7.19}$$

$$p_x(u) = \frac{1}{2\,x_{\max}}\,, \qquad -x_{\max} \leq u \leq +x_{\max} \tag{7.20}$$

$$\Delta x = \frac{2\,x_{\max}}{K} \tag{7.21}$$

the total quantization noise power results in

$$N_Q = 2\sum_{i=1}^{\frac{K}{2}} 2 \int_0^{+\frac{\Delta x}{2}} z^2 \, \frac{1}{2\,x_{\max}}\, \mathrm{d}z \tag{7.22-a}$$

$$= K\,\frac{1}{x_{\max}}\,\frac{1}{3}\,z^3\Big|_0^{\frac{\Delta x}{2}} \tag{7.22-b}$$

$$= \frac{\Delta x^2}{12} \tag{7.22-c}$$

and the signal power is

$$S = \int_{-x_{\max}}^{+x_{\max}} x^2\,\frac{1}{2\,x_{\max}}\,\mathrm{d}x \tag{7.22-d}$$

$$= \frac{1}{3}\,x_{\max}^2\,. \tag{7.22-e}$$

With $K = 2^w$ and (7.21), we obtain the signal-to-noise ratio

$$\frac{\mathrm{SNR}}{\mathrm{dB}} = 10\,\lg\left(\frac{S}{N_Q}\right) \tag{7.23-a}$$

$$= w\;20\,\lg(2) \tag{7.23-b}$$

$$\approx 6\,w\,. \tag{7.23-c}$$

This is the so-called *6-dB-per-bit rule*, which, however, is only accurate for this special case.

For the general case, an approximation can be provided. If the quantizer stepsize is sufficiently small ($\Delta x \ll x_{\max}$) and if $p_x(u)$ is sufficiently smooth, the probability density function can be approximated by its value in the middle of the interval as in

$$p_x(z + \hat{x}_i) \approx p_x(\hat{x}_i) \qquad \text{for } -\frac{\Delta x}{2} \leq z < +\frac{\Delta x}{2}\,. \tag{7.24}$$

Then, we find that

$$N_Q \approx \frac{\Delta x^2}{12} \tag{7.25}$$

approximately holds too, independently of the probability density function $p_x(u)$. This approximation corresponds to the assumption that within the quantization intervals the signal is uniformly distributed. Thus, the quantization noise $e(k)$ arises with uniform distribution $p_e(u)$ in the interval

$$-\frac{\Delta x}{2} < e \leq +\frac{\Delta x}{2} \tag{7.26}$$

and its power

$$N_Q = \mathrm{E}\left\{e^2\right\} = \int_{-\infty}^{+\infty} u^2 \, p_e(u) \, du \tag{7.27}$$

is also given by (7.25). In conclusion, (7.25) can be applied even in the case of a non-uniform probability density function $p_x(u)$, if the quantization is sufficiently small, i.e., the word length w is sufficiently large, $p_x(u)$ is sufficiently smooth, and overload effects are negligibly small.

If, furthermore, the signal power S is normalized to the squared overload amplitude $x^2_{\max}$ of the quantizer input by the use of the *form factor* F

$$S = F \, x^2_{\max} , \tag{7.28}$$

the signal-to-noise ratio is approximately described by

$$\frac{\mathrm{SNR}}{\mathrm{dB}} = 10 \lg \left(\frac{S}{N_Q}\right) \approx w \, 20 \lg(2) + 10 \lg(3F) \tag{7.29}$$

$$= w \, 6.02 + 10 \lg(3) + 10 \lg(F) . \tag{7.30}$$

The form factor F, which can be considered as normalized power, is depending on the shape of the probability density function.

Figure 7.4 depicts the general behavior of the signal-to-noise ratio as a function of the form factor F (normalized power). The signal-to-noise ratio is a linear function of the signal level, i.e., of the logarithmic signal power

$$10 \lg \left(\frac{S}{x^2_{\max}}\right) = 10 \lg(F) . \tag{7.31}$$

In case of overload, the signal-to-noise ratio drops rapidly with increasing signal power as indicated in Fig. 7.4. By scaling the amplitude of $x(k)$ or the overload

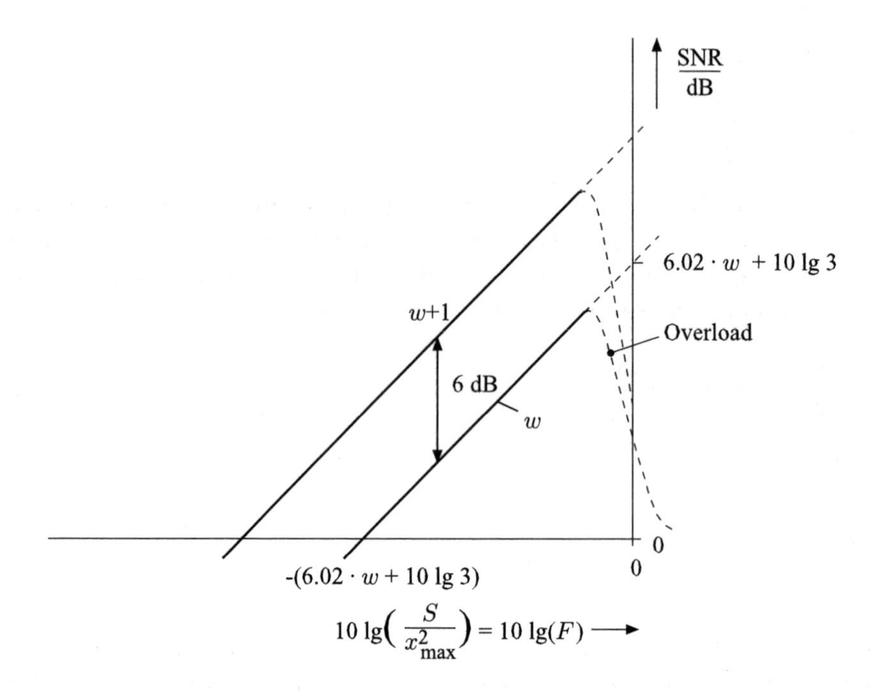

Figure 7.4: Signal-to-noise ratio for uniform quantization as a function of form factor F

amplitude $\pm x_{\max}$ of the quantizer, the overload effects can be kept small. In principle the quantizer may be derived in such a way that in (7.14-a) the contributions caused by quantization and overload are balanced

$$N_Q = N_O . \tag{7.32}$$

This case will not be discussed further here; for more details see, e.g., [Jayant, Noll 1984].

The impact of form factor F is shown in Table 7.1 for different full range signals (no quantizer overload, a), b), c)), as well as for signals with small overload probability ($P = 0.001$, d), e)).

Finally, it should be noted that for uniform quantization the *6-dB-per-bit rule* (7.23-c) is generally a good approximation. However, a constant penalty in the signal-to-noise ratio can be experienced, due to the shape of the probability density function. If small signal values occur much more often than large ones, or if the signal amplitude is so low that only a part of the dynamic range of the quantizer is exploited, the penalty can be substantial.

Table 7.1: Influence of the form factor F on the signal-to-noise ratio for uniform quantization (see also (7.29))
Overload of the quantizer with probability $P = 0.001$

Probability distribution $p_u(x)$	F	$10 \lg(3F)$
a) Uniform distribution	1/3	0
b) Distribution of a sinusoidal signal	1/2	+1.76
c) Triangular probability density function	1/6	−3.01
d) Laplace probability density function	$\approx 1/24$	≈ -9
e) Measurement of speech signals	1/300 to 1/20	−20 to −8

7.3 Non-uniform Quantization

For uniform quantization, the signal-to-noise ratio according to (7.29) is proportional to the signal level, hence it becomes smaller with decreasing signal power. However, especially in speech signals, small sample values are particularly frequent, corresponding to a probability density function (PDF), which can be approximated, for instance, by a Laplacian PDF, a gamma PDF, or by spherically invariant models, e.g., [Brehm, Stammler 1987], [Jayant, Noll 1984] (see Section 5.10). In this case, the resulting low signal-to-noise ratios can be improved by using a quantizer with a non-uniform amplitude resolution, which reduces the width of the quantization intervals in the low-amplitude region and allows larger intervals otherwise. A corresponding approach using *signal compression* is depicted in Fig. 7.5.

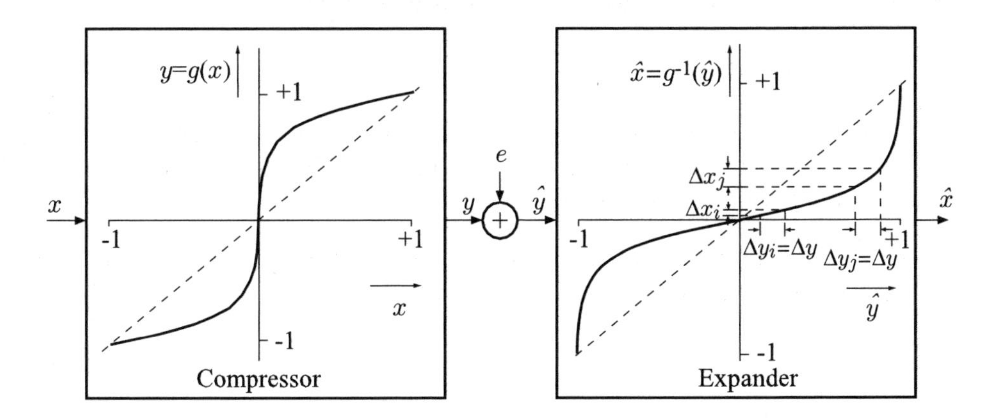

Figure 7.5: Principle of quantization with companding ($x_{\max}=y_{\max}=1$)

Prior to the actual quantization, the values $x(k)$ are non-linearly transformed with a compressor characteristic

$$y = g(x) \tag{7.33}$$

in such a way that for an unchanged amplitude range ($\pm y_{\max} = \pm x_{\max}$) small signal values are amplified more than large ones. Subsequently, uniform quantization with $K = 2^w$ quantization levels $\hat{y}_i$, $i = 1, 2, \dots, K$, is applied as in Section 7.2, which is modeled by adding a uniformly distributed, spectrally white noise signal $e(k)$ of power

$$N_Q = \frac{\Delta y^2}{12}\,; \qquad \Delta y = \frac{2\,y_{\max}}{K} \tag{7.34}$$

to the compressed signal y. Without loss of generality, we will in the following assume a normalized signal representation with

$$x_{\max} = y_{\max} = 1\,. \tag{7.35}$$

The non-linear distortion caused by the compressor characteristic should be removed by the inverse characteristic of the expander

$$\hat{x} = g^{-1}(\hat{y})\,, \tag{7.36}$$

before the quantized values $\hat{y}(k)$ are applied to any signal processing algorithm such as digital filtering.

The compressor characteristics with its instantaneous amplification is reversed as can be seen in Fig. 7.5. Thus the lower quantization levels $\hat{y}_i$ are attenuated more than the higher ones. This yields a non-uniform amplitude resolution of the values $\hat{x}(k)$, where the effective quantization levels are given by

$$\hat{x}_i = g^{-1}(\hat{y}_i)\,. \tag{7.37}$$

As a result the magnitude of the effective quantization noise $e(k)$ depends strongly on the amplitude of the input signal $x(k)$. The combination of compressor and expander is commonly called *compander*.

The relation between the different signals is illustrated in Fig. 7.6. The values $y(k)$ are quantized uniformly using a midrise quantizer

$$\hat{y} = f(y) \tag{7.38}$$

according to Fig. 7.3-b with word length of $w = 5$ and $w = 8$. The relatively fine amplitude resolution near the origin and the coarser resolution for larger signal values can clearly be seen.

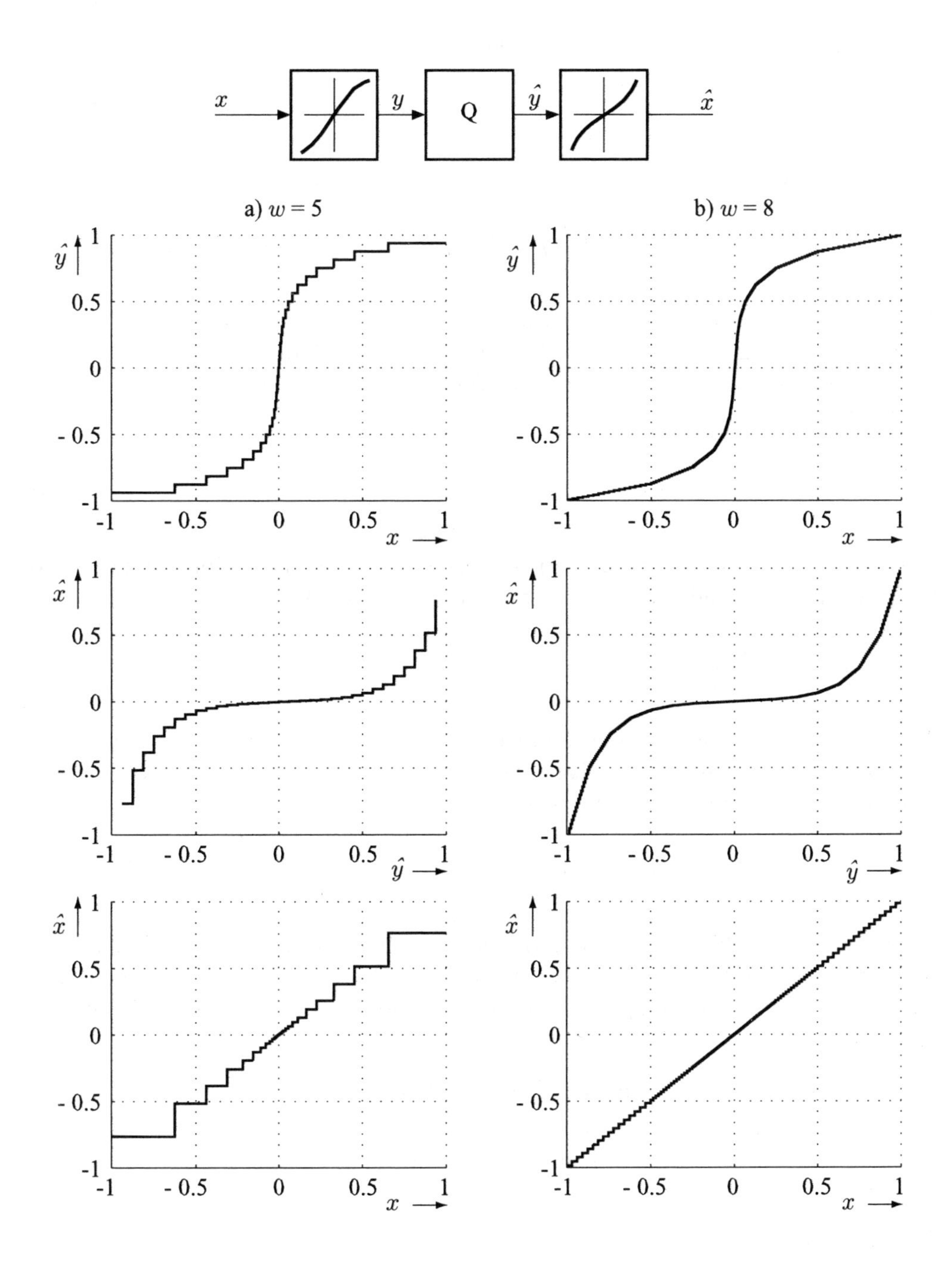

Figure 7.6: Companding quantization with piecewise linear approximation of the compressor and the expander characteristic (A-law characteristic, see also Fig. 7.9)

For the quantized values $\hat{y}$ and $\hat{x}$, we have

$$\hat{y} = \hat{y}_i \qquad \forall \qquad \hat{y}_i - \frac{\Delta y}{2} \leq y < \hat{y}_i + \frac{\Delta y}{2} \tag{7.39-a}$$

or, respectively,

$$\hat{x} = g^{-1}(\hat{y}_i) = \hat{x}_i \quad \forall \quad g^{-1}\left(\hat{y}_i - \frac{\Delta y}{2}\right) \leq x < g^{-1}\left(\hat{y}_i + \frac{\Delta y}{2}\right). \tag{7.39-b}$$

The quantization stepsizes Δx_i of the effective quantization intervals now depend on the magnitude of x. By simple geometrical considerations it can be shown that the different quantization stepsizes Δx_i are approximately determined by the gradient

$$\frac{\mathrm{d}\left(g^{-1}(\hat{y})\right)}{\mathrm{d}\,\hat{y}} = \left(g^{-1}(\hat{y})\right)' \tag{7.40}$$

of the inverse characteristic $g^{-1}(\hat{y})$ or, respectively, by the reciprocal $1/g'(\hat{x})$ of the gradient $g'(\hat{x})$ and the constant quantizer stepsize Δy:

$$\Delta x_i \approx \frac{\Delta y}{g'(\hat{x}_i)}. \tag{7.41}$$

A suitable criterion for the development of a compressor characteristic is the requirement for a constant relative quantization error. The quantization stepsize should therefore be proportional to the signal magnitude,

$$\Delta x(x) \approx \frac{\Delta y}{g'(x)} \sim |x|, \tag{7.42}$$

as far as this is achievable with K quantization intervals. For simplicity, we first consider only positive values of x and extend the resulting characteristic $g(x)$ later on to the negative range, such that a symmetric characteristic is obtained as illustrated in Fig. 7.5.

According to (7.42) the gradient $g'(x)$ should obey

$$\frac{1}{g'(x)} \stackrel{!}{=} c\,x; \qquad c = \text{const}; \qquad x > 0. \tag{7.43-a}$$

Thus, we get

$$g(x) = \int \frac{1}{c\,x}\,\mathrm{d}x = c_1 + c_2 \ln(x) \tag{7.43-b}$$

with appropriate constants c_1 and c_2. The desired compressor characteristic $g(x)$ is a logarithmic function, the expander characteristic an exponential function.

However, the function $\ln(x)$ is only defined for positive values and diverges for $x \to 0$. Consequently, the purely logarithmic compressor characteristic (7.43-b) is not practical. For this reason, the characterstic defined by (7.43-b) needs some pragmatic modifications. Two approximations to logarithmic quantization have found wide use as international standards providing almost a constant relative quantization error.

For the fixed (wire-line) digital telephone networks in Europe, the so-called *A-law characteristic* was defined as

$$g_A(x) = \begin{cases} \text{sign}(x) \cdot \dfrac{1 + \ln(A\,|x|)}{1 + \ln(A)} & \frac{1}{A} < |x| \leq +1 \\ \\ \dfrac{A\,x}{1 + \ln(A)} & -\frac{1}{A} \leq x \leq +\frac{1}{A}\,. \end{cases} \tag{7.44}$$

Near the origin, in the range

$$-\frac{1}{A} \leq x \leq +\frac{1}{A}\,, \tag{7.45}$$

the A-law compressor characteristic is based on a linear characteristic and is logarithmic beyond that. At $x = 1/A$ both characteristics meet smoothly without discontinuity. In the negative range the characteristic is mirrored.

For the linear part of the quantization characteristic a gradient of

$$g'_A(0) = 16 \tag{7.46}$$

is chosen, resulting in a parameter value $A = 87.56$. In accordance with (7.41), the effective quantization stepsizes Δx_i in the linear region

$$-\frac{1}{A} \leq x \leq +\frac{1}{A}\,, \tag{7.47}$$

i.e., for *small signals*, are thus reduced by the factor 2^{-4}. This corresponds to an increase of the signal-to-noise ratio in the linear region by

$$\Delta\text{SNR} = 20\,\lg(2^4) = 24.082\,\text{dB}\,. \tag{7.48}$$

According to the *6-dB-per-bit rule* a uniform quantizer would need $\Delta w = 4$ additional bits for the same signal-to-noise ratio for *small signals*.

In the digital telephone systems of North America and Japan, the so-called *μ-law characteristic* is utilized to approximate (7.43-b) in a slightly different way. This compressor characteristic is described by a single continuous function as

$$g_\mu(x) = \text{sign}(x)\,\frac{\ln(1 + \mu\,|x|)}{\ln(1 + \mu)} \quad \text{with } \mu = 255\,. \tag{7.49}$$

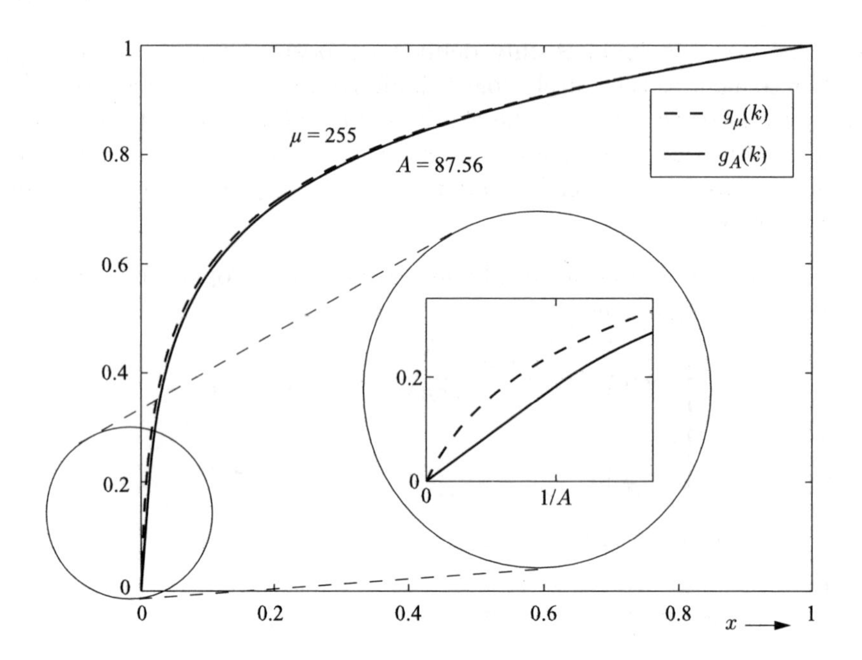

Figure 7.7: Comparison of A-law and μ-law compressor characteristics

For relatively small signals $x(k)$ or low values μ the compressor characteristic evolves as a linear function according to $\ln(1+\mu|x|) \approx \mu|x|$. For $\mu|x| \gg 1$, it is logarithmic since $\ln(1+\mu|x|) \approx \ln(\mu|x|)$. The gradient of the compressor characteristic at the origin is

$$g'_\mu(x=0) = \frac{\mu}{\ln(1+\mu)} = 45.99\,. \tag{7.50}$$

For the relatively small signals, the effective signal-to-noise ratio increases by

$$\Delta\text{SNR} = 20\,\lg(45.99) = 33.25\,\text{dB}\,. \tag{7.51}$$

The A-law characteristic and the μ-law characteristic are very similar. They are illustrated in Fig. 7.7.

The logarithmic companding results in a signal-to-noise ratio which is to a large extent independent of the signal level or the power of the sequence $x(k)$, respectively.

We now calculate the achievable signal-to-noise ratio. Presuming a uniform distribution of x within each quantization interval, the individual quantization noise power is $\Delta x_i^2/12$. The contribution of the i-th quantization interval to the total power of the quantization noise is

$$N_{Qi} = \frac{\Delta x_i^2}{12}\,P_i\,, \tag{7.52}$$

where $x(k)$ takes a value in the i-th quantization interval with a probability P_i. Under the assumption of a symmetric PDF $p_x(+u) = p_x(-u)$, the total quantization noise power N_Q can be calculated with (7.41) and (7.43-a):

$$N_Q = 2 \sum_{i=1}^{\frac{K}{2}} N_{Qi} \tag{7.53-a}$$

$$= 2 \sum_{i=1}^{\frac{K}{2}} \frac{\Delta y^2}{12} \; c^2 \; \hat{x}_i^2 \; P_i \tag{7.53-b}$$

$$\approx \frac{\Delta y^2}{12} \; c^2 \; S \, . \tag{7.53-c}$$

Hence, the noise power is proportional to the signal power. With

$$\Delta y \;=\; \frac{2 \; y_{\text{max}}}{2^w} \;=\; 2^{-(w-1)} \, , \qquad y_{\text{max}} \;=\; 1 \, , \tag{7.54}$$

the signal-to-noise-ratio

$$\frac{\text{SNR}}{\text{dB}} = 10 \; \text{lg} \left(\frac{S}{N_Q} \right) \;\approx\; w \; 20 \; \text{lg}(2) \;+\; 10 \; \text{lg}(3) \;-\; 20 \; \text{lg}(c) \tag{7.55}$$

$$= w \; 6.02 + 10 \; \text{lg} \, 3 - 20 \; \text{lg} \, c \tag{7.56}$$

becomes independent of the signal power. It now contains the constant c instead of the form factor F in (7.29). Applying (7.43-a), constant c amounts to

$$c_A \;=\; (1 \;+\; \ln A) \;\approx\; 5.47 \tag{7.57}$$

for the A-law characteristic in the range $1/A \leq |x| \leq 1$, and to

$$c_\mu \;\approx\; \ln(1 + \mu) \;\approx\; 5.55 \tag{7.58}$$

for the μ-law characteristic for $\mu \, x \gg 1$. With both characteristics a similar signal-to-noise ratio is thus obtained using (7.55):

$$\text{SNR}_A \approx 6 \cdot w \;-\; 9.99 \, \text{dB} \, , \tag{7.59-a}$$

$$\text{SNR}_\mu \approx 6 \cdot w \;-\; 10.11 \, \text{dB} \, . \tag{7.59-b}$$

The signal-to-noise ratio again satisfies a *6-dB-per-bit rule* which is almost independent of the signal level. This independence "costs" approximately 10 dB compared to (7.23-c) obtained for uniform quantization, i.e., for signals with uniform distribution and matching the peak-to-peak range of the quantizer. However, for uniform quantization the signal-to-noise ratio depends on the signal level. With uniform quantization substantial reductions must be expected in practice, according to

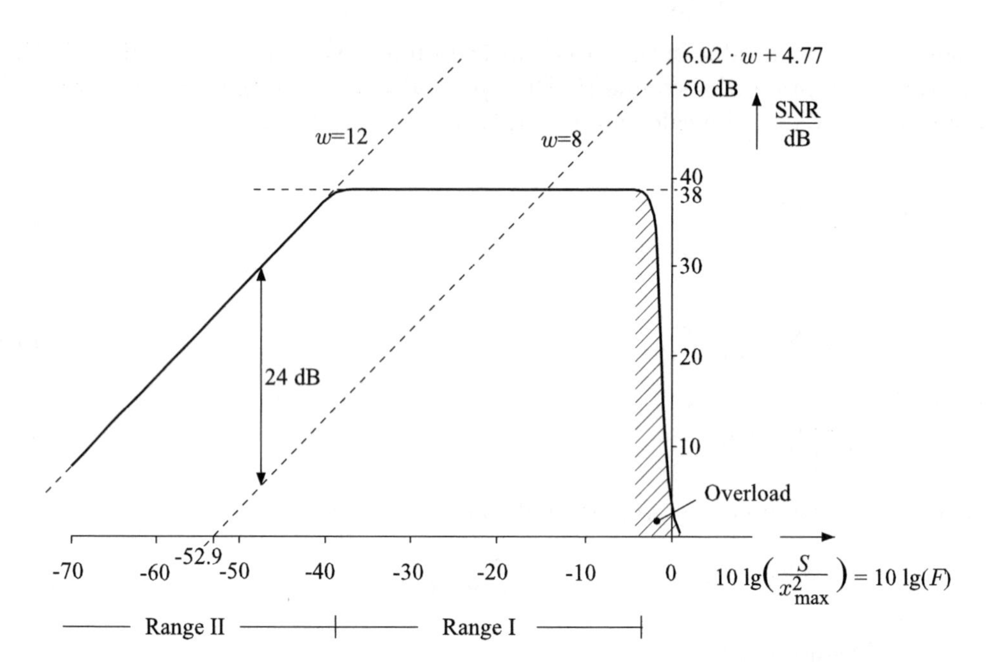

Figure 7.8: Signal-to-noise ratio for companding with the A-law characteristic $A = 87.56$ and a word length of $w = 8$

level variations corresponding to the form factor F. With the same peak-to-peak range, the logarithmic compander therefore proves indeed to be superior to the uniform quantizer over a wide range of amplitudes (see Fig. 7.8).

The approximations (7.59-a,b) do not apply to very small values of $|x|$. Close to the origin, we have a uniform quantization with an effective quantization stepsize of

$$\Delta x = \frac{\Delta y}{g'(x=0)} \tag{7.60}$$

and with the gradients given in (7.46) and (7.50). For the A-law characteristic, Fig. 7.8 shows the signal-to-noise ratio as a function of the signal power S for a word length of $w = 8$, which is common in the digital telecommunication network. In range I, the signal-to-noise ratio amounts to approximately 38 dB as derived in (7.59-a). In range II with $|x| \leq 1/A \approx 0.011$, i.e., for a range of approximately 1% of the maximum amplitude, the following expression applies in analogy to (7.29) with consideration of (7.48):

$$10 \lg \left(\frac{S}{N_Q} \right) \approx w\ 20 \lg(2) + 10 \lg(3F) + 24 \text{ dB}. \tag{7.61}$$

In this range, the signal-to-noise ratio is comparable to that of a uniform quantizer,

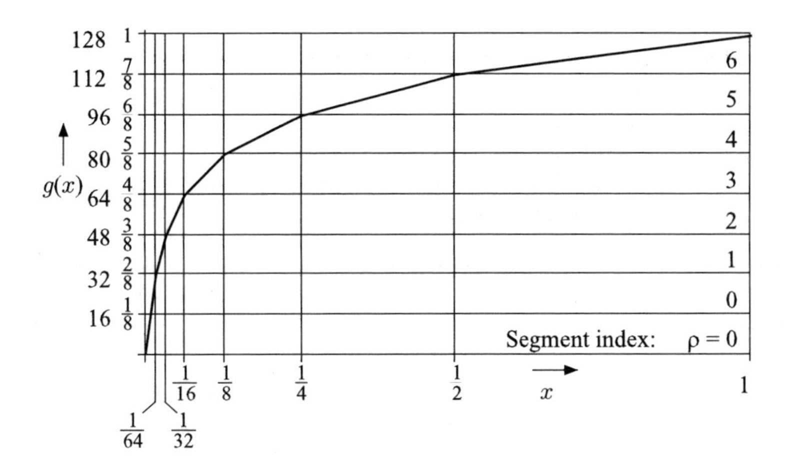

Figure 7.9: The 13-segment A-law characteristic

which, however, has a word length of $w = 12$. The corresponding improvement by 24 dB is also termed *compander gain*.

In practice, the A-law characteristic as well as the μ-law characteristic are each realized by a piecewise linear approximation, which for the A-law characteristic is depicted in Fig. 7.9.

The range $-1 \leq g \leq +1$ is divided into 16 equally spaced intervals, in which the A-law characteristic is approximated by straight lines. In the four innermost intervals ($0 < |x| < 1/64$) the A-law characteristic is almost linear (7.44). Thus in these four intervals the compressor characteristic can be approximated by one line. As a result the overall characteristic is approximated by $16 - 3 = 13$ segments of distinct slope. Due to the increasing length by a factor of 2 for each successive segment, the slopes of adjacent segments differ by a factor of 2.

For quantization with a word length of $w = 8$ bits, the first bit denotes the algebraic sign, the next 3 bits encode the respective segment, and the last 4 bits indicate the quantization level within the segment.

The effective quantization stepsize in the lowest segment amounts to

$$\Delta x_{\min} = \frac{1/128}{2^{-4}} = 2^{-11} \tag{7.62}$$

and in the highest segment to

$$\Delta x_{\max} = \frac{1/2}{2^{-4}} = 2^{-5} \, . \tag{7.63}$$

The resulting non-uniform quantization characteristic has already been depicted in Fig. 7.6-b.

Table 7.2: Coding law of the 13-segment A-law characteristic
a: number of leading zeros following the sign bit
b: binary-code of $7 - a$
c: the last four digits, if $a = 7$
the first four digits behind the leading 1, if a < 7
d: neglected digits

Segment index	Range	Binary representation $\|x\|$ (a, c, d)	Binary representation $\|\hat{y}\|$ (b, c)
ρ			
0	$0 \leq \|x\| < 2^{-7}$	0000000....	000....
0	$2^{-7} \leq \|x\| < 2^{-6}$	0000001....	001....
1	$2^{-6} \leq \|x\| < 2^{-5}$	000001....-	010....
2	$2^{-5} \leq \|x\| < 2^{-4}$	00001....--	011....
3	$2^{-4} \leq \|x\| < 2^{-3}$	0001....---	100....
4	$2^{-3} \leq \|x\| < 2^{-2}$	001....----	101....
5	$2^{-2} \leq \|x\| < 2^{-1}$	01....-----	110....
6	$2^{-1} \leq \|x\| < 2^{-0}$	1....------	111....

Quantization according to the 13-segment characteristic can also be achieved by uniform pre-quantization with $w_0 \geq w + 4$ and subsequent code conversion to the word length w. This coding law is summarized in Table 7.2 for $w_0 = 12$ and $w = 8$.

The 13-segment A-law coding rule can be derived on a bit level if we start with a 12 bit sign–magnitude representation

$$x = \text{sign}\{x\} \cdot |x| \,. \tag{7.64}$$

We denote by $0 \leq a \leq 7$ the number of leading zeros of the binary representation of $|x|$ and by c the next 4 bits as indicated in Table 7.2. The remaining d bits are neglected. Finally, the binary representation of $\hat{y}$ for $w = 8$ is obtained as shown in Table 7.3.

Table 7.3: Bit allocation of 13-segment A-law quantization

	1 bit	3 bits	4 bits
$\hat{y}$:	sign$\{x\}$	$7 - a$	c

This form of signal quantization which fulfills the international standard ITU G.711 [ITU-T Rec. G.711 1972] is the basis for digital speech transmission in the European digitized telecommunication networks with a bit rate of $B = w \cdot f_s = 64$ kbit/s per voice channel (ISDN, Integrated Services Digital Network).

7.4 Optimal Quantization

The quantizers discussed so far work with uniform or non-uniform stepsizes. The quantizer levels $\hat{x}_i$ are in the middle or at the edge of the intervals (see Fig. 7.3). In principle, the *interval limits* or *decision levels* x_i and the *quantizer representation levels* $\hat{x}_i$ for $i \in \{1, 2, \ldots, 2^w\}$ can be chosen arbitrarily as indicated in Fig. 7.10. In particular, they can be determined such that for a given signal PDF $p_x(u)$ the maximal SNR is obtained. In other words, the (scalar) optimal quantizer minimizes the power N_Q of the quantization error.

For non-uniformly distributed signals, a non-uniform resolution of the amplitude is to be expected; for signals like speech finer quantization of small amplitudes and coarser quantization of large amplitudes is desired. The characteristic should thus indeed show similarities to that of logarithmic companding. However, it might differ, as it is generated from a different minimization criterion and does not aim for PDF-independent SNR. The underlying optimization problem was solved in [Lloyd 1982] and [Max 1960]. This solution is called the *Lloyd–Max quantizer*.

In analogy to (7.17), the power of the quantization noise amounts to

$$N = \sum_{i=1}^{2^w} \int_{x_{i-1}}^{x_i} (\hat{x}_i - u)^2 \, p_x(u) \, du \,. \tag{7.65}$$

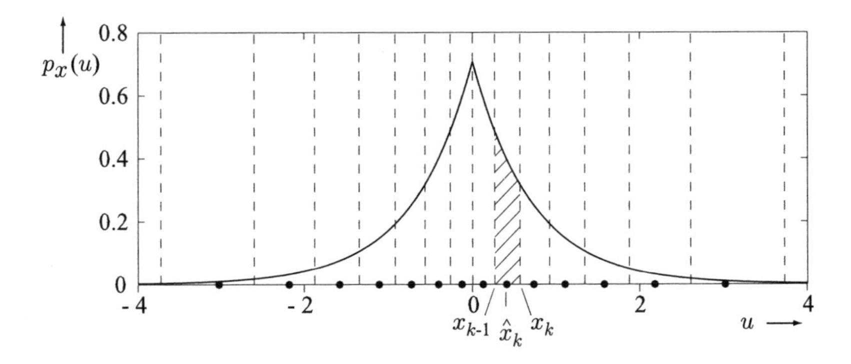

Figure 7.10: Relation between decision levels x_k and quantizer representation levels $\hat{x}_k$

Necessary conditions for determining the interval limits x_i and the representation levels $\hat{x}_i$ $(i = 0, 1, \ldots, 2^w - 1)$ are provided by partial derivatives. Taking into account that the outer limits x_0 and x_K $(K = 2^w)$ have to be treated separately, we obtain for $k = 1, 2, \ldots, 2^w - 1$

$$\frac{\partial N}{\partial x_k} = (\hat{x}_k - x_k)^2\, p_x(x_k) - (\hat{x}_{k+1} - x_k)^2\, p_x(x_k) \overset{!}{=} 0\,. \qquad (7.66\text{-a})$$

This yields

$$x_k = \frac{\hat{x}_k + \hat{x}_{k+1}}{2}, \qquad (7.66\text{-b})$$

and secondly for $k = 1, 2, \ldots, 2^w - 1$,

$$\frac{\partial N}{\partial \hat{x}_k} = 2 \int_{x_{k-1}}^{x_k} (\hat{x}_k - u)\, p_x(u)\, du \overset{!}{=} 0 \qquad (7.66\text{-c})$$

results in

$$\hat{x}_k = \frac{\int_{x_{k-1}}^{x_k} u\, p_x(u)\, du}{\int_{x_{k-1}}^{x_k} p_x(u)\, du}\,. \qquad (7.66\text{-d})$$

Hence, the optimal interval representatives $\hat{x}_i$ correspond to the centers of gravity of the quantization intervals. The optimal interval limits x_k are located midway between two adjacent representation levels with the two outer interval limits x_0 and x_K being exceptions. The latter are given by the range of x, for example, $x_0 = -\infty$ and $x_K = +\infty$. The conditions (7.66-b) and (7.66-d) can be numerically solved for arbitrary PDFs $p_x(u)$. The achievable improvement of the SNR (see also examples from Table 7.5) depends on the shape of the PDF $p_x(u)$ (e.g., [Jayant, Noll 1984]).

7.5 Adaptive Quantization

An alternative for reducing the dependency of the SNR on the (instantaneous) quantizer load is to use a uniform quantizer with K quantizer representation levels but with dynamical adaptation of the quantizer stepsize Δx.

Two basic solutions exist, which are designated in Fig. 7.11 as quantization with forward adaptation (AQF: *adaptive quantization forward*) or with backward adaptation (AQB: *adaptive quantization backward*) [Jayant, Noll 1984]. In both cases

$$\hat{x}(k) = \text{sign}(x(k))\; Z(k)\, \frac{\Delta x(k)}{2}, \qquad Z(k) \in \{1, 3, 5, \ldots\} \qquad (7.67)$$

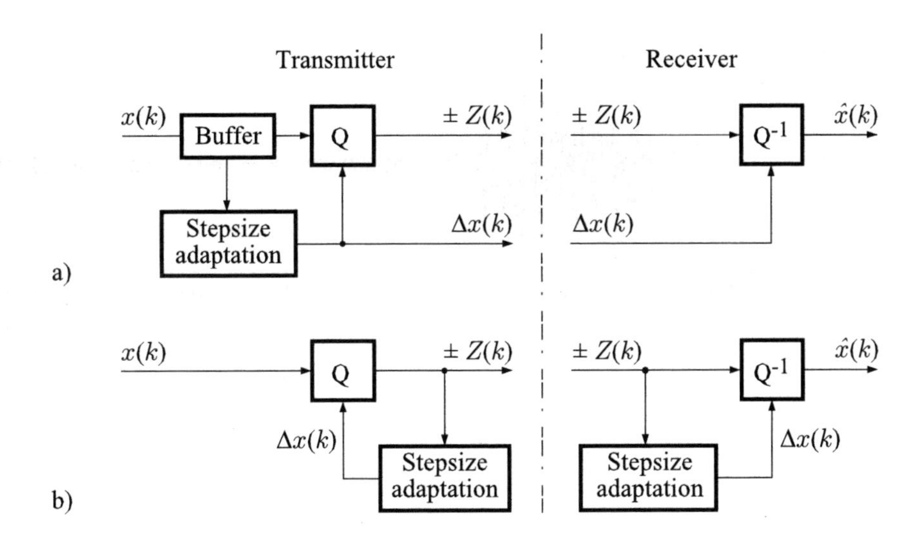

Figure 7.11: Adaptive quantization
a) Adaptive quantization with forward estimation (AQF)
b) Adaptive quantization with backward estimation (AQB)

applies to the quantized values, if a symmetric quantization characteristic according to Fig. 7.3-a is assumed.

With the AQF method, $\Delta x(k)$ is adjusted blockwise and transmitted (or, respectively, stored) as additional side information. Because of the extra required bit rate a relatively large block length of, for instance, $N = 128$ at $f_s = 8\,\text{kHz}$ is chosen. With the AQB method, there is no side information, as the quantizer stepsize is derived from $Z(k-1)$, which for undisturbed transmission is also available at the receiving side.

With both methods, the instantaneous power of $x(k)$ or $\hat{x}(k)$ is estimated and the stepsize is adjusted proportionally to the estimated standard deviation $\hat{\sigma}_x(k)$:

$$\Delta x(k) \;=\; c\,\hat{\sigma}_x(k)\,, \qquad c \;=\; \text{const}\,. \tag{7.68}$$

With the AQF method, the variance estimation is performed on blocks of N samples according to

$$\hat{\sigma}_x^2(k) \;=\; \frac{1}{N}\sum_{i=0}^{N-1} x^2(k_0+i) \quad \forall \;\; k \;=\; k_0, k_0+1, \ldots, k_0+N-1\,. \tag{7.69}$$

In the AQB method, however, σ_x^2 is estimated recursively using the already available quantized value $\hat{x}(k-1)$

$$\hat{\sigma}_x^2(k) \;=\; \alpha\,\hat{\sigma}_x^2(k-1) \;+\; (1-\alpha)\,\hat{x}^2(k-1)\,, \qquad 0 \;<\; \alpha \;<\; 1\,. \tag{7.70}$$

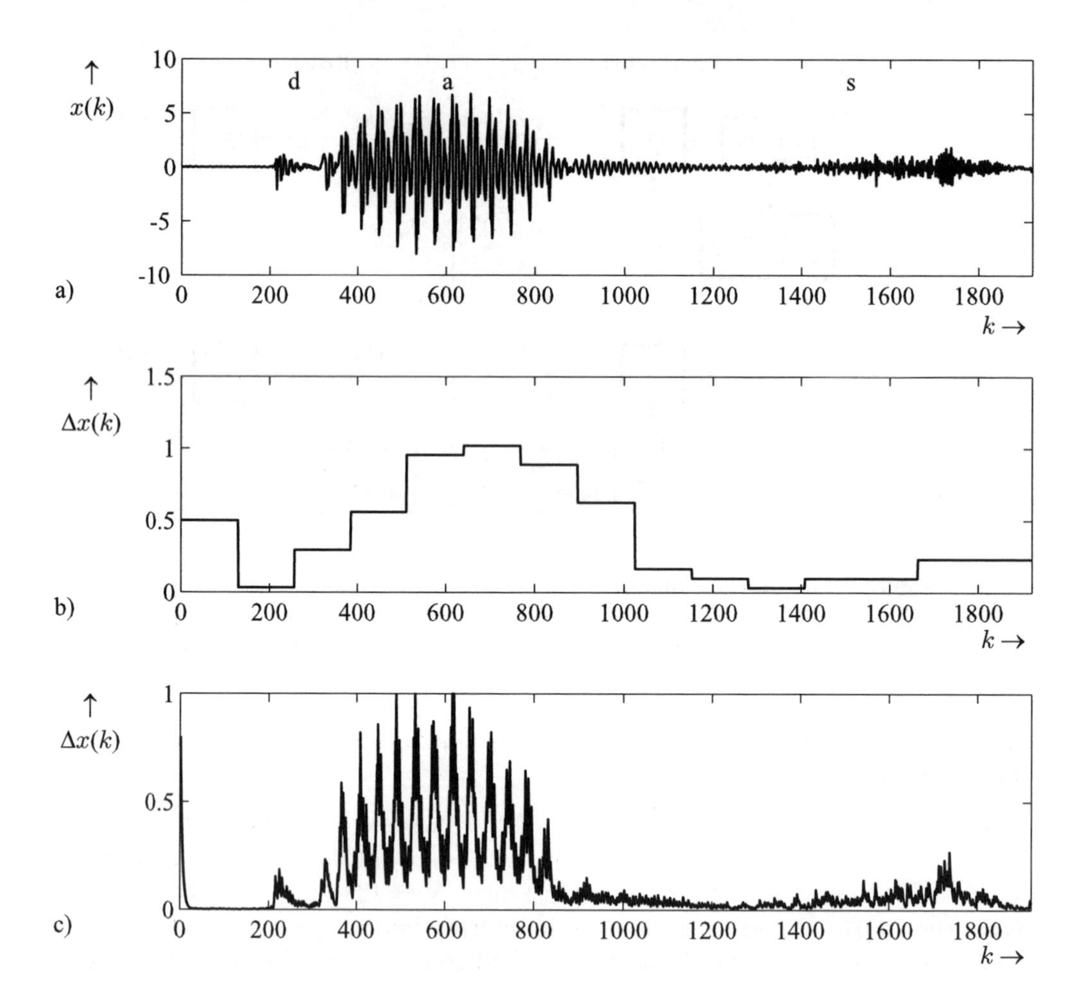

Figure 7.12: Example of the adaptation of the stepsize $\Delta x(k)$
a) Speech signal "das" (German)
b) Stepsize with forward adaptation (AQF)
c) Stepsize with backward adaptation (AQB)

The stepsize Δx can be adjusted more frequently than with the AQF method. This is illustrated by the example in Fig. 7.12.

For the AQB method an algorithm has been proposed in [Jayant 1973] which can be realized very efficiently. Due to (7.68), the following expression holds:

$$\frac{\Delta x(k)}{\Delta x(k-1)} = \frac{\hat{\sigma}_x(k)}{\hat{\sigma}_x(k-1)} \doteq M(k-1). \tag{7.71}$$

The term $M(k-1)$ will be called the *stepsize multiplier* from now on. Equation (7.71) in combination with (7.70), (7.68), and (7.67) yields

$$M^2(k-1) = \frac{\hat{\sigma}_x^2(k)}{\hat{\sigma}_x^2(k-1)} \tag{7.72-a}$$

$$= \alpha + (1-\alpha)\, Z^2(k-1)\, \frac{c^2}{4} \tag{7.72-b}$$

or,

$$M(k-1) = \sqrt{\alpha + (1-\alpha)\, Z^2(k-1)\, \frac{c^2}{4}}\,. \tag{7.72-c}$$

Hence, the stepsize multiplier depends only on Z. As Z takes only 2^{w-1} different values, $M(k-1)$ can be computed according to (7.72-c) in advance and stored in a table.

The computational steps required at time instant k at the transmit and receive sides are summarized below:

1. Computation of the new stepsize

$$\Delta x(k) = M(k-1) \cdot \Delta x(k-1)\,. \tag{7.73-a}$$

2. Quantization of $x(k)$ or, respectively, determination of Z according to

$$\hat{x}(k) = \operatorname{sign}\big(x(k)\big) \cdot Z(k) \cdot \frac{\Delta x(k)}{2}\,, \qquad Z(k) \in \{1, 3, 5, \dots\} \tag{7.73-b}$$

with

$$Z(k) = 2 \cdot \operatorname{int}\left(\frac{x(k)}{\Delta x(k)}\right) + 1 \tag{7.73-c}$$

(see also (7.9-a)).

3. Determination of the stepsize multiplier for $k+1$ by selecting the corresponding value

$$M(k) = f\big(Z(k)\big) \tag{7.73-d}$$

from a table.

Table 7.4: Stepsize multipliers $M = f(Z(k))$ for adaptive quantization of speech signals [Jayant 1973], [Jayant, Noll 1984]

		$Z =$	1	3	5	7	9	11	13	15
PCM	$w = 2$	$M =$	0.60	2.20						
	3		0.85	1.00	1.00	1.50				
	4		0.80	0.80	0.80	0.80	1.20	1.60	2.00	2.40
DPCM	2		0.80	1.60						
	3		0.90	0.90	1.25	1.75				
	4		0.90	0.90	0.90	0.90	1.20	1.60	2.00	2.40

Stepsize multipliers optimized for the adaptive quantization of speech signals are listed in Table 7.4 for speech (PCM). It should be noted that the optimization for prediction error signals (DPCM) gives somewhat different values [Jayant 1973].

In conclusion, the different quantization methods are represented comparatively for a short speech signal in Table 7.5 which shows the mean SNR and the segmental SNR[1] values obtained for the example from Fig. 7.13. The form factor according to (7.28) is $F = 0.0177$ in this case (see also Table 7.1).

With SNR values of 13.5 dB or 13.3 dB, respectively, the quantizer for companding with the A-law or μ-law characteristic provides distinctively better results in comparison to uniform quantization. Compared to the quantization process

Table 7.5: SNRs for quantization with $w = 4$ for the example given in Fig. 7.13

Quantization	SNR/dB	SNRseg/dB
Uniform	11.34	2.42
A-law characteristic	13.52	11.53
μ-law characteristic	13.34	12.02
Optimal quantizer	18.34	7.64
AQF	19.49	18.26
AQB	20.15	19.41

[1] The segmental signal-to-noise ratio SNRseg is defined as the average of the short-term SNR

$$\mathrm{SNR}(k') = 10 \lg \left(\frac{\hat{\sigma}_x^2(k')}{\hat{\sigma}_e^2(k')} \right) ,$$

where $\hat{\sigma}_x^2(k')$ and $\hat{\sigma}_e^2(k')$ are determined as short-term powers of the signal $x(k)$ and the quantization noise $e(k)$ for blocks of length N, while k' denotes the block index.

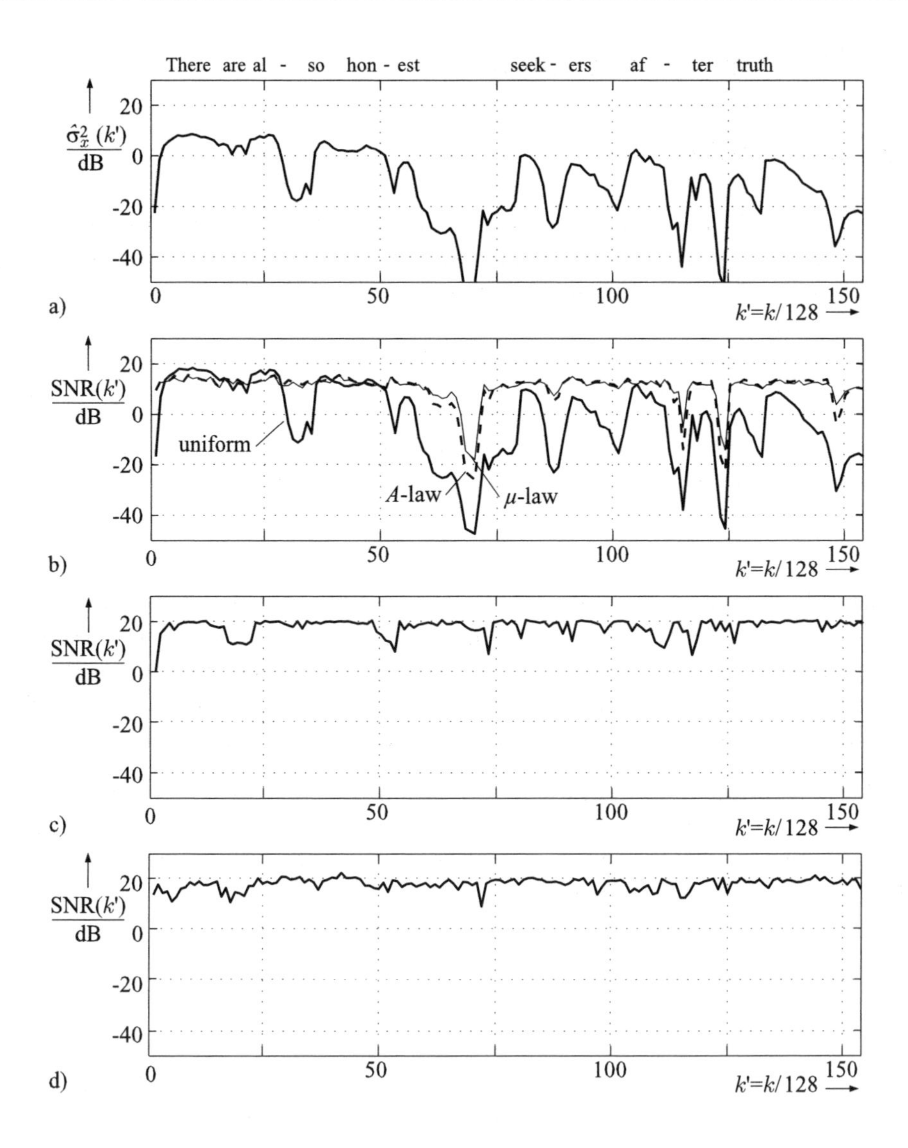

Figure 7.13: Short-term power and SNR for various quantizers with $w = 4$
a) Short-term power of the signal (Eq. (7.69), $N = 128$, $|x(k)|_{\max} = 8$)
b) Blockwise computed SNR for uniform quantization and for companding with the A-law or μ-law characteristic
c) Blockwise computed SNR for AQF
d) Blockwise computed SNR for AQB

with companding, the fixed optimal quantizer adjusted to the PDF of this signal segment achieves a mean SNR which is further improved by approximately 5 dB. The best result is obtained by adaptive quantization with backward estimation (AQB) with an SNR value of approximately 20 dB.

7.6 Vector Quantization

7.6.1 Principle

So far we have discussed *scalar quantization*. For the individual distributions of signal or parameter values x the suitable quantization intervals and quantizer reproduction levels $\hat{x}_i$ were identified. This procedure can be generalized: L values are combined to an L-dimensional *vector*

$$\mathbf{x} = (x_1, x_2, \ldots, x_L)^T, \tag{7.74}$$

are allocated to one of K possible L-dimensional *quantization cells*, and are replaced by a corresponding *quantizer representation vector*

$$\hat{\mathbf{x}}_i = (\hat{x}_{i,1}, \hat{x}_{i,2}, \ldots, \hat{x}_{i,L})^T. \tag{7.75}$$

This procedure is called *vector quantization* (VQ) [Gersho, Gray 1992]. With $L = 1$ the scalar quantization is included as a special case. The allocation of $\mathbf{x}$ to a suitable quantization cell, the *Voronoi cell*, is addressed by the cell index i. The corresponding representation vector is indexed by i in the *code book* consisting of K *code vectors* $\hat{\mathbf{x}}_i$ (quantizer reproduction vector, $i = 1, 2, \ldots, K = 2^w$).

In analogy to scalar quantization, vector quantization can be realized in the L-dimensional vector space with uniform as well as with non-uniform resolution. For the two-dimensional case, two vector quantizers with uniform and non-uniform resolution and $K = 25$ are depicted in Fig. 7.14.

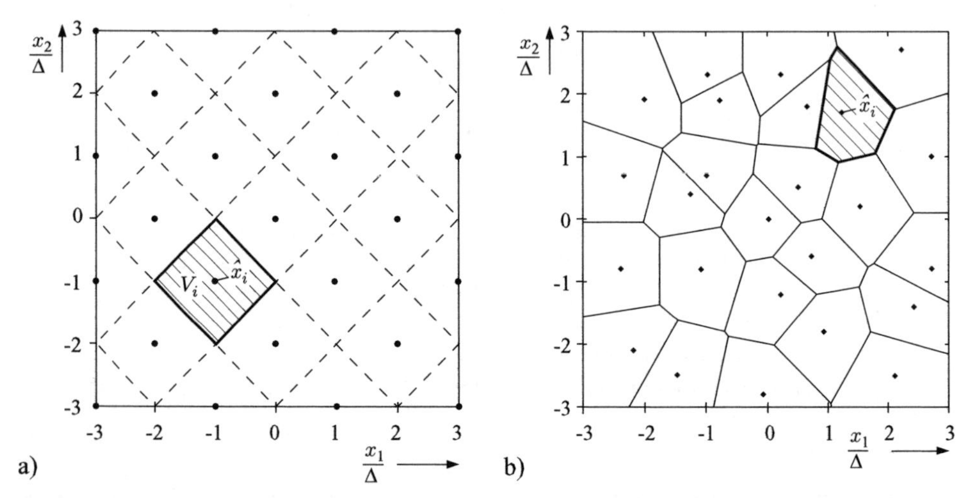

Figure 7.14: Vector quantization: example with $K = 25$ vectors of dimension $L = 2$
a) Uniform resolution (D_2-lattice)
b) Non-uniform resolution

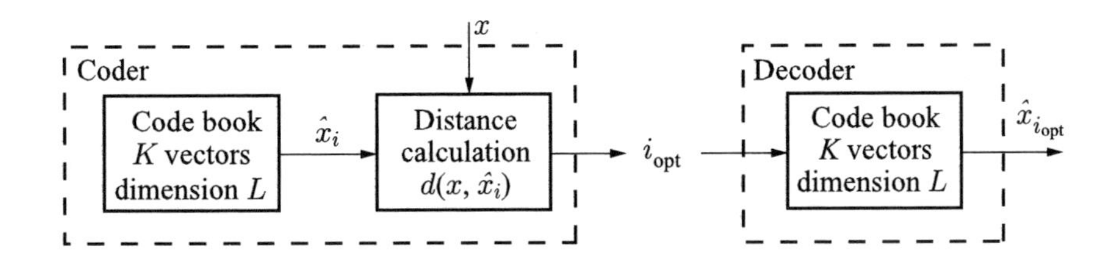

Figure 7.15: Principle of vector quantization

With a given code book, the vector quantization task is to replace an input vector $\mathbf{x}$ by the *most similar* vector $\hat{\mathbf{x}} = \hat{\mathbf{x}}_{i_{\text{opt}}}$. The choice is based on a distance or error measure $d(\mathbf{x}, \hat{\mathbf{x}})$ such that the condition

$$d(\mathbf{x}, \hat{\mathbf{x}}_{i_{\text{opt}}}) = \min_i d(\mathbf{x}, \hat{\mathbf{x}}_i) \tag{7.76}$$

is fulfilled. Thus, the boundaries of the Voronoi cells are implicitly determined.

Since the code book is known at the receiver, only the code book index i_{opt}, not the quantized vector $\hat{\mathbf{x}}_{i_{\text{opt}}}$, is transmitted. The basic procedure is illustrated in Fig. 7.15.

If the code book includes $K = 2^w$ vectors $\hat{\mathbf{x}}_i$ of dimension L, the selected index i_{opt} and hence, indirectly, the chosen vector can be coded with

$$w = \text{ld}\,(K) \text{ bits}\,. \tag{7.77}$$

With respect to a single component x_λ of vector $\mathbf{x}$, an effective word length of

$$\overline{w} = \frac{\text{ld}\,(K)}{L} \quad \text{[bits per component } x_\lambda] \tag{7.78}$$

results. With $K = 2^{10} = 1024$ and $L = 40$, which are typical dimensions in prediction error signal coding (see Section 8.5.3), only $1/4$ bit per value x_λ has to be transmitted.

Regarding the choice of the distance measure $d(\mathbf{x}, \hat{\mathbf{x}}_i)$, different possibilities exist. Note that the vector $\mathbf{x}$ may contain, e.g., either speech samples or some model-based codec parameters. If the speech signal is reconstructed from the quantized vectors $\hat{\mathbf{x}}$, the error vectors

$$\mathbf{e} = \hat{\mathbf{x}} - \mathbf{x} \tag{7.79}$$

affect the subjective speech quality in both cases differently. Therefore, different distance measures should be applied which should preferably correspond to the psychoacoustic perception.

For quantizing *signal* vectors, frequently the *squared error distortion measure* (Euclidian norm)

$$d(\mathbf{x}, \hat{\mathbf{x}}_i) = \frac{1}{L}\,(\mathbf{x} - \hat{\mathbf{x}}_i)^T(\mathbf{x} - \hat{\mathbf{x}}_i) \tag{7.80-a}$$

$$= \frac{1}{L}\sum_{\mu=1}^{L}(x_\mu - \hat{x}_{i\mu})^2\,, \qquad i = 1, 2, \ldots, K \tag{7.80-b}$$

is minimized. This corresponds to selecting the nearest neighbor $\hat{\mathbf{x}}_i$ of $\mathbf{x}$ in the L-dimensional vector space.

Alternatively, the *weighted mean square error*

$$d(\mathbf{x}, \hat{\mathbf{x}}_i) = \frac{1}{L}\,(\mathbf{x} - \hat{\mathbf{x}}_i)^T \cdot \mathbf{W} \cdot (\mathbf{x} - \hat{\mathbf{x}}_i) \tag{7.81}$$

is applied, with $\mathbf{W}$ representing a symmetric, positive-definite matrix of dimension $L \times L$. The errors of the individual vector components can, for instance, be weighted differently by a diagonal matrix $\mathbf{W}$.

For quantizing coefficient sets of linear predictors, usually different distance measures are utilized, e.g., the Itakura–Saito distance [Itakura, Saito 1968], which is defined as

$$d(\mathbf{x}, \hat{\mathbf{x}}_i) = \frac{(\mathbf{x} - \hat{\mathbf{x}}_i)^T\;\mathbf{R}^{(n+1)}\;(\mathbf{x} - \hat{\mathbf{x}}_i)}{\mathbf{x}^T\;\mathbf{R}^{(n+1)}\;\mathbf{x}}\,. \tag{7.82}$$

Here, vector $\mathbf{x}$ of dimension $L = n + 1$ includes the non-quantized predictor coefficients a_λ and vector $\hat{\mathbf{x}}_i$ the quantized representatives $\hat{a}_{i\lambda}$ according to

$$\mathbf{x} = (1, -a_1, -a_2, \ldots, -a_n)^T \tag{7.83-a}$$

$$\hat{\mathbf{x}}_i = (1, -\hat{a}_{i1}, -\hat{a}_{i2}, \ldots, -\hat{a}_{in})^T\,. \tag{7.83-b}$$

$\mathbf{R}^{(n+1)}$ denotes the squared auto-correlation matrix of dimension $(n+1) \times (n+1)$ of the signal segment for which the optimal predictor coefficients a_λ have been computed.

7.6.2 The Complexity Problem

Vector quantization might be computationally very intensive, as the input vector $\mathbf{x}$ must be compared to all K code vectors $\hat{\mathbf{x}}_i$ in order to minimize a distance measure according to (7.76). This case is called *full search*.

The required computational effort shall be estimated for the squared error distortion measure. In the search of the nearest neighbor $\hat{\mathbf{x}}_i$ to input vector $\mathbf{x}$ in (7.80-a) the division by L can be omitted since it is constant for every code book entry. Per distance computation, L differences, L squares, and $(L-1)$ additions must then be computed. This results in $3L-1$ operations per vector $\hat{\mathbf{x}}_i$, i.e., in total

$$\text{Op} = (3L-1)\,K \tag{7.84-a}$$

$$= (3L-1)\,2^{L\overline{w}}\,. \tag{7.84-b}$$

The computational complexity increases exponentially with the effective word length $\overline{w}$ according to (7.78). For real-time implementations of VQ encoders with the sampling rate $f_s = \frac{1}{T}$, these operations must be performed in the time period

$$\tau = L\,T = \frac{L}{f_s}\,. \tag{7.85}$$

This leads to the computational complexity

$$\text{CC} = \frac{\text{Op}}{\tau} = \frac{3L-1}{L}\,K\,f_s \approx 3\,K\,f_s\,. \tag{7.86}$$

For the typical dimensions $K = 1024$ and $f_s = 8\,\text{kHz}$, this results in

$$\text{CC} \approx 24.6\,\text{MOPS (Mega Operations Per Second)}\,. \tag{7.87}$$

Taking the computational capacity of state-of-the-art signal processors into account, this value is already a substantial load. This restricts the application of vector quantizers with complete search. For example, $K = 1024$ and $L = 40$ (i.e., $\tau = 5\,\text{ms}$ at $f_s = 8\,\text{kHz}$) yields $\overline{w} = 0.25$ bits per sample $x(k)$. The signal-to-noise ratio which can be achieved with such low bit rates depends, as will be shown below, on the statistical properties of the sequence $x(k)$.

Because of the complexity problem, the implementation of larger vector code books requires modifications which allow a reduction of the computational complexity. For this, fast search algorithms, e.g., with tree topology, structured code books, or cascaded vector quantizers, have been proposed (e.g., [Gray 1984], [Makhoul et al. 1985]), which, however, in general only provide suboptimal results. Theoretical bounds of hierarchical, i.e., cascaded, vector quantizers are discussed in [Erdmann 2004], [Erdmann, Vary 2004].

7.6.3 Lattice Quantization

With respect to complexity, *lattice quantizers* – a special class of structured vector quantization encoders – are of particular interest. Their code vectors are given by regular grid or lattice points in the L-dimensional vector space.

The simple example of the D_2-*lattice* has already been depicted in Fig. 7.14-a. The positions of the code vectors can be analytically described so that there is no need to store a code book. Furthermore, fast algorithms which render a full search superfluous can be developed. As an example, the D_L-lattice will be discussed here. The points of this type of lattice fulfill the two conditions (see also Fig. 7.14-a) that all vector components are integer multiples of a smallest unit Δ and that additionally the sum of the components is an even multiple of Δ:

$$\hat{x}_\mu = i\,\Delta \qquad ; \quad i \in \mathbb{Z} \tag{7.88-a}$$

$$\sum_{\mu=1}^{L} \hat{x}_\mu = 2\,m\,\Delta \qquad ; \quad m \in \mathbb{Z}\,. \tag{7.88-b}$$

Because of these conditions, the vector quantization can be reduced to simple-component, scalar rounding operations.

First, all components of the signal vector $\mathbf{x}$ are mathematically rounded to integer multiples of Δ. If the resulting component sum is even, the quantized vector $\hat{\mathbf{x}}$ has already been found. If the component sum is odd, the component which shows the biggest rounding error is rounded in the "wrong direction". Thus the condition of the even component sum is fulfilled.

The advantage of easy realization, however, is offset by the disadvantage that lattice quantizers are only optimal for uniformly distributed vectors $\mathbf{x}$. Lattice quantizers can be considered as vector generalizations of uniform scalar quantizers and can analogously be combined with companding.

A detailed presentation of the theory of lattice vector quantization can be found, for example, in [Conway, Sloane 1988].

7.6.4 Design of Optimal Vector Code Books

Optimal vector quantization is the L-dimensional generalization of the Lloyd–Max quantizer discussed in Section 7.4.

The scalar series $x(k)$ is replaced by the vector series $\mathbf{x}(k)$, which is described by the L-dimensional joint probability density $p_{\mathbf{x}}(\mathbf{u}) = p_{\mathbf{x}}(u_1, u_2, \dots, u_L)$. In analogy to (7.65) the K code vectors $\hat{\mathbf{x}}_i$ must be chosen such that the expected error value

$$\mathrm{E}\{d(\mathbf{x}, \hat{\mathbf{x}})\} = \sum_{i=1}^{K} \int_{V_i} d(\mathbf{u}, \hat{\mathbf{x}}_i)\; p_{\mathbf{x}}(\mathbf{u})\; d\mathbf{u} \tag{7.89}$$

becomes minimal.

The partial differentiation with respect to the representation vector $\hat{\mathbf{x}}_k$ provides a necessary condition in analogy to Section 7.4. When choosing the squared error distortion measure according to (7.80-a), the optimal representation vectors $\hat{\mathbf{x}}_k$ correspond to the centers of gravity (*centroids*) of the Voronoi cells.

In general, a mathematical relation between the K code vectors $\hat{\mathbf{x}}_i$ and the L-dimensional PDF cannot be formulated. But there is an elegant iterative design procedure, the *Linde–Buzo–Gray (LBG) algorithm* [Linde et al. 1980], which exists in two alternative versions (A) and (B).

Algorithm (A)

Code book optimization (A) begins with a random start code book and improves this iteratively by means of *training vectors* $\mathbf{x}$, until the decrease of the average distortion is below a certain limit or has reached a minimum. The algorithm consists of the following steps:

Step 0: a) Choose a start code book consisting of K random vectors $\hat{\mathbf{x}}_i$ (or "uniform" lattice vectors) of dimension L.
b) Set $m = 1$.

Step 1: a) Quantize the training sequence $\mathbf{x}(k)$, $k = 1, 2, \ldots, K_T$, with $K_T \gg K$ and compute the average distortion

$$D_m = \frac{1}{K_T} \sum_{k=1}^{K_T} d\big(\mathbf{x}(k), \hat{\mathbf{x}}_{i_{\text{opt}}}\big) . \tag{7.90}$$

b) Terminate the iteration if the relative difference between D_m and its previous value D_{m-1} is sufficiently small

$$\frac{|D_{m-1} - D_m|}{D_m} < \epsilon . \tag{7.91}$$

Step 2: a) Replace the old code vectors $\hat{\mathbf{x}}_i$ by the centroids of those training vectors $\mathbf{x}(k)$ which have been allocated to the old vectors $\hat{\mathbf{x}}_i$ (in generalization of (7.66-d)).
b) $m = m + 1$. Go to step 1.

The algorithm (A) will generally not deliver the best code book but find the local minimum of the quantization noise power. The choice of the start code book determines which local minimum will be achieved. This is why an alternative procedure has been proposed.

Algorithm (B)

The second design algorithm (B) starts with a single representation vector $\hat{\mathbf{x}}_1$ which is the centroid of K_T training vectors $\mathbf{x}$. Within each iteration the code book is splitted and applied to algorithm (A) as start code book. The aim of the special "splitting" procedure is to obtain a better start code book for algorithm (A) in the *last* iteration. However, it will still not guarantee that the global minimum of the quantization noise power is achieved.

Step 1: Set $\kappa = 1$ and determine the center of gravity of all K_T training vectors $\mathbf{x}$.
$\rightarrow \{\hat{\mathbf{x}}_i \mid i = 1\}$

Step 2: Split $\{\hat{\mathbf{x}}_i \mid i = 1, \ldots, \kappa\}$ using a small difference vector Δ which is chosen arbitrarily.
$\rightarrow \{\hat{\mathbf{x}}_i - \Delta\,;\ \hat{\mathbf{x}}_i + \Delta \mid i = 1, \ldots, \kappa\}$

Step 3: a) Run algorithm (A) with $\{\hat{\mathbf{x}}_i - \Delta\,;\ \hat{\mathbf{x}}_i + \Delta \mid i = 1, \ldots, \kappa\}$ as start code book resulting in 2κ optimized representation vectors.
$\rightarrow \{\hat{\mathbf{x}}_i \mid i = 1, \ldots, 2\kappa\}$

b) Set $\kappa = 2\kappa$.

Step 4: a) If $\kappa < K$ return to step 2,

b) otherwise the optimized vector code book is obtained.
$\rightarrow \{\hat{\mathbf{x}}_i \mid i = 1, \ldots, \kappa\}$

However, the split algorithm (B) has a distinct advantage if the code book indices have to be transmitted over a channel with bit errors [Goertz 1999]. For the transmission of any selected index a bit pattern of length w has to be assigned ($K = 2^w$). A single bit error on the transmission link might produce the index of a different code book entry $\hat{\mathbf{x}}_j$, which might have a very large distance to the desired entry $\hat{\mathbf{x}}_i$. Therefore, robust index assignment is needed. We get this robustness if we combine the index assignment with the splitting procedure. The distinction between the first code book which has only two entries $\hat{\mathbf{x}}_1$ and $\hat{\mathbf{x}}_2$ is made by one bit. As soon as we split one vector we add another address bit. Thus we can guarantee that a single bit error will produce an index corresponding to the neighborhood relations.

Examples of code book optimization with the LBG algorithm are presented in Fig. 7.16. For a vector length of $L = 2$, code books with $K = 256$ vectors, respec-

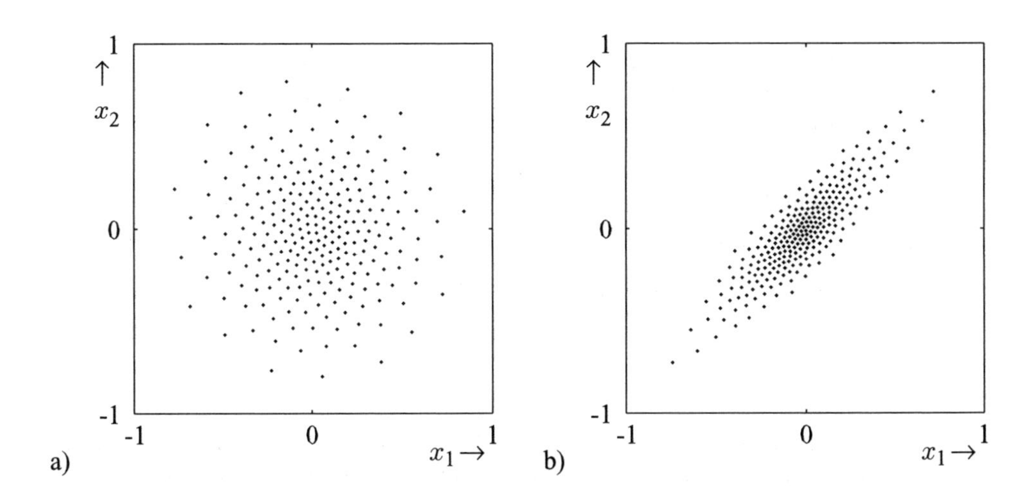

Figure 7.16: Example of a code book design with the LBG algorithm ($L = 2$, $K = 256$ or, respectively, $\overline{w} = 4$)
a) Gaussian source
b) First-order Gauss–Markov source ($a = 0.9$)

tively Voronoi cells, were designed. The first case (Fig. 7.16-a) is based on a non-correlated sequence $x(k)$ with Gaussian distribution, the second on a first-order Gauss–Markov source, i.e., an AR process (see also Section 6.1) with a first-order recursive filter. The feedback coefficient was $a = 0.9$. The code books were trained with 100000 vectors each, with the termination criterion set to a value of 10^{-6} for the difference for D from one iteration to the next according to step 1 b).

It can clearly be seen in Fig. 7.16-a that the cell sizes are adjusted to the probability density function and that in Fig. 7.16-b the correlation leeds to a higher cell density in the neighborhood of the diagonal of the (x_1, x_2)-plane. This is also underlined by the SNR, which results in 20.87 dB for the Gaussian source and in 24.05 dB for the correlated Gauss–Markov source. Evidently, the vector quantizer is able to exploit the correlation for an improvement of the SNR.

For $L\overline{w} = 10$, Table 7.6 shows the SNRs obtained with LBG code books, with variations of the effective word length in the range of

$$0.25 \leq \overline{w} \leq 5 \,. \tag{7.92}$$

Again, the uncorrelated Gaussian source and the correlated first-order Gauss–Markov source ($a = 0.9$) are compared.

The table shows that the vector quantizer exploits the correlation for an improvement of the SNR. Overall, however, for the given dimensions, relatively low SNR values are obtained.

Table 7.6: SNRs for vector quantization with LBG code books, uncorrelated and first-order correlated Gaussian source ($a = 0.9$), 100000 training vectors, $w = L\,\overline{w} = 10$

Vector dimension L	2	5	10	20	40
Effective word length $\overline{w}$	5	2	1	0.5	0.25
Gaussian source	25.93 dB	10.18 dB	4.94 dB	2.41 dB	1.17 dB
Gauss–Markov source	29.56 dB	15.99 dB	11.54 dB	8.68 dB	6.34 dB

For this reason, vector quantizers are generally not used for the direct signal quantization. Rather, they are utilized within model-based codec algorithms, in order to quantize prediction error signals (see Section 8.5.1) or other sets of parameters.

7.6.5 Gain–Shape Vector Quantization

For the vector quantization discussed so far, we assumed that the code book consists of quantizer representation vectors which are representative signal wave forms or representative sets of parameters. In signal quantization it may happen that the same signal shapes can occur with different amplitudes, e.g., if the volume of the speech signal is changed. Hence, the code book would have to contain vectors $\hat{\mathbf{x}}_i$ with the same *shape* and a different *gain*. This might possibly lead to a considerable increase in the size and complexity of the code book.

One possible solution consists of normalizing each input vector by means of a scaling factor derived from $\mathbf{x}$ (e.g., the biggest vector component x_μ). Per input vector, one additional scaling factor must then be transmitted.

Better results can be obtained if each code book vector $\hat{\mathbf{x}}_i$ is adjusted to each respective input vector $\mathbf{x}$ with an individually optimized *gain factor* g_i derived from $\mathbf{x}$ and $\hat{\mathbf{x}}_i$. This method is termed *gain–shape* vector quantization.

Utilizing the squared error distortion measure, the gain factor g_i for each arbitrary but fixed vector $\hat{\mathbf{x}}_i$ can be computed by minimizing the mean square error

$$d_i = d(\mathbf{x}, g_i\hat{\mathbf{x}}_i) = \frac{1}{L}\|\mathbf{x} - g_i\,\hat{\mathbf{x}}_i\|^2 \tag{7.93-a}$$

$$= \frac{1}{L}\sum_{\mu=1}^{L}(x_\mu - g_i\,\hat{x}_{i\mu})^2 \stackrel{!}{=} \min\,, \qquad i = 1, 2, \ldots, K\,. \tag{7.93-b}$$

After setting the partial derivative with respect to the unknown g_i to zero

$$\frac{\partial\, d_i}{\partial\, g_i} = -\frac{2}{L}\sum_{\mu=1}^{L}(x_\mu - g_i\, \hat{x}_{i\mu})\, \hat{x}_{i\mu} \stackrel{!}{=} 0\,, \tag{7.94}$$

the solution

$$g_{i,\text{opt}} = \frac{\sum\limits_{\mu=1}^{L} x_\mu\, \hat{x}_{i\mu}}{\sum\limits_{\mu=1}^{L} \hat{x}_{i\mu}^2} = \frac{\mathbf{x}^T\hat{\mathbf{x}}_i}{\|\hat{\mathbf{x}}_i\|^2} \tag{7.95}$$

results. If we insert the optimum gain factor (7.95) in (7.93-a), the minimum mean square error can be calculated explicitly for each code book vector $\hat{\mathbf{x}}_i$:

$$d_i = \frac{1}{L}\|\mathbf{x} - g_{i,\text{opt}}\, \hat{\mathbf{x}}_i\|^2 \tag{7.96-a}$$

$$= \frac{1}{L}\left\|\mathbf{x} - \frac{\mathbf{x}^T\hat{\mathbf{x}}_i}{\|\hat{\mathbf{x}}_i\|^2}\, \hat{\mathbf{x}}_i\right\|^2 \tag{7.96-b}$$

$$= \frac{1}{L}\left[\|\mathbf{x}\|^2 - \frac{(\mathbf{x}^T\hat{\mathbf{x}}_i)^2}{\|\hat{\mathbf{x}}_i\|^2}\right]. \tag{7.96-c}$$

Hence, we could in a first step evaluate (7.96-c) for each index i and identify the best code book vector $\hat{\mathbf{x}}_i$ which minimizes this expression. As the squared norm of $\mathbf{x}$ is constant, (7.96-c) can be minimized by maximizing the second term:

$$\frac{(\mathbf{x}^T\hat{\mathbf{x}}_i)^2}{\|\hat{\mathbf{x}}_i\|^2} \stackrel{!}{=} \max\,. \tag{7.97}$$

Then, in a second step, the corresponding optimum gain factor $g_{i,\text{opt}}$ has to be calculated according to (7.95) only for the selected best code book vector. Prior to transmission, this gain factor must be quantized.

Bibliography

Brehm, H.; Stammler, W. (1987). Description and Generation of Spherically Invariant Speech-Model Signals, *Signal Processing*, vol. 12, pp. 119–141.

Conway, J. H.; Sloane, N. J. A. (1988). *Sphere packings, lattices and groups*, Springer Verlag, New York.

Erdmann, C. (2004). *Hierarchical Vector Quantization: Theory and Application to Speech Coding*, PhD thesis. *Aachener Beiträge zu digitalen Nachrichtensystemen*, vol. 19, P. Vary (ed.), RWTH Aachen University.

Erdmann, C.; Vary, P. (2004). Performance of Multistage Vector Quantization in Hierarchical Coding, *European Transactions on Telecommunications*, vol. 15, no. 4, pp. 363–372.

Gersho, A.; Gray, R. M. (1992). *Vector Quantization and Signal Compression*, Kluwer Academic, Boston, Dordrecht, London.

Goertz, N. (1999). *Aufwandsarme Qualitätsverbesserungen bei der gestörten Übertragung codierter Sprachsignale*, PhD thesis, University of Kiel. (in German).

Gray, R. M. (1984). Vector Quantization, *IEEE ASSP Magazine*, vol. 1, no. 2, April, pp. 4–29.

Itakura, F.; Saito, S. (1968). Analysis Synthesis Telephony Based on the Maximum Likelihood Method, *Proceedings of the 6th International Congress of Acoustics*, Tokyo, Japan.

ITU-T Rec. G.711 (1972). Pulse Code Modulation (PCM) of Voice Frequencies, International Telecommunication Union (ITU).

Jayant, N. S. (1973). Adaptive Quantization with a One Word Memory, *Bell System Technical Journal*, vol. 52, no. 7, September, pp. 1119–1144.

Jayant, N. S.; Noll, P. (1984). *Digital Coding of Waveforms*, Prentice Hall, Englewood Cliffs, New Jersey.

Linde, Y.; Buzo, A.; Gray, R. M. (1980). An Algorithm for Vector Quantizer Design, *IEEE Transactions on Communications*, vol. 28, no. 1, January, pp. 84–95.

Lloyd, S. P. (1982). Least Squares Quantization in PCM, *IEEE Transactions on Information Theory*, vol. 28, March, pp. 129–136.

Makhoul, J.; Roucos, S.; Gish, H. (1985). Vector Quantization in Speech Coding, *IEEE Proceedings*, vol. 73, no. 11, November, pp. 1551–1588.

Max, J. (1960). Quantizing for Minimum Distortion, *IRE Transactions on Information Theory*, vol. 6, March, pp. 7–12.

Vary, P.; Heute, U.; Hess, W. (1998). *Digitale Sprachsignalverarbeitung*, B. G. Teubner, Stuttgart (in German).

8

Speech Coding

In telephone networks, speech signals are generally limited to the frequency band of 0.3 to 3.4 kHz. This specification of telephone bandwidth dates back to the analog age when frequency division multiplexing by single side band modulation and sub-audio signaling was used to realize, e.g., 10800 telephone channels on coaxial cables in the frequency band 4–60 MHz [ITU-T Rec. G.333 1988]. Even in ISDN networks the upper frequency limit is still 3.4 kHz, whereas the lower limit might be below 300 Hz. The speech signal is sampled at a rate of $f_s = 8$ kHz and quantized non-uniformly at 64 kbit/s (A-law characteristic, $w = 8$ bit/sample, see Section 7.3).

For economical reasons, this bit rate is not available for wireless communication systems such as cordless telephones and cellular radio networks, which allow on average only $\overline{w} = 0.5$–2 bit/sample. Moreover, there is an increasing demand for transmitting *wideband speech* (0.05 to 7 kHz, $f_s = 16$ kHz) in fixed and mobile networks without increasing the bit rates. Thus, the target bit rates for narrowband and wideband speech coding are in the range of $B = \overline{w} \cdot f_s = 4$–32 kbit/s. The majority of coding algorithms for these requirements is based on a model of speech production (see Chapter 2) and relies on masking properties of the auditory system only to a limited extent.

In contrast to this, algorithms for the coding of music signals are based mainly on models of the human auditory system, since appropriate source models do not exist. High audio quality with a bandwidth of about 16 kHz (i.e., $f_s = 32$ kHz) can be obtained with effectively $\overline{w} = 2$–4 bit/sample or $B = \overline{w} \cdot f_s = 64$–192 kbit/s.

Digital Speech Transmission: Enhancement, Coding and Error Concealment
Peter Vary and Rainer Martin

In this chapter, we will develop a comprehensive and unified description of the most important speech coding algorithms. The basic concepts, such as LPC *vocoder*, *differential pulse code modulation* (DPCM), and *code excited linear prediction* (CELP), as well as the relevant advanced coding techniques, will be discussed in detail. Selected codec standards are presented in Appendix A.

8.1 Classification of Speech Coding Algorithms

In public telephone networks, the audio bandwidth is mostly limited according to the IRS or the MIRS narrowband characteristic ((*modified*) *intermediate reference system*, [ITU-T Rec. P.48 1993], [ITU-T Rec. P.830 1996]) as shown in Fig. 8.1-a.

The fundamental frequency of adult speakers, which is below 200 Hz, is actually not transmitted. The average intelligibility of meaningless syllables is only about 91%, but the comprehensibility of words and sentences is considered to be sufficient. Nevertheless, everyone knows from experience that we sometimes need to spell words or names on the phone. This problem no longer exists if wideband speech is transmitted with a frequency range of $0.05\,\text{kHz} \leq f \leq 7.0\,\text{kHz}$ as shown in Fig. 8.1-b [ITU-T Rec. P.341 1998]. In this case, we have to increase the sampling rate to $f_s = 16\,\text{kHz}$ and we need source coding to reduce the bit rate in fixed as well as in wireless systems. Appropriate standards for wideband speech coding are available for ISDN [ITU-T Rec. G.722 1988], as well as for digital cellular networks [3GPP TS 26.190 2001], [ITU-T Rec. G.722.2 2002].

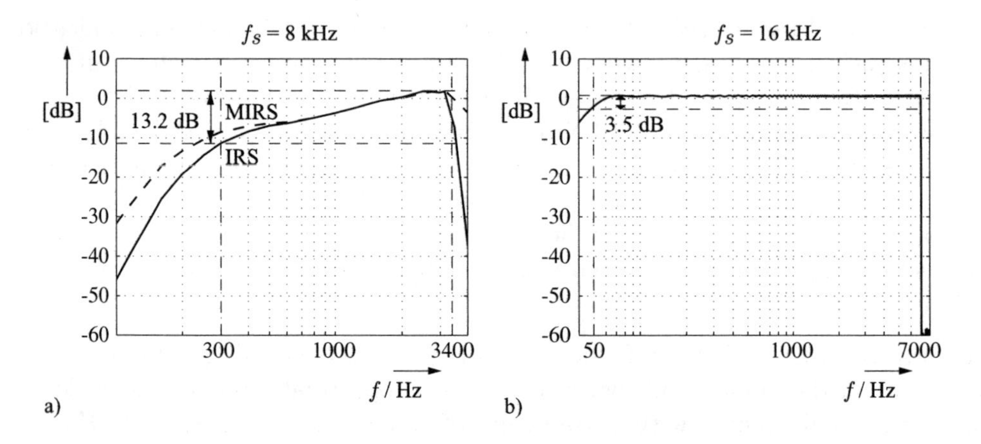

Figure 8.1: a) Magnitude response of the *intermediate reference systems* (IRS, MIRS) $f_s = 8\,\text{kHz}$ [ITU-T Rec. P.48 1993], [ITU-T Rec. P.830 1996]
b) Magnitude response for wideband handsfree telephony terminals $f_s = 16\,\text{kHz}$ [ITU-T Rec. P.341 1998]

Speech coding algorithms have to be optimized with regard to the following requirements:

- high speech quality
- low bit rate
- low complexity
- limited signal delay.

These criteria, some of which contradict each other, must be weighted differently according to the application. They are highly interrelated, as, for example, a constant speech quality may be obtained at a reduced data rate at the expense of increased complexity of the coding algorithms and/or an increased signal delay.

Speech coding algorithms can be subdivided, as depicted in Fig. 8.2, into three categories:

- waveform coding
- parametric coding
- hybrid coding.

In *waveform coding* [Jayant, Noll 1984], bit rate reduction is achieved through (fixed or adaptive) quantization of samples of the speech signal itself or some intermediate signal(s) such as a prediction error signal or subband signals. In comparison to plain quantization of speech samples, better results are achieved by applying a (fixed or adaptive) linear predictive filter (LP filter), which is adjusted according to the correlation properties of the signal. The LP filter can be considered to be a whitening filter (see Section 6.3). The resulting reduction of the signal dynamics in the time and the frequency domain can be quantified by the prediction gain. In time domain coding algorithms, quantization and predictive filtering are generally adaptive processes. At the receiver, the signal is reconstructed by applying the (quantized) residual signal to the synthesis filter. Both the synthesis filter as well as the inverse quantizer can be adjusted by backward or closed-loop adaptation. No side information about the quantizer and the synthesis filter needs to be transmitted. Only the quantized residual signal is transferred via the *signal channel* (Fig. 8.2-a). The effective quantization error can be spectrally shaped within certain limits in order to maximize the subjective speech quality, exploiting the psychoacoustic masking effect (see Section 8.3.3).

For $f_s = 8\,\text{kHz}$ and a target bit rate of $B = 32$ kbit/s, the predominant waveform coding scheme is *adaptive differential pulse code modulation* (ADPCM), which allows a reconstruction of the signal waveform with a signal-to-noise ratio of SNR = 30–35 dB. ADPCM (see Section 8.3.4) is used in digital cordless phones and in circuit multiplication equipment (two voice channels within 64 kbit/s).

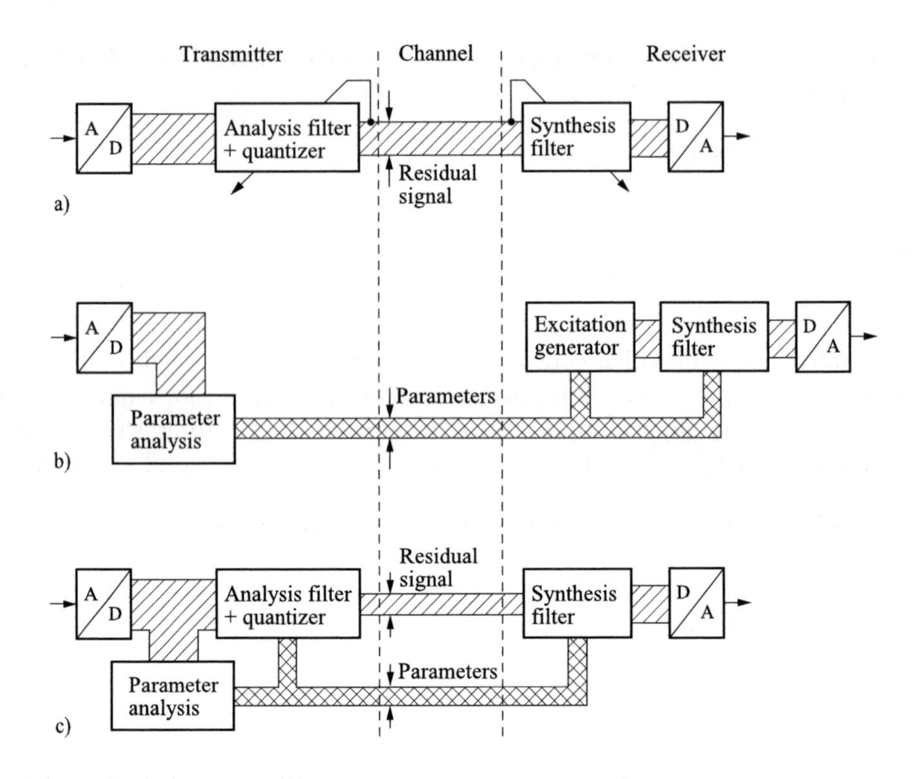

Figure 8.2: A classification of speech coding algorithms
a) Waveform coding
b) Parametric coding
c) Hybrid coding

In contrast to waveform coding, *parametric coders (vocoders)* (Fig. 8.2-b) do not encode the waveform, but a set of model parameters. This implies the realization of a speech production model as discussed in Section 6.1. The time-variant synthesis filter in the receiver can be interpreted as a model of the vocal tract. Its excitation signal (glottis signal) is delivered from a controlled generator. On the transmission side, the parameters of this model, i.e., the coefficients of the filter and the control parameters of the generator, are extracted from the speech samples by analysis procedures and transmitted in quantized form. At a typical bit rate of $B = 2.4\,\text{kbit/s}$ parametric coders produce a clearly intelligible but somewhat synthetic speech.

An interesting compromise between these two concepts is the *hybrid coding* approach (Fig. 8.2-c). As in the vocoder, the parameters of a time-variant synthesis filter are transmitted as side information, whereas the excitation signal is generated similarly to waveform coding by the quantized prediction error signal (residual signal). In consideration of the auditory system, it is possible to quantize the residual signal quite coarsely with respect to amplitude and time resolution. However, large quantization errors do not permit a derivation of the synthesis filter coefficients from the quantized residual signal at the receiver. Subjectively, telephone quality can almost be achieved with 0.75–1.5 bit/sample, whereas the measurable signal-to-noise ratio might only be 10 dB. The hybrid approach is widely used in digital mobile radio systems, e.g., [Kondoz 2004], [Hanzo et al. 2001], [Chu 2003], [Goldberg, Riek 2000], [Sluijter 2005] (see Section 8.5 and Appendix A).

A common feature of the three coding schemes is the time-varying synthesis filter which more or less approximates the vocal tract transfer function. The necessary processing at the transmitter is based on linear prediction (LP). Therefore, the generic term *linear predictive coding* (LPC) is widely used for any codecs of the three classes. The attribute LP is often used to characterize special variants of the hybrid approach such as CELP codecs (*code excited linear prediction*, see Section 8.5.3).

An alternative classification distinguishes between time domain algorithms using linear prediction and frequency domain algorithms based on short-term spectral analysis. The frequency domain algorithms, which rely more on auditory models, require at least 2 bit/sample and are especially suitable for music signals (e.g., [Rao, Hwang 1996], [Vary et al. 1998], [Brandenburg, Bosi 1997], [MPEG-2 1997]). Due to the underlying model of speech production, the predictive time domain algorithms are especially suitable for speech coding with effective bit rates of less than 2 bit/sample.

8.2 Model-Based Predictive Coding

For speech coding with medium to low bit rates, i.e., with effective word lengths of

$$\overline{w} = \frac{B}{f_s} \leq 2 \text{ bit/sample},$$

model-based time domain methods are widely used [Atal 1982]. The international codec standards for telecommunications almost exclusively rely on the simplifying model of speech production derived in Section 2.3. Some properties of the auditory system are exploited as well, especially spectral masking of quantization errors (see Section 2.4).

According to the classification given in Section 8.1, a common feature is an LP-analysis filter at the transmitter and a synthesis filter at the receiver. For public telephone applications the most interesting classes are waveform coding and hybrid coding. Both concepts are based on the model of speech production as illustrated in Fig. 8.3.

The transmitter processes speech samples $x(k)$, which are considered to be produced by an autoregressive process (AR process). It is assumed that these samples

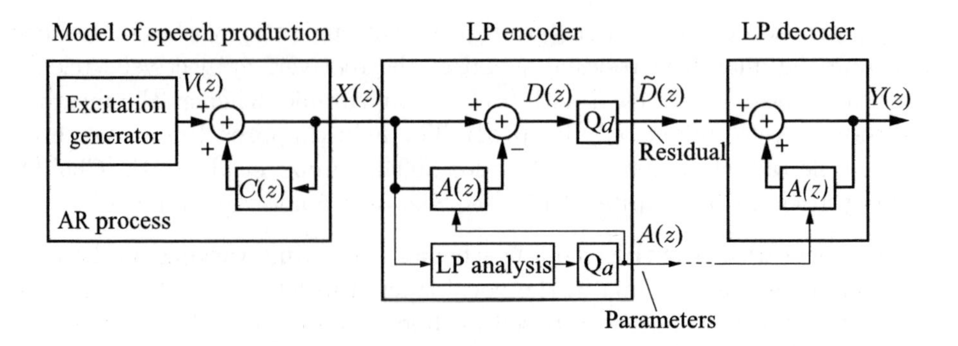

Figure 8.3: Basic principle of model-based predictive coding

have been generated from an excitation sequence $v(k)$ by a purely recursive time-variant filtering process (vocal tract filter).

In Chapter 6, it was shown that, if the model is strictly valid, optimal linear prediction in terms of the minimum mean square prediction error implicitly performs a system identification of the vocal tract filter. Hence, according to Fig. 8.3 the z-transform $D(z)$ of the residual signal $d(k)$ is described by

$$D(z) = \frac{1 - A(z)}{1 - C(z)} \, V(z) \, . \tag{8.1}$$

For perfect system identification, i.e., for $A(z) = C(z)$, the residual signal $d(k)$ is identical to the excitation signal $v(k)$ of the model filter. The analysis filter on the transmission side with the transfer function

$$G(z) = 1 - A(z) \tag{8.2}$$

is the inverse of the vocal tract filter.

The residual signal $d(k)$ is transmitted to the receiver in quantized form and is then used as the excitation signal for the synthesis filter

$$H(z) = \frac{1}{1 - A(z)} \, , \tag{8.3}$$

which corresponds to the vocal tract filter. Thus, the speech signal is resynthesized at the receiving side (decoding) according to the speech production model.

However, we must assume that

1. the utilized model of speech production is only approximately valid;
2. the estimated filter parameters are not accurate $(A(z) \approx C(z))$.

Nevertheless, with this approach the bit rate can be efficiently reduced. Even for inaccurate estimates of the filter parameters the frequency responses of the filters on the transmitting and receiving sides are exactly inverse to each other:

$$G(z) \cdot H(z) = 1 . \tag{8.4}$$

If we do not quantize the residual signal $d(k)$, the output signal and the input signal are identical:

$$y(k) = x(k) . \tag{8.5}$$

There is no signal delay. The key to bit rate reduction is that rather coarse quantization can be applied to the residual signal $d(k)$. If the quantized residual signal

$$\tilde{d}(k) = d(k) + \Delta(k) \tag{8.6}$$

is used to resynthesize the speech signal, the output signal $y(k)$ consists of the original $x(k)$ and a filtered version of the quantization noise:

$$y(k) = x(k) + r(k) . \tag{8.7}$$

The effective quantization noise $r(k)$ has (almost) the same spectral shape as the signal $x(k)$. Therefore, subjective perception is significantly improved by the psychoacoustic effect of masking.

The approach illustrated in Fig. 8.3 covers most of the predictive time domain concepts for speech coding. However, essential differences exist regarding the quantization of the residual signal (fixed or adaptive, scalar or vector quantization, error criterion) and the prediction type (sequential or block adaptation, with or without long-term prediction).

8.3 Differential Waveform Coding

8.3.1 First-Order DPCM

According to the classification given in Section 8.1, the *adaptive differential pulse code modulation*(ADPCM) can be attributed to the class of waveform coders.

The objective of linear prediction is bit rate reduction. The analog signal $x_a(t)$ will be digitized with the sampling rate f_s and a sufficient word length $w_0 \geq 12$. Thus, the initial bit rate is $B_0 = w_0 \cdot f_s$. With linear prediction we generate the residual signal $d(k)$. The subsequent quantization of $d(k)$ with shorter word length $w < w_0$ will lead to the reduced bit rate of $B = w \cdot f_s$.

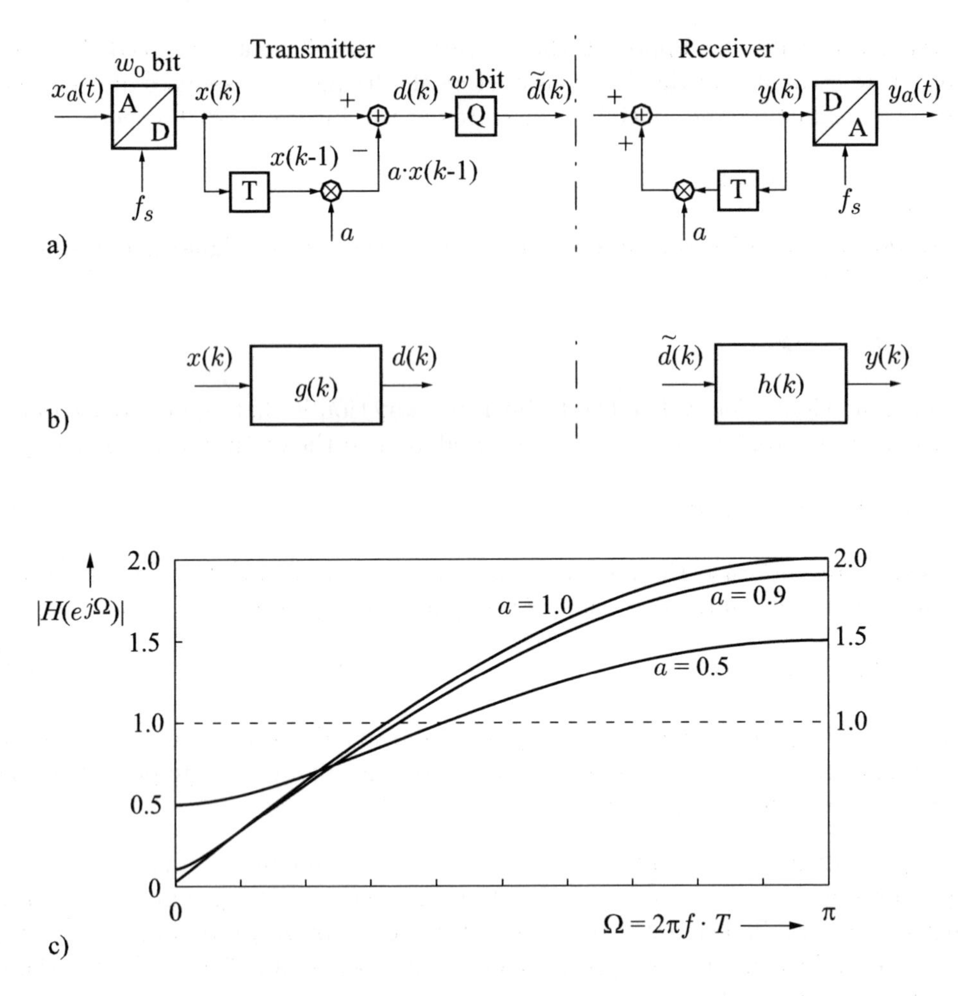

Figure 8.4: Differential pulse code modulation (DPCM) of first order
a) Block diagram
b) Equivalent filters
c) Magnitude response of the analysis filter ($f_s = 1/T$)

The simplest version is a *differential PCM system* (DPCM system) with a first-order predictor as depicted in Fig. 8.4. The current sample $x(k)$ is predicted by weighting the preceding $x(k-1)$ with coefficient a:

$$\hat{x}(k) = a \cdot x(k-1)\,. \tag{8.8}$$

The predictor coefficient a can be optimized as described in Chapter 6 by means of block adaptation (Section 6.3.1) or sequential adaptation using, for example, the LMS algorithm (Section 6.3.2). Here, we will first look at the block adaptive

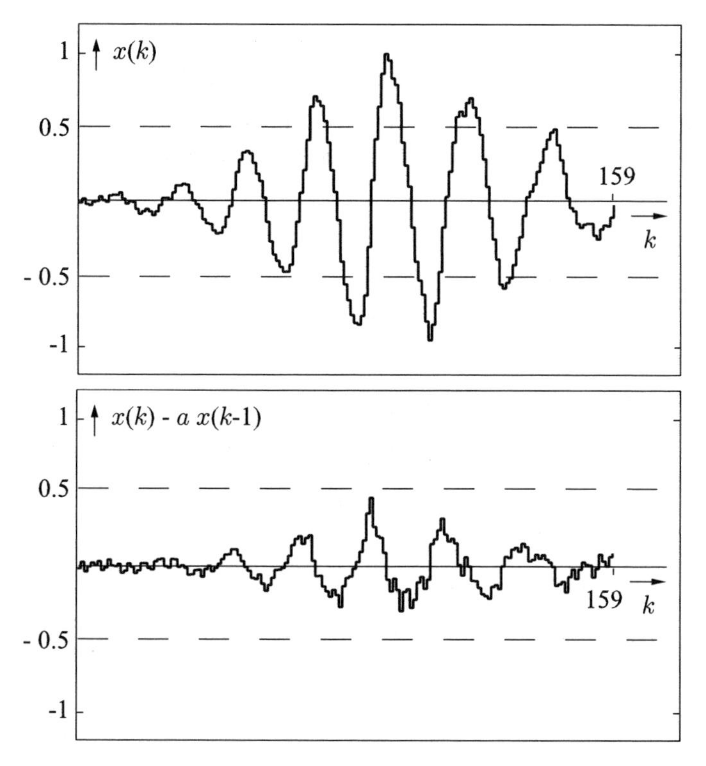

Figure 8.5: Example of block adaptive DPCM
($N = 160$, $f_s = 8$ kHz, $a = 0.9545$)

approach, which is also referred to as *adaptive predictive coding* (APC). The coefficient a, which is optimal in terms of the minimum mean square prediction error (see Eq. (6.31)), is

$$a_{\text{opt}} = \frac{\varphi_{xx}(1)}{\varphi_{xx}(0)} . \tag{8.9}$$

In practice, the auto-correlation values $\varphi_{xx}(i)$ are replaced by short-term estimates. The effect of a DPCM system with a first-order predictor which is optimal for a voiced signal segment of block length $N = 160$ ($N \cdot T = 20\,\text{ms}$) is depicted in Fig. 8.5. Compared to the input signal $x(k)$ the dynamic range of the residual signal $d(k)$ is distinctively reduced. The corresponding prediction gain (see Section 6.2)

$$G_p = \frac{\varphi_{xx}(0)}{\varphi_{dd}(0)} = \frac{\varphi_{xx}^2(0)}{\varphi_{xx}^2(0) - \varphi_{xx}^2(1)} \geq 1 \tag{8.10}$$

can be exploited for shortening the word length to $w < w_0$ under certain circumstances as shown in Section 8.3.3.

First, we will analyze the system performance in the frequency domain for the special case when no additional quantization is applied, i.e., $w = w_0$.

For a constant coefficient a (constant for the duration of the signal block), impulse responses can be determined for the transmitter and the receiver as suggested in Fig. 8.4:

$$g(k) = \begin{cases} 1 & k = 0 \\ -a & k = 1 \\ 0 & \text{otherwise} \end{cases} \tag{8.11-a}$$

$$h(k) = \begin{cases} a^k & k \geq 0 \\ 0 & k < 0 \,. \end{cases} \tag{8.11-b}$$

Through z-transformation of these impulse responses, we obtain the frequency responses

$$G(e^{j\Omega}) = 1 - a \cdot e^{-j\Omega} \tag{8.12-a}$$

or

$$|G(e^{j\Omega})| = \sqrt{1 + a^2 - 2\ a\ \cos\Omega}\,, \tag{8.12-b}$$

and for $|a| < 1$

$$H(e^{j\Omega}) = \sum_{k=0}^{\infty} a^k \cdot e^{-jk\Omega} \tag{8.13-a}$$

$$= \frac{1}{1 - a \cdot e^{-j\Omega}} = \frac{1}{G(e^{j\Omega})}\,. \tag{8.13-b}$$

Since the filter at the receiving side is inverse to the filter on the transmitting side, the output signal $y(k)$ and the input signal $x(k)$ are identical if $d(k)$ is not quantized, i.e., if $\tilde{d}(k) = d(k)$.

The magnitude response of the transmission filter for $a = 1$ is of special interest:

$$|G(e^{j\Omega})| = \sqrt{2 \cdot (1 - \cos\Omega)} \tag{8.14-a}$$

$$= 2 \cdot \left|\sin\left(\frac{\Omega}{2}\right)\right|\,. \tag{8.14-b}$$

The magnitude response outlined in Fig. 8.4-c for $a = 1$ is almost linear at low frequencies. This behavior approximates the magnitude response of the differentiator. The extreme values of the magnitude response are $|1 - a|$ and $|1 + a|$.

The logarithmic prediction gain of the example in Fig. 8.5 is

$$10\lg(G_p) = -10\lg(1 - a_{\text{opt}}^2) \approx 10.5\,\text{dB}\,.$$

If the achievable bit rate reduction increases with the prediction gain according to the 6-dB-per-bit rule, about 1.5 bit/sample could be saved for a gain of 10.5 dB compared to PCM. This interdependency will be studied in Section 8.3.3.

8.3.2 Open-Loop and Closed-Loop Prediction

So far we have discussed the simple DPCM system with a single coefficient a. In Chapter 6 it was shown that the prediction gain can be improved by increasing the filter order n and by adaptation of the predictor coefficients $a_i(k)$ $(i = 1, \dots, n)$ to the time-varying characteristics of speech.

The structure of the DPCM system with an n-th order prediction filter is shown in Fig. 8.6. As the prediction signal

$$\hat{x}(k) = \sum_{i=1}^{n} a_i(k) \cdot x(k-i) \tag{8.15}$$

is produced from preceding input samples $x(k-i)$, this structure is called *forward prediction* or *open-loop prediction.*

The time-variant coefficients of the transmitter must be known at the receiver. However, the transmission of the prediction coefficients requires a considerable part of the bit rate saved by quantization of $d(k)$.

Therefore, it would be attractive if the vector $\mathbf{a}(k) = \left(a_1(k), a_2(k), \dots, a_n(k)\right)^T$ of the predictor coefficients could be recalculated at the receiver without transmitting any *side information.* This can be achieved with the modified predictor structure of Fig. 8.7 (see Section 8.3.4), if the residual signal $d(k)$ is quantized with sufficient accuracy. The modification compared to Fig. 8.6 is that the estimated signal $\hat{x}(k)$ at the transmitter is derived in the same way as at the receiver. This variation is called *backward prediction* or *closed-loop prediction*, as the predictor and the quantizers are located within a loop.

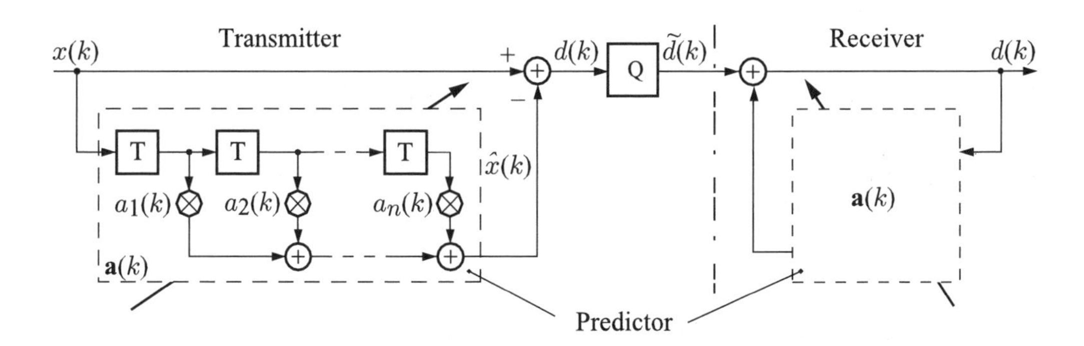

Figure 8.6: DPCM system with n-th order adaptive forward predictor (*open-loop prediction*)

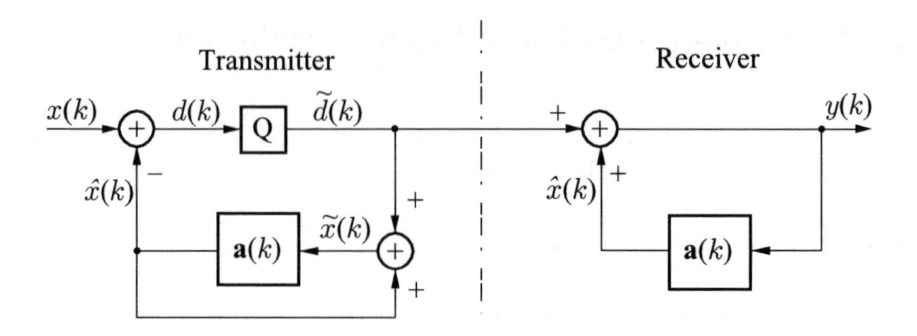

Figure 8.7: DPCM system with adaptive backward predictor (*closed-loop prediction*)

If the residual signal $d(k)$ is not quantized, then the open- and closed-loop structures produce exactly the same estimated signal $\hat{x}(k)$. In this case, we get

$$\tilde{d}(k) = d(k) = x(k) - \hat{x}(k) \tag{8.16-a}$$

and

$$\tilde{x}(k) = \tilde{d}(k) + \hat{x}(k) = x(k)\,. \tag{8.16-b}$$

Hence, the estimated signal for closed-loop prediction

$$\hat{x}(k) = \sum_{i=1}^{n} a_i(k) \cdot \tilde{x}(k-i) = \sum_{i=1}^{n} a_i(k) \cdot x(k-i) \tag{8.16-c}$$

is identical to the result produced by open-loop prediction (8.15) (see Fig. 8.6). Furthermore, with closed-loop prediction the decoder output signal $y(k)$ is identical to the intermediate signal $\tilde{x}(k)$ even if the residual signal $\tilde{d}(k)$ is quantized. However, if the residual signal $d(k)$ is quantized, open- and closed-loop prediction produce different signals $\hat{x}(k)$. Thus, in both cases quantization affects the output signal $y(k)$ differently. This will be discussed in the next section.

8.3.3 Quantization of the Residual Signal

In this section, we will show how the quantization of the residual signal affects the output signal $y(k)$ and how the prediction gain can be exploited for bit rate reduction. For reasons of simplicity, we assume error-free transmission.

8.3.3.1 Quantization with Open-Loop Prediction

The quantization of the residual signal $d(k)$ will be described by additive quantization noise $\Delta(k)$. Assuming uniform quantization with mathematical rounding, the quantization error can be expressed (see Section 7.2) as uniformly distributed noise in the range

$$-\frac{\Delta d}{2} < \Delta(k) \leq +\frac{\Delta d}{2}, \tag{8.17-a}$$

with the power

$$\sigma_\Delta^2 = \frac{(\Delta d)^2}{12}, \tag{8.17-b}$$

and a constant noise power spectral density

$$\Phi_{\Delta\Delta}(e^{j\Omega}) = \frac{(\Delta d)^2}{12}. \tag{8.17-c}$$

Here, Δd denotes the quantizer's stepsize.

According to Fig. 8.8-a, the quantized residual signal $\tilde{d}(k)$ consists of the two components $d(k)$ and $\Delta(k)$. Due to the linearity of the receiving filter, the output signal $y(k)$ consists of the sum of the filtered versions of these two components. Since the transmitting and the receiving filters are inverse to each other, we get the original signal $x(k)$ and a filtered version $r(k)$ of the noise $\Delta(k)$ according to

$$y(k) = x(k) + r(k). \tag{8.18}$$

The *reconstruction error* $r(k)$ is thus a spectrally shaped version of the white quantization noise $\Delta(k)$. As shown in Section 6.1, the optimization of the analysis filter

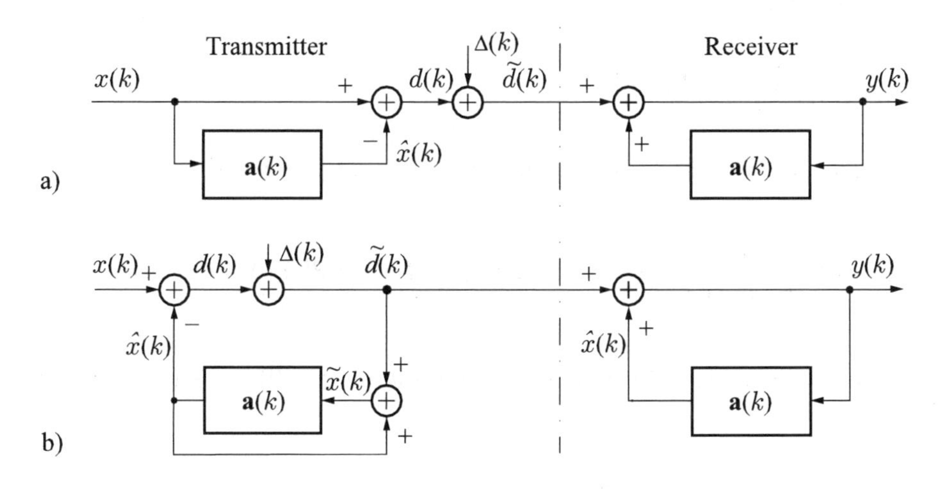

Figure 8.8: DPCM with quantization of the residual signal
a) Open-loop prediction (forward prediction)
b) Closed-loop prediction (backward prediction)

implies a system identification of the vocal tract. With adaptive linear prediction, the frequency response of the synthesis filter therefore approximates the instantaneous frequency response of the vocal tract filter. For this reason, the spectrum of the reconstruction error follows the spectral envelope of the speech signal. From a psychoacoustic point of view, this is advantageous as the quantization error is masked to a certain extent by the speech signal itself.

In this context, the significance of the prediction gain for the achievable signal-to-noise ratio and the required bit rate is of particular interest. In order to clarify this issue, we will look at the system in the frequency domain. We assume that the prediction is perfect in such a way that a spectrally flat residual signal $d(k)$ with constant power spectral density

$$\Phi_{dd}(e^{j\Omega}) = \text{const.} = \varphi_{dd}(0) \tag{8.19}$$

results. This assumption holds, as shown in Section 6.1, for unvoiced segments. With respect to the spectral envelope, and especially when utilizing a long-term predictor (see Section 6.4, Fig. 6.17), this assumption applies to voiced segments as well.

Furthermore, the relation between the power spectral densities of the residual signal and the input signal is given by

$$\Phi_{dd}(e^{j\Omega}) = \Phi_{xx}(e^{j\Omega}) \cdot |1 - A(e^{j\Omega})|^2 . \tag{8.20-a}$$

With (8.19) we get

$$\frac{\varphi_{dd}(0)}{|1 - A(e^{j\Omega})|^2} = \Phi_{xx}(e^{j\Omega}) . \tag{8.20-b}$$

The integration over the frequency interval $-\pi \leq \Omega \leq +\pi$ with

$$\varphi_{dd}(0) \, \frac{1}{2\pi} \int_{-\pi}^{\pi} \frac{1}{|1 - A(e^{j\Omega})|^2} \, \mathrm{d}\Omega = \frac{1}{2\pi} \int_{-\pi}^{\pi} \Phi_{xx}(e^{j\Omega}) \, \mathrm{d}\Omega = \varphi_{xx}(0) \tag{8.20-c}$$

yields the power $\varphi_{xx}(0)$ of signal $x(k)$. Dividing by $\varphi_{dd}(0)$ leads to the prediction gain

$$\frac{\varphi_{xx}(0)}{\varphi_{dd}(0)} = \frac{1}{2\pi} \int_{-\pi}^{\pi} \frac{1}{|1 - A(e^{j\Omega})|^2} \, \mathrm{d}\Omega = G_p . \tag{8.21}$$

Equation (8.21) integrates the squared magnitude response

$$|H(e^{j\Omega})|^2 = \frac{1}{|1 - A(e^{j\Omega})|^2} \tag{8.22}$$

of the synthesis filter.

The signal-to-noise ratio of the uniform quantizer is given by

$$\left(\frac{S}{N}\right)_{\tilde{d}} = \frac{\varphi_{dd}(0)}{\frac{(\Delta d)^2}{12}}. \tag{8.23}$$

The noise power $\varphi_{rr}(0)$ at the receiver output results from integration over the respective noise power spectral density using (8.17-c), (8.21), and (8.22)

$$\varphi_{rr}(0) = \frac{1}{2\pi} \int_{-\pi}^{\pi} \Phi_{\Delta\Delta}(e^{j\Omega}) \cdot |H(e^{j\Omega})|^2 \, d\Omega = \frac{(\Delta d)^2}{12} \cdot G_p. \tag{8.24}$$

Hence, with (8.21) and (8.23), the signal-to-noise ratio $\left(\frac{S}{N}\right)_y$ of the output signal is given by

$$\left(\frac{S}{N}\right)_y = \frac{\varphi_{xx}(0)}{\varphi_{rr}(0)} \tag{8.25-a}$$

$$= \frac{\varphi_{xx}(0)}{\frac{(\Delta d)^2}{12} \cdot G_p} = \frac{G_p \cdot \varphi_{dd}(0)}{\frac{(\Delta d)^2}{12} \cdot G_p} \tag{8.25-b}$$

$$= \frac{\varphi_{dd}(0)}{\frac{(\Delta d)^2}{12}} = \left(\frac{S}{N}\right)_{\tilde{d}}. \tag{8.25-c}$$

As a result, the signal-to-noise ratio is not improved by open-loop prediction. The prediction gain *cannot* be used to improve the *objective* signal-to-noise ratio. Due to the psychoacoustic masking effect, however, the *subjective* speech quality is significantly improved by spectral shaping of the quantization error, which can also be exploited for bit rate reduction.

8.3.3.2 Quantization with Closed-Loop Prediction

The analysis of the structure in Fig. 8.8-b is carried out in the time domain. Assuming an error-free transmission, we get

$$y(k) = \tilde{x}(k) = \hat{x}(k) + \tilde{d}(k) \tag{8.26-a}$$

$$= \hat{x}(k) + d(k) + \Delta(k) \tag{8.26-b}$$

$$= x(k) + \Delta(k). \tag{8.26-c}$$

In contrast to open-loop prediction, the reconstruction error $r(k)$ equals the spectrally white quantization noise $\Delta(k)$. Consequently, $r(k)$ appears as a spectrally white noise signal at the output of the receiver.

In this case, the prediction gain can be used to improve the signal-to-noise ratio

$$\left(\frac{S}{N}\right)_y = \frac{\varphi_{xx}(0)}{\varphi_{\Delta\Delta}(0)} \tag{8.27-a}$$

$$= \frac{\varphi_{xx}(0)}{\varphi_{dd}(0)} \cdot \frac{\varphi_{dd}(0)}{\varphi_{\Delta\Delta}(0)} \tag{8.27-b}$$

$$= G_p \cdot \left(\frac{S}{N}\right)_{\tilde{d}} . \tag{8.27-c}$$

Compared to the quantizer output, the signal-to-noise ratio SNR_y of the receiver output is increased by the prediction gain G_p (see (8.21))

$$\mathrm{SNR}_y = 10 \lg \left(\frac{S}{N}\right)_y = 10 \lg \left(\frac{S}{N}\right)_{\tilde{d}} + 10 \lg G_p . \tag{8.27-d}$$

This fact can be exploited for objective quality improvement, but also for bit rate reduction. In view of the 6-dB-per-bit rule (see Section 7.2), this implies that closed-loop prediction provides a $10 \lg G_p / 6$ bit advantage over PCM. Two views of the word length gain are equivalent:

1. SNR improvement for $w = w_0$:
   If we compare plain PCM and closed-loop DPCM both using the same word length w, the quantizer of the DPCM system has a lower peak-to-peak load by the residual signal ($\sigma_d < \sigma_x$). Therefore, we can adapt the quantization stepsize ($\Delta d < \Delta x$) and achieve a quantization noise power which is reduced according to the prediction gain.

2. Constant SNR but $w < w_0$:
   In practice, a certain target performance of SNR_y is required. Because of the relation

   $$\mathrm{SNR}_y = \mathrm{SNR}_{\tilde{d}} + 10 \lg G_p \tag{8.27-e}$$

   we design the quantizer for a target signal-to-noise ratio $\mathrm{SNR}_{\tilde{d}}$, which can be reduced by $\Delta\mathrm{SNR} = 10 \lg G_p$ compared to open-loop prediction. In accordance with (7.23-b) this implies a possible word length reduction by

   $$\Delta w_p = \frac{\Delta\mathrm{SNR}}{20 \lg 2} = \frac{10 \lg G_p}{20 \lg 2} = \frac{1}{2} \,\mathrm{ld}\, G_p . \tag{8.27-f}$$

Finally, it should be noted that white quantization noise is in principle less pleasant than the spectrally shaped noise. However, as the power of the white noise in closed-loop prediction may be significantly lower than that of the colored noise in open-loop prediction, a comparable subjective performance is achieved. Nevertheless, the objective performance of closed-loop prediction is better.

8.3.3.3 Spectral Shaping of the Quantization Error

As for the psychoacoustic aspects, open-loop and closed-loop prediction represent two extreme cases with respect to the reconstruction error. In the first case, the noise spectrum follows the spectral envelope of the speech signal, while the signal-to-noise ratio is not improved. The logarithmic distance between the power spectral densities of the speech signal and the reconstruction error is independent of the frequency. In the second case, a white quantization noise remains; however, the signal-to-noise power ratio is increased by the prediction gain. In order to exploit to some extent both the psychoacoustic masking effect and the prediction gain, a compromise between these two extremes is desirable. The signal-to-noise ratio should be improved to a certain extent. At the same time, the noise spectrum should be matched to the spectrum of the desired signal in such a way that, compared to the open-loop prediction, an increased distance results in the range of the formant frequencies, whereas a smaller distance is permissible in the "spectral valleys". This objective can be achieved by the technique of *noise shaping* [Schroeder et al. 1979].

The starting point for the derivation of such a structure is the closed-loop DPCM system according to Fig. 8.9-a. To highlight the functional relations between the input, the output and the quantization noise of the DPCM system, we prefer to use z-transform representations instead of power spectral densities. The z-transform of a finite segment of the quantization noise $\Delta(k)$ is denoted by $\Delta(z)$.

For the z-transform $\tilde{D}(z)$ according to Fig. 8.9-a we obtain

$$\tilde{D}(z) = X(z)\big(1 - A(z)\big) + \Delta(z)\big(1 - A(z)\big)\,. \tag{8.28-a}$$

Figure 8.9-b shows an alternative but equivalent structure, which is also described by (8.28-a).

The two noise components $\Delta(z)$ and $-\Delta(z) \cdot A(z)$ are added to the signal component

$$D_0(z) = X(z)\big(1 - A(z)\big)\,.$$

The first part results from quantizing $d(k)$, while the second one is obtained by filtering the difference between the input and the output of the quantizer. The behavior of closed-loop prediction (Fig. 8.9-a) can therefore be achieved by open-loop prediction and feedback of the filtered quantization error (Fig. 8.9-b). However,

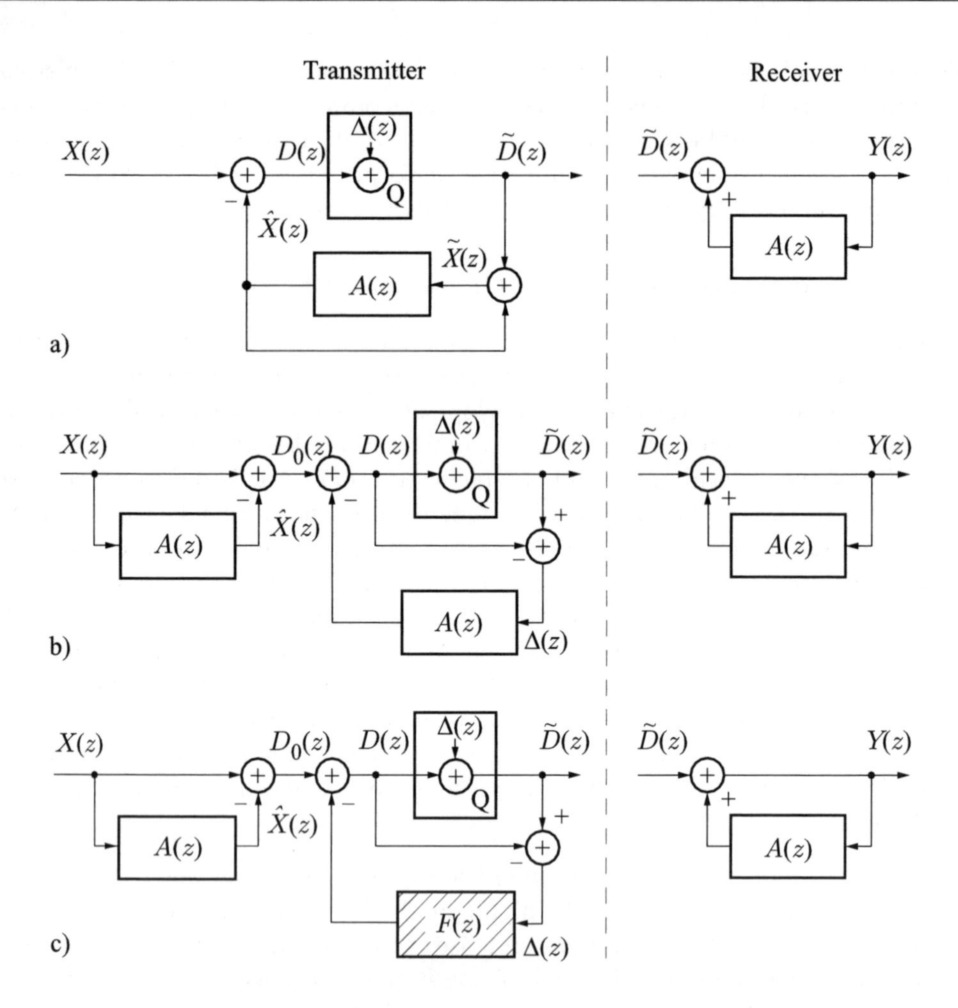

Figure 8.9: Spectral shaping of the quantization noise (*noise shaping*)
a) Closed-loop DPCM with quantization in the loop
b) Alternative but equivalent structure
c) Generalization of b)

this alternative structure has the advantage that the reconstruction error can be explicitly influenced to a certain extent. As shown in Fig. 8.9-c, a filter function $F(z)$ can be utilized for the noise feedback:

$$\tilde{D}(z) = X(z)\big(1 - A(z)\big) + \Delta(z)\big(1 - F(z)\big) . \tag{8.28-b}$$

Finally, we get for the receiver output

$$Y(z) = X(z) + \Delta(z)\,\frac{1 - F(z)}{1 - A(z)} . \tag{8.29}$$

For the choice of $F(z)$ it has to be considered that delay-less loops cannot be implemented.

A suitable function $F(z)$ can be derived from $A(z)$. According to [Schroeder et al. 1979], we choose

$$F(z) = A(z/\gamma) \quad \text{with} \quad 0 \leq \gamma \leq 1\,. \tag{8.30}$$

With the parameter γ the adaptation of the noise spectrum to the spectrum of the speech signal can be performed with respect to the desired compromise. The effect of this choice can be explained by the zeros z_{0i} of the LP-analysis filter. The product form

$$1 - A(z) = \frac{1}{z^n} \prod_{i=1}^{n} (z - z_{0i}) \tag{8.31-a}$$

leads to

$$1 - F(z) = 1 - A(z/\gamma) = \frac{1}{z^n} \prod_{i=1}^{n} (z - \gamma \cdot z_{0i})\,. \tag{8.31-b}$$

If a positive factor $\gamma < 1$ is chosen, the magnitude $|z_{0i}|$ of each zero is reduced, while the angles φ_{0i} are maintained:

$$\tilde{z}_{0i} = \gamma \cdot z_{0i} = \gamma \cdot |z_{0i}| \cdot e^{j\varphi_{0i}}\,. \tag{8.32}$$

The extrema of the resulting frequency response are less distinct, since the zeros which are inside the unit circle (see Chapter 6) are moved towards the origin of the z-plane. The special case of closed-loop prediction is covered with $\gamma = 1$, i.e., $F(z) = A(z)$, as shown in (8.31-b). The second extreme case of open-loop prediction is obtained for $\gamma = 0$, i.e., $1 - F(z) = \frac{1}{z^n} \cdot z^n = 1\,.$

The effective noise shaping is illustrated by an example in Fig. 8.10. Represented are the squared magnitude response of the synthesis filter for a finite signal segment

$$|H(e^{j\Omega})|^2 = \frac{1}{|1 - A(e^{j\Omega})|^2} \tag{8.33}$$

and the power spectral density of the reconstruction error

$$\Phi_{rr}(e^{j\Omega}) = \Phi_{\Delta\Delta}(e^{j\Omega}) \cdot \left| \frac{1 - A\left(\frac{1}{\gamma}\, e^{j\Omega}\right)}{1 - A(e^{j\Omega})} \right|^2 \tag{8.34}$$

for three values of γ. According to (8.17-c) with

$$\Phi_{\Delta\Delta}(e^{j\Omega}) = \frac{(\Delta d)^2}{12}\,, \tag{8.35}$$

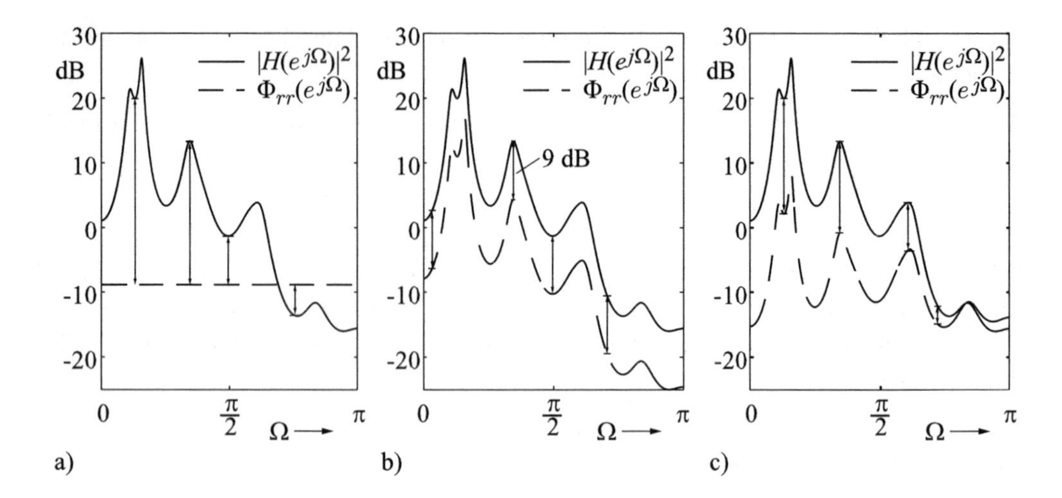

Figure 8.10: Noise shaping for the example of a vowel
a) Closed-loop prediction, $\gamma = 1$
b) Open-loop prediction, $\gamma = 0$
c) Noise shaping, $\gamma = 0.7$

we assumed a constant noise power spectral density due to uniform quantization, and for the purpose of illustration we exemplarily assume a quantizer with a signal-to-noise ratio of the quantized residual signal $\tilde{d}(k)$ of

$$\mathrm{SNR}_{\tilde{d}} = 10 \lg \left(\frac{\varphi_{dd}(0)}{\frac{(\Delta d)^2}{12}} \right) = 9\,\mathrm{dB} \quad \text{and} \quad \varphi_{dd}(0) = 1\,. \tag{8.36}$$

For $\gamma = 0$ (open-loop prediction, Fig. 8.10-b), we obtain $\mathrm{SNR}_y = \mathrm{SNR}_{\tilde{d}}$ and the noise spectrum follows the magnitude response with a constant distance given by the uniform quantization (in this case 9 dB).

For $\gamma = 1$ (closed-loop prediction, Fig. 8.10-a), and the same quantizer stepsize Δd, a constant white noise spectrum results. Compared to the open-loop prediction the signal-to-noise ratio is increased by the logarithmic prediction gain $10 \lg G_p$ (see (8.21)).

The DPCM system with noise shaping (Fig. 8.10-c) shows that in the lower formant frequency range the spectral signal-to-noise ratio has markedly increased in comparison to open-loop prediction. Due to the psychoacoustic masking effect the noise shaping provides the best *subjective* speech quality. Compared to closed-loop prediction, however, the objectively measured signal-to-noise ratio deteriorates.

The effect on the speech signal is depicted for a second example in Fig. 8.11. A fixed, $n = 8$-th order predictor was utilized in the structure according to Fig. 8.9-c. The

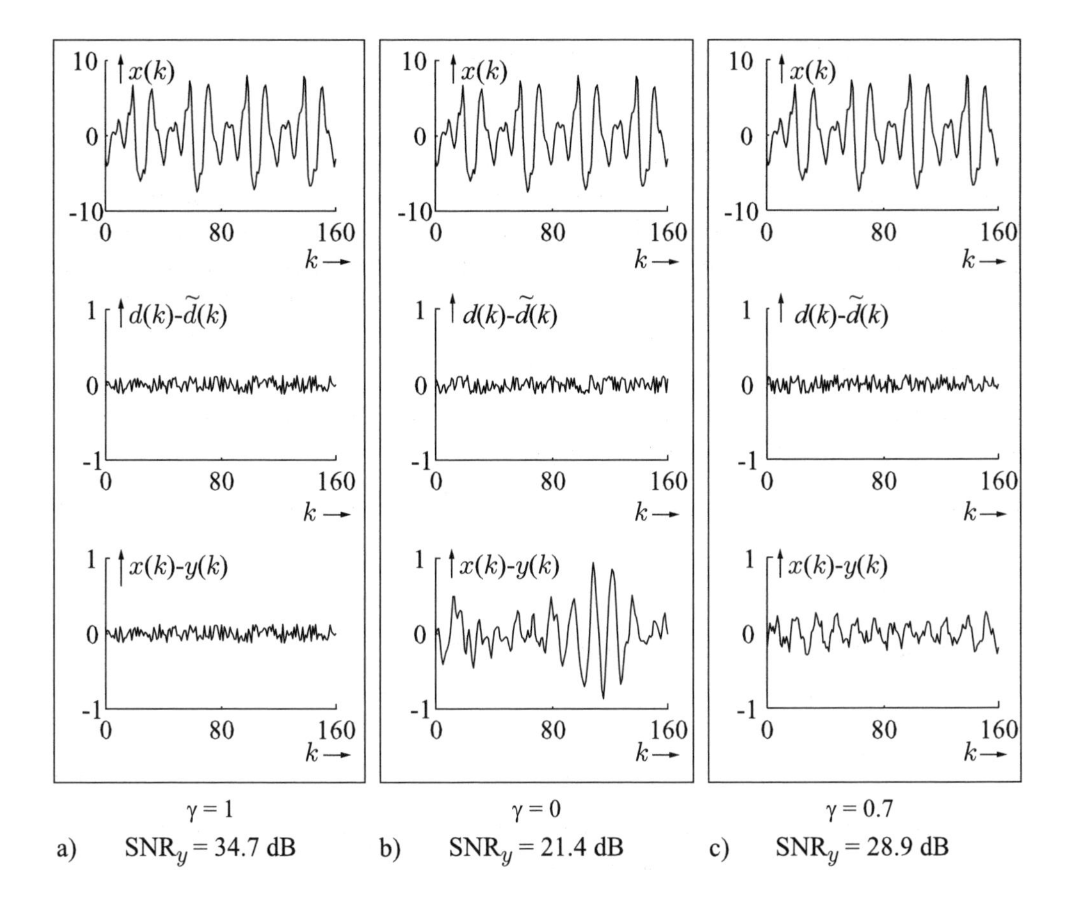

Figure 8.11: Example of differential coding according to Fig. 8.9-c
(predictor: $n = 8$, quantizer: $w = 5$, voiced speech)
a) Closed-loop prediction, $\gamma = 1$
b) Open-loop prediction, $\gamma = 0$
c) Noise shaping, $\gamma = 0.7$

quantizer with a word length of $w = 5$ was adjusted to the reduced dynamic range of the residual signal $d(k)$.

From Fig. 8.11 we can infer that the best objective match between the original signal $x(k)$ and output signal $y(k)$ is achieved with closed-loop prediction. The difference $x(k) - y(k)$ is the same as $d(k) - \tilde{d}(k)$ (see also (8.26-c)). The largest reconstruction error energy $\sum_k \big(x(k) - y(k)\big)^2$, i.e., the lowest SNR_y, results from open-loop prediction. However, the best auditory impression is obtained by quantization with noise shaping. Here, the energy of the reconstruction error is higher than with closed-loop prediction but smaller than with open-loop prediction. The parameter γ controls the noise shaping intensity and ranges typically in $0.6 \leq \gamma \leq 0.9$.

8.3.4 Adaptive Differential Pulse Code Modulation

Discussions in this section will refer to a system called *adaptive differential pulse code modulation* (ADPCM) with adaptive prediction and adaptive quantization. When applying the *least-mean-square* (LMS) algorithm (see Section 6.3.2) and adaptive quantization backwards (AQB, see Section 7.5), the adaptation of the predictor and the quantizer can be carried out recursively. Apart from the quantized residual signal $\tilde{d}(k)$, no additional information must be transmitted to the receiver. Furthermore, better results in terms of bit rate reduction and speech quality compared to DPCM are obtained.

We will look at the ADPCM structure according to Fig. 8.12 with closed-loop prediction and quantization within the loop.

For the derivation of the LMS algorithm we will in a first step assume that $d(k)$ is not quantized:

$$\tilde{d}(k) = d(k) . \tag{8.37}$$

We consider the *instantaneous power*

$$\hat{\sigma}_d^2(k) = d^2(k) \tag{8.38-a}$$

$$= \left(x(k) - \mathbf{a}^T(k)\, \tilde{\mathbf{x}}(k-1)\right)^2 \tag{8.38-b}$$

and calculate the instantaneous gradient $\hat{\nabla}(k)$ with respect to the coefficient vector $\mathbf{a}(k)$ in analogy to Section 6.3.2, (6.79-b), and (6.80-b):

$$\hat{\nabla} = \frac{\partial d^2(k)}{\partial \mathbf{a}(k)} = -2 \underbrace{\left(x(k) - \mathbf{a}^T(k)\, \tilde{\mathbf{x}}(k-1)\right)}_{d(k)} \tilde{\mathbf{x}}(k-1) \tag{8.39-a}$$

$$= -2\; d(k)\; \tilde{\mathbf{x}}(k-1) , \tag{8.39-b}$$

with the vector notation

$$\tilde{\mathbf{x}}(k-1) = \left(\tilde{x}(k-1), \tilde{x}(k-2), \ldots, \tilde{x}(k-n)\right)^T \tag{8.39-c}$$

$$\mathbf{a}(k) = \left(a_1(k), a_2(k), \ldots, a_n(k)\right)^T . \tag{8.39-d}$$

The new set of prediction coefficients $\mathbf{a}(k+1)$ is computed by updating the present vector $\mathbf{a}(k)$ in a direction opposite to that of the instantaneous gradient $\hat{\nabla}$. The corresponding simple recursive algorithm

$$\mathbf{a}(k+1) = \mathbf{a}(k) + 2\;\vartheta\; d(k)\; \tilde{\mathbf{x}}(k-1) \tag{8.40-a}$$

or, accordingly for the individual coefficient with index $i = 1, \ldots, n$,

$$a_i(k+1) = a_i(k) + 2\;\vartheta\; d(k)\; \tilde{x}(k-i) \tag{8.40-b}$$

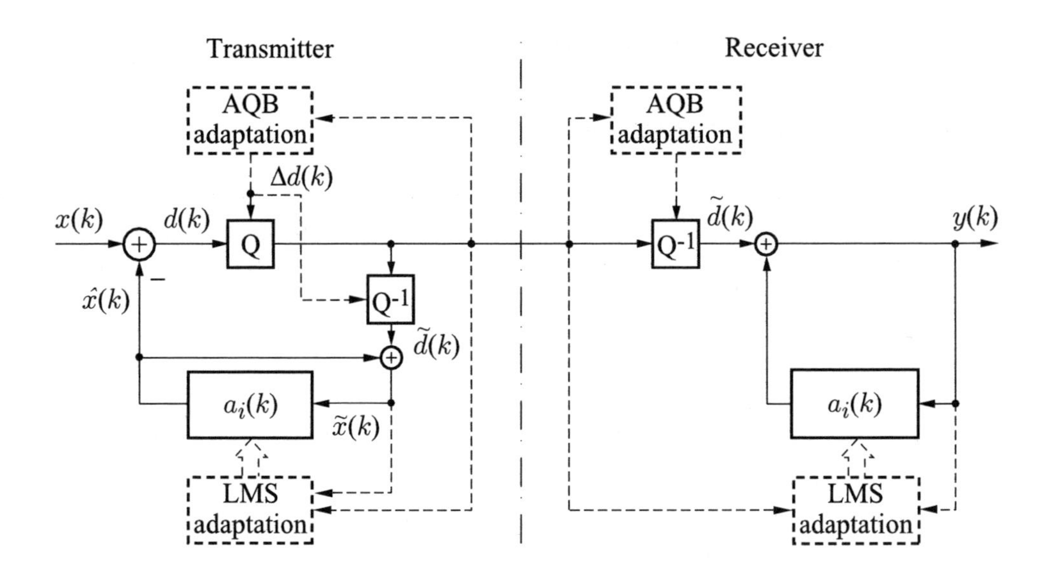

Figure 8.12: ADPCM with sequential adaptation

represents the LMS algorithm (see Section 6.3.2). The effective stepsize $2\,\vartheta$ controls the rate of adaptation and the stability (see (6.81-b)).

In the case of error-free transmission, the required information for the adaptation is also available at the receiver. Due to the structural correspondence,

$$\tilde{x}(k) = y(k) \tag{8.41}$$

is also valid if the residual signal $d(k)$ is quantized, i.e., we replace $d(k)$ by $\tilde{d}(k)$ at the transmitting side in (8.40-a). The predictor at the receiving side can be adjusted synchronously to the filter on the transmitting side, if in (8.40-a) we replace $d(k)$ by $\tilde{d}(k)$ and $\tilde{x}(k)$ by $y(k)$.

ADPCM with a word length of $w = 4$ provides almost the same speech quality as scalar logarithmic quantization of $x(k)$ with $w = 8$ (A-law PCM).

The speech codec used in digital cordless phones according to the *DECT standard* (see Appendix A.2, ITU-T/G.726) is based on the ADPCM system illustrated in Fig. 8.12. In this concept a prediction filter with poles and zeros is applied. The adaptation of the filter is carried out using the so-called *algebraic sign–LMS* algorithm for implementational simplicity. Each adjustment is proportional to the negative of an estimate of the gradient according to

$$a_i(k+1) = a_i(k) + 2\,\vartheta\ \mathrm{sign}\{d(k)\}\ \mathrm{sign}\{\tilde{x}(k-i)\}\,. \tag{8.42}$$

The adaptation does not require multiplications, which is advantageous for implementation in *application-specific integrated circuits* (ASICs).

8.4 Parametric Coding

8.4.1 Vocoder Structures

The second class of speech coding algorithms comprises the parametric coders, generally called *vocoders* (*vo*ice *coders*). Vocoders provide the greatest reduction of bit rate: effectively 0.1–0.5 bit/sample can be achieved. Although the speech sounds mostly synthetic, a sufficient intelligibility is obtained. "Speaker transparency" is only given to a limited degree. In contrast to waveform coding, an exact signal reproduction is not the main objective; in particular, phase information is not considered. Therefore, the perceived quality of the synthesized speech signal cannot be quantified by objective distortion measures such as the signal-to-noise ratio.

A common characteristic of the different time domain vocoders is a signal analysis procedure for extracting perceptually significant parameters from the speech signal. Using these parameters, which contain information about

- the instantaneous frequency response of the vocal tract filter with impulse response $h_0(k)$ and
- the excitation signal $v(k)$,

an output signal is synthesized in the decoder. In the synthesis procedure the excitation signal is built frame by frame, either by a noise signal or by a periodic impulse sequence, depending on whether the speech frame is classified as voiced or unvoiced. Some more advanced schemes provide also mixed voiced/unvoiced excitation modes. The fundamental differences of the vocoder variants lie in the structure of the synthesis filter and in the analysis of the filter parameters. Figure 8.13 shows a generic description of the synthesis procedure of a vocoder, i.e., the decoder at the receiving side.

The basic structure corresponds to the discrete time model of speech production according to Section 6.1. The time-variant synthesis filter is excited by $v(k)$ with either a noise-like or a periodic structure.

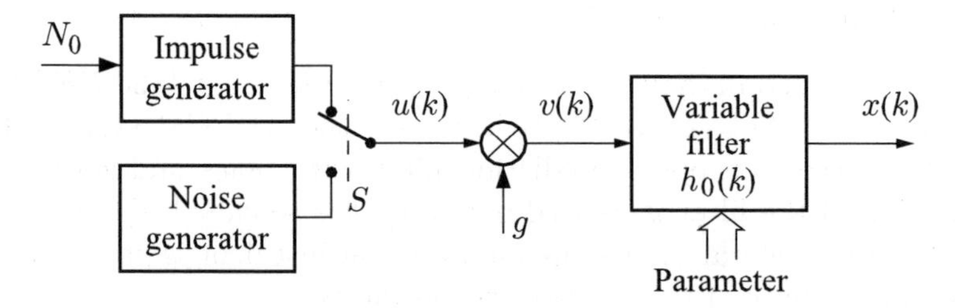

Figure 8.13: Parametric decoding of a vocoder
S : voiced/unvoiced switch
N_0: pitch period
g : gain

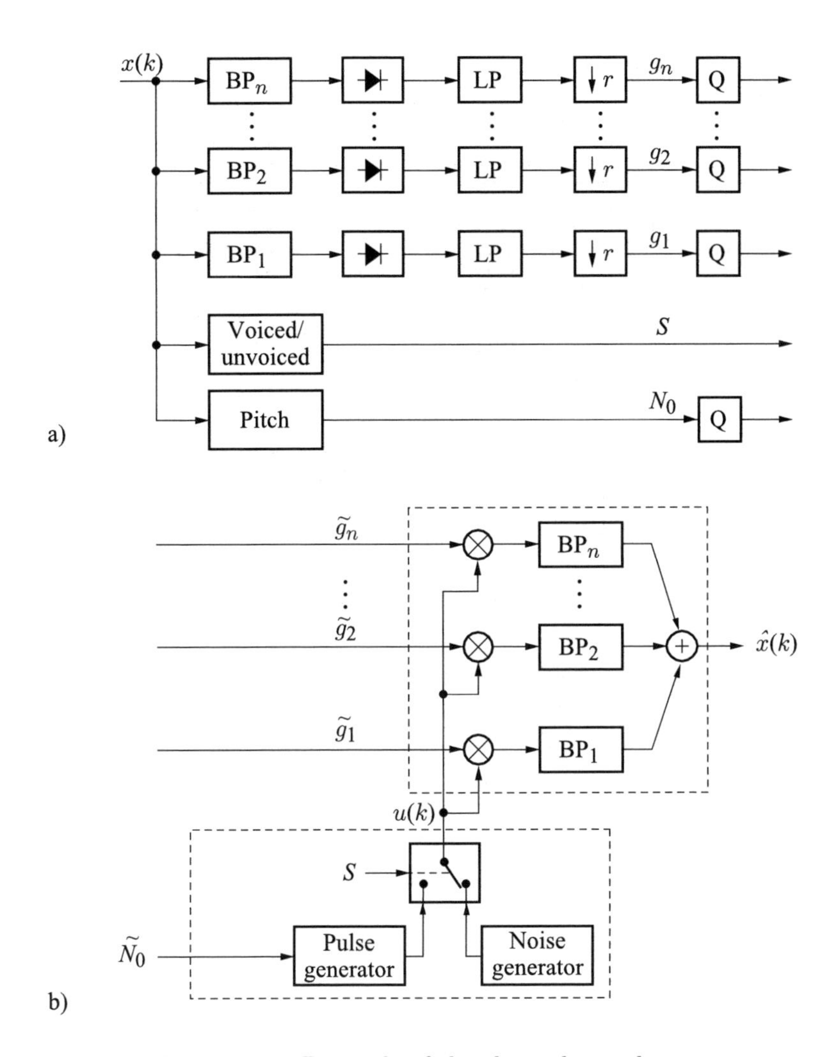

Figure 8.14: Principle of the channel vocoder
a) Transmitter
b) Receiver

The development of the different vocoders is closely linked to the technology available at the respective time. In what follows, we will first briefly discuss two early variants of the vocoder approaches. More detailed descriptions can be found in, for example, [Rabiner, Schafer 1978], [Sluijter 2005].

The oldest vocoder is the so-called *channel vocoder*, which was originally realized in analog technology with $n = 10$ channels [Dudley 1939], in order to transmit speech signals in analog form with a markedly reduced bandwidth. The basic principle is depicted in Fig. 8.14.

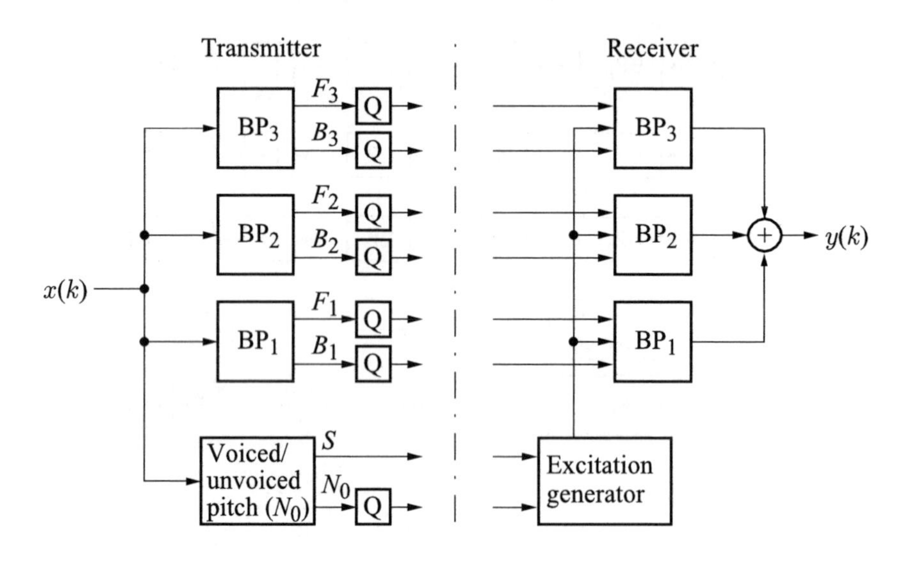

Figure 8.15: Principle of the formant vocoder

The synthesis filter consists of a parallel arrangement of bandpasses (Fig. 8.14-b). The corresponding output signals are added up to form signal $\hat{x}(k)$. All bandpasses are excited with the same signal $u(k)$, which is individually scaled by quantized time-variant factors $\tilde{g}_i$ $(i = 1, \ldots, n)$. The gain factors g_i are determined on the transmitting side by measuring the envelope of the short-term power in each frequency band (Fig. 8.14-a), while using the same bandpasses on the transmitting and receiving sides. Due to the relatively slow changes of the envelope, the scaling factors are transmitted with a highly reduced sampling frequency (subsampling by the factor r). If, at the receiving side, the quantized gain factors $\tilde{g}_i$ are combined with the bandpasses, the overall filter can be interpreted as an approximation of the vocal tract filter. Consequently, the desired frequency response is approximated by a parallel arrangement of n bandpasses with fixed center frequencies and bandwidths but different and variable gain factors $\tilde{g}_i$.

Another vocoder variant is the *formant vocoder* (e.g., [Rosenberg et al. 1971], [Rabiner, Schafer 1978]). In contrast to the channel vocoder, the spectral envelope does not result from measuring the energy in fixed frequency bands, but from an explicit determination of the formant frequencies F_i and formant bandwidths B_i. The synthesis filter can be realized as a cascade or, as depicted in Fig. 8.15, as a parallel connection of second-order filters.

With this concept bit rates of $B < 1$ kbit/s can be realized. However, this entails fundamental limitations on the naturalness of speech. With a correct analysis of the formant frequencies a better speech quality is achieved compared to the channel vocoder. The formant analysis is especially problematic if two formants lie close together. Linear prediction or cepstral analysis is suitable to determine the formants.

Apart from these time domain vocoders, related approaches exist, which represent frequency domain methods, such as the *phase vocoder* [Flanagan, Golden 1966] and the *cepstrum vocoder* (e.g., [O'Shaughnessy 2000], [Rabiner, Schafer 1978]), which will not be discussed here.

8.4.2 LPC Vocoder

*Linear predictive coding*is based on the all-pole model of speech production. The simplest variant is the *LPC vocoder*, which corresponds to the model of speech production derived in Section 2.3 and simplified in Section 6.1 (Fig. 6.1). The essential properties of this model are:

- the nasal tract is not considered, and the vocal tract is approximated by lossless tube segments or by a minimum-phase all-pole filter accordingly;
- based on a voiced/unvoiced classification of short signal frames of 20–30 ms duration, the excitation or the glottis signal is approximated by a periodic impulse sequence or a noise signal.

In contrast to the waveform coding of Section 8.3, the coding strategy depends on a dynamic signal classification. Often, a predictor or a synthesis filter of reduced filter order is utilized in unvoiced frames so that the mean bit rate can be reduced. The basic structure of the LPC vocoder is illustrated in Fig. 8.16.

A typical dimensioning will be explained by means of the so-called *LPC-10 algorithm* according to [Campbell et al. 1989] and [Tremain 1982].

The speech signal is sampled with a frequency of $f_s = 8\,\text{kHz}$ and segmented into frames of 22.5 ms duration ($N = 180$ samples). Using the covariance method (see Section 6.3) 10 reflection coefficients in the case of voiced frames and 4 reflection coefficients for unvoiced frames are computed.

The coefficients are quantized in the form of the *log area ratios* (see Section 8.4.3), where the first two coefficients are non-uniformly quantized with 5 bits each, while the remaining coefficients are represented with uniform resolution and word lengths of $w = 2, \ldots, 5$. The complete set of coefficients is coded with 41 bits or 20 bits in the case of voiced or unvoiced frames, respectively. For the pitch period and the voiced/unvoiced decision 7 bits are used, 5 bits for the logarithmic quantization of the gain factor and 1 bit for the synchronization. Thus, the bit rates for both voicing modes result in

$$\frac{54\text{ bits}}{22.5\,\text{ms}} = 2.4\text{ kbit/s}\quad(\text{voiced})\text{ and }\frac{33\text{ bits}}{22.5\,\text{ms}} = 1.47\text{ kbit/s}\quad(\text{unvoiced}).$$

The basic delay of this codec amounts to approximately 90 ms. Despite the reduced naturalness of the resynthesized speech, a relatively high intelligibility is achieved. The LPC-10 algorithm was primarily developed for encrypted transmission using modems in non-public analog telephone networks.

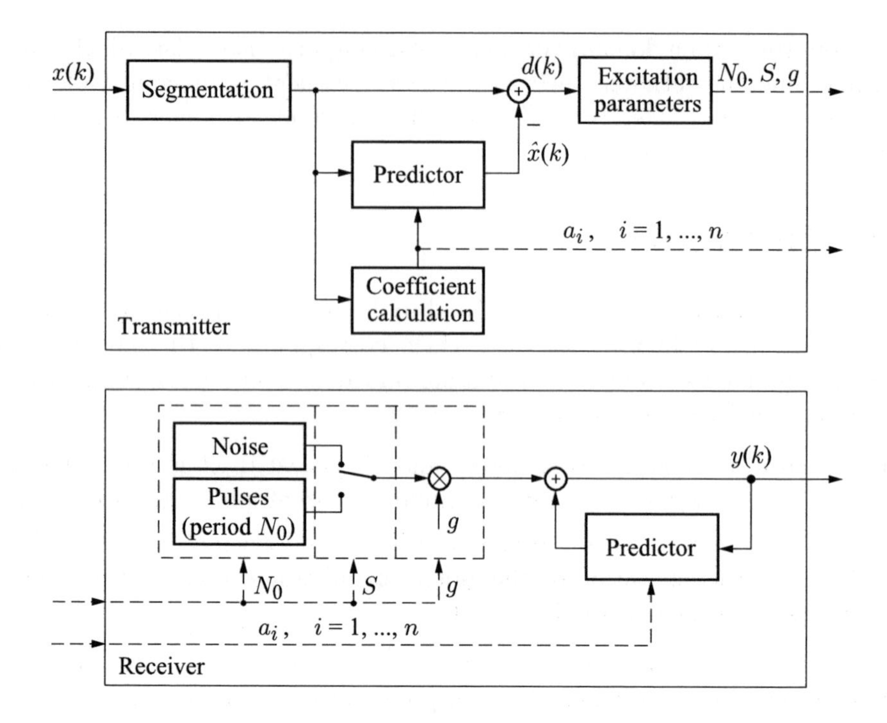

Figure 8.16: Principle of the LPC vocoder
(The parameter quantizers are omitted here for simplicity)

8.4.3 Quantization of the Predictor Coefficients

The speech quality of LPC vocoders also depends on how precisely the spectral envelope of the speech signal is matched by the frequency response of the synthesis filter, especially in the neighborhood of the formant frequencies.

This accuracy is determined by three factors: the LP-analysis algorithm (Chapter 6), the filter order, and finally the quantization of the coefficients. Generally, the predictor coefficients are computed with relatively high precision, e.g., in fixed point arithmetic with a word length of 16 bits. If the coefficients were directly transmitted as 16 bit numbers with a filter order of $n = 10$ and a block length of 20 ms, a bit rate of 160 bits/20 ms $=$ 8 kbit/s would be needed. Such a high accuracy is not required for the representation of the LP coefficients. They can be quantized at a significantly reduced bit rate. The effect of the quantization error on the frequency response and possibly on the stability of the synthesis filter strongly depends on the filter structure which is used. We can use various equivalent types, such as the direct structure, the lattice structure, the ladder structure, or the cascade of second-order filters. For the quantization of the filter coefficients, the scalar and vector methods discussed in Chapter 7 can in principle be used. As regards the bit rate reduction, a variety of specific solutions can be found in the literature, which differ in terms of complexity and consider the statistical properties of the coefficients of the underlying filter type in different ways.

A detailed overview, from which the following numerical examples are taken, can be found in [Kleijn, Paliwal 1995].

In order to evaluate the quality of an LPC quantizer, an objective measure or a distance measure, which is preferably independent of the chosen filter structure, is required. A common measure is the *mean spectral distortion* of the logarithmic frequency response of the synthesis filter. If the frequency responses of the synthesis filter for non-quantized and quantized coefficients are termed $H(e^{j\Omega})$ and $\hat{H}(e^{j\Omega})$, a spectral distance measure SD can be determined for each single speech frame as follows:

$$SD = \sqrt{\frac{1}{2\pi} \int_{-\pi}^{\pi} \left[10 \lg \left| H(e^{j\Omega}) \right|^2 - 10 \lg \left| \hat{H}(e^{j\Omega}) \right|^2 \right]^2 \mathrm{d}\Omega} \,. \tag{8.43}$$

Evaluating the spectral distortion SD for all frames in the test data and computing its average value over many frames gives the mean spectral distance $\overline{SD}$. Transparency is given for $\overline{SD} \leq 1\,\text{dB}$ (e.g., [Sugamura, Farvardin 1988], [Atal et al. 1989], [Kleijn, Paliwal 1995]).

The quantization methods to be discussed are of interest not only for the LPC vocoder, but also especially for the hybrid coding methods of Section 8.5.

8.4.3.1 Scalar Quantization of the LPC Coefficients

The coefficients a_i of a predictor in direct form must be represented very precisely in order to guarantee the stability of the synthesis filter. Besides this, each coefficient has to be quantized with the same precision, because each coefficient shows a similar impact on the frequency response. Using individual scalar optimal quantizers (see Section 7.4) with 6 bits per coefficient a_i, i.e., 60 bits per frame corresponding to a bit rate of $B = 3$ kbit/s, a mean spectral distortion $\overline{SD}$ of 1.83 dB has been determined in [Kleijn, Paliwal 1995] for a representative speech database. For 25% of the frames, an unstable synthesis filter resulted. Consequently, this type of quantization is not used in practice.

8.4.3.2 Scalar Quantization of the Reflection Coefficients

The reflection coefficients k_i of the acoustic tube model of speech production or of the corresponding digital lattice and ladder filters result from solving the normal equations (6.54-a), e.g., by means of the Levinson–Durbin algorithm (Section 6.3.1.3). An advantage of these structures is a guaranteed stability if the quantized parameters $\tilde{k}_i$ fulfill the condition

$$-1 < \tilde{k}_i < 1 .$$

In addition, not all the coefficients have to be represented with the same precision. A non-uniform bit allocation is allowed, where, for instance, the first reflection coefficient k_1 is quantized with 6 bits and the last coefficient k_{10} with only 2 bits. If individual optimal quantizers are adjusted to the probability density function of the respective reflection coefficient, the contribution of each single quantization interval to the mean square quantization error is the same. The effects of the quantization errors on the spectral distortion according to (8.43), however, strongly depend on the actual value of the reflection coefficient. This effect can be described by a U-shaped spectral sensitivity curve with its maximal values at $k_i = \pm 1$ (e.g., [Makhoul, Viswanathan 1975]). For this reason, large values of $|k_i|$ should be quantized more accurately. This aim is achieved by means of a non-linear transformation of each coefficient.

Two appropriate transformations are the inverse sine

$$S_i = \frac{2}{\pi} \arcsin(k_i) \tag{8.44}$$

and the inverse hyperbolic tangent

$$L_i = \text{arctanh}\ (k_i) \tag{8.45-a}$$

$$= \frac{1}{2} \ln\left(\frac{1+k_i}{1-k_i}\right); \quad |k_i| < 1 . \tag{8.45-b}$$

Both transformation characteristics are depicted in Fig. 8.17.

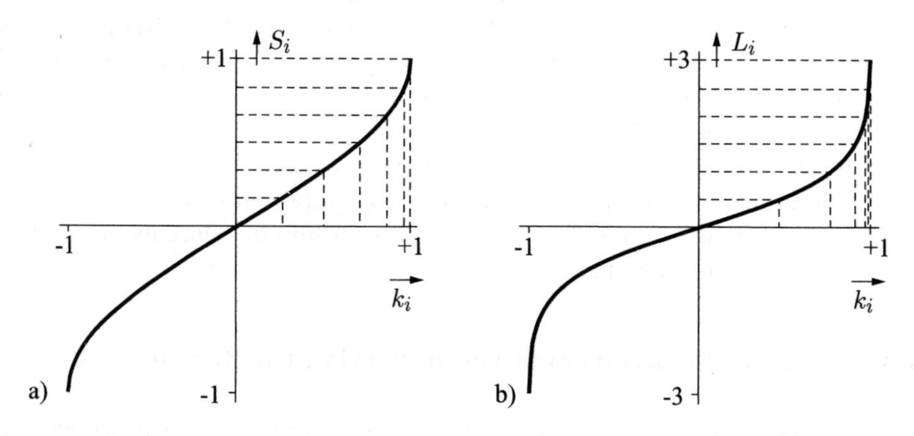

Figure 8.17: Non-linear transformation of the reflection coefficients
a) $S_i = \frac{2}{\pi} \arcsin(k_i)$
b) $L_i = \frac{1}{2} \ln\left(\frac{1+k_i}{1-k_i}\right)$

There is a direct relation between the reflection coefficient k_i and the cross-sectional areas A_i and A_{i+1} of two successive tube segments as discussed in Section 2.3:

$$k_i = \frac{A_i - A_{i+1}}{A_i + A_{i+1}} \, . \tag{8.45-c}$$

Note, that in comparison to Section 2.3, there is an index decrement by one ($k_{i-1} = r_i$).

Inserting (8.45-c) in (8.45-b) yields

$$L_i = \frac{1}{2} \ln \left(\frac{A_i}{A_{i+1}} \right) \, . \tag{8.45-d}$$

The transformed coefficient L_i is proportional to the logarithm of the quotient of the cross-sectional areas. Therefore, the coefficients L_i are named *log area ratios* (LARs). On the other hand, the inverse-sine transformation offers with S_i the *arcsine reflection coefficients* (ASRCs).

By means of uniform quantization of the transformed values S_i or L_i, the desired better resolution for large values of k_i is obtained. This represents a near-optimal but low-complexity alternative to optimal quantization. The second solution can be found in the *full-rate codec* of the GSM mobile radio communication system (see Appendix A). Here, the logarithm is approximated by linear segments. The $n = 8$ coefficients L_i are uniformly quantized with different word lengths of $w = 3$–6 (e.g., [Vary et al. 1988]). In total, the set of coefficients is coded with 36 bits per 20 ms referring to a bit rate of 1.8 kbit/s.

Table 8.1 shows the number of bits which is required to achieve a mean spectral distortion of $\overline{SD} \approx 1\,\text{dB}$ with optimal quantization of $n = 10$ reflection coefficients or transformed coefficients, respectively. Obviously, the use of the LAR or ASRC representation provides a gain of 2 bit/frame in comparison to the reflection coefficients.

Table 8.1: Scalar quantization of the reflection coefficients (according to [Kleijn, Paliwal 1995])

Coefficient	Mean spectral distortion [dB] $\overline{SD}$	Bits/frame (10 coefficients)
k_i	1.02	34
S_i	1.04	32
L_i	1.04	32

8.4.3.3 Scalar Quantization of the LSF Coefficients

The *line spectral frequencies* (LSFs) ([Itakura 1975], [Soong, Juang 1984], [Sugamura, Itakura 1986], [Kabal, Ramachandran 1986]) represent another method to ensure stability of the all-pole synthesis filter after LPC quantization. Furthermore, the LSF-coefficients have some favorable properties which can be exploited for quantization at low bit rates.

The basis of the LSF representation is a stability theorem for recursive digital filters and the decomposition of the function

$$G(z) = 1 - \sum_{i=1}^{n} a_i \, z^{-i} \tag{8.46}$$

$$= \sum_{i=0}^{n} \alpha_i \, z^{-i} \tag{8.47}$$

with

$$\alpha_0 = 1, \tag{8.48}$$

$$\alpha_i = -a_i \, ; \qquad i = 1, 2, \ldots, n \tag{8.49}$$

into a mirror polynomial

$$P(z) = G(z) + z^{-(n+1)} \, G(z^{-1}) \tag{8.50-a}$$

and an anti-mirror polynomial

$$Q(z) = G(z) - z^{-(n+1)} \, G(z^{-1}) \, . \tag{8.50-b}$$

The mirror and the anti-mirror properties are characterized by

$$P(z) = z^{-(n+1)} \, P(z^{-1}) \quad \text{and} \tag{8.51}$$

$$Q(z) = -z^{-(n+1)} \, Q(z^{-1}) \, , \quad \text{respectively.} \tag{8.52}$$

The polynomial $G(z)$ can be reconstructed as follows:

$$G(z) = \frac{1}{2}[P(z) + Q(z)] \, . \tag{8.53}$$

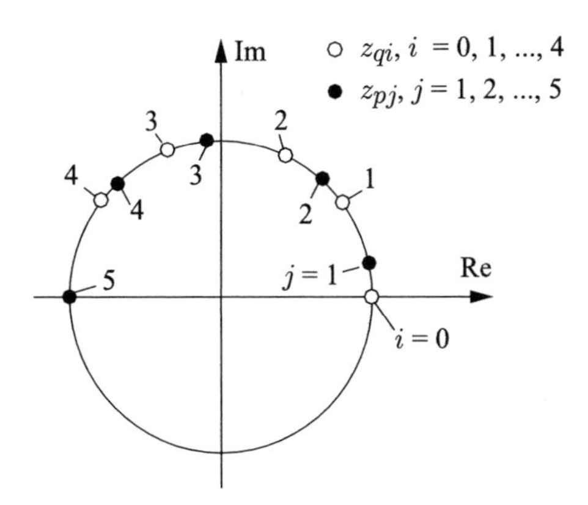

Figure 8.18: LSF parameters of an eighth order predictor

Thus, $G(z)$ can also be described by the zeros of $P(z)$ and $Q(z)$. It can be shown that $G(z)$ is minimum phase and thus the stability of the synthesis filter

$$H(z) = \frac{1}{G(z)} \tag{8.54}$$

is guaranteed ([Itakura 1975], [Schüssler 1994]), if the following hold:

- the zeros z_{p_i} and z_{q_i} of the polynomials $P(z)$ and $Q(z)$ are located on the unit circle of the z-plane

$$z_{p_i} = e^{j\omega_{p_i}} \qquad z_{q_i} = e^{j\omega_{q_i}}\,; \qquad i \in \{0, 1, \ldots, n\};$$

- the zero positions ω_{p_i} and ω_{q_i} of $P(z)$ and $Q(z)$, respectively, are interleaved on the unit circle as shown in Fig. 8.18 (see also (8.56)).

These two properties can be exploited to represent the zero positions ω_{p_i} and ω_{q_i} ($i \in \{0, 1, \ldots, n\}$) of the LSF parameters with a relatively low bit rate.

First, we assume an even filter order n. $P(z)$ has a zero at $z = e^{j\pi} = -1$ and $Q(z)$ one at $z = e^{j0} = +1$, which are denoted by $\omega_{p\frac{n}{2}+1} = \pi$ and $\omega_{q_0} = 0$. The locations of the remaining zeros depend on $G(z)$. These zeros occur in $\frac{n}{2}$ interleaved, complex conjugate pairs. The $2n+2$ zeros of $P(z)$ and $Q(z)$ fulfill the following condition:

$$P(z)\colon\; 0 < \omega_{p_1} < \omega_{p_2} < \cdots < \omega_{p\frac{n}{2}} < \omega_{p\frac{n}{2}+1} = \pi \tag{8.55-a}$$

$$Q(z)\colon\; 0 = \omega_{q_0} < \omega_{q_1} < \cdots < \omega_{q\frac{n}{2}} < \pi\,. \tag{8.55-b}$$

The complex conjugate zeros in the lower z-plane are not considered here, as they can be reconstructed from the others at the receiver. According to the relation

$$0 = \omega_{q_0} < \omega_{p_1} < \omega_{q_1} < \omega_{p_2} < \omega_{q_2} < \cdots < \omega_{q_{\frac{n}{2}}} < \omega_{p_{\frac{n}{2}+1}} = \pi \tag{8.56}$$

the following notation is introduced:

$$0 = \omega_0 < \omega_1 < \omega_2 < \cdots < \omega_n < \omega_{n+1} = \pi \,. \tag{8.57}$$

Since the first value $\omega_{q_0} = \omega_0 = 0$ and the last value $\omega_{p_{\frac{n}{2}+1}} = \omega_{n+1} = \pi$ are fixed, n LSF parameters $\omega_1, \ldots, \omega_n$ must be quantized. Instead of a straightforward scalar quantization of the LSF parameters, their sequential ordering can be exploited by scalar quantization of the differences of successive LSFs.

A more advanced vector quantization scheme which achieves a spectral distortion of only 1.04 dB for $n = 10$ and 28 bit/frame was proposed in [Xie, Adoul 1995]. The vector consisting of 10 LSF parameters $\omega_1 \ldots \omega_{10}$ is split into four groups:

$$\boldsymbol{\omega}_A = (\omega_3,\ \omega_7)$$

$$\boldsymbol{\omega}_B = \left(\frac{\omega_1}{\hat{\omega}_3},\ \frac{\hat{\omega}_3 - \omega_2}{\hat{\omega}_3}\right)$$

$$\boldsymbol{\omega}_C = \left(\frac{\omega_4 - \hat{\omega}_3}{\hat{\omega}_7 - \hat{\omega}_3},\ \frac{\omega_5 - \hat{\omega}_4}{\hat{\omega}_7 - \hat{\omega}_3},\ \frac{\hat{\omega}_7 - \omega_6}{\hat{\omega}_7 - \hat{\omega}_3}\right)$$

$$\boldsymbol{\omega}_D = \left(\frac{\omega_8 - \hat{\omega}_7}{\pi - \hat{\omega}_7},\ \frac{\omega_9 - \hat{\omega}_8}{\pi - \hat{\omega}_7},\ \frac{\pi - \omega_{10}}{\pi - \hat{\omega}_7}\right).$$

Four different quantizers are used. For the reference vector $\boldsymbol{\omega}_A$, an LBG-trained vector quantizer with a code book of size 64 (6 bits) is used which delivers $\hat{\omega}_3$ and $\hat{\omega}_7$. The normalized vectors $\boldsymbol{\omega}_B$, $\boldsymbol{\omega}_C$, and $\boldsymbol{\omega}_D$ are quantized with dedicated lattice quantizers with 5, 9, and 8 bits respectively, where the quantities $\hat{\omega}_4$ and $\hat{\omega}_8$ are obtained through a constraint of the lattice quantizer. For the details, the reader is referred to [Xie, Adoul 1995]. This quantizer is of practical significance as this concept has been adopted in various speech coding standards (see Appendix A).

A further reduction of the bit rate for the LSF coefficients can be achieved by exploiting the interframe correlation of sets of coefficients. In [Kataoka et al. 1996], with a spectral distortion of approx. 1.2 dB, this approach needs 18 bit/frame, i.e., an average of only 1.8 bit/coefficient.

8.5 Hybrid Coding

8.5.1 Basic Codec Concepts

The main application areas for source coding of speech signals with bit rates below the 32 kbit/s of standard ADPCM are digital mobile radio communication, speech storage, and "line multiplication", i.e., channel sharing by speech signal compression. Of particular interest are coders with 0.5–1.5 bits per sample, i.e., for telephone speech with a bit rate of $B = 4$–$12\,\text{kbit/s}$.

In these applications, hybrid speech codecs are used almost exclusively. As was shown in Fig. 8.2, these codecs take a position between waveform coding and parametric coding. As a common feature, the coefficients of a synthesis filter are transmitted as *side information* (parameter channel in Fig. 8.2), and the residual signal is approximated quite roughly with respect to the amplitude and/or time resolution. The literature shows a wide range of concepts, and a vast variety of different codec variants.

In this section, we will develop a uniform conceptual description of these approaches. The following sections will deal with typical variants, while some selected codec standards are outlined in Appendix A.

In hybrid codecs, short-term as well as long-term prediction are common; the decoder structure is shown in Fig. 8.19. The decoder consists of a cascade of a long-term prediction (LTP) and a linear prediction (LP) synthesis filter, which is excited by the quantized residual signal $\tilde{d}'(k)$. The predictors $A(z)$ and $P(z)$ are time variant and, as described in Chapter 6, adjusted blockwise. The coefficients are quantized as explained in Section 8.4.3. Since the filters are generally held constant for the length of a signal frame or subframe, respectively, their time variation will not be considered here for simplicity of the presentation. The typical frame duration is 20 ms. Frames may be divided into subframes of, e.g., 5 ms each.

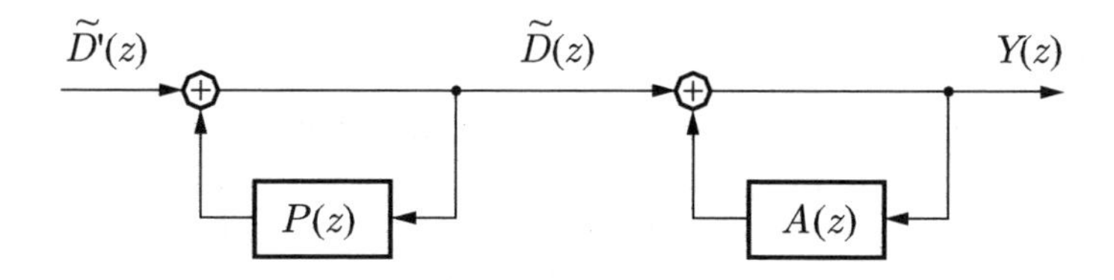

Figure 8.19: Structure of the hybrid decoder
$A(z)$: Short-term predictor (LP)
$P(z)$: Long-term predictor (LTP)

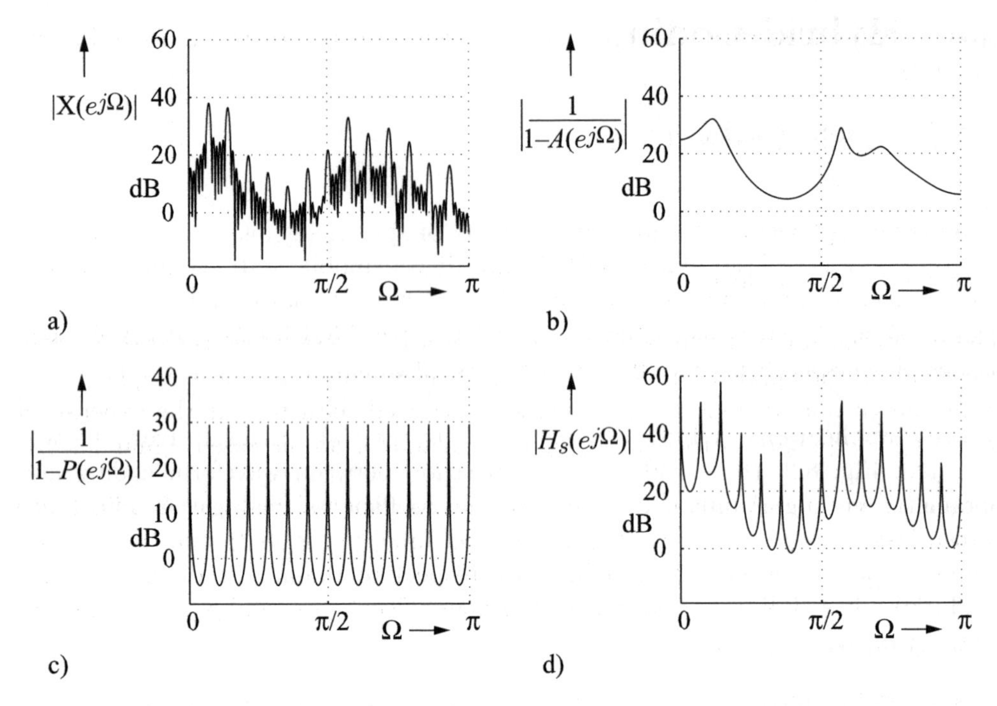

Figure 8.20: Functionality of the hybrid decoder
a) Short-term spectral analysis $|X(e^{j\Omega})|$ for a voiced segment (20 ms)
b) Magnitude response of the LP-synthesis filter $|1/(1-A(e^{j\Omega}))|$
c) Magnitude response of the LTP-synthesis filter $|1/(1-P(e^{j\Omega}))|$
d) Magnitude response of the cascade of both filters
$|H_s(e^{j\Omega})| = |1/(1-P(e^{j\Omega}))| \cdot |1/(1-A(e^{j\Omega}))|$

The basic functionality of the decoder will be illustrated by the following example: Fig. 8.20 shows the short-term spectral analysis of a (synthetic) voiced speech segment (20 ms), the magnitude responses of the LTP-synthesis filter, the LP-synthesis filter, and the overall filter. The first filter stage (LTP), which is adapted every subframe, shows a distinctive comb filter characteristic in voiced segments. Starting from a flat spectrum $\tilde{D}'(z)$ of the excitation signal, all spectral components located at multiples of the estimated fundamental frequency, i.e., at

$$\Omega_i = \frac{2\pi}{N_0} i \; ; \qquad i = (0), 1, 2, \ldots, \tag{8.58}$$

are amplified, and the components in between are attenuated. The result is signal $\tilde{d}(k)$ with an approximately spectrally flat envelope and a discrete, harmonic structure. Finally, with the second filter stage (LP) the spectral envelope of the speech segment is formed on $\tilde{d}(k)$ by the LP-synthesis filter.

The codec concepts to be discussed here basically differ in the predictor structures at the transmitter and in the representation of the residual signal by scalar or vector quantization.

8.5.1.1 Scalar Quantization of the Residual Signal

Basic schemes using scalar quantization are shown in Fig. 8.21. The effect of the quantizer will be described in the time domain by additive quantization noise $\Delta(k)$ according to

$$\tilde{d}'(k) = d'(k) + \Delta(k)\,. \tag{8.59}$$

Scalar quantization is applied to the residual signal $d'(k)$ either after two-stage open-loop prediction (Fig. 8.21-a) or inside a closed prediction loop (Fig. 8.21-b to Fig. 8.21-d). In all the cases, the quantized residual signal is termed $\tilde{d}'(k)$ and the same decoder of Fig. 8.19 is utilized.

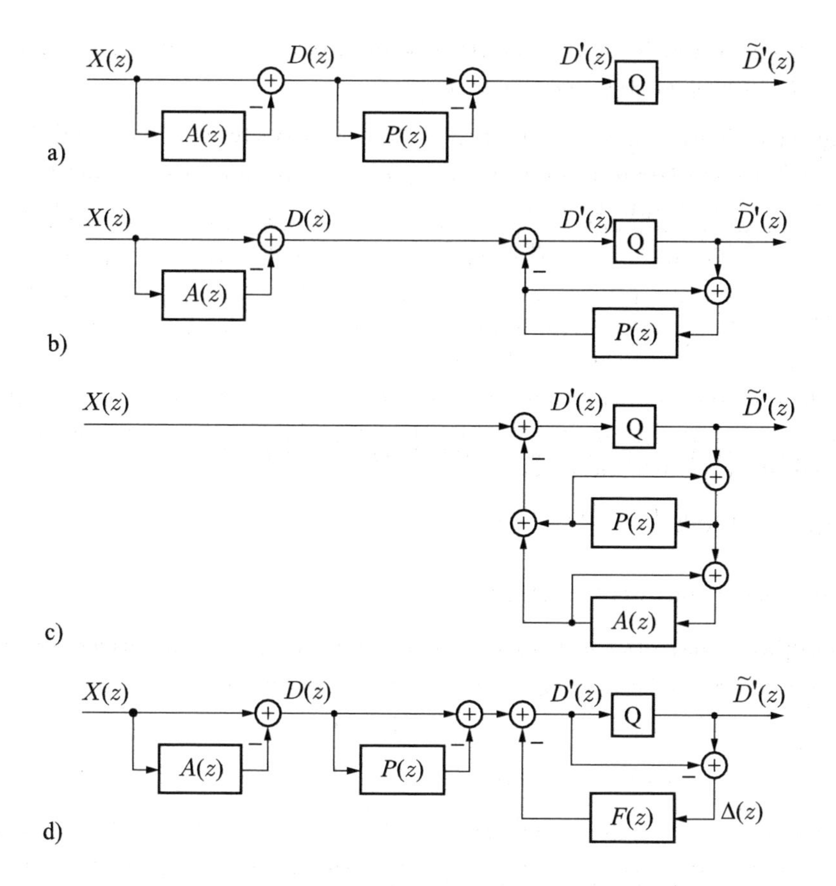

Figure 8.21: Hybrid coding with scalar quantization of the residual signal
a) LP and LTP open loop
b) LP open loop and LTP closed loop
c) LP and LTP closed loop
d) LP and LTP with noise shaping

If the quantizer is switched off, i.e., $\tilde{d}'(k) = d'(k)$, the three structures in Fig. 8.21-a to Fig. 8.21-c provide the same residual signal. Moreover, for identical LP and LTP coefficients on the transmitting and receiving side, the output signal $y(k)$ equals the input signal $x(k)$, since the filter operations are inverse to each other.

In contrast to this, with quantization of the residual signal $d'(k)$, marked differences in the reconstruction error occur due to the different positions of the quantizer.

As the predictors are adapted to signal frames of finite length, the following analysis can be performed for a signal $x(k)$ of limited duration, for which the z-transform $X(z)$ exists.

Without committing to a certain type of quantizer, the quantization error $\Delta(k)$ can thus be described by

$$\Delta(z) = \tilde{D}'(z) - D'(z) \tag{8.60}$$

in the z-domain. In analogy to the concept of noise shaping derived in Section 8.3.3, it can be shown that the performance of the block diagrams of Fig. 8.21-a to Fig. 8.21-c can be described exactly by the structure of Fig. 8.21-d with an adequate choice of the *noise shaping* filter $F(z)$. Of particular interest is the resulting reconstruction error in the decoder according to Fig. 8.19.

Open-Loop Short-Term and Long-Term Prediction

This case corresponds to the choice

$$F(z) = 0$$

resulting in

$$\tilde{D}'(z) = X(z)\,\big(1 - A(z)\big)\,\big(1 - P(z)\big) + \Delta(z)$$

$$Y(z) = X(z) + \frac{\Delta(z)}{\big(1 - P(z)\big)\,\big(1 - A(z)\big)}\,.$$

The spectrum of the quantization error is weighted with the frequency response of the cascaded synthesis filters. Applying psychoacoustic criteria, this noise shaping is advantageous, especially when the spectral envelope of the quantization error is flat. However, the prediction gain cannot be exploited to improve the signal-to-noise ratio, since both predictors are applied in the open-loop mode (see Section 8.3.3.1).

Open-Loop Short-Term and Closed-Loop Long-Term Prediction

This case corresponds to the choice

$$F(z) = P(z)$$

resulting in

$$\begin{aligned}
\tilde{D}'(z) &= X(z)\,\big(1 - A(z)\big)\,\big(1 - P(z)\big) + \Delta(z)\,\big(1 - P(z)\big) \\
Y(z) &= X(z) + \frac{\Delta(z)}{\big(1 - A(z)\big)}\,.
\end{aligned}$$

The spectrum of the quantization error is weighted with the spectral envelope of the input signal. Since the long-term predictor is inside a closed loop, only its corresponding part of the prediction gain can be used to improve the signal-to-noise ratio.

Closed-Loop Short-Term and Long-Term Prediction

This case corresponds to the choice

$$F(z) = A(z) + P(z) - A(z)\,P(z)$$

resulting in

$$\begin{aligned}
\tilde{D}'(z) &= X(z)\,\big(1 - A(z)\big)\,\big(1 - P(z)\big) \\
&\qquad + \;\Delta(z)\,\big(1 - A(z) - P(z) + A(z)\,P(z)\big) \\
&= X(z)\,\big(1 - A(z)\big)\,\big(1 - P(z)\big) + \Delta(z)\,\big(1 - A(z)\big)\,\big(1 - P(z)\big) \\
Y(z) &= X(z) + \Delta(z)\,.
\end{aligned}$$

Reconstruction and quantization error are identical. Due to the two-step closed-loop prediction, the total prediction gain can be used entirely to improve the signal-to-noise ratio. (see Section 8.3.3.2)

Short-Term and Long-Term Prediction with Noise Shaping

This is the most general case with

$$\begin{aligned}
\tilde{D}'(z) &= X(z)\,\big(1 - A(z)\big)\,\big(1 - P(z)\big) + \Delta(z)\,\big(1 - F(z)\big) \\
Y(z) &= X(z) + \Delta(z)\,\frac{1 - F(z)}{\big(1 - P(z)\big)\,\big(1 - A(z)\big)}\,.
\end{aligned}$$

Table 8.2: Effect of the noise shaping filter $F(z)$; $0 \leq (\gamma_1, \gamma_2) \leq 1$

$F(z)$	Reconstruction error
$A(z/\gamma_1)$	$\Delta(z)\ \dfrac{1}{1-P(z)}\ \dfrac{1-A(z/\gamma_1)}{1-A(z)}$
$P(z/\gamma_2)$	$\Delta(z)\ \dfrac{1-P(z/\gamma_2)}{1-P(z)}\ \dfrac{1}{1-A(z)}$
$A(z/\gamma_1)+P(z/\gamma_2)-A(z/\gamma_1)\ P(z/\gamma_2)$	$\Delta(z)\ \dfrac{1-P(z/\gamma_2)}{1-P(z)}\ \dfrac{1-A(z/\gamma_1)}{1-A(z)}$

In correspondence with spectral noise shaping according to Section 8.3.3.3, an auditory improvement can be achieved exploiting the masking effect of the human ear (see Section 2.5) by only partly using the prediction gain for the objective improvement. Different options for $F(z)$ and its effect on the reconstruction error are shown in Table 8.2.

With respect to psychoacoustic aspects, the choice of $\gamma_1 < 1$ and $\gamma_2 = 1$ is favorable. In this case, the spectral weighting corresponds to the conventional noise shaping (see Section 8.3.3.3), while additionally the prediction gain of the LTP loop is used for an objective improvement of the signal-to-noise ratio.

8.5.1.2 Vector Quantization of the Residual Signal

A target bit rate of effectively only 0.5–1.5 bits per sample suggests vector quantization of the residual signal. Note that due to prediction the residual signal is more or less decorrelated. For complexity reasons, only *gain–shape* vector quantization (Section 7.6.5) is applicable. Basically, code books for normalized residual signal vectors must be designed. The statistical analysis of speech material shows that the normalized residual signal vectors follow a multivariate Gaussian distribution to a good approximation. This does almost not depend on the speaker. Due to the non-uniform distribution, we can thus obtain better results by vector quantization than by scalar quantization despite decorrelation, as shown in Section 7.6.4. With the available low bit rates, however, only relatively low signal-to-noise ratios can be achieved. Therefore, the error criterion or the effective distance measure, respectively, is of great importance.

The introduction of noise shaping techniques (see Section 8.5.1.1) offers a considerable quality improvement with respect to psychoacoustic criteria. Here, the masking effect in speech perception is implicitly exploited. This approach can also

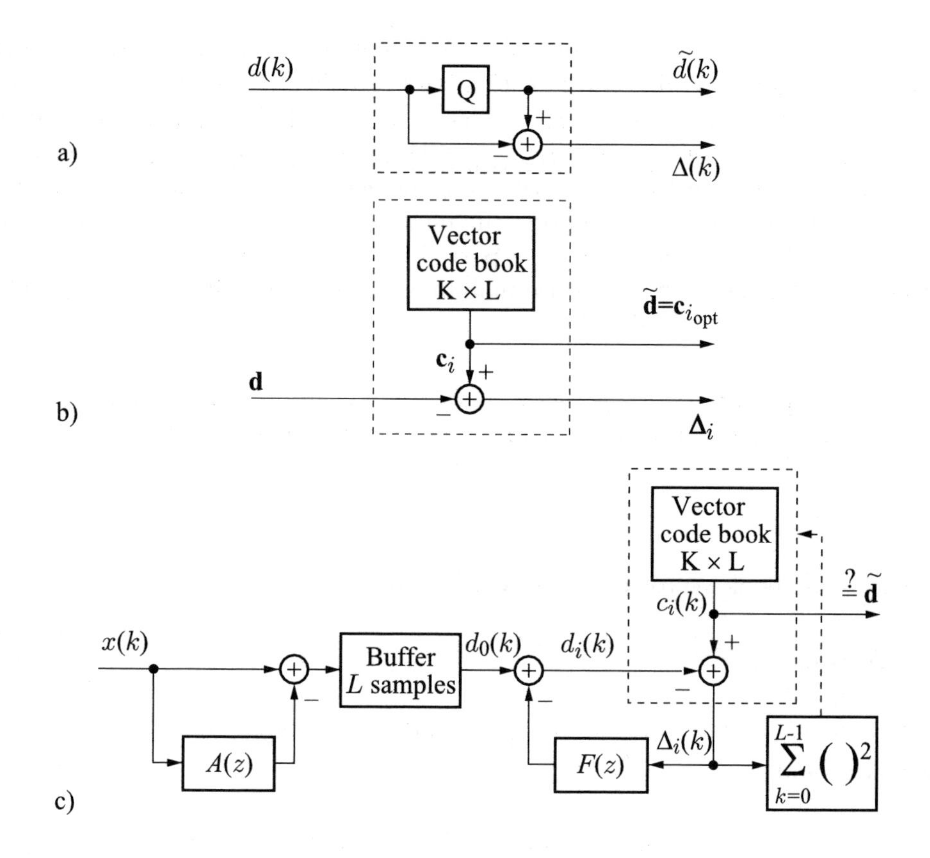

Figure 8.22: Vector quantization with spectral noise shaping
a) Scalar quantization of samples $d(k)$
b) Vector quantization of vector $\mathbf{d}$
c) Vector quantization with noise shaping
(Note: For reasons of simplicity, without long-term prediction)

be applied to vector quantization, resulting in what is called *analysis-by-synthesis* coding in the literature. It is the basis of most codec standards (see also Appendix A). In what follows, we will develop the approach, starting from scalar quantization with noise shaping according to Fig. 8.21-d.

In order to simplify the representation, we will disregard the long-term prediction in a first step and use plain vector quantization instead of gain–shape vector quantization. The starting point is Fig. 8.22-a. It shows the relation between the non-quantized scalar residual signal $d(k)$, the quantized value $\tilde{d}(k)$, and the resulting error

$$\Delta(k) = \tilde{d}(k) - d(k)\,. \tag{8.61}$$

For the vector approach, L samples of $d(k)$ are combined to a vector $\mathbf{d}$, which can be vector-quantized as described in Fig. 8.22-b. In simplified notation, we start with a segment of the sequence $x(k)$, which for $k = 0, 1, 2, \ldots, L-1$ corresponds to the respective vector elements and append outside this interval only zero values. Due to the linearity of the filters, the performance of the complete structure can be described correctly by segmenting the input sequence $x(k)$, separately filtering these segments, and superposing the partial reactions. The superposition results automatically if the filter states are maintained when going from one segment to the next.

The code book contains K vectors $\mathbf{c}_i$ $(i = 1, 2, \ldots, K)$ of dimension L. The best vector with respect to the smallest mean square error will be called

$$\tilde{\mathbf{d}} = \mathbf{c}_{i_{\text{opt}}} . \tag{8.62}$$

For each code book vector $\mathbf{c}_i$ an individual error vector

$$\Delta_i = \mathbf{c}_i - \mathbf{d} \tag{8.63}$$

can be calculated. This procedure applies to the case of open-loop prediction without noise shaping.

If we want to use noise shaping in combination with vector quantization of the residual, we have to apply the vector quantizer in analogy to Fig. 8.9 within the loop as shown in Fig. 8.22-c.

However, the vector $\mathbf{d}$ is not completely available in advance. Each element $d(k)$ of vector $\mathbf{d}$ depends on the preceding elements

$$\Delta_i(k-\kappa) = c_i(k-\kappa) - d(k-\kappa) , \qquad \kappa = 1, 2, \ldots \tag{8.64}$$

of the error vector Δ_i. For this reason, the vector $\mathbf{d}$ develops sample by sample for each $\mathbf{c}_i$ in the process of vector quantization. For each code book vector $\mathbf{c}_i$ a different vector $\mathbf{d}_i$ might ensue. For this reason the L input samples $d_0(k)$ must be buffered so that the filter loop can be processed for each code book vector $\mathbf{c}_i$, where the filter states of $F(z)$ have to be set to the original start values. Within the search procedure, the optimal excitation vector $\mathbf{c}_{i_{\text{opt}}}$ can be found among K entries by minimizing the mean square error. Obviously, the same mechanism is effective for the spectral shaping of the quantization error as for scalar quantization.

However, the practical realization of this principle is based on a different structure, which can be derived from Fig. 8.22-c by means of the following considerations.

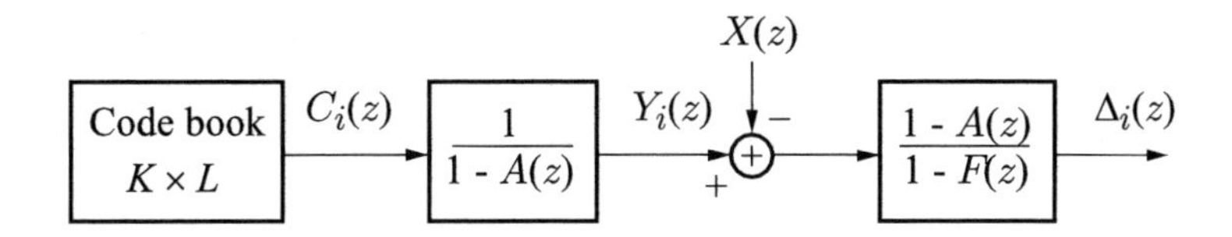

Figure 8.23: Principle of an analysis-by-synthesis coder

It is sufficient to look at only one segment to find the relations between $x(k)$, $d_0(k)$, $c_i(k)$, and $\Delta_i(k)$. According to Fig. 8.22-c with the z-transforms of the respective sequences, we have

$$\Delta_i(z) = -X(z)\,\frac{1-A(z)}{1-F(z)} + C_i(z)\,\frac{1}{1-F(z)} \tag{8.65-a}$$

$$= -X(z)\,\frac{1-A(z)}{1-F(z)} + C_i(z)\,\frac{1}{1-A(z)}\,\frac{1-A(z)}{1-F(z)} \tag{8.65-b}$$

$$= \left[Y_i(z) - X(z)\right]\,\frac{1-A(z)}{1-F(z)} \tag{8.65-c}$$

with

$$Y_i(z) = C_i(z)\,\frac{1}{1-A(z)}\,. \tag{8.66}$$

The result of (8.65-c) is depicted in Fig. 8.23.

With respect to a perceptually weighted spectral error criterion, the optimal vector quantization of the residual signal can be found within a search process. As a result we obtain an "optimal" replica $y_{i_{\text{opt}}}(k)$, which matches the speech signal $x(k)$ with minimum error. Note that the z-transform of the weighting filter refers to the inverse of the corresponding noise shaping filter.

This procedure, which is equivalent to Fig. 8.22-c, is defined in the literature by the generic term *analysis-by-synthesis* coding, and in this special form as CELP coding (*code excited linear prediction coding*, see also Section8.5.3) [Schroeder, Atal 1985].

This coding approach is computationally intensive since K different replicas $y_i(k)$ must be synthesized for each signal segment $x(k)$. In order to encode the residual signal with effectively 0.5 bits per sample corresponding to a bit rate of $B = 0.5$ bits $\cdot$ 8 kHz $= 4$ kbit/s, a code book with

$$\frac{\mathrm{ld}(K)}{L} = 0.5 \tag{8.67}$$

is required. For example, for $L = 20$ ($\hat{=}$ 2.5 ms), the size of the code book already amounts to $K = 2^{10} = 1024$.

Based on these concepts of scalar and vector quantization of the residual signal, the next sections will introduce different codec structures which provide comparable speech quality, and which differ considerably in terms of the bit rate and complexity. The concrete variants of the important codec standards are outlined in Appendix A.

8.5.2 Residual Signal Coding: RELP

The first concept to be dealt with follows the *adaptive predictive coding* (APC) scheme with block adaptation (see Section 8.3.1). The residual signal produced by short-term and long-term prediction is applied to a scalar quantizer (see Fig. 8.21-a). The core problem of this approach becomes evident in the following consideration. If the target bit rate is, for instance, $B = 15\,\text{kbit/s}$, only a bit rate of about 12 kbit/s is available for the coding of the residual signal, since the quantization of the LP and LTP filter parameters requires a bit rate of about 3 kbit/s (see Section 8.4.3 and 6.4). Thus, with a sampling frequency of $f_s = 8\,\text{kHz}$ only 1.5 bits per sample can be utilized. However, the power of the resulting quantization error $\Delta(k)$ would be so high that, even with noise shaping, no acceptable speech quality could be obtained.

A comparison with the LPC vocoder (see Section 8.4.2) reveals that it is not necessary to exactly reconstruct the residual signal at the receiver, as perception is insensitive to certain changes in this signal. Extracting and coding perceptually relevant aspects of the residual signal such as

- the correct temporal volume contour,
- the correct (quasi-)periodicity in voiced segments, and
- the noise-like character in unvoiced segments

provides communications-quality speech coders for the 2.4–8 kbit/s range.

Assuming that the lowest speech frequencies carry the highest perceptually important information, the relevant characteristics can mostly be reconstructed from a baseband of the residual signal extracted by a lowpass filter as shown in Fig. 8.24. This coding scheme is called the *baseband–RELP codec* [Un, Magill 1975]. For reasons of simplicity, the long-term predictor is not considered here.

The lowpass signal $d_{LP}(k)$ is decimated and transmitted in quantized form. In the receiver, the missing high frequencies are reconstructed by spectrally shifted versions of the baseband.

The prediction coefficients $a_i(k)$ $(i = 1, 2, \ldots, n)$ depend on time index k and are adapted blockwise every N samples, e.g., $N = 160$ $(\hat{=}\ 20\,\text{ms})$ (see Section 6.3.1).

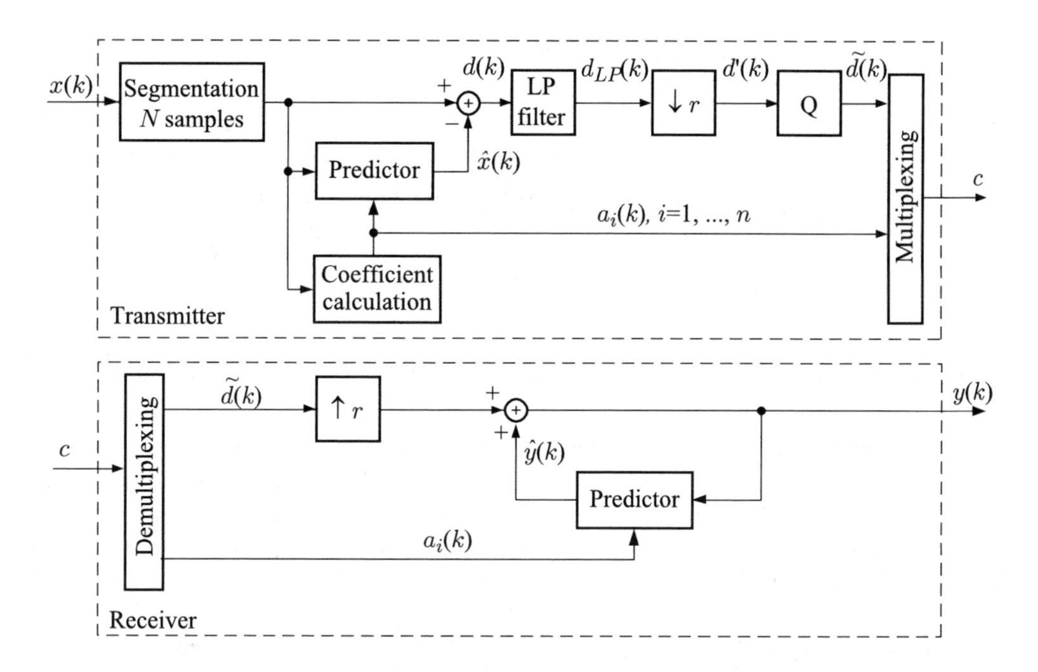

Figure 8.24: Principle of the baseband–RELP codec
(RELP: *Residual Excited Linear Prediction*)

The segmentation into frames is needed to buffer the speech samples for LP coefficient calculation and subsequent filtering. As a result, the codec introduces an *algorithmic delay* of at least one frame, i.e., N sampling intervals. This delay is increased by the time needed for the computation of the coefficients, the filter operations, and the signal delay caused by the lowpass. In a practical realization, such a codec will cause a total delay of 1.25 N to 2 N cycles (here 25–40 ms). To simplify the representation, we will assume that the coefficients $a_i(k)$ already exist in quantized form (see Section 8.4.3), i.e., the allocation of the quantized values to the bit patterns and the corresponding decoding at the receiver are not explicitly described here.

A generic feature of the baseband–RELP codec is that the prediction error signal $d(k)$ is applied to a lowpass filter with subsequent decimation by a factor r. The lowpass has a cutoff frequency of $\Omega_c = \pi/r$. Thus, the sampling frequency at the output can be reduced by factor r without spectral aliasing. The factor r is generally set to $r = 3$ or $r = 4$. At $f_s = 8\,\text{kHz}$, the baseband signal $d^{'}(k)$ has a bandwidth of $4/3\,\text{kHz}$ or $1\,\text{kHz}$, respectively, and a sampling frequency of $f_s^{'} = 8/3\,\text{kHz}$ or $f_s^{'} = 2\,\text{kHz}$, respectively. If, for example, a bit rate of $12\,\text{kbit/s}$ is available for coding the residual signal, each sample $d^{'}(k)$ can be encoded quite accurately with $12/8 \cdot r = 12/8 \cdot 3 = 4.5\,\text{bits}$ or $12/2 = 6\,\text{bits}$, respectively. The operation of lowpass filtering, downsampling and scalar quantization can be interpreted as joint time *and* amplitude quantization.

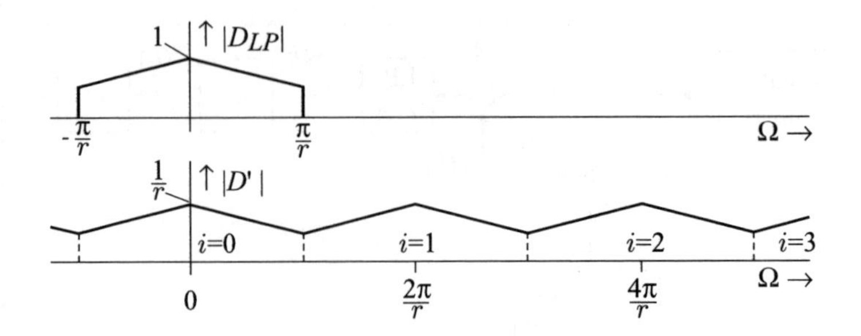

Figure 8.25: Generation of the excitation signal in the baseband–RELP decoder

At the receiving side, the sampling frequency is increased back to the original rate by replacing the missing samples with zeros. Thus, the baseband spectrum is mirrored $(r-1)$ times and overlapped. Disregarding the quantization (i.e., for $\tilde{d}(k) = d'(k)$) this process can be described exactly by introducing the decimation sequence $p(k)$ as follows:

$$\tilde{d}(k) = d'(k) = d_{LP}(k) \cdot p(k) = \begin{cases} d_{LP}(k) & k = \lambda \cdot r \\ 0 & k \neq \lambda \cdot r \end{cases} \qquad \lambda \in \mathbb{N}_0 \tag{8.68}$$

with

$$p(k) = \frac{1}{r} \sum_{i=0}^{r-1} e^{+j\frac{2\pi}{r}ik} = \begin{cases} 1 & k = \lambda \cdot r \\ 0 & k \neq \lambda \cdot r\,. \end{cases} \tag{8.69}$$

It is obviously not necessary to transmit the zero samples of $\tilde{d}(k)$. Because of the periodicity of the complex exponential function, $p(k)$ can be written as the superposition of r complex carrier signals with the frequencies $\Omega_i = \frac{2\pi}{r}i$ $(i = 0, 1, \ldots, r-1)$. According to the modulation theorem, the spectrum of $d'(k)$ results as a superposition of spectrally shifted versions of the baseband spectrum $D_{LP}(e^{j\Omega})$

$$D'(e^{j\Omega}) = \frac{1}{r} \sum_{i=0}^{r-1} D_{LP}(e^{j(\Omega - \frac{2\pi}{r}i)})\,. \tag{8.70}$$

The spectral interrelation is shown schematically in Fig. 8.25.

The result is a spectrally more or less flat broadband excitation signal $\tilde{d}(k)$, which is applied to the synthesis filter. This excitation signal essentially shows the perceptually relevant properties obtained if the predictor produces a spectrally flat residual signal $d(k)$. For unvoiced sounds, a broadband, noise-like residual signal emerges, while for periodic segments $\tilde{d}(k)$ shows a discrete line spectrum. Furthermore, in the baseband, i.e., for

$$0 \leq \Omega \leq \frac{\pi}{r}\,, \tag{8.71}$$

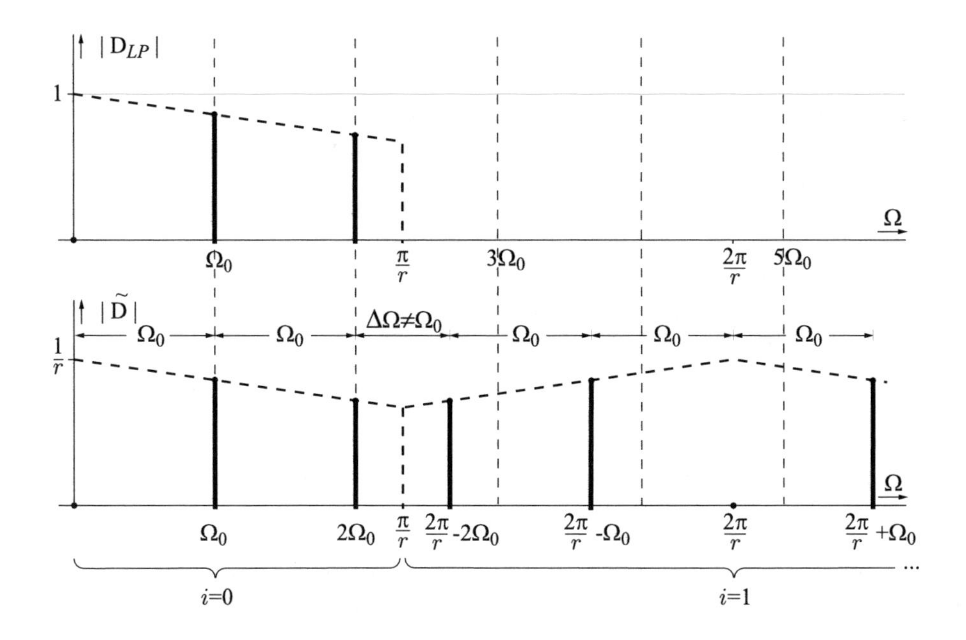

Figure 8.26: Spectral mirroring in the RELP concept

the codec provides transparent transmission except for the quantization noise $\Delta(k) = \tilde{d}(k) - d'(k)$.

Compared to the LPC vocoder, the RELP concept improves the speech quality considerably. The perceptually relevant aspects, which are essential for the naturalness of the speech signal, are, at least in the baseband, transmitted transparently. For the transition between different types of sounds, a mixed voiced/unvoiced excitation is also possible. However, especially for high-pitched voices of women and children, a disturbing metallic or vocoder-like sound becomes noticeable in voiced segments due to the lack of preserving the harmonic structure. In voiced segments, discrete line spectra are generated by the spectral mirroring, as shown in Fig. 8.26. As a result, the spectral components outside the baseband generally do not occur at multiples of the fundamental frequency Ω_0. This effect is less interfering with male voices, since most of the harmonics which hold the biggest part of the energy fall into the baseband. Due to the lowpass filtering, this codec principle is not suited for voiceband data and music signals.

The speech quality can be considerably improved if a long-term predictor is added. During the signal synthesis in voiced segments, the comb filter characteristic of the LTP-synthesis filter amplifies spectral components at multiples of the fundamental frequency and attenuates components lying in between.

Example: GSM Full-Rate Codec

The RELP concept has gained great significance for digital mobile radio communications [Sluijter 2005]. The *full-rate codec* of the European GSM (*Global System for Mobile communications*) standard is based on this principle as shown in Fig. 8.27.

The basic structure corresponds to the concept given in Fig. 8.21-b. Short-term prediction is implemented as an open-loop filter with order $n = 8$, while the long-term predictor (LTP) works in a closed loop. The baseband lowpass filter, the decimator, and the scalar quantizer are located within the LTP loop.

Thus, the prediction gain of the long-term prediction contributes to reducing the effective quantization noise power. Furthermore, the following details must be considered.

The second residual signal designated by $e(k)$ is processed blockwise. The lowpass filter is a linear-phase FIR filter with $m = 11$ coefficients. The filtering process is implemented as a *block filter*, where each $L = 40$ samples $e(k)$ are supplemented by $m - 1$ zeros and the filtered version $e_{LP}(k)$, consisting of $L + m - 1 = 50$ values, is calculated. In the block named RPE (*Regular Pulse Excitation*), an

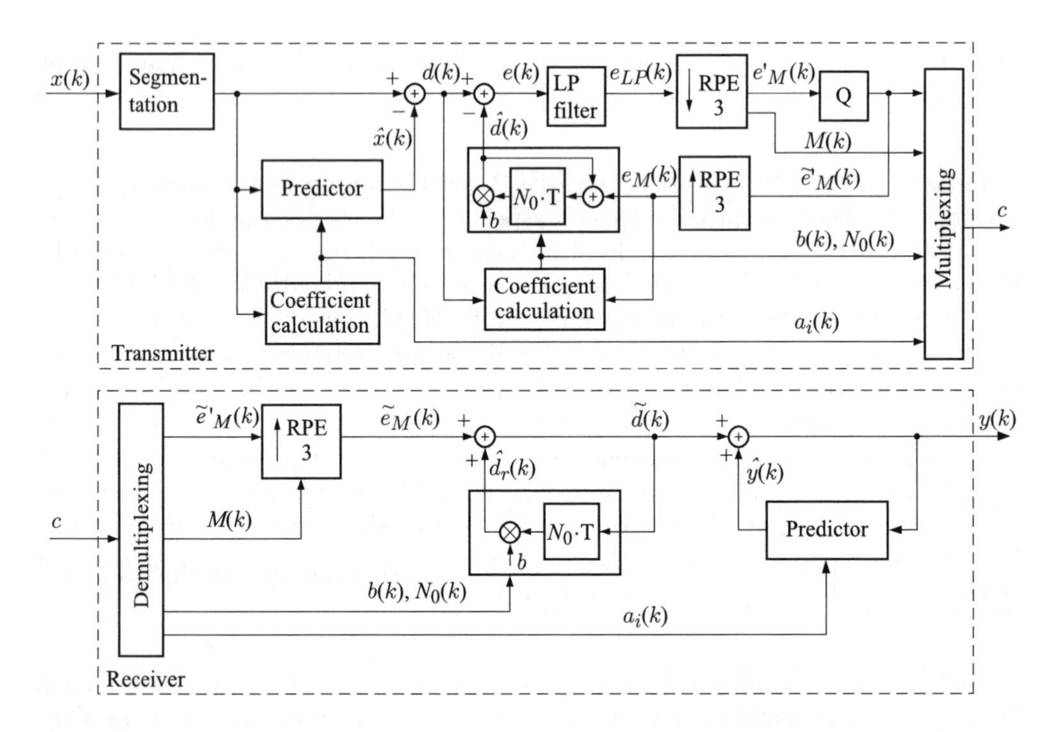

Figure 8.27: Block diagram of the GSM full-rate codec

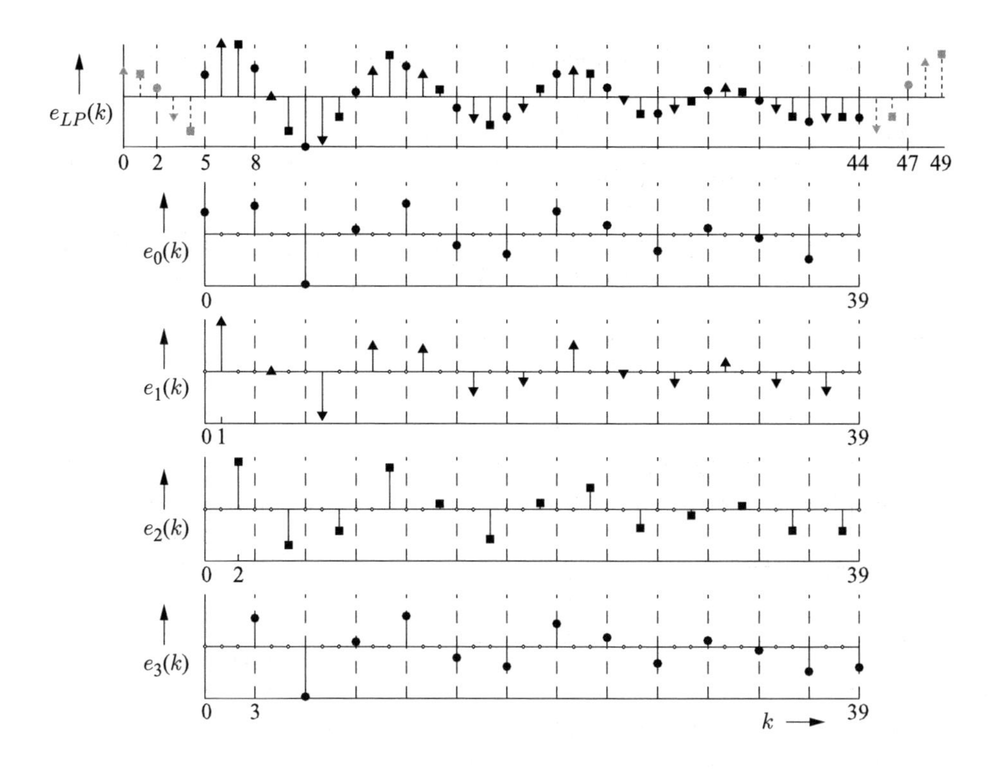

Figure 8.28: Adaptive decimation according to the RPE principle

adaptive downsampling by a factor $r = 3$ is applied.[1] As depicted in Fig. 8.28, the downsampling is adaptive in the sense that four different possibilities exist in principle, which vary in the decimation grid starting at $k = 0$, $k = 1$, $k = 2$, or $k = 3$. The sequence $e_{LP}(k)$ is divided into four subsequences $e_i(k)$, with the sequences $e_0(k)$ and $e_3(k)$ differing only in the first and in the last sample.

From the $L = 40$ central samples of $e_{LP}(k)$, $(k = 5, \dots, 44)$, 13 samples are selected with regard to the best subsequence according to the energy criterion

$$E_M = \max_i \sum_{\lambda=0}^{12} e_{LP}^2(i + \lambda \cdot 3 + 5)\,; \qquad i = 0, 1, 2, 3\,. \tag{8.72}$$

Besides the subsequence $e_M = e_{i_{\text{opt}}}$, the grid position $M = i_{\text{opt}}$ is determined every 5 ms as well. This quasi-random variation helps to avoid the tonal–metallic sound of the baseband–RELP codec. However, especially for high-pitched voices,

[1] This particular variation of the RELP structure with block filtering and adaptive downsampling results as a special case of the so-called RPE method [Kroon et al. 1986]. It is an analysis-by-synthesis coding process, where pulse-shaped excitation sequences are optimized on a regular grid (see Section 8.5.4).

a slight roughness of the reconstructed speech signal is produced [Sluyter et al. 1988].

The 13 selected samples are quantized with AQF (see Section 7.5) by normalizing and uniformly quantizing them with eight levels (3 bits). For the block maximum, a logarithmic quantizer with 2^6 levels is used, while the grid position is represented by 2 bits (see also Appendix A). For the residual signal, this leads to a bit rate of

$$B' = (13 \cdot 3 + 6 + 2)\ \text{bit}/5\,\text{ms} \tag{8.73}$$

$$= 9.4\ \text{kbit/s}\,. \tag{8.74}$$

The LP coefficients are coded as LARs with 36 bits/20 ms = 1.8 kbit/s (see Section 8.4.3) and the LTP parameters N_0 and b with $(7+2)$ bits/5 ms = 1.8 kbit/s. This results in a total bit rate of $B = 13.0$ kbit/s.

The example of Fig. 8.29 provides an insight into the mechanism of the codec. For the different intermediate signals of Fig. 8.27, a short-term spectral analysis was performed. A polyphase filter bank with a 3 dB channel bandwidth of 4 kHz/128 = 31.25 Hz and a 40 dB bandwidth of 62.5 Hz was utilized. The same linear scale was used for all magnitude responses.

Starting from the spectrum of the input signal x, we can clearly observe in the residual signal d the spectral whitening effect of the LP-analysis filter in Fig. 8.29-b. Regarding the spectrum of the LTP residual signal e (Fig. 8.29-c), the closed-loop structure containing a lowpass filter with a cutoff frequency of 4/3 kHz must be considered. Hence, a further prediction gain is noticeable only in the range up to $\approx$ 1.33 kHz. The LTP excitation signal e_M essentially contains the baseband of the residual signal e, which is transmitted to the receiver and broadened by time-varying spectral folding (Fig. 8.29-d).

The harmonic structure is regained in the residual $\tilde{d}$ by means of the LTP-synthesis filter (Fig. 8.29-e), and the spectral envelope is reconstructed through the LP-synthesis filtering (Fig. 8.29-f).

The comparison of Fig. 8.29-a and Fig. 8.29-f shows a transparent transmission in the range up to approx. 1.33 kHz, whereas the spectral amplitude characteristic of the higher-frequency components is only approximately reproduced. Due to the properties of human hearing, a relatively high quality of speech can be obtained, while the computational complexity is relatively small in comparison to the class of CELP codecs, which is the subject of the next section.

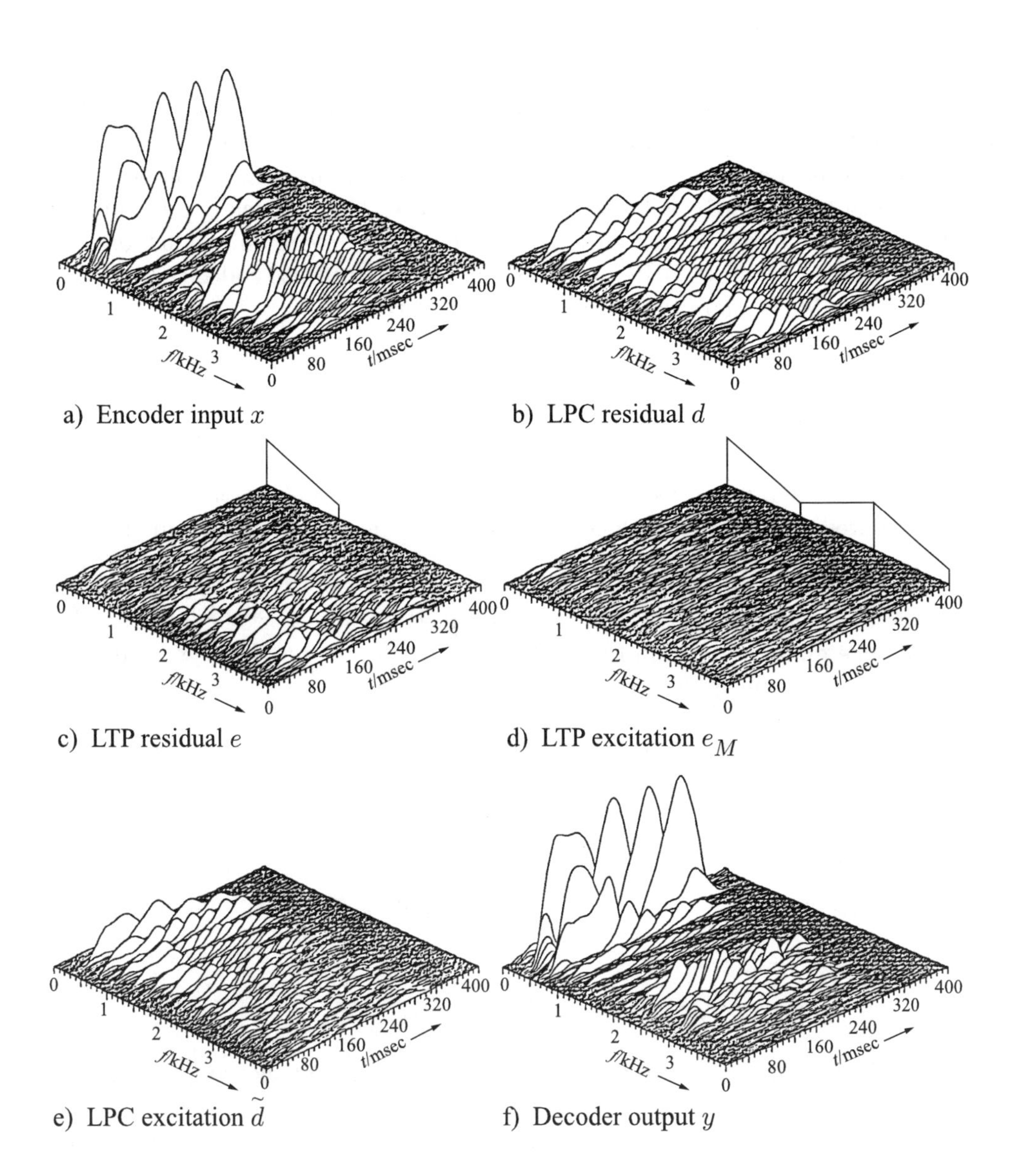

Figure 8.29: Example of the GSM full-rate codec; short-term spectral analysis for the syllable "De" (female voice, according to [Vary, Hofmann 1988])

8.5.3 Analysis by Synthesis: CELP

Code excited linear prediction (CELP) is the most important concept for medium to low bit rate speech codecs. Based on the original proposal [Schroeder, Atal 1985], many variations exist, e.g., [Atal et al. 1991], [Furui, Sondhi 1992], [Goldberg, Riek 2000], [Hanzo et al. 2001], [Chu 2003], and [Kondoz 2004], which can be found in most of the speech codec standards used today in telecommunications, e.g., the GSM half-rate codec (GSM-HR), the GSM enhanced full-rate codec (GSM-EFR), or the GSM/UMTS adaptive multi-rate codec (AMR).

8.5.3.1 Principle

The principle of analysis-by-synthesis coding was derived in Section 8.5.1 from the concept of predictive coding with noise shaping by replacing the scalar quantization of the residual signal $d(k)$ with vector quantization.

From these considerations the structure outlined in Fig. 8.23 resulted, which is depicted once more in detail in Fig. 8.30-a. For reasons of simplicity, we will again neglect the long-term prediction in the beginning.

The encoder on the transmission side contains a complete decoder. For each of the K code book vectors

$$\mathbf{c}_i = \big(c_i(0), c_i(1), \ldots, c_i(L-1)\big)^T ; \qquad i = 1, 2, \ldots, K \tag{8.75}$$

L values $y_i(\lambda)$, $\lambda = 0, 1, \ldots, L-1$, are provisionally synthesized. The reconstruction error

$$e(\lambda) = x'(\lambda) - y_i(\lambda) ; \qquad \lambda = 0, 1, \ldots, L-1 \tag{8.76}$$

is spectrally weighted so that the masking effect of human hearing is implicitly exploited (see also Section 8.5.1). The computationally intensive search of the best vector $\mathbf{c}_{i_{\text{opt}}}$ is based on minimizing the mean square spectrally weighted reconstruction error $e_w(\lambda)$.

8.5.3.2 Fixed Code Book

Since the vector code book according to Fig. 8.30 was originally obtained from a stochastic sequence with Gaussian distribution [Schroeder, Atal 1985], the terms *stochastic code book* or *fixed code book* are commonly used. The following explanations show how the computational complexity of the stochastic code book search

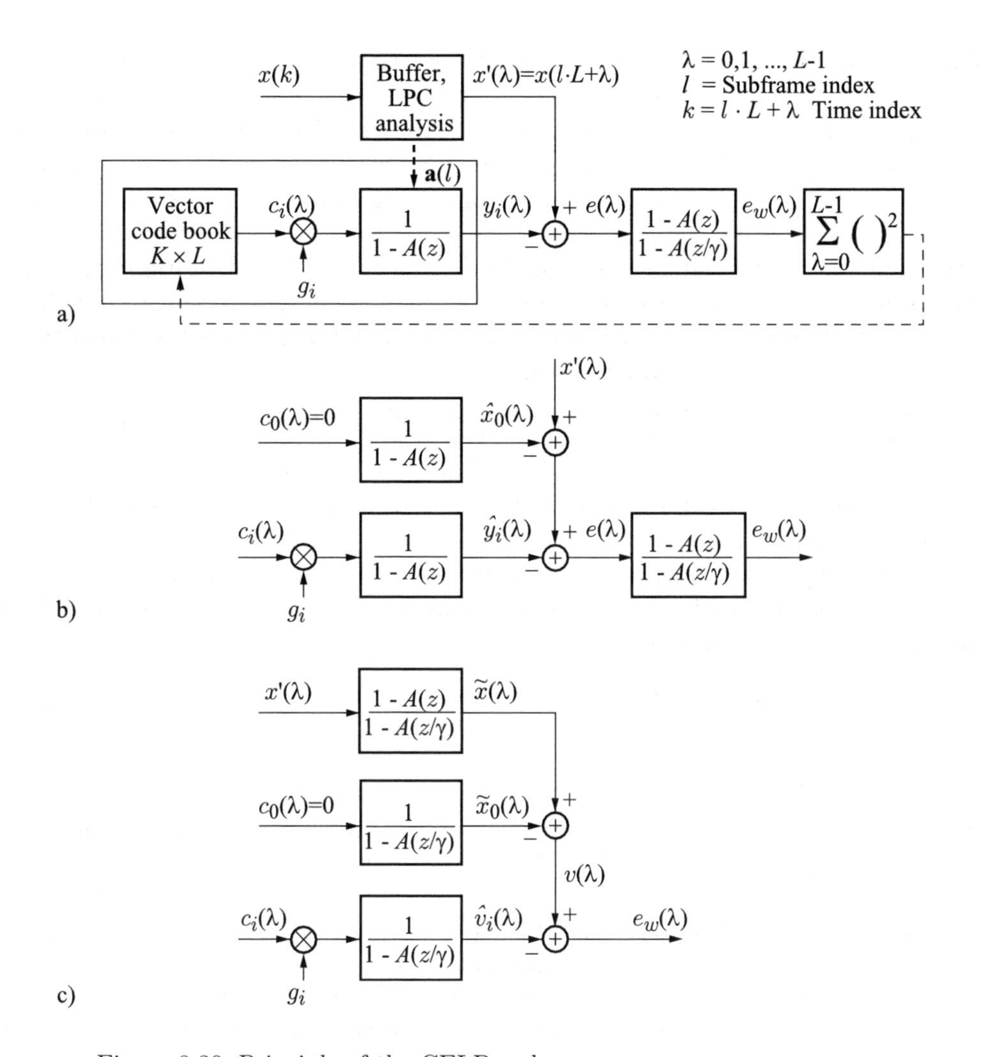

Figure 8.30: Principle of the CELP codec
a) Basic structure of the CELP concept
b) Blockwise processing ($\lambda = 0, 1, \ldots, L-1$) with ringing of the synthesis filter
c) Structure equivalent to b) with reduced complexity

and the respective gain factor g_i can be considerably reduced by modifying the structure of Fig. 8.30-a.

In CELP coders with the stochastic code book, short sequences of the residual waveform, i.e., blocks of L samples, are coded. At the output of the (time-variant) recursive synthesis filter with the transfer function

$$H(z) = \frac{1}{1 - A(z)}, \tag{8.77}$$

each selected optimal excitation vector $\mathbf{c}_{i_{\text{opt}}}$ contributes to the reconstructed signal y_i. Due to the ringing of the synthesis filter, this contribution is not limited to the time interval of length L. Therefore, when coding the l-th signal segment

$$x^{'}(\lambda) = x(l \cdot L + \lambda), \quad \lambda = 0, 1, \ldots, L-1, \tag{8.78}$$

the ringing of the preceding block must be taken into account. In Fig. 8.30-b, this contribution is denoted $\hat{x}_0(\lambda)$. It is generated by means of a second synthesis filter, which is excited by a zero sequence $c_0(\lambda)$. At the beginning of the l-th block its state variables are set to those which the synthesis filter achieved at the end of the $(l-1)$-th block for the best excitation vector. For the l-th block, the ringing contribution $\hat{x}_0(\lambda)$ can be precalculated. Then the modified sequence $x^{'}(\lambda) - \hat{x}_0(\lambda)$ is the target for the search of the new excitation vector $\mathbf{c}_i$.

Due to the linearity of the filter operations, the contribution of each excitation sequence $c_i(\lambda)$ to the weighted error sequence $e_w(\lambda)$ can be described by filtering with a cascade of two filters, i.e., the synthesis and the weighting filter. This yields the effective transmission function

$$\tilde{H}(z) = \frac{1}{1 - A(z)} \cdot \frac{1 - A(z)}{1 - A(z/\gamma)} \tag{8.79-a}$$

$$= \frac{1}{1 - A(z/\gamma)} = \frac{1}{1 - F(z)} . \tag{8.79-b}$$

The result is a marked reduction in computational complexity since now each excitation vector only needs to be filtered once.

With the same linearity argument, the contributions of the signal samples $x^{'}(\lambda)$ and the ringing contribution $\hat{x}_0(\lambda)$ included in $e_w(\lambda)$ can be determined. The result is depicted in Fig. 8.30-c. In principle, the weighting filter is shifted to the three signal branches via the two summation points.

The new target signal $v(\lambda)$, which is only computed once for each interval, must be approximated by the sequence $\hat{v}_i(\lambda)$ $(\lambda = 0, 1, \ldots, L-1)$ with respect to the minimal mean square error by choosing the best excitation vector $\mathbf{c}_i$ and the optimal scaling factor g_i.

Part of the problem can be solved in closed form. If according to (8.79-b) the impulse response of the weighted synthesis filter $1/(1 - A(z/\gamma))$ is denoted by $\tilde{h}(k)$, we have

$$\hat{v}_i(\lambda) = \sum_{\kappa=0}^{L-1} \tilde{h}(\lambda - \kappa) \cdot g_i \cdot c_i(\kappa); \quad \lambda = 0, 1, \ldots, L-1 . \tag{8.80}$$

For simplification the following vector/matrix notation is introduced:

$$\hat{\mathbf{v}}_i = (\hat{v}_i(0), \hat{v}_i(1), \ldots, \hat{v}_i(L-1))^T, \tag{8.81-a}$$

$$\mathbf{v} = (v(0), v(1), \ldots, v(L-1))^T, \tag{8.81-b}$$

$$\mathbf{H} = \begin{pmatrix} \tilde{h}(0) & 0 & 0 & \ldots & 0 \\ \tilde{h}(1) & \tilde{h}(0) & 0 & \ldots & 0 \\ \tilde{h}(2) & \tilde{h}(1) & \tilde{h}(0) & \ldots & 0 \\ \vdots & \vdots & \vdots & \ddots & \vdots \\ \tilde{h}(L-1) & \tilde{h}(L-2) & \tilde{h}(L-3) & \ldots & \tilde{h}(0) \end{pmatrix}. \tag{8.81-c}$$

Now the optimization criterion can be formulated as the squared norm of the reconstruction error vector:

$$\min_{i=1,\ldots,K} \sum_{\lambda=0}^{L-1} \left(v(\lambda) - \hat{v}_i(\lambda)\right)^2 = \min_{i=1,\ldots,K} \|\mathbf{v} - \hat{\mathbf{v}}_i\|^2 . \tag{8.82-a}$$

Hence, the term

$$E_i = \|\mathbf{v} - \hat{\mathbf{v}}_i\|^2 \tag{8.82-b}$$

$$= \|\mathbf{v} - g_i \mathbf{H}\mathbf{c}_i\|^2 \tag{8.82-c}$$

must be minimized.

By partial differentiation of E_i with respect to g_i, the optimal gain factor is determined for each excitation vector $\mathbf{c}_i$:

$$g_i = \frac{\mathbf{v}^T \mathbf{H}\mathbf{c}_i}{\|\mathbf{H}\mathbf{c}_i\|^2} . \tag{8.83}$$

Inserting the optimal gain factor g_i in (8.82-c) leads to an expression for the minimal error energy which can be achieved with each excitation vector $\mathbf{c}_i$

$$E_i = \|\mathbf{v}\|^2 - \frac{(\mathbf{v}^T \mathbf{H}\mathbf{c}_i)^2}{\|\mathbf{H}\mathbf{c}_i\|^2} . \tag{8.84}$$

In view of the optimization, the term $\|\mathbf{v}\|^2$ is constant so that the criterion reduces to the maximization of the fraction

$$\frac{(\mathbf{v}^T \mathbf{H}\mathbf{c}_i)^2}{\|\mathbf{H}\mathbf{c}_i\|^2} \stackrel{!}{=} \max . \tag{8.85}$$

The search for the maximum through variation of i leads to the best excitation vector $\mathbf{c}_{i_{\text{opt}}}$. Then the respective optimal gain factor $g_{i_{\text{opt}}}$ is explicitly calculated according to (8.83).

Table 8.3: Number of arithmetic operations for the code book search

a)	Numerator (8.85)	$L^2 + K(1 + L)$
b)	Denominator (8.85)	$K \cdot L^2$
c)	Division (8.85)	$K \cdot 16$
d)	Division (8.83)	$1 \cdot 16$
	Sum:	$K(17 + L + L^2) + L^2 + 16$

The computational complexity of the code book search can easily be estimated if, in view of a realization with programmable signal processors, a multiplication with subsequent addition is counted as one operation per instruction cycle (multiply–accumulate operation, MAC). For the division we assume a complexity of 16 instruction cycles.

Listed in Table 8.3 are the different numbers of arithmetic operations for the code book search according to (8.83) and (8.85).

For the estimation we assumed that the term $\mathbf{v}^T\mathbf{H}$ is only calculated once with L^2 operations, and that the numerator and denominator of (8.83), which occur as intermediate results when utilizing (8.85), are not recomputed.

Under these assumptions the required mean computational effort results in

$$\text{CE} = \frac{K(17 + L + L^2) + L^2 + 16}{L} f_s \tag{8.86}$$

operations per second. Hence, a typical dimensioning with $f_s = 8\,\text{kHz}$, $K = 256$ and $L = 40$ requires a high computational effort of approx. $85.2 \cdot 10^6$ operations per second.

Numerous approaches to reduce this computational complexity have been proposed in the literature. One successful approach is the choice of code vectors with only a few non-zero components. If, for example, each vector $\mathbf{c}_i$ contains only four non-zero pulses, the number of arithmetic operations for the denominator in (8.85) (see Table 8.3 b)) decreases from $K \cdot L^2$ to $K \cdot 4 \cdot L$. Thus, for $L = 40$ the dominating part of the computational complexity for the code book search is reduced by a factor of 10 (e.g. [Adoul et al. 1987]).

A widely used approach is the choice of a structured code book, which is called the *algebraic code book* (ACELP) [Laflamme et al. 1990]. A typical version can be found in the ITU codec G.729 ([ITU-T Rec. G.729 1996], [Salami et al. 1997]). The excitation sequence $c_i(\lambda)$ of length $L = 40$ contains only four non-zero pulses in the positions $\lambda_0, \lambda_1, \lambda_2$, and λ_3

$$c_i(\lambda) \in \{+1, 0, -1\}\,; \quad \lambda = 0, 1, \ldots, 39\,. \tag{8.87}$$

Table 8.4: ACELP code book: positions for non-zero pulses

Pulse positions	Sign	Tracks	Number of bits
λ_0	s_0	$0, 5, 10, 15, 20, 25, 30, 35$	3+1
λ_1	s_1	$1, 6, 11, 16, 21, 26, 31, 36$	3+1
λ_2	s_2	$2, 7, 12, 17, 22, 27, 32, 37$	3+1
λ_3	s_3	$3, 8, 13, 18, 23, 28, 33, 38$ $4, 9, 14, 19, 24, 29, 34, 39$	4+1

The 40 possible positions for non-zero pulses are divided into four tracks with 8 or 16 positions as indicated in Table 8.4.

With the signs $s_0, \dots, s_3$ of the four selected pulses we construct the excitation sequence

$$c_i(\lambda) = s_0\,\delta(\lambda - \lambda_0) + s_1\,\delta(\lambda - \lambda_1) + s_2\,\delta(\lambda - \lambda_2) + s_3\,\delta(\lambda - \lambda_3) \tag{8.88}$$

$$\delta(\lambda) = \begin{cases} 1 & \lambda = 0 \\ 0 & \text{else}\,. \end{cases}$$

This sparse ternary code book allows an efficient search procedure with significantly reduced complexity.

8.5.3.3 Long-Term Prediction, Adaptive Code Book

So far, the principle of CELP coding has been discussed without consideration of long-term prediction. To exploit the highly periodic nature of speech signals, occurring especially during voiced speech segments, the introduction of a long-term predictor is essential. Figure 8.31-a shows the CELP codec of Fig. 8.30-a extended by an LTP-synthesis filter. For the excitation sequence of the LP-synthesis filter we have

$$u^{'}(\lambda) = u(l \cdot L + \lambda) = g_i \cdot c_i(\lambda) + b \cdot u^{'}(\lambda - N_0)\,. \tag{8.89-a}$$

As shown in Fig. 8.32, we combine L consecutive samples of $u(k)$ according to

$$u^{'}(\lambda - j) = u(l{\cdot}L + \lambda - j) \doteq u_j(\lambda)\,;\; j = \text{const.},\; \lambda = 0, 1, \dots, L\text{–}1 \tag{8.89-b}$$

to a vector

$$\mathbf{u}_j = (u_j(0), u_j(1), \dots, u_j(L-1))^T \tag{8.89-c}$$

so that (8.89-a) yields

$$\mathbf{u}^{'} \equiv \mathbf{u}_0 = g_i\ \mathbf{c}_i + b\ \mathbf{u}_{N_0}\,. \tag{8.89-d}$$

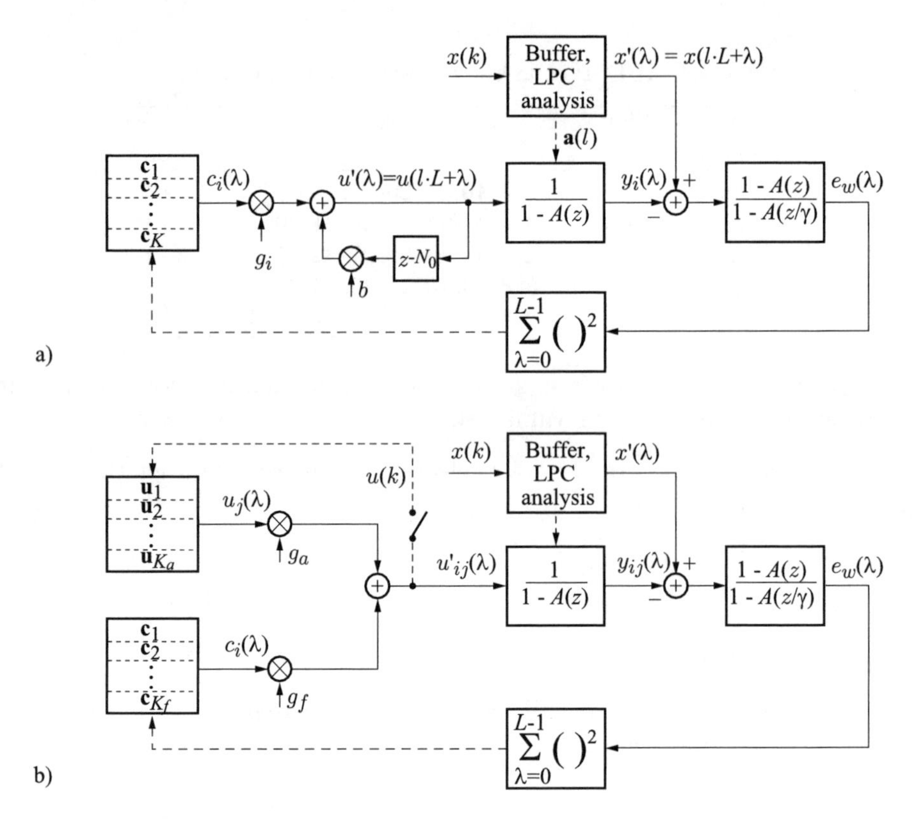

Figure 8.31: CELP codec with short-term and long-term prediction
a) Conventional realization of the LTP-synthesis filter
b) Realization of the LTP loop by means of an adaptive code book

This vector superposition is illustrated in Fig. 8.31-b. A generalization that still needs to be justified is made by substituting the pitch parameter N_0 with an index j ($j = 1, 2, \ldots, K_a$). In this concept, the LTP-synthesis filter is replaced by a pseudo code book containing vectors $\mathbf{u}_j$ of length L, which are overlapping segments of the recent past of the LP excitation signal. In order to distinguish the two partial contributions $g_i\,\mathbf{c}_i$ and $b\,\mathbf{u}_j$ according to (8.89-d), double indexing is introduced according to u'_{ij}, indicating the two contributions c_i and u_j. Furthermore, the weighting factors are replaced by $g_a = b$ and $g_f = g_i$. With the vector components $u_0(\lambda)$, $c_i(\lambda)$, and $u_j(\lambda)$ from (8.89-d), the excitation sequence $u'_{ij}(\lambda)$ of Fig. 8.31 is generated:

$$u'_{ij}(\lambda) \equiv u_0(\lambda) = g_i\; c_i(\lambda) + b\; u_j(\lambda) \tag{8.89-e}$$

$$= g_f\; c_i(\lambda) + g_a\; u_j(\lambda) \tag{8.89-f}$$

with $\lambda = 0, 1, \ldots, L-1$.

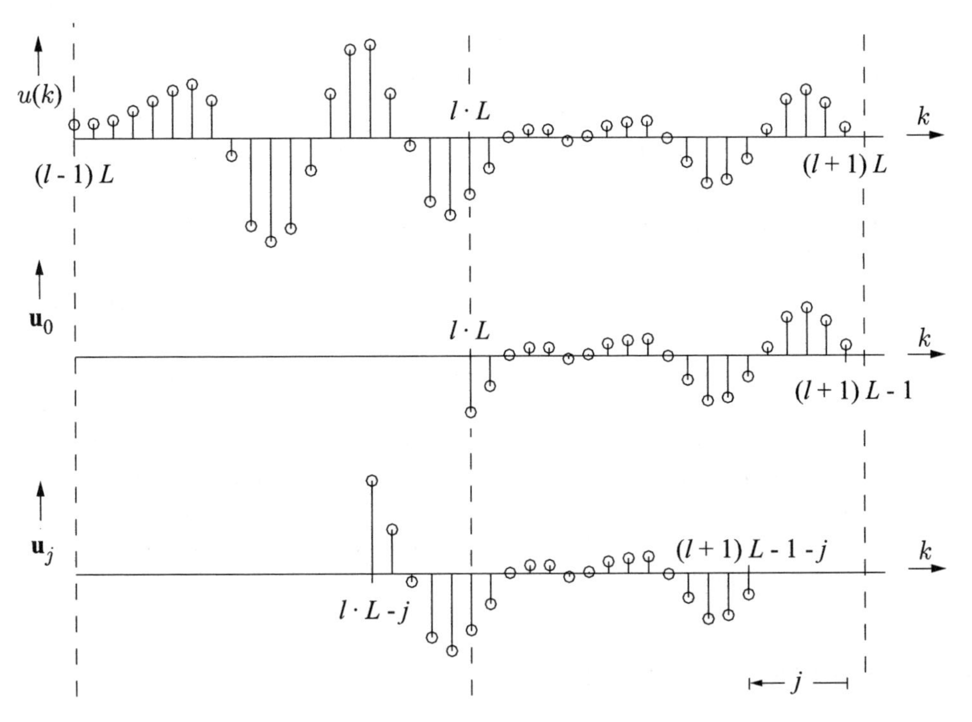

Figure 8.32: Definition of the vectors $\mathbf{u}_j$ of the adaptive code book ($L = 14$, $j = 0$, and $j = 4$)

For $j = N_0$ the terms (8.89-a) and (8.89-e) obviously coincide. The advantage of the modified structure is that the contribution of the LTP loop can be treated in the same way as that of the fixed code book. In particular, in an analysis-by-synthesis procedure, the LTP parameters b and N_0 or g_a and j_{opt}, respectively, can be determined just like the parameters $g_f = g_i$ and i_{opt}. This procedure is called *closed-loop* LTP. Since the content of the second code book changes depending on the signal, it is called the *adaptive code book*. Due to the *closed-loop* search of the LTP parameters g_a and j_{opt}, considerably better speech quality and a higher SNR of the synthesized speech signal $y(k)$ by 2–5 dB are achieved [Singhal, Atal 1984].

Note:

- In order to reach the optimal result, a complete search of all combinations of the two excitation vectors $\mathbf{u}_j$ and $\mathbf{c}_i$ with the respective optimal weighting factors g_a and g_f would have to be performed. Since this procedure has an extremely high computational complexity for the common code book dimensions, a suboptimal solution is chosen by a sequential search. First, in analogy to the criterion given in (8.85) and (8.83), the optimal adaptive code book entry and gain are determined. For the search we can make use of the fact that two successive vectors $\mathbf{u}_j$ and $\mathbf{u}_{j+1}$ consist of the same elements except for the first and last element of $\mathbf{u}_j$ and $\mathbf{u}_{j+1}$, respectively.

The selected code book entry is scaled and filtered by the LP-synthesis filter and subtracted from the speech signal. The resulting modified target vector is used for the stochastic code book search by applying (8.85) and (8.83).

- While searching for the best contribution $\mathbf{u}_j$, the adaptive code book is not altered for the current subframe of length L. Only after completing the search for the best vector $\mathbf{c}_i$ from the fixed code book is the adaptive codebook updated.
- Strictly speaking, the equivalence between Fig. 8.31-a and 8.31-b is only given for $j_{\text{opt}} = N_0 \geq L$. Due to the block processing with subframe length L and the sequential search in the adaptive and fixed code book, depending on the constellation of j and λ no definite samples $u(k)$ are available for $N_0 < L$ when searching the adaptive code book, since the contribution from the fixed code book is not yet known. In this case, the missing entries of the adaptive code book are generated, for example, from the periodic repetition of the last N_0 samples of the excitation sequence $u(k)$ of the preceding frame (subframe index $l-1$).

Examples: EFR and AMR Codecs

Most of the recent standards in speech coding are based on the concept of CELP. Two representative examples are the GSM-enhanced full-rate codec (EFR) and the GSM-adaptive multi-rate codec (AMR). In comparison to the first GSM full-rate codec as described in Section 8.5.2, the EFR codec ([ETSI Rec. GSM 06.60 1996], [Järvinen et al. 1997]) gives a substantially better quality which is almost equivalent to the speech quality of ADPCM at 32 kbit/s [ITU-T Rec. G.726 1990].

A simplified block diagram of the EFR codec is given in Fig. 8.33. The codec structure contains the basic CELP elements of a fixed and an adaptive code book with individual gain factors g_f and g_a. In addition, the decoded output signal $\hat{s}(k)$ is processed by an adaptive postfilter to improve the subjectively perceived speech quality (see Section 8.6). The frame length is 20 ms, and twice per frame 10 LP coefficients are calculated, which are transformed for quantization into the LSF representation. A dedicated vector matrix quantization scheme is applied for each set of 20 LSF coefficients [ETSI Rec. GSM 06.60 1996].

The EFR codec is based on a fixed algebraic code book (ACELP) [Salami et al. 1997b] with a subframe length of $L = 40$ consisting of effectively $K = 2^{35}$ vectors of dimension 40 with 10 non-zero values $s_\mu \in \{+1, \ -1\}$ each:

$$c_i(\lambda) \; = \; \sum_{\mu=0}^{9} s_\mu \cdot \delta(\lambda - \lambda_\mu) \, ; \qquad s_\mu \in \{+1, \ -1\} \, . \tag{8.90}$$

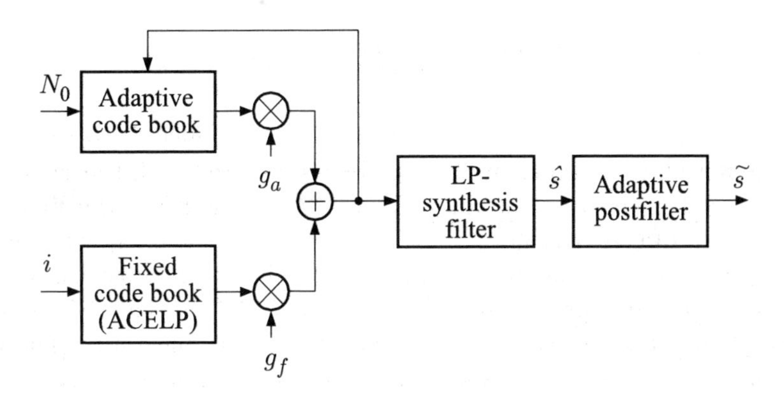

Figure 8.33: Simplified block diagram of the enhanced full-rate codec (GSM-EFR)

The code vectors are constructed, as shown in Table 8.5, from five tracks of eight interleaved positions, with two pulses per track. The signs of the two pulses are encoded with only one bit. This bit indicates the sign of the first pulse whereas the sign of the second pulse is depending on its position relative to the first pulse.

The bit allocation is summarized in Table 8.6.

The second example is the adaptive multi-rate (AMR) codec ([ETSI Rec. GSM 06.90 1998], [Bruhn et al. 1999], [Ekudden et al. 1999], [Järvinen 2000]), which was designed for GSM and UMTS. It can be considered as an extension of the EFR codec. The AMR codec has eight different bit rates from 4.75 kbit/s to 12.2 kbit/s. The mode of the highest bit rate is identical to the EFR codec. The overall structure is very similar to the block diagram of Fig. 8.33. The different bit rates are primarily achieved by using different ACELP code books (see also Appendix A). The objective of the AMR codec is to increase error robustness for better speech quality in adverse channel conditions. In GSM, the codec can be used in the half-rate and the full-rate channel. In the latter case, the gross bit rate including channel coding is 22.8 kbit/s. The bit rate allocation between speech coding and channel coding is controlled dynamically by the network. The codec mode can be switched

Table 8.5: ACELP code book of the EFR codec

Pulse position	Sign	Positions	Bits
$\lambda_0\,,\lambda_5$	s_0, s_5	$0, 5, 10, 15, 20, 25, 30, 35$	$2 \times 3 + 1$
$\lambda_1\,,\lambda_6$	s_1, s_6	$1, 6, 11, 16, 21, 26, 31, 36$	$2 \times 3 + 1$
$\lambda_2\,,\lambda_7$	s_2, s_7	$2, 7, 12, 17, 22, 27, 32, 37$	$2 \times 3 + 1$
$\lambda_3\,,\lambda_8$	s_3, s_8	$3, 8, 13, 18, 23, 28, 33, 38$	$2 \times 3 + 1$
$\lambda_4\,,\lambda_9$	s_4, s_9	$4, 9, 14, 19, 24, 29, 34, 39$	$2 \times 3 + 1$

Table 8.6: Bit allocation of the EFR codec

Parameter	Subframes 1&2 (5 ms each)	Subframes 3&4 (5 ms each)	Bits per frame 20 ms
2×10 LPC coefficients (*line spectral frequencies*)			38
Adaptive code book			
• delay N_0	9	6	30
• gain g_a	4	4	16
Fixed code book (ACELP)			
• pulse positions and signs	35	35	140
• gain g_s	5	5	20
Bits every 20 ms			244

every 20 ms. If the channel gets worse, the bit rate for channel coding is increased at the expense of the bit rate for speech coding. In the GSM full-rate channel, the AMR codec extends the lower C/I limit (Carrier-to-Interference limit) of the EFR codec of C/I $=$ 9 dB down to about C/I $\geq$ 4–7 dB. The introduction of this codec into existing GSM networks improves significantly the coverage, especially in buildings.

A simulation example is shown in Fig. 8.34.

The AMR codec was adopted by the 3GPP (3rd Generation Partnership Project) as the default speech codec for 3G wideband CDMA systems (UMTS, CDMA2000) [3GPP TS 26.090 2001].

A further development of the AMR concept is the adaptive multi-rate wideband speech codec (AMR-WB) ([Bessette et al. 2002], [3GPP TS 26.190 2001]) which extends the audio bandwidth to 7 kHz (see also Appendix A). The AMR-WB codec is also based on the ACELP code book and uses artificial wideband extension to synthesize the signal beyond 6.4 kHz as proposed in [Paulus 1996]. The codec has nine different bit rates from 23.85 kbit/s down to 6.6 kbit/s. In the case of clean speech without background noise and without transmission errors, the six highest modes (23.85–14.25 kbit/s) offer a speech quality which is equal to or better than that of the wire-line wideband codec G.722 ([ITU-T Rec. G.722 1988], split band ADPCM, 48, 56, and 64 kbit/s). The speech quality of the 12.65 kbit/s mode is at least equal to G.722 at 56 kbit/s, while the 8.85 kbit/s mode is still comparable to G.722 at 48 kbit/s. The two lowest modes (8.85 kbit/s and 6.6 kbit/s) are used only for adverse channel conditions or during network congestion. The AMR-WB codec has also been adopted as an ITU-T standard for multimedia applications [ITU-T Rec. G.722.2 2002].

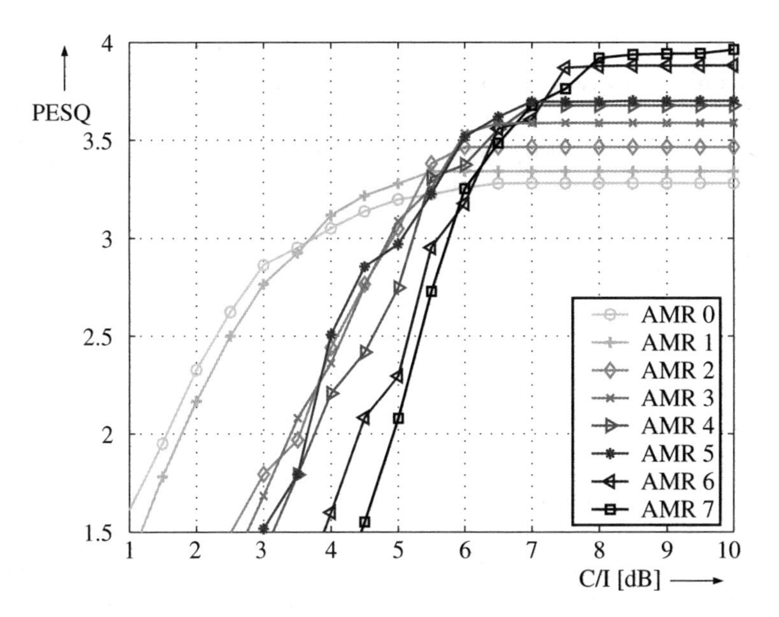

Figure 8.34: Speech quality improvement in adverse radio channel conditions by using the AMR codec; GSM channel TU50, average PESQ values for 70 seconds of speech
PESQ: *Perceptual Evaluation of Speech Quality* (see also Appendix B)
C/I: Carrier-to-Interference ratio

8.5.4 Analysis by Synthesis: MPE, RPE

In this section, two special variations of predictive analysis-by-synthesis coding are introduced, which can be considered to be intermediate solutions in the evolution of the CELP concept in retrospect. The methods in question are the MPE (*Multi-Pulse Excitation*) coding [Atal, Remde 1982], which has only gained a limited practical significance, and, derived from it, the general form of the RPE (*Regular Pulse Excitation*) coding [Kroon et al. 1986], which, in highly simplified form, is the basis for the GSM full-rate codec (see also Section 8.5.2).

8.5.4.1 MPE

For the description of this method, we reconsider the block diagram in Fig. 8.30, again disregarding the LTP filter in order to simplify the presentation of the general relations.

In contrast to CELP, no given vector code book is available for the excitation of the synthesis filter in the MPE approach (Fig. 8.30-a). Instead of the excitation vectors $g_i\,\mathbf{c}_i$, $i = 1, 2, \ldots, K$, $K \gg L$, used in a CELP codec, only one single excitation vector $\mathbf{c}$ exists in the MPE codec, which is constructed for each subframe of length L. This excitation vector contains elements other than zero only in $M < L$ positions λ_i, $i = 1, 2, \ldots, M$. This structure can be obtained through weighted superposition of M elementary vectors $\mathbf{e}_\mu$, which only show one unit pulse differing from zero. The multi-pulse excitation is given by

$$\mathbf{c} = \sum_{\mu=1}^{M} p_\mu \cdot \mathbf{e}_\mu \tag{8.91}$$

with

$$e_\mu(\lambda) = \begin{cases} 1 & \lambda = \lambda_\mu \\ 0 & \lambda \neq \lambda_\mu \end{cases} \quad \lambda = 0, 1, \ldots, L-1\,. \tag{8.92}$$

The optimization problem is to determine the optimal M pulse positions λ_μ and the optimal pulse amplitudes p_μ with respect to the smallest energy of the spectrally weighted error signal $e_w(\lambda)$ $(\lambda = 0, 1, \ldots, L-1)$. There is no closed-form analytic solution to the problem and a complete analysis of the countable, finite possibilities for quantization of the amplitudes p_μ must be discarded because of the computational effort. Therefore, a suboptimal, iterative solution is chosen, which searches the optimal pulses and amplitudes sequentially one pulse at a time in M steps.

In the first step we set

$$\mathbf{c} = p_1\, \mathbf{e}_1\,, \tag{8.93}$$

with p_1 and λ_1 still being undetermined. Thus, in analogy to the search in the CELP codec (Eqs. (8.81-c) to (8.85)), the optimal amplitude p_1 can be written in its general form according to

$$p_1 = \frac{\mathbf{v}^T \mathbf{H} \mathbf{e}_1}{\|\mathbf{H}\mathbf{e}_1\|^2} \tag{8.94}$$

(see also Fig. 8.30).

For this amplitude, depending on the L possible pulse positions λ_1, the error energy can be determined (see (8.84)) so that, according to (8.85), the criterion reduces to

$$\frac{(\mathbf{v}^T \mathbf{H} \mathbf{e}_1)^2}{\|\mathbf{H}\mathbf{e}_1\|^2} \stackrel{!}{=} \max\,. \tag{8.95}$$

The respective computational effort is very small, since vector $\mathbf{e}_1$ has only one single non-zero component.

As in the CELP approach, first vector $\mathbf{e}_1$ or pulse position λ_1 is determined according to (8.95). The pulse amplitude p_1 is calculated with (8.94).

In the second iteration step, the contribution of the first pulse $\hat{\mathbf{v}}_1 = p_1 \mathbf{H} \mathbf{e}_1$ is subtracted from the original target vector $\mathbf{v}$ (see Fig. 8.30-c)

$$\mathbf{v}_2 = \mathbf{v} - \hat{\mathbf{v}}_1 \tag{8.96}$$

before the next pulse is determined. The pulse positions λ_2 and the pulse amplitude p_2 are then determined with an altered target vector in analogy to λ_1 and p_1.

The remaining parameters λ_μ and p_μ are treated accordingly, leading to the complete vector according to (8.91) after M iteration steps.

Depending on the required speech quality, $M = 3 - 8$ pulses per 4 ms ($L = 32$) are needed. Logarithmic quantization of the amplitudes p_μ with (only) 5 bits each, and coding of the pulse positions λ_μ with 5 bits each, yield a data rate of 7.5 to 20 kbit/s. The bit rate for the LP filter and, if applicable, for the LTP filter has to be added. Hence, the minimal bit rate ranges in the region of 10 kbit/s.

8.5.4.2 RPE

One disadvantage of the MPE method is that a significant part of the available bit rate is used for coding the pulse positions, thus strongly limiting the number M of pulses. An alternative solution was proposed in [Kroon et al. 1986] which allows only a few, regular spaced pulse positions but increases the number of pulses significantly (see also [Sluijter 2005]). For the pulse positions K uniform grids are given as outlined in Fig. 8.35. K different excitation vectors $\mathbf{c}_i$ ($i = 1, 2, \ldots, K$) must be optimized which have only non-zero values on the grid positions.

Only the excitation vector $\mathbf{c}_{i_{\text{opt}}}$, leading to the minimal perceptually weighted error between the original and reconstructed signal, is transmitted. For a typical

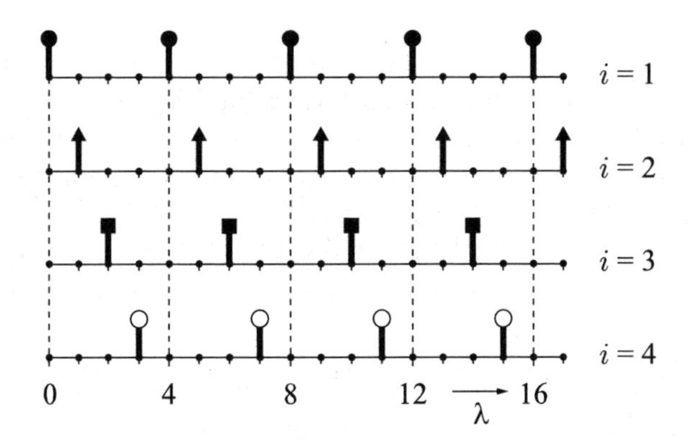

Figure 8.35: Definition of the RPE grids ($K = 4$)

dimension of $K = 4$, only 2 bits are required for coding the grid positions. Thus, the bit rate available for the stochastic excitation can be used to a large extent for scalar or vector quantization of the pulse amplitudes $p_i(\lambda)$.

The search for the optimal excitation vector

$$\mathbf{c}_{i_{\text{opt}}} \in \{\mathbf{c}_1, \mathbf{c}_2, \ldots, \mathbf{c}_K\} \tag{8.97}$$

consists of computing for each possible grid position i (see Fig. 8.35) with

$$c_i(\lambda) = \begin{cases} p_i(\lambda) & \lambda = \rho \cdot K + i - 1 \\ 0 & \lambda \neq \rho \cdot K + i - 1 \end{cases} \quad i = 1, 2, \ldots, K\,,\ \rho \in \mathbb{Z} \tag{8.98}$$

the amplitudes $p_i(\lambda)$ and the reconstruction error. In contrast to the MPE method, here the optimization can in principle be performed by a closed-form analytical solution.

The $m = L/K$ non-zero amplitudes $p_i(\lambda)$ of the vectors $\mathbf{c}_i$ are combined to vectors $\mathbf{g}_i$ and the condition (8.98) is described with a matrix $\mathbf{B}_i$ which has m columns, L rows, and its elements only take the values 0 or 1:

$$\mathbf{c}_i = \mathbf{B}_i \cdot \mathbf{g}_i\,. \tag{8.99}$$

The error energy E_i caused by vector $\mathbf{c}_i$ is, in analogy to (8.82-c) and Fig. 8.30,

$$E_i = \|\mathbf{v} - \mathbf{H}\mathbf{c}_i\|^2 \tag{8.100-a}$$

$$= \|\mathbf{v} - \mathbf{H}\mathbf{B}_i\mathbf{g}_i\|^2\,. \tag{8.100-b}$$

Partial differentiation with respect to the unknown vector $\mathbf{g}_i$ yields

$$\frac{\partial E_i}{\partial \mathbf{g}_i} = -2\ (\mathbf{H}\mathbf{B}_i)^T \left(\mathbf{v} - \mathbf{H}\mathbf{B}_i\mathbf{g}_i\right) \stackrel{!}{=} \mathbf{0} \tag{8.101}$$

or, accordingly, the general solution

$$\mathbf{g}_i = \left(\mathbf{B}_i^T\mathbf{H}^T\mathbf{H}\mathbf{B}_i\right)^{-1}(\mathbf{H}\mathbf{B}_i)^T\mathbf{v}\,. \tag{8.102}$$

Inserting (8.102) in (8.100-b) leads to a term for the error energy as a function of the grid positions $i = 1, 2, \ldots, K$.

Therefore, in the practical application, the best grid position i is determined first by minimizing the energy, and then the optimal excitation vector is established with (8.102) and (8.99).

Since the amplitudes $p_i(\lambda)$ are computed in non-quantized form, the subsequently performed amplitude quantization, and thus the quantization error, is not part of the optimization process.

The structure of the GSM full-rate codec (see Section 8.5.2) can be derived from this approach as a special case.

8.6 Adaptive Postfiltering

Predictive speech coding with effectively 0.5 to 2 bits per sample is the key element of many commercial applications such as mobile communication. This can be largely attributed to the technique of noise shaping or spectral error weighting, respectively, where at the speech encoder the reconstruction error is shaped. This spectral masking improves the perceptual speech quality significantly. However, admitting larger reconstruction errors in the region of the formants leads to an increased error power. The signal-to-noise ratio over frequency takes its highest values in the intervals with spectral peaks, while it is significantly worse in the spectral valleys (see also Fig. 8.10). Typically, the noise around the spectral peaks is below the masking threshold, while in valley regions it is not. Consequently, at low bit rates spectral error weighting alone is not sufficient to completely mask the noise so that an audible speech-dependent noise remains.

At the output of the speech decoder the effect of the non-masked quantization noise can be reduced to a certain extent through an adaptive postfilter (see Fig. 8.36). Since most of the perceived noise components come from spectral valleys, the postfilter attenuates the frequency components between pitch harmonics, as well as the components between formants. In speech perception, the formants and local spectral peaks are much more important than spectral valley regions. Therefore, by attenuating the components in spectral valleys, the postfilter only introduces minimal distortion in the speech signal, while a substantial noise reduction can be achieved.

In the literature, various postfilters which exploit this effect have been proposed. A unifying presentation, on which the following considerations are based, can be found in [Chen, Gersho 1995]. Different variants of this general approach are contained in various codec standards (see Appendix A).

The desired characteristics of the instantaneous frequency response of an adaptive postfilter are depicted for a voiced signal segment in Fig. 8.37.

The frequency response of the postfilter follows the spectral envelope and the spectral fine structure in such a way that the formants and the local maxima are preferably not altered, whereas the spectral valleys between the formants and pitch

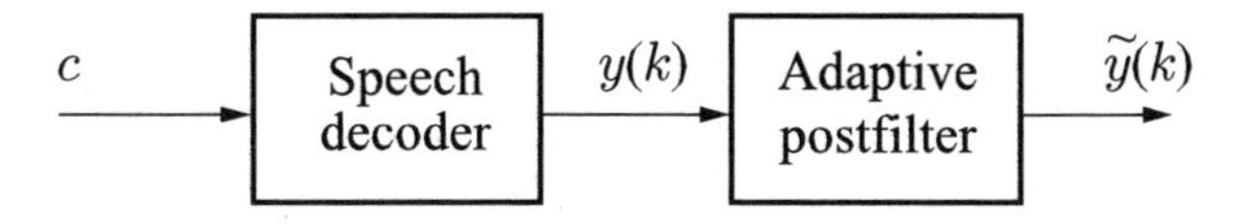

Figure 8.36: Adaptive postfiltering of the decoded speech signal

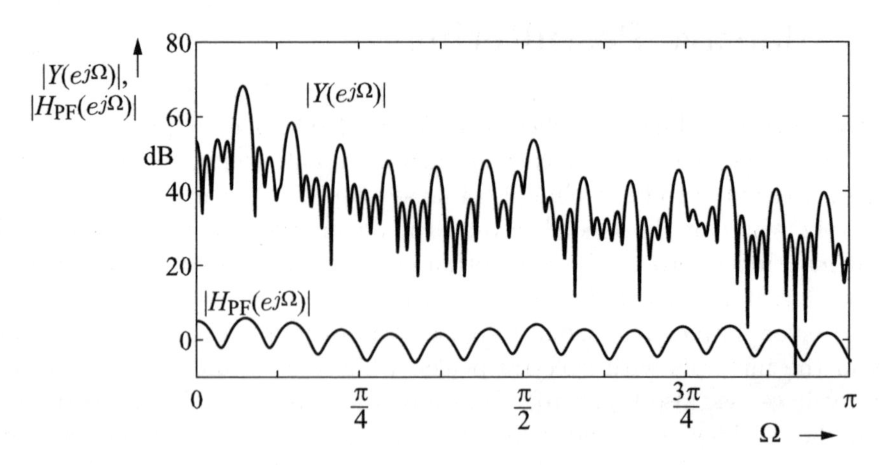

Figure 8.37: Signal spectrum $|Y(e^{j\Omega})|$ and magnitude response $|H_{\mathrm{PF}}(e^{j\Omega})|$ of an adaptive postfilter

harmonic peaks are attenuated. The filter parameters can generally be derived from the coefficients of the speech decoder's LTP- and LP-synthesis filters, or it can be determined by signal $y(k)$.

The required attenuation of the spectral valleys can be performed separately for the spectral envelope and for the spectral fine structure. Consequently, a short-term and a long-term postfilter with the transfer functions $H_A(z)$ and $H_P(z)$ are cascaded to achieve the desired characteristics:

$$H_{\mathrm{PF}}(z) = H_A(z) \cdot H_P(z)\,. \tag{8.103}$$

For the construction of $H_A(z)$, the pole radii of the LP-synthesis filter are scaled with a factor α according to

$$\tilde{H}_{\mathrm{LP}}(z) = \frac{1}{1 - A(z/\alpha)}\,; \qquad 0 \leq \alpha \leq 1\,. \tag{8.104}$$

As a result, the poles are moved radially towards the origin of the z-plane. The effect on the magnitude response is shown in Fig. 8.38.

With decreasing α the resonances become less pronounced as well. However, the magnitude responses also show a tilt towards higher frequencies, i.e., the relative intensity of the formants will change due to the postfilter, which is not desired.

This tilt can be largely avoided by means of the following transfer function:

$$H_A(z) = g_A \cdot \frac{1 - A(z/\beta)}{1 - A(z/\alpha)} \cdot (1 - \gamma \cdot z^{-1})\,; \qquad 0 \leq \alpha,\, \beta,\, \gamma \leq 1\,. \tag{8.105}$$

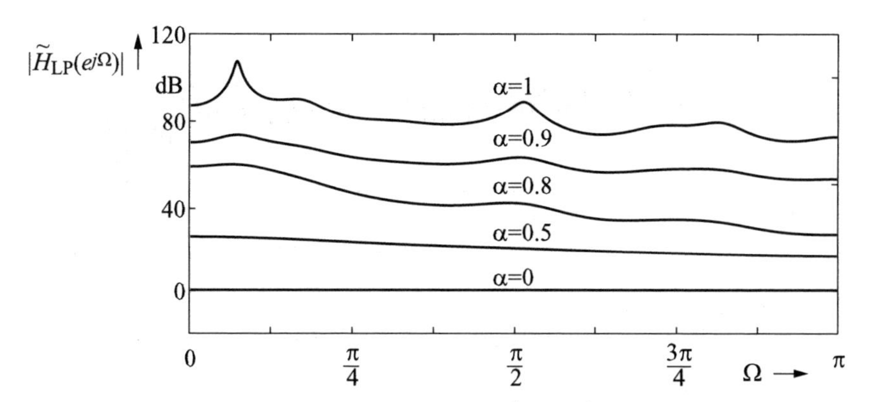

Figure 8.38: Magnitude response of $\tilde{H}_{LP}(e^{j\Omega})$ for different α; adjacent curves are raised by 20 dB each

Here g_A denotes a (time-variant) scaling factor which ensures that the power of the signal remains unchanged by the filtering process.

On a logarithmic scale, the relation between the terms $(1 - A(z/\beta))$ and $(1 - A(z/\alpha))$ appears as the difference between two curves of Fig. 8.38. Thus, the spectral tilt is already partly compensated. With a fixed (or adaptive) coefficient γ, the last product term $(1 - \gamma \cdot z^{-1})$ performs the remaining tilt compensation.

Figure 8.39-a,b exemplarily depicts the speech spectrum and the resulting magnitude response $|H_A(e^{j\Omega})|$ of the short-term postfilter.

The long-term postfilter $H_P(z)$ is either derived from the long-term predictor $P(z)$ or obtained by means of a renewed LTP analysis of the decoded signal $y(k)$. The latter approach is preferable if the LTP parameters (gain factor b or g_a, respectively, and delay N_0) on the transmission side are determined by an analysis-by-synthesis procedure. Due to the error criterion, the delay N_0 does not necessarily match the instantaneous pitch period of the signal in this case.

The periodic fine structure of Fig. 8.37 can be realized using the following approach:

$$H_P(z) = g_P \cdot \frac{1 + \varepsilon \cdot z^{-N_0}}{1 - \eta \cdot z^{-N_0}}\,; \qquad 0 \leq \varepsilon, \eta \leq 1\,, \tag{8.106}$$

with the scaling factor g_P. The transfer function $H_P(z)$ of the pole–zero postfilter has its poles at

$$z_{\infty i} = \rho \cdot e^{j\Omega_{\infty i}} \quad \text{with} \;\; \rho = \frac{1}{\sqrt[N_0]{\eta}}\,; \quad \Omega_{\infty_i} = \frac{2\pi}{N_0}\, i \tag{8.107}$$

and its zeros at

$$z_{0i} = \zeta \cdot e^{j\Omega_{0i}} \quad \text{with} \;\; \zeta = \frac{1}{\sqrt[N_0]{\varepsilon}}\,; \quad \Omega_{0_i} = \frac{\pi}{N_0}\,(2i+1) \tag{8.108}$$

for $i = 0, 1, \ldots, N_0 - 1$.

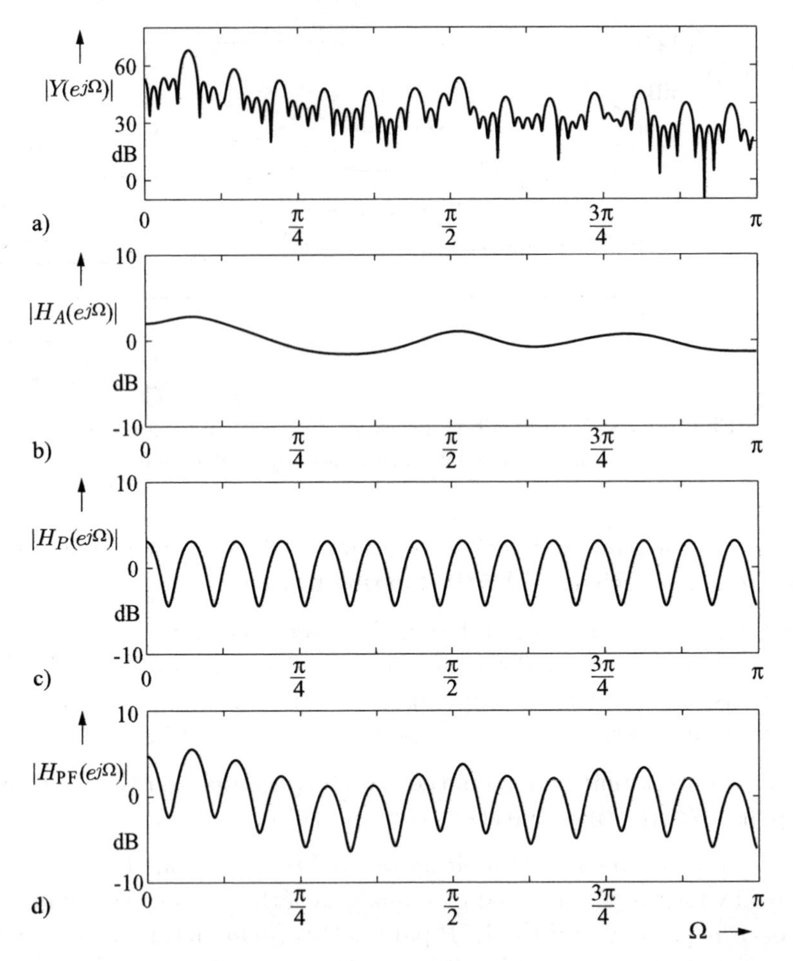

Figure 8.39: a) Magnitude spectrum of speech $|Y(e^{j\Omega})|$
b) Magnitude response of the short-term postfilter $|H_A(e^{j\Omega})|$
c) Magnitude response of the long-term postfilter $|H_P(e^{j\Omega})|$
d) Magnitude response of the complete postfilter $|H_{PF}(e^{j\Omega})|$
($\alpha = 0.8$, $\beta = 0.5$, $\gamma = 0.2$, $\varepsilon = 0.4$, $\eta = 0.05$)

[Chen, Gersho 1995] propose to adjust the coefficients ε and η depending on the LTP parameter b (see e.g. (6.83-a)) as follows:

$$\varepsilon = c_1 \cdot f(b)\,,\ \eta = c_2 \cdot f(b)\,; \quad 0 \leq c_1, c_2 < 1\,; \quad c_1 + c_2 = 0.5 \tag{8.109}$$

with

$$f(b) = \begin{cases} 0 & b < c_3 \\ b & c_3 \leq b \leq 1 \\ 1 & 1 < b \end{cases} \qquad 0 < c_3 < 1\,. \tag{8.110}$$

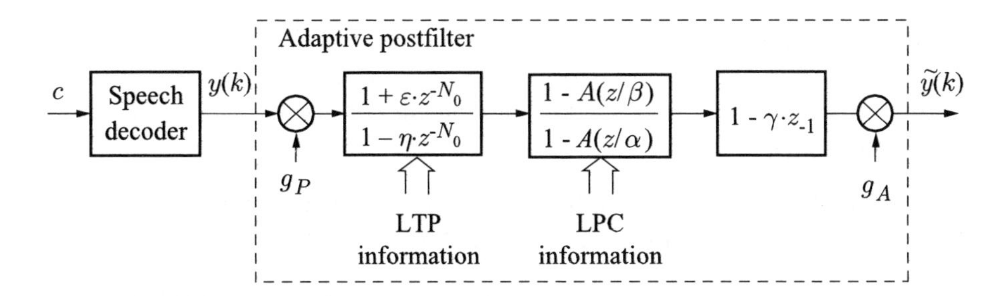

Figure 8.40: Speech decoder with adaptive postfilter

In unvoiced segments, in which b or g_a takes very small values, the long-term postfilter is turned off. The parameters $\alpha, \beta, \gamma, \varepsilon, \eta, c_1, c_2$, and c_3 must be experimentally adjusted to the respective codec. In general, c_2 is set very small or even zero so that the recursive part in (8.106) is canceled. A typical choice is: $\alpha = 0.8$, $\beta = 0.5$, $c_1 = 0.5$, $c_2 = 0.0$, and $c_3 = 0.6$.

The scaling factor g_P in (8.106) is adjusted in such a way that the short-term power of $y(k)$ is approximately not altered by the filtering with $H_P(z)$. The factor g_P is determined within relatively small time intervals, e.g., according to [Chen, Gersho 1995],

$$g_P = \frac{1 - \eta/b}{1 + \varepsilon/b} \,. \tag{8.111}$$

The second scaling factor g_A (see also Fig. 8.40) ensures that the powers of $y(k)$ and $\tilde{y}(k)$ match with a time resolution that lies in the range of the duration of the syllable.

All in all, this concept of adaptive postfiltering provides a further, in some circumstances significant, reduction of the audible quantization distortions. However, this postfiltering can cause considerable signal distortions if the signal runs through a chain of two or more codecs (tandem operation). In this case, it is better to switch off the postfilter.

Bibliography

3GPP TS 26.090 (2001). Adaptive Multi-Rate (AMR) Speech Codec; Transcoding Functions, 3GPP.

3GPP TS 26.190 (2001). AMR Wideband Speech Codec; Transcoding Functions, 3GPP.

Adoul, J. P.; Mabilleau, P.; Delprat, M.; Morisette, S. (1987). Fast CELP Coding Based on Algebraic Codes, *Proceedings of the IEEE International Conference on Acoustics, Speech, and Signal Processing (ICASSP)*, Dallas, Texas, USA, pp. 1957–1960.

Atal, B. S. (1982). Predictive Coding of Speech at Low Bit Rates, *IEEE Transactions on Communications*, vol. 30, April, pp. 600–614.

Atal, B. S.; Cox, R.; Kroon, P. (1989). Spectral Quantization and Interpolation for CELP Coders, *Proceedings of the IEEE International Conference on Acoustics, Speech, and Signal Processing (ICASSP)*, Glasgow, Scotland, pp. 69–72.

Atal, B. S.; Cuperman, V.; Gersho, A. (eds.) (1991). *Advances in Speech Coding*, Kluwer Academic, Boston, Massachusetts.

Atal, B. S.; Remde, J. R. (1982). A New Model of LPC Excitation for Producing Natural-sounding Speech at Low Bit Rates, *Proceedings of the IEEE International Conference on Acoustics, Speech, and Signal Processing (ICASSP)*, Paris, France, pp. 614–617.

Bessette, B.; Salami, R.; Lefebvre, R.; Jelinek, M.; Rotola-Pukkila, J.; Vainio, J.; Mikkola, H.; Järvinen, K. (2002). The Adaptive Multirate Wideband Speech Codec (AMR-WB), *IEEE Transactions on Speech and Audio Processing*, vol. 10, no. 8, November, pp. 620–636.

Brandenburg, K.; Bosi, M. (1997). Overview of MPEG Audio: Current and Future Standards for Low Bit-rate Audio Coding, *Journal of the Audio Engineering Society*, vol. 45, pp. 4–21.

Bruhn, S.; Blöcher, P.; Hellwig, K.; Sjöberg, J. (1999). Concepts and Solutions for Link Adaptation and Inband Signaling for the GSM AMR Speech Coding Standard, *IEEE Vehicular Technology Conference*, Houston, TX, USA, vol. 3, pp. 2451–2455.

Campbell, J. P.; Welch, V. C.; Tremain, T. E. (1989). The New 4800 bps Voice Coding Standard, *Proceedings of the Military Speech Techology*, pp. 64–70.

Chen, J. H.; Gersho, A. (1995). Adaptive Postfiltering for Quality Enhancement of Coded Speech, *IEEE Transactions on Speech and Audio Processing*, vol. 3, no. 1, January, pp. 59–71.

Chu, W. C. (2003). *Speech Coding Algorithms*, Wiley-Interscience, New York.

Dudley, H. (1939). The Vocoder, *Bell Labs Record*, vol. 17, pp. 122–126.

Ekudden, E.; Hagen, R.; Johansson, I.; Svedberg, J. (1999). The AMR Speech Coder, *Proceedings of the IEEE Workshop on Speech Coding*, Porvoo, Finland, pp. 117–119.

ETSI Rec. GSM 06.60 (1996). *Digital Cellular Telecommunications System (Phase 2+); Enhanced Full Rate (EFR) Speech Transcoding*, European Telecommunications Standards Institute.

ETSI Rec. GSM 06.90 (1998). *Digital Cellular Telecommunications System (Phase 2+); Adaptive Multi-Rate (AMR) Speech Transcoding*, European Telecommunications Standards Institute.

Flanagan, J. L.; Golden, R. (1966). Phase Vocoder, *Bell System Technical Journal*, vol. 45, pp. 1493–1509.

Furui, S.; Sondhi, M. M. (1992). *Advances in Speech Signal Processing*, Marcel Dekker, New York.

Goldberg, R.; Riek, L. (2000). *A Practical Handbook of Speech Coders*, CRC Press, New York.

Hanzo, L.; Somerville, F. C. A.; Woodard, J. P. (2001). *Voice Compression and Communications*, Wiley-Interscience, New York.

Itakura, F. (1975). Line Spectral Representation of Linear Prediction Coefficients of Speech Signals, *Journal of the Acoustical Society of America*, vol. 57, no. 1, pp. S35.

ITU-T Rec. G.333 (1988). 60 MHz Systems on standardized 2.6/9.5 mm Coaxial Cable Pairs, International Telecommunication Union (ITU).

ITU-T Rec. G.722 (1988). 7 kHz Audio Coding within 64 kbit/s, *Recommendation G.722*, International Telecommunication Union (ITU), pp. 269–341.

ITU-T Rec. G.722.2 (2002). Wideband Coding of Speech at Around 16 kbits/s Using Adaptive Multi-Rate Wideband (AMR-WB), International Telecommunication Union (ITU).

ITU-T Rec. G.726 (1990). 40, 32, 24, 16 kbit/s Adaptive Differential Pulse Code Modulation (ADPCM), *Recommendation G.726*, International Telecommunication Union (ITU).

ITU-T Rec. G.729 (1996). Coding of Speech at 8 kbit/s Using Conjugate-Structure Algebraic-Code-Excited Linear-Prediction (CS-ACELP), *Recommendation G.729*, International Telecommunication Union (ITU).

ITU-T Rec. P.341 (1998). Transmission characteristics for wideband (150-7000,Hz) digital hands-free telephony terrminals, *Recommendation P.341*, International Telecommunication Union (ITU).

ITU-T Rec. P.48 (1993). Specification for an Intermediate Reference System, *Recommendation P.48*, International Telecommunication Union (ITU).

ITU-T Rec. P.830 (1996). Subjective Performance Assessment of Telephone-Band and Wideband Digital Codecs. Annex D – Modified IRS Send and Receive Characteristics, *Recommendation P.830*, International Telecommunication Union (ITU).

Järvinen, K. (2000). Standardization of the Adaptive Multi-Rate Codec, *Proceedings of the European Signal Processing Conference (EUSIPCO)*, Tampere, Finland.

Järvinen, K.; Vainio, J.; Kapanen, P.; Salami, R.; Laflamme, C.; Adoul, J. P. (1997). GSM Enhanced Full Rate Speech Codec, *Proceedings of the IEEE International Conference on Acoustics, Speech, and Signal Processing (ICASSP)*, Munich, Germany, pp. 771–774.

Jayant, N. S.; Noll, P. (1984). *Digital Coding of Waveforms*, Prentice Hall, Englewood Cliffs, New Jersey.

Kabal, P.; Ramachandran, R. P. (1986). The Computation of Line Spectral Frequencies Using Chebyshev Polynomials, *IEEE Transactions on Acoustics, Speech and Signal Processing*, vol. 34, no. 6, December, pp. 1419–1426.

Kataoka, A.; Moriya, T.; Hayashi, S. (1996). An 8 kb/s Conjugate Structure CELP (CS-CELP) Speech Coder, *IEEE Transactions on Speech and Audio Processing*, vol. 4(6), pp. 401–411.

Kleijn, W. B.; Paliwal, K. K. (eds.) (1995). *Speech Coding and Synthesis*, Elsevier, Amsterdam.

Kondoz, A. M. (2004). *Digital Speech*, John Wiley & Sons, Inc., New York.

Kroon, P.; Deprettere, E. F.; Sluyter, R. (1986). Regular-Pulse Excitation - A Novel Approach to Effective and Efficient Multipulse Coding of Speech, *IEEE Transactions on Acoustics, Speech and Signal Processing*, vol. 34, no. 5, October, pp. 1054–1063.

Laflamme, C.; Adoul, J.-P.; Su, H.; Morisette, S. (1990). On Reducing Computational Complexity of Codebook Search in CELP Coder Through the Use of Algebraic Codes, *Proceedings of the IEEE International Conference on Acoustics, Speech, and Signal Processing (ICASSP)*, Albuquerque, New Mexico, USA.

Makhoul, J.; Viswanathan, R. (1975). Quantization Properties of Transmission Parameters in Linear Predictive Systems, *IEEE Transactions on Acoustics, Speech and Signal Processing*, vol. 23, no. 3, pp. 309–321.

MPEG-2 (1997). Advanced Audio Coding, AAC International Standard IS 13818-7, ISO/IEC JTC1/SC29 WG11, MPEG.

O'Shaughnessy, D. (2000). *Speech Communications*, IEEE Press, New York.

Paulus, J. (1996). *Codierung breitbandiger Sprachsignale bei niedriger Datenrate*, PhD thesis. *Aachener Beiträge zu digitalen Nachrichtensystemen*, vol. 6, P. Vary (ed.), RWTH Aachen University (in German).

Rabiner, L. R.; Schafer, R. W. (1978). *Digital Processing of Speech Signals*, Prentice Hall, Englewood Cliffs, New Jersey.

Rao, K. R.; Hwang, J. J. (1996). *Techniques & Standards for Image, Video & Audio Coding*, Prentice Hall, Upper Saddle River, New Jersey.

Rosenberg, A. E.; Schafer, R. W.; Rabiner, L. R. (1971). Effects of smoothing and quantizing the parameters of formant-coded voiced speech, *Journal of the Acoustical Society of America*, vol. 50, no. 6, pp. 1532–1538.

Salami, R.; Laflamme, C.; Bessette, B.; Adoul, J. P. (1997a). Description of ITU-T Rec. G.729 Annex A: Reduced Complexity 8 kbit/s CS-ACELP Coding, *Proceedings of the IEEE International Conference on Acoustics, Speech, and Signal Processing (ICASSP)*, Munich, Germany.

Salami, R.; Laflamme, C.; Bessette, B.; Adoul, J. P. (1997b). ITU-T G.729 Annex A: Reduced Complexity 8 kb/s CS-ACELP Codec for Digital Simultaneous Voice and Data, *IEEE Communications Magazine*, vol. 35, no. 9, September, pp. 56–63.

Schroeder, M. R.; Atal, B. S. (1985). Code-Excited Linear Prediction (CELP): High-Quality Speech at Very Low Bit Rates, *Proceedings of the IEEE International Conference on Acoustics, Speech, and Signal Processing (ICASSP)*, Tampa, Florida, USA, pp. 937–940.

Schroeder, M. R.; Atal, B. S.; Hall, J. (1979). Optimizing Digital Speech Coders by Exploiting Masking Properties of the Human Ear, *Journal of the Acoustical Society of America*, vol. 66, pp. 1647–1652.

Schüssler, H. W. (1994). *Digitale Signalverarbeitung I*, Springer Verlag, Berlin (in German).

Singhal, S.; Atal, B. S. (1984). Improving Performance of Multi-pulse LPC Coders at Low Bit Rates, *Proceedings of the IEEE International Conference on Acoustics, Speech, and Signal Processing (ICASSP)*, San Diego, USA, pp. 9–12.

Sluijter, R. J. (2005). *The Development of Speech Coding and the First Standard Coder for Public Mobile Telephony*, PhD thesis, Technical University Eindhoven.

Sluyter, R. J.; Vary, P.; Hofmann, R.; Hellwig, K. (1988). A Regular-pulse Excited Linear Predictive Codec, *Speech Communication*, vol. 7, no. 2, pp. 209–215.

Soong, F.; Juang, B. (1984). Line Spectrum Pair (LSP) and Speech Data Compression, *Proceedings of the IEEE International Conference on Acoustics, Speech, and Signal Processing (ICASSP)*, San Diego, USA, pp. 1.10.1–1.10.4.

Sugamura, N.; Farvardin, N. (1988). Quantizer Design in LSP Speech Analysis and Synthesis, *Proceedings of the IEEE International Conference on Acoustics, Speech, and Signal Processing (ICASSP)*, New York, USA, pp. 398–401.

Sugamura, N.; Itakura, F. (1986). Speech Analysis and Synthesis Methods Developed at ECL in NTT – From LPC to LSP, *Speech Communication*, vol. 5, pp. 199–215.

Tremain, T. E. (1982). The Government Standard Linear Predictive Coding Algorithm: LPC-10, *Speech Technology*, vol. 1, pp. 40–49.

Un, C. K.; Magill, D. T. (1975). The Residual-Excited Linear Prediction Vocoder with Transmission Rate Below 9.6 kbit/s, *IEEE Transactions on Communications*, vol. 23, pp. 1466–1474.

Vary, P.; Hellwig, K.; Hofmann, R.; Sluyter, R. J.; Galand, C.; Rosso, M. (1988). Speech Codec for the European Mobile Radio System, *Proceedings of the IEEE International Conference on Acoustics, Speech, and Signal Processing (ICASSP)*, New York, USA, pp. 227–230.

Vary, P.; Heute, U.; Hess, W. (1998). *Digitale Sprachsignalverarbeitung*, B. G. Teubner, Stuttgart (in German).

Vary, P.; Hofmann, R. (1988). Sprachcodec für das Europäische Funkfernsprechnetz, *Frequenz*, vol. 42, pp. 85–93. (in German).

Xie, M.; Adoul, J. (1995). Fast and Low Complexity LSF Quantization Using Algebraic Vector Quantizer, *Proceedings of the IEEE International Conference on Acoustics, Speech, and Signal Processing (ICASSP)*, Detroit, USA, pp. 716–719.

9

Error Concealment and Soft Decision Source Decoding

Digital speech, audio, and video communication over noisy channels usually comprises source and channel coding. The source encoder delivers source parameters such as A-law or μ-law coded speech samples, filter coefficients of the digital vocal tract model, or gain factors of subband signals. The achievable speech, audio, or video quality is determined by the quantizers and the resulting net bit rate of the coding algorithm. For error protection, channel coding is applied to the corresponding bit patterns of these parameters to preserve the quality level over a wide range of channel characteristics. Nevertheless, even channel coding cannot prevent the occurrence of residual bit errors in the case of (temporarily) adverse channel conditions that may lead to a severe degradation of the signal quality. These annoying effects can be reduced or even eliminated by means of *error concealment* (e.g., [Gerlach 1993], [Feldes 1993], [Feldes 1994], [Skoglund, Hedelin 1994], [Gerlach 1996], [Fingscheidt et al. 1998], [Fingscheidt 1998], [Fingscheidt, Vary 2001]).

A similar situation occurs in packet voice transmission via the Internet. Within the real-time constraints frames or packets of bits may arrive too late and have to be declared lost or erased (e.g., [Jayant, Christensen 1981], [Goodman et al. 1986], [Valenzuela, Animalu 1989], [Sereno 1991], [Erdöl et al. 1993], [Stenger et al. 1996], [Clüver 1996], [Martin et al. 2001]).

In either application, erased bits, lost packets of bits, or disturbed codec parameters have to be substituted to reduce the subjectively annoying effects. The reasoning

Digital Speech Transmission: Enhancement, Coding and Error Concealment
Peter Vary and Rainer Martin

behind error concealment is that most source codecs for speech, audio, and video transmission are not perfect. Due to practical reasons such as delay and complexity constraints, residual redundancy can be observed within the codec parameters in most cases. As Shannon has pointed out, this source coding suboptimality should be exploited at the receiver "to combat noise" [Shannon 1948].

In this chapter, we will first discuss by way of examples some standard error concealment techniques as they are applied in the GSM (*Global System for Mobile communications*) and the UMTS (*Universal Mobile Telecommunications System*) mobile radio systems. Then, a generic approach will be presented which is called *soft decision source decoding* [Fingscheidt 1998], [Fingscheidt, Vary 2001]. It can be applied to any parametric source decoder as it relies on residual source redundancy, reliability information from the channel (or channel decoder), and optimum parameter estimation (e.g., [Gerlach 1993], [Feldes 1993], [Feldes 1994], [Gerlach 1996]).

The subjective speech (audio or video) quality can be significantly enhanced with a muting mechanism or graceful degradation behavior in the case of a degrading channel.

Finally, the concepts of joint and iterative source–channel decoding will be introduced, which allow further improvements at the expense of an increased complexity. Iterative source–channel decoding is a derivative of the so-called turbo principle known from channel decoding.

9.1 Hard Decision Source Decoding

In speech, audio, and video source coding at medium and low bit rates, the signals are usually processed block by block or frame by frame. In speech and audio processing, frames may overlap in time. In speech coding a frame or block consists of typically $N = 160$ samples $s(\lambda)$, and covers a duration of 20 ms at a sample rate of $f_s = 8$ kHz. In video coding a block typically consists of $N \times N$ (e.g., $N = 8$ or $N = 16$) pixels $s(\lambda)$, while a frame denotes a complete picture.

In this chapter we will deal with one-dimensional signals such as speech and audio. However, the results can easily be extended to two-dimensional processing.

The signal s is segmented into blocks of N samples $s_k(\lambda)$ $(\lambda = 1, \ldots, N)$, where k is denoting the time or block index. For each block of samples $s_k(\lambda)$, a set of *codec parameters* such as predictor coefficients, gain factors, pitch lags, etc., is calculated by means of parameter analysis. For simplicity, we will consider one scalar parameter $\tilde{v}_k \in \mathbb{R}$ from this set here, as indicated in Fig. 9.1. The parameter $\tilde{v}_k$ is applied to the quantizer Q with 2^w quantizer reproduction levels. The 2^w levels can be addressed by w bits. Therefore, the quantized version $v_k \in \{v^{(i)} \in \mathbb{R},\ i = 0, 1, \ldots, 2^w - 1\}$ of each parameter sample $\tilde{v}_k$ is mapped

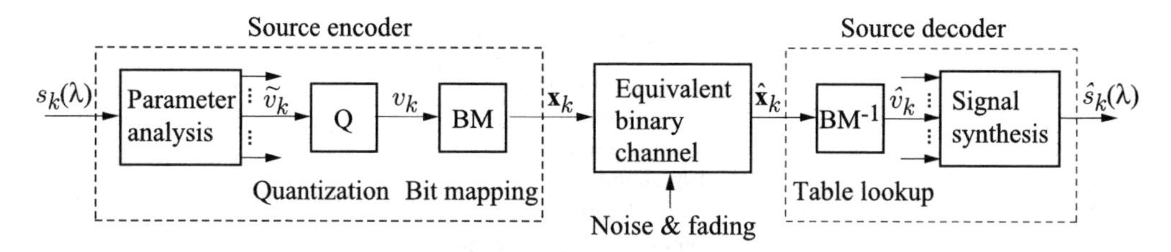

Figure 9.1: Conventional hard decision (HD) speech decoding

k	:	block or frame index (time)
$s_k(\lambda)$	:	speech sample at time $k \cdot N + \lambda$
$\tilde{v}_k$	:	real valued source codec parameter in frame k
v_k	:	quantized parameter $v_k \in \{v^{(i)},\ i = 0, 1, \ldots 2^w - 1\}$
$\mathbf{x}_k$	:	bit pattern (vector) of length w
$\hat{\mathbf{x}}_k$	:	received bit vector of length w
Q	:	quantization
BM	:	bit mapping

to a corresponding bit pattern $\mathbf{x}_k \in \{\mathbf{x}^{(i)},\ i = 0, 1, \ldots, 2^w - 1\}$ of length w (bit mapping), which is transmitted over the noisy channel.

For reasons of simplicity, we consider scalar parameters and scalar quantizers. All the concepts to be discussed below can easily be extended to vector quantization. The transmission of any quantized codec parameter v_k over the noisy channel will be described here by the equivalent binary symmetric channel with input vectors $\mathbf{x}_k$ and output vectors $\hat{\mathbf{x}}_k$. The w bits $x_k(\kappa)$ (bit index $\kappa = 1, \ldots, w$) of each vector $\mathbf{x}_k$ are transmitted sequentially.

The equivalent channel might consist of any combination of the noisy analog channel with channel (de)coding, (de)modulation, and equalization. Due to the channel noise, the received bit combination $\hat{\mathbf{x}}_k$ is possibly not identical to the transmitted one. In the conventional decoding scheme of Fig. 9.1 the received bit combination $\hat{\mathbf{x}}_k$ is applied to table decoding (*inverse bit mapping* (BM^{-1}) scheme). Thus, in the case of residual bit error(s) a wrong table entry is selected. The decoded parameter $\hat{v}_k$ is finally used within the synthesis algorithm to reconstruct signal samples $\hat{s}(\lambda)$. In what follows, this conventional solution will be called *hard decision (HD) source decoding*.

9.2 Conventional Error Concealment

Conventional error concealment is based on hard decision source decoding. Signal frames are converted by source encoding into bit frames, as shown in Fig. 9.2. In contrast to Fig. 9.1, the bit frame does not consist of one single parameter, but of the complete set of the quantized parameters of frame number k. If necessary, the

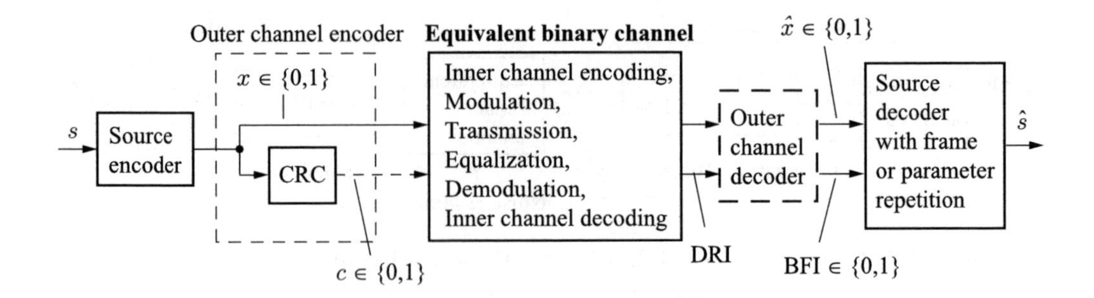

Figure 9.2: Transmission with conventional error concealment
DRI: Decoder Reliability Indicator (inner channel decoder)
BFI: Bad Frame Indicator (outer channel decoder)

bit frame may be subdivided into groups or vectors $\mathbf{x}$ of single bits x. For reasons of simplicity, we are omitting the frame, parameter, and bit indices here.

The bit frames are applied to two stages of channel coding: the *outer channel encoder* and the *inner channel encoder.* The inner channel code is used for error correction, while the outer code helps to control the error concealment at the receiving end by detecting residual errors after inner channel decoding.

In mobile radio systems such as GSM and UMTS the inner channel codec consists of a convolutional encoder and a trellis decoder (Viterbi algorithm and its derivatives). The outer channel encoder is a systematic block code, which adds *cyclic redundancy check* (CRC) bits c to groups of the subjectively most important bits.

As convolutional error protection by the inner channel coder is not of interest here, we will consider the *equivalent channel* for convenience, which can consist of any combination of inner channel encoding, modulation, noisy transmission, equalization, demodulation, and inner channel decoding.

A soft decision inner channel decoder which delivers *decoder reliability information* (DRI) is assumed: a binary flag, an integer number, or a real-valued quality measure per frame of samples s, per group $\mathbf{x}$ of bits, or even per individual bit x or c. At the receiver, a binary *bad frame indicator* (BFI) is produced by the outer channel decoder, which takes into account the reliability information from the inner decoder, e.g., the metric of the Viterbi decoder, the CRC check, and possibly other quality parameters such as the received field strength. The BFI, which is valid for the duration of a complete speech frame, is exploited within the error concealment and source decoding process to reduce the subjectively annoying effects of residual bit errors.

The GSM standards on substitution and muting of lost frames [Rec. GSM 06.11 1992], [Rec. GSM 06.21 1995], and [Rec. GSM 06.61 1996] propose simple mechanisms such as parameter repetition and step-by-step muting. They are driven by

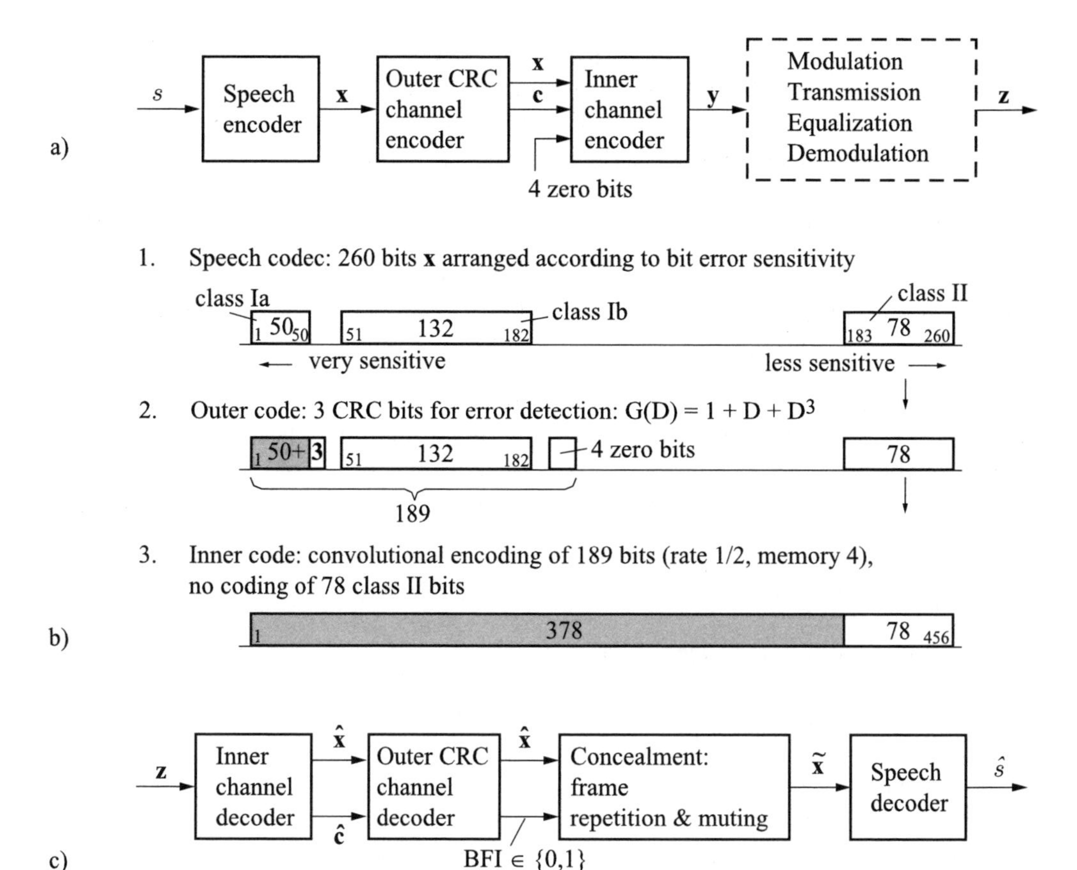

Figure 9.3: Standard error concealment of the GSM full-rate speech codec GSM 06.10
a) Block diagram of the encoder
b) Bit classes and error protection
c) Block diagram of the decoder

the binary BFI that marks the current received frame as good or bad. It can be seen as very coarse reliability information that may initiate the substitution of a complete frame, even if only a few bits have been disturbed. Conversely, the BFI may declare a frame reliable although some bits are incorrect.

As an example of conventional error concealment, we will briefly discuss this standard solution [Rec. GSM 06.11 1992]. In GSM the source and the channel encoding process of the so-called full-rate channel (22.8 kbit/s) is organized at the transmitter as shown in the block diagram of Fig. 9.3-a.

The standard full-rate speech encoder (see also Chapter 8 and Appendix A on codec standards) produces frames of 260 bits x every 20 ms. These bits are grouped, as illustrated in Fig. 9.3-b, according to their auditive sensitivity with respect to bit errors into $50 + 132 = 182$ class I bits and 78 class II bits. The 50 subjectively most important (class Ia) bits are protected by three CRC bits. The class Ia bits usually belong to the most important bits of various speech codec parameters. A systematic cyclic block code with the generator polynomial

$$G_{CRC}(D) = 1 + D + D^3 \tag{9.1}$$

is used, where D denotes the delay operator. Thus, the CRC and the resulting BFI give a measure of frame reliability rather than parameter or bit reliability. Class II comprises mostly the numerically least significant bits (LSBs).

The inner encoder performs feed-forward convolutional encoding of the 50 class Ia bits, the three parity check bits, and the remaining 132 class Ib bits; and no coding of the remaining 78 class II bits, which are not protected at all. The generator polynomials of the convolutional encoder of rate $r = 1/2$ and memory 4 are given by

$$\begin{aligned} G_1(D) &= 1 + D^3 + D^4 \\ G_2(D) &= 1 + D + D^3 + D^4 . \end{aligned} \tag{9.2}$$

According to the degree of the polynomials (9.2) and to the length of the encoder memory, four 0 bits are appended to the 182 class I bits and the three CRC bits to drive the encoder into the zero state. In total, 189 bits are applied to the convolutional encoder, which delivers $2 \cdot 189 = 378$ bits. Thus, the inner encoder produces $378 + 78 = 456$ bits every 20 ms. The gross bit rate is 456 bits/20 ms = 22.8 kbit/s.

Following the original standard [Rec. GSM 06.11 1992], error concealment is controlled by the binary BFI, as indicated in Fig. 9.3-c, while the BFI is produced by the outer channel decoder. The parity check bits are recalculated from the received 50 class Ia bits $\hat{x}$ which possibly have residual bit errors. Then they are compared to the received parity bits $\hat{\mathbf{c}}$. If these do not coincide, residual bit errors are assumed within the class Ia bits, and the error concealment procedure is activated via the BFI flag. If one single frame is marked bad, the complete frame of 260 bits $\hat{x}$ is replaced by the previous bit frame. If several consecutive bad frames occur, the last good frame is repeated while the following speech frames are gradually attenuated and muted after 320 ms (16 frames). In practice it is not sufficient to derive the BFI flag just from the CRC check, as multiple bit errors might occur which cannot be detected by this simple mechanism. It is up to the manufacturer of the mobile phone to use additional information for generating the BFI information, such as the metric of the inner channel decoder (DRI), the received field strength, etc. There are some proposals to enhance the reliability information (e.g., [Sereno 1991], [Su, Mermelstein 1992], [Heikkila et al. 1993], [Järvinen, Vainio 1994]). Alternatively, BFIs are not generated explicitly. In [Minde et al. 1993] weighting factors are computed to perform parameter substitution by weighted summation over previous frames.

More sophisticated concealment techniques have been designed for the GSM enhanced full-rate codec [Rec. GSM 06.61 1996], and the GSM adaptive multi-rate speech codec (AMR) [ETSI Rec. GSM 06.90 1998], which will also be used in the UMTS cellular system. In both cases the BFI flags of the present and the previous frame are considered. They control the concealment process, which is based on a state machine with seven states. Depending on the state, certain parameters of the codec are replaced by attenuated counterparts from the previous frame or by averaged values.

All these empirical algorithms improve the perceived speech quality significantly in adverse channel conditions and they are a major part of the whole error protection concept. If the error concealment were switched off, the speech quality would not be acceptable in many everyday-life transmission conditions.

This is the motivation to derive a theoretical framework leading to optimal error concealment techniques in terms of parameter estimation or soft decision source decoding. The optimum estimators should systematically exploit the residual redundancy of the source encoder and all kinds of quality information which can be made available at the receiver.

9.3 Softbits and L-values

The key to enhanced error concealment is to exploit reliability information within the source decoding process. This information may be available for groups $\hat{\mathbf{x}}$ of received bits or even for each individual bit $\hat{x}$. For reasons of simplicity, we omit in this section the frame index k and we consider the sequential transmission of bits $x(\kappa)$, where κ denotes the bit index which is in this case identical to the time index.

9.3.1 Binary Symmetric Channel (BSC)

For the representation of softbits and L-values, it is convenient to introduce a bipolar bit representation, i.e., the logical "0" is replaced by the real value $+1.0$ and the logical "1" by the real value -1.0.

A side effect of the bipolar bit representation is that the non-linear modulo-2 addition is replaced by the multiplication. Thus, we can describe the equivalent binary symmetric channel (BSC) by means of the two representations of Fig. 9.4.

For the time being, we assume that the equivalent channel is characterized by its constant bit error probability p_0 and by the error sequence $e(\kappa) \in \{+1, -1\}$. The channel output is given by

$$\hat{x}(\kappa) = x(\kappa) \cdot e(\kappa) \tag{9.3}$$

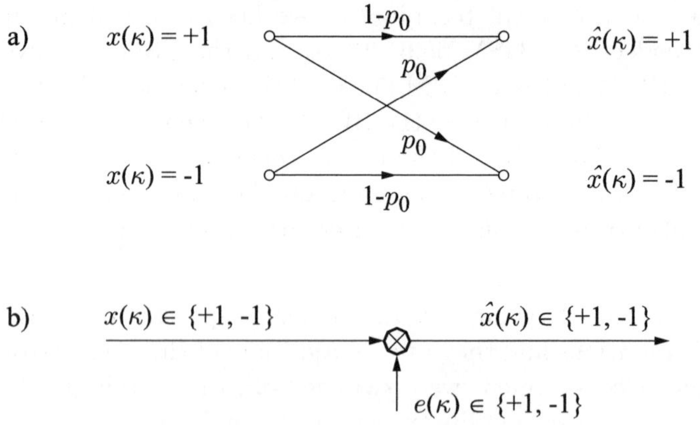

Figure 9.4: Binary symmetric channel for bipolar bit representation
p_0 : bit error probability of the channel
a) Transition diagram
b) Transmission model: multiplication of bits $x(\kappa)$ with the bit error sequence $e(\kappa)$

and a transmission error occurs if $e(\kappa) = -1$. The bit error probability p_0 is the same as the probability that $e(\kappa)$ takes the value -1.

A *reliability measure* would be the conditional probability that the received value $\hat{x}(\kappa) = +1$ or $\hat{x}(\kappa) = -1$ at time κ is correct. Alternatively, the *unreliability* could be expressed by the conditional probability that the observation is not correct.

Therefore, we may consider as a measure of unreliability the *instantaneous* conditional error probabilities

$$P(x(\kappa) \neq \hat{x}(\kappa) \mid \hat{x}(\kappa))\,; \quad \hat{x}(\kappa) \in \{+1, -1\} \tag{9.4-a}$$

and as a measure of reliability

$$P(x(\kappa) = \hat{x}(\kappa) \mid \hat{x}(\kappa))\,; \quad \hat{x}(\kappa) \in \{+1, -1\}\,. \tag{9.4-b}$$

The same quality information about the observed $\hat{x}(\kappa)$ at the channel output can obviously be quantified on different scales (reliability or unreliability).

Note A: The conditional error probabilities as defined in (9.4-a) must not necessarily be the same as the *bit error probability* p_0 of the channel. Only if $x(\kappa) = +1$ and $x(\kappa) = -1$ have the same probability of occurence, i.e., $P(x(\kappa) = +1) = P(x(\kappa) = -1) = 1/2$, the conditional error probabilities are the same as the bit error probability p_0 of the channel.

Note B: In data transmission it is usually assumed that both polarities of the bits $x(\kappa)$ have the same probability $P(x = +1) = P(x = -1) = 1/2$.

However, in speech, audio, and video coding the quantized parameters $v \in \{v^{(i)},\ i = 0, 1, \ldots, 2^w - 1\}$ and thus the corresponding bit patterns $\mathbf{x} \in \{\mathbf{x}^{(i)},\ i = 0, 1, \ldots, 2^w - 1\}$ (see Fig. 9.2) exhibit residual redundancy as a result of, e.g., non-symmetric parameter distributions or specific bit mapping rules (see block BM Fig. 9.1). Thus, the events $x = +1$ and $x = -1$ may *not* necessarily be equiprobable. This fact can be exploited within the error concealment process. If this information is not available or if it cannot be taken into consideration due to complexity constraints, then the assumption $P(x = +1) = P(x = -1) = 1/2$ will be used, too. It should be noted that if the parameter distribution and the bit mapping are symmetric, all bits x have the same probabilities for ± 1. However, in this case there is still redundancy on the parameter level due to non-uniform parameter distributions and parameter correlation, which can also be used for error concealment.

As basic *soft information* about the BSC from which any specific measure can be derived, we consider the joint information of a received hardbit and its instantaneous *conditional* error probability

$$\textit{BSC soft information} = \{\hat{x}(\kappa),\ P(x(\kappa) \neq \hat{x}(\kappa)|\hat{x}(\kappa))\}\,. \tag{9.5}$$

In the literature, there are equivalent measures for different purposes, such as the *softbit* and the *L-value* which are derived from the same soft information [Skoglund, Hedelin 1994], [Hagenauer 1995], [Hagenauer 1997], [Vary, Fingscheidt 1998], [Huber 2002].

The common root of softbits and L-values is the Bayes theorem:

$$P(x,\, \hat{x}) = P(x \mid \hat{x}) \cdot P(\hat{x}) = P(\hat{x} \mid x) \cdot P(x)\,. \tag{9.6}$$

For notational convenience, the index κ will be omitted insofar as it is not necessary to distinguish between different indices. With (9.6) the conditional error probabilities (9.4-a) and their complements (9.4-b) are given by

$$\begin{aligned} P(x \neq \hat{x} \mid \hat{x}) &= \frac{P(\hat{x} \mid x \neq \hat{x}) \cdot P(x \neq \hat{x})}{P(\hat{x})}\,; \quad \hat{x} \in \{+1, -1\} \\ P(x = \hat{x} \mid \hat{x}) &= \frac{P(\hat{x} \mid x = \hat{x}) \cdot P(x = \hat{x})}{P(\hat{x})}\,; \quad \hat{x} \in \{+1, -1\}\,. \end{aligned} \tag{9.7}$$

The expressions (9.7) are also called the *a posteriori probabilities. A posteriori* means we have received a bit $\hat{x}$ (with index κ), i.e., the receiver has already taken the decision $+1$ or -1 and we are interested in the probability that this decision was right or wrong. For calculating these *a posteriori* probabilities (9.7), we need access to the following quantities:

- $P(x = \pm 1)$: the *a priori probabilities*
- $P(\hat{x} = \pm 1 \mid x = \pm 1)$: the *transition probabilities*
- $P(\hat{x} = \pm 1)$: the *observation probabilities.*

The *a priori* probabilities $P(x = +1)$ and $P(x = -1)$ or the *a priori knowledge on bit level* can be measured in advance if representative coded signals and/or bit sequences $\mathbf{x}$ are available.

With regard to the transition probabilities, we have to distinguish between the different ± 1 cases and find easily the dependency on the bit error probability p_0 of the equivalent BSC channel:

$$P(\hat{x} = +1 \mid x = +1) = P(\hat{x} = -1 \mid x = -1) \quad = \quad 1 - p_0 \tag{9.8-a}$$

$$P(\hat{x} = +1 \mid x = -1) = P(\hat{x} = -1 \mid x = +1) \quad = \quad p_0 \, . \tag{9.8-b}$$

Finally, the observation probabilities $P(\hat{x} = +1)$ and $P(\hat{x} = -1)$ can be calculated from the quantities already known:

$$P(\hat{x} = +1) = P(x = +1) \cdot (1 - p_0) \; + \; P(x = -1) \cdot p_0 \tag{9.9-a}$$

$$P(\hat{x} = -1) = P(x = -1) \cdot (1 - p_0) \; + \; P(x = +1) \cdot p_0 \, . \tag{9.9-b}$$

Softbits and Binary Channel

The BSC receiver takes a preliminary hard decision $\hat{x}(\kappa) = +1$ or $\hat{x}(\kappa) = -1$. Based on this decision, i.e., for a fixed $\hat{x}(\kappa)$, the two conditional probabilities $P(x = +1 \mid \hat{x})$ and $P(x = -1 \mid \hat{x})$ can be computed according to (9.7), (9.8), and (9.9). They can be interpreted as the probabilities that either of the alternative decisions $x = +1$ or $x = -1$ is correct. Therefore, we are able to replace the first hard decision $\hat{x}(\kappa)$ by a second *soft estimate* or *softbit* $\tilde{x}(\kappa)$, which makes a compromise between $+1$ and -1 according to the two conditional probabilities. An optimal solution consists of minimizing the mean square estimation error. It can be shown (see Chapter 5) that the mean square error is minimized by the conditional expectation:

$$\tilde{x} = \mathrm{E}\{x \mid \hat{x}\} \tag{9.10}$$

$$= +1 \cdot P(x = +1 \mid \hat{x}) \; - \; 1 \cdot P(x = -1 \mid \hat{x}) \tag{9.11}$$

$$= \hat{x}\,(1 - 2 \cdot P(x \neq \hat{x} | \hat{x})) \, . \tag{9.12}$$

With the abbreviations

$$p_+ \;=\; P(x=+1) \qquad \text{and} \qquad p_- = P(x=-1)$$

we find with (9.7), the transition probabilities (9.8), and the observation probabilities (9.9) the explicit solutions for the two polarities of $\hat{x}$:

a) $\hat{x} = +1$:

$$\tilde{x} = \frac{p_+ \cdot (1-p_0) \;-\; p_- \cdot p_0}{p_+ \cdot (1-p_0) \;+\; p_- \cdot p_0} = \frac{p_+ \;-\; p_0}{p_+ \;-\; 2\,p_+ p_0 \;+\; p_0} \tag{9.13-a}$$

b) $\hat{x} = -1$:

$$\tilde{x} = \frac{p_+ \cdot p_0 \;-\; p_- \cdot (1-p_0)}{p_+ \cdot p_0 \;+\; p_- \cdot (1-p_0)} = -\frac{p_- \;-\; p_0}{p_- \;-\; 2\,p_- p_0 \;+\; p_0}\,. \tag{9.13-b}$$

In the general case of (9.13-a) and (9.13-b), the softbit is determined by the channel quality associated with p_0 and the *a priori* knowledge p_+ or $p_- = 1 - p_+$. There is no strict separation of the channel-related and the source-related contributions. Nevertheless, the sign of the softbit corresponds to the best hard decision $\hat{x}_{\text{opt}} = \text{sign}(\tilde{x})$ and the magnitude is a measure of reliability

$$0 \;\leq\; |\tilde{x}| \;\leq\; 1\,. \tag{9.14}$$

Two extreme cases are of special interest:

Case I $p_+ = p_- = 1/2$; no *a priori* knowledge about $x(\kappa)$ (no redundancy on bit level)

Case II: $p_0 = 1/2$; total disturbance of the channel (no transinformation, i.e., the channel output is independent from the input)

Case I: If $x = +1$ and $x = -1$ have the same probabilities $p_+ = p_- = 1/2$, then the two levels of $\hat{x}$ also occur with the same probability and the denominators in (9.7), (9.13-a), and (9.13-b) take the value $P(\hat{x} = \pm 1) = 1/2$. Thus, the two equations (9.13-a) and (9.13-b) can be combined to the single formula

$$\tilde{x} \;=\; \hat{x} \cdot (1 \;-\; 2 \cdot p_0)\,. \tag{9.15}$$

The *softbit* $\tilde{x}$ has the same sign as the hardbit $\hat{x}$, while the magnitude is determined by the transmission error probability p_0. If the transmission is free of error ($p_0 = 0$), then the softbit is the same as the hardbit $\hat{x}$. If the transmission error probability increases, the magnitude of the softbit decreases. For total disturbance ($p_0 = 1/2$) the softbit is muted to $\tilde{x} = 0$. There is no transinformation or mutual information in terms of information theory.

A useful feature of the softbit (9.15) is the preservation of the modulo-2 operation and/or the equivalent multiplication. Let us consider the transmission of two sequences of bits x_1 and x_2 which are received as hardbits $\hat{x}_1$ and $\hat{x}_2$ over two independent channels with corresponding error probabilities p_1 and p_2. The two hardbits $\hat{x}_1$ and $\hat{x}_2$ are combined by a modulo-2 operation, i.e., multiplication, to obtain

$$\hat{x}_3 = \hat{x}_1 \cdot \hat{x}_2 . \tag{9.16}$$

A bit error in $\hat{x}_3$ occurs only if we have a bit error either in $\hat{x}_1$ *or* in $\hat{x}_2$, i.e.,

$$p_3 = (1 - p_1) \cdot p_2 + (1 - p_2) \cdot p_1 . \tag{9.17}$$

If both bits are disturbed, the errors compensate each other and $\hat{x}_3$ is free of error.

If we define the corresponding softbit

$$\tilde{x}_3 = \hat{x}_3 \cdot (1 - 2 \cdot p_3) , \tag{9.18}$$

it can easily be shown that the following relation is valid:

$$\tilde{x}_3 = \tilde{x}_1 \cdot \tilde{x}_2 = \hat{x}_1 \cdot (1 - 2 \cdot p_1) \cdot \hat{x}_2 \cdot (1 - 2 \cdot p_2) \tag{9.19}$$

$$= \hat{x}_3 \cdot (1 - 2 \cdot p_3) . \tag{9.20}$$

As the magnitudes of the softbits $\tilde{x}_1$ and $\tilde{x}_2$ are not larger than $+1$, the magnitude of the combined softbit $\tilde{x}_3$, and thus its reliability $|\tilde{x}_3|$, cannot be larger than the smallest magnitude of either $\tilde{x}_1$ or $\tilde{x}_2$. In general, the combined bit will be less reliable.

Case II: The *a priori* probabilities are different, i.e., $p_+ \neq p_-$, but we have total disturbance ($p_0 = 1/2$). Then we get with (9.13-a) and (9.13-b)

$$\tilde{x} = p_+ - p_- . \tag{9.21}$$

The softbit is determined only by the *a priori* knowledge. If, for example, $x = +1$ occurs more often than $x = -1$, $\tilde{x}$ will be positive. If both polarities have the same probability $p_+ = p_-$, we get $\tilde{x} = 0$ as we do not have any *a priori* knowledge.

L-values and Binary Channel

The *conditional L-value* is an alternative logarithmic reliability measure which is widely used in soft-input, soft-output channel decoding [Hagenauer et al. 1996], [Hagenauer 1995], [Hagenauer 1980]. Instead of the conditional error probabilities (9.4-a) or their complements (9.4-a), we consider

$$L(x \mid \hat{x}) = \ln \left(\frac{P(x = +1 \mid \hat{x})}{P(x = -1 \mid \hat{x})} \right) . \tag{9.22-a}$$

According to the logarithmic function, the range of values is

$$-\infty < L(x \mid \hat{x}) < +\infty \,. \tag{9.22-b}$$

The extreme points are:

- $P(x = +1 \mid \hat{x}) = 1,\ P(x = -1 \mid \hat{x}) = 0 \quad \Rightarrow \quad L(x \mid \hat{x}) \rightarrow +\infty$
- $P(x = +1 \mid \hat{x}) = P(x = -1 \mid \hat{x}) = 1/2 \quad \Rightarrow \quad L(x \mid \hat{x}) = 0$
- $P(x = +1 \mid \hat{x}) = 0,\ P(x = -1 \mid \hat{x}) = 1 \quad \Rightarrow \quad L(x \mid \hat{x}) \rightarrow -\infty \,.$

The sign of the L-value is the optimum decision $\hat{x}_{\text{opt}}$ taking the bit error probability p_0 as well as the *a priori* knowledge p_+ and p_- into account. The larger the magnitude of $L(x \mid \hat{x})$, the larger is the reliability.

Equation (9.22-a) can be evaluated explicitly as a function of p_0, p_+, and p_-. By inserting (9.7) and (9.8) into (9.22-a) and by considering the two cases $\hat{x} = \pm 1$, we find

$$L(x \mid \hat{x}) = \hat{x} \cdot \ln\left(\frac{1 - p_0}{p_0}\right) + \ln\left(\frac{p_+}{p_-}\right) \tag{9.22-c}$$

$$= \hat{x} \cdot L_e + L_x \,. \tag{9.22-d}$$

Here, we have a clear separation of the channel-related contribution $\hat{x} \cdot L_e$ and the source-related contribution L_x (*a priori* knowledge). The term L_x is also called the L-value of the sequence $x(\kappa)$:

$$L_x = \ln\left(\frac{p_+}{p_-}\right) = \ln\left(\frac{1 - p_-}{p_-}\right) . \tag{9.23}$$

The channel quality is quantified by L_e, which is also called the L-value of the bit error sequence $e(\kappa)$

$$L_e = \ln\left(\frac{1 - p_0}{p_0}\right) . \tag{9.24}$$

It can easily be shown that p_0 can be recalculated from L_e

$$p_0 = \frac{1}{1 + e^{+L_e}} \,. \tag{9.25}$$

Generally speaking, the L-value of any binary sequence is the logarithm of the quotient of the probabilities that the sequence takes the values $+1$ and -1, whereas in the conditional L-value, conditional probabilities have to be inserted, which depend on the *a priori* knowledge p_+, p_- and the bit error probability p_0.

The conditional L-value $L(x|\hat{x})$ and the softbit $\tilde{x} = \hat{x} \cdot (1 - 2 \cdot P(x \neq \hat{x}|\hat{x}))$ are different representations of the same information.

The relation between the softbit $\tilde{x}$ and the conditional L-value $L(x|\hat{x})$ can also be derived by considering the different cases with $x = \pm 1$ and $\hat{x} = \pm 1$. Using (9.22-a) we obtain a closed expression for the conditional error probabilities

$$P(x \neq \hat{x}|\hat{x}) = \frac{1}{1 + e^{\hat{x} \cdot L(x|\hat{x})}} \qquad \hat{x} \in \{+1, -1\}\,, \tag{9.26}$$

and the softbit

$$\tilde{x} = \hat{x} \cdot (1 \;-\; 2 \cdot P(x \neq \hat{x}|\hat{x})) = \hat{x}\,\frac{e^{\hat{x} \cdot L(x|\hat{x})} - 1}{e^{\hat{x} \cdot L(x|\hat{x})} + 1} \tag{9.27-a}$$

$$= \hat{x}\,\frac{e^{\frac{1}{2}\hat{x} \cdot L(x|\hat{x})} - e^{-\frac{1}{2}\hat{x} \cdot L(x|\hat{x})}}{e^{\frac{1}{2}\hat{x} \cdot L(x|\hat{x})} + e^{-\frac{1}{2}\hat{x} \cdot L(x|\hat{x})}} \tag{9.27-b}$$

$$= \hat{x}\,\tanh\left(\hat{x}\frac{L(x|\hat{x})}{2}\right) \tag{9.27-c}$$

$$= \tanh\left(\frac{L(x|\hat{x})}{2}\right) \tag{9.27-d}$$

The optimal decision is given by

$$\hat{x}_{\text{opt}} = \text{sign}\{\tilde{x}\} = \text{sign}\{L(x|\hat{x})\} \tag{9.28}$$

and the bitwise decoder reliability can be expressed with different scaling either as

$$\text{DRI}_{\tilde{x}} = |\tilde{x}| \tag{9.29}$$

or

$$\text{DRI}_L = |L(x|\hat{x})|\,. \tag{9.30}$$

Note that the sign of $\hat{x}_{\text{opt}}$ may be different from the sign of $\hat{x}$ if we have a bias in (9.22) due to non-zero *a priori* knowledge $p_+ \neq p_-$.

Finally it should be mentioned that the two reliability measures (9.29), (9.30) can easily be extended to groups of bits.

With respect to the modulo-2 addition, we can derive a combination rule for L-values as well. Let us consider two independent sequences with bits x_1 and x_2 with L-values

$$L_{x1} = \ln\left(\frac{1 - p_1}{p_1}\right); \qquad L_{x2} = \ln\left(\frac{1 - p_2}{p_2}\right). \tag{9.31}$$

For calculating the L-value of the combined sequence

$$x_3 \;=\; x_1 \cdot x_2 \tag{9.32}$$

we need to compute the probabilities $1 - p_3$ and p_3 that x_3 takes the values $+1$ and -1, respectively. By using the relation (9.25) for $p_i,\ i = 1, \ldots, 3$ we find the combination rule which is called the *box-plus operation* [Hagenauer et al. 1996]:

$$L_{x3} = \ln\left(\frac{1 \,+\, e^{L_{x1}} \,\cdot\, e^{L_{x2}}}{e^{L_{x1}} \,+\, e^{L_{x2}}}\right) \tag{9.33-a}$$

$$= 2 \cdot \operatorname{arctanh}\left(\prod_{i=1}^{2} \tanh\left(\frac{L_{xi}}{2}\right)\right) \tag{9.33-b}$$

$$= L_{x1} \boxplus L_{x2} \tag{9.33-c}$$

$$\approx \operatorname{sign}(L_{x1}) \cdot \operatorname{sign}(L_{x2}) \cdot \min\left(|L_{x1}|,\ |L_{x2}|\right) . \tag{9.33-d}$$

Note: The approximation (9.33-d) is based on the assumption that the magnitude of one of the two L-values is very high [Hagenauer et al. 1996]. If L_{x1} and L_{x2} denote conditional L-values, this would mean that one of the bits is very realiable.

9.3.2 Fading–AWGN Channel

We consider the transmission over a fading channel with additive white Gaussian noise (AWGN) using binary phase shift keying (BPSK) and coherent demodulation, but without (inner) channel coding. The option of inner channel coding will be discussed in the next section. The continuous-time baseband model of this transmission scheme is illustrated in Fig. 9.5-a. The main difference to the binary channel (BSC) is that we have access not only to the received hardbit

$$\hat{x} \;=\; \operatorname{sign}(z) \tag{9.34}$$

but also to the *real-valued* output samples $z(\kappa)$ of the matched filter receiver. Furthermore, we assume that the fading/attenuation factor a and the power $\sigma_n^2 = N_0/2$ of the white noise are known. Therefore, an instantaneous bit error rate $p_0(\kappa)$ can be calculated (see Fig. 9.5-b).

The bit duration is T_b, and the continuous transmitter signal $x_T(t)$ consists of the concatenated stream of bits $x(\kappa)$. The transmitter signal $x_T(t)$ can be described by applying a train of weighted Dirac impulses to a transmit filter with rectangular impulse response $g(t)$:

$$x_T(t) = \left[\sum_{\kappa} x(\kappa) \cdot \delta(t - \kappa \cdot T_b)\right] \;*\; g(t) \tag{9.35-a}$$

$$= x(k)\,, \quad \text{if} \quad k \cdot T_b \;\leq\; t \;<\; (k+1) \cdot T_b \tag{9.35-b}$$

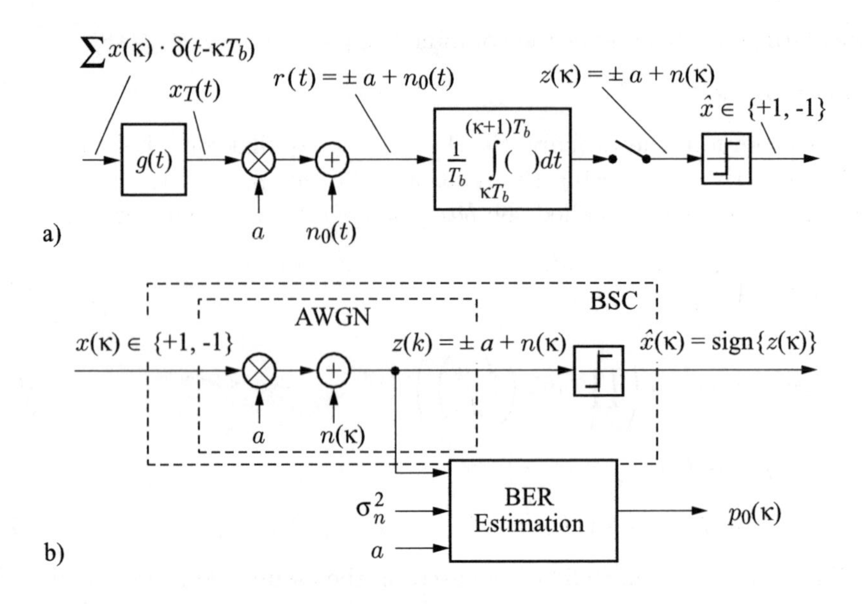

Figure 9.5: Fading–AWGN channel model for coherent BPSK
a) Continuous time baseband model
$n_0(t)$: white noise with power spectral density $N_0/2$
a : fading/attenuation factor
b) Discrete time models for AWGN and BSC
$p_0(\kappa)$: instantaneous bit error rate
σ_n^2 : power of the noise samples $n(\kappa)$

with

$$g(t) = \begin{cases} 1 & 0 \leq t < T_b \\ 0 & \text{else}. \end{cases} \tag{9.36}$$

The transmitter signal $x_T(t)$ is attenuated by a constant or slowly time-varying *fading factor*

$$0 < a(\kappa). \tag{9.37}$$

The fading/attenuation factor a is considered to be a constant during a bit interval

$$\kappa \cdot T_b \leq t < (\kappa + 1) \cdot T_b .$$

On the transmission link AWGN $n_0(t)$ with constant power spectral density $N_0/2$ is added to $x_T(t)$, and the coherent matched filter receiver averages the received baseband signal

$$\begin{aligned} r(t) &= x_T(t) + n_0(t) \\ &= \pm a + n_0(t) \end{aligned} \tag{9.38}$$

by integration over the κ-th bit interval. The real-valued sample

$$z(\kappa) = \frac{1}{T_b} \int\limits_{\kappa T_b}^{(\kappa+1)T_b} r(t)\,\mathrm{d}t = \pm a + n(\kappa) \tag{9.39}$$

taken at the end of the bit interval consists of the signal component $\pm a$ and a sample $n(\kappa)$ of the averaged (filtered) noise $n_0(t)$. Finally, the hard decision $\hat{x} \in \{+1, -1\}$ is taken. Bit errors occur in the case of $x(\kappa) \cdot a = -a$ and $n(\kappa) > +a$ or in the case of $x(\kappa) \cdot a = +a$ and $n(\kappa) < -a$. Thus, we can use the equivalent discrete-time block diagram of Fig. 9.5-b.

For the following considerations we need the energy E_b per transmitted bit $x(\kappa)$ and the power σ_n^2 of the noise samples $n(\kappa)$. If the amplitude of the transmitter signal $x_T(t)$ is considered dimensionless, the energy per bit $x(\kappa)$ is given by

$$E_b = \int\limits_{kT_b}^{(k+1)T_b} x_T^2(t)\,\mathrm{d}t = x^2(\kappa) \cdot T_b = T_b\,. \tag{9.40}$$

The receiving matched filter (averaging integrator) has the rectangular impulse response

$$h(t) = \begin{cases} 1/T_b & 0 \leq t < T_b \\ 0 & \text{else} \end{cases} \tag{9.41}$$

and a frequency response $H(\omega)$. As the noise signal $n_0(t)$ has a Gaussian probability density function (PDF), the PDF of the samples $n(\kappa)$ has the same shape but a different power. The power of the noise samples $n(\kappa)$ is that of the noise component $n_c(t)$ at the filter output as indicated in Fig. 9.6 and can be calculated by means of Parseval's theorem (e.g., [Papoulis 1968]):

$$\begin{aligned} \sigma_n^2 &= \frac{1}{2\pi} \int\limits_{-\infty}^{+\infty} \frac{N_0}{2} \cdot |H(\omega)|^2\,\mathrm{d}\omega \\ &= \frac{N_0}{2} \cdot \int\limits_{-\infty}^{+\infty} h^2(t)\,\mathrm{d}t \\ &= \frac{N_0}{2} \cdot \frac{1}{T_b}\,. \end{aligned} \tag{9.42}$$

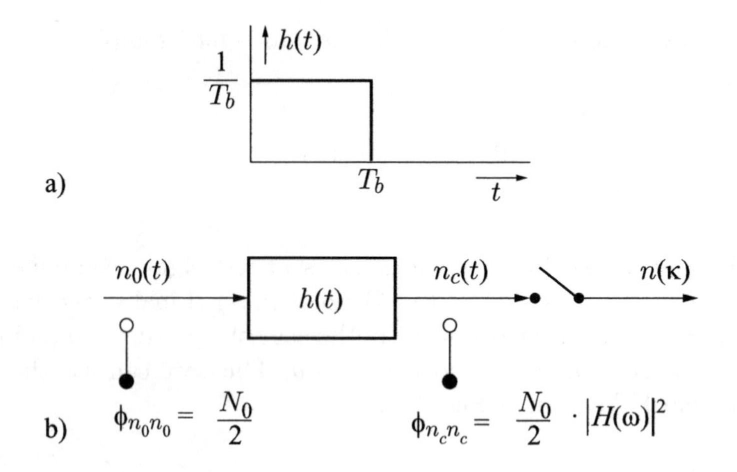

Figure 9.6: Processing of the white noise $n_0(t)$: matched filtering and sampling
a) Impulse response of the matched filter
b) Time domain and frequency domain relations

The PDF of the noise samples $n(\kappa)$ is therefore given by

$$p_n(u) = \frac{1}{\sqrt{2\pi}\sigma_n} \cdot \exp\left(-\frac{u^2}{2\sigma_n^2}\right) \tag{9.43}$$

$$= \frac{1}{\sqrt{2\pi}\sigma_n} \cdot \exp\left(-\frac{E_b}{N_0}u^2\right). \tag{9.44}$$

The definition of the conditional L-value is basically the same as for the binary channel, apart from the fact that we replace in (9.22-a) the binary variable $\hat{x}$ by the real-valued sample z

$$L(x \mid z) \; = \; \ln\left(\frac{P(x = +1 \mid z)}{P(x = -1 \mid z)}\right). \tag{9.45}$$

In contrast to $\hat{x}$, which is bipolar, i.e., discrete, the samples z take real values. Therefore, we use Bayes' theorem in *mixed form* here, e.g., [Proakis 1995];

$$P(x \mid z) \cdot p(z) \; = \; p(z \mid x) \cdot P(x)\,, \tag{9.46}$$

where $P(\cdot)$ denotes a discrete probability and $p(\cdot)$ a PDF. The conditional PDF $p(z \mid x)$ can easily be derived because of the deterministic relations

$$z(\kappa) \; = \; a \cdot x(\kappa) \; + \; n(\kappa)\,,$$

$$n(\kappa) \; = \; z(\kappa) \; - \; a \cdot x(\kappa)$$

and thus

$$p_n(u = z - a \cdot x) \;=\; \frac{1}{\sqrt{2\pi}\sigma_n} \cdot \exp\left(-\frac{E_b}{N_0}(z - a \cdot x)^2\right) . \tag{9.47}$$

With (9.45), (9.46), and (9.47) we find the conditional L-value for the fading-AWGN channel

$$L(x \mid z) = \ln\left(\frac{P(x = +1 \mid z)}{P(x = -1 \mid z)}\right) \tag{9.48-a}$$

$$= \ln\left(\frac{p(z \mid x = +1) \cdot P(x = +1)}{p(z \mid x = -1) \cdot P(x = -1)}\right) \tag{9.48-b}$$

$$= \ln\left(\frac{p(z \mid x = +1)}{p(z \mid x = -1)}\right) \;+\; \ln\left(\frac{P(x = +1)}{P(x = -1)}\right) \tag{9.48-c}$$

$$= z \cdot 4a\frac{E_b}{N_0} \;+\; L_x \tag{9.48-d}$$

$$= z \cdot 2a\frac{S}{N} \;+\; L_x \tag{9.48-e}$$

$$= z \cdot L_c \;+\; L_x \tag{9.48-f}$$

with

$$L_c = 4a\frac{E_b}{N_0} \;=\; 2a\frac{S}{N} \tag{9.49}$$

$$S = E_b \cdot T_b \qquad ; \qquad N = \sigma_n^2 = \frac{N_0}{2\,T_b}$$

$$\frac{S}{N} = \frac{E_b}{N_0/2} .$$

The result is similar to the L-value (9.22-d) obtained for the binary channel. However, the product $\hat{x} \cdot L_e$ has been replaced by $z \cdot L_c$.

The optimal decision, taking both the knowledge about the channel and the *a priori* knowledge into consideration, is given by:

$$\hat{x}_{\text{opt}} \;=\; \text{sign}\left(L(x \mid z)\right) \;=\; \text{sign}\left(z \cdot L_c \;+\; L_x\right), \tag{9.50}$$

and the reliability measure is

$$\text{DRI} \;=\; |L(x \mid z)| \;=\; |z \cdot L_c \;+\; L_x| \, . \tag{9.51}$$

A comparison of the two channel models discussed so far is shown in Fig. 9.7.

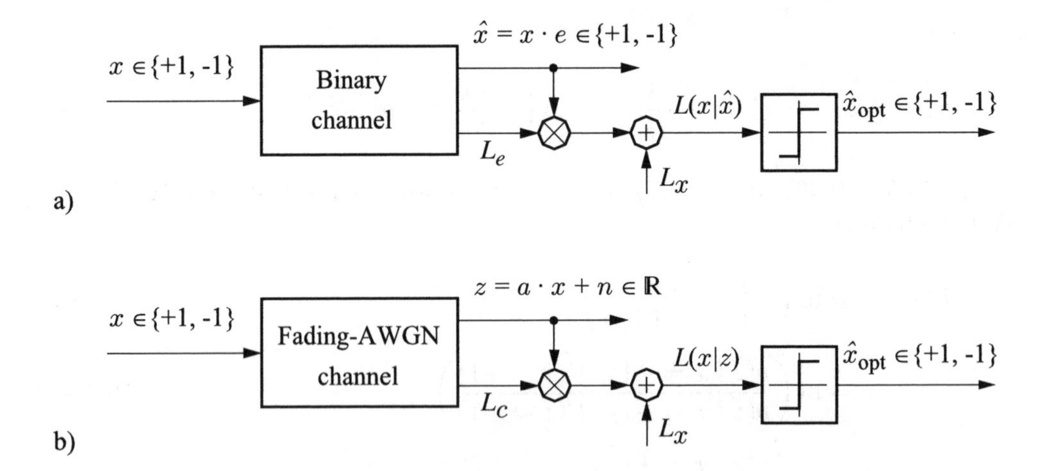

Figure 9.7: Conditional L-values and optimal decisions $\hat{x}_{\text{opt}}$
a) Binary channel
b) Fading-AWGN channel without inner channel coding

The main difference between the two models is that in (9.22-d) the magnitude $|\hat{x} \cdot L_e|$ in (9.22-d), and thus the reliability information about the binary channel, is a constant, whereas the corresponding quantity $|z(\kappa) \cdot L_c|$ in (9.48-f) may change from sample to sample. Thus, we obtain a measure of the *instantaneous channel* quality. In analogy to (9.25) and (9.27), we can derive expressions for the *instantaneous bit error rate* [Hagenauer 1980], [Hagenauer 1995] (9.48)

$$p_0(\kappa) = \frac{1}{1 + e^{+|z(\kappa) \cdot L_c|}} . \tag{9.52}$$

The softbit $\tilde{x}$ corresponding to (9.15) can be represented as a function of the conditional L-value

$$\tilde{x}(\kappa) = \tanh\left(\frac{L(x(\kappa)|z(\kappa))}{2}\right) \tag{9.53}$$

$$= \tanh\left(\frac{|z(\kappa) \cdot L_c + L_x|}{2}\right) . \tag{9.54}$$

The *channel state information* (CSI) L_c comprises the fading factor a as well as the signal-to-noise ratio S/N at the output of the matched filter. Both values have to be estimated at the receiver as indicated in Fig. 9.5-b. From (9.52) it is obvious how an individual instantaneous bit error probability can be assigned to each received value $z(\kappa)$ and each optimal decision $\hat{x}_{\text{opt}}(\kappa)$.

9.3.3 Channel with Inner SISO Decoding

If the equivalent fading–AWGN channel includes inner channel coding, the required reliability information can be provided by a soft-input, soft-output (SISO) channel decoder. The corresponding block diagram is given in Fig. 9.8, where we have re-introduced the time or frame index k.

Note: A block usually consists of several parameters, e.g., LP-coefficients, gain factors, etc. For simplicity, we consider here only a single parameter v_k or the corresponding bit pattern $\mathbf{x}_k$ as illustrated in Fig. 9.9. Each parameter value v_k is transmitted as bit vector $\mathbf{x}_k = (x_k(1), x_k(2), \ldots, x_k(w))$ of length w. The stream of bit vectors $\mathbf{x}_k$ is applied to the inner convolutional channel encoder which delivers a bit sequence $y(\kappa)$. The best choice for a convolutional SISO decoder is the channel decoder of Bahl et al. [Bahl et al. 1974] because it is able to yield conditional log-likelihood values (L-values)

$$L(x_k(\kappa) \mid \mathbf{z}) = \ln\left(\frac{P(x_k(\kappa) = +1 \mid \mathbf{z})}{P(x_k(\kappa) = -1 \mid \mathbf{z})}\right) \tag{9.55}$$

where $\mathbf{z}$ may denote the complete received sequence of real-valued samples $z(\kappa)$, a vector $\mathbf{z}_k$ consisting of w samples, or even a single sample $z(\kappa)$. The hardbit is given by

$$\hat{x}_k(\kappa) = \operatorname{sign}\left\{L(x_k(\kappa) \mid \mathbf{z})\right\} . \tag{9.56}$$

The conditional L-value $L(x_k(\kappa) \mid \mathbf{z})$ may or may not comprise *a priori* information, given by L_x according to (9.23). Note that in practice the less complex soft-output Viterbi algorithm (SOVA) [Hagenauer, Hoeher 1989] or a comparable solution, e.g., [Huber, Rüppel 1990], is often used, which yields approximations to the L-values of Eq. (9.55).

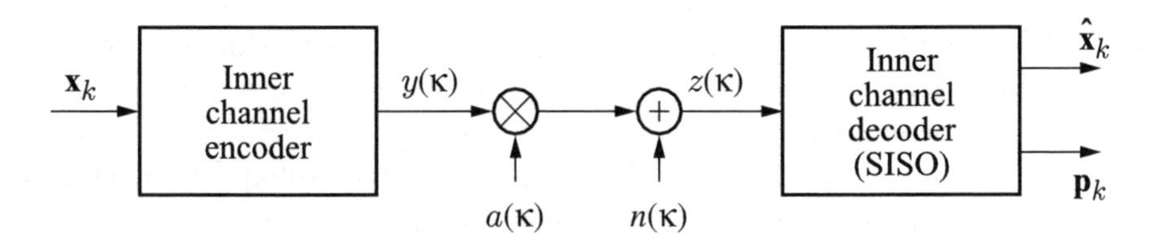

Figure 9.8: Equivalent channel with inner SISO decoding
SISO : Soft-Input, Soft-Output
$\mathbf{x}_k$, $\hat{\mathbf{x}}_k$: binary (bipolar) vectors of length w at time or frame index k
$y(\kappa)$: output sequence of the convolutional inner channel encoder
$z(k)$: sequence of the real-valued received samples
$\mathbf{p}_k$: vector of w instantaneous bit error rates (decoder reliabilities)

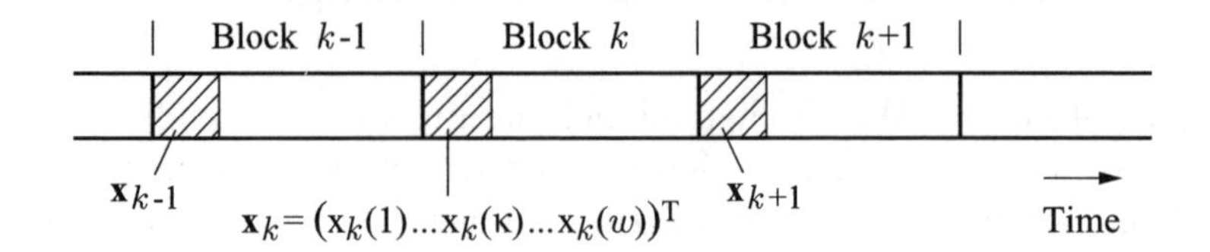

Figure 9.9: Block structure and sequential transmission of bit patterns $\mathbf{x}_k$
k = block or time index
$\mathbf{x}_k = (x_k(1), \ldots, x_k(\kappa), \ldots, x_k(w))^T$

Some of these SISO decoder algorithms include in the calculation of $\mathbf{p}_k$ *a priori* knowledge L_x or p_+ and p_-. For clarity of presentation it is assumed here that $L(x_k(\kappa)|\mathbf{z})$ comprises only the channel related quality. Thus, the conditional L-value can be translated into the instantaneous bit error probabilities $p_k(\kappa)$ according to (9.52) (Note: The instantaneous bit error probabilities may now also depend on the frame index k).

Then the SISO decoder transforms the received sequence of real-valued samples $z(\kappa)$ into hardbit vectors $\hat{\mathbf{x}}_k$ and bit error probability vectors $\mathbf{p}_k$, both with dimension w.

9.4 Soft Decision (SD) Source Decoding

In this section we will derive a concept for exploiting soft information within the source decoding process. The *softbit source decoding* (SBSD) approach, in short *soft decision* (SD) source decoding (e.g., [Fingscheidt, Vary 2001], [Fingscheidt, Vary 1997a]), should be compatible with existing transmission systems so that we do not have to modify the transmitter. In the literature, there are various proposals for error concealment exploiting reliability and/or *a priori* information (e.g., [Gerlach 1993], [Gerlach 1996], [Feldes 1993], [Feldes 1994], [Wong et al. 1984], [Sundberg 1978]). The presentation given here follows [Fingscheidt, Vary 2001]. As a reference we consider the conventional solution of Fig. 9.1 with hardbit or hard decision (HD) decoding at the receiver by table lookup.

In the soft decision approach, we replace the table lookup module by a parameter estimator as illustrated in Fig. 9.10. The soft information on the bit level in terms of softbits $\tilde{x}_k(\kappa)$, conditional L-values $L(x_k(\kappa)\,|\,\mathbf{z})$, or hardbits $\hat{x}_k(\kappa)$ plus (instantaneous) bit error rate $p_k(\kappa)$ and *a priori* knowledge is transformed to the

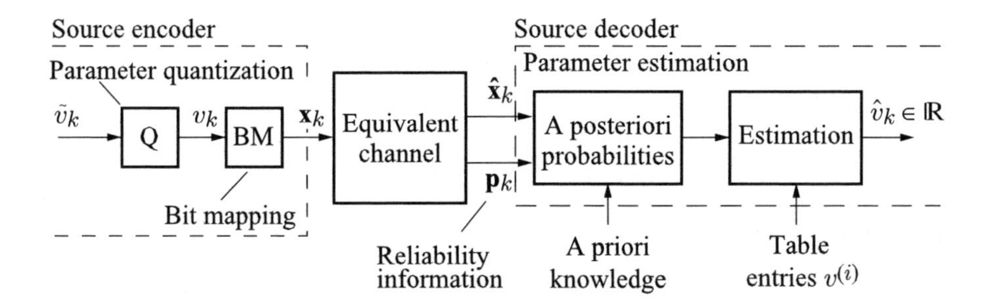

Figure 9.10: Soft decision (SD) source decoding
$\tilde{v}_k$: real-valued source codec parameter at time (frame or subframe) k
v_k : quantized parameter $v_k \in \{v^{(i)},\ i = 0, 1, \ldots, 2^w - 1\}$
$\mathbf{x}_k$: bit pattern, $\mathbf{x}_k \in \{\mathbf{x}^{(i)},\ i = 0, 1, \ldots, 2^w - 1\}$, vector of length w
$\hat{\mathbf{x}}_k$: received bit vector of length w
$\mathbf{p}_k$: vector of instantaneous bit error rates at time k
Q : quantization
BM: bit mapping

parameter level in such way that the codec parameters v_k can be estimated according to some optimization criterion, while the following information is taken into consideration:

- received hardbits $\hat{\mathbf{x}}_k$
- channel quality on bit level $\mathbf{p}_k = (p_k(1), \ldots, p_k(w))$
- *a priori* knowledge on parameter level: $P(v_k)$, $P(v_k \mid v_{k-1})$, ... (probability distribution, correlation, ...).

The codec parameter $\tilde{v}_k$ at time instant k is quantized according to $\mathbf{Q}[\tilde{v}_k] = v_k$ with $v_k \in \{v^{(i)},\ i = 0, 1, \ldots, 2^w - 1\}$ (Quantization Table) and can be represented by the quantization table index i. At the time instant k, a bit combination

$$\mathbf{x}_k = (x_k(1),\ x_k(2),\ \ldots,\ x_k(\kappa),\ \ldots,\ x_k(w)) \tag{9.57}$$

consisting of w bits is assigned via bit mapping (BM) to each quantized parameter v_k (or quantization table index i). The bits are assumed to be bipolar, i.e., $x_k(\kappa) \in \{-1,\ +1\}$. Due to the channel noise the received bit combination $\hat{\mathbf{x}}_k$ is possibly not identical to the transmitted one. In the conventional hard decision (HD) decoding scheme of Fig. 9.1, the received bit combination $\hat{\mathbf{x}}_k$ is applied to table decoding (inverse bit mapping (BM^{-1}) scheme). Thereafter, the decoded parameter $\hat{v}_k$ is used within the specific source decoder algorithm to reconstruct samples $\hat{s}(k \cdot N + \lambda) = \hat{s}_k(\lambda)$ of the speech signal.

The concept of error concealment by soft decision (SD) source decoding as depicted in Fig. 9.10 requires reliability information in terms of estimated instantaneous bit error probabilities

$$\mathbf{p}_k = (p_k(1),\, p_k(2),\, \ldots,\, p_k(\kappa),\, \ldots,\, p_k(w)) \tag{9.58}$$

of the hardbit combination $\hat{\mathbf{x}}_k$.

The kernel of the soft decision (SD) decoding algorithm consists of

- step 1: calculation of 2^w *a posteriori* probabilities
  $$P(\mathbf{x}^{(i)} \mid \hat{\mathbf{x}}_k) \text{ with } i \in \{0, 1, \ldots, 2^w - 1\}$$
- step 2: estimation of a real-valued parameter $\hat{v}_k$.

These steps will be described below. We start with the description of the second step.

9.4.1 Parameter Estimation

Let us assume that the *a posteriori* terms $P(\mathbf{x}^{(i)} \mid \hat{\mathbf{x}}_k)$ can be computed from the information on the bit level, which is available at the soft output of the equivalent channel which might or might not include inner channel decoding. If we have received a certain bit pattern $\hat{\mathbf{x}}_k$, then the probability $P(\mathbf{x}^{(i)} \mid \hat{\mathbf{x}}_k)$ quantifies the reliability of the decision that the pattern $\mathbf{x}^{(i)}$, and thus the quantized parameter value $v^{(i)}$, was transmitted at time k.

The *a posteriori* probabilities can be extended as shown below to take into consideration previous and/or future bit patterns $\hat{\mathbf{x}}_n$ with $n = k, k-1, \ldots, k-N$ and/or with $n = k+1, \ldots, k+K$ which are received later. The latter solution requires the introduction of some delay into the decoding process. The corresponding general *a posteriori* probability reads

$$P(\mathbf{x}_k = \mathbf{x}^{(i)} \mid \hat{\mathbf{x}}_k, \ldots) = P\big(\mathbf{x}_k = \mathbf{x}^{(i)} \mid \hat{\mathbf{x}}_{k+K}, \hat{\mathbf{x}}_{k+K-1}, \ldots, \hat{\mathbf{x}}_{k+1}, \hat{\mathbf{x}}_k, \hat{\mathbf{x}}_{k-1}, \hat{\mathbf{x}}_{k-2}, \ldots, \hat{\mathbf{x}}_{k-N}\big)\,. \tag{9.59}$$

Due to its computational complexity and the achievable quality improvements, the most relevant cases are $N = 0$ and $N = 1$, both with $K = 0$.

Once all 2^w (or in the general case $2^{w\cdot(N+1+K)}$) *a posteriori* terms have been computed for the present received pattern $\hat{\mathbf{x}}_k$, we can estimate the transmitted parameter value using different optimization criteria. The estimation error criterion should reflect the impact of parameter errors on the subjective quality of the soft-decoded signal.

For most of the speech, audio, and video codec parameters the minimum mean square (MS) error criterion is appropriate. In the case of speech these parameters may be PCM samples, spectral coefficients, gain factors, etc. However, for some parameters, such as the pitch period, the maximum *a posteriori* (MAP) estimator should be applied.

9.4.1.1 MAP Estimation

The MAP estimator follows the criterion

$$\hat{v}_k = v^{(\nu)} \quad \text{with} \quad \nu = \arg\max_i P\big(\mathbf{x}^{(i)} \mid \hat{\mathbf{x}}_k, \ldots\big) \,. \tag{9.60}$$

MAP estimation minimizes the probability of an erroneously decoded parameter [Melsa, Cohn 1978]. The decoded parameter $\hat{v}_k$ equals one of the code book/quantization table entries. In the case of error-free transmission, only one of the 2^w *a priori* probabilities takes the value 1, all the others are zero. In this situation the MAP decoder selects the same table entry as the hard decision decoder.

9.4.1.2 MS Estimation

The optimum decoded parameter in a minimum mean square error sense (see also Chapter 5) equals

$$\hat{v}_k = \sum_{i=0}^{2^w-1} v^{(i)} \cdot P\big(\mathbf{x}_k^{(i)} \mid \hat{\mathbf{x}}_k, \ldots\big) \,. \tag{9.61}$$

According to the orthogonality principle of linear mean square estimation (see, e.g., [Melsa, Cohn 1978]), the variance of the estimation error $e_0 = \hat{v}_k - v_k$ is $\sigma_e^2 = \sigma_v^2 - \sigma_{\hat{v}}^2$ with σ_v^2 being the variance of the undisturbed parameter v_k and $\sigma_{\hat{v}}^2$ denoting the variance of the estimated parameter $\hat{v}_k$. Because $\sigma_e^2 \geq 0$ we can state that the variance of the estimated parameter is smaller than or equals the variance of the error-free parameter.

For the worst-case channel with $p_0 = 0.5$, the *a posteriori* probabilities simplify to $P(\mathbf{x}^{(i)} \mid \hat{\mathbf{x}}_k, \ldots) = P(\mathbf{x}^{(i)})$. If in this case the unquantized parameter $\tilde{v}_k$ as well as the quantization table entries $v_k^{(i)}$ are distributed symmetrically around zero, the MS estimated parameter according to Eq. (9.61) is attenuated to zero (by weighted averaging, i.e., conditional expectation of v). Such symmetries are often found for gain factors in speech and audio encoders. Thus, the MS estimation of gain factors results in an inherent muting mechanism providing a graceful degradation of the signal quality with decreasing channel quality. This is one of the main advantages of soft decision source decoding.

On the other hand, if the channel is free of errors ($p_e = 0$) and $\mathbf{x}^{(\kappa)}$ has been transmitted, then all the parameter transition probabilities are zero except $P(\hat{\mathbf{x}}_k \mid \mathbf{x}^{(\kappa)}) = 1$. This yields $P(\mathbf{x}^{(\kappa)} \mid \hat{\mathbf{x}}_k, \ldots) = 1$ while all other *a posteriori* probabilities become zero. As a consequence, the MS estimator also yields the correct parameter value $\hat{v}_k = v_k$. This is equivalent to bit exactness in clear channel situations.

Finally, it should be mentioned that all of the algorithms discussed above can be used in the case of vector quantization. The only difference to scalar quantization lies in the estimation step. Let us consider a P-tuple of codec parameters $\tilde{\mathbf{v}}_k \in \mathbb{R}^P$ which is coded by w bits. The quantized parameter vector is then $\mathbf{Q}[\tilde{\mathbf{v}}_k] = \mathbf{v}_k$ with $\mathbf{v}_k \in$ CB (code book). MAP estimation simply yields the most probable parameter vector $\hat{\mathbf{v}}_k \in \{\mathbf{v}^{(i)},\ i = 0, 1, \ldots, 2^w - 1\}$ instead of a scalar, whereas MS estimation can be formulated as

$$\hat{\mathbf{v}}_k = \sum_{i=0}^{2^w-1} \mathbf{v}^{(i)} \cdot P\big(\mathbf{x}^{(i)} \mid \hat{\mathbf{x}}_k, \ldots\big) . \tag{9.62}$$

The MS estimator is to be considered as a weighted sum of the code book entries.

9.4.2 The *A Posteriori* Probabilities

For the estimation of a source codec parameter at the receiver, *a posteriori* probabilities providing information about any possibly transmitted bit combination $\mathbf{x}_k \in \{\mathbf{x}^{(i)},\ i = 0, \ldots, 2^w - 1\}$ are required (step 1 of the SD algorithm).

In the calculation described here we will only consider the most recent bit vector $\hat{\mathbf{x}}_k$ within the estimation process. Any earlier or later bit patterns will be neglected, i.e., $N = 0$ and $K = 0$ in (9.59),

We have to apply Bayes' theorem in order to obtain the *a posteriori* probabilities $P(\mathbf{x}_k = \mathbf{x}^{(i)} \mid \hat{\mathbf{x}}_k)$ for $i \in \{0, 1, \ldots, 2^w - 1\}$:

$$P(\mathbf{x}_k,\, \hat{\mathbf{x}}_k) \;=\; P(\mathbf{x}_k \mid \hat{\mathbf{x}}_k) \cdot P(\hat{\mathbf{x}}_k) \;=\; P(\hat{\mathbf{x}}_k \mid \mathbf{x}_k) \cdot P(\mathbf{x}_k) . \tag{9.63}$$

The probabilities $P(\hat{\mathbf{x}}_k)$ are not known. They depend on the *a priori* probabilities $P(\mathbf{x}_k = \mathbf{x}^{(k)})$ and on the transition probabilities $P(\hat{\mathbf{x}}_k \,|\, \mathbf{x}_k = \mathbf{x}^{(k)})$. Thus, these probabilities can be computed via the marginal distribution as follows:

$$P(\hat{\mathbf{x}}_k) = \sum_{\kappa=0}^{2^w-1} P\big(\hat{\mathbf{x}}_k \mid \mathbf{x}_k = \mathbf{x}^{(\kappa)}\big) \cdot P\big(\mathbf{x}_k = \mathbf{x}^{(\kappa)}\big). \tag{9.64}$$

Finally, we find the solution for the *a posteriori* probabilities

$$P\big(\mathbf{x}_k = \mathbf{x}^{(i)} \mid \hat{\mathbf{x}}_k\big) = \frac{P\big(\hat{\mathbf{x}}_k \mid \mathbf{x}_k = \mathbf{x}^{(i)}\big) \cdot P\big(\mathbf{x}_k = \mathbf{x}^{(i)}\big)}{\sum_{\kappa=0}^{2^w-1} P\big(\hat{\mathbf{x}}_k \mid \mathbf{x}_k = \mathbf{x}^{(\kappa)}\big) \cdot P\big(\mathbf{x}_k = \mathbf{x}^{(\kappa)}\big)}. \tag{9.65}$$

For the calculation of (9.64) and (9.65) we need the following probabilities:

- *a priori* probabilities $P\big(\mathbf{x}_k = \mathbf{x}^{(\kappa)}\big)$
- transition probabilities $P\big(\hat{\mathbf{x}}_k \mid \mathbf{x}_k = \mathbf{x}^{(\kappa)}\big)$, $\kappa = 0, 1, \ldots, 2^w - 1$

which will be discussed below.

9.4.2.1 The *A Priori* Knowledge

As regards the specification of the required *a priori* knowledge, there is a certain degree of freedom.

For the computation of (9.65), we need *a priori* knowledge about the quantized parameter in terms of the 2^w probabilities $P(\mathbf{x}^{(i)}) = P(v^{(i)})$, $i = 0, 1, \ldots, 2^w - 1$, i.e., the histogram of the quantized parameter.

In the general case, we model the quantized parameter as a Markov process of $(N + K)$-th order (N past values and K future values). For reasons of simplicity we will restrict our considerations to the case $K = 0$ here according to

$$P\left(\mathbf{x}_k \mid \mathbf{x}_{k+K}, \ldots, \mathbf{x}_{k+1}, \mathbf{x}_{k-1}, \ldots, \mathbf{x}_{k-N}\right) = P\left(\mathbf{x}_k \mid \mathbf{x}_{k-1}, \ldots, \mathbf{x}_{k-N}\right). \tag{9.66}$$

To find an appropriate Markov order, it is convenient to measure terms for different N such as $P(\mathbf{x}_k)$, $P(\mathbf{x}_k \mid \mathbf{x}_{k-1})$, and $P(\mathbf{x}_k, \mathbf{x}_{k-1})$ or even higher-order conditional and joint probabilities. This can be achieved by applying a large signal database to the source encoder and by counting the occurrences of the different quantizer output symbols, or different pairs of output symbols. We call $P(\mathbf{x}_k)$ *zeroth-order* a priori *knowledge* ($N = 0$, AK0) because it gives a statistical description of a zero-order Markov process, i.e., a memoryless process. Accordingly, we call $P(\mathbf{x}_k \mid \mathbf{x}_{k-1})$

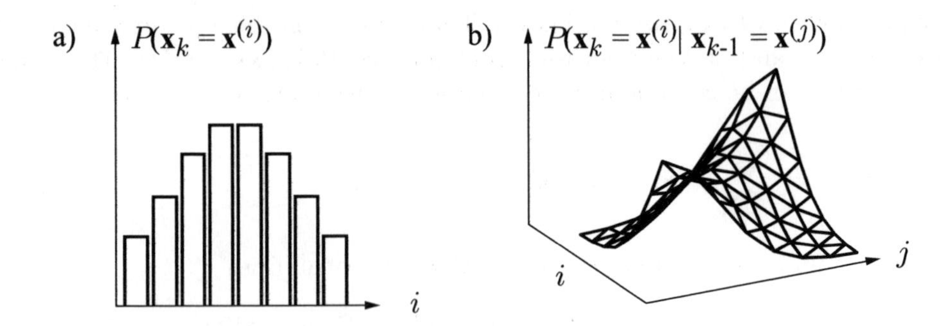

Figure 9.11: *A priori* parameter knowledge
a) AK0: zeroth order (N=0), 1D parameter/bit pattern histogram
b) AK1: first order (N=1), 2D parameter/bit pattern histogram

or $P(\mathbf{x}_k, \mathbf{x}_{k-1})$ first-order *a priori* knowledge ($N = 1$, AK1) because it refers to a first-order Markov process. The decision on which model should be taken is a matter of the

- observed redundancy
- allowed complexity of the soft decision source decoder
- trade-off between performance and complexity.

The two cases AK0 and AK1 are illustrated by the histograms of Fig. 9.11.

If we model a parameter as a zeroth-order Markov process, 2^w probabilities $P(\mathbf{x}_k = \mathbf{x}^{(i)})$ with $i \in \{0, 1, \ldots, 2^w - 1\}$ have to be stored in the decoder. With the entropy defined as

$$H(\mathbf{x}_k) \;=\; -\sum_{i=0}^{2^w-1} P\big(\mathbf{x}^{(i)}\big) \log_2 P\big(\mathbf{x}^{(i)}\big) \,, \qquad (9.67)$$

the redundancy of $\Delta R = M - H(\mathbf{x}_k)$ can be used for error concealment.

If a parameter is modeled as a first order Markov process, the 2^{2w} probabilities $P(\mathbf{x}_k = \mathbf{x}^{(i)} \mid \mathbf{x}_{k-1} = \mathbf{x}^{(j)})$ with $i, j \in \{0, \ldots, 2^w - 1\}$ have to be stored in the decoder. Then, a redundancy of $\Delta R = M - H(\mathbf{x}_k \mid \mathbf{x}_{k-1})$ can be used for error concealment, with the conditional entropy $H(\mathbf{x}_k \mid \mathbf{x}_{k-1})$ (see, e.g., [Cover, Thomas 1991]). This concept can be extended to even higher Markov orders N, while the storage requirements are $2^{w(N+1)}$ words.

9.4.2.2 The Parameter Transition Probabilities

Knowing the (instantaneous) bit error probabilities $p_k(\kappa)$, $\kappa = 1, \ldots, w$, of the known received bit $\hat{x}_k(\kappa)$, we get the *bit* transition probabilities for any transmitted bit $x_k(\kappa)$ as

$$P\big(\hat{x}_k(\kappa) \mid x_k(\kappa) = x^{(i)}(\kappa)\big) = \begin{cases} 1 - p_k(\kappa) & \text{if } \hat{x}_k(\kappa) = x^{(i)}(\kappa) \\ p_k(\kappa) & \text{if } \hat{x}_k(\kappa) \neq x^{(i)}(\kappa)\,. \end{cases} \tag{9.68}$$

If we consider an equivalent channel with independent bit errors (memoryless), the *parameter* transition probability reads

$$P\big(\hat{\mathbf{x}}_k \mid \mathbf{x}_k = \mathbf{x}^{(i)}\big) = \prod_{\kappa=0}^{w-1} P\big(\hat{x}_k(\kappa) \mid x^{(i)}(\kappa)\big)\,. \tag{9.69}$$

This term includes the channel characteristics and provides the probability of a transition from any possibly transmitted bit combination $\mathbf{x}^{(i)}$, $i \in \{0, 1, \ldots, 2^w - 1\}$, at time k, to the known received bit combination $\hat{\mathbf{x}}_k$.

In real-life applications, the assumption of a memoryless equivalent channel can be a coarse approximation, even if an interleaving scheme is employed within inner channel coding. However, the achievable error concealment based on Eq. (9.69) might still be very effective.

For the clarification of (9.69), we consider the calculation of the parameter transition probability by way of example.

Example:

$w = 3$,

$\mathbf{p}_k = (p_0,\ p_0,\ p_0)$,

$\hat{\mathbf{x}}_k = (+1,\ +1,\ -1)$ and

$i = 2$, i.e., $\mathbf{x}_k = \mathbf{x}^{(2)} = (+1,\ -1,\ +1)$

$P\big(\hat{\mathbf{x}}_k \mid \mathbf{x}_k = \mathbf{x}^{(2)}\big) = (1 - p_0) \cdot p_0^2\,.$

The bit error rate is assumed to be constant during the transmission of the 3 bits. We are interested in the probability that at time instant k the bit pattern $\mathbf{x}_k^{(2)} = \mathbf{x}^{(2)} = (+1,\ -1,\ +1)$, i.e., $i = 2$, is transmitted and that we receive the bit pattern $\hat{\mathbf{x}}_k = (+1,\ +1,\ -1)$. In this case, we would obviously receive one bit without error and two erroneous bits. The corresponding parameter transition probability for the quantization table entry $i = 2$ is therefore given by

$$P\big(\hat{\mathbf{x}}_k \mid \mathbf{x}_k = \mathbf{x}^{(2)}\big) = (1 - p_0) \cdot\ p_0^2\,. \tag{9.70}$$

In case of $p_+ = p_- = 1/2$, i.e., $L_x = 0$, the parameter transition probabilities can be derived from the softbits (9.15) by considering the different cases $x^{(i)}(\kappa) = \pm 1$ and $\hat{x}(\kappa) = \text{sign}\{\tilde{x}(\kappa)\} = \pm 1$ as follows:

$$P\left(\hat{\mathbf{x}}|\mathbf{x}^{(i)}\right) = \prod_{\kappa=0}^{w-1} \frac{\left|x^{(i)}(\kappa) + \tilde{x}(\kappa)\right|}{2} . \tag{9.71}$$

9.4.2.3 Approaches with Different Degrees of *A Priori* Knowledge

No *A Priori* Knowledge (NAK)

If there is no *a priori* knowledge available about the regarded parameter, we have to assume that the quantizer output symbols are uncorrelated and equally likely with

$$P(v^{(i)}) = P(\mathbf{x}^{(i)}) = \frac{1}{2^w} . \tag{9.72}$$

In this case only the channel-dependent soft information can be exploited. The required probabilities (9.65) are

$$P\big(\mathbf{x}_k = \mathbf{x}^{(i)} \mid \hat{\mathbf{x}}_k\big) = \frac{P\big(\hat{\mathbf{x}}_k \mid \mathbf{x}_k = \mathbf{x}^{(i)}\big)}{\sum\limits_{m=0}^{2^w-1} P\big(\hat{\mathbf{x}}_k|\mathbf{x}_k = \mathbf{x}^{(m)}\big)} . \tag{9.73}$$

Zeroth-Order *A Priori* Knowledge (AK0)

It should be noted that in practical coding schemes the assumption of equally likely quantizer outputs is usually not met. The widely used Lloyd–Max quantizers [Max 1960] (see also Section 7.4) yield, for instance, identical quantization error variance contributions of each quantization interval i, but no identical probabilities $P(\mathbf{x}_k = \mathbf{x}^{(i)})$.

If there is zeroth-order *a priori* knowledge available ($N = 0$), the Bayes rule yields according to (9.65) *a posteriori* probabilities

$$P\big(\mathbf{x}_k = \mathbf{x}^{(i)} \mid \hat{\mathbf{x}}_k\big) = \frac{P\big(\hat{\mathbf{x}}_k \mid \mathbf{x}_k = \mathbf{x}^{(i)}\big) \cdot P\big(\mathbf{x}_k = \mathbf{x}^{(i)}\big)}{\sum\limits_{m=0}^{2^w-1} P\big(\hat{\mathbf{x}}_k \mid \mathbf{x}_k = \mathbf{x}^{(m)}\big) \cdot P\big(\mathbf{x}_k = \mathbf{x}^{(m)}\big)} , \tag{9.74}$$

which take the parameter distribution, i.e., $P\left(v_k = v^{(i)}\right) = P\left(\mathbf{x}_k = \mathbf{x}^{(i)}\right)$, into account.

First-Order *A Priori* Knowledge (AK1)

The *a posteriori* term in Eq. (9.74) assumes successive bit combinations to be independent. If there is a residual correlation of first order, i.e., if

$H(\mathbf{x}_k) > H(\mathbf{x}_k \mid \mathbf{x}_{k-1})$, the *a posteriori* probabilities can be extended to take this correlation into account as well. The maximum information that is available at the decoder consists of the complete sequence of the bit combinations already received

$$\hat{\mathbf{x}}_k,\, \hat{\mathbf{x}}_{k-1},\, \hat{\mathbf{x}}_{k-2},\, \ldots\, ,\, \hat{\mathbf{x}}_{k-N} = \hat{\mathbf{x}}_k,\, \hat{\mathbf{X}}_{-1}\,,$$

where $\hat{\mathbf{X}}_{-1}$ includes the complete history of the received bit combinations until the previous time instant $k-1$. Given the first-order *a priori* knowledge $P(\mathbf{x}_k = \mathbf{x}^{(i)} \mid \mathbf{x}_{k-1} = \mathbf{x}^{(j)})$, the *a posteriori* probabilities $P(\mathbf{x}_k = \mathbf{x}^{(i)} \mid \hat{\mathbf{x}}_k, \hat{\mathbf{X}}_{-1})$ exploiting this complete history can be computed by means of a recursion. The MS estimation can be carried out as given in (9.61). The derivation of the calculation of the AK1, ... , AKN *a posteriori* probabilities can be found in [Fingscheidt 1998], [Fingscheidt, Vary 2001].

9.5 Application to Model Parameters

In this section, we will demonstrate the capabilities of soft decision source decoding by applying it to artificial Gaussian parameters. We will focus on a single codec parameter rather than on signal reconstruction. The parameter $\tilde{v}_k$, which is taken from a first-order autoregressive process with correlation factor of $\rho_{\tilde{v}\tilde{v}} = 0.9$, for example, is quantized by a scalar Lloyd–Max quantizer (LMQ) [Max 1960] using w bits. Correlation values of this magnitude can be found, for instance, in gain factors of speech and audio codecs. As we are not interested in the optimization of the bit mapping scheme here, we choose the natural binary code (NBC) [Jayant, Noll 1984]. We employ BPSK modulation over an AWGN channel and coherent demodulation, i.e., Fig. 9.5 with $a = 1$ (no fading). The reliability information in terms of instantaneous bit error probabilities $p_k(\kappa)$ or $|z(\kappa) \cdot L_c|$ in Eq. (9.48) is assumed to be ideally known, and thus the results presented in Figs. 9.12–9.16 can be interpreted as upper bounds for implementations with estimated reliability.

As a measure of quality we choose the global signal-to-noise ratio on the parameter level

$$\mathrm{SNR} = 10 \log_{10} \frac{\mathrm{E}\left\{\tilde{v}^2\right\}}{\mathrm{E}\left\{(\hat{v} - \tilde{v})^2\right\}} \tag{9.75}$$

which is henceforth called the *parameter SNR*.

For typical speech, audio, and video coding schemes the SNR of the most sensitive parameters seems to be a reasonable measure of parameter quality even for sporadic but extreme parameter errors.

9.5.1 Soft Decision Decoding without Channel Coding

In the first simulation, we do not use any channel coding. Thus, the input bit $x_k(\kappa)$ and the output bit $\hat{x}_k(\kappa)$ of the channel belong to the bit combinations of the quantized codec parameters v_k and $\hat{v}_k$, respectively. The soft decision (SD) decoding techniques use softbits or instantaneous bit error probabilities and *a priori* probabilities. In the example of Fig. 9.12, we choose a quantizer with $w = 3$, i.e., $2^w = 8$, quantizer reproduction levels. Under clear channel conditions (no channel noise), the SNR due to quantization is about 15 dB.

While the parameter SNR of hard decision (HD) decoding decreases rapidly with decreasing channel quality (E_b/N_0 ratio), soft decision decoding allows gains depending on the amount of parameter *a priori* knowledge which is used. By exploiting only the soft information at the channel output (i.e., no *a priori* knowledge, NAK, only the instantaneous bit error rates $p_k(\kappa)$), the SNR performance is slightly improved in comparison to hard decision decoding. A further improvement can be gained by using the zeroth-order *a priori* knowledge (AK0) which corresponds to the best case if the correlation is not taken into consideration or if the parameter is uncorrelated (i.e., $\rho_{\tilde{v}\tilde{v}} = 0$). However, if we exploit the correlation of the model parameter according to $\rho_{\tilde{v}\tilde{v}} = 0.9$, then the first order *a priori* knowledge (AK1) allows significant additional gains.

The conclusion from the experiment is that in the case of a highly correlated parameter, soft decision decoding by mean-sqare estimation (MS) exploiting first

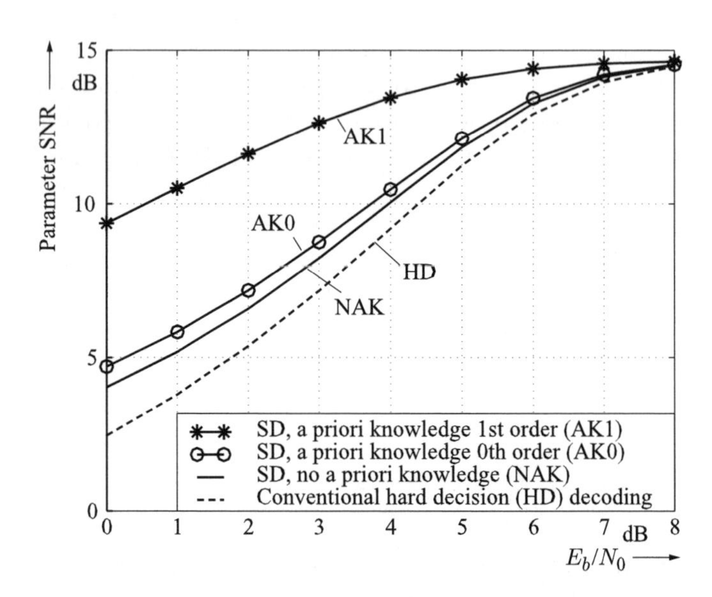

Figure 9.12: SNR performance of soft decision (SD) decoding without channel coding (Gaussian source s, Lloyd–Max quantizer (LMQ) with $w = 3$ bits, $\rho_{\tilde{v}\tilde{v}} = 0.9$, MS estimation)

order *a priori* knowledge (AK1) allows improvements of the parameter SNR of up to 6–10 dB or corresponding E_b/N_0 gains of up to 3–6 dB in comparison to hard decision decoding.

The PCM standard G.711 [ITU-T Rec. G.711 1972] provides a single parameter – the speech sample itself – quantized by $w = 8$ bits. Soft decision demodulation for PCM speech, where the PDF of speech samples is exploited as *a priori* knowledge was also investigated by Sundberg et al. ([Wong et al. 1984], [Sundberg 1978]).

The A-law PCM transmission over an AWGN channel was simulated as described before. The channel quality in terms of E_b/N_0 is again assumed to be ideally known. We measured entropy values of $H(\mathbf{x}_k) = 7.83$ bits and $H(\mathbf{x}_k|\mathbf{x}_{k-1}) = 6.5$ bits if pause segments were removed from the speech database. Including speech pauses, both values are much lower, depending on the percentage of speaker activity. This indicates that the usage of at least zeroth-order *a priori* knowledge is recommendable. The difference between entropy and conditional entropy reflects the amount of redundancy due to correlation that can be used by first-order *a priori* knowledge.

Figure 9.13 shows four different cases in terms of speech SNR (here: equal to parameter SNR) as a function of the E_b/N_0 ratio. If *a priori* knowledge is ex-

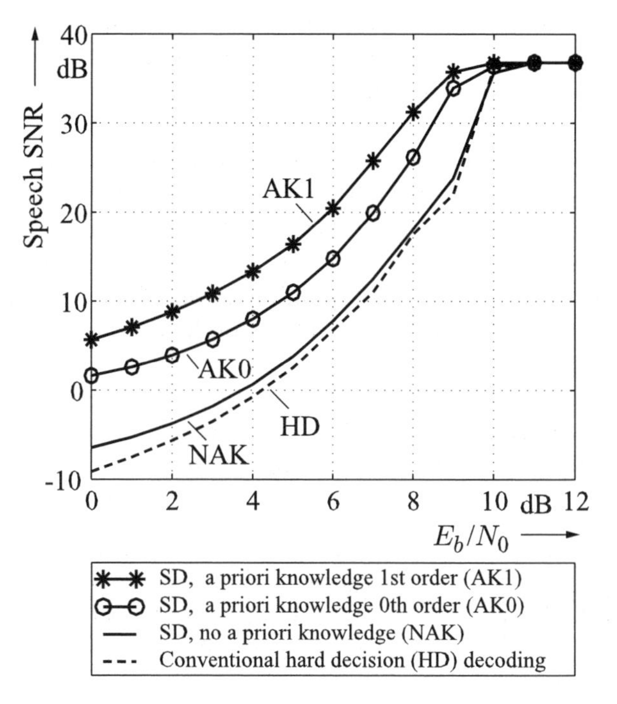

Figure 9.13: SNR performance of soft decision (SD) decoding without channel coding (A-law PCM, 64 kbit/s, MS estimation) [Fingscheidt, Vary 2001]; © 2001 IEEE

ploited, the quality of the MS estimated speech degrades asymptotically to 0 dB with decreasing E_b/N_0, i.e., the inherent muting mechanism of MS estimation. As in Fig. 9.12, the shape of the curves strongly depends on the order of exploited *a priori* knowledge. In comparison to hard decision (HD) decoding, soft decision (SD) decoding by MS estimation without *a priori* knowledge (SD, NAK) leads to a small speech SNR gain of about 1–2 dB, while the exploitation of *a priori* knowledge allows speech SNR gains of up to 10 dB (zero order) and up to about 15 dB (first order), respectively. This corresponds to a significant enhancement of the perceived speech quality although the model of the speech as a Markov process of first order is actually too simple. Further improvements can be gained by increasing the model order [Fingscheidt, Vary 1997a], [Fingscheidt, Vary 1997b], [Fingscheidt et al. 1998].

9.5.2 Soft Decision Decoding with Channel Coding

In this section, soft decision decoding in connection with channel coding will be investigated. In a first step, we apply an outer parity check code to each parameter individually and exploit the added redundancy in the soft decision decoding process. We do not explicitly implement an outer channel decoder, as indicated in Fig. 9.14, but we use the additional artificial redundancy to support parameter estimation by improved *a priori* knowledge.

As a consequence of the even (or odd) parity check bit, transmit vectors $\mathbf{y}_k \in \{\mathbf{y}^{(i)},\ i = 0, \ldots, 2^w - 1\}$ with an odd (or even) number of bits do not exist. If such bit patterns $\hat{\mathbf{y}}_k$ occur at the receiver due to bit errors, the corresponding *a posteriori* probabilities $P(\mathbf{x}_k = \mathbf{x}^{(i)} \mid \hat{\mathbf{y}}_k)$ or conditional L-values $L(\mathbf{x}_k \mid \mathbf{y}_k)$ can easily be calculated with slight modifications of the approach described in Section 9.4.2 [Fingscheidt et al. 1999]. The code has the rate $r = w/(w+1)$.

The equivalent channel of Fig. 9.14 may include inner channel coding and decoding. In any case, the availability of reliability information in terms of the instantaneous bit error probabilities is assumed at the channel output.

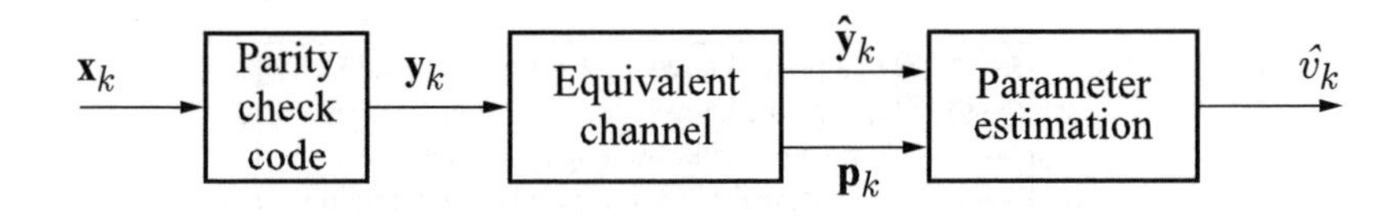

Figure 9.14: Support of soft decision decoding by an outer channel block code (e.g., parity check per parameter)

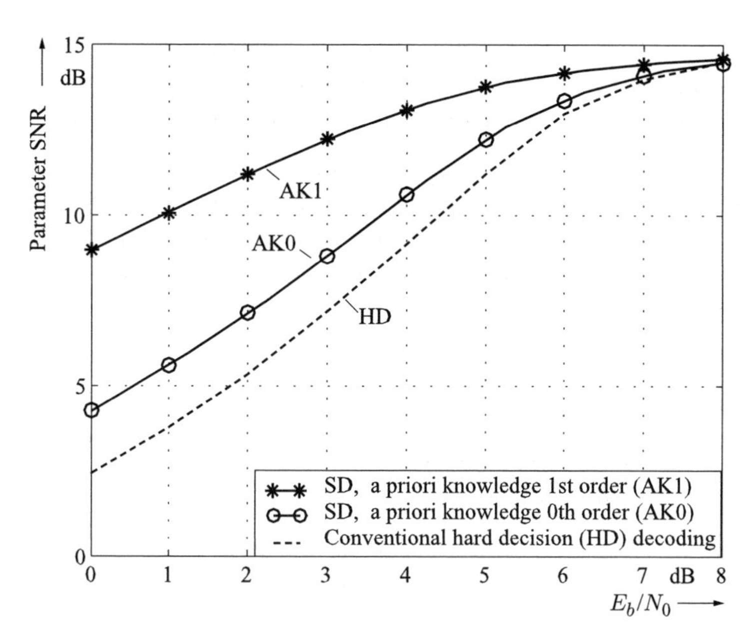

Figure 9.15: SNR performance of soft decision source decoding with outer parity check code (sequence of Gaussian parameters $\tilde{v}_k$, LMQ, bit patterns x_k with $w = 3$ bits, $\rho_{\tilde{v}\tilde{v}} = 0.9$, one parity check bit per bit pattern x_k, MS estimation)

An example is shown in Fig. 9.15. A sequence of Gaussian parameters $\tilde{v}_k$ with unit variance is applied to a $w = 3$ bit scalar Lloyd–Max quantizer. The bit mapping is natural binary coding (NBC), and the equivalent channel is AWGN with BPSK modulation and coherent demodulation. The soft output of the channel is computed as given in (9.48) and (9.52). The Gaussian source is a first-order Markov process with correlation $\rho = 0.9$. Within the soft decoding process we can exploit this correlation (first-order *a priori* knowledge) or we can disregard it (zeroth-order *a priori* knowledge).

The comparison of the achievable parameter SNR as a function of the E_b/N_0 ratio of the channel shows the worst performance for conventional hard decision decoding via table lookup (dashed line). In this case, no outer channel coding is applied, and the total transmission power is spent on the transmission of the uncoded bits $\mathbf{x}_k$. The term E_b quantifies in any case the energy which is spent to transmit one *information* bit $x_k(\kappa)$. However, if a parity check bit is added to each group $\mathbf{x}_k$ while keeping the same total transmit power per parameter as before, the parameter quality can be enhanced significantly by soft decision source decoding (solid lines), especially if we exploit the parameter correlation (solid line with stars). The assumed correlation of $\rho = 0.9$ is comparable to the measured correlation of parameters of various codec standards for speech and audio, such as gain factors in CELP speech codecs and subband audio codecs.

In a second step, we take inner channel coding into consideration and realize the equivalent channel of Fig. 9.10 as a combination of a convolutional channel encoder, an AWGN channel, and a channel decoder with soft-decision output.

We choose $w = 2$ and a convolutional encoder with a coding rate of $r = 1/2$ and constraint length $L = 5$. A sequential realization of the channel decoder according to Bahl et al. [Bahl et al. 1974] is used. In the clear channel condition the parameter SNR provided by a 2 bit Lloyd–Max quantization (LMQ) is 9.3 dB. We consider the transmission model of Fig. 9.8 with the options

- hard decision source decoding, using only $\hat{\mathbf{x}}_k$ (no *a priori* knowledge)
- soft decision source decoding, using $\hat{\mathbf{x}}_k$, $\mathbf{p}_k$, and $P(v_k = v^{(i)}) = P(\mathbf{x}^{(i)})$ (*a priori* knowledge of zero order)
- soft decision source decoding, using $\hat{\mathbf{x}}_k$, $\mathbf{p}_k$, $P(v^{(i)})$, and $P(v_k = v^{(i)} \mid v_{k-1} = v^{(j)})$ (*a priori* knowledge of first order).

Figure 9.16 shows that soft decision source decoding is able to reduce the gradient of quality loss beyond the typical threshold of soft input channel decoding with

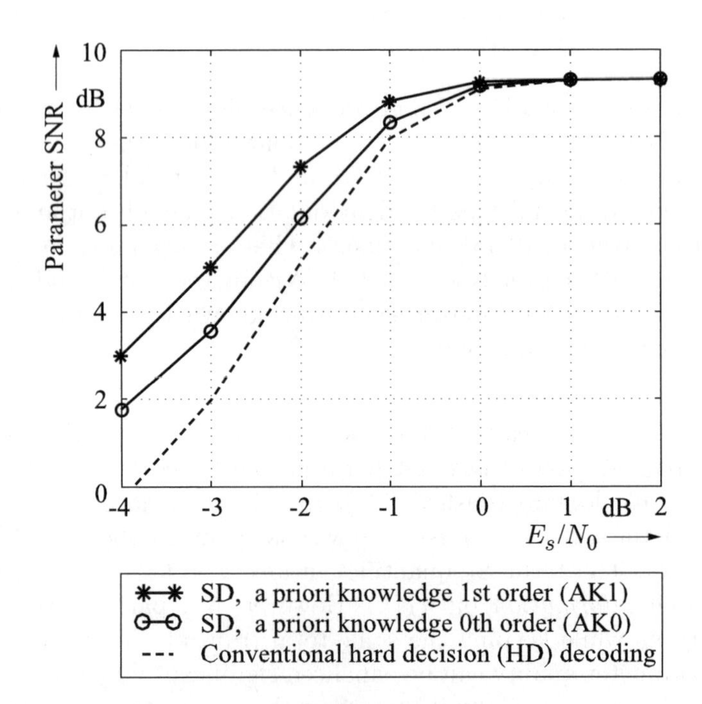

Figure 9.16: SNR performance of soft decision decoding in connection with rate $r = 1/2$ convolutional channel coding using a sequential channel decoder; LMQ, $w = 2$, $\rho_{\tilde{v}\tilde{v}} = 0.9$, MS estimation, $E_s = rE_b = \frac{1}{2}E_b$ [Fingscheidt, Vary 2001]; © 2001 IEEE

Table 9.1: Number of bits w, entropy $H(\mathbf{x}_k)$, conditional entropy $H(\mathbf{x}_k \mid \mathbf{x}_{k-1})$ of the parameters of the GSM full-rate codec [Fingscheidt, Scheufen 1997a], [Rec. GSM 06.10 1992], [Fingscheidt, Scheufen 1997b]

$\left[\frac{\text{bit}}{\text{parameter}}\right]$	LAR no.							
	1	2	3	4	5	6	7	8
w	6	6	5	5	4	4	3	3
$H(\mathbf{x}_k)$	5.43	4.88	4.75	4.53	3.73	3.76	2.84	2.88
$H(\mathbf{x}_k \mid \mathbf{x}_{k-1})$	4.46	4.29	4.18	4.09	3.37	3.39	2.49	2.46

$\left[\frac{\text{bit}}{\text{parameter}}\right]$	LTP		RPE		
	Lag	Gain	Grid	Max.	Pulse
w	7	2	2	6	3
$H(\mathbf{x}_k)$	6.31	1.88	1.96	5.39	2.86
$H(\mathbf{x}_k \mid \mathbf{x}_{k-1})$	5.75	1.74	1.96	4.29	2.86

hard decision source decoding. However, if the soft output of the channel decoder in terms of L-values or instantaneous bit error rates $\mathbf{p}_k$ is exploited within the source decoder, the soft decision source decoding scheme based on the first-order model allows E_s/N_0 (E_s: energy per coded bit) or E_b/N_0 (E_b: energy per information bit) gains of 1 dB or more.

In a final example the concept of soft decision speech decoding is applied to the GSM full-rate speech decoder [Rec. GSM 06.10 1992], [Fingscheidt, Scheufen 1997a], [Fingscheidt, Scheufen 1997b]. In Table 9.1 some codec parameters and their corresponding entropy values are listed. It is obvious that for most of the parameters first-order *a priori* knowledge will be helpful. Exceptions are the RPE grid and the RPE pulses (see also Chapter 8 and Appendix A). For reasons of simplicity, each codec parameter is modeled as a first-order Markov process.

For example, LAR no. 1 (Log Area Ratio, see Chapter 8), which is one of the subjectively most important parameters, provides a redundancy of 1.54 bits. This is more than 25%. Even speech codecs with lower bit rates than the GSM full-rate codec provide a very high amount of residual redundancy within the spectral parameters: e.g., [Alajaji et al. 1996] found about 29% of redundancy for the line spectral frequencies (LSFs) of the FS 1016 CELP [FS 1016 CELP 1992] due to non-uniform distribution (zeroth-order *a priori* knowledge concerning time) and due to intraframe correlation (first-order *a priori* knowledge concerning correlation to LSFs of the same frame).

It turns out that the non-integer GSM codec parameters can be well estimated using an MS estimator. In contrast to this, the estimation of a pitch period (LTP lag) or the RPE grid position should be performed by an MAP estimator.

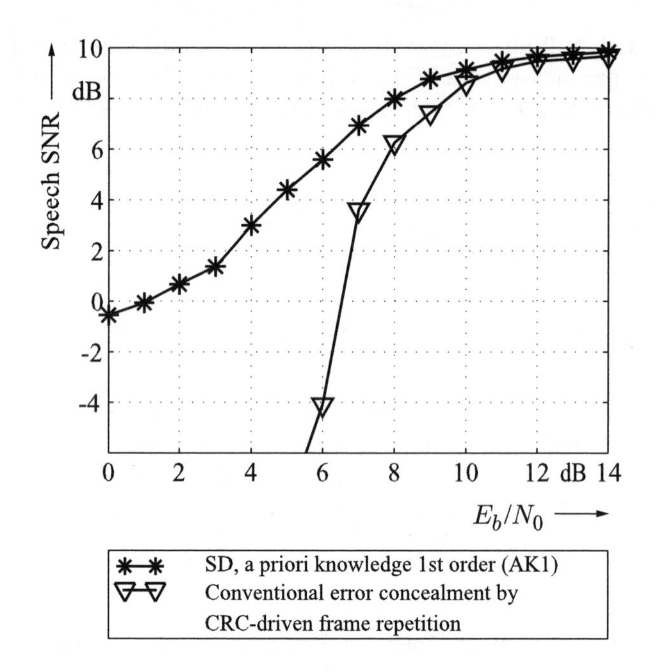

Figure 9.17: SNR performance of GSM full-rate soft decision speech decoding, TU50 (MS estimation) [Fingscheidt, Vary 2001]; © 2001 IEEE

Depicted in Fig. 9.17 are the results of a complete GSM simulation using the COSSAP GSM library [COSSAP 1995] with speech and channel (de)coding, (de)interleaving, (de)modulation, a channel model, and equalization. The channel model represents a typical case of an urban area (TU) with six characteristic propagation paths [Rec. GSM 05.05 1992] and a user speed of 50 km/h (TU50). Soft-output channel decoding is carried out by the algorithm of Bahl et al. [Bahl et al. 1974]. The conventional reference GSM decoder performs error concealment by a frame repetition (FR) algorithm as proposed in [Rec. GSM 06.11 1992]. The bad frame indicator (BFI) is simply set by the evaluation of the CRC (Cyclic Redundancy Check).

The SNR is surely not the optimum measure for speech quality. However, informal listening tests reveal the superiority of the soft decision speech decoder in comparison to the conventional decoding scheme in all situations of vehicle speeds and C/I ratios. Error concealment by soft decision speech decoding provides quite a good subjective speech quality down to C/I = 6 dB, whereas conventional frame repetition produces severe distortions at C/I = 7 dB. Even long error bursts caused by a low vehicle speed can be decoded sufficiently. In the soft decision decoding simulation, the annoying clicks of hard decision decoding in the case of CRC failures and the synthetic sounds of the frame repetition disappear completely and turn into slightly noisy or modulated speech. This enhances the perceived speech quality significantly.

A very attractive feature of this soft decision source decoding technique is the fact that it can be applied to any source coding algorithm without modifications at the transmitter side. In clear channel conditions, bit exactness is always preserved.

9.6 Further Improvements

Further improvements are obtained if we replace the outer parity check code in Fig. 9.14 by a more powerful block code [Fingscheidt et al. 1999] or even by a non-linear block code, e.g., [Heinen, Vary 2000], [Hindelang et al. 2000], [Heinen 2001], [Heinen, Vary 2005], which is not intended for explicit channel decoding at the receiver but for increasing the *a priori* knowledge in support of the soft decision parameter estimation process.

This combination of soft decision source and channel decoding is actually an attractive alternative to conventional error concealment based on parity check coding. Furthermore, if the residual correlation of the codec parameters exceeds a certain value, soft decision source decoding with support by outer channel block coding can outperform conventional channel coding [Hindelang et al. 2000].

In any case, this concept is also advantageous in combination with conventional inner channel coding, even if the residual parameter correlation is low.

Finally, it should be mentioned that the concept of soft decision source decoding opens up possibilities for iterative source–channel decoding, e.g., [Görtz 2000], [Farvardin 1990], [Adrat et al. 2002], [Perkert et al. 2001]. This approach of joint and iterative source-channel decoding is called *turbo error–concealment* [Adrat 2003]. The decoding process is based on the turbo principle [Berrou, Glavieux 1996], as illustrated in Fig. 9.18. One of the two component decoders is a channel decoder, the other is a soft decision source decoder. The inner SISO channel decoder provides *extrinsic information* to the soft decision source decoder which itself extracts extrinsic information on the bit level from the parameter *a posteriori* probabilities

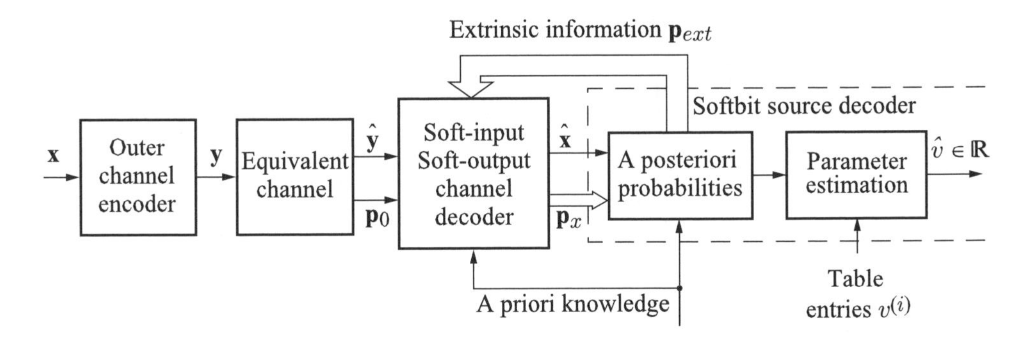

Figure 9.18: The concept of iterative source–channel decoding ("Turbo error-concealment" [Adrat 2003])

and feeds it back to the channel decoder. After terminating the iterations, the final step consists of estimating the codec parameter as described in Section 9.4.1 using the resulting reliability information (bit error probabilities) on the parameter level.

The improved capabilities of the iterative source–channel decoding are demonstrated by the example of Fig. 9.19. Gauss–Markov source parameters $\tilde{v}_k$ with unit variance and correlation $\rho = 0.95$ are applied to an optimum Lloyd–Max quantizer (LMQ) with $w = 3$ bit/parameter. This degree of correlation can be found in certain parameters of some standardized speech and audio codecs. The bit mapping rule is natural binary coding. No outer channel encoding is used here. The inner channel decoder is a rate $r = 1/2$ recursive non-systematic convolutional channel code (RNSC) with constraint length $L = 7$ and generator polynomials

$$G_1(D) = \frac{1 + D^2 + D^3 + D^5 + D^6}{1 + D + D^2 + D^3 + D^4 + D^6} \tag{9.76-a}$$

$$G_2(D) = \frac{1 + D + D^4 + D^6}{1 + D + D^2 + D^3 + D^4 + D^6} \, . \tag{9.76-b}$$

A target parameter SNR of 13.5 dB is assumed.

In comparison to non-iterative decoding schemes, further significant improvements of the error robustness can obviously be achieved. For further details, the reader is referred to, for example, [Adrat 2003], [Adrat, Vary 2004], [Adrat et al. 2005], [Adrat, Vary 2005].

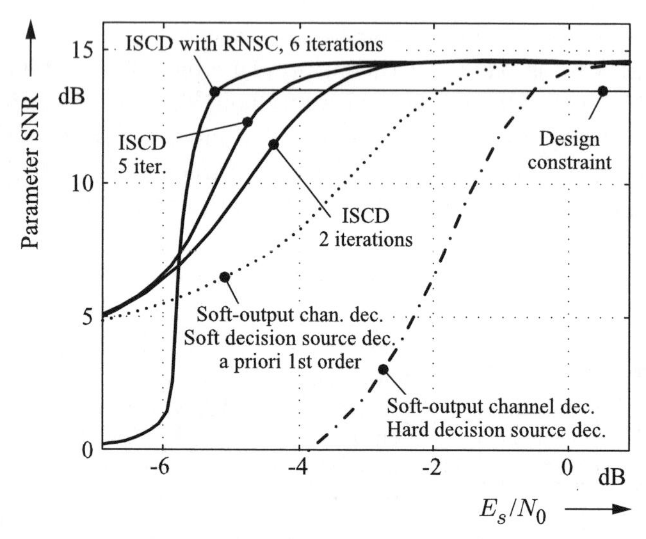

Figure 9.19: Example of iterative source–channel decoding (ISCD):
Gauss–Markov parameter with $\rho = 0.95$, LMQ with $w = 3$ bit/parameter
RNSC: Recursive Non-Systematic Convolutional code, $r = 1/2$, $L = 7$
Soft decision source decoding with *a priori* knowledge of first order
("Turbo error concealment" [Adrat 2003])

Bibliography

Adrat, M. (2003). *Iterative Source-Channel Decoding for Digital Mobile Communications*, PhD thesis. *Aachener Beiträge zu digitalen Nachrichtensystemen*, vol. 16, P. Vary (ed.), RWTH Aachen University.

Adrat, M.; Brauers, J.; Clevorn, T.; Vary, P. (2005). The EXIT-Characteristics of Softbit-Source Decoders, *IEEE Communication Letters*, vol. 9, no. 6, June, pp. 540–542.

Adrat, M.; Hänel, R.; Vary, P. (2002). On Joint Source-Channel Decoding for Correlated Sources, *Proceedings of the IEEE International Conference on Acoustics, Speech, and Signal Processing (ICASSP)*, Orlando, Florida, USA.

Adrat, M.; Vary, P. (2004). Iterative Source-Channel Decoding with Code Rates near $r = 1$, *IEEE International Conference on Communications*, Paris, France.

Adrat, M.; Vary, P. (2005). Iterative Source-Channel Decoding: Improved System Design Using EXIT-Charts, *EURASIP Journal on Applied Signal Processing (Special Issue: Turbo Processing)*, vol. 2005, no. 6, May, pp. 928–941.

Alajaji, F. I.; Phamdo, N. C.; Fuja, T. E. (1996). Channel Codes that Exploit the Residual Redundancy in CELP-Encoded Speech, *IEEE Transactions on Speech and Audio Processing*, vol. 4, no. 5, September, pp. 325–336.

Bahl, L. R.; Cocke, J.; Jelinek, F.; Raviv, J. (1974). Optimal Decoding of Linear Codes for Minimizing Symbol Error Rate, *IEEE Transactions on Information Theory*, vol. 20, March, pp. 284–287.

Berrou, C.; Glavieux, A. (1996). Near Optimum Error Correcting Coding and Decoding: Turbo-Codes, *IEEE Transactions on Communications*, vol. 44, October, pp. 1261–1271.

Clüver, K. (1996). An ATM Speech Codec With Improved Reconstruction of Lost Cells, *Signal Processing VIII, Proceedings of the European Signal Processing Conference (EUSIPCO)*, Trieste, Italy, pp. 1641–1643.

COSSAP (1995). *Model Libraries Volume 3*, Synopsys Inc., Mountain View, California.

Cover, T. M.; Thomas, J. A. (1991). *Elements of Infomation Theory*, John Wiley & Sons, Inc., New York.

Erdöl, N.; Castelluccia, C.; Zilouchian, A. (1993). Recovery of Missing Speech Packets Using the Short-Time Energy and Zero-Crossing Measurements, *IEEE Transactions on Speech and Audio Processing*, vol. 1, no. 3, July, pp. 295–303.

ETSI Rec. GSM 06.90 (1998). *Digital Cellular Telecommunications System (Phase 2+); Adaptive Multi-Rate (AMR) Speech Transcoding*, European Telecommunications Standards Institute.

Farvardin, N. (1990). A Study of Vector Quantization for Noisy Channels, *IEEE Transactions on Information Theory*, vol. 36, no. 4, July, pp. 799–809.

Feldes, S. (1993). Enhancing Robustness of Coded LPC-Spectra to Channel Errors by Use of Residual Redundancy, *Proceedings of EUROSPEECH*, Berlin, Germany, pp. 1147–1150.

Feldes, S. (1994). *Restredundanzbasierte Rekonstruktionsverfahren bei der digitalen Sprachübertragung in Mobilfunksystemen*, PhD thesis, TH Darmstadt, Germany (in German).

Fingscheidt, T. (1998). *Softbit-Sprachdecodierung in digitalen Mobilfunksystemen*, PhD thesis. *Aachener Beiträge zu digitalen Nachrichtensystemen*, vol. 9, P. Vary (ed.), RWTH Aachen University (in German).

Fingscheidt, T.; Heinen, S.; Vary, P. (1999). Joint Speech Codec Parameter and Channel Decoding of Parameter Individual Block Codes (PIBC), *IEEE Workshop on Speech Coding*, Porvoo, Finland, pp. 75–77.

Fingscheidt, T.; Scheufen, O. (1997a). Error Concealment in the GSM System by Softbit Speech Decoding, *Proceedings of the 9th Aachen Symposium "Signaltheorie"*, Aachen, Germany, pp. 229–232.

Fingscheidt, T.; Scheufen, O. (1997b). Robust GSM Speech Decoding Using the Channel Decoder's Soft Output, *Proceedings of EUROSPEECH*, Rhodes, Greece, pp. 1315–1318.

Fingscheidt, T.; Vary, P. (1997a). Robust Speech Decoding: A Universal Approach to Bit Error Concealment, *Proceedings of the IEEE International Conference on Acoustics, Speech, and Signal Processing (ICASSP)*, Munich, Germany, vol. 3, pp. 1667–1670.

Fingscheidt, T.; Vary, P. (1997b). Speech Decoding with Error Concealment Using Residual Source Redundancy, *Proceedings of the IEEE Workshop on Speech Coding*, Pocono Manor, Pennsylvania, USA, pp. 91–92.

Fingscheidt, T.; Vary, P. (2001). Softbit Speech Decoding: A New Approach to Error Concealment, *IEEE Transactions on Speech and Audio Processing*, vol. 9, no. 3, pp. 240–250.

Fingscheidt, T.; Vary, P.; Andonegui, J. A. (1998). Robust Speech Decoding: Can Error Concealment Be Better Than Error Correction?, *Proceedings of the IEEE International Conference on Acoustics, Speech, and Signal Processing (ICASSP)*, Seattle, Washington, USA, vol. 1, pp. 373–376.

FS 1016 CELP (1992). Details to Assist in Implementation of Federal Standard 1016 CELP, US Department of Commerce – National Technical Information Service.

Gerlach, C. G. (1993). A Probabilistic Framework for Optimum Speech Extrapolation in Digital Mobile Radio, *Proceedings of the IEEE International Conference on Acoustics, Speech, and Signal Processing (ICASSP)*, Minneapolis, Minnesota, vol. 2, pp. 419–422.

Gerlach, C. G. (1996). *Beiträge zur Optimalität in der codierten Sprachübertragung*, PhD thesis. *Aachener Beiträge zu digitalen Nachrichtensystemen*, vol. 5, P. Vary (ed.), RWTH Aachen University (in German).

Goodman, D. J.; Lockhart, G. B.; Wasem, O. J.; Wong, W.-C. (1986). Waveform Substitution Techniques for Recovering Missing Speech Segments in Packet Voice Communications, *IEEE Transactions on Acoustics, Speech and Signal Processing*, vol. 34, no. 6, December, pp. 1440–1448.

Görtz, N. (2000). Iterative Source-Channel Decoding using Soft-In/Soft-Out Decoders, *Proceedings of the International Symposium on Information Theory (ISIT)*, Sorrento, Italy, p. 173.

Hagenauer, J. (1980). Viterbi Decoding of Convolutional Codes for Fading- and Burst-Channels, *Proceedings of the 1980 Zurich Seminar on Digital Communications*, Zurich, Switzerland, pp. G2.1–G2.7.

Hagenauer, J. (1995). Source-Controlled Channel Decoding, *IEEE Transactions on Communications*, vol. 43, no. 9, September, pp. 2449–2457.

Hagenauer, J. (1997). Vom Analogwert zum Bit und zurück, *Frequenz*, vol. 51, no. 9, September, pp. 211–227 (in German).

Hagenauer, J.; Hoeher, P. (1989). A Viterbi Algorithm with Soft-Decision Outputs and its Applications, *Proceedings of GLOBECOM*, Dallas, Texas, USA, pp. 1680–1686.

Hagenauer, J.; Offer, E.; Papke, L. (1996). Iterative Decoding of Binary Block and Convolutional Codes, *IEEE Transactions on Information Theory*, vol. 42, no. 2, March, pp. 429–445.

Heikkila, I.; Jokinen, H.; Ranta, J. (1993). A Signal Quality Detecting Circuit and Method for Receivers in the GSM System, European Patent # 0648032 A 1.

Heinen, S. (2001). *Quellenoptimierter Fehlerschutz für digitale Übertragungskanäle*, PhD thesis. *Aachener Beiträge zu digitalen Nachrichtensystemen*, vol. 14, P. Vary (ed.), RWTH Aachen University (in German).

Heinen, S.; Vary, P. (2000). Source Optimized Channel Codes (SOCCs) for Parameter Protection, *Proceedings of the International Symposium on Information Theory (ISIT)*, Sorrento, Italy.

Heinen, S.; Vary, P. (2005). Source Optimized Channel Coding for Digital Transmission Channels, *IEEE Transactions on Communications*, vol. 53, no. 4, April, pp. 592–600.

Hindelang, T.; Heinen, S.; Vary, P.; Hagenauer, J. (2000). Two Approaches to Combined Source-Channel Coding: A Scientific Competition in Estimating Correlated Parameters, *International Journal of Electronics and Communications (AEÜ)*, vol. 54, no. 6, December, pp. 364–378.

Huber, J. (2002). Grundlagen der Wahrscheinlichkeitsrechnung für iterative Decodierverfahren, *Elektrotechnik und Informationstechnik, Heft 11*, pp. 386–394 (in German).

Huber, J.; Rüppel, A. (1990). Zuverlässigkeitsschätzung für die Ausgangssymbole von Trellis-Decodern, *International Journal of Electronics and Communications (AEÜ)*, vol. 44, pp. 8–21 (in German).

ITU-T Rec. G.711 (1972). Pulse Code Modulation (PCM) of Voice Frequencies, International Telecommunication Union (ITU).

Järvinen, K.; Vainio, J. (1994). Detection of Defective Speech Frames in a Receiver of a Digital Speech Communication System, World Patent # WO 96/09704.

Jayant, N. S.; Christensen, S. W. (1981). Effects of Packet Losses in Waveform Coded Speech and Improvements due to an Odd-Even Sample-Interpolation Procedure, *IEEE Transactions on Communications*, vol. 29, no. 2, February, pp. 101–109.

Jayant, N. S.; Noll, P. (1984). *Digital Coding of Waveforms*, Prentice Hall, Englewood Cliffs, New Jersey.

Martin, R.; Hoelper, C.; Wittke, I. (2001). Estimation of Missing LSF Parameters Using Gaussian Mixture Models, *Proceedings of the IEEE International Conference on Acoustics, Speech, and Signal Processing (ICASSP)*, Salt Lake City, Utah, USA.

Max, J. (1960). Quantizing for Minimum Distortion, *IRE Transactions on Information Theory*, vol. 6, March, pp. 7–12.

Melsa, J. L.; Cohn, D. L. (1978). *Decision and Estimation Theory*, McGraw-Hill Kogakusha, Tokyo.

Minde, T. B.; Mustel, P.; Nilsson, H.; Lagerqvist, T. (1993). Soft Error Correction in a TDMA Radio System, World Patent # WO 95/16315.

Papoulis, A. (1968). *Systems and Transforms with Applications in Optics*, McGraw-Hill, New York.

Perkert, R.; Kaindl, M.; Hindelang, T. (2001). Iterative Source and Channel Decoding for GSM, *Proceedings of the IEEE International Conference on Acoustics, Speech, and Signal Processing (ICASSP)*, Salt Lake City, Utah, USA, pp. 2649–2652.

Proakis, J. G. (1995). *Digital Communications*, 3rd edn, McGraw-Hill, New York.

Rec. GSM 05.05 (1992). Recommendation GSM 05.05 Radio Transmission and Reception, ETSI TC-SMG.

Rec. GSM 06.10 (1992). Recommendation GSM 06.10 GSM Full Rate Speech Transcoding, ETSI TC-SMG.

Rec. GSM 06.11 (1992). Recommendation GSM 06.11 Substitution and Muting of Lost Frames for Full Rate Speech Traffic Channels, ETSI TC-SMG.

Rec. GSM 06.21 (1995). European Digital Cellular Telecommunications System Half Rate Speech Part 3: Substitution and Muting of Lost Frames for Half Rate Speech Traffic Channels (GSM 06.21), ETSI TM/TM5/TCH-HS.

Rec. GSM 06.61 (1996). Digital Cellular Telecommunications System: Substitution and Muting of Lost Frames for Enhanced Full Rate (EFR) Speech Traffic Channels (GSM 06.61), ETSI TC-SMG.

Sereno, D. (1991). Frame Substitution and Adaptive Post-Filtering in Speech Coding, *Proceedings of EUROSPEECH*, Genoa, Italy, pp. 595–598.

Shannon, C. E. (1948). A Mathematical Theory of Communication, *Bell System Technical Journal*, vol. 27, July, pp. 379–423.

Skoglund, M.; Hedelin, P. (1994). Vector Quantization Over a Noisy Channel Using Soft Decision Decoding, *Proceedings of the IEEE International Conference on Acoustics, Speech, and Signal Processing (ICASSP)*, Adelaide, Australia, vol. 5, pp. 605–608.

Stenger, A.; Younes, K. B.; Reng, R.; Girod, B. (1996). A New Error Concealment Technique for Audio Transmission with Packet Loss, *Signal Processing VIII, Proceedings of the European Signal Processing Conference (EUSIPCO)*, Trieste, Italy, pp. 1965–1968.

Su, H.-Y.; Mermelstein, P. (1992). Improving the Speech Quality of Cellular Mobile Systems Under Heavy Fading, *Proceedings of the IEEE International Conference on Acoustics, Speech, and Signal Processing (ICASSP)*, San Francisco, California, USA, pp. II 121–124.

Sundberg, C.-E. (1978). Soft Decision Demodulation for PCM Encoded Speech Signals, *IEEE Transactions on Communications*, vol. 26, no. 6, June, pp. 854–859.

Valenzuela, R. A.; Animalu, C. N. (1989). A New Voice-Packet Reconstruction Technique, *Proceedings of the IEEE International Conference on Acoustics, Speech, and Signal Processing (ICASSP)*, Glasgow, UK, pp. 1334–1336.

Vary, P.; Fingscheidt, T. (1998). From Soft Decision Channel Decoding to Soft Decision Speech Decoding, *Proceedings of the 2. ITG-Fachtagung "Codierung für Quelle, Kanal und Übertragung"*, VDE-Verlag, Aachen, pp. 155–162.

Wong, W. C.; Steele, R.; Sundberg, C.-E. (1984). Soft Decision Demodulation to Reduce the Effect of Transmission Errors in Logarithmic PCM Transmitted over Rayleigh Fading Channels, *AT&T Bell Laboratories Technical Journal*, vol. 63, no. 10, December, pp. 2193–2213.

10

Bandwidth Extension (BWE) of Speech Signals

The characteristic sound of *telephone speech* with its restricted audio quality is mainly caused by the limitation of the transmitted bandwidth to the frequency range of the old analog telephone. In the conversion process from analog to digital transmission technology, the frequency bandwidth from 300 Hz to 3.4 kHz [ITU-T Rec. G.712 1988] has been retained for reasons of compatibility.

In the long run, true wideband transmission with a bandwidth of 7 kHz will be introduced into the telephone networks. However, this will require a very long transition period in which many or, in the beginning, even most terminals and parts of the network have not yet been equipped with the wideband capability. In this situation, the speech quality may be improved at the receiving end by means of *artificial bandwidth extension* (BWE).

Bandwidth extension of telephone speech or audio signals is a very recent and attractive research topic. The objective is to increase the frequency bandwidth and to improve the perceived speech or audio quality by artificially adding some spectral components. These components are generated within the decoding process using information which is completely extracted from the transmitted narrowband signal in the extreme case. Alternatively, the network or the encoder may allow the transmission of a small amount of side information to support the process of bandwidth extension.

Digital Speech Transmission: Enhancement, Coding and Error Concealment
Peter Vary and Rainer Martin

In this chapter, it will be shown how some specialized versions of BWE with side information have already been used in the context of speech coding and how the purely receiver-based extension may work.

10.1 Narrowband versus Wideband Telephony

The minimum demands on the telephone bandwidth were specified in the CCITT Red Book from 1961: at the lower and the upper cutoff frequencies of $f_L = 300\,\text{Hz}$ and $f_U = 3.4\,\text{kHz}$, the analog filters in the transmission path may attenuate the signal by no more than 10 dB with regard to the level at the frequency of 800 Hz ([ITU-T Rec. G.132 1988], [ITU-T Rec. G.151 1988]). At that time, the reasons for the bandwidth limitation were the use of analog frequency division multiplex (FDM) transmission with a frequency grid of 4 kHz and the optional use of sub-audio telegraphy for out-of-band signaling. The specification of the cutoff frequencies of 300 Hz and 3.4 kHz and of the filter characteristics was based on subjective listening tests. According to this specification, the intelligibility of (meaningless) syllables is about 91% and the comprehension of sentences is about 99% [Terhardt 1998], [Brosze et al. 1992], [Schmidt, Brosze 1967]. However, listening experiments have shown that a certain increase of the acoustic bandwidth significantly improves not only the perceived speech quality ([Krebber 1995], [Voran 1997]) but also the intelligibility of, for example, unvoiced and fricative speech sounds.

Since then, the public telephone networks have been converted almost completely to digital transmission techniques. According to the international standards [ITU-T Rec. G.711 1972] and [ITU-T Rec. G.712 1988], the speech signals are sampled at a sampling rate of $f_s = 8\,\text{kHz}$ and the samples are quantized using non-linear companding according to the A-law or μ-law characteristic (see Section 7.3) with 8 bits per sample, resulting in the bit rate of 64 kbit/s. A minimum stopband attenuation of at least 25 dB has to be achieved above 4.6 kHz. The attenuation tolerance scheme for PCM transmission is given in Fig. 10.1-a.

For cellular phone systems, a further limitation of the frequency range is specified in order to reduce the amount of disturbing low-frequency background noise. In the GSM system, for instance, the sensitivity of both the sending and the receiving terminal should provide an attenuation of at least 12 dB below 100 Hz [ETSI Rec. GSM 03.50 1998].

As we know from everyday life, speech intelligibility on the phone seems to be sufficient, at least for a normal conversation, although the fundamental frequency of speech is not transmitted in most cases. We become aware of the limited intelligibility of syllables if we are forced to understand unknown words or names. In these cases we often need to spell a word, especially to distinguish between certain

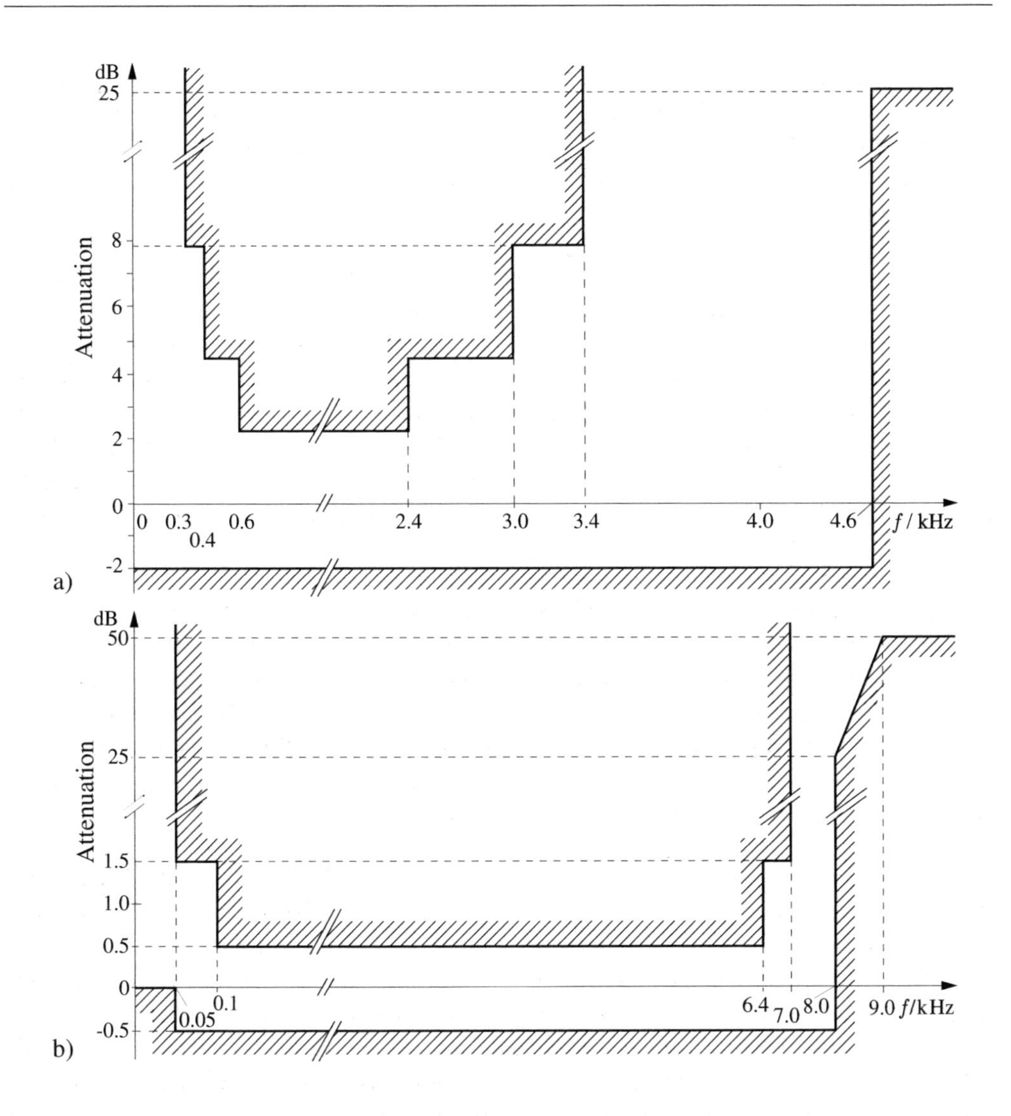

Figure 10.1: Tolerance scheme for the magnitude frequency response of the PCM anti-aliasing filter according to:
a) ITU-T G.712: narrowband speech
b) ITU-T G.722: wideband speech

unvoiced or plosive utterances, such as /f/ and /s/ or /p/ and /t/. A second restriction of the subjectively perceived speech quality is that some speaker-specific features are not retained on the phone. Speaker transparency is frequently limited.

These restrictions can be overcome with the introduction of *wideband speech transmission*, which is characterized by the extended frequency band from 50 Hz to 7.0 kHz, as indicated in Fig. 10.1-b by the magnitude tolerance scheme of the corresponding anti-aliasing input filter according to [ITU-T Rec. G.722 1988]. The sampling frequency has to be increased from $f_s = 8\,\text{kHz}$ to $f_s^{'} = 16\,\text{kHz}$, and the required resolution for A/D conversion has to be improved from $w = 8$ bits (non-uniform A-law or μ-law) to $w' = 14$ bits (uniform) per sample. In what follows, conventional telephone speech will be called *narrowband speech* (as opposed to *wideband speech*).

An example of unvoiced speech with significant frequency content beyond 3.4 kHz is given in Fig. 10.2, which shows a spectral comparison of the original speech with the corresponding *narrowband* and *wideband* versions. A closer look at Fig. 10.2 reveals that narrowband speech may lack significant parts of the spectrum and that the difference between wideband speech and original speech is still clearly noticeable. The frequency range of wideband telephone speech is comparable to the bandwidth of AM radio broadcasting, which provides excellent speech quality and intelligibility of syllables but is nowadays no longer considered to be adequate for music.

Here, we will only discuss transmission of speech. The introduction of wideband transmission in a telephone network requires at least new terminals with better electro-acoustic front-ends, improved A/D converters, and new speech codecs. In addition, signaling procedures are needed for detection and activation of the wideband capability. In cellular radio networks, expensive modifications are necessary, since speech-codec-specific error protection (channel coding) is implemented in the base stations and not in the centralized switching centers.

Several wideband speech codecs have been standardized in the past. In 1985, a first wideband speech codec for ISDN and teleconferencing with bit rates of 64, 56 and 48 kbit/s was specified by CCITT ([ITU-T Rec. G.722 1988], see also Appendix A on codec standards). For the sake of compatibility with the existing terminals, this standard comprises a dedicated signaling procedure for the automatic detection of the capabilities of the far-end terminal and a fall-back mode to narrowband ISDN (A-law, μ-law), if necessary. However, this codec has not yet been introduced in public telephone networks. It is mainly used in the context of radio broadcast stations by external reporters using special terminals and ISDN connections from outside to the studio.

In 1999, a second wideband codec [ITU-T Rec. G.722.1 1999] was introduced which produces almost comparable speech quality at reduced bit rates of 32 and 24 kbit/s.

The breakthrough will probably be the recently standardized *adaptive multi-rate wideband* (AMR-WB) speech codec, specified by the 3GPP initiative (3rd Generation Partnership Project) and by ETSI for CDMA cellular networks such as UMTS. This 7 kHz codec has eight different modes with bit rates from 6.6 kbit/s up to 23.85 kbit/s with increasing quality [3GPP TS 26.190 2001], [Salami et al. 1997]. This codec has been adopted by ITU [ITU-T Rec. G.722.2 2002]. A further extension has been specified by 3GPP recently with the AMR-WB+ codec, which

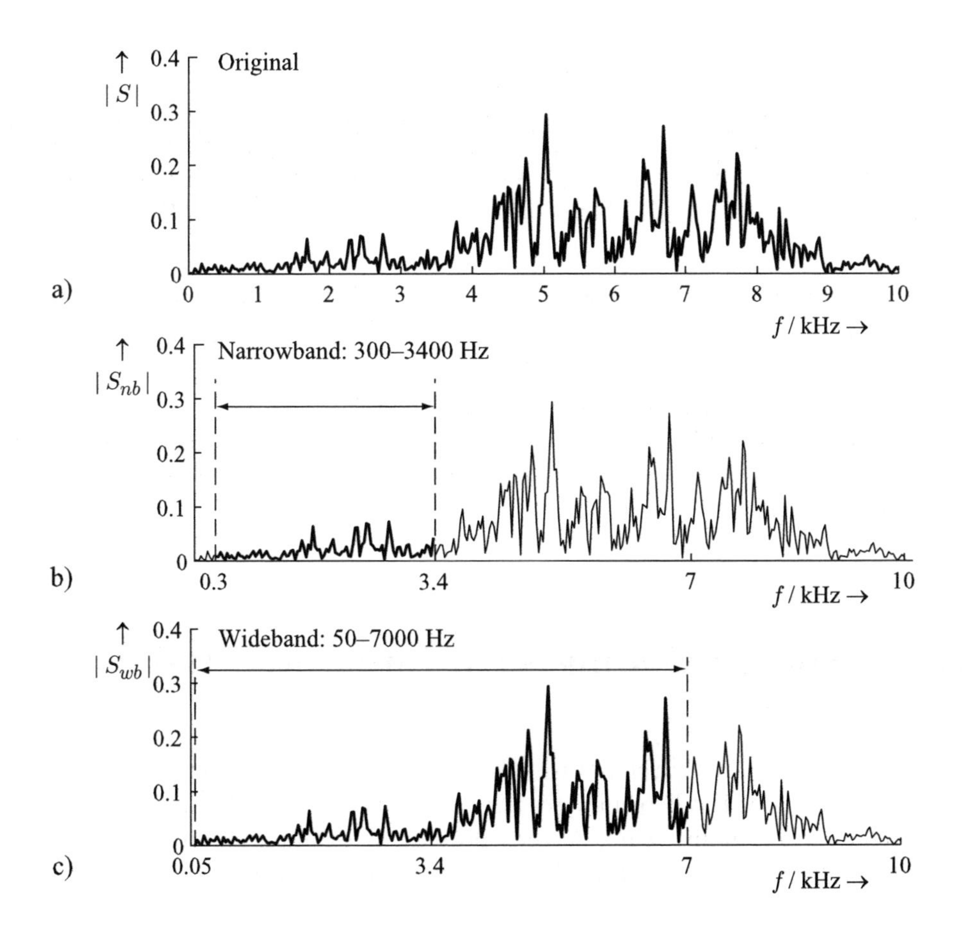

Figure 10.2: Example of a short-term spectrum of an unvoiced utterance
a) S: original speech
b) S_{nb}: narrowband telephone speech
c) S_{wb}: wideband telephone speech

is an extension of the AMR-WB codec in terms of the frequency range up to 16 kHz and bit rates up to 32 kbit/s. The aforementioned AMR-WB codec is part of the AMR-WB+ standard [3GPP TS 26.290 2005] (see also Appendix A).

Although cellular phones are replaced by new models much more often than fixed line telephones, there will be a long transitional period with both narrowband and wideband terminals in use in both cellular and fixed networks. During this period, *artificial bandwidth extension* will be a very attractive feature to increase the acceptance of the new wideband terminals. As indicated in Fig. 10.3, there may be a narrowband terminal at the far end and narrowband transmission over the network, while the near-end terminal already has the wideband capabilities. For reasons of compatibility, the narrowband codec has to be used for bidirectional transmission.

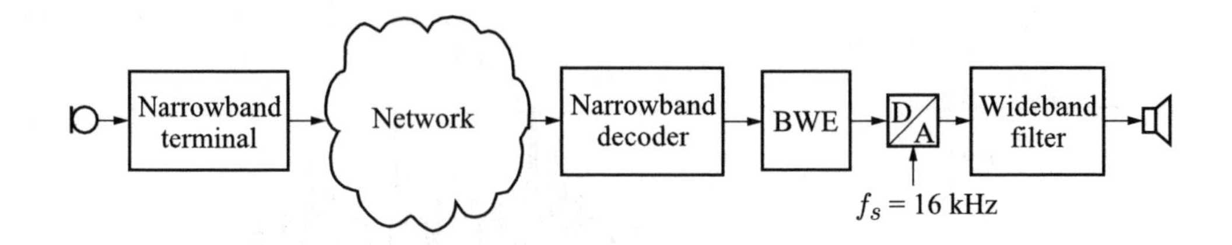

Figure 10.3: Artificial bandwidth extension (BWE) at the receiving end

However, due to the increased audio bandwidth of the near-end terminal (sampling rate 16 kHz), artificial bandwidth extension (BWE) can be applied to improve the received speech signal. This approach does not require any modification of the sending terminal and the network.

10.2 Speech Coding with Integrated BWE

Artificial BWE is closely related to speech coding. In fact, some very special and effective variants of BWE techniques have been used as an integral part of various speech codecs for many years. A very prominent example in this respect is the GSM full-rate codec [Rec. GSM 06.10 1992] (see also Chapter 8).

Most of the BWE algorithms proposed in the literature are based on the source–filter model of speech production. The extension of the source signal (excitation) and of the frequency response of the synthesis filter (spectral envelope) can be treated separately [Carl 1994], [Carl, Heute 1994]. The latter is much more challenging because the ear is rather insensitive with respect to coarse quantization or approximation of the excitation signal. Therefore, BWE can be implemented with great success if the information on the complete spectral envelope is transmitted, while the extension of the excitation is performed at the receiver without additional side information.

This idea has been used for coding *narrowband telephone speech* for quite a long time to achieve bit rates below 16 kbit/s with moderate computational complexity. The excitation signal $d(k)$ is transmitted with a bandwidth even smaller than the standard telephone bandwidth by applying lowpass filtering and sample rate decimation. In this way, more bits can be assigned to each of the residual samples which are transmitted. The basic concept, which was originally proposed by [Makhoul, Berouti 1979], is called *baseband RELP* (Residual Excited Linear Prediction). It is illustrated in Fig. 10.4 (see also Section 8.5.2).

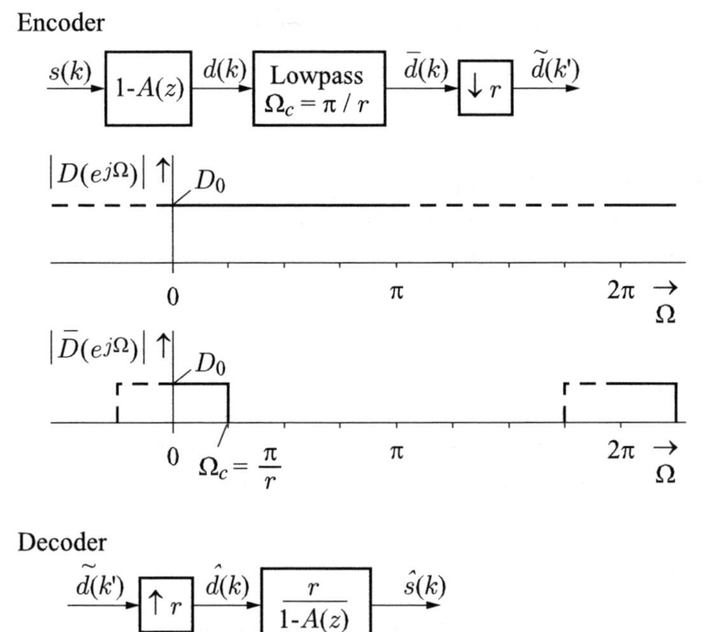

Figure 10.4: Basic principle of the RELP codec, example $r = 4$

The prediction error signal $d(k)$ or *residual* signal is obtained by linear prediction using the analysis filter

$$1 - A(z) = 1 - \sum_{\nu=1}^{n} a(\nu) \cdot z^{-\nu}$$

to produce the residual signal

$$d(k) \;=\; x(k) \;-\; \sum_{\nu=1}^{n} a(\nu) \cdot x(k-\nu)\,.$$

Due to the limited bit rate, the bandwidth is reduced by applying a lowpass filter with a normalized cutoff frequency of $\Omega_c = \pi/r$ to $d(k)$. The decimation factor $r \in \{2, 3, \ldots\}$ is fixed and used for the sample rate decimation of the lowpass filtered residual $\bar{d}(k)$ according to

$$\tilde{d}(k') = \bar{d}(k = k' \cdot r) .$$

The number of samples which have to be quantized and coded is therefore reduced by r (see also Section 8.5.2) and more bits are available for the scalar quantization of the remaining residual samples $\tilde{d}(k')$. At the receiving end, the missing samples are replaced by zeros

$$\hat{d}(k) = \begin{cases} \tilde{d}(k') & k = k' \cdot r \\ 0 & \text{else} \end{cases}$$

and thus the baseband of the residual signal, which occupies the frequency range $-\pi/r \leq \Omega \leq \pi/r$, is repeated r times and scaled by $1/r$ with center frequencies

$$\Omega_i = \frac{2\pi}{r} \cdot i \quad ; \qquad i = 0, 1, 2, \ldots, r-1 . \tag{10.1}$$

The bandwidth extended excitation signal is given by

$$\hat{D}(e^{j\Omega}) = \frac{1}{r} \cdot \sum_{i=0}^{r-1} D\left(e^{j(\Omega - \Omega_i)}\right) . \tag{10.2}$$

This method yields reasonable results due to the insensitivity of the human ear at higher frequencies, especially for unvoiced sounds. In this case, the regenerated excitation signal $\hat{d}(k)$ is a noise signal with a flat spectral envelope. If the input signal is voiced, the spectral repetitions according to (10.2) will deliver a spectrum with discrete components and a flat envelope. However, the discrete components are not necessarily harmonics of the fundamental frequency. Therefore, this type of speech codec produces a slightly metallic sound, especially for female voices.

The extension of the excitation signal by spectral repetition or spectral folding as given by (10.2) is called *high-frequency regeneration* [Makhoul, Berouti 1979] which may be considered as artificial BWE of the excitation signal. The overall spectral envelope of the speech segment is reconstructed from the spectrally flat excitation by applying the synthesis filter $1/(1 - A(z))$ which covers the whole frequency range $0 \leq \Omega \leq \pi$. In contrast to true artificial BWE, there is no need to estimate the spectral envelope in the *extension band* $\frac{\pi}{r} \leq \Omega \leq \pi$. The transmission of the coefficients of $a(\nu)$ with $\nu = 1, 2, \ldots, n$ may be considered as the transmission of *auxiliary information* for the construction of the decoded signal in the extension band.

This concept of the baseband RELP was later refined for different standardized speech codecs. A prominent example is the basic full-rate speech codec of the GSM system, e.g., [Rec. GSM 06.10 1992], [Vary et al. 1988], [Sluijter 2005].

More recently, BWE has been applied in the context of wideband speech coding, e.g., [Paulus, Schnitzler 1996], [Paulus 1996], [Taori et al. 2000], [Schnitzler 1999], [Erdmann et al. 2001], [3GPP TS 26.171 2001]. In these approaches, CELP coding (see also Section 8.5.3) is applied to the speech components of up to about 6 kHz, and artificial BWE is used to synthesize a supplementary signal for the narrow frequency range from 6 to 7 kHz. The extension is supported by transmitting different amounts of side information which controls the spectral envelope and the level of noise excitation in the extension band (see also Appendix A). Similar approaches have also been introduced in the context of MPEG audio coding as *spectral band replication* (SBR) [Dietz et al. 2002], [Gröschel et al. 2003]. SBR has been used to enhance the coding efficiency of MP3 (MP3pro) and *advanced audio coding* (AACPlus), e.g., [Dietz et al. 2002], [Ehret et al. 2003], [Gröschel et al. 2003], [Homm et al. 2003].

10.3 BWE without Auxiliary Transmission

The most challenging application of BWE is the improvement of narrowband telephone speech at the receiving end without transmitting any auxiliary information, as shown in Fig. 10.3. This configuration will be required for a long time before the conversion of the telephone networks from narrowband transmission to true wideband transmission will be completed.

10.3.1 Basic Approaches and Classification

The early proposals for BWE used simple (analog or digital) signal processing techniques without taking the model of speech production into account. Probably the first proposal for BWE was made as early as 1933 by [Schmidt 1933], who tried to extend the speech bandwidth by non-linear processing. In 1972, a first application was conducted by the BBC [Croll 1972], aiming at the improvement of telephone speech contributions to broadcast programs. Two different signal processing techniques were used for the regeneration of the lower frequencies (80–300 Hz) and the higher frequencies (3.4–7.6 kHz). Artificial low-frequency components were produced by rectification of the narrowband signal and by applying a lowpass filter with a cutoff frequency of $f_c = 300\,\text{Hz}$. The high-frequency components were inserted as bandpass noise, which was modulated by the spectral speech components in the range 2.4–3.4 kHz. However, these methods were too simple to produce consistently improved speech as the level of the low-frequency content is often too high or too low and the high-frequency extension introduces noise disturbance.

A crucial signal processing approach was made in 1983 by P. J. Patrick [Patrick 1983]. He made a distinction between unvoiced and voiced sounds in several of his experiments. During voiced segments a synthetic component at the fundamental frequency was added and a copied and scaled version of the spectrum between 300 Hz and 3.4 kHz was inserted by FFT techniques beyond 3.4 kHz. These pure signal processing concepts also produce artifacts due to a level and frequency mismatch of the synthetic components, and they are not able to deliver consistently natural sounding speech.

In the context of audio transmission using ADPCM coding with a sample rate of 16 kHz and 4 bits per sample, [Dietrich 1984] has found that the subjective quality can be improved by allowing the output filter following the D/A conversion to deliberately leave some aliasing, due to a wide transition region of the output lowpass filter from 8 to 12 kHz. Advanced implementations based on this idea have also been proposed in the context of audio coding [Larsen et al. 2002], [Larsen, Aarts 2004]. It should be noted that these simple techniques work quite well at frequencies above 8 kHz but that they fail in the interesting 3.4–7.0 kHz *gap of the telephone frequencies.*

A breakthrough was made in 1994 by the proposals of H. Carl and U. Heute [Carl 1994], [Carl, Heute 1994], Y. M. Cheng, D. O'Shaughnessy, and P. Mermelstein [Cheng et al. 1994], and Y. Yoshida and M. Abe [Yoshida, Abe 1994]. They explicitly took the model of speech production into consideration.

These approaches exploit the redundancy of the speech production mechanism, as well as specific properties of auditory perception. This pioneering work should be regarded as the starting point of a series of further investigations, e.g., [Iyengar et al. 1995], [Epps, Holmes 1999], [Enbom, Kleijn 1999], [Avendano et al. 1995], [Park, Kim 2000], [Fuemmeler et al. 2001], [Kornagel 2001], [Jax, Vary 2000], [Jax, Vary 2002], [Jax 2002], [Jax, Vary 2003], [Jax 2004].

The common concept of the artificial BWE algorithms for speech is to exploit the redundancy of the linear source–filter model. This model consists of an *autoregressive* (AR) filter (corresponding to the vocal tract) and a source producing a spectrally flat excitation. According to this model, BWE can be separated into two separate tasks [Carl 1994]:

- *extension of the spectral envelope* of the speech signal and
- *extension of the excitation signal.*

The crucial point is the estimation of the spectral envelope in the extension bands (subscript *eb*, frequency range below 300 Hz and beyond 3.4 kHz), while the extension of the excitation signal is far less critical, as shown, for example, by Carl [Carl 1994], [Carl, Heute 1994]. This is also known from speech coding with integrated BWE (see also Section 10.2).

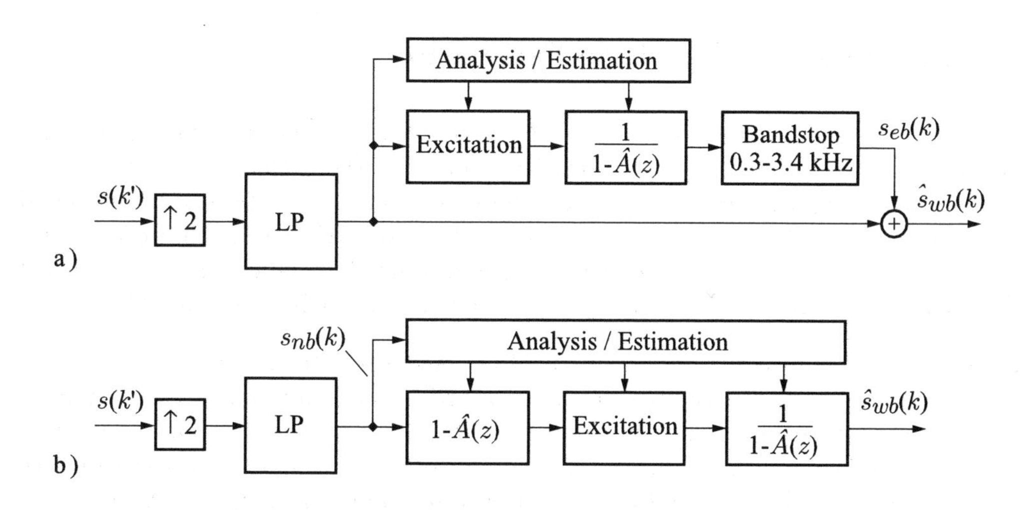

Figure 10.5: Classification of BWE algorithms
a) Parallel extension: explicit superposition of an extension signal $s_{eb}(k)$
b) Serial extension: filtering and modification of the narrowband signal $s_{nb}(k)$

The model-based algorithms for BWE proposed in the literature so far can be classified as illustrated in Fig. 10.5. The two classes, both of which extend the spectral envelope and the excitation separately, may be denoted as *parallel extension* and *serial extension*. In the first case, an artificial signal $s_{eb}(k)$ consisting of the (highpass and lowpass) extension components is added to the telephone speech signal, which has been interpolated to $f_s = 16\,\text{kHz}$. In the serial approach, the interpolated telephone speech signal is passed through a cascade of a wideband analysis filter and the corresponding inverse synthesis filter, while the excitation signal is extended in between. As the analysis and the synthesis filter are exactly inverse to each other, the output signal $\hat{s}_{wb}(k)$ contains the narrowband input signal $s_{nb}(k)$ (in the case of a "narrowband transparent" excitation extension).

Most of the algorithms proposed in the literature derive the spectral envelope of the wideband speech signal in a *first step.*

In a *second step*, the baseband of the excitation signal (0.3–3.4 kHz) is obtained from the narrowband speech signal by means of linear prediction (LP).

Finally, in a *third step*, the artificial wideband speech signal is produced by applying the extended excitation signal (see next section) to the extended AR filter in the context of either the parallel or the serial BWE approach of Fig. 10.5.

Before describing specific solutions, a general estimation framework [Vary 2004] will be discussed in the next section, which covers most of the BWE algorithms as proposed in the literature.

10.3.2 Spectral Envelope Estimation

In this section, the procedure of conditional estimation is described in general terms using vector notation (see also Chapter 5 for the scalar case). The speech enhancement algorithms are based on the conditional estimation of the spectral envelope of speech, which may be described by the AR coefficients of the AR model of speech production, the corresponding reflection coefficients, the cepstral coefficients, or any other related representation, see also Chapter 2 and Chapter 6.

The generic set-up is illustrated in Fig. 10.6. According to this model, a vector $\mathbf{a}$ of parameters, e.g., AR coefficients, is obtained by an analysis procedure A from the original (wideband) speech signal s. In the real application, the parameters $\mathbf{a}$ are not accessible; instead, we have some disturbed/degraded observations $\mathbf{b}$, which are gained by a second analysis procedure B of a signal y, which has been disturbed or degraded (e.g., by the telephone channel). The analysis algorithms A and B must not necessarily be identical. Analysis is carried out by block or frame processing. The two procedures A and B might include quantization so that the vectors $\mathbf{a}$ and $\mathbf{b}$ only take a finite number of different values.

It is, then, the task of the conditional estimator to form an estimate $\hat{\mathbf{a}}$ for the parameter vector $\mathbf{a}$ of the present frame by using the present disturbed observation $\mathbf{b}$ and *a priori* knowledge in terms of the statistics of $\mathbf{a}$. These statistics are either discrete probabilities $P(\mathbf{a})$ if $\mathbf{a}$ is quantized, or probability density functions (PDFs) $p(\mathbf{a})$ if $\mathbf{a}$ is not quantized. In addition, statistical knowledge about the disturbance/degradation in terms of transition probabilities $P(\mathbf{b} \mid \mathbf{a})$ or conditional PDFs $p(\mathbf{b} \mid \mathbf{a})$ may be taken into consideration. If information about $\mathbf{a}$ has been lost due to the disturbance, the original values cannot be reconstructed without errors. Thus, the estimation relies on finding the best possible estimate $\hat{\mathbf{a}}$ in a statistical sense, e.g., by minimizing the average estimation error. For this purpose

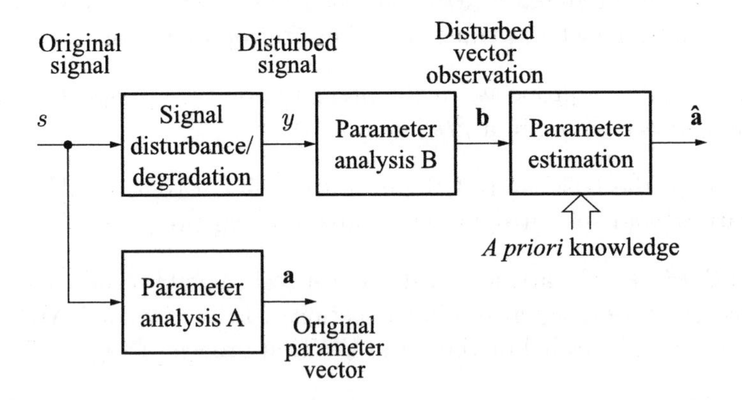

Figure 10.6: Conditional estimation in a parameter domain using *a priori* knowledge

the *a posteriori* PDF $p(\mathbf{a}\,|\,\mathbf{b})$ of the original vector $\mathbf{a}$ conditioned on the instantaneous observation $\mathbf{b}$ is exploited. A cost function $C(\mathbf{a},\hat{\mathbf{a}})$ is introduced [Melsa, Cohn 1978], which assigns a cost value or weight to each combination of undisturbed $\mathbf{a}$ and estimated $\hat{\mathbf{a}}$ and thus weights the estimation error for each given pair $(\mathbf{a},\,\hat{\mathbf{a}})$.

An estimation rule $\hat{\mathbf{a}} = f(\mathbf{b})$ which minimizes the expectation of the cost function is required. Let $n = \dim \mathbf{a}$ and $m = \dim \mathbf{b}$. Then the average costs, i.e., expectation of $C(\mathbf{a},\hat{\mathbf{a}})$, can be formulated by integration over the joint PDF[1] of the undisturbed and disturbed value

$$\rho_0 = \mathrm{E}\{C(\mathbf{a},\hat{\mathbf{a}})\} = \int\limits_{\mathbf{R}^m}\int\limits_{\mathbf{R}^n} C(\mathbf{a},\hat{\mathbf{a}})\cdot p(\mathbf{a},\mathbf{b})\,\mathrm{d}\mathbf{a}\,\mathrm{d}\mathbf{b}\,. \tag{10.3}$$

The estimation rule $\hat{\mathbf{a}} = f(\mathbf{b})$ can be found by minimizing ρ_0. After applying Bayes' theorem, equation (10.3) can be converted as follows:

$$\rho_0 = \int\limits_{\mathbf{R}^m}\left(\int\limits_{\mathbf{R}^n} C(\mathbf{a},\hat{\mathbf{a}})\cdot p(\mathbf{a}|\mathbf{b})\,\mathrm{d}\mathbf{a}\right)p(\mathbf{b})\,\mathrm{d}\mathbf{b}\,. \tag{10.4}$$

As $p(\mathbf{b})$ is non-negative, the minimum of ρ_0 can be found by minimizing the inner integral for every possible observation $\mathbf{b}$ [Melsa, Cohn 1978]

$$\rho_1 = \mathrm{E}\{C(\mathbf{a},\hat{\mathbf{a}})|\mathbf{b}\} = \int\limits_{\mathbf{R}^n} C(\mathbf{a},\hat{\mathbf{a}})\cdot p(\mathbf{a}|\mathbf{b})\,\mathrm{d}\mathbf{a}\,. \tag{10.5}$$

Conditional Minimum Mean Square Error Estimation

Choosing the Euclidean error norm as cost function, i.e., $C(\mathbf{a},\hat{\mathbf{a}}) = (\mathbf{a}-\hat{\mathbf{a}})^T(\mathbf{a}-\hat{\mathbf{a}})$, the minimization of ρ_1 with respect to $\hat{\mathbf{a}}$

$$\nabla_{\hat{\mathbf{a}}}\left[\int\limits_{\mathbf{R}^n}(\mathbf{a}-\hat{\mathbf{a}})^T(\mathbf{a}-\hat{\mathbf{a}})\cdot p(\mathbf{a}|\mathbf{b})\,\mathrm{d}\mathbf{a}\right] = -\int\limits_{\mathbf{R}^n} 2(\mathbf{a}-\hat{\mathbf{a}})\cdot p(\mathbf{a}|\mathbf{b})\,\mathrm{d}\mathbf{a} \overset{!}{=} 0 \tag{10.6}$$

and $\int p(\mathbf{a}|\mathbf{b})\,\mathrm{d}\mathbf{a} = 1$ lead to the *minimum mean square error* (MMSE) or conditional mean estimator:

$$\hat{\mathbf{a}} = \mathrm{E}\{\mathbf{a}|\mathbf{b}\} = \int\limits_{\mathbf{R}^n} \mathbf{a}\cdot p(\mathbf{a}|\mathbf{b})\,\mathrm{d}\mathbf{a}\,. \tag{10.7}$$

[1] For notational convenience, the following compact definition is introduced:

$$\text{Let } \mathbf{x} = (x_1, x_2, \dots, x_p)^T \in \mathbb{R}^p\text{, then } \int\limits_{\mathbf{R}^p} f(\mathbf{x})\,\mathrm{d}\mathbf{x} \doteq \int\limits_{-\infty}^{\infty}\cdots\int\limits_{-\infty}^{\infty} f(\mathbf{x})\,\mathrm{d}x_1\cdots\mathrm{d}x_p\,.$$

The *a posteriori* probability density $p(\mathbf{a} \,|\, \mathbf{b})$ is unknown, but by using Bayes' theorem once more, (10.7) can be rewritten as

$$\hat{\mathbf{a}} = \frac{\int\limits_{\mathbf{R}^n} \mathbf{a} \cdot p(\mathbf{b}|\mathbf{a}) \cdot p(\mathbf{a}) \, d\mathbf{a}}{p(\mathbf{b})} = \frac{\int\limits_{\mathbf{R}^n} \mathbf{a} \cdot p(\mathbf{b}|\mathbf{a}) \cdot p(\mathbf{a}) \, d\mathbf{a}}{\int\limits_{\mathbf{R}^n} p(\mathbf{b}|\mathbf{a}) \cdot p(\mathbf{a}) \, d\mathbf{a}} . \tag{10.8}$$

If $\mathbf{a}$ and $\mathbf{b}$ take discrete values (e.g., due to quantization), corresponding discrete formulations of (10.7) and (10.8) can be derived. The integrals have to be replaced by summations and the PDFs by discrete probabilities. Even a mixed form is possible where the statistics of only one quantity are discrete. In this case, we need the "mixed form" of Bayes' theorem.

Conditional Maximum *A Posteriori* Estimation

Another useful function to weight the estimation error for (10.4) is the uniform cost

$$C(\mathbf{a}, \hat{\mathbf{a}}) = \begin{cases} 0 & |\mathbf{a} - \hat{\mathbf{a}}| < \epsilon \\ 1 & \text{else} . \end{cases} \tag{10.9}$$

To minimize the integral of (10.5) with this cost function, the maximum of $p(\mathbf{a}, \mathbf{b})$ must be in the area where $C = 0$. Thus, the estimate $\hat{\mathbf{a}}$ is obtained by searching the maximum of the *a posteriori* PDF

$$\hat{\mathbf{a}} = \arg\max_{\mathbf{a}} p(\mathbf{a}, \mathbf{b}) , \tag{10.10}$$

which can also be reformulated with Bayes' rule:

$$\hat{\mathbf{a}} = \arg\max_{\mathbf{a}} \frac{p(\mathbf{b}|\mathbf{a}) \cdot p(\mathbf{a})}{p(\mathbf{b})} . \tag{10.11}$$

If the *a posteriori* probability density is symmetric and unimodal, the MMSE estimate equals the *maximum a posteriori* (MAP) estimate (see Chapter 5 or, e.g., [Van Trees 1968]).

Applications

In the literature, different proposals of estimating the wideband spectral envelope have been made which may be considered as applications and specializations of the described estimation framework.

In [Carl 1994], the wideband LP coefficients are obtained by *code book mapping* between the coefficient set of the narrowband speech (observation $\mathbf{b}$) and a pre-trained wideband code book (quantized target vectors $\mathbf{a}$). The selection of the most probable entry of the code book, i.e., the most probable wideband coefficient, is to be considered as a special case of conditional MAP estimation.

In [Park, Kim 2000], the required PDFs are approximated by *Gaussian mixture models* (GMMs) [Reynolds, Rose 1995], [Vaseghi 1996].

The solution proposed in [Jax, Vary 2000] is also based on GMMs but in addition a statistical *hidden Markov state model* (HMM) [Rabiner 1989], [Papoulis 1991] is used to identify typical speech sounds. The disturbed observation **b** is represented by a feature vector **X**, which is derived from the interpolated narrowband speech signal, while a pre-trained code book (vectors **a**) contains the cepstral coefficients of wideband speech. The conditional estimate is calculated according to the MMSE criterion (see also Section 10.3.4).

10.3.3 Extension of the Excitation Signal

According to the simplifying linear model of speech production, the excitation signal $d(k)$ is spectrally flat: for voiced sounds, it contains sinusoids at multiples of the fundamental (pitch) frequency of the speech segment where all harmonics have almost the same amplitude; during unvoiced sounds, the excitation is more or less white noise.

The frequencies of the narrowband signal are limited to the telephone frequency band 300 Hz $\leq f \leq$ 3.4 kHz. If the narrowband signal is interpolated to 16 kHz and the wideband analysis filter $1 - \hat{A}(z)$ is applied as indicated in Fig. 10.7-a, b, the residual signal $\hat{d}_{nb}(k)$ is almost flat within the telephone band and zero outside.

Due to these properties the missing high-frequency components of the excitation signal can be produced by modulation, i.e., a frequency shift of Ω_{M} [Carl 1994], [Fuemmeler et al. 2001], [Kornagel 2001]

$$\tilde{d}(k) = \hat{d}_{nb}(k)\, 2\cos(\Omega_{\mathrm{M}} k)\,, \tag{10.12}$$

where the choice of Ω_{M} is discussed subsequently. The real-valued modulation produces *two* shifted components

$$\tilde{D}(e^{j\Omega}) = \hat{D}_{nb}(e^{j(\Omega-\Omega_{\mathrm{M}})}) + \hat{D}_{nb}(e^{j(\Omega+\Omega_{\mathrm{M}})})\,, \tag{10.13}$$

and a highpass filter with impulse response $h_{HP}(k)$ may be applied (see Fig. 10.7-c) to produce the excitation in the extension band:

$$\hat{d}_{eb}(k) = \tilde{d}(k) * h_{HP}(k)\,. \tag{10.14}$$

The estimated wideband excitation signal is finally obtained by the superposition of the narrowband excitation $\hat{d}_{nb}(k)$ and the excitation $\hat{d}_{eb}(k)$ of the extension band. The algorithmic delay k_0 of the highpass filter has to be compensated for in the path of the baseband signal:

$$\hat{d}_{wb}(k) \;=\; \hat{d}_{nb}(k-k_0) \;+\; \hat{d}_{eb}(k)\,. \tag{10.15}$$

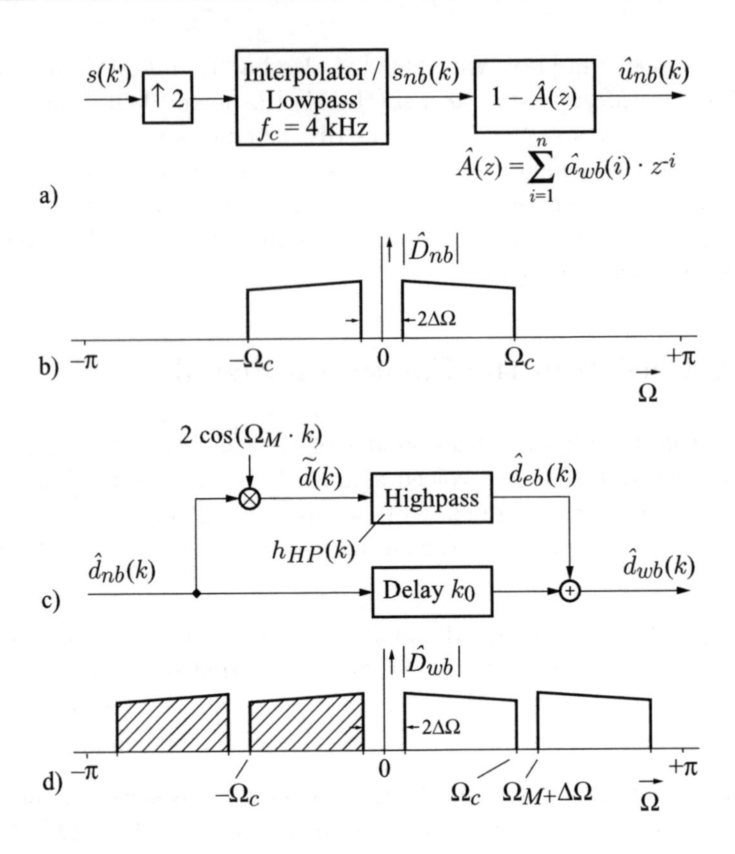

Figure 10.7: Narrowband and wideband excitation signals
a) Interpolation and analysis filtering of the narrowband signal
b) Spectral shape of the narrowband residual signal $\hat{d}_{nb}(k)$
c) Extension of the excitation signal by modulation according to (10.14) and (10.15)
d) Spectral shape of an extended residual signal $\hat{d}_{wb}(k)$ for $\Omega_c - \Delta\Omega \leq \Omega_M \leq \pi - \Omega_c$

By selecting the modulation frequency Ω_M, several modulation schemes can be chosen:

- Modulation with the *Nyquist* frequency, i.e., $\Omega_\mathrm{M} = \pi$ and $\cos(\Omega_\mathrm{M} k) = (-1)^k$, corresponds to the method of *spectral mirroring* as proposed in [Makhoul, Berouti 1979]. In this special case the two shifted copies of the baseband spectrum coincide, so that the highpass filter from Fig. 10.7-c is not needed.

  However, due to the cutoff frequency of the narrowband input signal, there is a spectral gap in $\tilde{d}_{wb}(k)$ between 3.4 and 4.6 kHz with a width of 1.2 kHz ($\pi - 2 \cdot \Omega_c$, see Fig 10.7). Furthermore, the discrete spectral components of the extended frequency band are in general not harmonics of the fundamental frequency.

- To prevent this spectral gap, the modulation frequency can be chosen in such a way that the shifted spectrum is a seamless continuation of the baseband spectrum

$$\Omega_{\mathrm{M}} = \Omega_{3.4} \quad \text{with} \quad \Omega_{3.4} = 2\pi \frac{3.4\,\mathrm{kHz}}{f_{\mathrm{s}}}, \tag{10.16}$$

  where f_{s} denotes the sampling rate. In general there is a misalignment of discrete spectral components in the extension band during voiced sounds.

- A further possibility to control Ω_{M} takes the pitch frequency Ω_{p} of the current speech frame into account: the modulation frequency is adapted in such a way that it is always an integer multiple of the estimated pitch frequency $\tilde{\Omega}_{\mathrm{p}}$ [Fuemmeler et al. 2001], e.g.,[2]

$$\Omega_{\mathrm{M}} = \left\lceil \frac{\Omega_{3.4}}{\tilde{\Omega}_{\mathrm{p}}} \right\rceil \tilde{\Omega}_{\mathrm{p}}. \tag{10.17}$$

  This method guarantees that the harmonics in the extended frequency band will always match the harmonic structure of the baseband. The pitch-adaptive modulation reacts quite sensitively to small errors of the estimate of the pitch frequency, because these are significantly enlarged by the factor $\lceil \Omega_{3.4}/\tilde{\Omega}_{\mathrm{p}} \rceil$. Therefore, a reliable pitch estimator is needed if the pitch-adaptive method is to give rise to improvements in comparison to the two fixed modulation schemes.

Informal listening tests have shown that – assuming that the BWE of the spectral envelope works well – the human ear is surprisingly insensitive to distortions of the excitation signal at frequencies above 3.4 kHz. For example, spectral gaps of moderate width as produced by bandstop filters are almost inaudible. Furthermore, a misalignment of the harmonic structure of speech at frequencies beyond 3.4 kHz does not significantly degrade the subjective quality of the enhanced speech signal. A good compromise between subjective quality and computational complexity is the modulation with the fixed frequency of $\Omega_{\mathrm{M}} = \Omega_{3.4}$.

It should be noted that a couple of different other proposals for the extension of the excitation signal can be found in the literature. These techniques include, for example, non-linearities [Valin, Lefebvre 2000], [Kornagel 2003] or synthetic multiband excitation (MBE vocoder) [Chan, Hui 1996].

10.3.4 Example BWE Algorithm

The task of artificial BWE is decribed in Fig. 10.8 in the context of conditional estimation of Fig. 10.6-a, with two different specific analysis procedures A and B.

[2] $\lceil x \rceil$ denotes the smallest integer larger than x.

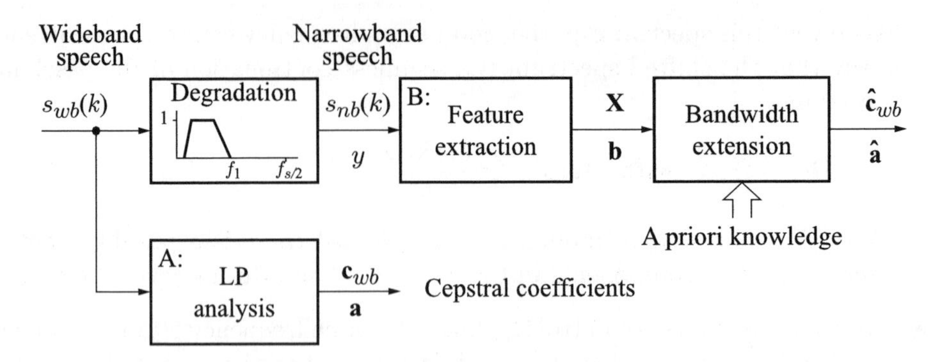

Figure 10.8: Model of BWE by conditional estimation
Correspondence with respect to Fig. 10.6-a: $\mathbf{c}_{wb} = \mathbf{a}$, $\mathbf{X} = \mathbf{b}$, $\hat{\mathbf{c}}_{wb} = \hat{\mathbf{a}}$ (vectors); wb=wideband; nb=narrowband

In analysis A, a vector of cepstral coefficients $\boldsymbol{c}_{\text{wb}}$ of the wideband speech signal $s_{\text{wb}}(k)$ is calculated, whereas analysis B delivers a feature vector $\mathbf{X}$.

As a representative example, the serial BWE approach, illustrated in Fig. 10.9, [Jax 2002], [Jax, Vary 2003], will be described here. For the representation of the wideband spectral envelope the cepstral coefficients $\hat{\mathbf{c}}_{wb}$ are used, since the corresponding logarithmic magnitude frequency response shows strong correlation with human perception.

The estimation of the cepstral coefficients is based on a feature vector $\mathbf{X}$ and an underlying state model of speech production. Each speech frame of 20 ms with time or frame index k' can be characterized by a state S_i $(i = 1, \ldots, N_s)$, the typical vector of cepstral coefficients $\hat{\mathbf{c}}_i$, and the "measured" feature vector $\mathbf{X}$.

The estimated cepstral coefficients $\hat{\mathbf{c}}_{wb}$ are converted to the wideband LP coefficients (see Section 3.7.2) of the analysis filter $1 - \hat{A}(z)$ and of the corresponding all-pole (vocal tract) filter $1/(1 - \hat{A}(z))$ (Section 6.1).

By applying the FIR analysis filter $1 - \hat{A}(z)$ to the narrowband input signal $s_{nb}(k)$ which has been interpolated previously to a sample rate of $f_s = 16$ kHz, an estimate $\hat{d}_{nb}(k)$ of the narrowband excitation signal (prediction residual) is derived. The *extension of the excitation signal* converts the narrowband excitation signal $\hat{d}_{nb}(k)$ to an extended version $\hat{d}_{wb}(k)$ by modulation, exploiting the spectral flatness. The extended wideband excitation signal $\hat{d}_{wb}(k)$ is fed into the wideband all-pole synthesis filter $1/(1 - \hat{A}(z))$ to synthesize the enhanced output speech $\hat{s}_{wb}(k)$.

A comparison of the original spectrum, the narrowband spectrum, and the artificially extended spectrum is shown in Fig. 10.10.

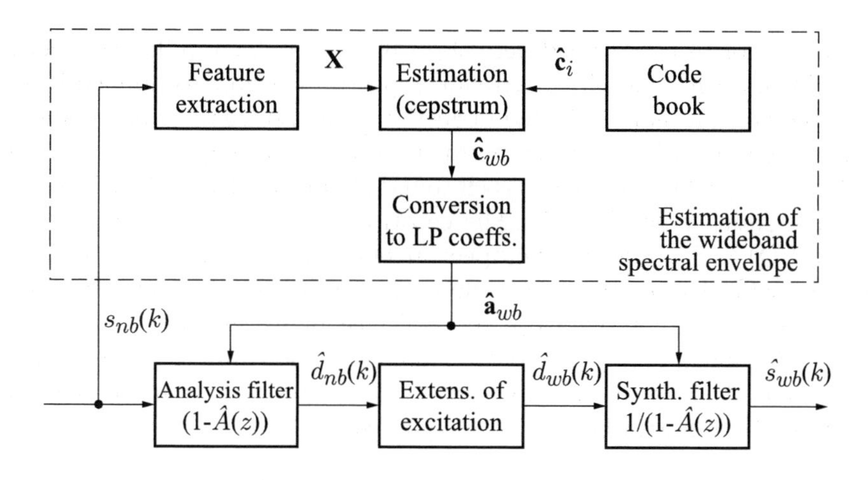

Figure 10.9: Block diagram of a serial BWE algorithm

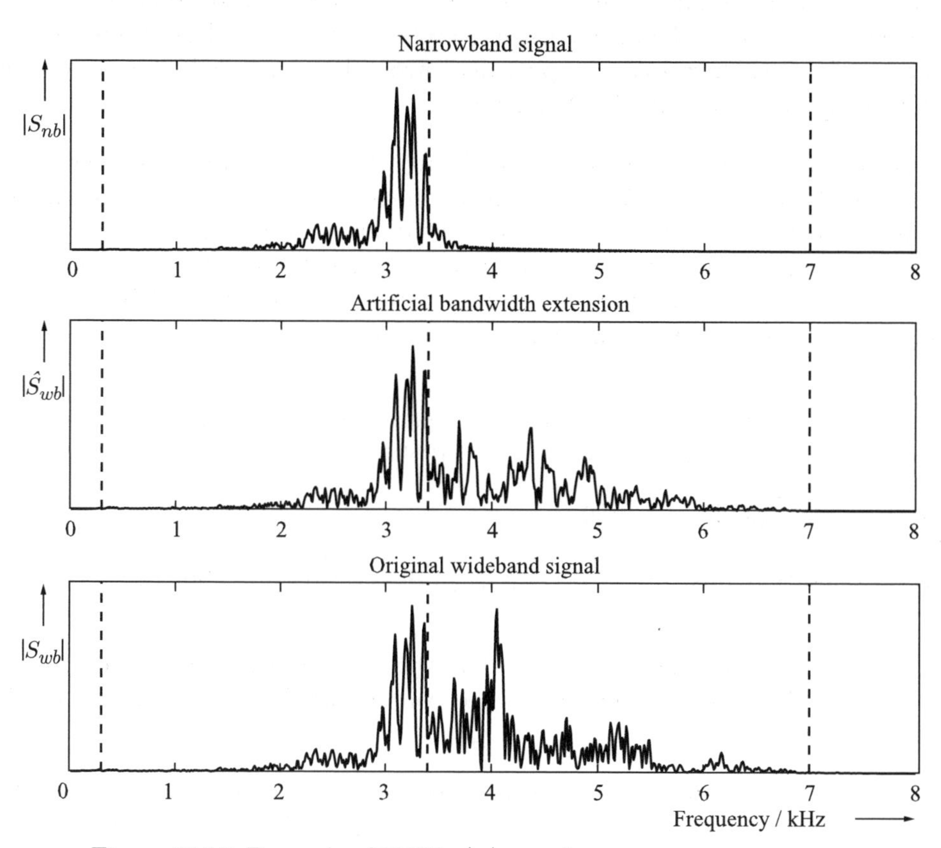

Figure 10.10: Example of BWE: /s/ sound
Modulation with fixed frequency $f_M = 3.4\,\text{kHz}$

In the BWE algorithm described here [Jax 2002], [Jax, Vary 2003], the method of the conditional estimation of the extended spectral envelope is applied in a more sophisticated version than explained in Section 10.3.2, since the *a priori* knowledge is based on a state model of speech production. Each state S_i, $i = 1, 2, \ldots, N_S$, of the model is assigned to a typical speech sound (frame of 20 ms) which is associated with a representative envelope $\hat{\mathbf{c}}_i$. The cepstral vectors $\hat{\mathbf{c}}_i$ are entries of a *vector quantizer* (VQ) code book of the spectral envelope representation $\mathbf{c}_{wb}$ (vector of cepstral coefficients of the true wideband speech signal). Each centroid $\hat{\mathbf{c}}_i$ of the vector quantizer represents the spectral envelope of a typical speech sound. However, wideband speech s_{wb} is only available in the training phase, whereas in the application phase of the BWE algorithm the states S_i have to be identified by classification of the narrowband speech signal s_{nb}.

For each signal frame, a vector $\mathbf{X}$ of features, which is chosen to deliver maximum information about the state S_i, is extracted from the narrowband signal. The vector $\mathbf{X}$ contains features like the normalized auto-correlation function (or LP coefficients of the narrowband signal), zero crossing rate, normalized frame energy, gradient index, local kurtosis, and spectral centroid. For a detailed description refer to [Jax 2002], [Jax, Vary 2003], [Jax 2004].

The connection between the observations $\mathbf{X}$ and the states S_i (and thus the corresponding code book entries $\hat{\mathbf{c}}_i$) is established by a state-specific statistical model. For each state S_i the features $\mathbf{X}$, as well as the unknown spectral envelope $\mathbf{c}_{wb}$, exhibit characteristic statistical relations. The following statistical quantities can be measured during an offline training process with representative wideband speech signals $s_{wb}(k)$ and corresponding narrowband signals $s_{nb}(k)$:

- the code book entries $\hat{\mathbf{c}}_i$ of the vector quantizer (e.g., by using the standard LBG training algorithm [Linde et al. 1980]),
- the state probabilities $P(S_i)$,
- the conditional feature PDFs $p(\mathbf{X}|S_i)$ (observation probabilities).

The wideband speech is needed to calculate the true state sequence, and the narrowband speech is used to determine the conditional observation PDFs of the feature vectors $\mathbf{X}$.

As the *observation PDF* is conditioned on the state S_i, a separate PDF $p(\mathbf{X}|S_i)$ exists for each state. A common way to model measured high-dimensional PDFs is the approximation with *Gaussian mixture models* (GMMs; see, e.g., [Reynolds, Rose 1995], [Vaseghi 1996]). According to the definition of the state model, it is assumed that the observation $\mathbf{X}$ for each frame only depends on this particular frame.

By the MMSE estimation rule (10.7), a continuous estimation of the parameter vector $\mathbf{c}_{wb}$ of dimension n can be performed with the *a posteriori* PDF $p(\mathbf{a}|\mathbf{b}) = p(\mathbf{c}|\mathbf{X})$.

Thus, the *minimum mean square error* (MMSE) estimator for the cepstral coefficient vector is given by[3]

$$\hat{\mathbf{c}}_{\text{MMSE}} = \mathrm{E}\{\mathbf{c}|\mathbf{X}\} = \int\limits_{\mathbf{R}^n} \mathbf{c} \cdot p(\mathbf{c}|\mathbf{X})\, \mathrm{d}\mathbf{c}\,. \tag{10.18}$$

As a model of the conditional PDF $p(\mathbf{c}|\mathbf{X})$ does not exist in closed form, this quantity has to be expressed indirectly via the states of the model

$$p(\mathbf{c}|\mathbf{X}) = \sum_{i=1}^{N_S} p(\mathbf{c}, S_i|\mathbf{X})\,. \tag{10.19}$$

Insertion of $p(\mathbf{c}, S_i|\mathbf{X}) = p(\mathbf{c}|S_i, \mathbf{X}) \cdot P(S_i|\mathbf{X})$ into (10.19) and (10.18) yields

$$\hat{\mathbf{c}}_{\text{MMSE}} = \sum_{i=1}^{N_S} P(S_i|\mathbf{X}) \cdot \int\limits_{\mathbf{R}^n} \mathbf{c}\, p(\mathbf{c}|S_i, \mathbf{X})\, \mathrm{d}\mathbf{c}\,, \tag{10.20}$$

which can be written as

$$\hat{\mathbf{c}}_{\text{MMSE}} = \sum_{i=1}^{N_S} \hat{\mathbf{c}}_i\, P(S_i|\mathbf{X})\,. \tag{10.21}$$

Hence, the estimated coefficient vector $\hat{\mathbf{c}}_{\text{MMSE}}$ is calculated by a weighted sum of the individual code book entries $\hat{\mathbf{c}}_i$. The weights are the respective *a posteriori* probabilities of the states. The described MMSE estimator can be seen as as a *soft classification* which is comparable to the error concealment algorithm described in Section 9.4.1.2. A block diagram of this estimation procedure is given in Fig. 10.11.

The *a posteriori* probability $P(S_i|\mathbf{X})$ can be formulated in terms of the measured state probabilities $P(S_i)$ and the measured conditional feature PDFs $p(\mathbf{X}|S_i)$ as follows:

$$P(S_i|\mathbf{X}) = \frac{p(S_i, \mathbf{X})}{p(\mathbf{X})} = \frac{p(\mathbf{X}|S_i) \cdot P(S_i)}{\sum_{j=1}^{N_S} p(\mathbf{X}|S_j) \cdot P(S_j)}\,. \tag{10.22}$$

In the denominator of (10.22) the hardly tractable PDF $p(\mathbf{X})$ of the observation sequence has been replaced by a summation over the joint PDF $p(S_j, \mathbf{X}) = p(\mathbf{X}|S_j) \cdot P(S_j)$.

[3]See footnote 1 on page 373

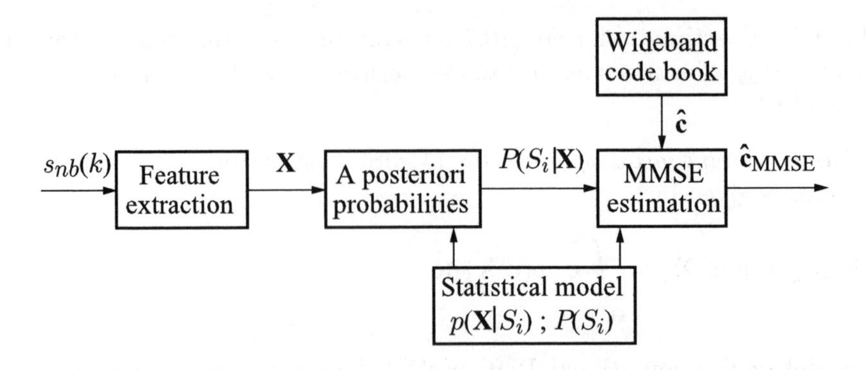

Figure 10.11: Conditional estimation of the extended spectral envelope (cepstral coefficients) taking a state model into account

Note: A more elaborate hidden Markov state model (HMM) can be used as proposed in [Jax 2002] and [Jax, Vary 2003], where the state transition probabilities $P(S_i|S_j)$ $(i, j = 1, 2, \ldots, N_s)$ are also taken into consideration. For more details the reader is referred to [Jax 2004].

In informal listening tests, it was found that the enhanced speech exhibits a significantly improved quality. By additionally exploiting the transition probabilities $P(S_i|S_j)$ of the HMM, the occurrence of artifacts is reduced considerably. The quality of the enhanced speech may even further be increased by using speaker-dependent instead of speaker-independent modeling [Jax 2002].

Bibliography

3GPP TS 26.171 (2001). AMR Wideband Speech Codec; General Description, 3GPP.

3GPP TS 26.190 (2001). AMR Wideband Speech Codec; Transcoding Functions, 3GPP.

3GPP TS 26.290 (2005). Extended Adaptive Multi-Rate Wideband (AMR-WB+) Codec, 3GPP.

Avendano, C.; Hermansky, H.; Wan, E. A. (1995). Beyond Nyquist: Towards the Recovery of Broad-Bandwidth Speech from Narrow-Bandwidth Speech, *Proceedings of EUROSPEECH*, Madrid, Spain, vol. 1, pp. 165–168.

Brosze, O.; Schmidt, K. O.; Schmoldt, A. (1992). Der Gewinn an Verständlichkeit beim "Fernsprechen", *Nachrichtentechnische Zeitschrift (NTZ)*, vol. 15, no. 7, pp. 349–352 (in German).

Carl, H. (1994). *Untersuchung verschiedener Methoden der Sprachkodierung und eine Anwendung zur Bandbreitenvergrößerung von Schmalband-Sprachsignalen*, PhD thesis. *Arbeiten über Digitale Signalverarbeitung*, vol. 4, U. Heute (ed.), Ruhr-Universität Bochum (in German).

Carl, H.; Heute, U. (1994). Bandwidth Enhancement of Narrow-Band Speech Signals, *Proceedings of the European Signal Processing Conference (EUSIPCO)*, Edinburgh, Scotland, vol. 2, pp. 1178–1181.

Chan, C.-F.; Hui, W.-K. (1996). Wideband Enhancement of Narrowband Coded Speech Using MBE Re-Synthesis, *Proceedings of the International Conference on Signal Processing (ICSP)*, vol. 1, pp. 667–670.

Cheng, Y.; O'Shaughnessy, D.; Mermelstein, P. (1994). Statistical Recovery of Wideband Speech from Narrowband Speech, *IEEE Transactions on Speech and Audio Processing*, vol. 2, no. 4, October, pp. 544–548.

Croll, M. G. (1972). Sound-Quality Improvement of Broadcast Telephone Calls, Technical Report 1972/26, The British Broadcasting Corporation (BBC).

Dietrich, M. (1984). Performance and Implementation of a Robust ADPCM Algorithm for Wideband Speech Coding with 64kbit/s, *Proceedings of the International Zürich Seminar on Digital Communications*, Zurich, Switzerland.

Dietz, M.; Liljeryd, L.; Kjorling, K.; Kunz, O. (2002). Spectral Band Replication: a Novel Approach in Audio Coding, *112th Convention of the Audio Engineering Society (AES)*, Munich, Germany.

Ehret, A.; Dietz, M.; Kjorling, K. (2003). State-of-the-art Audio Coding for Broadcasting and Mobile Applications, *114th Convention of the Audio Engineering Society (AES)*, Amsterdam, The Netherlands.

Enbom, N.; Kleijn, W. B. (1999). Bandwidth Expansion of Speech Based on Vector Quantization of the Mel Frequency Cepstral Coefficients, *Proceedings of the IEEE Workshop on Speech Coding*, Porvoo, Finland, pp. 171–173.

Epps, J.; Holmes, W. H. (1999). A New Technique for Wideband Enhancement of Coded Narrowband Speech, *Proceedings of the IEEE Workshop on Speech Coding*, Porvoo, Finland, pp. 174–176.

Erdmann, C.; Vary, P.; Fischer, K.; Xu, W.; Marke, M.; Fingscheidt, T.; Varga, I.; Kaindl, M.; Quinquis, C.; Kovesi, B.; Massaloux, D. (2001). A Candidate Proposal for a 3GPP Adaptive Multi-Rate Wideband Speechcodec, *Proceedings of the IEEE International Conference on Acoustics, Speech, and Signal Processing (ICASSP)*, Salt Lake City, Utah, USA, pp. 757–760.

ETSI Rec. GSM 03.50 (1998). *Digital Cellular Telecommunications System (Phase 2+); Transmission Planning Aspects of the Speech Service in the GSM Public Land Mobile (PLMN) System, Version 8.1.0*, European Telecommunications Standards Institute.

Fuemmeler, J. A.; Hardie, R. C.; Gardner, W. R. (2001). Techniques for the Regeneration of Wideband Speech from Narrowband Speech, *EURASIP Journal on Applied Signal Processing*, vol. 4, December, pp. 266–274.

Gröschel, A.; Schug, M.; Beer, M.; Henn, F. (2003). Enhancing Audio Coding Efficiency of MPEG Layer-2 with Spectral Band Replication for DigitalRadio (DAB) in a Backwards Compatible Way, *114th Convention of the Audio Engineering Society (AES)*, Amsterdam, The Netherlands.

Homm, D.; Ziegler, T.; Weidner, R.; Bohm, R. (2003). Implementation of a DRM Audio Decoder (aacPlus) on ARM Architecture, *114th Convention of the Audio Engineering Society (AES)*, Amsterdam, The Netherlands.

ITU-T Rec. G.132 (1988). ITU-T Recommendation G.132, Attenuation Performance, in Blue Book, vol. Fascicle III.1, General Characteristics of International Telephone Connections and Circuits, International Telecommunication Union (ITU).

ITU-T Rec. G.151 (1988). General Performance Objectives Applicable to all Modern International Circuits and National Extension Circuits, in Blue Book, vol. Fascicle III.1, General Characteristics of International Telephone Connections and Circuits, International Telecommunication Union (ITU).

ITU-T Rec. G.711 (1972). Pulse Code Modulation (PCM) of Voice Frequencies, International Telecommunication Union (ITU).

ITU-T Rec. G.712 (1988). Performance Characteristics of PCM Channels Between 4-wire Interfaces at Voice Frequencies, International Telecommunication Union (ITU).

ITU-T Rec. G.722 (1988). 7 kHz Audio Coding within 64 kbit/s, *Recommendation G.722*, International Telecommunication Union (ITU), pp. 269–341.

ITU-T Rec. G.722.1 (1999). Coding at 24 and 32 kbit/s for Hands-free Operation in Systems with Low Frame Loss, International Telecommunication Union (ITU).

ITU-T Rec. G.722.2 (2002). Wideband Coding of Speech at Around 16 kbits/s Using Adaptive Multi-Rate Wideband (AMR-WB), International Telecommunication Union (ITU).

Iyengar, V.; Rabipour, R.; Mermelstein, P.; Shelton, B. R. (1995). Speech Bandwidth Extension Method and Apparatus, US patent no. 5455888.

Jax, P. (2002). *Enhancement of Bandlimited Speech Signals: Algorithms and Theoretical Bounds*, PhD thesis. *Aachener Beiträge zu digitalen Nachrichtensystemen*, vol. 15, P. Vary (ed.), RWTH Aachen University.

Jax, P. (2004). Bandwidth Extension for Speech, *in* E. Larsen; R. M. Aarts (eds.), *Audio Bandwidth Extension*, John Wiley & Sons, Ltd., Chichester, chapter 6, pp. 171–236.

Jax, P.; Vary, P. (2000). Wideband Extension of Telephone Speech Using a Hidden Markov Model, *Proceedings of the IEEE Workshop on Speech Coding*, Delavan, Wisconsin, USA, pp. 133–135.

Jax, P.; Vary, P. (2002). An Upper Bound on the Quality of Artificial Bandwidth Extension of Narrowband Speech Signals, *Proceedings of the IEEE International Conference on Acoustics, Speech, and Signal Processing (ICASSP)*, Orlando, Florida, USA, vol. 1, pp. 237–240.

Jax, P.; Vary, P. (2003). On Artificial Bandwidth Extension of Telephone Speech, *Signal Processing*, vol. 83, no. 8, August, pp. 1707–1719.

Kornagel, U. (2001). Spectral Widening of the Excitation Signal for Telephone-Band Speech Enhancement, *Proceedings of the International Workshop on Acoustic Echo and Noise Control (IWAENC)*, Darmstadt, Germany, pp. 215–218.

Kornagel, U. (2003). Improved Artificial Low-Pass Extension of Telephone Speech, *Proceedings of the International Workshop on Acoustic Echo and Noise Control (IWAENC)*, Kyoto, Japan, pp. 107–110.

Krebber, W. (1995). *Sprachübertragunsqualität von Fernsprech-Handapparaten*, PhD thesis, RWTH Aachen University (in German).

Larsen, E. R.; Aarts, R. M.; Danessis, M. (2002). Efficient High-Frequency Bandwidth Extension of Music and Speech, *112th Convention of the Audio Engineering Society (AES)*, Munich, Germany.

Larsen, E. R.; Aarts, R. M. (eds.) (2004). *Audio Bandwidth Extension - Application of Psychoacoustics, Signal Processing and Loudspeaker Design*, John Wiley & Sons, Ltd., Chichester.

Linde, Y.; Buzo, A.; Gray, R. M. (1980). An Algorithm for Vector Quantizer Design, *IEEE Transactions on Communications*, vol. 28, no. 1, January, pp. 84–95.

Makhoul, J.; Berouti, M. (1979). High-Frequency Regeneration in Speech Coding Systems, *Proceedings of the IEEE International Conference on Acoustics, Speech, and Signal Processing (ICASSP)*.

Melsa, J. L.; Cohn, D. L. (1978). *Decision and Estimation Theory*, McGraw-Hill Kogakusha, Tokyo.

Papoulis, A. (1991). *Probability, Random Variables, and Stochastic Processes*, 3rd edn, McGraw-Hill, New York.

Park, K. Y.; Kim, H. S. (2000). Narrowband to Wideband Conversion of Speech Using GMM-Based Transformation, *Proceedings of the IEEE International Conference on Acoustics, Speech, and Signal Processing (ICASSP)*, Istanbul, Turkey, vol. 3, pp. 1847–1850.

Patrick, P. J. (1983). *Enhancement of Bandlimited Speech Signal*, PhD thesis, Loughborough University of Technology.

Paulus, J. (1996). *Codierung breitbandiger Sprachsignale bei niedriger Datenrate*, PhD thesis. *Aachener Beiträge zu digitalen Nachrichtensystemen*, vol. 6, P. Vary (ed.), RWTH Aachen University (in German).

Paulus, J.; Schnitzler, J. (1996). 16 kbit/s Wideband Speech Coding Based on Unequal Subbands, *Proceedings of the IEEE International Conference on Acoustics, Speech, and Signal Processing (ICASSP)*, Atlanta, Georgia, USA, pp. 651–654.

Rabiner, L. R. (1989). A Tutorial on Hidden Markov Models and Selected Applications in Speech Recognition, *Proceedings of the IEEE*, vol. 77, no. 2, February, pp. 257–286.

Rec. GSM 06.10 (1992). Recommendation GSM 06.10 GSM Full Rate Speech Transcoding, ETSI TC-SMG.

Reynolds, D. A.; Rose, R. C. (1995). Robust Text-Independent Speaker Identification Using Gaussian Mixture Speaker Models, *IEEE Transactions on Speech and Audio Processing*, vol. 3, no. 1, January, pp. 72–83.

Salami, R.; Laflamme, C.; Bessette, B.; Adoul, J. P. (1997). Description of ITU-T Rec. G.729 Annex A: Reduced Complexity 8 kbit/s CS-ACELP Coding, *Proceedings of the IEEE International Conference on Acoustics, Speech, and Signal Processing (ICASSP)*, Munich, Germany.

Schmidt, K. O. (1933). Neubildung von unterdrückten Sprachfrequenzen durch ein nichtlinear verzerrendes Glied, *Telegraphen- und Fernsprech-Technik*, vol. 22, no. 1, January, pp. 13–22 (in German).

Schmidt, K. O.; Brosze, O. (1967). *Fernsprech-Übertragung*, Fachverlag Schiele und Schön, Berlin (in German).

Schnitzler, J. (1999). *Breitbandige Sprachcodierung: Zeitbereichs- und Frequenzbereichskonzepte*, PhD thesis. *Aachener Beiträge zu digitalen Nachrichtensystemen*, vol. 12, P. Vary (ed.), RWTH Aachen University (in German).

Sluijter, R. J. (2005). *The Development of Speech Coding and the First Standard Coder for Public Mobile Telephony*, PhD thesis, Technical University Eindhoven.

Taori, R.; Sluijter, R. J.; Gerrits, A. J. (2000). Hi-BIN: An Alternative Approach to Wideband Speech Coding, *Proceedings of the IEEE International Conference on Acoustics, Speech, and Signal Processing (ICASSP)*, Istanbul, Turkey, vol. 2, pp. 1157–1160.

Terhardt, E. (1998). *Akustische Kommunikation: Grundlagen mit Hörbeispielen*, Springer Verlag, Berlin.

Valin, J.-M.; Lefebvre, R. (2000). Bandwidth Extension of Narrowband Speech for Low Bit-Rate Wideband Coding, *IEEE Workshop on Speech Coding*, Delavan, Wisconsin, USA, pp. 130–132.

Van Trees, H. L. (1968). *Detection, Estimation, and Modulation Theory, Part I*, John Wiley & Sons, Inc., New York.

Vary, P. (2004). Advanced Signal Processing in Speech Communication, *Proceedings of the European Signal Processing Conference (EUSIPCO)*, Vienna, Austria, pp. 1449–1456.

Vary, P.; Hellwig, K.; Hofmann, R.; Sluyter, R.; Galand, C.; Rosso, M. (1988). Speech Codec for the European Mobile Radio System, *Proceedings of the IEEE International Conference on Acoustics, Speech, and Signal Processing (ICASSP)*, New York, USA, pp. 227–230.

Vaseghi, S. V. (1996). *Advanced Signal Processing and Digital Noise Reduction*, John Wiley & Sons, Ltd., Chichester, and B. G. Teubner, Stuttgart.

Voran, S. (1997). Listener Ratings of Speech Passbands, *Proceedings of the IEEE Workshop on Speech Coding*, Pocono Manor, Pennsylvania, USA, pp. 81–82.

Yoshida, Y.; Abe, M. (1994). An Algorithm to Reconstruct Wideband Speech from Narrowband Speech Based on Codebook Mapping, *Proceedings of the International Conference on Spoken Language Processing (ICSLP)*, Yokohama, Japan, pp. 1591–1594.

11

Single and Dual Channel Noise Reduction

This chapter is concerned with algorithms for reducing additive noise. We will primarily focus on methods which use a single microphone, but also present systems with two microphones. More general multi-microphone beamforming techniques will be discussed in Chapter 12. The first part of this chapter gives an introduction to the basic principles and implementation aspects. We will introduce the Wiener filter and the "spectral subtraction" technique, as well as noise power spectral density estimation techniques. The second part is more advanced and dedicated to non-linear optimal estimators. These estimators explicitly use the probability density function of short time spectral coefficients and are thus better able to utilize information available *a priori*. We will derive maximum likelihood, maximum *a posteriori*, and minimum mean square estimators for the estimation of the complex DFT coefficients, as well as for functions of the spectral amplitude. We will conclude this chapter with a discussion of the spatial correlation properties of sound fields, two-microphone noise cancellation, and speech enhancement approaches.

Digital Speech Transmission: Enhancement, Coding and Error Concealment
Peter Vary and Rainer Martin

11.1 Introduction

When a speech communication device is used in environments with high levels of ambient noise, the noise picked up by the microphone will significantly impair the quality and the intelligibility of the transmitted speech signal. The quality degradations can be very annoying, especially in mobile communications where hands free devices are frequently used in noisy environments such as cars. Compared to a close talking microphone, the use of a hands free device can lower the SNR of the microphone signal by more than 20 dB. Thus, even when the noise levels are moderate, the SNR of a hands free microphone signal is rarely better than 25 dB. Single microphone speech enhancement systems can improve the quality of speech signals and help to reduce listener fatigue. When the signal is transmitted via a low bit rate speech coder, such as the *mixed excitation linear prediction* (MELP) coder, or is used in conjunction with a cochlear implant, a noise reduction pre-processor can improve not only the quality [Kang, Fransen 1989] but also the intelligibility of the transmitted signal [Collura 1999], [Dörbecker 1998]. Furthermore, dual channel systems typically outperform single channel approaches [Greenberg, Zurek 1992], [Van Hoesel, Clark 1995], [Wouters, Vanden Berghe 2001].

To illustrate the performance of single channel systems, Fig. 11.1 plots the time domain waveforms of a clean speech signal, a noisy signal, and an enhanced signal vs. the sampling index k. The noise is non-stationary car noise. The single channel noise reduction algorithm which was used in this example, significantly reduces the level of the disturbing noise on average. However, due to the difficulty of tracking fast variations of the background noise, the short noise bursts around $k = 40000$ and $k = 10000$ are not removed.

Given the large diversity of acoustic environments and noise reduction applications and the resulting, sometimes conflicting performance requirements for noise reduction algorithms, it is apparent that there cannot be only one single "optimal" algorithm. Hence, a large variety of algorithms have been developed which have proved to be beneficial in certain noise environments or certain applications. Most of these more successful algorithms use statistical considerations for the computation of the enhanced signal. Most of them work in the frequency or some other transformation domain, and some of them also include models of human hearing [Tsoukalas et al. 1993], [Virag 1999], [Gustafsson et al. 1998]. So far there are no international (ITU or ETSI) standards for noise reduction algorithms, although noise reduction algorithms have become part of speech coding systems recently. In fact, a noise reduction pre-processor [Ramabadran et al. 1997] is part of the *Enhanced Variable Rate Codec* (EVRC, TIA/EIA IS-127) standard. A set of minimum requirements for noise reduction pre-processors has been defined in conjunction with the ETSI/3GPP *adaptive multi-rate* (AMR) codec [3GPP TS 122.076 V5.0.0 2002], [3GPP TR 126.978, V4.0.0 2001].

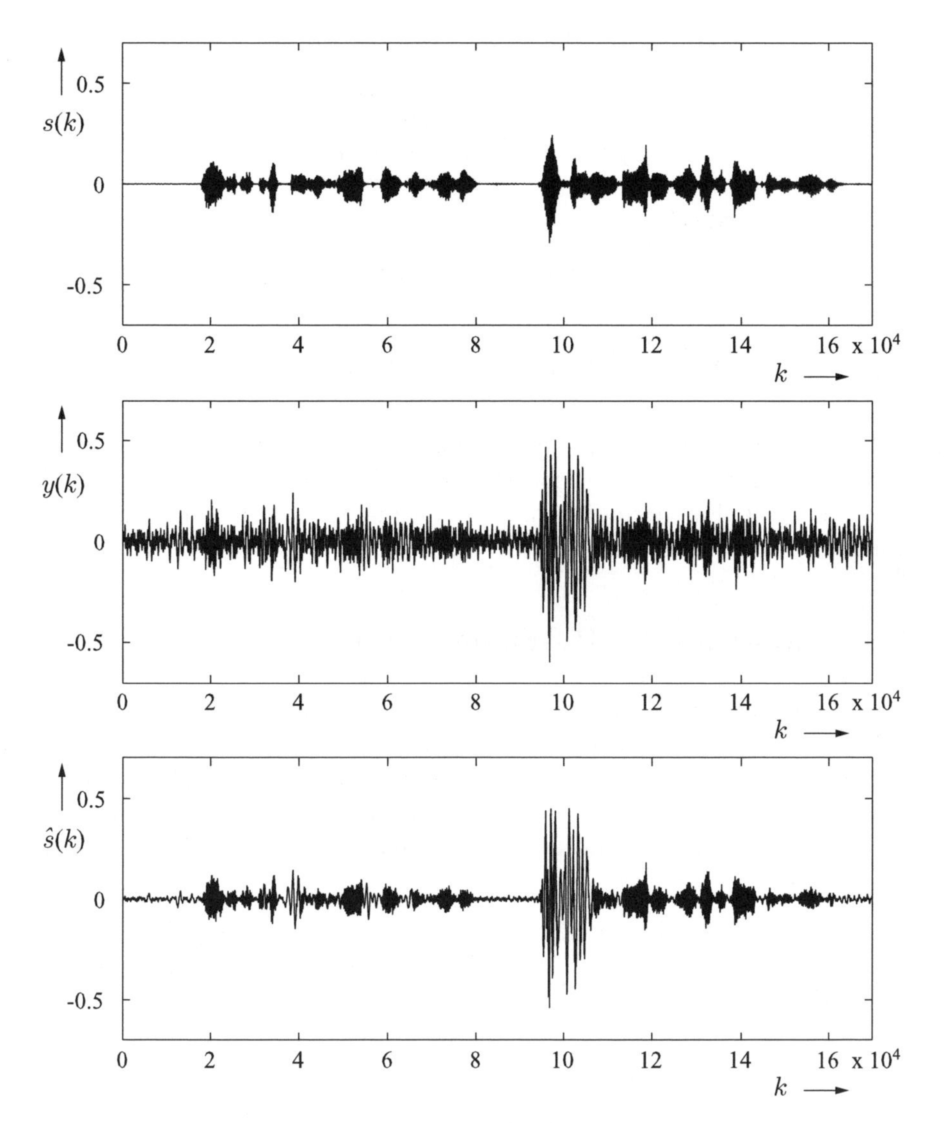

Figure 11.1: Time domain waveforms of a clean (top), a noisy (middle), and an enhanced (bottom) speech signal

In this chapter, we will give a general introduction to noise reduction algorithms and then discuss some of the more important topics in detail. We will look especially at statistical models for the estimation of speech signals and their spectral coefficients and at methods for the estimation of the noise power spectral density. Throughout this chapter, we will focus exclusively on additive noise, i.e., consider disturbed signals $y(k)$ which are a sum of the speech signal $s(k)$ and the noise signal $n(k)$, $y(k) = s(k) + n(k)$, as shown in Fig. 11.2. Furthermore, $s(k)$ and $n(k)$ are assumed to be statistically independent, which also implies $\mathrm{E}\{s(k)n(i)\} = 0 \,\forall\, k, i$. The enhanced signal will be denoted by $\hat{s}(k)$.

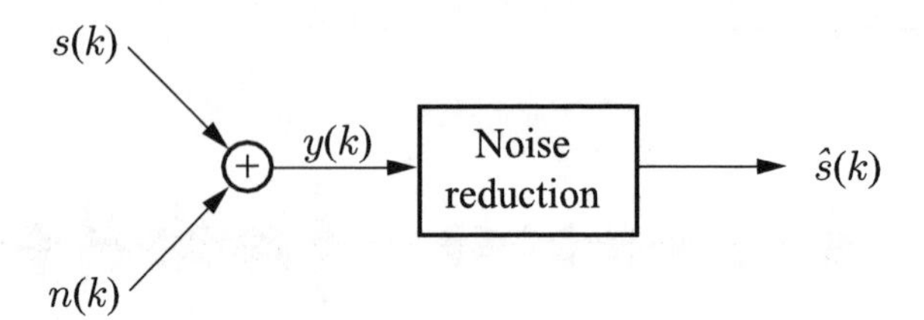

Figure 11.2: The noise reduction filter

11.2 Linear MMSE Estimators

Linearly constrained minimum mean square error (MMSE) estimators play an important role in signal estimation theory as they are relatively easy to develop and to implement. Furthermore, they are optimal for jointly Gaussian signals (see Section 5.12.3) in the unconstrained case.

11.2.1 Non-causal IIR Wiener filter

Our discussion is based on the signal model depicted in Fig. 11.3. Adding the noise signal $n(k)$ to the undisturbed desired signal $s(k)$ yields a noisy signal $y(k)$. The enhanced signal $\hat{s}(k)$ is compared to a reference signal $d(k)$. The resulting error signal $e(k)$ is used to compute the filter coefficients $h(k)$. For the time being, we assume that all signals are wide sense stationary and zero mean random processes.

The optimal linear filter is, then, time invariant and can be characterized by its impulse response $h(k)$. In the general case, the filter with impulse response $h(k)$ is neither a causal nor an FIR filter. Thus, the output signal $\hat{s}(k)$ of the linear time-invariant IIR filter is given by

$$\hat{s}(k) = \sum_{\kappa=-\infty}^{\infty} h(\kappa)\, y(k-\kappa)\,. \tag{11.1}$$

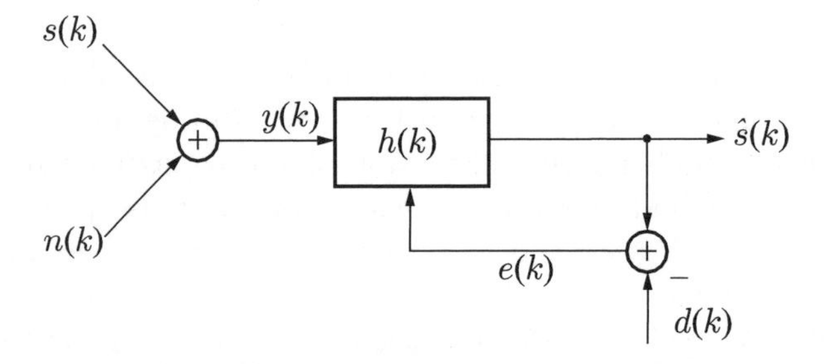

Figure 11.3: The linear filter problem for reducing additive noise

For computing the impulse response of the optimal filter, we define a target signal $d(k) = s(k)$ and minimize the mean square error

$$\mathrm{E}\left\{e^2(k)\right\} = \mathrm{E}\left\{\left(\hat{s}(k) - d(k)\right)^2\right\} = \mathrm{E}\left\{\left(\hat{s}(k) - s(k)\right)^2\right\}. \tag{11.2}$$

The partial derivative of the mean square error $\mathrm{E}\left\{e^2(k)\right\}$ with respect to a coefficient $h(i)$ leads to the condition

$$\frac{\partial \mathrm{E}\left\{e^2(k)\right\}}{\partial h(i)} = \frac{\partial \mathrm{E}\left\{\left(\sum\limits_{\kappa=-\infty}^{\infty} h(\kappa)\, y(k-\kappa) - s(k)\right)^2\right\}}{\partial h(i)} = 0 \tag{11.3}$$

for all $i \in \mathbb{Z}$ which may be written as

$$\sum_{\kappa=-\infty}^{\infty} h(\kappa)\, \varphi_{yy}(i-\kappa) = \varphi_{ys}(i)\,, \; \forall i \in \mathbb{Z}\,. \tag{11.4}$$

$\varphi_{yy}(\lambda) = \mathrm{E}\left\{y(k)y(k+\lambda)\right\}$ and $\varphi_{ys}(\lambda) = \mathrm{E}\left\{y(k)s(k+\lambda)\right\}$ are the auto-correlation function of the noisy signal $y(k)$ and the cross-correlation function of the noisy signal $y(k)$ and the clean speech signal $s(k)$, respectively.

Equation (11.4) is recognized as a convolution in the correlation function domain and can therefore be easily solved in the Fourier domain using power spectral densities $\varphi_{yy}(\lambda) \overset{\mathcal{F}}{\circ\!\!-\!\!\bullet} \Phi_{yy}(e^{j\Omega})$ and $\varphi_{ys}(\lambda) \overset{\mathcal{F}}{\circ\!\!-\!\!\bullet} \Phi_{ys}(e^{j\Omega})$. For $\Phi_{yy}(e^{j\Omega}) \neq 0$ we obtain

$$H(e^{j\Omega}) = \frac{\Phi_{ys}(e^{j\Omega})}{\Phi_{yy}(e^{j\Omega})} \quad \text{and} \quad h(k) = \frac{1}{2\pi} \int_{-\pi}^{\pi} H(e^{j\Omega}) e^{j\Omega k}\, \mathrm{d}\Omega\,. \tag{11.5}$$

For the above additive noise model with $\Phi_{yy}(e^{j\Omega}) = \Phi_{ss}(e^{j\Omega}) + \Phi_{nn}(e^{j\Omega})$, it follows that

$$H(e^{j\Omega}) = \frac{\Phi_{ss}(e^{j\Omega})}{\Phi_{ss}(e^{j\Omega}) + \Phi_{nn}(e^{j\Omega})} = \frac{\Phi_{ss}(e^{j\Omega})/\Phi_{nn}(e^{j\Omega})}{1 + \Phi_{ss}(e^{j\Omega})/\Phi_{nn}(e^{j\Omega})} \tag{11.6}$$

where $\Phi_{ss}(e^{j\Omega})$ and $\Phi_{nn}(e^{j\Omega})$ are the power spectral densities of the clean speech and the noise signal, respectively. The estimated signal spectrum may then be written as

$$\widehat{S}(e^{j\Omega}) = H(e^{j\Omega}) Y(e^{j\Omega})\,. \tag{11.7}$$

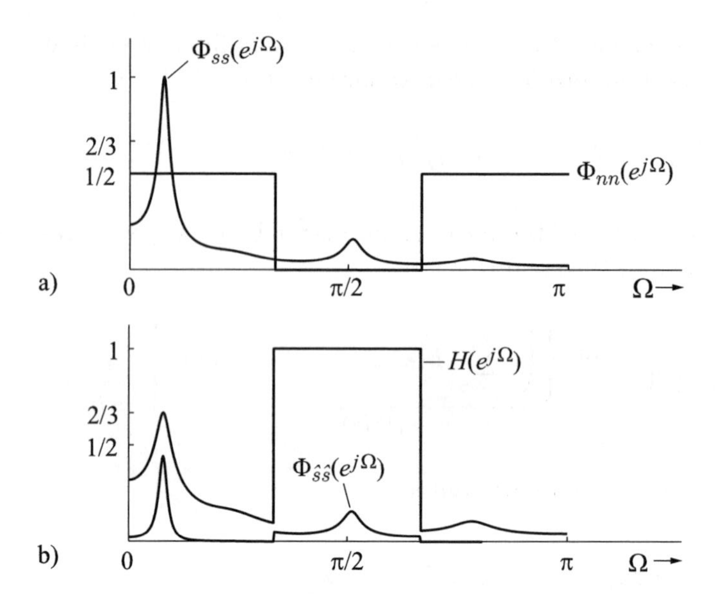

Figure 11.4: Wiener filter: principle of operation
a) Power spectral densities of the target signal and of the noise signal
b) Resulting frequency response and the power spectral density of the output signal

The non-causal IIR *Wiener filter* (11.6) evaluates the signal-to-noise ratio $\Phi_{ss}(e^{j\Omega})/\Phi_{nn}(e^{j\Omega})$ at a given frequency Ω. This is illustrated in Fig. 11.4 where the power spectral densities of the clean speech and the noise are shown in the upper plot and the resulting frequency response of the Wiener filter in the lower plot. When the SNR is large at a given frequency Ω, $H(e^{j\Omega})$ approaches unity and the corresponding frequency component will be passed on without being attenuated. For low SNR conditions, we have $\Phi_{ss}(e^{j\Omega}) \ll \Phi_{nn}(e^{j\Omega})$ and $H(e^{j\Omega}) \approx 0$. The corresponding frequency component of the input signal will be attenuated. Therefore, the noise reduction task will be most effectively accomplished if the speech signal and the noise do not occupy the same frequency bands. In the case of overlapping frequency bands, the noise reduction will also result in an attenuation of the desired speech signal.

For the additive noise model the power spectral densities in (11.6) are real valued and even symmetric. Therefore, the frequency response and the impulse response are also real valued and even symmetric. Since the convolution with an even symmetric impulse response does not alter the phase spectrum of the input signal, the non-causal IIR Wiener filter is a "zero-phase" filter.

In a practical implementation, the non-causal Wiener solution (11.6) requires further modifications. Spectral factorization techniques allow us in principle to derive causal optimal IIR filters [Kailath 1981]. However, also in this case, a truncation of the impulse response may be necessary. It is therefore important to derive the optimal FIR Wiener filter as well.

11.2.2 The FIR Wiener Filter

We now restrict the impulse response $h(k)$ of the optimal filter to be of finite and even order N, and to be causal, i.e.,

$$h(k) = \begin{cases} \text{arbitrary} & 0 \le k \le N \\ 0 & \text{otherwise}. \end{cases} \tag{11.8}$$

To obtain a linear-phase solution for the additive noise model, we use a delayed reference signal $d(k) = s(k-N/2)$ in the derivation of the optimal filter in Fig. 11.3. From (11.4) we find immediately a set of $N+1$ equations

$$\sum_{\kappa=0}^{N} h(\kappa)\, \varphi_{yy}(i-\kappa) = \varphi_{ss}(i-N/2), \quad i = 0 \dots N, \tag{11.9}$$

which can be stacked in vector/matrix notation as

$$\begin{pmatrix} \varphi_{yy}(0) & \varphi_{yy}(1) & \cdots & \varphi_{yy}(N) \\ \varphi_{yy}(1) & \varphi_{yy}(0) & \cdots & \varphi_{yy}(N-1) \\ \vdots & \vdots & \ddots & \vdots \\ \varphi_{yy}(N) & \varphi_{yy}(N-1) & \cdots & \varphi_{yy}(0) \end{pmatrix} \begin{pmatrix} h(0) \\ h(1) \\ \vdots \\ h(N) \end{pmatrix} = \begin{pmatrix} \varphi_{ss}(-N/2) \\ \varphi_{ss}(-N/2+1) \\ \vdots \\ \varphi_{ss}(N/2) \end{pmatrix} \tag{11.10}$$

where we used $\varphi_{yy}(\lambda) = \varphi_{yy}(-\lambda)$. Equation (11.10) may be written as

$$\mathbf{R}_{yy}\, \mathbf{h} = \boldsymbol{\varphi}_{ss} \tag{11.11}$$

where

$$\mathbf{R}_{yy} = \begin{pmatrix} \varphi_{yy}(0) & \varphi_{yy}(1) & \cdots & \varphi_{yy}(N) \\ \varphi_{yy}(1) & \varphi_{yy}(0) & \cdots & \varphi_{yy}(N-1) \\ \vdots & \vdots & \ddots & \vdots \\ \varphi_{yy}(N) & \varphi_{yy}(N-1) & \cdots & \varphi_{yy}(0) \end{pmatrix} \tag{11.12}$$

is the auto-correlation matrix and

$$\mathbf{h} = \begin{pmatrix} h(0) \\ h(1) \\ \vdots \\ h(N) \end{pmatrix} \quad \text{and} \quad \boldsymbol{\varphi}_{ss} = \begin{pmatrix} \varphi_{ss}(-N/2) \\ \varphi_{ss}(-N/2+1) \\ \vdots \\ \varphi_{ss}(N/2) \end{pmatrix} \tag{11.13}$$

are the coefficient vector and the auto-correlation vector, respectively. If $\mathbf{R}_{yy}$ is invertible, the coefficient vector is given by

$$\mathbf{h} = \mathbf{R}_{yy}^{-1}\ \boldsymbol{\varphi}_{ss}\,. \tag{11.14}$$

Since $\mathbf{R}_{yy}$ is a symmetric Toeplitz matrix, the system of equations in (11.11) can be efficiently solved using the Levinson–Durbin recursion. Also, we note that the correlation vector $\boldsymbol{\varphi}_{ss}$ is symmetric, and therefore the vector of filter coefficients,

$$h(N/2 - i) = h(N/2 + i)\,, \qquad i \in \{-N/2, \ldots, N/2\}\,, \tag{11.15}$$

is symmetric as well with respect to $i = N/2$. When the reference signal is delayed by $N/2$ samples, the solution to the optimal filter problem is a linear-phase FIR filter. Of course, the filter can also be optimized without a delay in the reference signal. In general, the resulting filter is still causal but does not have the linear-phase property in this case.

To account for the time-varying statistics of speech and noise signals the application of the Wiener filter to speech processing requires the use of short-term correlation functions. This can be done either by using a block processing approach or by approximating the Wiener filter with, for instance, a stochastic gradient algorithm such as the *normalized least-mean-square* (NLMS) algorithm [Haykin 1996]. In both cases the resulting filter will be time variant.

11.3 Speech Enhancement in the DFT Domain

The IIR and the FIR Wiener filters discussed so far are based on time domain linear filtering of stationary signals. However, the IIR solution to the optimization problem has led us quite naturally into the frequency domain. For stationary signals, approximations to the IIR Wiener filter may easily be realized in the frequency domain by using either Fourier transform or filter bank techniques. When the input signal is non-stationary, however, the filter coefficients must be continuously adapted. The filter will then be linear only on short, quasi-stationary segments of the signal.

Before we investigate these implementation aspects in detail, we would like to approach the noise reduction problem from a different point of view and treat it directly as an estimation problem of short, quasi-stationary speech segments (or speech frames) in the discrete Fourier transform (DFT) domain. We compute DFT frames using a perfect reconstruction analysis–synthesis system as shown in Fig. 11.5. The frame-based processing approach segments the incoming noisy signal into short frames of typically 5–30 ms duration. Each of these frames is transformed into the DFT domain, enhanced, inverse DFT transformed, and added to the previously processed signal with some overlap to smooth out discontinuities at the

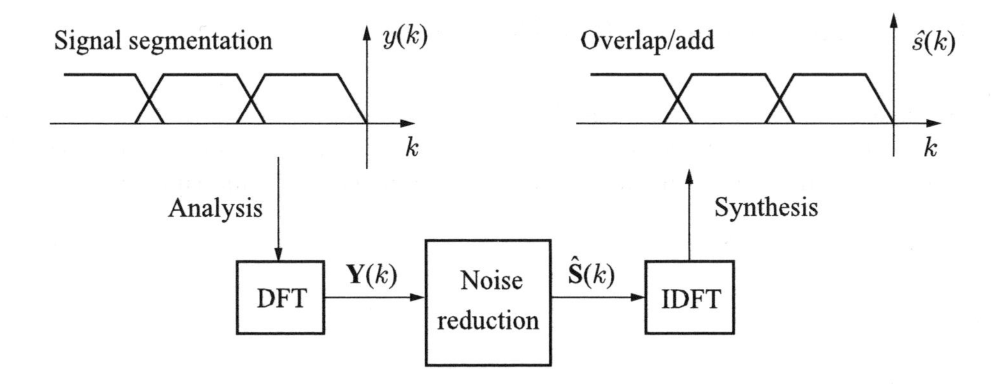

Figure 11.5: DFT domain implementation of the noise reduction filter

frame boundaries. If we assume that our analysis–synthesis system is sufficiently benign when the modified frames are overlap-added to construct the enhanced signal [Griffin, Lim 1984], we can focus on the enhancement of a *single* quasi-stationary frame of speech.

To this end, we define signal vectors of size M for the noisy speech $y(k)$, the clean speech $s(k)$, and the noise $n(k)$,

$$\mathbf{y}(k) = (y(k-M+1), y(k-M+2), \ldots, y(k))^T \tag{11.16}$$

$$\mathbf{s}(k) = (s(k-M+1), s(k-M+2), \ldots, s(k))^T \tag{11.17}$$

$$\mathbf{n}(k) = (n(k-M+1), n(k-M+2), \ldots, n(k))^T . \tag{11.18}$$

When k is the index of the current sample, each of these vectors holds a frame of the M most recent signal samples. Prior to computing the DFT, these vectors are weighted with an analysis window $\mathbf{w} = (w(0), w(1), \ldots, w(M-1))^T$,

$$\mathbf{Y}(k) = \text{DFT}\{\mathbf{w} \otimes \mathbf{y}(k)\} \tag{11.19}$$

$$\mathbf{S}(k) = \text{DFT}\{\mathbf{w} \otimes \mathbf{s}(k)\} \tag{11.20}$$

$$\mathbf{N}(k) = \text{DFT}\{\mathbf{w} \otimes \mathbf{n}(k)\} , \tag{11.21}$$

where $\otimes$ denotes an element-by-element multiplication of two vectors or matrices. The DFT domain vectors at sample index k are written in terms of their frequency components as

$$\mathbf{Y}(k) = (Y_0(k), \ldots, Y_\mu(k), \ldots, Y_{M-1}(k))^T \tag{11.22}$$

$$\mathbf{S}(k) = (S_0(k), \ldots, S_\mu(k), \ldots, S_{M-1}(k))^T \tag{11.23}$$

$$\mathbf{N}(k) = (N_0(k), \ldots, N_\mu(k), \ldots, N_{M-1}(k))^T . \tag{11.24}$$

11.3.1 The Wiener Filter Revisited

The correspondence between time domain convolution and Fourier domain multiplication suggests the definition of the DFT $\widehat{\mathbf{S}}(k) = (\widehat{S}_0(k), \widehat{S}_1(k), \ldots, \widehat{S}_{M-1}(k))^T$ of the enhanced signal frame as the result of an elementwise multiplication

$$\widehat{\mathbf{S}}(k) = \mathbf{H}(k) \otimes \mathbf{Y}(k) \tag{11.25}$$

of a weight vector

$$\mathbf{H}(k) = (H_0(k), H_1(k), \ldots, H_{M-1}(k))^T \tag{11.26}$$

and the DFT vector $\mathbf{Y}(k)$. Assuming that the DFT coefficients are (asymptotically) independent, we may minimize the mean square error

$$\mathrm{E}\left\{\left|S_\mu(k) - \widehat{S}_\mu(k)\right|^2\right\} \tag{11.27}$$

independently for each frequency bin μ. The partial derivative of

$$\mathrm{E}\left\{|S_\mu(k) - \widehat{S}_\mu(k)|^2\right\} = \mathrm{E}\left\{(S_\mu(k) - H_\mu(k)Y_\mu(k))\,(S_\mu(k) - H_\mu(k)Y_\mu(k))^*\right\}$$

with respect to the real part of $H_\mu(k)$ yields the condition

$$\frac{\partial \mathrm{E}\left\{\left|S_\mu(k) - \widehat{S}_\mu(k)\right|^2\right\}}{\partial \mathrm{Re}\left\{H_\mu(k)\right\}} = 0 \tag{11.28}$$

and hence

$$\mathrm{Re}\left\{H_\mu(k)\right\} = \frac{\mathrm{E}\left\{|S_\mu(k)|^2\right\}}{\mathrm{E}\left\{|Y_\mu(k)|^2\right\}} = \frac{\mathrm{E}\left\{|S_\mu(k)|^2\right\}}{\mathrm{E}\left\{|S_\mu(k)|^2\right\} + \mathrm{E}\left\{|N_\mu(k)|^2\right\}}\,. \tag{11.29}$$

For the imaginary part we obtain from

$$\frac{\partial \mathrm{E}\left\{|S_\mu(k) - \widehat{S}_\mu(k)|^2\right\}}{\partial \mathrm{Im}\left\{H_\mu(k)\right\}} = 0 \tag{11.30}$$

the result

$$\mathrm{Im}\left\{H_\mu(k)\right\} = 0\,. \tag{11.31}$$

Therefore,

$$H_\mu(k) = \frac{\mathrm{E}\left\{|S_\mu(k)|^2\right\}}{\mathrm{E}\left\{|S_\mu(k)|^2\right\} + \mathrm{E}\left\{|N_\mu(k)|^2\right\}} \tag{11.32}$$

is a real valued weight which does not modify the phase of the noisy coefficients. $H_\mu(k)$ is defined in terms of statistical expectations as a function of time and is therefore in general time varying. There are numerous ways (non-parametric or model based) to compute approximations to these expectations. In general, some averaging (over time or frequency) is necessary. For a single microphone system, the most prominent estimation procedure is based on time-averaged modified periodograms [Welch 1967]. In multi-microphone systems, we might also smooth over microphone channels and thus reduce the amount of smoothing over time.

It is quite instructive to juxtapose this result to the Wiener filter in (11.6). For stationary signals, $\mathrm{E}\left\{|S_\mu(k)|^2\right\}$ and $\mathrm{E}\left\{|N_\mu(k)|^2\right\}$ are estimates of the power spectra $\Phi_{ss}(e^{j\Omega_\mu})$ and $\Phi_{nn}(e^{j\Omega_\mu})$. For a properly normalized window $\mathbf{w}$, these estimates are asymptotically ($M \to \infty$) unbiased. On the one hand, (11.32) is therefore closely related to the Wiener filter. On the other hand, it is conceptually quite different from the Wiener filter as we did not start from a *time domain linear* filtering problem. While a frequency domain approximation of (11.6) implies the implementation of a linear filter, e.g., by means of a fast convolution, the estimation of DFT coefficients as in (11.25) and (11.32) leads to a cyclic convolution in the time domain. However, there is no principal reason why the latter should not work and the cyclic effects known from linear theory must not be interpreted as a disturbance in case of optimal frequency domain estimators. However, we should use a perfect reconstruction overlap-add scheme, tapered analysis and synthesis windows, and sufficient overlap between frames to control estimation errors at the frame edges.

In Fig. 11.6 we illustrate the principle of the DFT-based noise reduction algorithm. The magnitude squared DFT coefficients of a noisy, voiced speech sound and the estimated noise floor are shown in the left hand plot. The right hand plot depicts the magnitude squared DFT coefficients of the enhanced speech sound, as well as the estimated noise *power spectral density* (PSD) of the noisy speech. We observe that the harmonic (high SNR) peaks of the speech sound are well reproduced, while the valleys in between these sounds, where the noise is predominant, are attenuated. As a result the global SNR of the speech sound is improved.

Despite the apparently straightforward solution to the estimation problem, one critical question remains: (11.6) as well as (11.32) require knowledge of the PSD of the clean speech or of the SNR of the noisy signal at a given frequency Ω_μ. However, neither the clean speech PSD nor the SNR are readily available. To arrive at more practical solutions, we discuss another (however closely related) approach to noise reduction which is known as *spectral subtraction*.

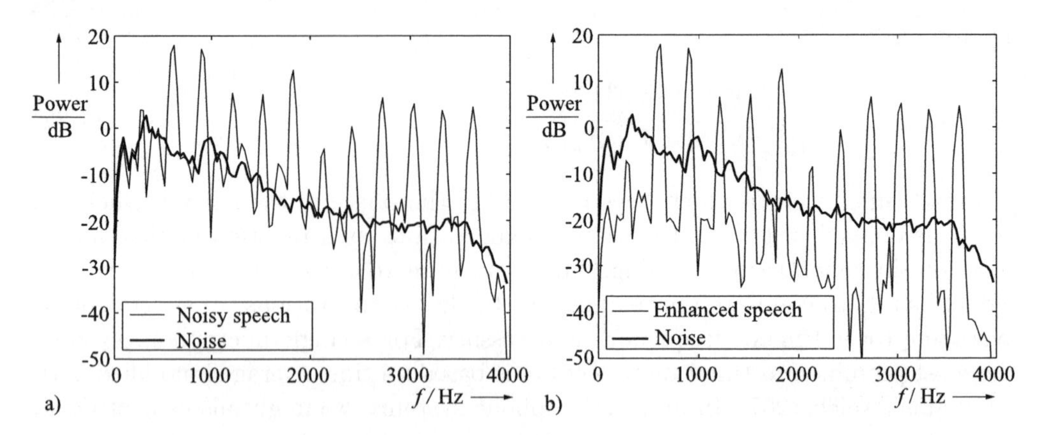

Figure 11.6: Principle of DFT-based noise reduction
a) Short-time spectrum of noisy signal and the estimated noise PSD
b) Short-time spectrum of the enhanced signal and the estimated noise PSD

11.3.2 Spectral Subtraction

The basic idea of spectral subtraction [Boll 1979] and related proposals [Preuss 1979], [Berouti et al. 1979] is to subtract an estimate of the noise floor from an estimate of the spectrum of the noisy signal. Since the speech signal is non-stationary, this has to be done on a short-time basis, preferably using a DFT-based analysis–synthesis system.

Using the frame-wise processing approach of Section 11.3 and an appropriately normalized analysis window $\mathbf{w}$, the power of the clean speech signal at the discrete frequencies $\Omega_\mu = \frac{\mu 2\pi}{M}$ is given by

$$\begin{aligned}\mathrm{E}\left\{|S_\mu(k)|^2\right\} &= \mathrm{E}\left\{|Y_\mu(k)|^2\right\} - \mathrm{E}\left\{|N_\mu(k)|^2\right\} \\ &= \mathrm{E}\left\{|Y_\mu(k)|^2\right\}\left[1 - \frac{\mathrm{E}\left\{|N_\mu(k)|^2\right\}}{\mathrm{E}\left\{|Y_\mu(k)|^2\right\}}\right] \\ &= \mathrm{E}\left\{|Y_\mu(k)|^2\right\} |\tilde{H}_\mu(k)|^2 \,. \end{aligned} \tag{11.33}$$

Since a sliding window DFT may be interpreted as an analysis filter bank (see Section 4.1.6), $\mathrm{E}\left\{|S_\mu(k)|^2\right\}$ represents the power of a complex-valued subband signal $S_\mu(k)$ for any fixed μ.

In analogy to the relation of input and output power in linear systems [Papoulis, Unnikrishna Pillai 2001], the spectral subtraction method may be interpreted as a time-variant filter with magnitude frequency response

$$\tilde{H}_\mu(k) = \sqrt{1 - \frac{\mathrm{E}\left\{|N_\mu(k)|^2\right\}}{\mathrm{E}\left\{|Y_\mu(k)|^2\right\}}} = \sqrt{\frac{\mathrm{E}\left\{|S_\mu(k)|^2\right\}}{\mathrm{E}\left\{|Y_\mu(k)|^2\right\}}} = \sqrt{H_\mu(k)} \;, \tag{11.34}$$

which is the square root of (11.32). Since we are subtracting in the PSD domain, this approach is called *power subtraction.*

When the spectral subtraction in (11.33) is used in conjunction with the Wiener filter (11.32), we obtain

$$H_\mu(k) = \frac{\mathrm{E}\left\{|Y_\mu(k)|^2\right\} - \mathrm{E}\left\{|N_\mu(k)|^2\right\}}{\mathrm{E}\left\{|Y_\mu(k)|^2\right\}} = 1 - \frac{\mathrm{E}\left\{|N_\mu(k)|^2\right\}}{\mathrm{E}\left\{|Y_\mu(k)|^2\right\}} . \tag{11.35}$$

This and other variations of the spectral gain function, such as *magnitude subtraction*

$$\sqrt{\mathrm{E}\left\{|Y_\mu(k)|^2\right\}} - \sqrt{\mathrm{E}\left\{|N_\mu(k)|^2\right\}} = \sqrt{\mathrm{E}\left\{|Y_\mu(k)|^2\right\}} \left[1 - \frac{\sqrt{\mathrm{E}\left\{|N_\mu(k)|^2\right\}}}{\sqrt{\mathrm{E}\left\{|Y_\mu(k)|^2\right\}}}\right] \tag{11.36}$$

are subsumed in a generalized, magnitude squared spectral gain function

$$|\tilde{H}_\mu(k)|^2 = \left[1 - \left(\frac{\mathrm{E}\left\{|N_\mu(k)|^2\right\}}{\mathrm{E}\left\{|Y_\mu(k)|^2\right\}}\right)^\beta\right]^\alpha . \tag{11.37}$$

Using magnitude squared DFT spectra and short-time estimates $\widehat{|N_\mu(k)|^2}$ and $\widehat{|Y_\mu(k)|^2}$, the above spectral subtraction rule may then be cast into a more practical heuristic form

$$\widehat{|S_\mu(k)|^2} = |Y_\mu(k)|^2 \left[1 - \left(\frac{\widehat{|N_\mu(k)|^2}}{\widehat{|Y_\mu(k)|^2}}\right)^\beta\right]^\alpha \tag{11.38}$$

where we have to make sure that the result is real valued and not negative. The parameters α and β can be either kept fixed or adapted to the characteristics of the speech and the noise signals [Hansen 1991]. The gain function of the generalized subtraction filter (11.37) is shown in Fig. 11.7 for three different parameter settings as a function of the ratio $\overline{\gamma}_\mu(k) = \mathrm{E}\left\{|Y_\mu(k)|^2\right\} / \mathrm{E}\left\{|N_\mu(k)|^2\right\}$.

Thus, the formulation of the subtraction approach in terms of a multiplicative spectral gain function not only unifies the various noise reduction approaches, but is also helpful for implementing these algorithms since $0 \leq |\tilde{H}(e^{j\Omega})| \leq 1$ is a normalized quantity.

The noise PSD, which is a necessary ingredient of all of these spectral weighting rules, can be estimated using the *minimum statistics* [Martin 1994], [Martin 2001a] approach, soft-decision estimators [Sohn, Sung 1998], or using a *voice activity detector* (VAD) [McAulay, Malpass 1980].

The "pure" spectral subtraction approach is not very robust with respect to estimation errors in the spectral gain function. As a result, practical implementations suffer from speech distortions and a fluctuating residual noise, which is commonly known as "musical noise".

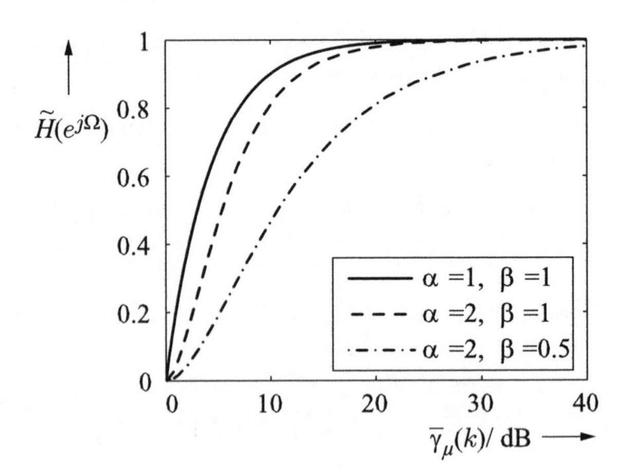

Figure 11.7: Spectral gain functions $\widetilde{H}(e^{j\Omega})$ as a function of $\overline{\gamma}_\mu(k) = \mathrm{E}\left\{|Y_\mu(k)|^2\right\} / \mathrm{E}\left\{|N_\mu(k)|^2\right\}$: power subtraction ($\alpha = \beta = 1$), Wiener filter ($\alpha = 2,\ \beta = 1$), and magnitude subtraction ($\alpha = 2,\ \beta = 0.5$)

11.3.3 Estimation of the *A Priori* SNR

In the previous section, we have seen that the spectral subtraction technique can be used on its own or in conjunction with the Wiener filter. In the latter case, it supplies an estimate for the unknown PSD of the clean speech. Another approach to the realization of the Wiener filter [Scalart, Vieira Filho 1996] is to express it in terms of the *a priori* SNR

$$\eta(e^{j\Omega}) = \frac{\Phi_{ss}(e^{j\Omega})}{\Phi_{nn}(e^{j\Omega})} \quad \text{or} \quad \eta_\mu(k) = \frac{\mathrm{E}\left\{|S_\mu(k)|^2\right\}}{\mathrm{E}\left\{|N_\mu(k)|^2\right\}} \tag{11.39}$$

and to estimate the *a priori* SNR instead of the clean speech PSD. Since for most noise reduction applications the *a priori* SNR ranges over an interval of −20–30 dB and is invariant with respect to the signal scaling, it is easier to deal with the SNR in (e.g., fixed point) implementations than using PSDs. Another reason for using the *a priori* SNR lies in the availability of a simple yet powerful estimation algorithm, which will be discussed below.

The estimation of the *a priori* SNR is based on the *a posteriori* SNR $\gamma_\mu(k)$, which is defined as the ratio of the periodogram of the noisy signal and the noise power

$$\gamma(e^{j\Omega}) = \frac{|Y(e^{j\Omega})|^2}{\Phi_{nn}(e^{j\Omega})} \quad \text{or} \quad \gamma_\mu(k) = \frac{|Y_\mu(k)|^2}{\mathrm{E}\left\{|N_\mu(k)|^2\right\}}\,. \tag{11.40}$$

If an estimate of the noise PSD is available, the *a posteriori* SNR is easily measurable.

The *a priori* SNR can now be expressed as

$$\eta_\mu(k) = \frac{\mathrm{E}\{|S_\mu(k)|^2\}}{\mathrm{E}\{|N_\mu(k)|^2\}} = \mathrm{E}\{\gamma_\mu(k) - 1\} \tag{11.41}$$

where the "decision-directed" approach [Ephraim, Malah 1984] may be used to estimate $\eta_\mu(k)$. To apply this approach to the estimation of the *a priori* SNR at time $k+r$, an estimate $|\widehat{S_\mu(k)}|$ of the clean speech amplitudes $|S_\mu(k)|$ at time k must be available. Furthermore, we assume that $|S_\mu(k+r)| \approx |S_\mu(k)|$, which holds for quasi-stationary speech sounds, but is less valid for transient sounds. Then, the *a priori* SNR $\widehat{\eta}_\mu(k+r)$ of the frame at $k+r$ is computed as a linear combination of an estimate based on the signal frame at time k and of an instantaneous realization of (11.41),

$$\widehat{\eta}_\mu(k+r) = \alpha_\eta \frac{|\widehat{S_\mu(k)}|^2}{\mathrm{E}\{|N_\mu(k)|^2\}} + (1-\alpha_\eta)\max(\gamma_\mu(k+r) - 1, 0)\,, \tag{11.42}$$

where α_η is typically in the range $0.9 \leq \alpha_\eta \leq 0.99$.

It has frequently been argued [Cappé 1994], [Scalart, Vieira Filho 1996] that this estimation procedure contributes considerably to the subjective quality of the enhanced speech, especially to the reduction of "musical noise". Therefore, it is rather useful in combination with almost any of the above noise reduction methods.

Since $\widehat{\eta}_\mu(k+r)$ as defined above does not fully account for speech absence, the *a priori* SNR may also be computed with a speech power estimate $|\widehat{S_\mu(k)}|^2$ which is conditioned on the presence of speech [Cohen, Berdugo 2001], i.e.,

$$\widehat{\eta}_\mu(k+r) = \alpha_\eta \frac{|\widehat{S_\mu(k)}|^2{}_{|\text{speech is present}}}{\mathrm{E}\{|N_\mu(k)|^2\}} + (1-\alpha_\eta)\max(\gamma_\mu(k+r) - 1, 0)\,. \tag{11.43}$$

Now, $\widehat{\eta}_\mu(k+r)$ is implicitly scaled with the *a priori* probability of speech presence. Various estimators may be used to compute the conditional speech power estimate $|\widehat{S_\mu(k)}|^2{}_{|\text{speech is present}}$ (see Section 11.4). To conclude we note that there are other ways to exploit the idea of recursive estimation, e.g., [Linhard, Haulick 1999], [Beaugeant, Scalart 2001], which generally lead to smoother estimates than the standard methods.

11.3.4 Musical Noise and Countermeasures

The "musical noise" phenomenon is as easy to explain as it is difficult to avoid. The term "musical noise" describes the randomly fluctuating noise floor which is frequently observed in noise-reduced signals. It is especially noticeable for the

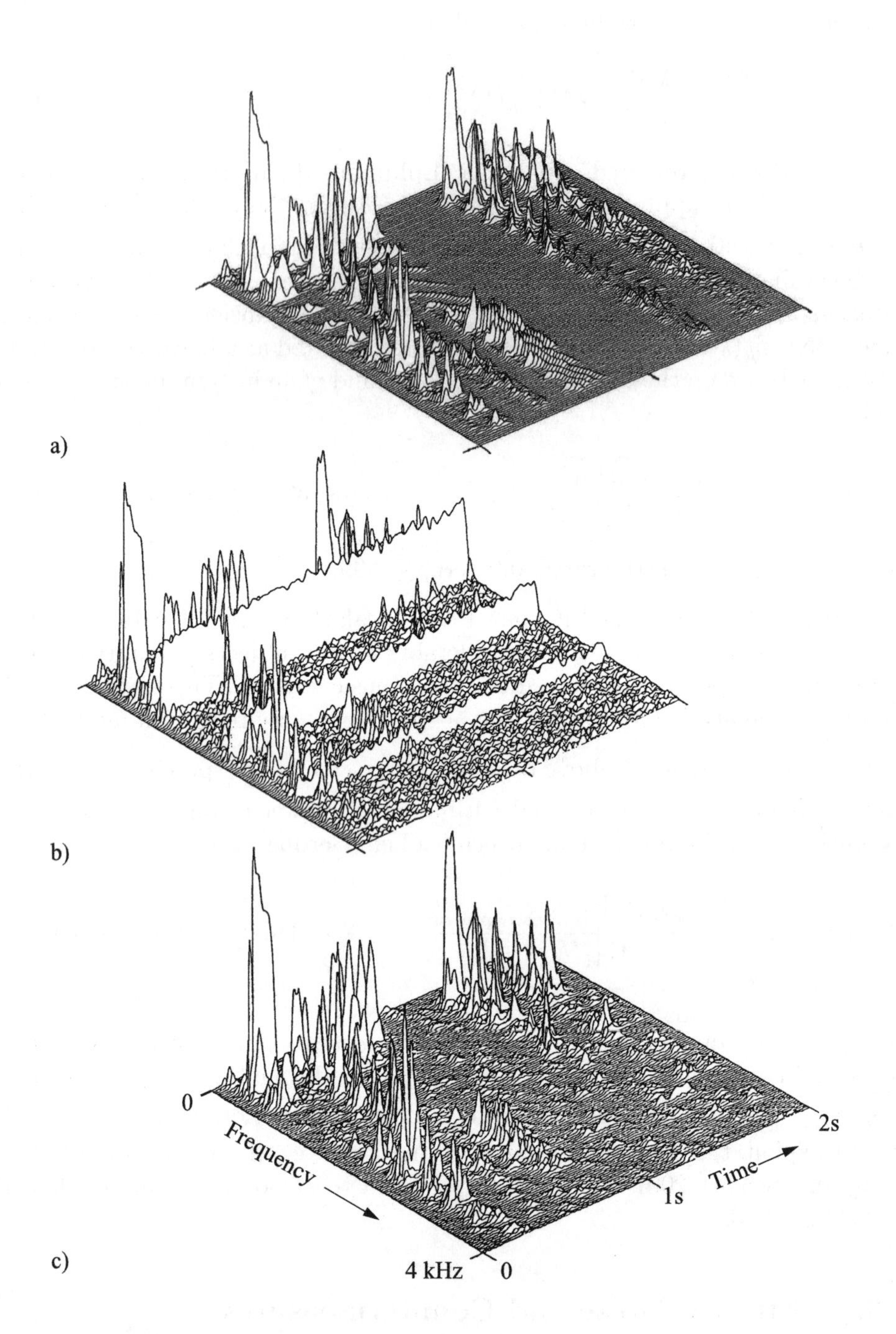

Figure 11.8: Short-term magnitude spectra vs. time and frequency
a) of a clean speech signal,
b) of the clean signal with additive white noise and harmonic tones,
c) of the enhanced signal using magnitude subtraction

simple spectral weighting methods as outlined in the previous sections. The musical noise phenomenon can be explained by estimation errors in the frequency domain which lead to spurious peaks in the spectral representation of the enhanced signal. When the enhanced signal is reconstructed via the IFFT and overlap-add, these peaks correspond to a sinusoidal excitation whose frequency varies randomly from frame to frame.

Figure 11.8 depicts the time evolution of the short-term magnitude spectrum of a clean (a), a noisy (b), and the corresponding enhanced signal (c) [Vary 1985]. The speech signal is disturbed by white Gaussian noise and three stationary harmonic tones at 1, 2, and 3 kHz. The enhanced spectrum is computed using the magnitude subtraction rule (11.36). Clearly, the stationary tones can be completely removed. Also, the level of the white noise is significantly reduced. However, some of the low-power speech components are lost and random spectral peaks, especially during the speech pause, remain.

A statistical evaluation of these estimation errors is given in [Vary 1985]. Quite a substantial effort in speech enhancement research has been spent on techniques which avoid the "musical noise" phenomenon. We will briefly describe two commonly used countermeasures.

11.3.4.1 Noise Oversubtraction and Spectral Floor

The subtraction of the average noise PSD from the time-varying magnitude squared DFT coefficients of the noisy signal leads to residual spectral peaks, which in turn result in the "musical noise" phenomenon. The musical noise is especially disturbing during speech pauses. The noise power overestimation increases the estimated noise PSD in order to reduce the amplitude of these random spectral peaks after subtraction [Berouti et al. 1979]. Furthermore, a "spectral floor" is applied to the resultant spectral coefficients, preventing them from falling below a preset minimum level. Hence, we obtain spectral components which fluctuate less during speech pauses.

In practice, the estimated noise level is increased by a factor of $O_S = 1 \ldots 2$,

$$\mathrm{E}\{|\widehat{N_\mu(k)}|^2\} = O_S \cdot \mathrm{E}\left\{|N_\mu(k)|^2\right\} . \tag{11.44}$$

The spectral floor imposes a lower limit $\beta \, \mathrm{E}\left\{|N_\mu(k)|^2\right\}$ on the magnitude squared DFT coefficients. For the spectral gain functions discussed so far, β cannot be much smaller than 0.1 without giving rise to noticeable "musical noise". On the other hand, it cannot be much larger without rendering the noise reduction algorithm useless.

During speech pauses or low SNR conditions, the overestimated noise PSD estimate helps to cover up the fluctuations in the DFT coefficients. During speech activity, however, it may lead to distortions of the speech signal. With a large oversubtraction factor the enhanced speech may sound muffled or even clipped. It is therefore not completely obvious how large the overestimation factor should be. In fact, one single constant overestimation factor does not improve the speech quality for all SNR conditions. Some authors have studied more general noise "oversubtraction" models [Hansen 1991], [Lockwood, Boudy 1991], [Lockwood et al. 1991], [Lockwood, Boudy 1992] and also explored the benefits of underestimating the (unbiased) noise PSD estimate for speech recognition [Kushner et al. 1989] and speech transmission [Händel 1995] purposes. We will investigate this issue in some more detail below.

Figure 11.9 plots the measured quality of the enhanced speech versus the oversubtraction factor for a frame SNR ranging between 0 dB and 30 dB. The quality criterion is the *mean square distance* between the enhanced and the clean speech in the *line spectral frequency* (LSF, see Chapter 8) domain [Paliwal, Atal 1991]

$$\Delta^2_{\mathrm{LSF}} = \frac{1}{10L} \sum_{\lambda=1}^{L} \sum_{i=1}^{10} c_i (\tilde{f}_{i,\lambda} - f_{i,\lambda})^2 \tag{11.45}$$

where $f_{i,\lambda}$ and $\tilde{f}_{i,\lambda}$ denote the i-th LSF parameter of the λ-th signal frame of the clean and the enhanced noisy speech signal, respectively. A total of 10 spectral parameters is used. L denotes the total number of speech signal frames, and c_i is

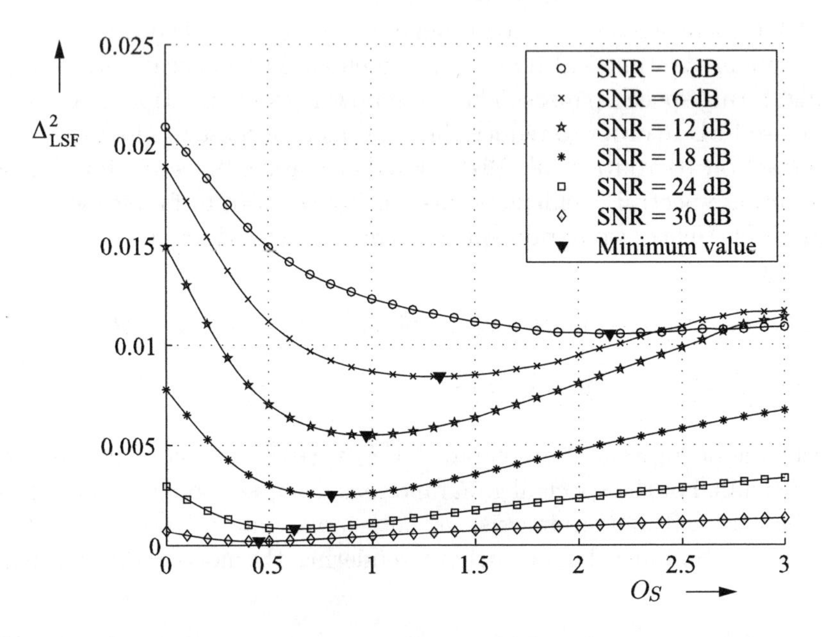

Figure 11.9: Error Δ^2_{LSF} for $0 \leq O_S \leq 3$ and several SNR classes

Table 11.1: Weights for the LSF distortion measure [Paliwal, Atal 1991]

LSF index	1–8	9	10
c_i	1.0	0.64	0.16

given in Table 11.1. According to this table, errors at higher LSFs do not contribute as much to the total error as errors at lower frequencies [Paliwal, Atal 1991]. The evaluation was performed using 60 seconds of phonetically balanced female and 60 seconds of phonetically balanced male speech which were recorded at a sampling rate of 8 kHz. White Gaussian noise was added at six different levels to the clean speech and the noisy speech was processed with the soft-decision MMSE-LSA noise reduction (see Section 11.4.4.3) algorithm. The processing was repeated for a number of oversubtraction factors O_S, while all other parameters were kept constant. The SNR of every single input frame was computed and assigned to one of six SNR classes, each of them corresponding to SNR intervals of 6 dB (see Table 11.2).

For each SNR class the average distance between the clean and the enhanced LSF parameters was computed. Finally, the distance data was interpolated by means of cubic splines. The optimal value of the examined parameter was found as the minimum of the interpolated data.

We find that for low SNR conditions an oversubtraction factor between 1.3 and 2 should be employed. For high SNR conditions *underestimating* the noise PSD is preferable. Similar findings were reported in [Kushner et al. 1989] for spectral and cepstral subtraction in the context of speech recognition. Nevertheless, a constant oversubtraction factor between 1.3 and 2 appears to be a good choice, as in this case, and for the high SNR cases the distortions are not greater than those obtained for the noisy original (oversubtraction of zero).

Table 11.2: Upper and lower SNR limits of SNR classes used for the evaluation of spectral parameters

SNR class	Lower class limit	Upper class limit
0 dB	−3 dB	3 dB
6 dB	3 dB	9 dB
12 dB	9 dB	15 dB
18 dB	15 dB	21 dB
24 dB	21 dB	27 dB
30 dB	27 dB	33 dB

11.3.4.2 Limitation of the *A Priori* SNR

[Cappé 1994], [Malah et al. 1999] proposed to apply a lower limit to the estimated *a priori* SNR in order to reduce the annoying musical tones. The lower limit has the greatest effect on frequency bins which do *not* contain speech. Thus, it is of great importance for the perceived quality during speech pauses but also for the overall shape of the speech spectrum during speech activity. The limit on the *a priori* SNR limits the maximum attenuation. This can be easily seen when using the Wiener filter but also holds for the more complicated non-linear estimators [Martin et al. 2000]. For the Wiener filter in (11.6) and low SNR conditions, we obtain

$$H(e^{j\Omega}) = \frac{\Phi_{ss}(e^{j\Omega})/\Phi_{nn}(e^{j\Omega})}{1 + \Phi_{ss}(e^{j\Omega})/\Phi_{nn}(e^{j\Omega})} \approx \frac{\Phi_{ss}(e^{j\Omega})}{\Phi_{nn}(e^{j\Omega})} . \tag{11.46}$$

Thus, by applying a lower bound to the *a priori* SNR, the maximum attenuation of the filter is limited.

11.3.5 Aspects of Spectral Analysis/Synthesis

The noise reduction methods presented so far were developed for spectral domain processing. Besides the DFT, other spectral transforms as well as filter banks are in principle suited for this task. Since spectral transforms and filter banks are discussed in detail in Chapter 3 and 4, we summarize here issues which are specifically related to noise reduction.

One of the most important questions to be answered concerns the spectral resolution which should be provided by the spectral transform. The design can be guided by the characteristics of the source signal and/or the properties of the receiver, the human ear, or a speech recognizer. While the former approach suggests a high resolution in order to separate the distinct peaks of the pitch harmonics and to attenuate the noise in between the peaks, the latter allows for less resolution, especially at higher frequencies. Both approaches have been advocated in the literature. We will briefly discuss two typical implementations, the first based on DFT processing and overlap-add, the second based on filter banks.

11.3.5.1 DFT-Based Analysis/Synthesis

Since perfect reconstruction is an important issue for high SNR conditions only, we must design our analysis–synthesis system such that it is optimal also in the presence of spectral modifications. This problem has been treated, for example, in [Allen 1977], [Griffin, Lim 1984], and [Cappé 1995]. In what follows, we will outline a widely used approach based on the sliding window DFT.

Assuming an additive, independent noise model, the noisy signal $y(k)$ is given by $s(k) + n(k)$, where $s(k)$ denotes the clean speech signal and $n(k)$ the noise. All signals are sampled at a sampling rate of f_s. We apply a short-time Fourier analysis to the input signal by computing the DFT of each overlapping windowed frame,

$$Y_\mu(\lambda) = \sum_{\ell=0}^{M-1} y(\lambda r + \ell)\, w(\ell)\, e^{-j\frac{2\pi\mu\ell}{M}} . \tag{11.47}$$

Here, r denotes the frame shift, $\lambda \in \mathbb{Z}$ is the frame index, $\mu \in \{0, 1, \ldots, M-1\}$ is the frequency bin index, which is related to the normalized center frequency $\Omega_\mu = \mu 2\pi / M$, and $w(\ell)$ denotes the window function.

The analysis–synthesis system must balance conflicting requirements of sufficient spectral resolution, little spectral leakage, smooth transitions between signal frames, low delay, and low complexity. Delay and complexity constraints limit the overlap of the signal frames. However, the frame shift must not be too large so as not to degrade the quality of the enhanced signal. Typical implementations of DFT-based noise reduction algorithms use a Hann window with a 50% overlap ($r/M = 0.5$) or a Hamming window with a 75% overlap ($r/M = 0.25$) for spectral analysis, and a rectangular window for synthesis.

Frequently, the speech enhancement algorithm is used in conjunction with a speech coder. The segmentation of the input signal into frames and the selection of an analysis window is therefore closely linked to the frame alignment of the speech coder and the admissible algorithmic delay [Martin, Cox 1999], [Martin et al. 1999]. The total algorithmic delay of a joint enhancement and coding system is minimized when the frame shift of the noise reduction pre-processor is adjusted in such a way that $l(M - M_O) = lr = M_C$ with $l \in \mathbb{N}$ and where M_C and M_O denote the frame length of the speech coder and the length of the overlapping portions of the pre-processor frames, respectively. This situation is depicted in Fig. 11.10.

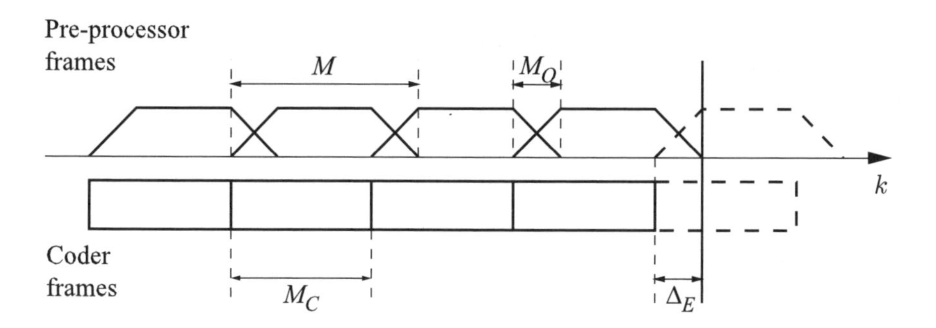

Figure 11.10: Frame alignment of enhancement pre-processor and speech coder ($r = M_C$) [Martin et al. 1999]

The additional delay Δ_E is equal to M_O due to the enhancement pre-processor. For example, for the MELP coder and its frame length of $M_C = 180$, we use an FFT length of $M = 256$ and have $M_O = 76$ overlapping samples between adjacent signal frames. This requires the use of a *flat-top* (*Tukey*) window.

Reducing the number of overlapping samples M_O, and thus the delay of the joint system, has several effects. First, with a flat-top analysis window, the side-lobe attenuation (or spectral leakage) during spectral analysis is decreased, which leads to increased crosstalk between frequency bins. This might complicate the speech enhancement task as most enhancement algorithms assume independent frequency bins and do not exploit correlation between bins. Secondly, as the overlap between frames is reduced, transitions between adjacent frames of the enhanced signal become less smooth. Discontinuities arise because the analysis window attenuates the input signal mostly at the edges of a frame, while estimation errors which occur during the processing of the frame in the spectral domain tend to spread evenly over the whole frame. This leads to larger relative estimation errors at the frame edges. The resulting discontinuities, which are most notable in low SNR conditions, may lead to pitch estimation errors and other speech coder artifacts.

These discontinuities are greatly reduced if we use not only a tapered analysis window but also a tapered window for spectral synthesis. A tapered synthesis window is beneficial when the overlap M_O is less than 40% of the DFT length M. In this case, the square root of the Tukey window

$$h(k) = \begin{cases} \sqrt{0.5\left(1-\cos\left(\frac{\pi k}{M_O}\right)\right)} & 1 \le k \le M_O \\ 1 & M_O + 1 \le k \le M - M_O - 1 \\ \sqrt{0.5\left(1-\cos\left(\frac{\pi (M-k)}{M_O}\right)\right)} & M - M_O \le k \le M \end{cases} \tag{11.48}$$

can be used as an analysis and synthesis window. It results in a perfect reconstruction system if the signal is not modified between analysis and synthesis. Note that the use of a tapered synthesis window is also in line with the results of Griffin and Lim [Griffin, Lim 1984] for the MMSE reconstruction of modified short-time spectra. As outlined in Chapter 4, this may be considered as an analysis–synthesis filter bank.

Filter bank implementations are especially attractive as they can be adapted to the spectral and the temporal resolution of the human ear. Also, there are many applications, such as hearing aids, where the computational complexity is of utmost importance. Filter banks with non-uniform spectral resolution allow a reduction in the number of channels without sacrificing the quality of the enhanced signal. In principle, the filter bank can be realized by discrete filters or by polyphase filter banks with uniform or non-uniform spectral resolution [Doblinger 1991], [Doblinger, Zeitlhofer 1996], [Kappelan et al. 1996], [Gülzow, Engelsberg 1998], [Gülzow et al. 2003], [Vary 2005]. Chapter 4 dicusses typical realizations in detail.

11.4 Optimal Non-linear Estimators

We return to the issue of how to estimate the clean speech DFT coefficients given the noisy coefficients. In contrast to earlier sections, we now aim at developing optimal *non-linear* estimators. Unlike the optimal linear estimators, these estimators require knowledge of the probability density functions of the speech and noise DFT coefficients. Similar to the exposition in Section 11.3, we do not assume stationary signals but develop estimators for a single, quasi-stationary frame of speech. To enhance the readability of the derivations, we will drop the frame index λ and denote all quantities as functions of the frequency bin index μ only. Thus, for the signal frame under consideration, the DFT coefficients of the clean speech and the noisy signal will be denoted by

$$\begin{aligned} S_\mu &= A_\mu e^{j\alpha_\mu} \quad \text{and} \\ Y_\mu &= R_\mu e^{j\theta_\mu} , \end{aligned} \tag{11.49}$$

respectively. Since we assume zero mean and mutually independent real and imaginary parts of equal variance, the power of the μ-th spectral component may be written as

$$\begin{aligned} \sigma_{S,\mu}^2 &= \mathrm{E}\left\{\mathrm{Re}\{S_\mu\}^2\right\} + \mathrm{E}\left\{\mathrm{Im}\{S_\mu\}^2\right\} = \mathrm{E}\left\{|S_\mu|^2\right\} = \mathrm{E}\left\{A_\mu^2\right\} , \\ \sigma_{N,\mu}^2 &= \mathrm{E}\left\{\mathrm{Re}\{N_\mu\}^2\right\} + \mathrm{E}\left\{\mathrm{Im}\{N_\mu\}^2\right\} = \mathrm{E}\left\{|N_\mu|^2\right\} , \end{aligned} \tag{11.50}$$

and

$$\begin{aligned} \mathrm{E}\left\{\mathrm{Re}\{Y_\mu\}^2\right\} + \mathrm{E}\left\{\mathrm{Im}\{Y_\mu\}^2\right\} &= \mathrm{E}\left\{|Y_\mu|^2\right\} = \mathrm{E}\left\{R_\mu^2\right\} \\ &= \sigma_{S,\mu}^2 + \sigma_{N,\mu}^2 . \end{aligned} \tag{11.51}$$

Most of the derivations employ the complex Gaussian model for the PDF of the DFT coefficients of speech and noise signals, as outlined in Section 5.11. This model is valid when the DFT length is significantly larger than the span of correlation in the signal.

The resulting estimators are frequently written as a multiplication of a spectral gain function and the DFT coefficients of the noisy signal, and we will follow this convention to a certain degree. However, this notation is not always appropriate, since the estimators are highly non-linear. In most cases the spectral gain function itself is a function of the DFT coefficients of the noisy signal.

Although we will primarily discuss estimators which are optimal with respect to the MMSE, we will begin with maximum likelihood (ML) estimation. We show that for the Gaussian model the maximum likelihood estimate is closely related to the spectral subtraction technique.

11.4.1 Maximum Likelihood Estimation

The *maximum likelihood* (ML) estimate of the speech power $\sigma_{S,\mu}^2$ in the μ-th DFT bin maximizes the joint probability density of the observed spectral coefficient Y_μ with respect to $\sigma_{S,\mu}^2$. The joint probability function conditioned on the unknown parameter $\sigma_{S,\mu}^2$ is given by (5.113), i.e.,

$$p(\mathrm{Re}\{Y_\mu\}, \mathrm{Im}\{Y_\mu\} \mid \sigma_{S,\mu}^2) = \frac{1}{\pi(\sigma_{S,\mu}^2 + \sigma_{N,\mu}^2)} \exp\left(-\frac{|Y_\mu|^2}{\sigma_{S,\mu}^2 + \sigma_{N,\mu}^2}\right). \quad (11.52)$$

If we set the first derivative of (11.52) with respect to $\sigma_{S,\mu}^2$ to zero, we obtain the estimated speech power [McAulay, Malpass 1980],

$$\widehat{\sigma}_{S,\mu}^2 = |Y_\mu|^2 - \sigma_{N,\mu}^2 = R_\mu^2 - \sigma_{N,\mu}^2 \,. \quad (11.53)$$

Since the second derivative is negative this is indeed a maximum. The ML estimator is an unbiased estimator of $\sigma_{S,\mu}^2$ since the expected value of $\widehat{\sigma}_{S,\mu}^2$ is $\mathrm{E}\left\{\widehat{\sigma}_{S,\mu}^2\right\} = \sigma_{Y,\mu}^2 - \sigma_{N,\mu}^2 = \sigma_{S,\mu}^2$. In a practical application of this estimator, we must ensure, however, that the estimate of the variance is always non-negative. Using the phase of the noisy input, an estimate of the spectral coefficient of clean speech is given by

$$\widehat{S}_\mu = \sqrt{1 - \frac{\sigma_{N,\mu}^2}{|Y_\mu|^2}}|Y_\mu|e^{j\theta_\mu} = \sqrt{1 - \frac{\sigma_{N,\mu}^2}{|Y_\mu|^2}}Y_\mu = G_{\mathrm{ML}} Y_\mu \quad (11.54)$$

where, again, we must make sure that the argument of the square root is non-negative. Note that although we have written the result as a product of the noisy coefficient and a gain function, the estimator is non-linear. The gain function depends on the DFT coefficient. $|Y_\mu|^2$ in the denominator may be smoothed over frequency or, if the signal is short-time stationary, over time. Using (11.50), equation (11.54) is recognized as an approximation to the power subtraction rule.

Similarly, (11.53) may be used in conjunction with the MMSE filter (11.32). Replacing the unkown speech power by its ML estimate, we obtain

$$\widehat{S}_\mu = \frac{|Y_\mu|^2 - \sigma_{N,\mu}^2}{\mathrm{E}\left\{|Y_\mu|^2\right\}} Y_\mu = G_{\mathrm{WML}} Y_\mu \quad (11.55)$$

where the gain function in (11.55) is, after smoothing the numerator, an approximation to (11.32).

An alternative solution to the ML signal estimation problem can be based on the conditional joint density (5.119) of noisy spectral coefficients, given the spectral amplitude A_μ [McAulay, Malpass 1980]. In this case, the clean speech spectral amplitude A_μ is the unknown parameter and the clean speech phase α_μ is assumed to be uniformly distributed between 0 and 2π. The ML estimate of the spectral amplitude is obtained from (5.119) by averaging over the phase. More specifically, we maximize

$$p(Y_\mu \mid A_\mu) = \int_0^{2\pi} p(Y_\mu \mid A_\mu, \alpha_\mu) p(\alpha_\mu)\, \mathrm{d}\alpha_\mu$$

$$= \frac{1}{\pi\sigma_{N,\mu}^2} \exp\left(-\frac{|Y_\mu|^2 + A_\mu^2}{\sigma_{N,\mu}^2}\right) \frac{1}{2\pi} \int_0^{2\pi} \exp\left(\frac{2A_\mu \mathrm{Re}\{Y_\mu e^{-j\alpha_\mu}\}}{\sigma_{N,\mu}^2}\right) \mathrm{d}\alpha_\mu$$

with respect to A_μ. The integral in the above equation is known as the modified Bessel function of the first kind. For $\frac{2A_\mu |Y_\mu|}{\sigma_{N,\mu}^2} \geq 3$ it can be approximated by

$$\frac{1}{2\pi} \int_0^{2\pi} \exp\left(\frac{2A_\mu \mathrm{Re}\{Y_\mu e^{-j\alpha_\mu}\}}{\sigma_{N,\mu}^2}\right) \mathrm{d}\alpha_\mu \approx \frac{1}{\sqrt{2\pi \frac{2A_\mu |Y_\mu|}{\sigma_{N,\mu}^2}}} \exp\left(\frac{2A_\mu |Y_\mu|}{\sigma_{N,\mu}^2}\right).$$

Differentiation of

$$p(Y_\mu | A_\mu) \approx \frac{1}{2\pi\sigma_{N,\mu}\sqrt{\pi A_\mu |Y_\mu|}} \exp\left(-\frac{(A_\mu - |Y_\mu|)^2}{\sigma_{N,\mu}^2}\right) \tag{11.56}$$

with respect to A_μ leads to the approximate ML estimate of the spectral magnitude

$$\widehat{A}_\mu = \frac{|Y_\mu|}{2}\left(1 \pm \sqrt{1 - \frac{\sigma_{N,\mu}^2}{R_\mu^2}}\right). \tag{11.57}$$

Retaining the phase of the noisy signal, we obtain for the complex enhanced coefficient

$$\widehat{S}_\mu = \left(0.5 + 0.5\sqrt{1 - \frac{\sigma_{N,\mu}^2}{R_\mu^2}}\right) Y_\mu = G_{\mathrm{MML}} Y_\mu \tag{11.58}$$

where, again, we have to make sure that the argument of the square root is non-negative. Note that this spectral gain function provides for a maximum of 6 dB of noise reduction.

11.4.2 Maximum *A Posteriori* Estimation

The maximum *a posteriori* estimator finds the clean speech coefficients which maximize $p(S_\mu \mid Y_\mu)$. It allows explict modeling of the *a priori* density of the speech coefficients and noise, since, using Bayes' theorem, we might as well maximize

$$p(S_\mu \mid Y_\mu)\, p(Y_\mu) = p(Y_\mu \mid S_\mu)\, p(S_\mu)\,. \tag{11.59}$$

For the Gaussian signal model, it is easily verified that the MAP estimator is identical to the Wiener filter (11.32).

More interesting solutions arise when we estimate the magnitude and phase, i.e., solve

$$\begin{aligned} p(A_\mu, \alpha_\mu \mid Y_\mu)\, p(Y_\mu) &= p(Y_\mu \mid A_\mu, \alpha_\mu)\, p(A_\mu, \alpha_\mu) \\ &= \frac{2A_\mu}{2\pi^2 \sigma_{N,\mu}^2 \sigma_{S,\mu}^2} \exp\left(-\frac{|Y_\mu - A_\mu e^{j\alpha_\mu}|^2}{\sigma_{N,\mu}^2}\right) \exp\left(-\frac{A_\mu^2}{\sigma_{S,\mu}^2}\right). \end{aligned}$$

The MAP estimate of the clean speech phase α_μ is simply the noisy phase θ_μ [Wolfe, Godsill 2001], whereas the MAP estimate of the amplitude A_μ yields

$$\widehat{A}_\mu = \frac{\eta_\mu + \sqrt{\eta_\mu^2 + 2(1+\eta_\mu)\frac{\eta_\mu}{\gamma_\mu}}}{2(1+\eta_\mu)} R_\mu\,. \tag{11.60}$$

MAP estimation procedures have also been developed in the context of supergaussian speech models [Lotter, Vary 2003], [Lotter, Vary 2005], [Dat et al. 2005].

11.4.3 MMSE Estimation

We will now turn to the MMSE estimation of the DFT coefficients under the Gaussian assumption. When the real and the imaginary parts of the DFT coefficients are statistically independent, the general non-linear solution is identical to the Wiener filter. In this case, the MMSE estimate of the complex coefficients separates into two independent estimators for the real and the imaginary parts, i.e.,

$$\begin{aligned} \widehat{S}_\mu &= \mathrm{E}\{S_\mu \mid Y_\mu\} \\ &= \int_{-\infty}^{\infty} \mathrm{Re}\{S_\mu\}\, p(\mathrm{Re}\{S_\mu\} \mid \mathrm{Re}\{Y_\mu\})\, \mathrm{dRe}\{S_\mu\} \\ &\quad + j \int_{-\infty}^{\infty} \mathrm{Im}\{S_\mu\}\, p(\mathrm{Im}\{S_\mu\} \mid \mathrm{Im}\{Y_\mu\})\, \mathrm{dIm}\{S_\mu\} \\ &= \mathrm{E}\{\mathrm{Re}\{S_\mu\} \mid \mathrm{Re}\{Y_\mu\}\} + j\,\mathrm{E}\{\mathrm{Im}\{S_\mu\} \mid \mathrm{Im}\{Y_\mu\}\}\,. \end{aligned} \tag{11.61}$$

Using the result of Section 5.12.3 with Gaussian densities for the noisy coefficients and for the speech coefficients as defined in (5.111),

$$p_{\mathrm{Re}\{Y_\mu\}}(u) = \frac{1}{\sqrt{\pi(\sigma_{S,\mu}^2 + \sigma_{N,\mu}^2)}} \exp\left(-\frac{u^2}{\sigma_{S,\mu}^2 + \sigma_{N,\mu}^2}\right), \tag{11.62}$$

the estimate of the real part evaluates to

$$\begin{aligned} \mathrm{E}\left\{\mathrm{Re}\{S_\mu\} \mid \mathrm{Re}\{Y_\mu\}\right\} &= \frac{1}{\sigma_{Y,\mu}^2 \sigma_{S,\mu}^2 \pi p\left(\mathrm{Re}\{Y_\mu\}\right)} \exp\left(-\frac{\mathrm{Re}\{Y_\mu\}^2}{\sigma_{N,\mu}^2}\right) \\ &\quad \cdot \int_{-\infty}^{\infty} u \exp\left(-\frac{\sigma_{S,\mu}^2 + \sigma_{N,\mu}^2}{\sigma_{S,\mu}^2 \sigma_{N,\mu}^2} u^2 + \frac{2\mathrm{Re}\{Y_\mu\}}{\sigma_{N,\mu}^2} u\right) du \\ &= \frac{\sigma_{S,\mu}^2}{\sigma_{S,\mu}^2 + \sigma_{N,\mu}^2} \mathrm{Re}\{Y_\mu\}. \end{aligned} \tag{11.63}$$

For the imaginary part we obtain with

$$p_{\mathrm{Im}\{Y_\mu\}}(u) = \frac{1}{\sqrt{\pi \sigma_{Y,\mu}^2}} \exp\left(-\frac{u^2}{\sigma_{Y,\mu}^2}\right) \tag{11.64}$$

the result

$$\begin{aligned} \mathrm{E}\left\{\mathrm{Im}\{S_\mu\} \mid \mathrm{Im}\{Y_\mu\}\right\} &= \frac{1}{\sigma_{Y,\mu}^2 \sigma_{S,\mu}^2 \pi p\left(\mathrm{Im}\{Y_\mu\}\right)} \exp\left(-\frac{\mathrm{Im}\{Y_\mu\}^2}{\sigma_{N,\mu}^2}\right) \\ &\quad \cdot \int_{-\infty}^{\infty} u \exp\left(-\frac{\sigma_{S,\mu}^2 + \sigma_{N,\mu}^2}{\sigma_{S,\mu}^2 \sigma_{N,\mu}^2} u^2 + \frac{2\mathrm{Im}\{Y_\mu\}}{\sigma_{N,\mu}^2} u\right) du \\ &= \frac{\sigma_{S,\mu}^2}{\sigma_{S,\mu}^2 + \sigma_{N,\mu}^2} \mathrm{Im}\{Y_\mu\} \end{aligned}$$

which, combined with the solution for the real part, yields

$$\mathrm{E}\left\{S_\mu \mid Y_\mu\right\} = \frac{\sigma_{S,\mu}^2}{\sigma_{S,\mu}^2 + \sigma_{N,\mu}^2} Y_\mu . \tag{11.65}$$

For the Gaussian signal model the optimal filter is therefore a zero-phase filter. The phase of the noisy signal is (except for a possible overall delay of the signal due to spectral analysis) not modified. For supergaussian speech models, estimators for the real and the imaginary parts can be developed as well [Martin 2002], [Martin, Breithaupt 2003], [Martin 2005b], [Martin 2005c]. In contrast to the estimators based on Gaussian speech models, they also lead to a modification of the short-time phase.

11.4.4 MMSE Estimation of Functions of the Spectral Magnitude

In speech enhancement, the estimation of the magnitude of the short-time Fourier coefficients of clean speech is much easier to achieve than the estimation of the short-time phase. When the short-time spectral amplitudes can be accurately estimated, additional (and independently derived) information on the phase is of little use [Wang, Lim 1982]. An important class of MMSE estimators [Porter, Boll 1984], [Ephraim, Malah 1984] therefore estimates the spectral magnitudes and uses the short-time phase of the noisy input signal for reconstruction. It turns out that these estimators deliver a fairly natural sounding residual noise without the unnatural fluctuations which are known as "musical noise".

The MMSE amplitude estimator minimizes the quadratic error in the spectral amplitudes $\mathrm{E}\left\{(A_\mu - \widehat{A}_\mu)^2\right\}$. As before, the conditional expectation

$$\widehat{A}_\mu = \mathrm{E}\{A_\mu | Y_\mu\} = \int_0^\infty A_\mu \, p(A_\mu | Y_\mu) \, \mathrm{d}A_\mu$$

$$= \int_0^\infty \int_0^{2\pi} A_\mu \, p(A_\mu, \alpha_\mu | Y_\mu) \, \mathrm{d}\alpha_\mu \, \mathrm{d}A_\mu \tag{11.66}$$

is the solution to this problem. With the help of Bayes' theorem

$$p(A_\mu, \alpha_\mu | Y_\mu) = \frac{p(Y_\mu \mid A_\mu, \alpha_\mu) \, p(A_\mu, \alpha_\mu)}{\int_0^\infty \int_0^{2\pi} p(Y_\mu \mid A_\mu, \alpha_\mu) \, p(A_\mu, \alpha_\mu) \, \mathrm{d}\alpha_\mu \, \mathrm{dA}_\mu} \tag{11.67}$$

the optimal estimate can be written as

$$\widehat{A}_\mu = \frac{\int_0^\infty \int_0^{2\pi} A_\mu \, p(Y_\mu | A_\mu, \alpha_\mu) \, p(A_\mu, \alpha_\mu) \, \mathrm{d}\alpha_\mu \, \mathrm{d}A_\mu}{\int_0^\infty \int_0^{2\pi} p(Y_\mu | A_\mu, \alpha_\mu) \, p(A_\mu, \alpha_\mu) \, \mathrm{d}\alpha_\mu \, \mathrm{d}A_\mu} \, . \tag{11.68}$$

This result can obviously be generalized to arbitrary deterministic and invertible functions of A_μ [Porter, Boll 1984]. We denote this function by $c\{\cdot\}$ and its inverse by $c^{-1}\{\cdot\}$. In the general case, the optimal estimator is then given by

$$\widehat{A}_\mu = c^{-1} \left\{ \frac{\int_0^\infty \int_0^{2\pi} c\{A_\mu\} \, p(Y_\mu | A_\mu, \alpha_\mu) \, p(A_\mu, \alpha_\mu) \, \mathrm{d}\alpha_\mu \, \mathrm{d}A_\mu}{\int_0^\infty \int_0^{2\pi} p(Y_\mu | A_\mu, \alpha_\mu) \, p(A_\mu, \alpha_\mu) \, \mathrm{d}\alpha_\mu \, \mathrm{d}A_\mu} \right\} . \tag{11.69}$$

Assuming a Gaussian noise model, the conditional probability (5.119) can be substituted into (11.68). Furthermore, with the assumptions of a uniform distribution for the clean speech phase and the independence of the clean speech magnitude and phase, the phase can be integrated to yield

$$\widehat{A}_\mu = c^{-1}\left\{\frac{\int_0^\infty c\{A_\mu\}\, p(A_\mu) \int_0^{2\pi} \exp\left(-\frac{|Y_\mu - A_\mu e^{j\alpha_\mu}|^2}{\sigma_{N,\mu}^2}\right) \mathrm{d}\alpha_\mu \,\mathrm{d}A_\mu}{\int_0^\infty p(A_\mu) \int_0^{2\pi} \exp\left(-\frac{|Y_\mu - A_\mu e^{j\alpha_\mu}|^2}{\sigma_{N,\mu}^2}\right) \mathrm{d}\alpha_\mu \,\mathrm{d}A_\mu}\right\} \tag{11.70}$$

$$= c^{-1}\left\{\frac{\int_0^\infty c[A_\mu]\, p(A_\mu) \exp\left(-\frac{(R_\mu^2 + A_\mu^2)}{\sigma_{N,\mu}^2}\right) I_0\left(\frac{2R_\mu A_\mu}{\sigma_{N,\mu}^2}\right) \mathrm{d}A_\mu}{\int_0^\infty p(A_\mu) \exp\left(-\frac{(R_\mu^2 + A_\mu^2)}{\sigma_{N,\mu}^2}\right) I_0\left(\frac{2R_\mu A_\mu}{\sigma_{N,\mu}^2}\right) \mathrm{d}A_\mu}\right\}$$

where we used

$$\frac{1}{2\pi}\int_0^{2\pi} \exp\left(-\frac{|Y_\mu - A_\mu e^{ju}|^2}{\sigma_{N,\mu}^2}\right) \mathrm{d}u = \exp\left(-\frac{R_\mu^2 + A_\mu^2}{\sigma_{N,\mu}^2}\right) I_0\left(\frac{2R_\mu A_\mu}{\sigma_{N,\mu}^2}\right).$$

By averaging over a large speech database, the above estimator can be evaluated numerically and tabulated as a function of the *a posteriori* SNR γ_μ and the *a priori* SNR η_μ [Porter, Boll 1984]. For the Gaussian speech model,

$$p(A_\mu, \alpha_\mu) = \frac{A_\mu}{\pi\sigma_{S,\mu}^2} \exp\left(-\frac{A_\mu^2}{\sigma_{S,\mu}^2}\right), \tag{11.71}$$

it can be evaluated analytically. While the numeric approach allows us to take the actual distribution of the clean speech magnitudes into account, it may also lead to large lookup tables and is not very practical without further simplifications. The parametric (Gaussian) approach leads to parametric closed form solutions. In what follows, three cases will be considered.

11.4.4.1 MMSE Magnitude Estimation

In the case of *MMSE amplitude* estimation (also known as the *MMSE short-time spectral amplitude* or *MMSE-STSA* estimator) we have $c\{A_\mu\} = A_\mu$. The optimal estimator can be expressed in terms of the complete Γ-function and the confluent hypergeometric function $F_1(\cdot,\cdot,\cdot)$ [Ephraim, Malah 1984], [Gradshteyn, Ryzhik 1994]. After some (tedious) computations, we obtain from (11.70)

$$\widehat{A}_\mu = \frac{R_\mu}{\gamma_\mu}\sqrt{v_\mu}\,\Gamma(1.5)\,F_1(-0.5,1,-v_\mu) \tag{11.72}$$

where v_μ is defined as

$$v_\mu = \frac{\eta_\mu}{1+\eta_\mu}\gamma_\mu\,. \tag{11.73}$$

Therefore, using the phase of the noisy coefficient, we have

$$S_\mu = \frac{\sqrt{v_\mu}}{\gamma_\mu}\,\Gamma(1.5)\,F_1(-0.5,1,-v_\mu)Y_\mu = G_{\mathrm{STSA}}Y_\mu\,. \tag{11.74}$$

11.4.4.2 MMSE Magnitude Squared Estimation

As a second example we consider $c\{A_\mu\} = A_\mu^2$. The *MMSE magnitude squared* estimator then minimizes $\mathrm{E}\left\{(A_\mu^2 - \widehat{A}_\mu^2)^2\right\}$. Again, the optimal estimator is the conditional expectation which factors in the real and the imaginary parts,

$$\begin{aligned}
\widehat{A}_\mu^2 = \mathrm{E}\left\{A_\mu^2|Y_\mu\right\} &= \int_0^\infty\int_0^{2\pi} A_\mu^2\,p(A_\mu,\alpha_\mu|Y_\mu)\,\mathrm{d}\alpha_\mu\,\mathrm{d}A_\mu \\
&= \int_{-\infty}^{\infty}\int_{-\infty}^{\infty} \left(\mathrm{Re}\left\{S_\mu\right\}^2 + \mathrm{Im}\left\{S_\mu\right\}^2\right) \\
&\qquad \cdot\, p\left(\mathrm{Re}\left\{S_\mu\right\},\mathrm{Im}\left\{S_\mu\right\} \mid \mathrm{Re}\left\{Y_\mu\right\},\mathrm{Im}\left\{Y_\mu\right\}\right)\,\mathrm{dRe}\left\{S_\mu\right\}\,\mathrm{dIm}\left\{S_\mu\right\} \\
&= \int_{-\infty}^{\infty} \mathrm{Re}\left\{S_\mu\right\}^2 p\left(\mathrm{Re}\left\{S_\mu\right\} \mid \mathrm{Re}\left\{Y_\mu\right\}\right)\,\mathrm{dRe}\left\{S_\mu\right\} \\
&\quad + \int_{-\infty}^{\infty} \mathrm{Im}\left\{S_\mu\right\}^2 p\left(\mathrm{Im}\left\{S_\mu\right\} \mid \mathrm{Im}\left\{Y_\mu\right\}\right)\,\mathrm{dIm}\left\{S_\mu\right\}\,.
\end{aligned} \tag{11.75}$$

In this case, the estimator is much easier to compute. It is related to the Wiener filter [Accardi, Cox 1999],

$$\widehat{A}_\mu^2 = \left(\frac{\sigma_{S,\mu}^2}{\sigma_{S,\mu}^2+\sigma_{N,\mu}^2}\right)^2 |Y_\mu|^2 + \frac{\sigma_{S,\mu}^2\sigma_{N,\mu}^2}{\sigma_{S,\mu}^2+\sigma_{N,\mu}^2}\,. \tag{11.76}$$

For supergaussian speech models the corresponding estimator is developed in [Breithaupt, Martin 2003].

11.4.4.3 MMSE Log Amplitude Estimation

Small speech signal amplitudes are very important for speech intelligibility. It is therefore sensible to use an error measure which places more emphasis on small signal amplitudes, e.g., a compressive type of function $c\{A_\mu\} = \log(A_\mu)$.

The *MMSE log spectral amplitude* (MMSE-LSA) estimator minimizes the mean square error of the logarithmically weighted amplitudes $\mathrm{E}\left\{(\log(A_\mu) - \log(\widehat{A}_\mu))^2\right\}$ and thus improves the estimation of small amplitudes.

The MMSE-LSA estimate is given by [Ephraim, Malah 1985]

$$\widehat{A}_\mu = \exp\left(\mathrm{E}\left\{\ln(A_\mu)|Y_\mu\right\}\right) , \tag{11.77}$$

provided that DFT bins are uncorrelated. The solution can be expressed as

$$\widehat{A}_\mu = \frac{\xi_\mu}{1+\xi_\mu} \exp\left(\frac{1}{2}\int_{v_\mu}^{\infty} \frac{\exp\{-t\}}{t}\,\mathrm{d}t\right) |Y_\mu| \tag{11.78}$$

where for a practical implementation, the exponential integral function in (11.78) can be tabulated as a function of v_μ. Therefore,

$$S_\mu = \frac{\xi_\mu}{1+\xi_\mu} \exp\left(\frac{1}{2}\int_{v_\mu}^{\infty} \frac{\exp\{-t\}}{t}\,\mathrm{d}t\right) Y_\mu = G_{\mathrm{LSA}} Y_\mu \,. \tag{11.79}$$

11.5 Joint Optimum Detection and Estimation of Speech

The optimal estimators of the previous sections implicitly assume that speech is present in the noisy signal. This is, of course, not always the case and the estimator for clean speech conditioned on the presence of speech is not optimal when speech is absent. Also, to improve the subjective listening quality, it can be advantageous to explicitly distinguish the two cases "speech present" and "speech absent". For example, the listening quality can be improved when the speech enhancement algorithm applies a constant, frequency-independent attenuation $G_{\min}$ to the noisy signal during speech pauses [Yang 1993], [Malah et al. 1999]. The MMSE estimator can be extended to account for speech presence or absence. Below, the resulting joint optimum detection and estimation problem [Middleton, Esposito 1968], as outlined in Section 5.12.4, is solved in the context of amplitude estimation.

To this end, we introduce the two hypotheses of "speech is present in DFT bin μ" and "speech is absent in DFT bin μ" and denote these two hypotheses by $H_\mu^{(1)}$ and $H_\mu^{(0)}$, respectively. All statistical quantities will now be conditioned on these hypotheses. The two hypotheses $H_\mu^{(1)}$ and $H_\mu^{(0)}$ can be stated in terms of the DFT coefficients as

$$H_\mu^{(0)} : Y_\mu = N_\mu \quad \text{and} \tag{11.80}$$

$$H_\mu^{(1)} : Y_\mu = S_\mu + N_\mu \,. \tag{11.81}$$

We denote the *a priori* probability that speech is present and the *a priori* probability that speech is absent in bin μ with $p_\mu = P(H_\mu^{(1)})$ and $q_\mu = P(H_\mu^{(0)}) = 1 - p_\mu$, respectively.

In the case of quadratic cost functions, the joint optimal detector and estimator is a linear combination of two terms as in (5.157). For example, the estimator for the spectral magnitudes may be decomposed into

$$\widehat{A}_\mu = G^{(1)}\,\mathrm{E}\left\{A_\mu \mid Y_\mu, H_\mu^{(1)}\right\} + G^{(0)}\,\mathrm{E}\left\{A_\mu | Y_\mu, H_\mu^{(0)}\right\}, \tag{11.82}$$

where $\mathrm{E}\left\{A_\mu | Y_\mu, H_\mu^{(1)}\right\}$ and $\mathrm{E}\left\{A_\mu | Y_\mu, H_\mu^{(0)}\right\}$ are the optimal estimators under the hypotheses $H_\mu^{(1)}$ and $H_\mu^{(0)}$, respectively, and the multiplicative weighting factors are given by

$$G^{(1)} = \frac{\Lambda_\mu(Y_\mu)}{1 + \Lambda_\mu(Y_\mu)} = \frac{p_\mu p(Y_\mu \mid H_\mu^{(1)})}{q_\mu p(Y_\mu \mid H_\mu^{(0)}) + p_\mu p(Y_\mu \mid H_\mu^{(1)})} \tag{11.83}$$

$$G^{(0)} = \frac{1}{1 + \Lambda_\mu(Y_\mu)} = \frac{q_\mu p(Y_\mu(\lambda) \mid H_\mu^{(0)}(\lambda))}{q_\mu p(Y_\mu \mid H_\mu^{(0)}) + p_\mu p(Y_\mu \mid H_\mu^{(1)})} \,. \tag{11.84}$$

These *soft-decision* weights are functions of the generalized likelihood ratio

$$\Lambda_\mu(Y_\mu) = \frac{p(Y_\mu \mid H_\mu^{(1)})\, p_\mu}{p(Y_\mu \mid H_\mu^{(0)})\, q_\mu} \,. \tag{11.85}$$

In the case of speech absence, we strive for a frequency-independent attenuation $\mathrm{E}\left\{A_\mu | Y_\mu, H_\mu^{(0)}\right\} = G_{\min} R_\mu$ of the noisy input magnitude. Thus, we may simplify the above estimator, i.e.,

$$\widehat{A}_\mu = \frac{\Lambda_\mu(Y_\mu)}{1+\Lambda_\mu(Y_\mu)} \mathrm{E}\left\{A_\mu \mid Y_\mu, H_\mu^{(1)}\right\} + \frac{1}{1+\Lambda_\mu(Y_\mu)} G_{\min} R_\mu \,. \tag{11.86}$$

Another solution may be obtained when the conditional density $p(Y_\mu \mid A_\mu, \alpha_\mu)$ is replaced by the conditional density of the magnitude $p(R_\mu \mid A_\mu, \alpha_\mu)$. A completely analogous derivation leads to the soft-decision estimation as proposed in [McAulay, Malpass 1980],

$$\widehat{A}_\mu = \mathrm{E}\left\{A_\mu \mid R_\mu, H_\mu^{(1)}\right\} P(H_\mu^{(1)} \mid R_\mu) + \mathrm{E}\left\{A_\mu \mid R_\mu, H_\mu^{(0)}\right\} P(H_\mu^{(0)} \mid R_\mu) \,.$$

In [McAulay, Malpass 1980], $\mathrm{E}\left\{A_\mu \mid R_\mu, H_\mu^{(1)}\right\}$ is replaced by the ML estimate (11.58) and $\mathrm{E}\left\{A_\mu \mid R_\mu, H_\mu^{(0)}\right\}$ is set to zero. In this case, the likelihood ratio Λ_k is based on the conditional PDF (5.122)

$$\Lambda_\mu(R_\mu) = \frac{p(R_\mu \mid H_\mu^{(1)})\, p_\mu}{p(R_\mu \mid H_\mu^{(0)})\, q_\mu} \,. \tag{11.87}$$

We note that the soft-decision approach can be used in conjunction with other estimators, e.g., the MMSE-LSA estimator, and other cost functions as well. In general, however, the factoring property of (11.82) will be lost.

As a final example, we consider the estimation of $\ln(A_\mu)$ and obtain

$$\widehat{A}_\mu = \exp\left(\mathrm{E}\left\{\ln(A_\mu) \mid Y_\mu, H_\mu^{(1)}\right\}\right)^{\frac{\Lambda_\mu(Y_\mu)}{1+\Lambda_\mu(Y_\mu)}} (G_{\min} R_\mu)^{\frac{1}{1+\Lambda_\mu(Y_\mu)}} \tag{11.88}$$

where we make use of the MMSE-LSA estimator as outlined in Section 11.4.4.3 and the fixed attenuation $G_{\min}$ during speech pause. This estimator [Cohen, Berdugo 2001] results in larger improvements in the segmental SNR than the estimator in (11.86) but also in more speech distortions. This less desirable behavior (see [Ephraim, Malah 1985]) is improved when the soft-decision modifier is multiplicatively combined with the MMSE-LSA estimator [Malah et al. 1999],

$$\widehat{A}_\mu = \frac{\Lambda_\mu(Y_\mu)}{1+\Lambda_\mu(Y_\mu)} \exp\left(\mathrm{E}\left\{\ln(A_\mu) \mid Y_\mu, H_\mu^{(1)}\right\}\right) . \tag{11.89}$$

This estimator achieves less noise reduction than the estimator in (11.88).

11.6 Computation of Likelihood Ratios

For the Gaussian model, the PDFs of Y_μ conditioned on the hypotheses $H_\mu^{(0)}$ or $H_\mu^{(1)}$ follow from (5.113),

$$p(Y_\mu \mid H_\mu^{(0)}) = \frac{1}{\pi\sigma_{N,\mu}^2} \exp\left(-\frac{R_\mu^2}{\sigma_{N,\mu}^2}\right) \tag{11.90}$$

and

$$p(Y_\mu \mid H_\mu^{(1)}) = \frac{1}{\pi\left(\sigma_{N,\mu}^2 + \mathrm{E}\left\{A_\mu^2 \mid H_\mu^{(1)}\right\}\right)} \exp\left(-\frac{R_\mu^2}{\sigma_{N,\mu}^2 + \mathrm{E}\left\{A_\mu^2 \mid H_\mu^{(1)}\right\}}\right).$$

The PDF of complex Gaussians depends only on the magnitude R_μ and not on the phase. The expectation of the speech power, $\mathrm{E}\left\{A_\mu^2 \mid H_\mu^{(1)}\right\}$, is now explicitly conditioned on the presence of speech and thus excludes speech pauses. The likelihood ratio is then given by

$$\begin{aligned}\Lambda_\mu &= \frac{1-q_\mu}{q_\mu} \frac{\sigma_{N,\mu}^2}{\pi\left(\sigma_{N,\mu}^2 + \mathrm{E}\left\{A_\mu^2 \mid H_\mu^{(1)}\right\}\right)} \\ &\quad \cdot \exp\left(-\frac{R_\mu^2}{\sigma_{N,\mu}^2 + \mathrm{E}\left\{A_\mu^2 \mid H_\mu^{(1)}\right\}} + \frac{R_\mu^2}{\sigma_{N,\mu}^2}\right) \\ &= \frac{1-q_\mu}{q_\mu} \frac{1}{1+\xi_\mu} \exp\left(\gamma_\mu \frac{\xi_\mu}{1+\xi_\mu}\right)\end{aligned} \tag{11.91}$$

where ξ_μ is the *a priori* SNR conditioned on the presence of speech

$$\xi_\mu = \frac{\mathrm{E}\left\{A_\mu^2 \mid H_\mu^{(1)}\right\}}{\sigma_{N,\mu}^2}. \tag{11.92}$$

According to [Ephraim, Malah 1984], ξ_μ can be expressed as a function of the unconditional expectation $\mathrm{E}\left\{A_\mu^2\right\} = \mathrm{E}\left\{|S_\mu|^2\right\}$

$$\frac{\mathrm{E}\left\{A_\mu^2 \mid H_\mu^{(1)}\right\}}{\sigma_{N,\mu}^2} = \frac{\mathrm{E}\left\{A_\mu^2\right\}}{\sigma_{N,\mu}^2} \frac{1}{1-q_\mu} = \eta_\mu \frac{1}{1-q_\mu} \tag{11.93}$$

with

$$\eta_\mu = \frac{\mathrm{E}\left\{A_\mu^2\right\}}{\sigma_{N,\mu}^2}. \tag{11.94}$$

Other solutions are discussed in [Cohen, Berdugo 2001].

11.7 Estimation of the *A Priori* Probability of Speech Presence

The computation of the likelihood ratio requires knowledge of the *a priori* probability p_μ of speech presence in each frequency bin μ. These probabilities do not just reflect the proportion of speech spurs to speech pauses. They should also take into account that, during voiced speech, most of the speech energy is concentrated in the frequency bins which correspond to the speech harmonics. The frequency bins in between the speech harmonics contain mostly noise. Consequently, there has been some debate as to which value would be most appropriate for p_μ. In the literature *a priori* probabilities for speech presence range between 0.5 and 0.8 [McAulay, Malpass 1980], [Ephraim, Malah 1984].

However, a fixed *a priori* probability p_μ can only be a compromise since the location of the speech harmonics varies with the fundamental frequency of the speaker. Tracking the probability of speech presence individually in all frequency bins should therefore result in an improved performance. This requires an estimation procedure for the *a priori* probabilities which we outline below [Soon et al. 1999], [Malah et al. 1999].

11.7.1 A Hard-Decision Estimator Based on Conditional Probabilities

A hard-decision approach can be developed using the likelihood ratio

$$\widetilde{\Lambda}_\mu(R_\mu) = \frac{p(R_\mu \mid H_\mu^{(1)})}{p(R_\mu \mid H_\mu^{(0)})} = \frac{\exp(-(R_\mu^2 + A_\mu^2)/\sigma_{N,\mu}^2)\, I_0(2R_\mu A_\mu/\sigma_{N,\mu}^2)}{\exp(-R_\mu^2/\sigma_{N,\mu}^2)}$$
$$= \exp(-\xi_\mu)\, I_0(2\sqrt{\xi_\mu \gamma_\mu}) \quad (11.95)$$

where we assumed equal *a priori* probabilities $p_\mu = q_\mu = 0.5$ and used the conditional densities in (5.119) and (5.114). We introduce an index function $\mathfrak{I}_\mu(\lambda)$ which denotes the result of the test for the λ signal frame

$$\mathfrak{I}_\mu(\lambda) = \begin{cases} 1 & \widetilde{\Lambda}_\mu(R_\mu(\lambda)) > 1 \\ 0 & \widetilde{\Lambda}_\mu(R_\mu(\lambda)) \leq 1\,. \end{cases} \quad (11.96)$$

An estimate $\widehat{p}_\mu(\lambda)$ for the probability of speech presence for the λ-th signal frame and the μ-th frequency bin can then be obtained by computing a recursive average of the index function

$$\widehat{p}_\mu(\lambda) = (1 - \beta_p)\, \widehat{p}_\mu(\lambda - 1) + \beta_p\, \mathfrak{I}_\mu(\lambda) \quad (11.97)$$

where β_p is typically set to 0.1.

11.7.2 Soft-Decision Estimation

The probability of speech presence can also be obtained from Bayes' theorem as

$$P(H_\mu^{(1)} \mid R_\mu) = \frac{p(R_\mu \mid H_\mu^{(1)})\, p_\mu}{p(R_\mu \mid H_\mu^{(1)})\, p_\mu + p(R_\mu \mid H_\mu^{(0)})\, q_\mu} = \frac{\Lambda_\mu(R_\mu)}{1+\Lambda_\mu(R_\mu)}$$

$$= \frac{p(R_\mu \mid H_\mu^{(1)})}{p(R_\mu \mid H_\mu^{(1)}) + p(R_\mu \mid H_\mu^{(0)})} \tag{11.98}$$

where the last identity again assumes equal *a priori* probabilities, $p_\mu = q_\mu = 0.5$. Substituting the conditional densities, we obtain

$$P(H_\mu^{(1)}|R_\mu) = \frac{\exp(-\xi_\mu)\, I_0(2\sqrt{\xi_\mu \gamma_\mu})}{1+\exp(-\xi_\mu)\, I_0(2\sqrt{\xi_\mu \gamma_\mu})} \tag{11.99}$$

which is recursively smoothed to yield an estimate of the probability of speech presence

$$\widehat{p}_\mu(\lambda) = (1-\beta_p)\, \widehat{p}_\mu(\lambda-1) + \beta_p\, P(H_\mu^{(1)} \mid R_\mu)\,. \tag{11.100}$$

A similar estimator [Malah et al. 1999] can be devised by substituting the conditional densities of the complex coefficients Y_μ into the likelihood ratio

$$P(H_\mu^{(1)}|Y_\mu) = \frac{\widehat{\Lambda}_\mu(Y_\mu)}{1+\widehat{\Lambda}_\mu(Y_\mu)} \tag{11.101}$$

with

$$\widetilde{\Lambda}_\mu(Y_\mu) = \frac{\exp(\gamma_\mu \xi_\mu/(1+\xi_\mu))}{1+\xi_\mu}\,. \tag{11.102}$$

11.7.3 Estimation Based on the *A Posteriori* SNR

The probability for speech presence can also be obtained from a test on the *a priori* SNR η_μ where the *a priori* SNR is compared to a preset threshold $\eta_{\min}$. Since the *a priori* SNR parameterizes the statistics of the *a posteriori* SNR in terms of the exponential density, a test can be devised which relies exclusively on the *a posteriori* SNR [Malah et al. 1999]. In this test we compare the *a posteriori* SNR to a threshold γ_q. When speech is present, the decisions are smoothed over time for each frequency bin using a first-order recursive system

$$\widehat{p}_\mu(\lambda) = \alpha_p\, \widehat{p}_\mu(\lambda-1) + (1-\alpha_p)\, I_\mu(\lambda) \tag{11.103}$$

where $I_\mu(\lambda)$ again denotes an index function with

$$I_\mu(\lambda) = \begin{cases} 1 & \gamma_\mu(\lambda) > \gamma_q \\ 0 & \gamma_\mu(\lambda) \leq \gamma_q \,. \end{cases} \tag{11.104}$$

During speech pauses, the probability p_μ is set to a fixed value $p_\mu \gtrapprox 0$. Good results are obtained using $\gamma_q = 0.8$ and $\alpha_q = 0.95$ [Malah et al. 1999].

11.8 VAD and Noise Estimation Techniques

All noise suppression methods discussed in the previous sections require knowledge of the PSD of the disturbing noise. Noise power estimation is therefore an important (and sometimes neglected) component of speech enhancement algorithms. The noise PSD is in general time varying and not known *a priori*. It must be estimated and updated during the execution of the noise reduction algorithm. When the noise is non-stationary, it is not sufficient to sample the noise in a speech pause prior to a speech phrase and to keep this estimate fixed during speech activity. On the contrary, we must frequently update the noise estimate in order to track the noise PSD with sufficient accuracy. Unlike methods built upon microphone arrays, where averaging over the various microphone signals is possible, single microphone algorithms can obtain the noise PSD only by smoothing the noisy signal over time or frequency. However, since the noise is non-stationary, we cannot smooth over arbitrarily large amounts of noisy data. We must balance the error variance of the noise PSD estimate and the ability to track changes in the noise PSD.

Among the many approaches to noise PSD estimation, the most prominent are certainly based on voice activity detectors (VAD). Voice activity detection works well for moderate to high SNRs. Besides, for noise estimation, VADs are also used for controlling *discontinuous transmission* (DTX) in mobile voice communication systems [Freeman et al. 1989], [Srinivasan, Gersho 1993], [3GPP TS 126.094, V4.0.0 2001] and for detecting speech spurs in speech recognition applications. As we will see, a single algorithm will not be perfect for all applications. The VAD algorithm needs to be optimized for the specific task.

In high levels of possibly non-stationary noise, however, most VADs do not perform well. Furthermore, for noise reduction applications, we need an estimate of the noise floor rather than binary voice activity decisions. Methods based on "soft" estimation criteria are therefore preferable since they provide for an update of the noise power during speech activity. Besides VAD-based methods, we therefore also consider techniques which are based on energy histograms, on the probability of speech presence and soft-decision updates, and on minimum power tracking, also known as *minimum statistics* [Martin 1994], [Martin 2001a].

11.8.1 Voice Activity Detection

Over the years, many algorithms for voice activity detection have been proposed, e.g., [McAulay, Malpass 1980], [Van Compernolle 1989]. A VAD can be described by means of a finite state machine with at least two states, "speech is present" and "speech is absent". Sometimes, additional states are used to cope with transitions between these two basic states.

When speech is absent, the noise PSD can be estimated using a first-order recursive system with $0 < \alpha < 1$,

$$\widehat{\sigma}^2_{Y,\mu}(\lambda) = \alpha\, \widehat{\sigma}^2_{Y,\mu}(\lambda - 1) + (1 - \alpha)|Y_\mu(\lambda)|^2 \,. \tag{11.105}$$

The key to the successful design of a VAD is to use features of the speech signal which either are – at least in principle – independent of the noise power level or can be normalized on the noise power. Much ingenuity is required to find the "right" set of features. Once these features have been selected, the problem can be treated with classical detection theory.

A comparison of standard VAD algorithms is presented, for example, in [Beritelli et al. 2001]. A simple approach is to use an estimated SNR [Martin 1993] and to compare the estimated SNR to one or several fixed thresholds. This might work well for high SNR conditions, but for low SNR conditions we will encounter a substantial amount of erroneous detections. Typical VAD implementations therefore use several features. Some of the more common methods are discussed below.

11.8.1.1 Detectors Based on the Subband SNR

An SNR-based VAD may be made more robust by computing the instantaneous SNR in frequency subbands and by averaging the SNR over all of these bands. The spectral analysis may be realized as a filter bank with uniform or non-uniform resolution or by a sliding window DFT. We will now examine this concept in greater detail and consider a detection algorithm based on the *a posteriori* SNR of DFT coefficients. To develop the detection algorithm, we introduce two hypotheses

$$H^{(0)}(\lambda): \quad \text{speech is absent in signal frame } \lambda \text{, and}$$

$$H^{(1)}(\lambda): \quad \text{speech is present in signal frame } \lambda \,.$$

Assuming Gaussian PDFs of equal variance for the real and the imaginary parts of the complex M-point DFT coefficients and mutual statistical independence of all DFT bins, we may write the joint PDF for a vector $\mathbf{Y}(\lambda)$ of $M/2-1$ complex noisy DFT coefficients when speech is present as

$$p_{\mathbf{Y}(\lambda)|H^{(1)}(\lambda)}\left(Y_1(\lambda),\ldots,Y_{M/2-1}(\lambda)\mid H^{(1)}(\lambda)\right) = \prod_{\mu=1}^{M/2-1}\frac{1}{\pi\left(\sigma_{N,\mu}^2+\sigma_{S,\mu}^2\right)}\exp\left(-\frac{|Y_\mu(\lambda)|^2}{\sigma_{N,\mu}^2+\sigma_{S,\mu}^2}\right). \tag{11.106}$$

The DC and Nyquist frequency bins are not included in the product. $\sigma_{S,\mu}^2$ and $\sigma_{N,\mu}^2$ denote the power of the speech and the noise coefficients in the μ-th frequency bin, respectively. When speech is absent, we have

$$p_{\mathbf{Y}(\lambda)|H^{(0)}(\lambda)}\left(Y_1(\lambda),\ldots,Y_{M/2-1}(\lambda)\mid H^{(0)}(\lambda)\right) = \prod_{\mu=1}^{M/2-1}\frac{1}{\pi\,\sigma_{N,\mu}^2}\exp\left(-\frac{|Y_\mu(\lambda)|^2}{\sigma_{N,\mu}^2}\right). \tag{11.107}$$

Speech can now be detected by comparing the log-likelihood ratio [Van Trees 1968], [Sohn et al. 1999]

$$\begin{aligned}\Lambda\left(\mathbf{Y}(\lambda)\right) &= \frac{1}{M/2-1}\log\left(\frac{p_{\mathbf{Y}(\lambda)|H^{(1)}(\lambda)}(Y_1(\lambda),\ldots,Y_{M/2-1}(\lambda)\mid H^{(1)}(\lambda))}{p_{\mathbf{Y}(\lambda)|H^{(0)}(\lambda)}(Y_1(\lambda),\ldots,Y_{M/2-1}(\lambda)\mid H^{(0)}(\lambda))}\right)\\ &= \frac{1}{M/2-1}\left[\sum_{\mu=1}^{M/2-1}\left(\frac{|Y_\mu(\lambda)|^2}{\sigma_{N,\mu}^2}-\frac{|Y_\mu(\lambda)|^2}{\sigma_{N,\mu}^2+\sigma_{S,\mu}^2}-\log\frac{\sigma_{N,\mu}^2+\sigma_{S,\mu}^2}{\sigma_{N,\mu}^2}\right)\right]\end{aligned}$$

to a threshold $\mathfrak{L}_{thr}$

$$\Lambda\left(\mathbf{Y}(\lambda)\right) \underset{H^{(1)}(\lambda)}{\overset{H^{(0)}(\lambda)}{\lessgtr}} \mathfrak{L}_{thr}\,. \tag{11.108}$$

The test can be further simplified if the unknown speech variances $\sigma_{S,\mu}^2$ are replaced by their ML estimates (11.53)

$$\widehat{\sigma}_{S,\mu}^2 = |Y_\mu(\lambda)|^2 - \sigma_{N,\mu}^2\,. \tag{11.109}$$

In this case we obtain

$$\Lambda\left(\mathbf{Y}(\lambda)\right) = \frac{1}{M/2-1}\left[\sum_{\mu=1}^{M/2-1}\left(\frac{|Y_\mu(\lambda)|^2}{\sigma_{N,\mu}^2} - \log\frac{|Y_\mu(\lambda)|^2}{\sigma_{N,\mu}^2} - 1\right)\right] \quad (11.110)$$

$$\Lambda\left(\mathbf{Y}(\lambda)\right) \underset{H^{(1)}(\lambda)}{\overset{H^{(0)}(\lambda)}{\lessgtr}} \mathfrak{L}_{thr}\ , \quad (11.111)$$

which is recognized as a discrete approximation of the Itakura–Saito distortion measure between the magnitude squared signal spectrum $|Y_\mu(\lambda)|^2$ and the noise power spectrum $\sigma_{N,\mu}^2$ [Markel, Gray 1976, chapter 6].

If we retain only the first term in the summation, the detection test simplifies to a test on the average *a posteriori* SNR (11.40). In fact, as observed by several authors [Häkkinen, Väänänen 1993], [Malah et al. 1999], the *a posteriori* SNR γ_μ can be used to build reliable VADs. Since γ_μ is normalized on the noise PSD, its expectation is equal to unity during speech pause. A VAD is therefore obtained by comparing the average *a posteriori* SNR $\overline{\gamma}(\lambda)$ to a fixed threshold γ_{thr},

$$\overline{\gamma}(\lambda) = \frac{1}{M/2-1}\sum_{\mu=1}^{M/2-1}\gamma_\mu(\lambda) \underset{H^{(1)}(\lambda)}{\overset{H^{(0)}(\lambda)}{\lessgtr}} \gamma_{thr}\,. \quad (11.112)$$

For the Gaussian model, the variance of $\gamma_\mu(\lambda)$ is equal to one during speech pause. If we assume that all frequency bins are mutually independent, the variance of the average *a posteriori* SNR is, during speech pause, given by

$$\mathrm{var}\{\overline{\gamma}\} = \frac{1}{M/2-1}\,. \quad (11.113)$$

As a consequence, the speech detection threshold can be set to $\gamma_{thr} = 1+a\sqrt{\mathrm{var}\{\overline{\gamma}\}}$ where a is in the range $2 \leq a \leq 4$. The examples in Fig. 11.11 show the typical performance of this detector for a high SNR and a low SNR signal. The noise in this experiment is computer-generated white Gaussian noise. As we can see, this detector works almost perfectly for high SNR conditions. When the SNR is low, speech activity is not always properly flagged. For low SNR conditions, a more advanced algorithm is required. We note that this approach may be varied in several ways. For example, [Yang 1993] employs a test

$$\frac{1}{M/2-1}\sum_{k=1}^{M/2-1}\max\left(\frac{R_\mu^2-\sigma_{N,\mu}^2}{R_\mu^2},0\right) \underset{H^{(1)}(\lambda)}{\overset{H^{(0)}(\lambda)}{\lessgtr}} \mathfrak{T}_{thr}\,. \quad (11.114)$$

Also, the input signal may be decomposed by means of a filter bank as outlined below.

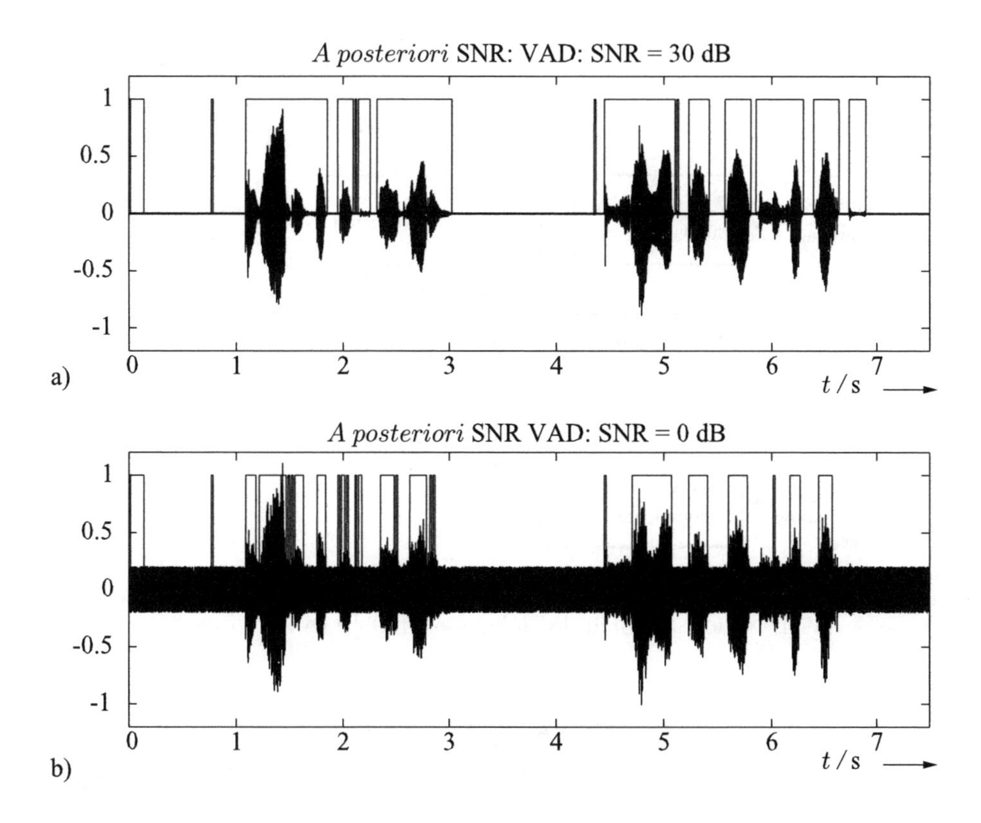

Figure 11.11: Speech signals and VAD decision for the algorithm based on the *a posteriori* SNR (11.112)
a) 30 dB segmental SNR
b) 0 dB segmental SNR

11.8.1.2 VAD (Option 1) of the Adaptive Multi-rate Coder

A state-of-the-art implementation of a VAD for speech transmission (DTX) purposes is shown in Fig. 11.12. This VAD is standardized as one of two different VAD algorithms for the ETSI *adaptive multi-rate* speech coder [3GPP TS 126.094, V4.0.0 2001], [Vähätalo, Johansson 1999]. It is typical in its use of several speech features, a background noise estimator, a detection unit, and a hangover generator. An intermediate VAD decision is based on the instantaneous SNR obtained from the outputs of a nine-channel filter bank as outlined in the previous section. The detection threshold, to which the sum of the channel SNRs is compared, is adapted to the level of the noise power. When the noise power is large, the threshold is lowered to allow for reliable detection of speech at the expense of "false alarms" when no speech is present. The purpose of the hangover generator is to prevent

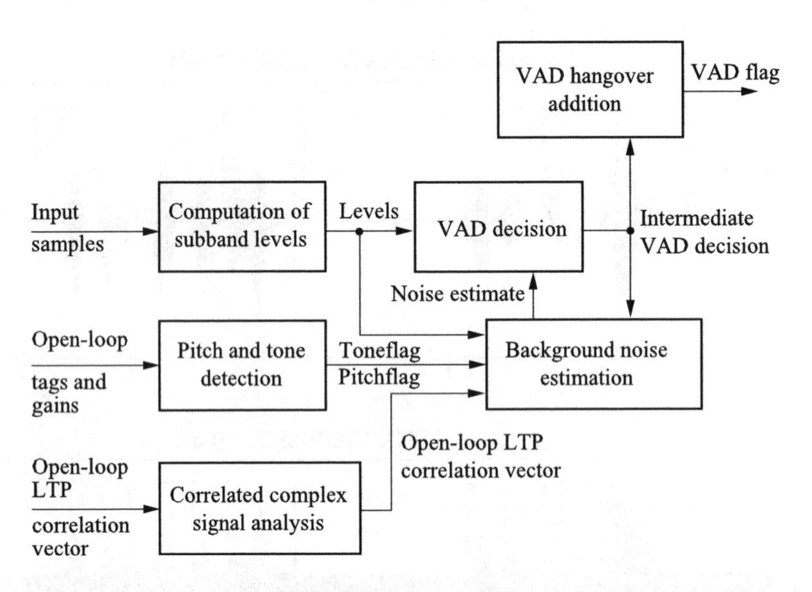

Figure 11.12: Block diagram of VAD algorithm of the adaptive multi-rate speech codec (VAD option 1 [Vähätalo, Johansson 1999]); © 1999 IEEE

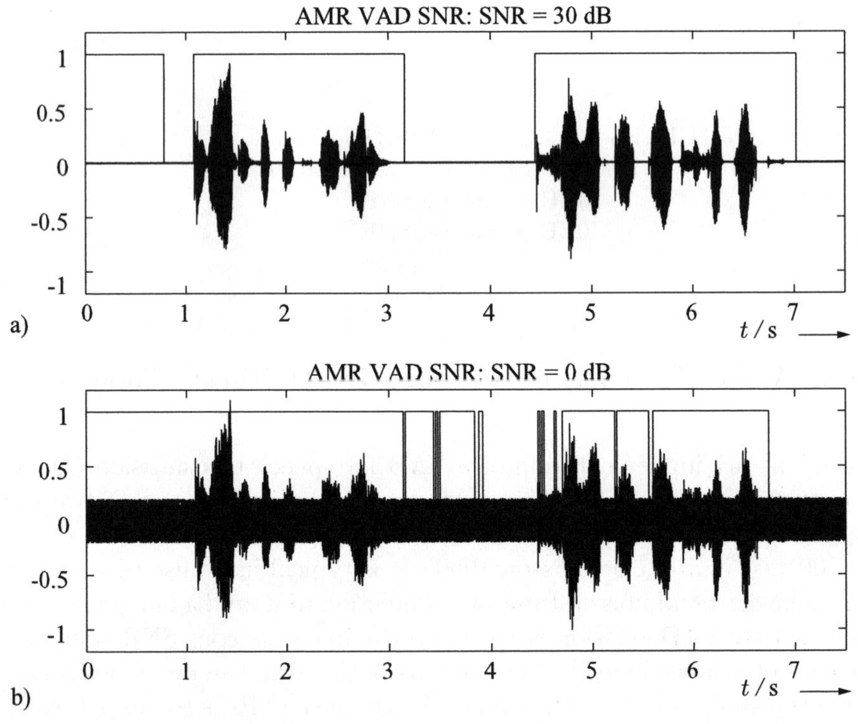

Figure 11.13: Speech signals and VAD decision for the AMR VAD 1
a) 30 dB segmental SNR
b) 0 dB segmental SNR

early returns from a "speech is present" state to a "speech is absent" state. This is especially useful for protecting low-energy speech sounds at the end of words. However, the hangover generator evaluates other criteria as well and modifies the final VAD decision accordingly [3GPP TS 126.094, V4.0.0 2001].

The performance of the AMR VAD is shown in Fig. 11.13 for the same two speech samples as in Fig. 11.11. For both SNR cases, the hangover can be clearly observed. For the low SNR condition, the speech onsets are not always properly detected. Because of the large amount of hangover the AMR VAD is not an ideal candidate for noise reduction applications. In a noise reduction application, we want to sample the noise also in the brief periods between words or syllables.

11.8.1.3 Noise Estimation Based on Frame Energy Histograms

The PDF of the short-term energy or the logarithmic short-term energy of clean speech has a distinct bimodal structure [McAulay, Malpass 1980], [Van Compernolle 1989]. One mode is located at zero and corresponds to speech pauses. The other represents speech. When noise is added to the speech signal, the mode which corresponds to speech pause reflects the energy of the noise.

This bimodal structure can be exploited for voice activity detection and noise power estimation. Using the histogram and the cumulative histogram of frame energies, an adaptive energy threshold parameter $\mathfrak{H}_{thr}$ can be determined [McAulay, Malpass 1980]. The energy of a given frame of noisy speech is compared to this threshold and the frames which are classified to contain only noise are averaged to obtain a noise power estimate. In a slightly modified implementation of this procedure [Van Compernolle 1989], Gaussian densities are fitted to the histogram of the logarithmic frame energies and the threshold is set in such a way that the *a posteriori* probabilities of speech absence and speech presence are equal. Thus, the probability of error is minimized [Melsa, Cohn 1978].

The histogram method may be used to estimate the noise power adaptively in frequency bands. By picking the most frequent histogram bin in each frequency band an estimate of the noise power is obtained [Hirsch, Ehrlicher 1995]. This works well when speech pauses are predominant. However, if the histogram is computed for $M/2+1$ bins of an M-point DFT and each frequency bin is resolved into H_M histogram bins, the memory requirements for this histogram-based noise power

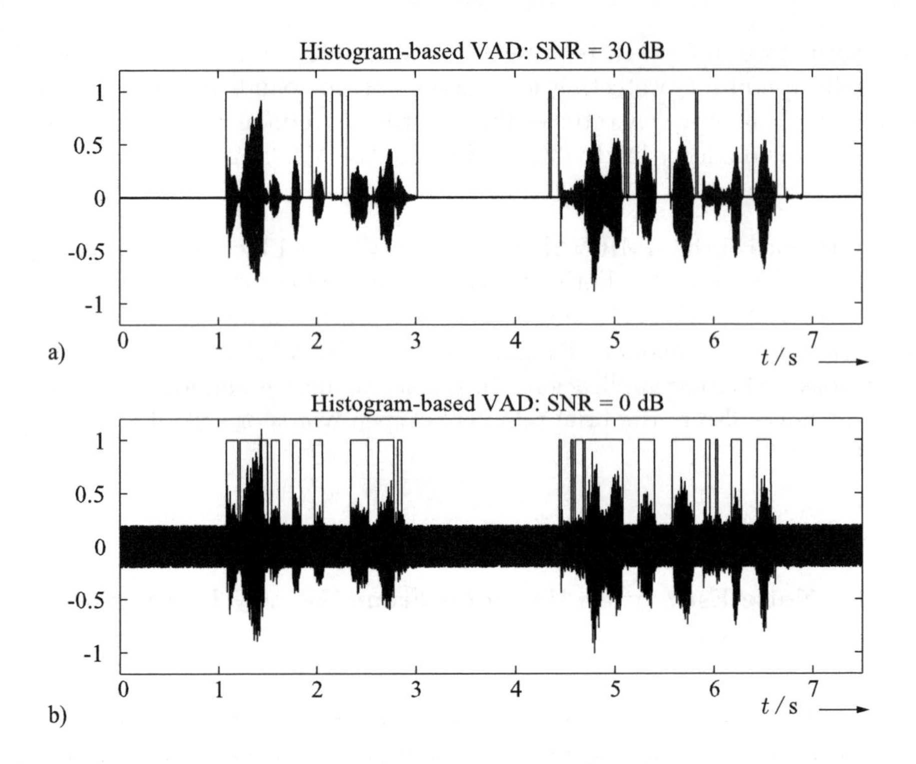

Figure 11.14: Speech signals and VAD decision for the histogram-based VAD
a) 30 dB segmental SNR
b) 0 dB segmental SNR

estimator are $(M/2+1)H_M$ memory words. The advantage of this procedure is that no explicit VAD is necessary. For the speech files used, for example, in Fig. 11.13 the resulting VAD decisions are shown in Fig. 11.14.

11.8.2 Noise Estimation Using a Soft-Decision Detector

Using a Gaussian model as outlined in Section 11.8.1.1, the probability of speech presence or absence can be determined. These probabilities can also be used to estimate and/or update the background noise power. A noise estimation algorithm based on these probabilities was proposed in [Sohn, Sung 1998] for the purpose of robust voice activity detection. This noise estimate can also be used directly for speech enhancement purposes without employing a VAD [Beaugeant 1999]. We will briefly outline the method.

Under the assumption that DFT coefficients are mutually independent, the MMSE estimate $\widehat{\sigma}^2_{N,\mu}(\lambda) = \mathrm{E}\left\{\sigma^2_{N,\mu}(\lambda) \mid Y_\mu(\lambda)\right\}$ of the background noise power can be written as

$$\begin{aligned}\widehat{\sigma}^2_{N,\mu}(\lambda) = &\ \mathrm{E}\left\{\sigma^2_{N,\mu}(\lambda) \mid Y_\mu(\lambda), H^{(0)}_\mu(\lambda)\right\} P\left(H^{(0)}_\mu(\lambda) \mid Y_\mu(\lambda)\right) \\ &+ \mathrm{E}\left\{\sigma^2_{N,\mu}(\lambda) \mid Y_\mu(\lambda), H^{(1)}_\mu(\lambda)\right\} P\left(H^{(1)}_\mu(\lambda) \mid Y_\mu(\lambda)\right) \end{aligned} \tag{11.115}$$

where we now define the hypotheses $H^{(0)}_\mu(\lambda)$ and $H^{(1)}_\mu(\lambda)$ for speech absence and presence, respectively, individually for each frequency bin. In analogy to Section 11.5, the probabilities $P(H^{(0)}_\mu(\lambda) \mid Y_\mu(\lambda))$ and $P(H^{(1)}_\mu(\lambda) \mid Y_\mu(\lambda))$ can be written as a function of the generalized likelihood ratio

$$\Lambda_\mu(\lambda) = \frac{p(Y_\mu(\lambda) \mid H^{(1)}_\mu(\lambda))P(H^{(1)}_\mu(\lambda))}{p(Y_\mu(\lambda) \mid H^{(0)}_\mu(\lambda))P(H^{(0)}_\mu(\lambda))}, \tag{11.116}$$

i.e.,

$$P(H^{(0)}_\mu(\lambda) \mid Y_\mu(\lambda)) = \frac{1}{1+\Lambda_\mu(\lambda)} \tag{11.117}$$

and

$$P(H^{(1)}_\mu(\lambda) \mid Y_\mu(\lambda)) = \frac{\Lambda_\mu(\lambda)}{1+\Lambda_\mu(\lambda)}\,. \tag{11.118}$$

The evaluation of the expectations in (11.115) is in general difficult. Therefore, simplified estimators are used. During speech activity the noise power estimate is not updated and the estimate is replaced by the estimate of the previous frame

$$\mathrm{E}\left\{\sigma^2_{N,\mu}(\lambda) \mid H^{(1)}_\mu(\lambda)\right\} \approx \widehat{\sigma}^2_{N,\mu}(\lambda-1)\,. \tag{11.119}$$

During speech pause the squared magnitude $|Y_\mu(k)|^2$ of the current frame is used as an estimate of the noise power, i.e., $\mathrm{E}\left\{\sigma^2_{N,\mu}(k) \mid H^{(0)}_\mu(\lambda)\right\} \approx |Y_\mu(k)|^2$. The estimator based on the probabilities of speech presence and absence can therefore be written as [Sohn, Sung 1998]

$$\widehat{\sigma}^2_{N,\mu}(\lambda) = \frac{1}{1+\Lambda_\mu(\lambda)}\,|Y_\mu(\lambda)|^2 + \frac{\Lambda_\mu(\lambda)}{1+\Lambda_\mu(\lambda)}\,\widehat{\sigma}^2_{N,\mu}(\lambda-1) \tag{11.120}$$

which is recognized as a first-order recursive (smoothing) filter with a time-varying and frequency-dependent smoothing parameter. The likelihood ratio can be estimated as outlined in Section 11.5 or be replaced by a frequency-independent average.

11.8.3 Noise Power Estimation Based on Minimum Statistics

In contrast to noise estimation based on voice activity detection, the *minimum statistics* algorithm does not use any explicit threshold to distinguish between speech activity and speech pause and is therefore more closely related to soft-decision methods than to traditional voice activity detection. Similar to soft-decision methods, it can also update the estimated noise PSD during speech activity. Speech enhancement based on minimum statistics was proposed in [Martin 1994] and improved in [Martin 2001a]. In [Meyer et al. 1997] it was shown that the minimum statistics algorithm [Martin 1994] performs relatively well in non-stationary noise. Our discussion of the *minimum statistics* approach follows closely [Martin 2001a], [Martin 2005a].

The minimum statistics method rests on two conditions, namely that the speech and the disturbing noise are statistically independent and that the power of a noisy speech signal frequently decays to the power level of the disturbing noise. It is therefore possible to derive an accurate noise PSD estimate by tracking the minimum of the noisy signal PSD. Since the minimum is smaller than or equal to the mean, the minimum tracking method requires a bias correction. It turns out that the bias is a function of the variance of the smoothed signal PSD and as such depends on the smoothing parameter of the PSD estimator.

To highlight some of the obstacles which are encountered when implementing a minimum tracking approach, we consider a simplified algorithm based on recursively smoothed magnitude squared DFT coefficients

$$\widehat{\sigma}^2_{Y,\mu}(\lambda) = \alpha\, \widehat{\sigma}^2_{Y,\mu}(\lambda - 1) + (1 - \alpha)\, |Y_\mu(\lambda)|^2 \tag{11.121}$$

where μ is the frequency bin index and λ is the frame index. Successive frames of the input signal are shifted by r samples in the time domain. Figure 11.15 plots the magnitude squared DFT coefficients $|Y_\mu(\lambda)|^2$, the smoothed signal power $\widehat{\sigma}^2_{Y,\mu}(\lambda)$, and the minimum of the smoothed power within a window of 96 consecutive power values as a function of the frame index λ and for a single frequency bin $\mu = 25$. The speech signal is degraded by a non-stationary vehicular noise with an overall SNR of approximately 10 dB. The window size is $M = 2r = 256$. The DFT coefficients are recursively smoothed with an equivalent (rectangular) window length of $T_{SM} = 0.2$ seconds which represents a good compromise between smoothing the noise and tracking the speech signal. By assuming independent DFT coefficients and equating the variance of $\widehat{\sigma}^2_{Y,\mu}(\lambda)$ to the variance of a moving-average estimator with window length T_{SM}, the smoothing parameter α in (11.121) can be computed as $\alpha = (T_{SM} f_s / r - 1)/(T_{SM} f_s / r + 1) \approx 0.85$. The noise power estimate is obtained by picking the minimum value within a sliding window of $D = 96$ consecutive values of $\widehat{\sigma}^2_{Y,\mu}(\lambda)$, regardless of whether speech is present or not.

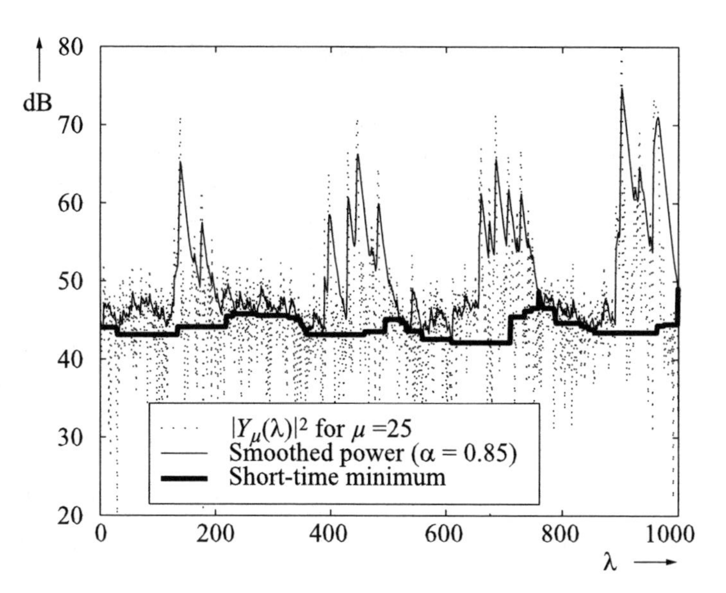

Figure 11.15: $|Y_\mu(\lambda)|^2$, smoothed power $\widehat{\sigma}^2_{Y,\mu}(\lambda)$ (11.121), and short-time minimum of the smoothed power for a noisy speech signal and a single frequency bin $\mu = 25$ [Martin 2001a]; © 2001 IEEE

Apparently, the minimum tracking provides a rough estimate of the noise power. However, further improvements are possible if the following issues are resolved:

- Smoothing with a fixed smoothing parameter α widens the peaks of speech activity of the smoothed PSD estimate $\widehat{\sigma}^2_{Y,\mu}(\lambda)$. This will lead to inaccurate noise estimates, as the sliding window for the minimum search might slip into broad peaks. Thus, we cannot use fixed smoothing parameters close to one and, as a consequence, the noise estimate will have a relatively large variance. This undesirable behavior can be circumvented with a time- and frequency-dependent smoothing parameter.

- The noise estimate as shown in Fig. 11.15 is biased towards lower values. Using results from extremal statistics theory, the bias can be computed and compensated.

- In the case of increasing noise power, the minimum tracking lags behind. A faster update of the noise power is possible if the variance of the smoothed power is taken into account.

In the following sections we will address these issues.

11.8.3.1 Derivation of the Smoothing Parameter

The key to a successful application of the *minimum statistics* principle is to use a time- and frequency-dependent smoothing parameter in (11.121). To derive an optimal smoothing parameter $\alpha_\mu(\lambda)$, we assume speech absence ($\sigma^2_{S,\mu}(\lambda) = 0$) and consider again the first-order smoothing equation for $\widehat{\sigma}^2_{Y,\mu}(\lambda)$,

$$\widehat{\sigma}^2_{Y,\mu}(\lambda) = \alpha_\mu(\lambda) \cdot \widehat{\sigma}^2_{Y,\mu}(\lambda - 1) + (1 - \alpha_\mu(\lambda)) \cdot |Y_\mu(\lambda)|^2 \,. \tag{11.122}$$

For speech pause, we want $\widehat{\sigma}^2_{Y,\mu}(\lambda)$ to be as close as possible to the noise PSD $\sigma^2_{N,\mu}(\lambda)$. Therefore, our objective is to minimize the conditional mean square error

$$\mathrm{E}\left\{(\widehat{\sigma}^2_{Y,\mu}(\lambda) - \sigma^2_{N,\mu}(\lambda))^2 \mid \widehat{\sigma}^2_{Y,\mu}(\lambda - 1)\right\} \tag{11.123}$$

from one iteration step to the next. After substituting $\widehat{\sigma}^2_{Y,\mu}(\lambda)$ in (11.123) and using $\mathrm{E}\left\{|Y_\mu(\lambda)|^2\right\} = \sigma^2_{N,\mu}(\lambda)$ and $\mathrm{E}\left\{|Y_\mu(\lambda)|^4\right\} = 2\sigma^4_{N,\mu}(\lambda)$, the mean square error is given by

$$\begin{aligned}
&\mathrm{E}\left\{(\widehat{\sigma}^2_{Y,\mu}(\lambda) - \sigma^2_{N,\mu}(\lambda))^2 \mid \widehat{\sigma}^2_{Y,\mu}(\lambda - 1)\right\} \\
&= \mathrm{E}\left\{\left(\alpha_\mu(\lambda) \cdot \widehat{\sigma}^2_{Y,\mu}(\lambda - 1) - \sigma^2_{N,\mu}(\lambda)\right)^2 \,\middle|\, \widehat{\sigma}^2_{Y,\mu}(\lambda - 1)\right\} \\
&\quad + \mathrm{E}\left\{(1 - \alpha_\mu(\lambda))^2 \cdot |Y_\mu(\lambda)|^4 \,\middle|\, \widehat{\sigma}^2_{Y,\mu}(\lambda - 1)\right\} \\
&\quad + 2\,\mathrm{E}\left\{\left(\alpha_\mu(\lambda) \cdot \widehat{\sigma}^2_{Y,\mu}(\lambda - 1) - \sigma^2_{N,\mu}(\lambda)\right) \cdot (1 - \alpha_\mu(\lambda)) \cdot |Y_\mu(\lambda)|^2 \,\middle|\, \widehat{\sigma}^2_{Y,\mu}(\lambda - 1)\right\} \\
&= \alpha^2_\mu(\lambda) \left[\widehat{\sigma}^2_{Y,\mu}(\lambda - 1) - \sigma^2_{N,\mu}(\lambda)\right]^2 + \sigma^4_{N,\mu}(\lambda) \cdot (1 - \alpha_\mu(\lambda))^2
\end{aligned} \tag{11.124}$$

where we have also assumed that successive signal frames are independent. Setting the first derivative with respect to $\alpha_\mu(\lambda)$ to zero yields

$$\alpha_\mu(\lambda)_{\langle\mathrm{opt}\rangle} = \frac{1}{1 + \left(\dfrac{\widehat{\sigma}^2_{Y,\mu}(\lambda - 1)}{\sigma^2_{N,\mu}(\lambda)} - 1\right)^2} \,. \tag{11.125}$$

Since the second derivative is non-negative, a minimum is achieved. The term $\widetilde{\gamma}_\mu(\lambda) = \widehat{\sigma}^2_{Y,\mu}(\lambda - 1)/\sigma^2_{N,\mu}(\lambda)$ in the denominator of (11.125) is recognized as a smoothed version of the *a posteriori* SNR.

Figure 11.16 plots the optimal smoothing parameter $\alpha_\mu(\lambda)_{\langle\mathrm{opt}\rangle}$ for $0 \leq \widetilde{\gamma}_\mu(\lambda) \leq 10$. Since the optimal smoothing parameter $\alpha_\mu(\lambda)_{\langle\mathrm{opt}\rangle}$ is between zero and one, a stable and non-negative power estimate $\widehat{\sigma}^2_{Y,\mu}(\lambda)$ is guaranteed.

Having assumed a speech pause in the above derivation does not pose any principal problems. The optimal smoothing procedure reacts to speech activity in the

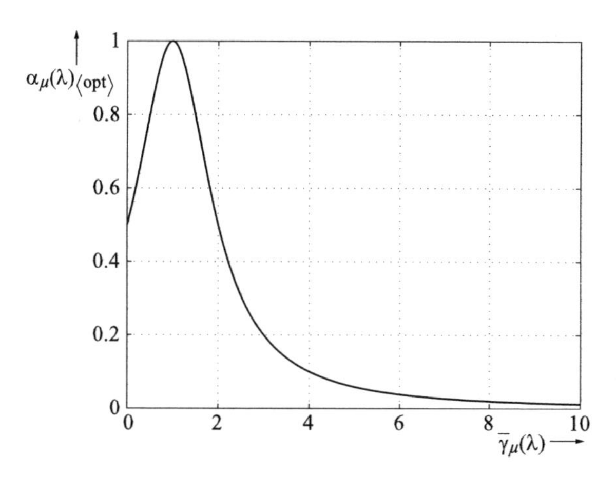

Figure 11.16: Optimal smoothing parameter $\alpha_\mu(\lambda)_{\langle\text{opt}\rangle}$ [Martin 2001a]; © 2001 IEEE

same way as to highly non-stationary noise. In the case of speech activity, the smoothing parameter is reduced to small values, which enables the PSD estimate $\widehat{\sigma}^2_{Y,\mu}(\lambda)$ to closely follow the time-varying PSD of the noisy speech signal. However, since in any implementation the true (and unknown) noise power $\sigma^2_{N,\mu}(\lambda)$ must be replaced by an estimate, additional measures are necessary to stabilize the smoothing procedure [Martin 2001a]. Finally, we note that for other methods of frequency analysis, e.g., a filter bank with non-uniform frequency bands, the optimal smoothing parameter must be adapted to the statistics of the subband signals [Martin, Lotter 2001].

11.8.3.2 Statistics of Minimum Power Estimates

The minimum tracking PSD estimation approach determines the minimum of the short-time PSD estimate within a finite window of length D. Since for non-trivial densities the minimum value of a set of random variables is smaller than their mean, the minimum noise estimate is necessarily biased.

The bias depends on the signal statistics and the smoothing method [Martin 2005a]. It can be computed analytically only if successive values of $\widehat{\sigma}^2_{Y,\mu}(\ell)$, $\ell \in \{\lambda, \ldots, \lambda - i, \ldots, \lambda - D + 1\}$ are independent, identically distributed (i.i.d.) random variables. Unless the sequence of successive $\widehat{\sigma}^2_{Y,\mu}(\lambda)$ values is subsampled, this is clearly not given. In this section we therefore consider the case of correlated short-term PSD estimates and discuss an approximate solution. To simplify notations, we restrict ourselves to the case of speech pause. All results carry over to the case of speech activity by replacing the noise variance by the variance of the noisy speech signal.

11.8.3.3 Mean of the Minimum of Correlated PSD Estimates

We consider the minimum $\sigma^2_{\min,\mu}(\lambda)$ of D successive short-term PSD estimates $\widehat{\sigma}^2_{Y,\mu}(\ell)$, $\ell \in \{\lambda, \ldots, \lambda - i, \ldots, \lambda - D + 1\}$. For an infinite sequence of signal frames and a fixed smoothing parameter $0 \leq \alpha < 1$, the short-term PSD estimate $\widehat{\sigma}^2_{Y,\mu}(\lambda)$ can be written as

$$\widehat{\sigma}^2_{Y,\mu}(\lambda) = (1 - \alpha) \sum_{i=0}^{\infty} \alpha^i |Y_\mu(\lambda - i)|^2 . \qquad (11.126)$$

For independent, exponentially, and identically distributed magnitude squared DFT coefficients $|Y_\mu(\lambda)|^2$, the characteristic function of the PDF of $\widehat{\sigma}^2_{Y,\mu}(\lambda)$ is then given by (see, e.g., [Johnson et al. 1994, chapter 18])

$$\Phi_{\sigma,Y}(\omega) = \prod_{i=0}^{\infty} \frac{1}{1 - j\omega\sigma^2_{N,\mu}(\lambda)\,(1 - \alpha)\,\alpha^i} . \qquad (11.127)$$

Since the PDF of $\widehat{\sigma}^2_{Y,\mu}(\lambda)$ is scaled by $\sigma^2_{N,\mu}(\lambda)$, the minimum statistics of the short-term PSD estimate are also scaled by $\sigma^2_{N,\mu}(\lambda)$ [David 1980, section 6.2]. Therefore, the mean $\mathrm{E}\left\{\sigma^2_{\min,\mu}(\lambda)\right\}$ of the minimum power is proportional to $\sigma^2_{N,\mu}(\lambda)$ and the variance is proportional to $\sigma^4_{N,\mu}(\lambda)$. Without loss of generality, it is thus sufficient to compute the mean and the variance for $\sigma^2_{N,\mu}(\lambda) = 1$. We introduce the bias $1 - \mathrm{E}\left\{\sigma^2_{\min,\mu}(\lambda)\right\}_{|\sigma^2_{N,\mu}(\lambda)=1}$ and determine the mean

$$B^{-1}_{\min,\mu}(\lambda) = \mathrm{E}\left\{\sigma^2_{\min,\mu}(\lambda)\right\}_{|\sigma^2_{N,\mu}(\lambda)=1} \qquad (11.128)$$

of the minimum of correlated variates $\sigma^2_{Y,\mu}(\lambda)$ as a function of the inverse normalized variance

$$Q_{\mathrm{eq},\mu}(\lambda) = \frac{2\sigma^4_{N,\mu}(\lambda)}{\mathrm{var}\{\sigma^2_{Y,\mu}(\lambda)\}} \qquad (11.129)$$

by generating exponentially distributed data with variance $\sigma^2_{N,\mu}(\lambda) = 1$ and by averaging minimum values for various values of D. $B^{-1}_{\min,\mu}(\lambda)$ is the factor by which the minimum is smaller than the mean. The inverse normalized variance $Q_{\mathrm{eq},\mu}(\lambda)$ is also called "equivalent degrees of freedom", since non-recursive (moving-average) smoothing of $Q_{\mathrm{eq},\mu}(\lambda)$ independent squared Gaussian variates will yield an estimate with the same variance. $Q_{\mathrm{eq},\mu}(\lambda)$ thus quantifies the amount of smoothing: when $\alpha_\mu(\lambda)$ is close to one, the estimated signal power $\widehat{\sigma}^2_{Y,\mu}(\lambda)$ is significantly smoother than $|Y_\mu(\lambda)|^2$. In this case, the variance of $\widehat{\sigma}^2_{Y,\mu}(\lambda)$ is small and $Q_{\mathrm{eq},\mu}(\lambda)$ large. The minimum of D such samples is, then, close to the mean.

The result of this evaluation is shown in Fig. 11.17. It depicts $B^{-1}_{\min}$ as a function of the length D of the minimum search window and as a function of the equivalent degrees of freedom $Q_{\mathrm{eq},\mu}$.

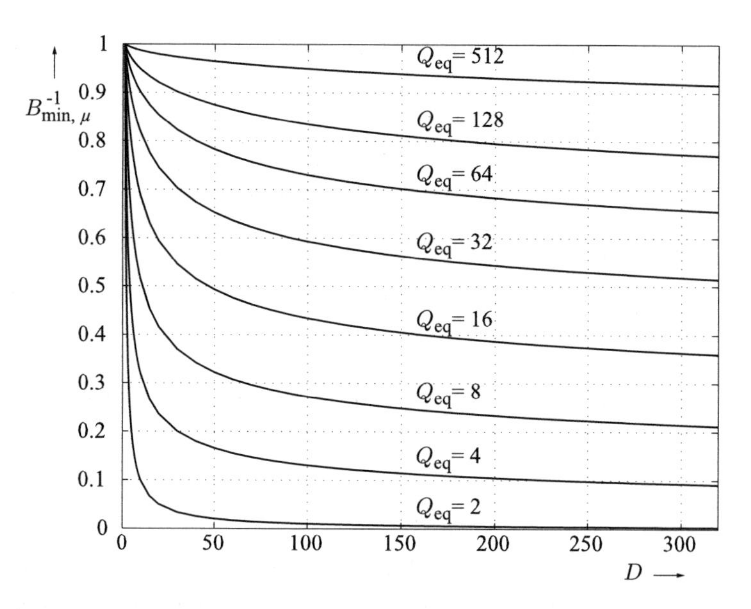

Figure 11.17: Mean of minimum of $1 \leq D \leq 320$ correlated and recursively averaged periodogram values for $\sigma_Y^2 = 1$ and Q_{eq} equivalent degrees of freedom. A Hann window with 50% overlap is used [Martin 2005a]

For software implementations it is practical to have a closed form approximation of the inverse mean $B_{\text{min},\mu}$, i.e., the bias correction factor. We note that $B_{\text{min},\mu} = D$ for $Q_{\text{eq},\mu}(\lambda) = 2$ and $B_{\text{min},\mu} = 1$ for $D = 1$. While the latter is obvious, we must prove the former. For $Q_{\text{eq},\mu}(\lambda) = 2$ the smoothed power exhibits only two degrees of freedom, which is true when no smoothing is in effect, i.e., for $\alpha_\mu(\lambda) = 0$. The probability density of the minimum $\sigma^2_{\text{min},\mu}(\lambda)$ of D i.i.d. random variables $\sigma^2_{Y,\mu}(\ell)$, $\ell \in \{\lambda, \ldots, \lambda - D + 1\}$, is given by

$$p_{\sigma^2_{\text{min},\mu}(\lambda)}(y) = D\,(1 - P_{\sigma^2_{Y,\mu}}(y))^{D-1}\, p_{\sigma^2_{Y,\mu}}(y) \tag{11.130}$$

where $P_{\sigma^2_{Y,\mu}}(y)$ denotes the probability distribution function of $\sigma^2_{Y,\mu}$. For $Q_{\text{eq},\mu}(\lambda) = 2$ and the Gaussian assumption, $\sigma^2_{Y,\mu}$ is exponentially distributed and

$$\begin{aligned} \mathrm{E}\left\{\sigma^2_{\text{min},\mu}(\lambda)\right\}_{|\sigma^2_{N,\mu}(\lambda)=1} = \frac{1}{B_{\text{min},\mu}} &= \int_0^\infty (1 - P_{\sigma^2_{Y,\mu}}(y))^D \, \mathrm{d}y \\ &= \frac{2}{Q_{\text{eq},\mu}(\lambda)} \int_0^\infty e^{-yD} \, \mathrm{d}y \,. \end{aligned} \tag{11.131}$$

Therefore, for $Q_{\text{eq},\mu}(\lambda) = 2$ we obtain $B_{\text{min},\mu} = D$.

Using an asymptotic result in [Gumbel 1958, section 7.2], we may approximate the inverse mean of the minimum by

$$B_{\min,\mu}(\lambda) \approx 1 + (D-1)\frac{2}{\widetilde{Q}_{\mathrm{eq},\mu}(\lambda)}\,\Gamma\left(1+\frac{2}{Q_{\mathrm{eq},\mu}(\lambda)}\right)^{f_2(D)} \tag{11.132}$$

where $\widetilde{Q}_{\mathrm{eq},\mu}(\lambda)$ is a scaled version of $Q_{\mathrm{eq},\mu}(\lambda)$

$$\widetilde{Q}_{\mathrm{eq},\mu}(\lambda) = \frac{Q_{\mathrm{eq},\mu}(\lambda) - 2f_1(D)}{1 - f_1(D)} \tag{11.133}$$

and $f_1(D)$ and $f_2(D)$ are functions of D. In most cases, the simplified approximation

$$B_{\min,\mu}(\lambda) \approx 1 + (D-1)\frac{2}{\widetilde{Q}_{\mathrm{eq},\mu}(\lambda)} \tag{11.134}$$

works equally well, since the additional term in (11.132) reduces the approximation error for small values of $Q_{\mathrm{eq},\mu}(\lambda)$ only. Small values occur predominantly when a significant amount of speech power is present. During speech activity, however, it is highly unlikely that $\widehat{\sigma}^2_{Y,\mu}(\lambda)$ attains a minimum.

Table 11.3 lists values for $f_1(D)$ and $f_2(D)$ as a function of D. Values in between can be obtained by linear interpolation. $\Gamma(\cdot)$ denotes the complete gamma function [Gradshteyn, Ryzhik 1994]. The approximation using (11.134) has a relative mean square error of less than 0.07 and a maximum relative error of less than 67%. The largest errors are obtained for small values of Q_{eq}. The approximation using (11.132) results in a relative mean square error of less than 0.001 and maximum relative error of less than 6%. In a real-time application with fixed window length D, $f_1(D)$ and $f_2(D)$ are determined from Table 11.3 and (11.133) and (11.134) are evaluated during runtime.

Table 11.3: Parameters for the approximation of the mean of the minimum (11.132) and (11.134) [Martin 2005a]

D	$f_1(D)$	$f_2(D)$	D	$f_1(D)$	$f_2(D)$
1	0	0	60	0.841	3.1
2	0.26	0.15	80	0.865	3.38
5	0.48	0.48	120	0.89	4.15
8	0.58	0.78	140	0.9	4.35
10	0.61	0.98	160	0.91	4.25
15	0.668	1.55	180	0.92	3.9
20	0.705	2.0	220	0.93	4.1
30	0.762	2.3	260	0.935	4.7
40	0.8	2.52	300	0.94	5

11.8.3.4 An Unbiased Noise Estimator Based on Minimum Statistics

As a result of the previous sections, we can now provide an *unbiased* estimator of the noise PSD $\sigma^2_{N,\mu}(\lambda)$,

$$\widehat{\sigma}^2_{N,\mu}(\lambda) = \frac{\sigma^2_{\min,\mu}(\lambda)}{\mathrm{E}\left\{\sigma^2_{\min,\mu}(\lambda)\right\}_{|\sigma^2_{N,\mu}(\lambda)=1}} = B_{\min,\mu}\left(D, Q_{\mathrm{eq},\mu}(\lambda)\right)\sigma^2_{\min,\mu}(\lambda)\,,$$

where we now emphasize the dependency of $B_{\min,\mu}$ on D and $Q_{\mathrm{eq},\mu}(\lambda)$. The unbiased estimator requires knowledge of the normalized variance

$$\frac{1}{Q_{\mathrm{eq},\mu}(\lambda)} = \frac{\mathrm{var}\{\widehat{\sigma}^2_{Y,\mu}(\lambda)\}}{2\,\sigma^4_{N,\mu}(\lambda)} \tag{11.135}$$

of the smoothed PSD estimate $\widehat{\sigma}^2_{Y,\mu}(\lambda)$ at any given time and frequency index.

To estimate the variance of the smoothed PSD estimate $\widehat{\sigma}^2_{Y,\mu}(\lambda)$, we use a first-order smoothing recursion for the approximation of the first moment, $\mathrm{E}\left\{\widehat{\sigma}^2_{Y,\mu}(\lambda)\right\}$, and the second moment, $\mathrm{E}\left\{\left(\widehat{\sigma}^2_{Y,\mu}(\lambda)\right)^2\right\}$, of $\widehat{\sigma}^2_{Y,\mu}(\lambda)$,

$$\begin{aligned}
\overline{P}_\mu(\lambda) &= \beta_\mu(\lambda)\,\overline{P}_\mu(\lambda-1) + (1-\beta_\mu(\lambda))\,P_\mu(\lambda)\\
\overline{P^2}_\mu(\lambda) &= \beta_\mu(\lambda)\,\overline{P^2}_\mu(\lambda-1) + (1-\beta_\mu(\lambda))\,P^2_\mu(\lambda)\\
\widehat{\mathrm{var}}\{\widehat{\sigma}^2_{Y,\mu}(\lambda)\} &= \overline{P^2}_\mu(\lambda) - \overline{P}^2_\mu(\lambda)
\end{aligned} \tag{11.136}$$

where $\overline{P}_\mu(\lambda)$ and $\overline{P^2}_\mu(\lambda)$ denote the estimated first and second moments, respectively. Good results are obtained by choosing the smoothing parameter $\beta_\mu(\lambda) = \alpha^2_\mu(\lambda)$ and by limiting $\beta_\mu(\lambda)$ to values less than or equal to 0.8 [Martin 2001a]. Finally, $1/Q_{\mathrm{eq},\mu}(\lambda)$ is estimated by

$$\frac{1}{Q_{\mathrm{eq},\mu}(\lambda)} \approx \frac{\widehat{\mathrm{var}}\{\widehat{\sigma}^2_{Y,\mu}(\lambda)\}}{2\,\widehat{\sigma}^4_{N,\mu}(\lambda-1)}\,, \tag{11.137}$$

and this estimate is limited to a maximum of 0.5 corresponding to $Q_{\mathrm{eq},\mu}(\lambda) = 2$. Since an increasing noise power can be tracked only with some delay, the minimum statistics estimator has a tendency to underestimate highly non-stationary noise. Furthermore, since the bias compensation (11.132) (or (11.134)) depends on the estimated normalized variance, the bias compensation factor is a random variable with a variance depending on the variance of $\sigma^2_{Y,\mu}(\lambda)$. It is therefore advantageous to increase the inverse bias $B_{\min,\mu}(\lambda)$ by a factor $B_c(\lambda) = 1 + a_v \sqrt{\overline{Q^{-1}}(\lambda)}$ proportional to the normalized standard deviation of the short-term estimate $\widehat{\sigma}^2_{Y,\mu}(\lambda)$. We compute the average normalized variance as

$$\overline{Q^{-1}}(\lambda) = \frac{1}{M} \sum_{\mu=0}^{M-1} \frac{1}{Q_{\mathrm{eq},\mu}(\lambda)} \tag{11.138}$$

and choose a_v typically as $a_v = 2.12$. This bias correction has an impact only when the short-term PSD estimate, and thus the estimated variance, has a large variance. Without the bias correction, the variations in $B_{\min,\mu}(D, Q_{\mathrm{eq},\mu}(\lambda))$ would push the minimum to values which are too low. For stationary noise this factor is close to one.

Although the minimum statistics approach was originally developed for a sampling rate of $f_s = 8000\,\mathrm{Hz}$ and a frame advance of 128 samples, it can easily be adapted to other sampling rates and frame advance schemes. The length D of the minimum search window must be set proportional to the frame rate. For a given sampling rate f_s and a frame advance r, the duration of the time window for minimum search, $D \cdot r / f_s$, should be equal to approximately 1.5 seconds. When a constant smoothing parameter [Martin 1994] is used in (11.122), the length D of the window for minimum search must be at least 50% larger than that for the adaptive smoothing algorithm. The minimum search itself is efficiently implemented by subdividing the search window of length D into subwindows. This, as well as a method to improve tracking of non-stationary noise, is explained in greater detail in [Martin 2001a].

11.8.3.5 The Initial Example Revisited

We demonstrate the performance of the adaptive smoothing and the bias compensation with a second look at the noisy speech file of Fig. 11.15. Figure 11.18 plots $|Y_\mu(\lambda)|^2$, the smoothed power $\widehat{\sigma}^2_{Y,\mu}(\lambda)$, the noise estimate $\widehat{\sigma}^2_{N,\mu}(\lambda)$, and the time-varying smoothing parameter $\alpha_\mu(\lambda)$ for the same noisy speech file and the same frequency bin as in Fig. 11.15. We see that the time varying smoothing parameter allows the estimated signal power to closely follow the variations of the speech signal. During speech pause the noise is well smoothed. Also, the bias compensation works very well, as the smoothed power and the estimated noise power follow each other closely during speech pause. We also note that the noise PSD estimate is updated during speech activity. This is a major advantage of the minimum statistics approach.

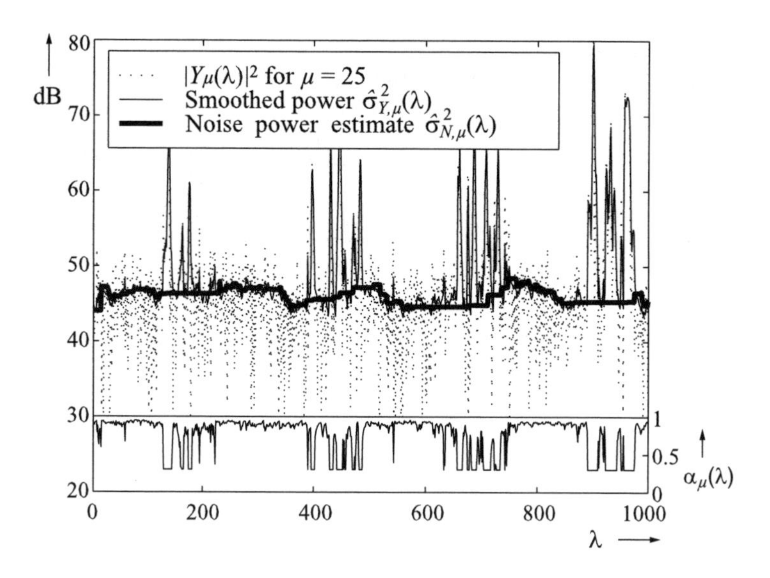

Figure 11.18: $|Y_\mu(\lambda)|^2$, smoothed power $\widehat{\sigma}^2_{Y,\mu}(\lambda)$ (11.122), and noise estimate $\widehat{\sigma}^2_{N,\mu}(\lambda)$ for a noisy speech signal and a single frequency bin $\mu = 25$. The time-varying smoothing parameter $\alpha_\mu(\lambda)$ is shown in the lower inset graph [Martin 2001a]; © 2001 IEEE.

11.9 Dual Channel Systems

Single microphone speech enhancement algorithms are favored in many applications because they are relatively easy to apply. Their performance, however, is limited, especially when the noise is non-stationary. The performance of noise reduction algorithms can be expected to improve, when more than one microphone is available. In this case, the spatial characteristics of the sound field can be exploited, e.g. for the estimation of *a priori* unknown statistical quantities.

Fig. 11.19 depicts the basic scenario with two microphones and a single, possibly adaptive filter with impulse response $h(k)$. This system differs from Fig 11.3 in that a second microphone is added and this second signal is taken as the target signal $d(k)$ in Fig 11.3. Furthermore, we also provide the error signal $e(k)$ as an additional output.

The computation of the impulse response $h(k)$ of the non-causal IIR Wiener filter for the dual channel case is analogous to the derivation in Section 11.2.1. For wide sense stationary signals, the minimization of

$$\mathrm{E}\left\{e^2(k)\right\} = \mathrm{E}\left\{(y_2(k) - \widehat{y}(k))^2\right\} \tag{11.139}$$

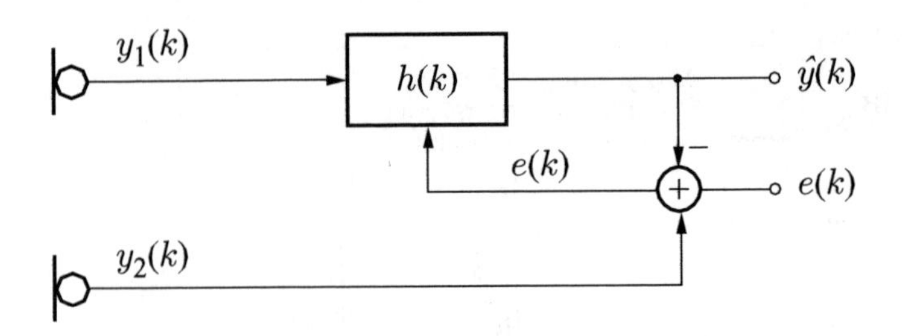

Figure 11.19: Dual channel noise reduction system

in the mean-square sense with

$$\widehat{y}(k) = \sum_{i=-\infty}^{\infty} h(i) y_1(k-i) \tag{11.140}$$

results in the necessary condition

$$\mathrm{E}\left\{(y_2(k) - \widehat{y}(k))\widehat{y}(k)\right\} = 0 \tag{11.141}$$

and in the frequency response

$$H(e^{j\Omega}) = \frac{\Phi_{y_1 y_2}(e^{j\Omega})}{\Phi_{y_1 y_1}(e^{j\Omega})} \tag{11.142}$$

of the optimal filter. In the general case of additive speech and noise signals,

$$\begin{aligned} y_1(k) &= s_1(k) + n_1(k) \\ y_2(k) &= s_2(k) + n_2(k)\,, \end{aligned} \tag{11.143}$$

and when the speech and the noise signals are statistically independent, the MMSE IIR filter is given by

$$H(e^{j\Omega}) = \frac{\Phi_{y_1 y_2}(e^{j\Omega})}{\Phi_{y_1 y_1}(e^{j\Omega})} = \frac{\Phi_{s_1 s_2}(e^{j\Omega}) + \Phi_{n_1 n_2}(e^{j\Omega})}{\Phi_{s_1 s_1}(e^{j\Omega}) + \Phi_{n_1 n_1}(e^{j\Omega})}\,. \tag{11.144}$$

The frequency response of the optimal filter may be decomposed into two independent optimal filters $H_s(e^{j\Omega})$ and $H_n(e^{j\Omega})$ for the estimation of the speech signal $s_2(k)$ and the noise signal $n_2(k)$, respectively,

$$H(e^{j\Omega}) = \frac{\Phi_{s_1 s_2}(e^{j\Omega})}{\Phi_{y_1 y_1}(e^{j\Omega})} + \frac{\Phi_{n_1 n_2}(e^{j\Omega})}{\Phi_{y_1 y_1}(e^{j\Omega})} = H_s(e^{j\Omega}) + H_n(e^{j\Omega})\,. \tag{11.145}$$

At the output of the optimal filter, we obtain the components of $y_1(k)$ which are correlated with the second channel $y_2(k)$, regardless whether they are speech or noise. Uncorrelated components are suppressed. Depending on the correlation properties of the speech and the noise signals, the optimal filter will act primarily as either a noise or a speech estimator.

For the optimal estimate $\widehat{y}_{\text{opt}}(k)$, the minimum error $\mathrm{E}\left\{e^2(k)\right\}_{|\min}$ is given with (11.141) by

$$\begin{aligned}\mathrm{E}\left\{e^2(k)\right\}_{|\min} &= \mathrm{E}\left\{(y_2(k)-\widehat{y}_{\text{opt}}(k))y_2(k)\right\} \\ &= \varphi_{y_2y_2}(0) - \sum_{i=-\infty}^{\infty} h_{\text{opt}}(i)\varphi_{y_1y_2}(i) \qquad (11.146) \\ &= \frac{1}{2\pi}\int_{-\pi}^{\pi} \Phi_{y_2y_2}(e^{j\Omega})\mathrm{d}\Omega - \frac{1}{2\pi}\int_{-\pi}^{\pi} H(e^{j\Omega})\Phi^*_{y_1y_2}(e^{j\Omega})\mathrm{d}\Omega\,,\end{aligned}$$

where Parseval's theorem (Table 3.2) was used in the last equality. The cross-PSD $\Phi_{y_1y_2}(e^{j\Omega})$ is the FTDS of $\varphi_{y_1y_2}(\ell) = \mathrm{E}\left\{(y_1(k)y_2(k+\ell))\right\}$. Using (11.142) in (11.146) and merging both integrals, we obtain

$$\begin{aligned}\mathrm{E}\left\{e^2(k)\right\}_{|\min} &= \frac{1}{2\pi}\int_{-\pi}^{\pi} \Phi_{y_2y_2}(e^{j\Omega})\left(1-\frac{\Phi_{y_1y_2}(e^{j\Omega})\Phi^*_{y_1y_2}(e^{j\Omega})}{\Phi_{y_1y_1}(e^{j\Omega})\Phi_{y_2y_2}(e^{j\Omega})}\right)\mathrm{d}\Omega \\ &= \frac{1}{2\pi}\int_{-\pi}^{\pi} \Phi_{y_2y_2}(e^{j\Omega})\left(1-\left|\gamma_{y_1y_2}(e^{j\Omega})\right|^2\right)\mathrm{d}\Omega \qquad (11.147)\end{aligned}$$

where $\left|\gamma_{y_1y_2}(e^{j\Omega})\right|^2$ denotes the *magnitude squared coherence* (MSC) function [Bendat, Piersol 1966], [Carter 1987] of the two microphone signals, i.e.,

$$\left|\gamma_{y_1y_2}(e^{j\Omega})\right|^2 = \frac{|\Phi_{y_1y_2}(e^{j\Omega})|^2}{\Phi_{y_1y_1}(e^{j\Omega})\Phi_{y_2y_2}(e^{j\Omega})}\,. \qquad (11.148)$$

The MSC constitutes a normalized, frequency-dependent measure of correlation with

$$0 \leq \left|\gamma_{y_1y_2}(e^{j\Omega})\right|^2 \leq 1\,. \qquad (11.149)$$

It indicates the linear relation between the two signals and, according to (11.147), gives an indication of how effective the Wiener filter is. Obviously, the effectiveness

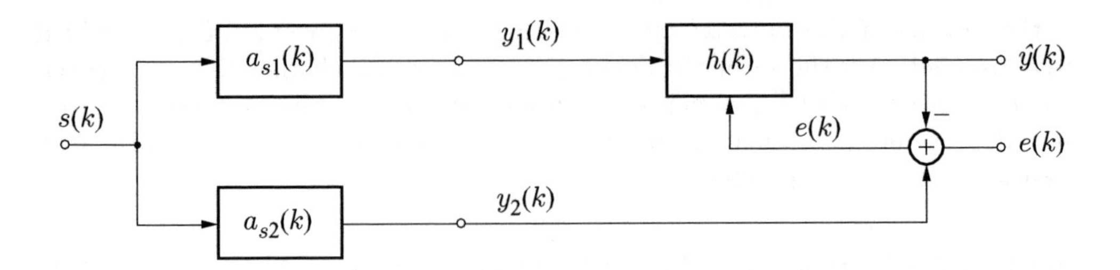

Figure 11.20: Single source signal model

of the Wiener filter depends on the signal model which the two microphone signals obey. In what follows, we consider three special cases:

- When the two microphone signals are uncorrelated, i.e. $\mathrm{E}\{y_1(k)y_2(\ell)\} = 0, \quad \forall k, \ell,$ the MSC is equal to zero and the error power is equal to the power of the signal $y_2(k)$. The impulse response of the optimal filter is identical to zero.

- When the two microphone signals are linearly related, the MSC is equal to one. In Fig. 11.20, the two microphone signals originate from a single source signal $s(k)$, i.e. $y_1(k) = a_{s1}(k) * s(k)$ and $y_2(k) = a_{s2}(k) * s(k)$. For this signal model we obtain with

$$\begin{aligned} \Phi_{y_1y_2}(e^{j\Omega}) &= A_{s1}^*(e^{j\Omega})A_{s2}(e^{j\Omega})\Phi_{ss}(e^{j\Omega}) \\ \Phi_{y_1y_1}(e^{j\Omega}) &= \left|A_{s1}(e^{j\Omega})\right|^2 \Phi_{ss}(e^{j\Omega}) \\ \Phi_{y_2y_2}(e^{j\Omega}) &= \left|A_{s2}(e^{j\Omega})\right|^2 \Phi_{ss}(e^{j\Omega}) \end{aligned} \tag{11.150}$$

  the MSC $\left|\gamma_{y_1y_2}(e^{j\Omega})\right|^2 = 1$. In general, it can be shown that the MSC is invariant with respect to linear transforms [Bendat, Piersol 1966].

- In many practical cases the correlation and thus the coherence varies with frequency. This situation occurs, when linearly related signals are disturbed by uncorrelated noise, or, when spatially distributed, mutually uncorrelated sources contribute to the microphone signals. An interesting special case is the *ideal diffuse sound field*. In such a sound field, the MSC of two microphone signals $y_1(k)$ and $y_2(k)$ obeys [Kuttruff 1990]

$$|\gamma_{y_1y_2}(\Omega)|^2 = \frac{\sin^2(\Omega\, f_s\, d\, c^{-1})}{(\Omega\, f_s\, d\, c^{-1})^2}\,, \tag{11.151}$$

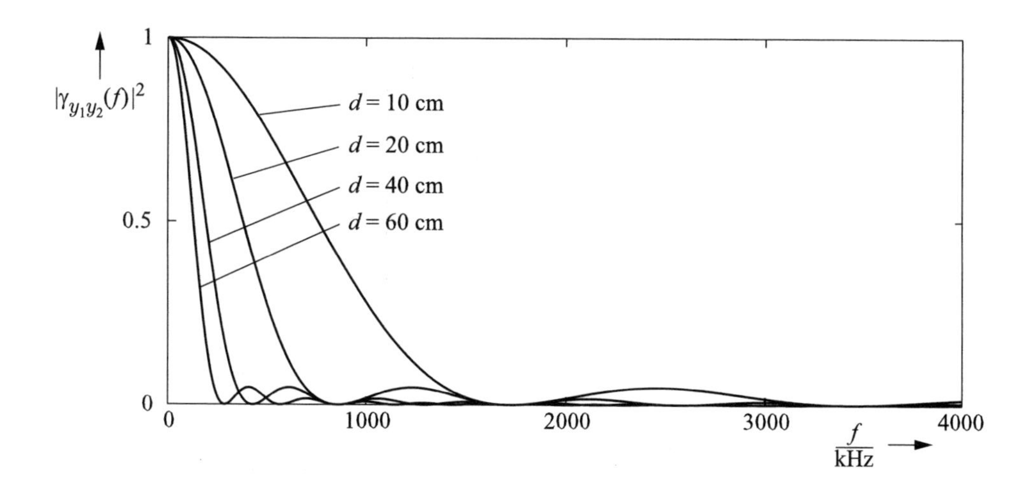

Figure 11.21: MSC of two microphone signals in an isotropic sound field for omnidirectional microphones and a distance $d = 0.1\,\text{m}$, $0.2\,\text{m}$, $0.4\,\text{m}$, and $0.6\,\text{m}$

where d, c, and f_s denote the distance between the *omnidirectional* microphones, the speed of sound, and the sampling rate, respectively. Figure 11.21 plots the MSC (11.151) as a function of frequency $f = \frac{\Omega\, f_s}{2\,\pi}$ for several inter-microphone distances d. The MSC of the ideal diffuse sound field attains its first zero at $f_c = \frac{c}{2d}$. Hence, for frequencies above $f_c = \frac{c}{2d}$, very little correlation is observed.

As an example, Fig. 11.22 shows the estimated MSC for stationary office noise and two omnidirectional microphones. Especially for low frequencies, the estimated coherence matches the MSC of the ideal diffuse sound field quite well. By contrast, for signals which originate from a single source a high degree of coherence is observed. Fig. 11.23 depicts the PSD (top) and the estimated coherence (bottom) of two signals which originate from a speaker in a car environment. For most frequencies an MSC above 0.9 is observed.

It is common practice to estimate the coherence on the basis of magnitude squared Fourier coefficients or the periodogram. The estimation of the coherence function necessarily requires averaging [Carter et al. 1973]. If we use the (cross-)periodogram without averaging as an approximation to the (cross-)PSD, we have

$$|\gamma_{y_1 y_2,\mu}(k)|^2 = \frac{|Y_{1,\mu}(k) Y_{2,\mu}^*(k)|^2}{|Y_{1,\mu}(k)|^2\,|Y_{2,\mu}(k)|^2} = 1 \tag{11.152}$$

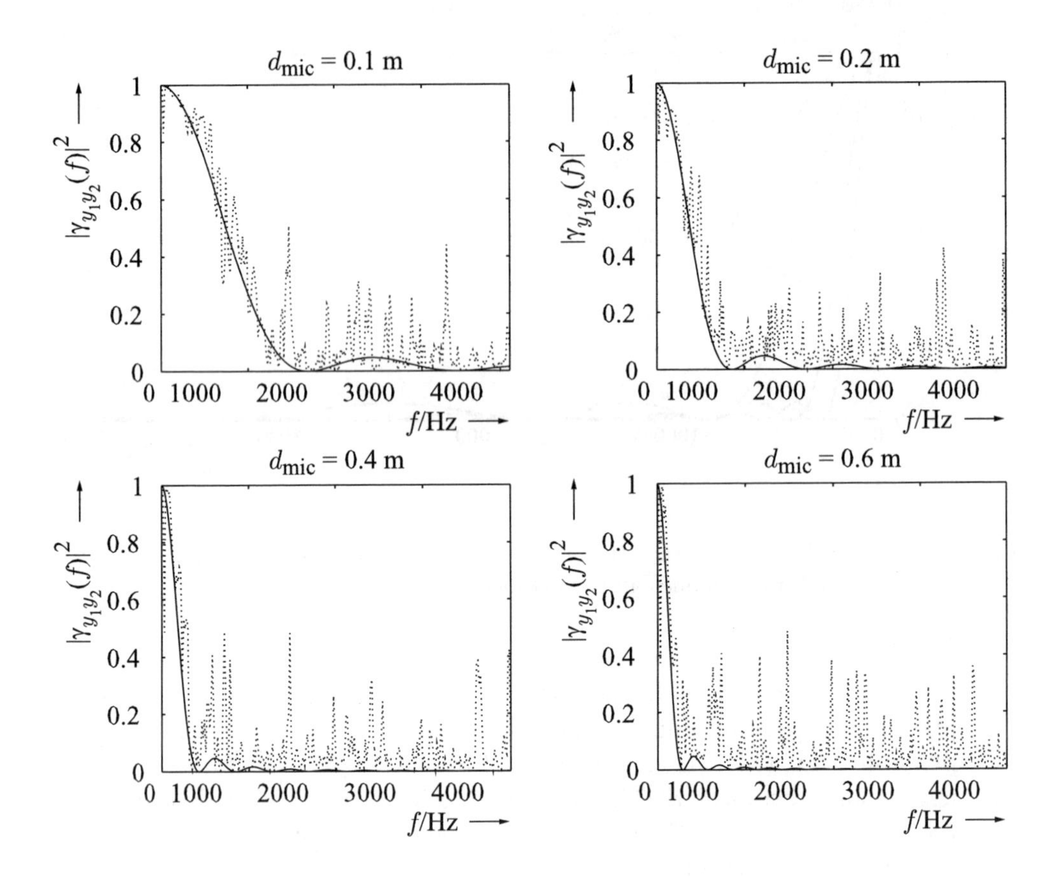

Figure 11.22: MSC of the ideal diffuse sound field and estimated MSC of office noise for omnidirectional microphones and a distance $d = 0.1$ m, 0.2 m, 0.4 m and 0.6 m [Martin 1995]

for frequency bin μ independent of the actual correlation properties. When reverberation comes into play, the estimation of the MSC using the DFT is not trivial, even when the signals are stationary and when sufficient averaging is applied. Long reverberation tails combined with DFT-based block processing might introduce a severe bias in the estimated MSC [Martin 1995].

As the effectiveness of the dual channel Wiener filter depends much on the correlation between the microphone signals, we will now consider several applications and the corresponding signal models. We distinguish two basic methods: *noise cancellation* based on a noise-only reference signal and *noise reduction* based on a symmetric dual channel signal model. In both cases the MSC of speech and noise is the key to analyzing and understanding the performance of these algorithms.

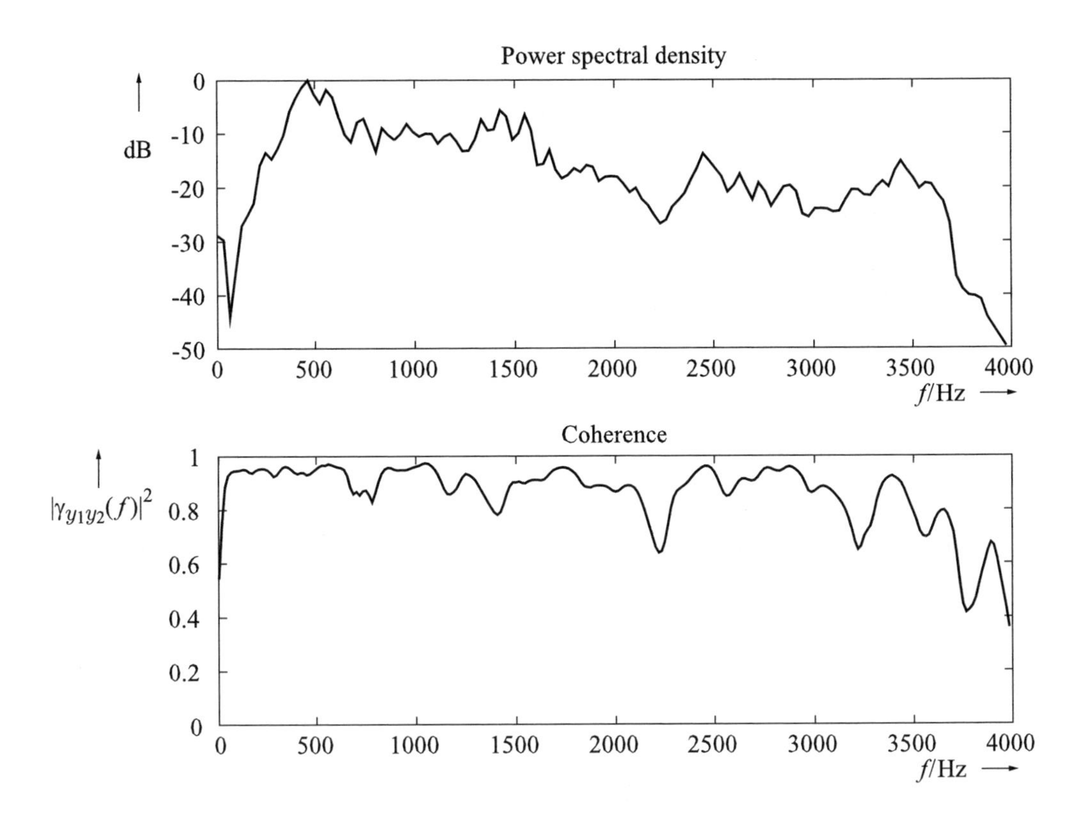

Figure 11.23: Average PSD (top) and measured MSC (bottom) of a speech signal in a reverberant room. The distance between the hypercardioid microphones is 0.2 m [Martin 1995]

11.9.1 Noise Cancellation

The noise cancellation technique can be applied when the noise is coherently received by the two microphones. This is the case, when the noise signals are linear transforms of a single noise source as shown in Fig. 11.24. We assume that the desired speech signal $s(k)$ is disturbed by additive noise which originates from a noise source $n(k)$ via the impulse response $a_{n2}(k)$ of a linear system. The noise and the speech signals are picked up by the second microphone. Therefore, $y_2(k) = a_{s2}(k) * s(k) + a_{n2}(k) * n(k)$. Furthermore, a noise reference is available, which originates from the same noise signal $n(k)$ and is picked up at the first microphone as $y_1(k) = a_{n1} * n(k)$. This noise reference signal is free of the speech signal. The Wiener filter is then used to estimate the noise which disturbs the desired signal $s(k)$.

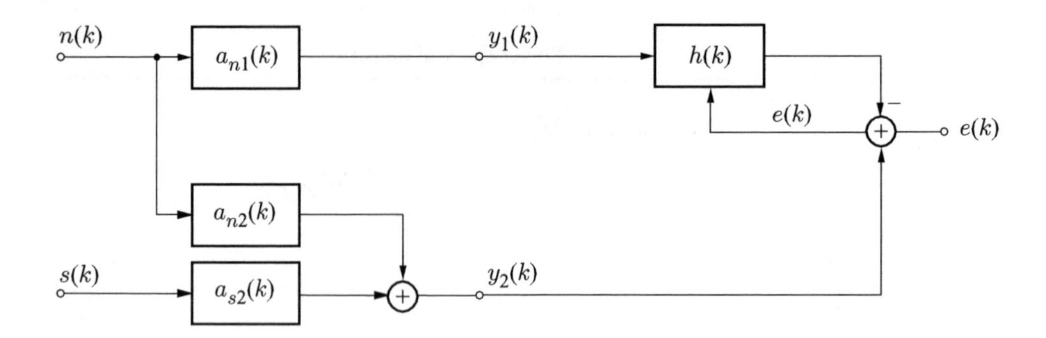

Figure 11.24: Signal model for noise cancellation

For this signal model we find that

$$\Phi_{y_1y_2}(e^{j\Omega}) = A_{n1}^*(e^{j\Omega})A_{n2}(e^{j\Omega})\Phi_{nn}(e^{j\Omega}) \tag{11.153}$$

and therefore

$$|\gamma_{y_1y_2}(e^{j\Omega})|^2 = \tag{11.154}$$

$$\frac{\left|A_{n1}^*(e^{j\Omega})A_{n2}(e^{j\Omega})\Phi_{nn}(e^{j\Omega})\right|^2}{|A_{n1}(e^{j\Omega})|^2\,\Phi_{nn}(e^{j\Omega})\left(|A_{s2}(e^{j\Omega})|^2\,\Phi_{ss}(e^{j\Omega}) + |A_{n2}(e^{j\Omega})|^2\,\Phi_{nn}(e^{j\Omega})\right)}$$

and

$$H(e^{j\Omega}) = \frac{A_{n1}^*(e^{j\Omega})A_{n2}(e^{j\Omega})}{|A_{n1}(e^{j\Omega})|^2} = \frac{A_{n2}(e^{j\Omega})}{A_{n1}(e^{j\Omega})}\,. \tag{11.155}$$

As the output of the Wiener filter is a noise estimate, the error signal $e(k)$ is the desired output which contains the speech signal

$$\begin{aligned} e(k) &= y_2(k) - h(k) * y_1(k) \\ &= a_{s2}(k) * s(k) + (a_{n2}(k) - a_{n1}(k) * h(k)) * n(k)\,. \end{aligned} \tag{11.156}$$

Thus, using the non-causal IIR Wiener filter, perfect noise cancellation is possible if $A_{n1}(e^{j\Omega}) \neq 0$. In a practical implementation using an FIR filter, an approximation to (11.155) must be used, which will in general reduce the amount of noise reduction. Furthermore, we notice from (11.154) that the speech signal reduces the coherence of the microphone signals and therefore increases the error. Thus, in the noise cancellation application, the speech signal acts as a disturbance! In an adaptive implementation of the noise canceller using, e.g., the LMS algorithm [Widrow et al. 1975], it is therefore advisable to adapt the noise estimation filter only, when little or no speech is present.

For the noise cancellation application, we define with (11.147) the normalized error power

$$R = -10\log_{10}\frac{\int_{-\pi}^{\pi}\Phi_{y_2y_2}(e^{j\Omega})\left(1-\left|\gamma_{y_1y_2}(e^{j\Omega})\right|^2\right)\mathrm{d}\Omega}{\int_{-\pi}^{\pi}\Phi_{y_2y_2}(e^{j\Omega})\mathrm{d}\Omega} \tag{11.157}$$

as a measure of performance. For example, a frequency-independent MSC of the noise signals of 0.9 provides a noise power reduction of 10 dB.

A successful application of the noise cancellation technique is acoustic echo cancellation (see Chapter 13). In acoustic echo cancellation the disturbing echo is available as a digital signal and can be fed directly into the canceller. A high degree of coherence is achieved. In the context of reducing additive acoustic noise, however, the requirements of the signal model in Fig. 11.24 are hard to fulfill. In the interior of a car, for example, we frequently encounter distributed noise sources and diffuse noise fields. Then, the microphones must be sufficiently close to achieve correlation over a large range of frequencies. This, however, leads inevitably to leakage of the speech signal into the reference channel [Armbrüster et al. 1986], [Degan, Prati 1988]. Therefore, it is difficult to obtain a noise-only reference which is free of the desired speech signal $s(k)$. In the diffuse noise field the noise cancellation approach can work only for low frequencies (and when very little speech leaks into the reference microphone!).

A decoupling of the two microphones with respect to the speech signal can be achieved by additional means, e.g. by using a facemask in an aircraft cockpit [Harrison et al. 1986], or by using a vibration sensor which is immune to air-borne sounds. In the former application, the information bearing signal $y_2(k)$ is picked up inside the mask while the noise reference, i.e. the input to the Wiener filter, is picked up outside.

11.9.1.1 Implementation of the Adaptive Noise Canceller

In practice, the filter $h(k)$ must be causal and also adaptive, since $a_{n1}(k)$ and $a_{n2}(k)$ are in general not fixed but time varying. A standard solution to the adaptation problem is to use an FIR filter $h(k)$ of order N and either a block adaptation of the coefficient vector according to

$$\mathbf{R}_{y_2y_2}\,\mathbf{h} = \boldsymbol{\varphi}_{y_1y_2} \tag{11.158}$$

or the NLMS algorithm for an iterative coefficient update [Widrow et al. 1975],

$$\mathbf{h}(k+1) = \mathbf{h}(k) + \beta(k)\,e(k)\,\mathbf{y}_2(k)\,. \tag{11.159}$$

Here, $\beta(k)$ denotes a possibly time-varying stepsize parameter. Since the speech signal of the primary channel disturbs the adaptation of the noise cancellation filter, the adaptation must be slowed down or completely stopped whenever the speaker becomes active. This is quite analogous to the echo cancellation problem which will be discussed in depth in Chapter 13. In the presence of speech, too large a stepsize will lead to distortions of the estimated noise. Since these distortions are correlated with the speech signal $s(k)$ they will be perceived as distortions of the speech signal. Too small a stepsize will slow down the adaptation of the adaptive filter. Thus, a good balance between averaging and tracking is desirable.

Furthermore, adaptive noise cancellers have been devised which exploit the short-term periodic structure of voiced speech [Sambur 1978]. Other approaches employ a cascade of cancellers to remove the speech in the reference signal and to remove the noise in the output signal [Faucon et al. 1989], or combine the two-microphone canceller with single channel approaches [Kroschel, Linhard 1988], [Gustafsson et al. 1999].

11.9.2 Noise Reduction

In many speech communication scenarios, it will not be possible to prevent the speech signal from leaking into the microphone signal $y_1(k)$. Thus, the speech signal will also be estimated by the adaptive filter and canceled to some extent. Also, the noise source is, in general, not a single acoustic point source but has some spatial distribution. Then, the correlation between the noise signals in both channels will be reduced. A typical example where this kind of problem prevails is speech pickup with two microphones in the ideal diffuse noise field. Diffuse noise fields arise in reverberant environments and are thus quite common in speech communication applications.

We therefore consider a scenario where both ambient noise and speech are picked up by the reference microphone as shown in Fig. 11.25. We assume that the speech signal originates from a point source and that the noise components in the two microphone signals exhibit a low degree of correlation. These requirements can be fulfilled in a ideal diffuse noise field, where above a cutoff frequency $f_c = c/(2d)$ the noise signals are mostly uncorrelated. The microphone signals can be therefore written as

$$y_1(k) = a_{s1}(k) * s(k) + n_1(k) \tag{11.160}$$

and

$$y_2(k) = a_{s2}(k) * s(k) + n_2(k) \; . \tag{11.161}$$

As before, we assume that the speech signals are not correlated with the noise signals. Thus, the linearly constrained MMSE IIR filter is given by

$$H(e^{j\Omega}) = \frac{\Phi_{y_1y_2}(e^{j\Omega})}{\Phi_{y_1y_1}(e^{j\Omega})} = \frac{A_{s1}^*(e^{j\Omega})A_{s2}(e^{j\Omega})\Phi_{ss}(e^{j\Omega}) + \Phi_{n_1n_2}(e^{j\Omega})}{\left|A_{s1}(e^{j\Omega})\right|^2 \Phi_{ss}(e^{j\Omega}) + \Phi_{n_1n_1}(e^{j\Omega})} \; . \tag{11.162}$$

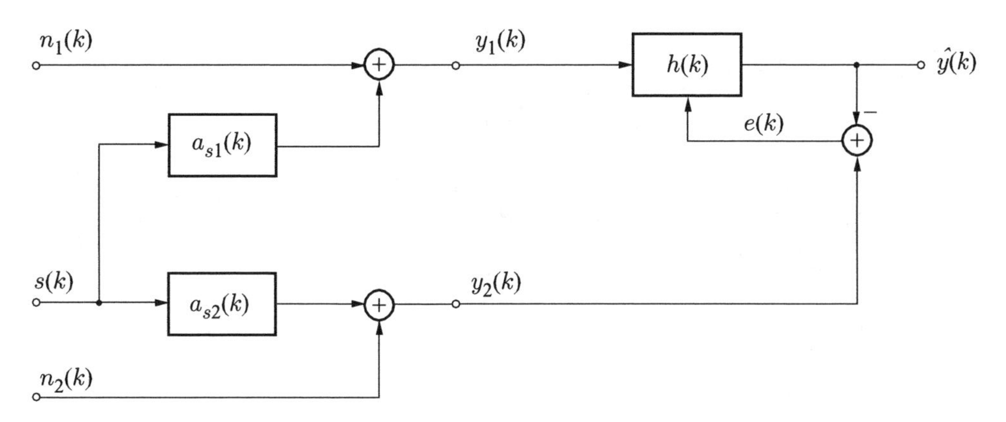

Figure 11.25: Signal model for noise reduction

The output signal of the dual channel noise reduction system is the signal $\widehat{y}(k)$. When the noise is uncorrelated above a cutoff frequency $f_c = \frac{c}{2d}$ and

$$A_{s1}(e^{j\Omega})A_{s2}^*(e^{j\Omega}) \approx \left|A_{s1}(e^{j\Omega})\right|^2 \approx 1\,, \tag{11.163}$$

the frequency response of the optimal filter

$$H(e^{j\Omega}) = \frac{\Phi_{y_1y_2}(e^{j\Omega})}{\Phi_{y_1y_1}(e^{j\Omega})} = \frac{A_{s1}^*(e^{j\Omega})A_{s2}(e^{j\Omega})\Phi_{ss}(e^{j\Omega})}{|A_{s1}(e^{j\Omega})|^2\,\Phi_{ss}(e^{j\Omega}) + \Phi_{n_1n_1}(e^{j\Omega})} \tag{11.164}$$

approaches the frequency response of the single channel Wiener filter. However, in contrast to the single channel Wiener filter, no *a priori* knowledge about the clean speech PSD is required. The filter is computed using the microphone signals only. However, for the above assumption to hold, we must place the microphones within the *critical distance* of the speech source. Within the critical distance the direct sound energy is larger than the energy of the reverberant sounds and thus a high degree of coherence is achieved.

11.9.3 Implementations of Dual Channel Noise Reduction Systems

In a practical realization of the two-channel speech enhancement system, we implement the adaptive filter either by means of block processing in the frequency domain or by means of the NLMS algorithm in the time domain [Martin, Vary 1992], [Martin 2001b]. Since the input signals $y_1(k)$ and $y_2(k)$ of the speech enhancement system in Fig. 11.25 may be interchanged, the algorithm itself may be symmetrised by using a second adaptive filter which uses the second channel $y_2(k)$ as its input and the first channel $y_1(k)$ as the reference signal. The resulting system, including further enhancements such as preemphasis and deemphasis filters,

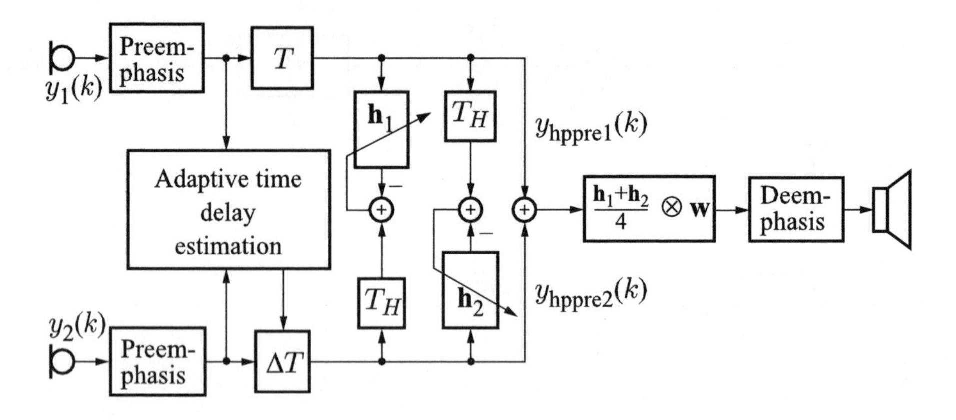

Figure 11.26: Symmetric dual microphone speech enhancement system. ΔT denotes the estimated time delay of arrival. The window coefficients in vector $\mathbf{w}$ are multiplied with the adaptive filter taps and thus smooth the frequency response

is shown in Fig. 11.26. This system requires a time delay compensation [Knapp, Carter 1976] to align the two input signals with respect to the speech signal. The adaptive filters may be implemented using a linear-phase version of the NLMS algorithm [Martin 2001b].

Other implementations of this principle use frequency domain methods and/or more than two microphones [Zelinski 1988], [Zelinski 1990]. In the context of microphone arrays, the filtering of the summed signal as in Fig. 11.26 has become known as the *postfilter* approach [Marro et al. 1998]. Also, the MSC function which gives an indication of which frequency bands contain correlated (i.e., useful) signal components has been used to dereverberate [Allen et al. 1977], [Marro et al. 1996] and to denoise speech [Ehrmann et al. 1995], [Le Bouquin-Jeannès et al. 1997].

11.9.4 Combined Single and Dual Channel Noise Reduction

To conclude this chapter we note that the single and dual channel approaches may be advantageously combined. An example of such a system is given in [Dörbecker, Ernst 1995]. Here, the Wiener filter as well as the noise PSD estimation algorithm uses the cross-correlation information of the two microphone channels. The resulting system is depicted in Fig. 11.27. Since this system delivers two output channels it can be used, for example, for a binaural hearing aid.

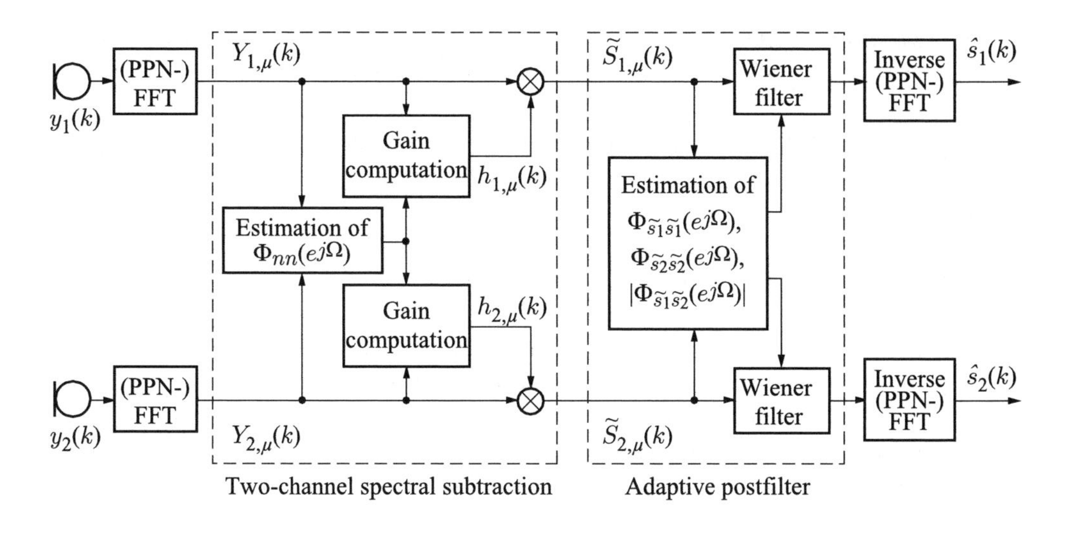

Figure 11.27: Two-channel spectral subtraction and Wiener filtering (PPN = Poly-Phase Network) [Dörbecker, Ernst 1995]

For the derivation of the two-channel noise PSD estimation we again define the two input signals of the algorithm as

$$\begin{aligned} y_1(k) &= a_{s1}(k) * s(k) + n_1(k) \\ y_2(k) &= a_{s2}(k) * s(k) + n_2(k) \end{aligned} \tag{11.165}$$

and assume that the magnitudes of the frequency responses $A_{s1}(e^{j\Omega})$ and $A_{s2}(e^{j\Omega})$ of filter $a_{s1}(k)$ and $a_{s2}(k)$, respectively, are equal,

$$|A_{s1}(e^{j\Omega})| \approx |A_{s2}(e^{j\Omega})| = |A_s(e^{j\Omega})| . \tag{11.166}$$

Furthermore, the two noise signals are assumed to be uncorrelated with each other and with the speech signal. We then have

$$\Phi_{y_1y_1}(e^{j\Omega}) = |A_s(e^{j\Omega})|^2\, \Phi_{ss}(e^{j\Omega}) + \Phi_{n_1n_1}(e^{j\Omega}) , \tag{11.167}$$

$$\Phi_{y_2y_2}(e^{j\Omega}) = |A_s(e^{j\Omega})|^2\, \Phi_{ss}(e^{j\Omega}) + \Phi_{n_2n_2}(e^{j\Omega}) \tag{11.168}$$

and

$$|\Phi_{y_1y_2}(e^{j\Omega})| \;=\; |A_s(e^{j\Omega})|^2\, \Phi_{ss}(e^{j\Omega}) . \tag{11.169}$$

If the two noise PSDs $\Phi_{n_1n_1}(e^{j\Omega})$ and $\Phi_{n_2n_2}(e^{j\Omega})$ are assumed to be equal, $\Phi_{n_1n_1}(e^{j\Omega}) = \Phi_{n_2n_2}(e^{j\Omega}) = \Phi_{nn}(e^{j\Omega})$, the geometric mean

$$\sqrt{\Phi_{y_1y_1}(e^{j\Omega})\Phi_{y_2y_2}(e^{j\Omega})} = |A_s(e^{j\Omega})|^2 \, \Phi_{ss}(e^{j\Omega}) + \Phi_{nn}(e^{j\Omega}) \tag{11.170}$$

leads to

$$\Phi_{nn}(e^{j\Omega}) = \sqrt{\Phi_{y_1y_1}(e^{j\Omega})\Phi_{y_2y_2}(e^{j\Omega})} - |\Phi_{y_1y_2}(e^{j\Omega})| \, . \tag{11.171}$$

This estimate may then be used in conjunction with a single channel noise reduction algorithm in each of the two microphone channels. In the second stage, a frequency-domain adaptive postfilter is employed which implements the dual channel noise reduction principle as outlined in Section 11.9.2. Thus, this method is suited, for example, for binaural hearing aids.

Bibliography

3GPP TR 126.978, V4.0.0 (2001). Universal Mobile Telecommunications System (UMTS); Results of the AMR noise suppression selection phase.

3GPP TS 122.076 V5.0.0 (2002). Digital cellular telecommunications system (Phase 2+); Universal Mobile Telecommunications System (UMTS); Noise Suppression for the AMR Codec; Service Description.

3GPP TS 126.094, V4.0.0 (2001). Universal Mobile Telecommunications System (UMTS); Mandatory Speech Codec Speech Processing Functions, AMR Speech Codec; Voice Activity Detector (VAD).

Accardi, A. J.; Cox, R. (1999). A Modular Approach to Speech Enhancement with an Application to Speech Coding, *Proceedings of the IEEE International Conference on Acoustics, Speech, and Signal Processing (ICASSP)*, Phoenix, Arizona, USA, vol. 1, pp. 201–204.

Allen, J. B. (1977). Short Term Spectral Analysis, Synthesis, and Modification by Discrete Fourier Transform, *IEEE Transactions on Acoustics, Speech and Signal Processing*, vol. ASSP-25, no. 3, June, pp. 235–238.

Allen, J. B.; Berkley, D. A.; Blauert, J. (1977). Multimicrophone Signal-Processing Technique to Remove Room Reverberation from Speech Signals, *Journal of the Acoustical Society of America*, vol. 62, no. 4, October, pp. 912–915.

Armbrüster, W.; Czarnach, R.; Vary, P. (1986). Adaptive Noise Cancellation with Reference Input - Possible Applications and Theoretical Limits, *Proceedings of the European Signal Processing Conference (EUSIPCO)*, pp. 391–394.

Beaugeant, C. (1999). *Réduction de Bruit et Contrôle d'Echo pour les Applications Radiomobiles*, PhD thesis, University of Rennes 1.

Beaugeant, C.; Scalart, P. (2001). Speech Enhancement Using a Minimum Least Square Amplitude Estimator, *Proceedings of the International Workshop on Acoustic Echo and Noise Control (IWAENC)*, Darmstadt, Germany, pp. 191–194.

Bendat, J. S.; Piersol, A. G. (1966). *Measurement and Analysis of Random Data*, John Wiley & Sons, Ltd., Chichester.

Beritelli, F.; Casale, S.; Ruggeri, G. (2001). Performance Evaluation and Comparison of ITU-T/ETSI Voice Activity Detectors, *Proceedings of the IEEE International Conference on Acoustics, Speech, and Signal Processing (ICASSP)*, Salt Lake City, Utah, USA, pp. 1425–1428.

Berouti, M.; Schwartz, R.; Makhoul, J. (1979). Enhancement of Speech Corrupted by Acoustic Noise, *Proceedings of the IEEE International Conference on Acoustics, Speech, and Signal Processing (ICASSP)*, Tulsa, Oklahoma, USA, pp. 208–211.

Boll, S. F. (1979). Suppression of Acoustic Noise in Speech Using Spectral Subtraction, *IEEE Transactions on Acoustics, Speech and Signal Processing*, vol. 27, pp. 113–120.

Breithaupt, C.; Martin, R. (2003). MMSE Estimation of Magnitude-Squared DFT Coefficients with Supergaussian Priors, *Proceedings of the IEEE International Conference on Acoustics, Speech, and Signal Processing (ICASSP)*, Hong Kong, China, vol. I, pp. 848–851.

Cappé, O. (1994). Elimination of the Musical Noise Phenomenon with the Ephraim and Malah Noise Suppressor, *IEEE Transactions on Speech and Audio Processing*, vol. 2, no. 2, April, pp. 345–349.

Cappé, O. (1995). Evaluation of Short-time Spectral Attenuation Techniques for the Restoration of Musical Recordings, *IEEE Transactions on Speech and Audio Processing*, vol. 3, no. 1, January, pp. 84–93.

Carter, G. C. (1987). Coherence and Time Delay Estimation, *Proceedings of the IEEE*, vol. 75, no. 2, February, pp. 236–255.

Carter, G. C.; Knapp, C. H.; Nuttall, A. H. (1973). Estimation of the Magnitude-Squared Coherence Function via Overlapped Fast Fourier Transform Processing, *IEEE Transactions on Audio and Electroacoustics*, vol. 21, no. 4, August, pp. 337–344.

Cohen, I.; Berdugo, B. (2001). Speech Enhancement for Non-Stationary Noise Environments, *Signal Processing*, vol. 81, pp. 2403–2418.

Collura, J. S. (1999). Speech Enhancement and Coding in Harsh Acoustic Noise Environments, *Proceedings of the IEEE Workshop on Speech Coding*, pp. 162–164.

Dat, T. H.; Takeda, K.; Itakura, F. (2005). Generalized Gamma Modeling of Speech and its Online Estimation for Speech Enhancement, *Proceedings of the IEEE International Conference on Acoustics, Speech, and Signal Processing (ICASSP)*, Philadelphia, Pennsylvania, USA, vol. IV, pp. 181–184.

David, H. A. (1980). *Order Statistics*, John Wiley & Sons, Ltd., Chichester.

Degan, N. D.; Prati, C. (1988). Acoustic Noise Analysis and Speech Enhancement Techniques for Mobile Radio Applications, *Signal Processing*, vol. 15, no. 1, pp. 43–56.

Doblinger, G. (1991). An Efficient Algorithm for Uniform and Nonuniform Digital Filter Banks, *IEEE International Symposium on Circuits and Systems (ISCAS)*, Raffles City, Singapore, pp. 646–649.

Doblinger, G.; Zeitlhofer, T. (1996). Improved Design of Uniform and Nonuniform Modulated Filter Banks, *Proceedings of the IEEE Nordic Signal Processing Symposium (NORSIG)*, Helsinki, Finland, pp. 327–330.

Dörbecker, M. (1998). *Mehrkanalige Signalverarbeitung zur Verbesserung akustisch gestörter Sprachsignale am Beispiel elektronischer Hörhilfen*, PhD thesis. *Aachener Beiträge zu digitalen Nachrichtensystemen*, vol. 10, P. Vary (ed.), RWTH Aachen University (in German).

Dörbecker, M.; Ernst, S. (1995). Combination of Two-Channel Spectral Subtraction and Adaptive Wiener Post-Filtering for Noise Reduction and Dereverberation, *Proceedings of the European Signal Processing Conference (EUSIPCO)*, pp. 995–998.

Ehrmann, F.; Le Bouquin-Jeannès, R.; Faucon, G. (1995). Optimization of a Two-sensor Noise Reduction Technique, *Signal Processing Letters*, vol. 2, pp. 108–110.

Ephraim, Y.; Malah, D. (1984). Speech Enhancement Using a Minimum Mean-Square Error Short-Time Spectral Amplitude Estimator, *IEEE Transactions on Acoustics, Speech and Signal Processing*, vol. 32, no. 6, December, pp. 1109–1121.

Ephraim, Y.; Malah, D. (1985). Speech Enhancement Using a Minimum Mean-Square Error Log-Spectral Amplitude Estimator, *IEEE Transactions on Acoustics, Speech and Signal Processing*, vol. 33, no. 2, April, pp. 443–445.

Faucon, G.; Mezalek, S. T.; Le Bouquin, R. (1989). Study and Comparison of Three Structures for Enhancement of Noisy Speech, *Proceedings of the IEEE International Conference on Acoustics, Speech, and Signal Processing (ICASSP)*, Glasgow, Scotland, pp. 385–388.

Freeman, D. K.; Cosier, G.; Southcott, C. B.; Boyd, I. (1989). The Voice Activity Detector for the Pan-European Digital Cellular Mobile Telephone Service,, *Proceedings of the IEEE International Conference on Acoustics, Speech, and Signal Processing (ICASSP)*, Glasgow, Schottland, pp. 369–372.

Gradshteyn, I. S.; Ryzhik, I. M. (1994). *Table of Integrals, Series, and Products*, 5th edn, Academic Press, San Diego, California.

Greenberg, J. E.; Zurek, P. M. (1992). Evaluation of an Adaptive Beamforming Method for Hearing Aids, *Journal of the Acoustical Society of America*, vol. 91, no. 3, mar, pp. 1662–1676.

Griffin, D. W.; Lim, J. S. (1984). Signal Estimation from Modified Short-Time Fourier Transform, *IEEE Transactions on Acoustics, Speech and Signal Processing*, vol. 32, no. 2, April, pp. 236–243.

Gülzow, T.; Engelsberg, A. (1998). Comparison of a Discrete Wavelet Transformation and a Nonuniform Polyphase Filterbank Applied to Spectral Subtraction Speech Enhancement, *Signal Processing*, vol. 64, no. 1, pp. 5–19.

Gülzow, T.; Ludwig, T.; Heute, U. (2003). Spectral-Substraction Speech Enhancement in Multirate Systems with and without Non-uniform and Adaptive Bandwidths, *Signal Processing*, vol. 83, pp. 1613–1631.

Gumbel, E. J. (1958). *Statistics of Extremes*, Columbia University Press, New York.

Gustafsson, H.; Nordholm, S.; Claesson, I. (1999). Spectral Subtraction Using Dual Microphones, *Proceedings of the International Workshop on Acoustic Echo and Noise Control (IWAENC)*, Pocono Manor, Pennsylvania, USA, pp. 60–63.

Gustafsson, S.; Jax, P.; Vary, P. (1998). A Novel Psychoacoustically Motivated Audio Enhancement Algorithm Preserving Background Noise Characteristics, *Proceedings of the IEEE International Conference on Acoustics, Speech, and Signal Processing (ICASSP)*, Seattle, Washington, USA, pp. 397–400.

Häkkinen, J.; Väänänen, M. (1993). Background Noise Suppressor for a Car Hands-Free Microphone, *Proceedings of the International Conference on Signal Processing Applications and Techniques (ICSPAT)*, Santa Clara, California, USA, pp. 300–307.

Händel, P. (1995). Low-Distortion Spectral Subtraction for Speech Enhancement, *Proceedings of the European Conference on Speech Communication and Technology (EUROSPEECH)*, Madrid, Spain, pp. 1549–1552.

Hansen, J. H. L. (1991). Speech Enhancement Employing Adaptive Boundary Detection and Morphological Based Spectral Constraints, *Proceedings of the IEEE International Conference on Acoustics, Speech, and Signal Processing (ICASSP)*, Toronto, Canada, pp. 901–905.

Harrison, W. A.; Lim, J. S.; Singer, E. (1986). A New Application of Adaptive Noise Cancellation, *IEEE Transactions on Acoustics, Speech and Signal Processing*, vol. 34, no. 1, February, pp. 21–27.

Haykin, S. (1996). *Adaptive Filter Theory*, 3rd edn, Prentice Hall, Englewood Cliffs, New Jersey.

Hirsch, H. G.; Ehrlicher, C. (1995). Noise Estimation Techniques for Robust Speech Recognition, *Proceedings of the IEEE International Conference on Acoustics, Speech, and Signal Processing (ICASSP)*, Detroit, Michigan, USA, vol. 1, pp. 153–156.

Johnson, N. L.; Kotz, S.; Balakrishnan, N. (1994). *Continuous Univariate Distributions*, John Wiley & Sons, Ltd., Chichester.

Kailath, T. (1981). *Lectures on Wiener and Kalman Filtering*, CISM Courses and Lectures No. 140, Springer Verlag, Berlin, Heidelberg, New York.

Kang, G. S.; Fransen, L. J. (1989). Quality Improvements of LPC-Processed Noisy Speech by Using Spectral Subtraction, *IEEE Transactions on Acoustics, Speech and Signal Processing*, vol. 37, no. 6, pp. 939–942.

Kappelan, M.; Strauß, B.; Vary, P. (1996). Flexible Nonuniform Filter Banks Using Allpass Transformation of Multiple Order, *Proceedings of Signal Processing VIII - Theories and Applications*, Trieste, Italy, pp. 1745–1748.

Knapp, C. H.; Carter, G. C. (1976). The Generalized Correlation Method for Estimation of Time Delay, *IEEE Transactions on Acoustics, Speech and Signal Processing*, vol. 24, August, pp. 320–327.

Kroschel, K.; Linhard, K. (1988). Combined Methods for Adaptive Noise Cancellation, *Proceedings of the European Signal Processing Conference (EUSIPCO)*, Grenoble, France, pp. 411–414.

Kushner, W. M.; Goncharoff, V.; Wu, C.; Nguyen, V.; Damoulakis, J. N. (1989). The Effects of Subtractive-Type Speech Enhancement/Noise Reduction Algorithms on Parameter Estimation for Improved Recognition and Coding in High Noise Environments, *Proceedings of the IEEE International Conference on Acoustics, Speech, and Signal Processing (ICASSP)*, Glasgow, Scotland, pp. 211–214.

Kuttruff, H. (1990). *Room Acoustics*, 3rd edn, Applied Science Publishers, Barking.

Le Bouquin-Jeannès, R.; Akbari Azirani, A.; Faucon, G. (1997). Enhancement of Speech Degraded by Coherent and Incoherent Noise using a Cross-spectral Estimator, *IEEE Transactions on Speech and Audio Processing*, vol. 5, pp. 484–487.

Linhard, K.; Haulick, T. (1999). Noise Subtraction with Parametric Recursive Gain Curves, *Proceedings of the European Conference on Speech Communication and Technology (EUROSPEECH)*, vol. 6, pp. 2611–2614.

Lockwood, P.; Baillargeat, C.; Gillot, J. M.; Boudy, J.; Faucon, G. (1991). Noise Reduction for Speech Enhancement in Cars: Non-Linear Spectral Subtraction/Kalman Filtering, *Proceedings of the European Conference on Speech Communication and Technology (EUROSPEECH)*, pp. 83–86.

Lockwood, P.; Boudy, J. (1991). Experiments with a Nonlinear Spectral Subtractor (NSS), Hidden Markov Models and the Projection, for Robust Speech Recognition in Cars, *Proceedings of the European Conference on Speech Communication and Technology (EUROSPEECH)*, pp. 79–82.

Lockwood, P.; Boudy, J. (1992). Experiments with a Nonlinear Spectral Subtractor (NSS), Hidden Markov Models and the Projection, for Robust Speech Recognition in Cars, *Speech Communication*, vol. 11, no. 2–3, pp. 215–228.

Lotter, T.; Vary, P. (2003). Noise Reduction by Maximum A Posteriori Spectral Amplitude Estimation with Supergaussian Speech Modeling, *Proceedings of the International Workshop on Acoustic Echo and Noise Control (IWAENC)*, Kyoto, Japan, pp. 83–86.

Lotter, T.; Vary, P. (2005). Speech Enhancement by MAP Spectral Amplitude Estimation Using a Super-Gaussian Speech Model, *EURASIP Journal on Applied Signal Processing*, vol. 2005, no. 7, May, pp. 1110–1126.

Malah, D.; Cox, R. V.; Accardi, A. J. (1999). Tracking Speech-Presence Uncertainty to Improve Speech Enhancement in Non-Stationary Noise Environments, *Proceedings of the IEEE International Conference on Acoustics, Speech, and Signal Processing (ICASSP)*, Phoenix, Arizona, USA, pp. 789–792.

Markel, J. D.; Gray, A. H. (1976). *Linear Prediction of Speech*, Springer Verlag, Berlin, Heidelberg, New York.

Marro, C.; Mahieux, Y.; Simmer, K. U. (1996). Performance of Adaptive Dereverberation Techniques Using Directivity Controlled Arrays, *Proceedings of the European Signal Processing Conference (EUSIPCO)*, Trieste, Italy, pp. 1127–1130.

Marro, C.; Mahieux, Y.; Simmer, K. U. (1998). Analysis of Noise Reduction and Dereverberation Techniques Based on Microphone Arrays with Postfiltering, *IEEE Transactions on Speech and Audio Processing*, vol. 6, no. 3, pp. 240–259.

Martin, R. (1993). An Efficient Algorithm to Estimate the Instantaneous SNR of Speech Signals, *Proceedings of the European Conference on Speech Communication and Technology (EUROSPEECH)*, Berlin, Germany, pp. 1093–1096.

Martin, R. (1994). Spectral Subtraction Based on Minimum Statistics, *Proceedings of the European Signal Processing Conference (EUSIPCO)*, Edinburgh, Scotland, pp. 1182–1185.

Martin, R. (1995). *Hands-free Telephones Based on Multi-Channel Echo Cancellation and Noise Reduction*, PhD thesis. *Aachener Beiträge zu digitalen Nachrichtensystemen*, vol. 3, P. Vary (ed.), RWTH Aachen University (in German).

Martin, R. (2001a). Noise Power Spectral Density Estimation Based on Optimal Smoothing and Minimum Statistics, *IEEE Transactions on Speech and Audio Processing*, vol. 9, no. 5, July, pp. 504–512.

Martin, R. (2001b). Small Microphone Arrays with Postfilters for Noise and Acoustic Echo Reduction, *in* M. Brandstein; D. Ward (eds.), *Microphone Arrays*, Springer Verlag, Berlin, Heidelberg, New York.

Martin, R. (2002). Speech Enhancement Using MMSE Short Time Spectral Estimation with Gamma Distributed Speech Priors, *Proceedings of the IEEE International Conference on Acoustics, Speech, and Signal Processing (ICASSP)*, Orlando, Florida, USA, vol. I, pp. 253–256.

Martin, R. (2005a). Bias Compensation Methods for Minimum Statistics Noise Power Spectral Density Estimation, *Signal Processing.* Special Issue on Speech and Audio Processing (to appear).

Martin, R. (2005b). Speech Enhancement based on Minimum Mean Square Error Estimation and Supergaussian Priors, *IEEE Transactions on Speech and Audio Processing*, vol. 13, no. 5, pp. 845–856.

Martin, R. (2005c). Statistical Methods for the Enhancement of Noisy Speech, *in* J. Benesty; S. Makino; J. Chen (eds.), *Speech Enhancement*, Springer Verlag, Berlin, Heidelberg, New York.

Martin, R.; Breithaupt, C. (2003). Speech Enhancement in the DFT Domain Using Laplacian Speech Priors, *Proceedings of the International Workshop on Acoustic Echo and Noise Control (IWAENC)*, Kyoto, Japan, pp. 87–90.

Martin, R.; Cox, R. V. (1999). New Speech Enhancement Techniques for Low Bit Rate Speech Coding, *Proceedings of the IEEE Workshop on Speech Coding*, Porvoo, Finland, pp. 165–167.

Martin, R.; Kang, H. G.; Cox, R. V. (1999). Low Delay Analysis/Synthesis Schemes for Joint Speech Enhancement and Low Bit Rate Speech Coding, *Proceedings of the European Conference on Speech Communication and Technology (EUROSPEECH)*, Budapest, Hungary, vol. 3, pp. 1463–1466.

Martin, R.; Lotter, T. (2001). Optimal Recursive Smoothing of Non-Stationary Periodograms, *Proceedings of the International Workshop on Acoustic Echo and Noise Control (IWAENC)*, Darmstadt, Germany, pp. 167–170.

Martin, R.; Vary, P. (1992). A Symmetric Two Microphone Speech Enhancement System – Theoretical Limits and Application in a Car Environment, *Proceedings of the Fifth IEEE Signal Processing Workshop*, Uttica, Illinois, USA, pp. 4.5.1–4.5.2.

Martin, R.; Wittke, J.; Jax, P. (2000). Optimized Estimation of Spectral Parameters for the Coding of Noisy Speech, *Proceedings of the IEEE International Conference on Acoustics, Speech, and Signal Processing (ICASSP)*, Istanbul, Turkey, vol. III, pp. 1479–1482.

McAulay, R. J.; Malpass, M. L. (1980). Speech Enhancement Using a Soft-Decision Noise Suppression Filter, *IEEE Transactions on Acoustics, Speech and Signal Processing*, vol. 28, no. 2, December, pp. 137–145.

Melsa, J. L.; Cohn, D. L. (1978). *Decision and Estimation Theory*, McGraw-Hill Kogakusha, Tokyo.

Meyer, J.; Simmer, K. U.; Kammeyer, K. D. (1997). Comparison of One- and Two-Channel Noise-Estimation Techniques, *Proceedings of the International Workshop on Acoustic Echo and Noise Control (IWAENC)*, London, UK, pp. 17–20.

Middleton, D.; Esposito, R. (1968). Simultaneous Optimum Detection and Estimation of Signals in Noise, *IEEE Transactions on Information Theory*, vol. 14, no. 3, pp. 434–444.

Paliwal, K. K.; Atal, B. (1991). Efficient Vector Quantization of LPC Parameters at 24 Bits/Frame, *Proceedings of the IEEE International Conference on Acoustics, Speech, and Signal Processing (ICASSP)*, Toronto, Canada, pp. 661–664.

Papoulis, A.; Unnikrishna Pillai, S. (2001). *Probability, Random Variables, and Stochastic Processes*, 4th edn, McGraw-Hill, New York.

Porter, J. E.; Boll, S. F. (1984). Optimal Estimators for Spectral Restoration of Noisy Speech, *Proceedings of the IEEE International Conference on Acoustics, Speech, and Signal Processing (ICASSP)*, San Diego, California, USA, pp. 18A.2.1–18A.2.4.

Preuss, R. D. (1979). A Frequency Domain Noise Cancelling Preprocessor for Narrowband Speech Communication Systems, *Proceedings of the IEEE International Conference on Acoustics, Speech, and Signal Processing (ICASSP)*, Washington, DC, USA, pp. 212–215.

Ramabadran, T. V.; Ashley, J. P.; McLaughlin, M. J. (1997). Background Noise Suppression for Speech Enhancement and Coding, *Proceedings of the IEEE Workshop on Speech Coding*, Pocono Manor, Pennsylvania, USA, pp. 43–44.

Sambur, M. R. (1978). Adaptive Noise Canceling for Speech Signals, *IEEE Transactions on Acoustics, Speech and Signal Processing*, vol. 26, no. 5, pp. 419–423.

Scalart, P.; Vieira Filho, J. (1996). Speech Enhancement Based on A Priori Signal to Noise Estimation, *Proceedings of the IEEE International Conference on Acoustics, Speech, and Signal Processing (ICASSP)*, Atlanta, Georgia, USA, pp. 629–632.

Sohn, J.; Kim, N. S.; Sung, W. (1999). A Statistical Model-Based Voice Activity Detector, *Signal Processing Letters*, vol. 6, no. 1, pp. 1–3.

Sohn, J.; Sung, W. (1998). A Voice Activity Detector Employing Soft Decision Based Noise Spectrum Adaptation, *Proceedings of the IEEE International Conference on Acoustics, Speech, and Signal Processing (ICASSP)*, Seattle, Washington, USA, vol. 1, pp. 365–368.

Soon, I. Y.; Koh, S. N.; Yeo, C. K. (1999). Improved noise suppression filter using self-adaptive estimator of probability of speech absence, *Signal Processing*, vol. 75, pp. 151–159.

Srinivasan, K.; Gersho, A. (1993). Voice Activity Detection for Cellular Networks, *Proceedings of the IEEE Workshop on Speech Coding*, St. Jovite, Canada, pp. 85–86.

Tsoukalas, D.; Paraskevas, M.; Mourjopoulos, J. (1993). Speech Enhancement using Psychoacoustic Criteria, *Proceedings of the IEEE International Conference on Acoustics, Speech, and Signal Processing (ICASSP)*, pp. 359–362.

Vähätalo, A.; Johansson, I. (1999). Voice Activity Detection for GSM Adaptive Multi-Rate Codec, *Proceedings of the IEEE Workshop on Speech Coding*, Porvoo, Finland, pp. 55–57.

Van Compernolle, D. (1989). Noise Adaptation in a Hidden Markov Model Speech Recognition System, *Computer Speech and Language*, vol. 3, pp. 151–167.

Van Hoesel, R. J. M.; Clark, G. M. (1995). Evaluation of a Portable Two-Microphone Adaptive Beamforming Speech Processor with Cochlear Implant Patients, *Journal of the Acoustical Society of America*, vol. 97, no. 4, pp. 2498–2503.

Van Trees, H. L. (1968). *Detection, Estimation, and Modulation Theory*, MIT Press, Cambridge, Massachusetts.

Vary, P. (1985). Noise Suppression by Spectral Magnitude Estimation - Mechanism and Theoretical Limits, *Signal Processing*, vol. 8, pp. 387–400.

Vary, P. (2005). An Adaptive Filterbank Equalizer for Speech Enhancement, *Signal Processing*. Special Issue on Speech and Audio Processing (to appear).

Virag, N. (1999). Single Channel Speech Enhancement Based on Masking Properties of the Human Auditory System, *IEEE Transactions on Speech and Audio Processing*, vol. 7, no. 2, pp. 126–137.

Wang, D. L.; Lim, J. S. (1982). The Unimportance of Phase in Speech Enhancement, *IEEE Transactions on Acoustics, Speech and Signal Processing*, vol. 30, no. 4, pp. 679–681.

Welch, P. D. (1967). The Use of Fast Fourier Transform for the Estimation of Power Spectra: A Method Based on Time Averaging Over Short, Modified Periodograms, *IEEE Transactions on Audio and Electroacoustics*, vol. 15, no. 2, pp. 70–73.

Widrow, B.; Glover, J. R.; McCool, J. M.; Kaunitz, J.; Williams, C. S.; Hearn, R. H.; Zeidler, J. R.; Dong, E.; Goodlin, R. C. (1975). Adaptive Noise Cancelling: Principles and Applications, *Proceedings of the IEEE*, vol. 63, no. 12, December, pp. 1692–1716.

Wolfe, P. J.; Godsill, S. J. (2001). Simple Alternatives to the Ephraim and Malah Suppression Rule for Speech Enhancement, *Proceedings of the eleventh IEEE Workshop on Statistical Signal Processing*, Singapore, vol. II, pp. 496–499.

Wouters, J.; Vanden Berghe, J. (2001). Speech Recognition in Noise for Cochlear Implantees with a Two-Microphone Monaural Adaptive Noise Reduction System, *Ear and Hearing*, vol. 22, no. 5, pp. 420–430.

Yang, J. (1993). Frequency Domain Noise Suppression Approaches in Mobile Telephone Systems, *Proceedings of the IEEE International Conference on Acoustics, Speech, and Signal Processing (ICASSP)*, Minneapolis, Minnesota, pp. 363–366.

Zelinski, R. (1988). A Microphone Array with Adaptive Post-Filtering for Noise Reduction in Reverberant Rooms, *Proceedings of the IEEE International Conference on Acoustics, Speech, and Signal Processing (ICASSP)*, New York City, USA, pp. 2578–2581.

Zelinski, R. (1990). Noise Reduction Based on Microphone Array with LMS Adaptive Post-Filtering, *Electronics Letters*, vol. 26, no. 24, November, pp. 2036–2037.

12

Multi-channel Noise Reduction

In this chapter we will study the question of how sound pickup in a noisy and reverberant environment can be improved by adding spatial diversity to the signal acquisition front-end. Spatially distributed receivers, i.e., microphone arrays, and multi-channel signal processing techniques allow the exploitation of spatial and statistical features of signals and the achievement of a performance which surpasses that of single channel systems.

12.1 Introduction

When more than one microphone is available for sound pickup the signal enhancement task may be facilitated by exploiting the multivariate deterministic and stochastic properties of the signals. From a deterministic viewpoint, the signals at the various microphones differ in that they arrive via different acoustic paths at the microphones and thus, also differ in their short-term amplitude and phase. From a stochastic perspective, multi-channel methods allow the evaluation of the second-order and higher-order statistics of the spatial sound field. Sources which are close to the array will generate mostly coherent microphone signals while distant and distributed sources lead to uncorrelated signals. Thus,

Digital Speech Transmission: Enhancement, Coding and Error Concealment
Peter Vary and Rainer Martin

short-term amplitude, short-term phase, and the statistics of the signals may be used to differentiate between sources and to perform source separation.

Array technology was developed for and has been used in radar and sonar systems for quite some time [Monzingo, Miller 1980], [Haykin 1985], [Gabriel 1992]. Very frequently, these systems are designed for narrowband signals. The application of array technology to speech signals can be challenging as speech is a wideband signal spanning several octaves. Furthermore, in many speech processing applications the environment is highly reverberant. As a consequence, the desired signal will arrive not only from one primary direction but also via reflections from the enclosing walls.

In this chapter we first develop the basic scenario and define signal models and performance measures. We will then consider microphone arrays and their properties for stationary environments, i.e., with fixed beam patterns. We explain typical design procedures for beamformers where we assume that the direction of incidence of the source signal is given and that the sound field is stationary. Finally, in Section 12.8, we briefly discuss postfilter techniques and adaptive beamforming approaches.

12.2 Sound Waves

Sound is a mechanical vibration that propagates through matter in the form of waves. Sound waves may be described in terms of a sound pressure field $p(\mathbf{r}, t)$ and a sound velocity vector field $\mathbf{u}(\mathbf{r}, t)$ which are both functions of a spatial vector $\mathbf{r}$ and time t. While the sound pressure characterizes the density variations (we do not consider the DC component), the sound velocity describes the velocity of dislocation of the physical particles which carry the waves.

In the context of our applications, the relation between the quantities of the sound field may be linearized. Then, for a wave propagating in just one spatial dimension $\mathfrak{r}$, these two quantities are related as (see also Section 2.3.1)

$$\frac{\partial p}{\partial \mathfrak{r}} = -\rho_0 \frac{\partial u}{\partial t} \quad \text{and} \quad \frac{\partial u}{\partial \mathfrak{r}} = -\frac{1}{\rho_0 c^2} \frac{\partial p}{\partial t} \tag{12.1}$$

where c and ρ_0 are the speed of sound and the density at rest, respectively. Both equations may be combined into a wave equation

$$\Delta p = \frac{\partial^2 p}{\partial \mathfrak{r}^2} = \frac{1}{c^2} \frac{\partial^2 p}{\partial t^2} \tag{12.2}$$

which, when using the appropriate definition of the Laplace operator Δp, is also valid for three-dimensional wave propagation. For example, in Cartesian coordinates $\mathfrak{x}$, $\mathfrak{y}$, and $\mathfrak{z}$ we have

$$\Delta p = \frac{\partial^2 p}{\partial \mathfrak{x}^2} + \frac{\partial^2 p}{\partial \mathfrak{y}^2} + \frac{\partial^2 p}{\partial \mathfrak{z}^2} \,. \tag{12.3}$$

For plane waves, the surfaces of constant sound pressure are planes which propagate in a given spatial direction. A harmonic plane wave which propagates in the $\mathfrak{x}$ direction may be written as

$$p(\mathfrak{x}, t) = \widehat{p} e^{j(\omega t - \widetilde{\beta}\mathfrak{x})} \tag{12.4}$$

where $\widetilde{\beta} = 2\pi/\lambda$, λ and $\widehat{p}$ are the *wave number*, the *wave length* and the *amplitude*, respectively. Using (12.1), the $\mathfrak{x}$ component of the sound velocity is then given by

$$u_{\mathfrak{x}}(\mathfrak{x}, t) = \frac{1}{\rho_0 c} p(\mathfrak{x}, t) \,. \tag{12.5}$$

Thus, for a plane wave, the sound velocity is proportional to the sound pressure.

Waves which have a constant sound pressure on concentrical spheres are also of interest. The wave equation (12.2) delivers the solution for such a *spherical wave* which propagates in radial direction r as

$$p(r, t) = \frac{1}{r} f(r - ct) \,, \tag{12.6}$$

where f is the propagating waveform. The amplitude of the sound wave diminishes with increasing distance from the center of the spheres. We may use the abstraction of a *point source* to explain the generation of such *spherical waves.*

An ideal point source may be represented by its acoustic volume velocity $v(t)$ [Kuttruff 2004]. Furthermore, with (12.1) we have

$$\frac{\partial u_r(r, t)}{\partial t} = -\frac{1}{\rho_0} \frac{\partial p(r, t)}{\partial r} = \frac{1}{\rho_0} \left(\frac{f(r - ct)}{r^2} - \frac{\mathrm{d}f(r - ct)/\mathrm{d}r}{r} \right) . \tag{12.7}$$

For an infinitesimally small sphere of radius r, the velocity vector may be integrated to yield $v(t) \approx 4\pi r^2 u_r(r, t)$. For $r \to 0$, the second term on the right hand side of (12.7) is smaller than the first. Therefore, with (12.7) we find

$$\frac{dv(t)}{dt} \approx \frac{4\pi}{\rho_0} f(r - ct)_{|r \to 0} \tag{12.8}$$

and, with (12.6),

$$p(r, t) = \frac{\rho_0}{4\pi r} \frac{dv(t - r/c)}{dt} \tag{12.9}$$

which characterizes, again, a spherical wave. The sound pressure is inversely proportional to the radial distance r from the point source. For a harmonic excitation

$$v(t) = \widehat{v}e^{j\omega t} \tag{12.10}$$

we find the sound pressure

$$p(r,t) = \frac{j\omega\rho_0\widehat{v}e^{j\omega(t-r/c)}}{4\pi r} = \frac{j\omega\rho_0\widehat{v}e^{j(\omega t-\widetilde{\beta}r)}}{4\pi r} \tag{12.11}$$

and hence, with (12.7) and an integration with respect to time, the sound velocity

$$u_r(r,t) = \frac{p(r,t)}{\rho_0 c}\left(1 + \frac{1}{j\widetilde{\beta}r}\right). \tag{12.12}$$

Clearly, (12.11) and (12.12) satisfy (12.7). Because of the second term in the parentheses in (12.12), sound pressure and sound velocity are not in phase. Depending on the distance of the observation point to the point source, the behaviour of the wave is distinctly different. When the second term cannot be neglected the observation point is in the *nearfield* of the source. For $\widetilde{\beta}r \gg 1$ the observation point is in the *farfield*. The transition from the nearfield to the farfield depends on the wave number $\widetilde{\beta}$ and, as such, on the wave length or the frequency of the harmonic excitation.

12.3 Spatial Sampling of Sound Fields

For our purposes, microphones may be modeled as discrete points in space at whose location the spatial sound pressure field is sampled. Unless explicitly stated, we assume that the microphones are omnidirectional (sound pressure) receivers, i.e., they have the same sensitivity regardless of the direction of the impinging sound.

Figure 12.1 illustrates the general scenario of a single sound source and a distributed microphone array. The position of the source and the positions of the N_M microphones with respect to a reference coordinate system are denoted by vectors $\mathbf{r}_s$ and $\mathbf{r}_\ell$, $\ell = 1 \ldots N_M$, respectively. $\mathbf{e}_x$, $\mathbf{e}_y$, and $\mathbf{e}_z$ denote orthogonal unit vectors spanning a Cartesian coordinate system.

In an anechoic environment (no reverberation or noise), the microphone signals $y_\ell(t)$, $\ell = 1, \ldots N_M$, are delayed and attenuated versions of the source signal $s_0(t)$,

$$y_\ell(t) = \frac{1}{||\mathbf{r}_\ell - \mathbf{r}_s||} s_0(t - \tau_\ell) \tag{12.13}$$

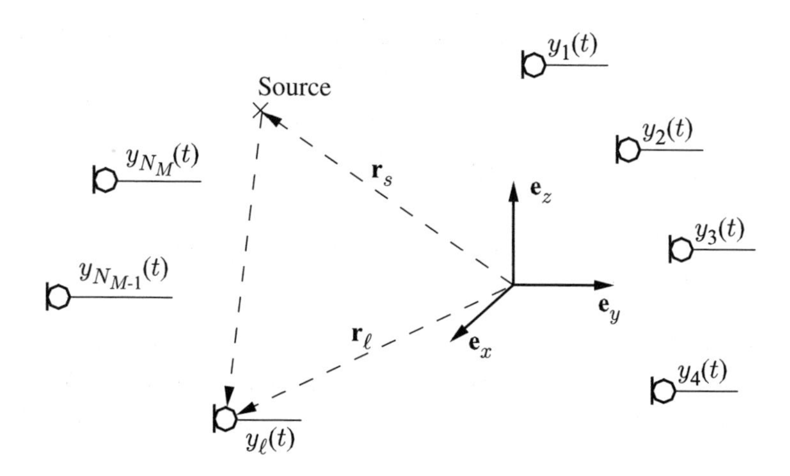

Figure 12.1: Microphone array in the nearfield of a single source. $\mathbf{e}_x$, $\mathbf{e}_y$, and $\mathbf{e}_z$ are orthogonal unit vectors spanning a Cartesian coordinate system

where $s_0(t)$ originates from a point source at $\mathbf{r}_s$ and $||\mathbf{r}_s||$ denotes the norm of vector $\mathbf{r}_s$. The absolute signal delay τ_ℓ of the ℓ-th microphone signal is given by

$$\tau_\ell = \frac{||\mathbf{r}_\ell - \mathbf{r}_s||}{c} . \tag{12.14}$$

We denote the signal which is received at the origin of the coordinate system by $s(t)$. τ_0 is the signal delay from the source to the origin. The origin of the coordinate system is henceforth referred to as the *reference point*. Then,

$$s(t) = \frac{1}{||\mathbf{r}_s||} s_0(t - \tau_0) \tag{12.15}$$

where the reference point may not coincide with the source location.

The reference point could be the geometric center of the array or, for convenience, the location of one of the microphones. Frequently, we are not interested in the absolute delay of the signal from the source to the microphones but rather in the delay of the signals relative to the signal which is received at the reference point. The *relative signal delay* $\Delta\tau_\ell$, i.e., the time delay difference between the received signal at the reference point and at the ℓ-th microphone, is then given by

$$\Delta\tau_\ell = \tau_0 - \tau_\ell = \frac{1}{c} \left(||\mathbf{r}_s|| - ||\mathbf{r}_\ell - \mathbf{r}_s|| \right) . \tag{12.16}$$

Thus, in an anechoic environment the ℓ-th microphone signal $y_\ell(t)$ may now be written as a function of the signal at the reference point,

$$y_\ell(t) = \frac{||\mathbf{r}_s||}{||\mathbf{r}_\ell - \mathbf{r}_s||} s\left(t + \Delta\tau_\ell\right) . \tag{12.17}$$

If the Fourier transform $S(j\omega)$ of $s(t)$ exists, the Fourier transform $Y_\ell(j\omega)$ of the microphone signals $y_\ell(t)$ may be written as

$$\begin{aligned} Y_\ell(j\omega) &= \frac{||\mathbf{r}_s||}{||\mathbf{r}_\ell - \mathbf{r}_s||} S(j\omega) \exp\left(j2\pi f \Delta\tau_\ell\right) \\ &= \frac{||\mathbf{r}_s||}{||\mathbf{r}_\ell - \mathbf{r}_s||} S(j\omega) \exp\left(j\widetilde{\beta}\left(||\mathbf{r}_s|| - ||\mathbf{r}_\ell - \mathbf{r}_s||\right)\right) \end{aligned} \quad (12.18)$$

where $\widetilde{\beta} = \frac{2\pi}{\lambda} = \frac{2\pi f}{c}$ is the *wave number* as before. In the above model no assumptions about the distance between the source and the array were made. The source may be arbitrarily close to the microphones. Therefore, this model is denoted as the *nearfield* model.

12.3.1 The Farfield Model

We now assume that the distance between the sound source and the microphone array is much larger than the largest dimension (the *aperture*) of the array and much larger than the wavelength, i.e., $\widetilde{\beta} r \gg 1$. In this case the sound waves which are picked up by the microphones may be modeled as plane waves. This *farfield* scenario is illustrated in Fig. 12.2. When the source is far from the array then

$$\frac{||\mathbf{r}_s||}{||\mathbf{r}_1 - \mathbf{r}_s||} \approx \frac{||\mathbf{r}_s||}{||\mathbf{r}_2 - \mathbf{r}_s||} \approx \cdots \approx \frac{||\mathbf{r}_s||}{||\mathbf{r}_{N_M} - \mathbf{r}_s||} . \quad (12.19)$$

The absolute and the relative attenuation of the source signal at the microphones are approximately the same for all microphones. The phase differences between the

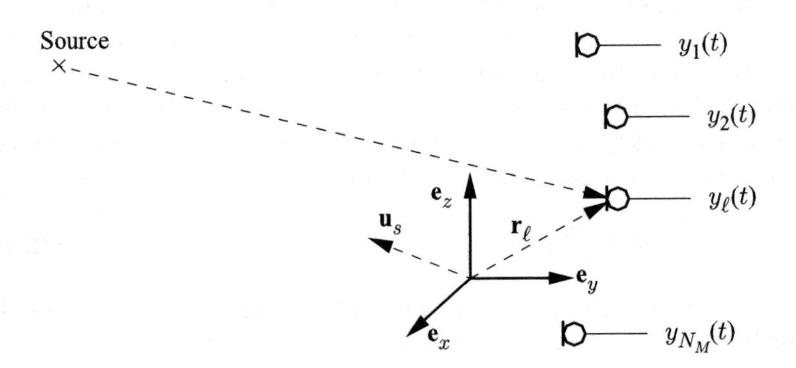

Figure 12.2: Microphone array in the farfield of a single source

microphone signals, however, depend on the distance between the microphones and the wavelength of the impinging wave.

For the single source farfield scenario, when the reference point is close to the microphones, $||\mathbf{r}_\ell|| \ll ||\mathbf{r}_s||$, we may write

$$y_\ell(t) \approx s(t + \Delta\tau_\ell) \approx s\left(t + \frac{\langle \mathbf{r}_\ell, \mathbf{u}_s \rangle}{c}\right) \tag{12.20}$$

where $\mathbf{u}_s$ is a unit vector which points from the reference point towards the source, i.e., $\mathbf{u}_s = \mathbf{r}_s/||\mathbf{r}_s||$, and $\langle \mathbf{r}_\ell, \mathbf{u}_s \rangle$ denotes the inner product of vectors $\mathbf{r}_\ell$ and $\mathbf{u}_s$. When the reference point is closer to the source than the microphones, as shown in Fig. 12.2, this inner product is negative. In Cartesian coordinates we have

$$y_\ell(t) \approx s\left(t + \frac{\mathbf{r}_\ell^T \mathbf{u}_s}{c}\right) . \tag{12.21}$$

The relative delays depend on the array geometry and the components of $\mathbf{r}_\ell$ in the direction of $\mathbf{u}_s$.

In the frequency domain, and for the farfield model, the microphone signals can then be expressed as

$$Y_\ell(j\omega) = S(j\omega)\exp(j2\pi f \Delta\tau_\ell) , \tag{12.22}$$

where we assume that the Fourier transform of signal $S(j\omega)$ exists. When the microphone signals are sampled with sampling rate f_S, we may write the signal spectra of the sampled microphone signals $y_\ell(k)$ as a function of the normalized frequency Ω as

$$Y_\ell(e^{j\Omega}) = S(e^{j\Omega})\exp\left(j\Omega f_S \Delta\tau_\ell\right) , \tag{12.23}$$

or, in vector notation,

$$\mathbf{Y}(e^{j\Omega}) = S(e^{j\Omega})\mathbf{a} . \tag{12.24}$$

with a vector of signal spectra

$$\mathbf{Y}(e^{j\Omega}) = \left(Y_1(e^{j\Omega}), \ldots , Y_{N_M}(e^{j\Omega})\right)^T \tag{12.25}$$

and the *propagation vector*

$$\mathbf{a} = \left(\exp\left(j\Omega f_s \Delta\tau_1\right), \exp\left(j\Omega f_s \Delta\tau_2\right), \ldots , \exp\left(j\Omega f_s \Delta\tau_{N_M}\right)\right)^T . \tag{12.26}$$

In the farfield scenario, $\mathbf{a}^H \mathbf{Y}(e^{j\Omega})$ will yield the sum of perfectly phase aligned microphone signals.

12.3.2 The Uniform Linear Array

Figure 12.3 depicts an important special case, the linear array with N_M uniformly spaced microphones in the farfield of the acoustic source. The microphones are symmetrically aligned with an inter-microphone distance d along the z-axis. The reference point is in the geometric center of the array. In this case it is convenient to introduce polar coordinates. We define an azimuth φ_s and an elevation θ_s which characterize the direction of the desired sound source. Due to the symmetry of the array, the relative signal delays and thus, the array response, do not depend on the azimuth φ_s.

In Cartesian coordinates, the relative delays are given by $\Delta\tau_\ell = \pm \mathbf{e}_z^T \mathbf{u}_s ||\mathbf{r}_\ell||/c$, or, using the source elevation θ_s, by

$$\Delta\tau_\ell = \frac{d}{c}\left(\frac{N_M+1}{2} - \ell\right)\cos(\theta_s)\,, \quad \ell = 1, 2, \ldots, N_M\,. \tag{12.27}$$

When the elevation θ_s is zero or π we have an *endfire* orientation of the array with respect to the source. Then, the *look direction* of the array coincides with the positive z-axis and $\Delta\tau_\ell$ attains its maximum absolute value. In *broadside* orientation the look direction is perpendicular to the z-axis, i.e., $\theta_s = \pi/2$ and $\Delta\tau_\ell = 0$ for all ℓ.

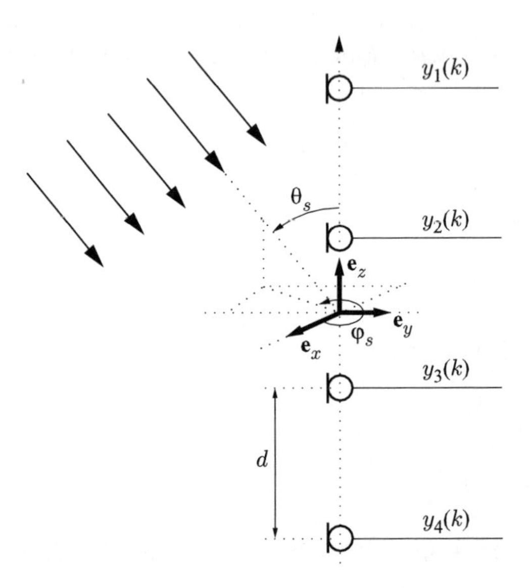

Figure 12.3: Uniform linear microphone array with $N_M = 4$

12.3.3 Phase Ambiguity and Coherence

The sound pressure field $p(\mathbf{r}, t)$ is a function of space and time. The microphone array samples this wave field at discrete points in space. In analogy to the sampling theorem of discrete time signals, the spatial sampling of the sound field must obey a spatial sampling relation if *spatial aliasing* is to be avoided. The analysis of a sound field by means of a microphone array can lead to ambiguous results if the spatial and temporal sampling constraints are not properly taken into account.

Since the complex exponential in (12.22) is periodic, integer increments in $f\Delta\tau_\ell$ lead to the same value of $\exp(j2\pi f\Delta\tau_\ell)$. Therefore, for a fixed relative delay $\Delta\tau_\ell$, the signal phase may vary quite significantly as a function of frequency.

For a harmonic plane wave and two closely spaced microphones located along the z-axis at

$$\mathbf{r}_1 = \frac{d}{2}\mathbf{e}_z \qquad \text{and} \qquad \mathbf{r}_2 = -\frac{d}{2}\mathbf{e}_z \tag{12.28}$$

where $\mathbf{e}_z$ is the unit vector in z-direction, the sampling condition can be derived from the phase term $\exp(j2\pi f\Delta\tau_\ell)$. If we denote the phase of the harmonic wave at the first and the second microphone by $\phi_1 = \phi + \widetilde{\beta}\langle\mathbf{e}_z, \mathbf{u}_s\rangle\, d/2$ and $\phi_2 = \phi - \widetilde{\beta}\langle\mathbf{e}_z, \mathbf{u}_s\rangle\, d/2$ respectively, the phase difference is given by

$$\Delta\phi = \phi_1 - \phi_2 = \widetilde{\beta} d\,\langle\mathbf{e}_z, \mathbf{u}_s\rangle = \widetilde{\beta} d\cos(\theta_s) \tag{12.29}$$

where θ_s denotes the angle between the positive z-axis and the direction of the sound source. The phase difference is zero when the wave impinges from the broadside direction ($\mathbf{u}_s \perp \mathbf{e}_z$). The phase difference is largest when the microphones are in endfire orientation ($\mathbf{u}_s \parallel \mathbf{e}_z$ or $\mathbf{u}_s \parallel -\mathbf{e}_z$). Obviously, ambiguous phase values are avoided when, for $\theta_s \in [0, \pi]$, the phase difference $\Delta\phi = \widetilde{\beta} d\cos(\theta_s)$ is in the range $-\pi < \Delta\phi \leq \pi$. Thus, a one-to-one mapping of $\theta_s \in [0, \pi]$ and $\Delta\phi$ is achieved for

$$d \leq \frac{\lambda}{2} \quad \text{or} \quad f \leq \frac{c}{2d}\,. \tag{12.30}$$

Figure 12.4 plots the phase difference between two microphones as a function of the angle of incidence θ_s and frequency. Clearly, for $\theta_s = \pm\pi/2$, the phase difference is zero. Additionally it can be seen that for a fixed frequency $f \geq c/(2d)$ there is no one-to-one mapping of $\theta_s \in [0, \pi]$ and $\Delta\phi$.

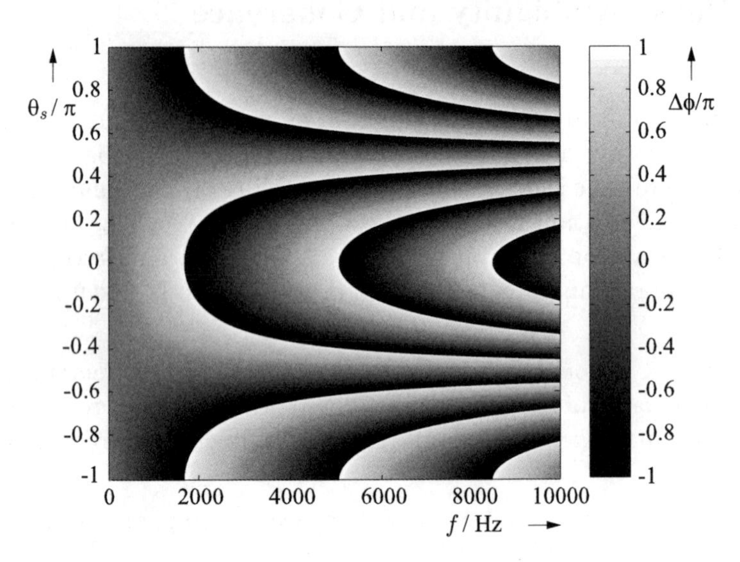

Figure 12.4: Phase difference $\Delta\phi/\pi$ for $-\pi \leq \theta_s \leq \pi$ of two microphone signals recorded at distance $d = 0.1\,\text{m}$ as a function of frequency f ($c = 340\,\text{m/s}$)

12.3.4 Spatial Correlation Properties of Acoustic Signals

Besides the deterministic phase differences, multi-channel signal processing techniques allow the exploitation of the statistical properties of the microphone signals, especially the spatial correlation of sound fields. In the context of speech enhancement, three cases are of special interest:

1. The microphone signals y_ℓ are uncorrelated.
   Fully uncorrelated signals originate from the microphones themselves as thermal self-noise. Although we may assume that microphones are selected such that this noise does not degrade the perceived quality of the microphone signals, we must be aware that self-noise might be amplified by the multi-channel signal processing algorithm.

2. The microphone signals exhibit a frequency-dependent correlation.
   A typical example of frequency-dependent correlation is the ideal diffuse sound field, see Section 11.9. In this sound field, sound energy impinges with equal power from all spatial directions onto the microphone array. For omnidirectional receivers, the magnitude squared coherence can then be written as a squared sinc function (11.151). This type of correlation is encountered in reverberant spaces with multiple, distributed noise sources such as in car compartments or in noisy offices.

3. The microphone signals are highly correlated.
   Highly correlated signals occur when the sound source is close to the microphones and the microphone signals contain little noise and reverberation. Then, the direct sound dominates and leads to high and frequency-independent correlation of the microphone signals.

12.4 Beamforming

The task of the beamforming algorithm is to combine the sampled microphone signals such that a desired (and possibly time-varying) spatial selectivity is achieved. In single source scenarios, this comprises the formation of a beam of high gain in the direction of the source.

12.4.1 Delay-and-Sum Beamforming

The simplest method to solve the signal combination problem is to combine the microphone signals such that the desired signals add up constructively. In an acoustic environment where direct sound is predominant and reverberation plays only a minor role, this phase alignment can be achieved by appropriately delaying the microphone signals. Then, the resulting phase-aligned signals are added to form a single output signal. This is known as the *delay-and-sum* beamformer.

The delay-and-sum beamformer as shown in Fig. 12.5 is simple in its implementation and provides for easy steering of the beam towards the desired source. When

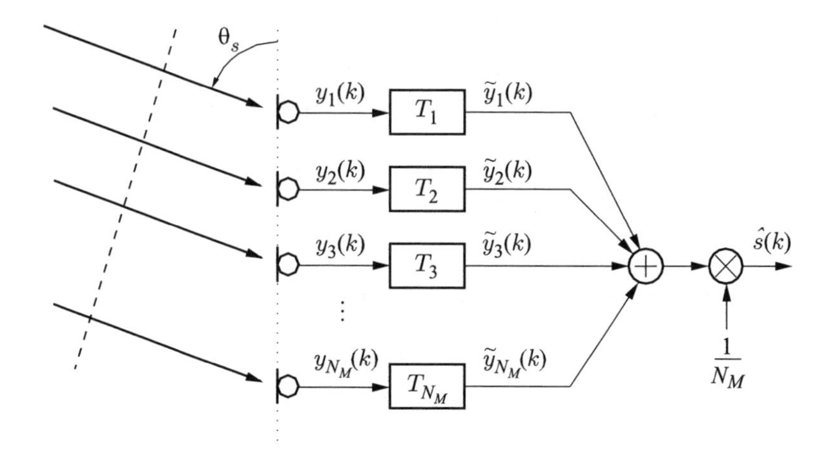

Figure 12.5: Delay-and-sum beamformer

written as a function of a continuous time argument t, the signals picked up by the individual microphones are

$$y_\ell(t) = s_\ell(t) + n_\ell(t) = s(t + \Delta\tau_\ell) + n_\ell(t)\,, \tag{12.31}$$

where $s(t)$ is the desired source signal at the reference position and $n_\ell(t)$ is the noise signal at the ℓ-th microphone. Again, we assume that the desired source is in the farfield and any signal attenuation is absorbed in $s(t)$. We denote the delayed signals by

$$\widetilde{y}_\ell(t) = \widetilde{s}_\ell(t) + \widetilde{n}_\ell(t) = s(t + \Delta\tau_\ell - T_\ell) + n_\ell(t - T_\ell) \tag{12.32}$$

where the delay T_ℓ is applied to the ℓ-th signal. Besides a relative delay which depends on ℓ, T_ℓ includes a channel-independent delay T_B such that $T_\ell \geq 0$ for all ℓ. For a digital implementation and since T_ℓ is, in general, not equal to an integer multiple of the sampling period, the accurate compensation of signal delays requires some form of signal interpolation. The interpolation may be implemented by fractional delay filters in the time or the frequency domain [Crochiere, Rabiner 1981], [Crochiere, Rabiner 1983], [Laakso et al. 1996]. In what follows, we will consider only sampled signals and digital beamformer implementations. Therefore, the channel-independent constant delay T_B is chosen to be equal to an integer multiple of the sampling period $1/f_s$.

The sampled output signal of the delay-and-sum beamformer may then be written as

$$\widehat{s}(kT) = \frac{1}{N_M}\sum_{\ell=1}^{N_M} \widetilde{y}_\ell(kT) = \frac{1}{N_M}\sum_{\ell=1}^{N_M} \widetilde{s}_\ell(kT) + \frac{1}{N_M}\sum_{\ell=1}^{N_M} \widetilde{n}_\ell(kT). \tag{12.33}$$

For a broadside orientation of the array with respect to the source, T_ℓ for all $\ell = 1 \ldots N_M$, and hence T_B, are set to zero. Then, the delay-and-sum beamformer comprises a scaled sum of the microphone signals only. For ease of notation, we omit the sampling period T in the sequel.

For the farfield scenario and when the delay from the source to the microphones is perfectly equalized, we have $T_\ell = T_B + \Delta\tau_\ell$ and with (12.33)

$$\widehat{s}(k) = s(k - T_B f_s) + \frac{1}{N_M}\sum_{\ell=1}^{N_M} \widetilde{n}_\ell(k)\,. \tag{12.34}$$

12.4.2 Filter-and-Sum Beamforming

A more general processing model is the *filter-and-sum* beamformer as shown in Fig. 12.6 where, before summation, each microphone signal is filtered,

$$\widetilde{y}_\ell(k) = \sum_{m=0}^{M} h_\ell(m) y_\ell(k - m)\,. \tag{12.35}$$

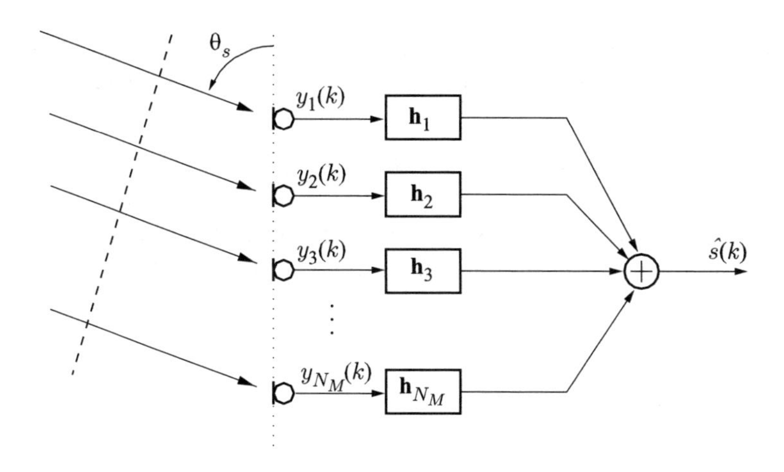

Figure 12.6: Filter-and-sum beamformer

We assume that FIR filters of order M are used. To simplify notations, we define a vector of signal samples

$$\mathbf{y}_\ell(k) = (y_\ell(k), y_\ell(k-1), \ldots, y_\ell(k-M))^T \tag{12.36}$$

and a vector of real-valued filter coefficients

$$\mathbf{h}_\ell = (h_\ell(0), h_\ell(1), \ldots, h_\ell(M))^T . \tag{12.37}$$

Then, the output signal of the filter-and-sum beamformer may be written as

$$\widehat{s}(k) = \sum_{\ell=1}^{N_M} \mathbf{h}_\ell{}^T \mathbf{y}_\ell(k) . \tag{12.38}$$

For the signal $s(k)$ of a single (desired) source and the farfield scenario, we obtain in the frequency domain (provided all Fourier transforms exist)

$$\begin{aligned}\widehat{S}(e^{j\Omega}) &= \sum_{\ell=1}^{N_M} H_\ell(e^{j\Omega}) Y_\ell(e^{j\Omega}) \\ &= S(e^{j\Omega}) \sum_{\ell=1}^{N_M} H_\ell(e^{j\Omega}) \exp\left(j\Omega f_s \Delta\tau_\ell\right) \\ &= S(e^{j\Omega}) H(e^{j\Omega}, \mathbf{u}_s) .\end{aligned} \tag{12.39}$$

$H_\ell(e^{j\Omega})$ denotes the frequency response of the ℓ-th FIR filter. Thus, for a fixed source position and fixed microphone positions $\mathbf{r}_\ell$ the array response

$$H(e^{j\Omega}, \mathbf{u}_s) = \sum_{\ell=1}^{N_M} H_\ell(e^{j\Omega}) \exp(j\Omega f_s \Delta\tau_\ell) = \mathbf{H}^T(e^{j\Omega}) \mathbf{a} \tag{12.40}$$

of the filter-and-sum beamformer depends on the propagation vector $\mathbf{a}$ (12.26) of the point source in the farfield and the vector of filter responses

$$\mathbf{H}(e^{j\Omega}) = \left(H_1(e^{j\Omega}), H_2(e^{j\Omega}), \ldots, H_{N_M}(e^{j\Omega})\right)^T . \tag{12.41}$$

When a delay-and-sum beamformer is steered towards the source in the farfield, we obtain a special case of the filter-and-sum beamformer with the filter coefficient vector

$$\mathbf{H}(e^{j\Omega}) = \frac{\exp(-j\Omega f_s T_B)}{N_M}\mathbf{a}^* = \frac{1}{N_M}\mathbf{e} \tag{12.42}$$

where

$$\mathbf{e} = \left(\exp\left(-j\Omega f_s T_1\right), \exp\left(-j\Omega f_s T_2\right), \ldots, \exp\left(-j\Omega f_s T_{N_M}\right)\right)^T . \tag{12.43}$$

is the *steering vector.*

In the case of random source signals, we obtain the power spectrum of the output signal as the Fourier transform of the autocorrelation function

$$\begin{aligned}
&\mathrm{E}\left\{\widehat{s}(k)\widehat{s}(k+k')\right\} \\
&\quad = \mathrm{E}\left\{\sum_{\ell=1}^{N_M}\sum_{\ell'=1}^{N_M}\sum_{m=0}^{M}\sum_{m'=0}^{M} h_\ell(m)y_\ell(k-m)y_{\ell'}(k-m'+k')h_{\ell'}(m')\right\} \\
&\quad = \sum_{\ell=1}^{N_M}\sum_{\ell'=1}^{N_M}\sum_{m=0}^{M}\sum_{m'=0}^{M} h_\ell(m)\varphi_{y_\ell y_{\ell'}}(k'-m'+m)h_{\ell'}(m') .
\end{aligned} \tag{12.44}$$

Hence,

$$\Phi_{\widehat{s}\widehat{s}}(e^{j\Omega}) = \mathbf{H}^H(e^{j\Omega})\mathbf{\Phi}_{yy}(e^{j\Omega})\mathbf{H}(e^{j\Omega}) \tag{12.45}$$

where

$$\mathbf{\Phi}_{yy}(e^{j\Omega}) = \begin{pmatrix} \Phi_{y_1y_1}(e^{j\Omega}) & \Phi_{y_1y_2}(e^{j\Omega}) & \cdots & \Phi_{y_1y_{N_M}}(e^{j\Omega}) \\ \Phi_{y_2y_1}(e^{j\Omega}) & \Phi_{y_2y_2}(e^{j\Omega}) & \cdots & \Phi_{y_2y_{N_M}}(e^{j\Omega}) \\ \vdots & \vdots & \ddots & \vdots \\ \Phi_{y_{N_M}y_1}(e^{j\Omega}) & \Phi_{y_{N_M}y_2}(e^{j\Omega}) & \cdots & \Phi_{y_{N_M}y_{N_M}}(e^{j\Omega}) \end{pmatrix} \tag{12.46}$$

is a matrix of cross-power spectra.

When the microphone signals originate from a single point source, we obtain the Fourier transform of the cross-correlation function of the microphone signals

$$\Phi_{y_\ell y_{\ell'}}(e^{j\Omega}) = \Phi_{ss}(e^{j\Omega}) e^{j\Omega f_s(\Delta\tau_\ell - \Delta\tau_{\ell'})} \tag{12.47}$$

and hence

$$\Phi_{\widehat{s}\widehat{s}}(e^{j\Omega}) = \Phi_{ss}(e^{j\Omega}) \mathbf{H}^H(e^{j\Omega}) \mathbf{a}\mathbf{a}^H \mathbf{H}(e^{j\Omega}) = \Phi_{ss}(e^{j\Omega}) |H(e^{j\Omega}, \mathbf{u}_s)|^2 . \tag{12.48}$$

For mutually uncorrelated microphone signals we have

$$\Phi_{\widehat{s}\widehat{s}}(e^{j\Omega}) = \sum_{\ell=1}^{N_M} |H_\ell(e^{j\Omega})|^2 \Phi_{y_\ell y_\ell}(e^{j\Omega}) . \tag{12.49}$$

The design of the filter-and-sum beamformer reduces to computing filter coefficient vectors $\mathbf{h}_\ell(m)$ such that a performance measure is optimized. Before we investigate these methods we introduce such measures.

12.5 Performance Measures and Spatial Aliasing

12.5.1 Array Gain and Array Sensitivity

Microphone arrays are designed to improve the SNR of a desired source signal. The *array gain* characterizes the performance of a microphone array as the ratio of the SNR at the output of the array with respect to the average SNR of the microphone signals. Using the matrix notation of the previous section and the trace operator Tr(), the average powers of the desired signals and of the noise signals at the microphones are given by $\mathrm{Tr}(\mathbf{\Phi}_{ss}(e^{j\Omega}))/N_M$ and $\mathrm{Tr}(\mathbf{\Phi}_{nn}(e^{j\Omega}))/N_M$, respectively, whereas the corresponding powers at the output are given by $\mathbf{H}^H(e^{j\Omega})\mathbf{\Phi}_{ss}(e^{j\Omega})\mathbf{H}(e^{j\Omega})$ and $\mathbf{H}^H(e^{j\Omega})\mathbf{\Phi}_{nn}(e^{j\Omega})\mathbf{H}(e^{j\Omega})$, respectively. Therefore, the frequency-dependent array gain may be defined as [Herbordt 2005]

$$G(e^{j\Omega}) = \frac{\mathrm{Tr}(\mathbf{\Phi}_{nn}(e^{j\Omega}))}{\mathrm{Tr}(\mathbf{\Phi}_{ss}(e^{j\Omega}))} \frac{\mathbf{H}^H(e^{j\Omega})\mathbf{\Phi}_{ss}(e^{j\Omega})\mathbf{H}(e^{j\Omega})}{\mathbf{H}^H(e^{j\Omega})\mathbf{\Phi}_{nn}(e^{j\Omega})\mathbf{H}(e^{j\Omega})} . \tag{12.50}$$

Assuming a farfield scenario with identical speech and identical noise power spectral densities at all microphones and mutually uncorrelated noise signals, we obtain for the delay-and-sum beamformer with (12.48) and (12.49),

$$G(e^{j\Omega}) = \frac{N_M \Phi_{nn}(e^{j\Omega})}{N_M \Phi_{ss}(e^{j\Omega})} \frac{\frac{1}{N_M^2} N_M^2 \Phi_{ss}(e^{j\Omega})}{\frac{1}{N_M^2} N_M \Phi_{nn}(e^{j\Omega})} = N_M . \tag{12.51}$$

Under the above assumptions, the gain of the delay-and-sum beamformer does not depend on frequency. For partially correlated noise signals $\widetilde{n}_\ell(k)$ the improvement can be significantly lower. For example, in the diffuse noise field, the gain is close to zero for frequencies $f < \frac{c}{2d}$ or $\lambda > 2d$ where d is the inter-microphone distance.

Furthermore, we introduce a performance measure which characterizes the *sensitivity* of the array with respect to spatially and temporally white noise. For instance, this noise can be thermal noise originating from the microphones or the amplifiers. It can also serve as a model for random phase or position errors due to the physical implementation of the array. For mutually uncorrelated noise signals we find

$$\Phi_{n_\ell n_{\ell'}}(e^{j\Omega}) = \begin{cases} \Phi_{n_\ell n_\ell}(e^{j\Omega}) & \ell = \ell' \\ 0 & \ell \neq \ell' \end{cases} . \tag{12.52}$$

Thus, when the noise signals are temporally white, the array gain is given by

$$G_W(e^{j\Omega}) = \frac{|\mathbf{H}^H(e^{j\Omega})\mathbf{a}|^2}{\mathbf{H}^H(e^{j\Omega})\mathbf{H}(e^{j\Omega})} . \tag{12.53}$$

The inverse of the white noise gain $G_W(e^{j\Omega})$ is called the *susceptibility* of the array. It characterizes the sensitivity of the array with respect to uncorrelated noise.

12.5.2 Directivity Pattern

The spatial selectivity of the array in the farfield of a source is characterized by its *directivity pattern*

$$\Psi(e^{j\Omega}, \mathbf{u}_s) = |H(e^{j\Omega}, \mathbf{u}_s)|^2 = \mathbf{H}^H(e^{j\Omega})\mathbf{a}\mathbf{a}^H\mathbf{H}(e^{j\Omega}) = |\mathbf{H}^H(e^{j\Omega})\mathbf{a}|^2 \tag{12.54}$$

where the vectors $\mathbf{u}_s$ and $\mathbf{a}$ denote a unit vector in the direction of the source and the propagation vector (12.26) of the impinging sound respectively. The directivity pattern depicts the attenuation of the sound energy from a given direction. It is a useful tool for array performance analysis especially when sounds propagate coherently and there is no or only little reverberation. Besides frequency and direction of arrival, $\Psi(e^{j\Omega}, \mathbf{u}_s)$ in (12.54) depends also on the inter-microphone distances and the filter coefficients.

As an example we consider the delay-and-sum beamformer and the farfield scenario. With (12.40) and (12.42), we obtain the directivity pattern of the uniform linear array

$$\Psi(e^{j\Omega}, \mathbf{u}_s) = \begin{cases} \dfrac{\sin^2\left(\pi(\cos(\theta) - \cos(\theta_s))N_M d/\lambda\right)}{N_M^2 \sin^2\left(\pi(\cos(\theta) - \cos(\theta_s))d/\lambda\right)} & \cos(\theta) \neq \cos(\theta_s) \\ 1 & \cos(\theta) = \cos(\theta_s) . \end{cases} \tag{12.55}$$

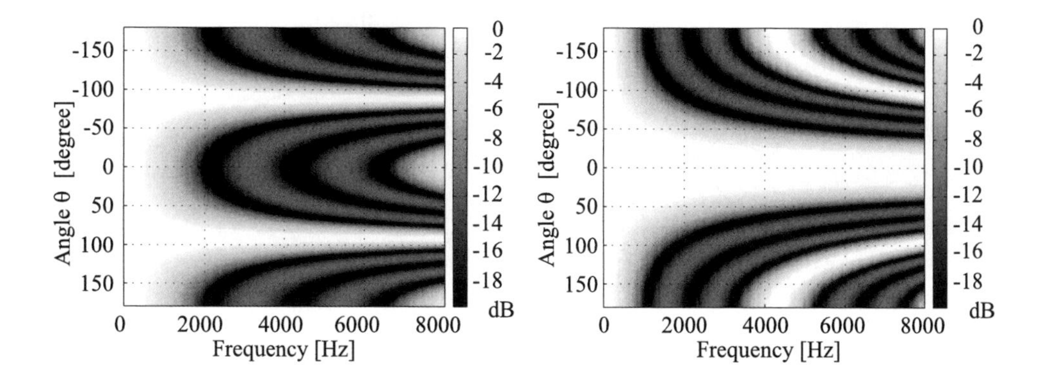

Figure 12.7: Directivity pattern of the delay-and-sum beamformer with $N_M = 4$ microphones and $d = 4\,\mathrm{cm}$ for the broadside (left) and endfire (right) orientation

θ_s denotes the direction of arrival and θ the look direction of the beamformer. The corresponding pattern is shown in Fig. 12.7 for the broadside (left) and the endfire (right) orientation. In this example, we use a uniform linear array with $d = 4\,\mathrm{cm}$. Then, $\lambda/d = 1$ corresponds to $f = 8500\,\mathrm{Hz}$. The beams in the respective look directions can be clearly recognized. Furthermore, we note that the directivity at low frequencies is not very pronounced and that for high frequencies spatial aliasing in the form of side lobes is visible.

The directivity pattern (12.55) is a periodic function of the difference $\vartheta = (\cos(\theta) - \cos(\theta_s))$ with period λ/d. For a broadside array ($\cos(\theta) = 0$), ϑ attains values between -1 and 1 when θ_s sweeps from 0 to π. Thus, spatial aliasing is avoided when $\frac{\lambda}{d} \geq 1$. For an endfire array the same sweep will result in values $0 \leq \vartheta \leq 2$ and therefore in this case $\frac{\lambda}{d} \geq 2$ will avoid spatial aliasing.

In general, $|\cos(\theta) - \cos(\theta_s)| \leq 1 + |\cos(\theta)|$ holds and the above relations can be cast into the more general form [Kummer 1992]:

$$\frac{d}{\lambda} \leq \frac{1}{1 + |\cos(\theta)|} . \tag{12.56}$$

Directivity patterns may also be represented in two- or three-dimensional polar directivity plots. Figure 12.8 depicts two- and three-dimensional polar plots for a filter-and-sum beamformer (see Section 12.6) and a uniform linear array with four microphones. These plots are computed for a frequency of 1000 Hz and for broadside and endfire orientations.

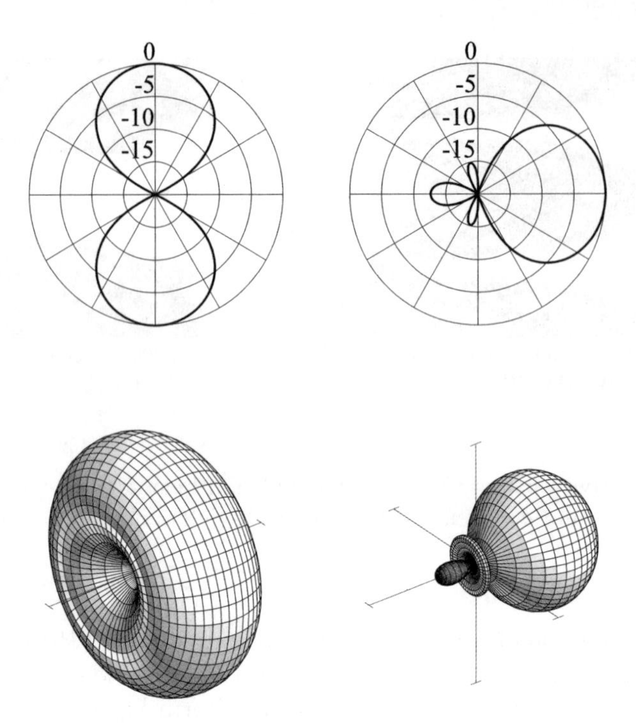

Figure 12.8: Directivity pattern in dB of a filter-and-sum beamformer with four omnidirectional microphones with $d = 4.25\,\text{cm}$ and broadside (left) and endfire (right) orientations ($f = 1000\,\text{Hz}$)

12.5.3 Directivity and Directivity Index

The directivity pattern characterizes the performance of the array when directional sources are present. When noise impinges from many different directions, an integral measure is better suited. The *directivity* is defined as the ratio of the directivity pattern in look direction and the directivity pattern averaged over all directions,

$$D(e^{j\Omega}) = \frac{\Psi(e^{j\Omega}, \mathbf{u})}{\frac{1}{4\pi}\int_{A_1} \Psi(e^{j\Omega}, \mathbf{u}_s)\,\mathrm{d}a}, \tag{12.57}$$

where A_1 denotes the surface of a unit sphere with the array at its center. In polar coordinates, we have $\mathrm{d}a = \sin(\theta)\mathrm{d}\theta\mathrm{d}\varphi$ and therefore

$$D(e^{j\Omega}) = \frac{\Psi(e^{j\Omega}, \mathbf{u})}{\frac{1}{4\pi}\int_0^{\pi}\int_0^{2\pi} \Psi(e^{j\Omega}, \mathbf{u}_s(\theta,\varphi))\sin(\theta)\,\mathrm{d}\theta\mathrm{d}\varphi} \quad . \tag{12.58}$$

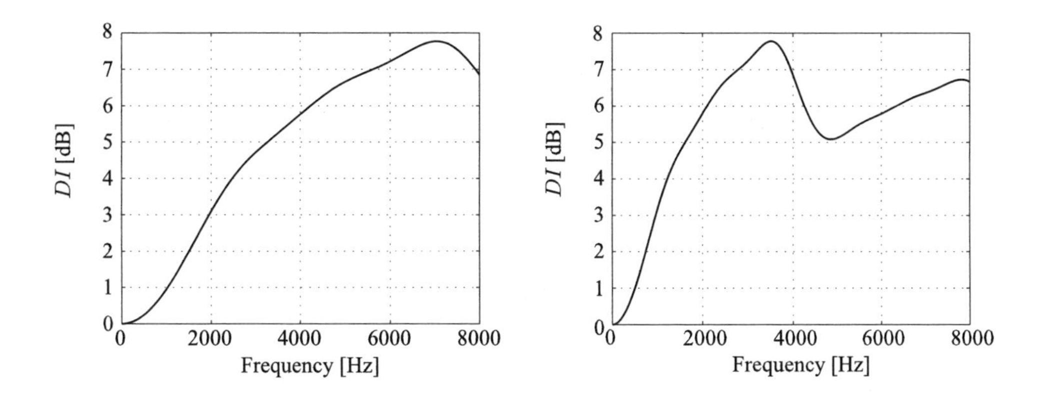

Figure 12.9: Directivity index of the delay-and-sum beamformer with $N_M = 4$ microphones and $d = 4\,\mathrm{cm}$ for the broadside (left) and endfire (right) orientation

The directivity thus characterizes the array performance for a single desired source and an ideal diffuse noise field. For the farfield scenario, the directivity can be expressed in terms of the array filters

$$D(e^{j\Omega}) = \frac{|\mathbf{H}^H(e^{j\Omega})\mathbf{a}|^2}{\sum_{\ell=1}^{N_M}\sum_{\ell'=1}^{N_M} H_\ell(e^{j\Omega})H_{\ell'}^*(e^{j\Omega})\gamma_{n_\ell n_{\ell'}}(e^{j\Omega})} \tag{12.59}$$

where $\gamma_{n_\ell n_{\ell'}}(e^{j\Omega})$ is the spatial coherence function of the noise signals n_ℓ and $n_{\ell'}$ in the ideal diffuse sound field with,

$$\gamma_{n_\ell n_{\ell'}}(e^{j\Omega}) = \frac{1}{4\pi}\int_{A_1} \exp(j\widetilde{\beta}\,\langle(\mathbf{r}_\ell - \mathbf{r}_{\ell'}), \mathbf{u}_s\rangle)\,\mathrm{d}a \tag{12.60}$$

$$= \begin{cases} \frac{\sin(\widetilde{\beta} d_{\ell,\ell'})}{\widetilde{\beta} d_{\ell,\ell'}} & \ell \neq \ell' \\ 1 & \ell = \ell' \end{cases} = \begin{cases} \frac{\sin(\Omega f_s d_{\ell,\ell'} c^{-1})}{(\Omega f_s d_{\ell,\ell'} c^{-1})} & \ell \neq \ell' \\ 1 & \ell = \ell' \end{cases}$$

where $d_{\ell,\ell'}$ represents the distance between the ℓ-th and the ℓ'-th microphone.

Figure 12.9 plots the *directivity index* $DI(e^{j\Omega}) = 10\log_{10}\left(D(e^{j\Omega})\right)$ of the same delay-and-sum beamformer as in Fig. 12.7 for the ideal diffuse noise field. For low frequencies, the directivity index of the beamformer in endfire orientation is significantly higher than for the broadside orientation. For the broadside orientation, the gain does not exceed 3 dB below 2000 Hz.

12.5.4 Example: Differential Microphones

Before we discuss more advanced beamformer designs we start out with the simple case of two closely spaced microphones ($d < \lambda/2$). The two microphone signals

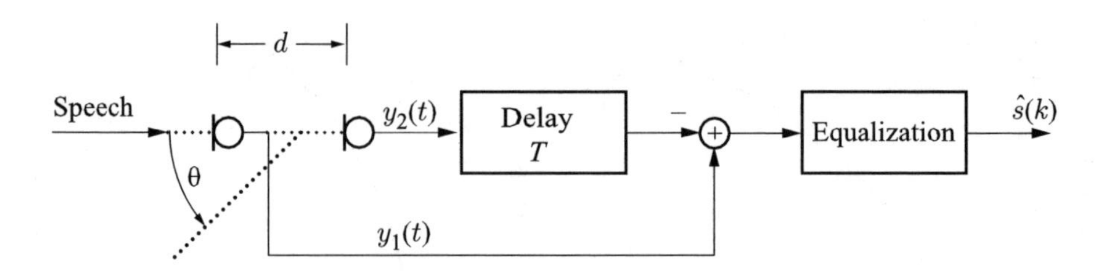

Figure 12.10: Differential microphones

$y_1(t)$ and $y_2(t)$ are combined in a simple delay-and-subtract operation as shown in Fig. 12.10. The delay T is sometimes termed *internal delay* as opposed to the *external delay* of the acoustic path between the two microphones. The array design and the resulting beam patterns are reminiscent of directional microphones, where the delay-and-subtract operation is achieved by means of the acoustic design. Typically, the differential microphone array is combined with an equalizer to compensate for undesirable effects of the subtraction on the frequency response.

We now compute the frequency response of the differential microphone array under the farfield assumption. When the reference point is selected to be the midpoint of the line joining the two microphones, the two microphone signals may be written as

$$y_1(t) = s\left(t + \frac{d}{2c}\cos(\theta)\right) \quad \text{and} \tag{12.61}$$

$$y_2(t) = s\left(t - \frac{d}{2c}\cos(\theta)\right) . \tag{12.62}$$

Thus, after summation and before equalization we obtain

$$\widetilde{s}(t) = s\left(t + \frac{d}{2c}\cos(\theta)\right) - s\left(t - \frac{d}{2c}\cos(\theta) - T\right) \tag{12.63}$$

and in the Fourier transform domain

$$\widetilde{S}(j\omega) = S(j\omega) \cdot \left(e^{j\omega\left(\frac{d}{2c}\cos(\theta)\right)} - e^{-j\omega\left(\frac{d}{2c}\cos(\theta)+T\right)}\right) \tag{12.64}$$

$$= S(j\omega) \cdot e^{j\omega\left(\frac{d}{2c}\cos(\theta)\right)} \cdot \left[1 - e^{-j\omega\frac{d}{c}\left(\cos(\theta)+\frac{cT}{d}\right)}\right] . \tag{12.65}$$

The magnitude of the frequency response can be written as

$$|H(j\omega)| = \left|\frac{\widetilde{S}(j\omega)}{S(j\omega)}\right| = 2 \cdot \left|\sin\left(\frac{\omega d}{2c}\left(\cos(\theta) + \frac{cT}{d}\right)\right)\right| . \tag{12.66}$$

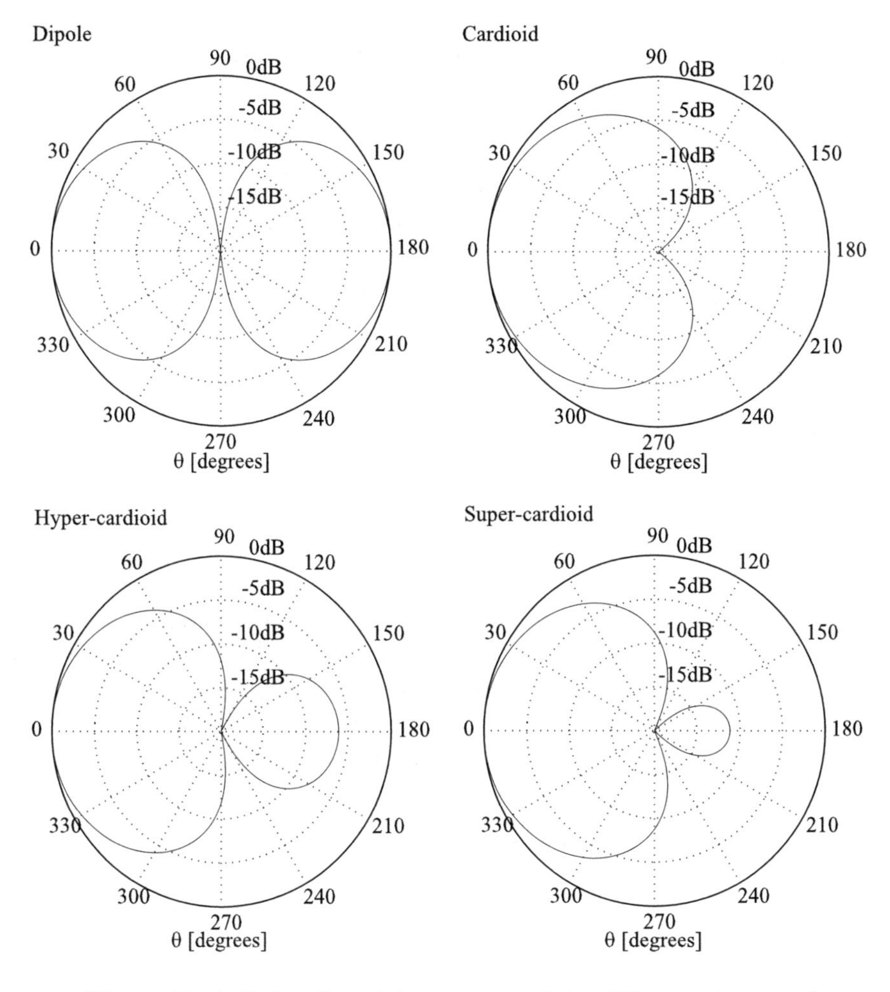

Figure 12.11: Polar directivity pattern of the differential arrays for $d = 0.015\,\mathrm{m}$, $f = 1\,\mathrm{kHz}$ various values of β

With the assumption that $\frac{\omega d}{2c} \ll \frac{\pi}{2}$, which is equivalent to $2\,d \ll \lambda = \frac{c}{f}$, and with $\frac{cT}{d} \lessapprox 1$ we may use the approximation $\sin(\theta) \approx \theta$ and therefore,

$$\left|\frac{\widetilde{S}(j\omega)}{S(j\omega)}\right| \approx 2 \cdot \left|\frac{\omega d}{2c}\left(\cos(\theta) + \frac{cT}{d}\right)\right| = \left|\frac{\omega d}{c}\left(\cos(\theta) + \frac{cT}{d}\right)\right| . \tag{12.67}$$

With the substitution

$$T = \frac{d}{c}\frac{\beta}{1-\beta} \quad \text{or} \quad \beta = \frac{T}{T + \frac{d}{c}} = \frac{Tc}{Tc + d} < 1 \tag{12.68}$$

Table 12.1: Parameters and average directivity index DI_{av} of the differential microphones for $d = 0.015\,\text{m}$

Characteristic	$T\frac{c}{d}$	β	DI_{av} / dB
Dipole	0	0	4.7
Cardioid	1	0.5	4.8
Super-cardioid	0.57	0.3631	5.7
Hyper-cardioid	0.34	0.2537	6.0

we may write the magnitude response as

$$\left|\frac{\widetilde{S}(j\omega)}{S(j\omega)}\right| = \left|\frac{\omega d}{c}\left(\cos(\theta) + \frac{\beta}{1-\beta}\right)\right| \tag{12.69}$$

$$= \frac{1}{1-\beta}\frac{d}{c} \cdot |\omega \cdot ((1-\beta)\cdot\cos(\theta) + \beta)| \,. \tag{12.70}$$

Since the magnitude of the approximate frequency response increases linearly with ω, the frequency response of the array corresponds to a first-order differentiator. Therefore, a first-order lowpass filter may be used for equalization. This, however, might amplify low frequency noise. Table 12.1 lists the parameter β for several directional characteristics and the resulting average directivity index.

The differential microphone array can easily be extended into an adaptive null-steering array. Assuming a look direction $\theta \in [-\pi/2, \pi/2]$ and combining two directional patterns, a null of the combined beampattern can be steered towards a disturbing signal source in the range $\theta \in [\pi/2, 3\pi/2]$ [Elko, Nguyen Pong 1997].

12.6 Design of Fixed Beamformers

If the wavelength is much larger than the array aperture, the delay-and-sum beamformer delivers only little directivity. By optimizing the coefficients of a filter-and-sum beamformer, the directivity of a beamformer at low frequencies can be significantly improved. In this section we will examine a frequently used beamformer design, the *minimum variance distortionless response* beamformer, in greater detail.

12.6.1 Minimum Variance Distortionless Response Beamformer

The *minimum variance distortionless response* (MVDR) beamformer minimizes the output noise variance while constraining the array to have unit gain in look direction. To derive the optimal beamformer coefficients, we write the noise power spectral density at the output of the filter-and-sum beamformer as

$$\Phi_{\tilde{n}\tilde{n}}(e^{j\Omega}) = \mathbf{H}^H(e^{j\Omega})\mathbf{\Phi}_{nn}(e^{j\Omega})\mathbf{H}(e^{j\Omega}) \tag{12.71}$$

where $\mathbf{\Phi}_{nn}(e^{j\Omega})$ denotes the power spectral density matrix of the noise signals $n_\ell(k)$, $\ell = 1, \ldots, N_M$. The distortionless response constraint is written as

$$\mathbf{H}^H(e^{j\Omega})\mathbf{a} = 1\,, \tag{12.72}$$

where $\mathbf{a}$ is the propagation vector as before. Here, we use the Hermitian transpose ^{H} for notational convenience. Furthermore, to improve the readability of the following derivations, we will drop the dependency on frequency Ω.

Since the source signal is not distorted, the MVDR beamformer also maximizes

$$\widetilde{D} = \frac{|\mathbf{H}^H\mathbf{a}|^2}{\mathbf{H}^H\mathbf{\Phi}_{nn}\mathbf{H}} \tag{12.73}$$

for a source in the farfield of the array and thus the directivity in an ideal diffuse noise field.

To solve the constraint optimization problem, we use Lagrange's method and minimize

$$\mathfrak{L}(\mathbf{H}, q) = \mathbf{H}^H\mathbf{\Phi}_{nn}\mathbf{H} + q(\mathbf{H}^H\mathbf{a} - 1)\,. \tag{12.74}$$

Computing the gradient with respect to $\mathbf{H}$ and q results in the necessary conditions

$$2\mathbf{\Phi}_{nn}\mathbf{H} + q\mathbf{a} = 0 \quad \text{and} \quad \mathbf{H}^H\mathbf{a} = 1\,. \tag{12.75}$$

Solving the first condition for $\mathbf{H}$ and using the result in the second condition we obtain

$$\mathbf{H} = -\frac{1}{2}q\mathbf{\Phi}_{nn}^{-1}\mathbf{a} \quad \text{and} \quad q = \frac{-2}{\mathbf{a}^H\mathbf{\Phi}_{nn}^{-1}\mathbf{a}}\,, \tag{12.76}$$

and finally

$$\mathbf{H}_{\text{MVDR}} = \frac{\mathbf{\Phi}_{nn}^{-1}\mathbf{a}}{\mathbf{a}^H\mathbf{\Phi}_{nn}^{-1}\mathbf{a}}\,. \tag{12.77}$$

Besides the dependency on frequency, the optimal solution is a function of the spectral correlation matrix of the noise and of the propagation vector (12.26).

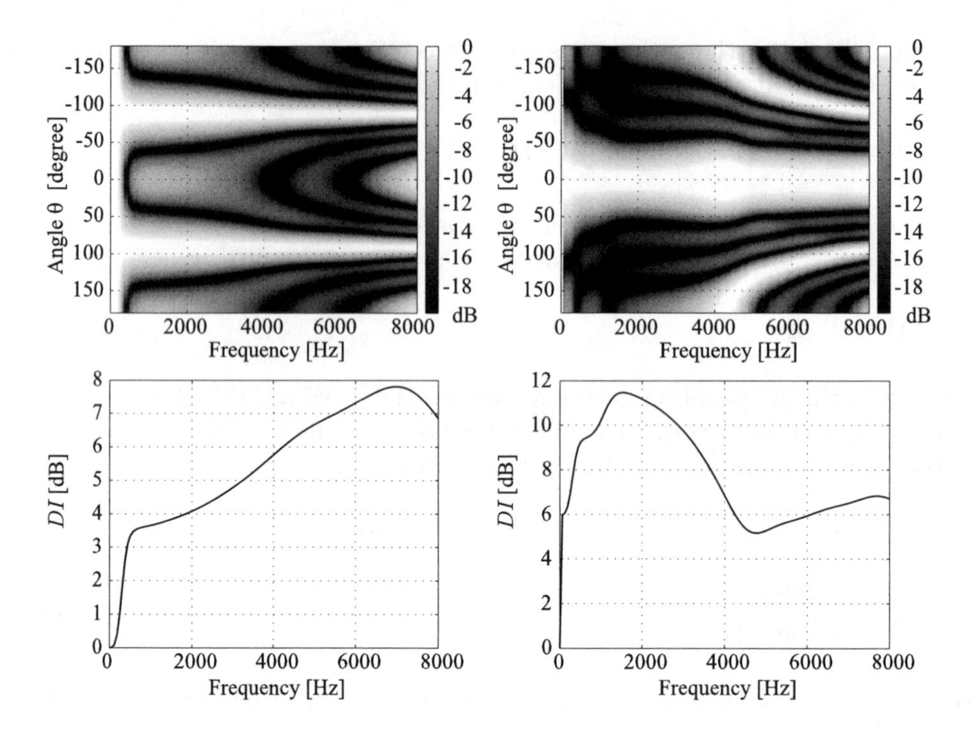

Figure 12.12: Directivity pattern and directivity index of the MVDR beamformer with $N_M = 4$ microphones and $d = 4\,\text{cm}$ for the broadside (left) and endfire (right) orientation optimized for the ideal diffuse noise field

Figure 12.12 shows the directivity pattern and the gain of an MVDR beamformer for the broadside (left plot) and endfire (right plot) orientation. The array geometry is the same as in Fig. 12.7. The array is optimized for the ideal diffuse noise field. We observe a significantly increased directivity for low frequencies. For high frequencies, the performance of the MVDR beamformer is close to the performance of the delay-and-sum beamformer.

As a special case, we consider mutually uncorrelated noise signals with identical PSDs. Then, the noise power spectral density matrix is a diagonal matrix and the optimal filter vector (12.77) evaluates to

$$\mathbf{H} = \frac{\mathbf{a}}{N_M}, \tag{12.78}$$

which is the delay-and-sum beamforming solution for the design condition (12.72) and when channel-independent delay T_B is neglected.

The high directivity of the MVDR beamformer at low frequencies also results in a high susceptibility to uncorrelated noise. In order to minimize the noise power in each frequency bin, the MVDR beamformer cancels out correlated noise components. The distortionless response, however, can be only maintained if the response of individual filters $H_\ell(e^{j\Omega})$ is much larger than unity for low frequencies. The high gain of the individual filters at low frequencies leads then to an amplification of uncorrelated noise in the microphone signals. For high frequencies and in an ideal diffuse noise field, the MVDR beamformer approaches the delay-and-sum solution. For the delay-and-sum beamformer we obtain

$$\frac{1}{G_W} = \mathbf{H}^H\mathbf{H} = \frac{1}{N_M}\,. \tag{12.79}$$

This is the smallest value that the susceptibility can attain under the distortionless response constraint [Dörbecker 1998].

Furthermore, the propagation vector $\mathbf{a}$ can be modified to suit the actual transmission conditions. For example, *head-related transfer functions* can be used to model head-related effects which is of importance in hearing aid applications [Lotter, Vary 2006].

12.6.2 MVDR Beamformer with Limited Susceptibility

The susceptibility of the MVDR beamformer can be controlled by including an additional constraint in the design procedure [Cox et al. 1986]. We require that the susceptibility attains a fixed value K_0,

$$\frac{1}{G_W} = \mathbf{H}^H\mathbf{H} = K_0\,. \tag{12.80}$$

Using Lagrange's method again we now have

$$\mathfrak{L}(\mathbf{H}, q, q') = \mathbf{H}^H\mathbf{\Phi}_{nn}\mathbf{H} + q(\mathbf{H}^H\mathbf{a} - 1) + q'(\mathbf{H}^H\mathbf{H} - K_0)\,. \tag{12.81}$$

Computing the gradient, we obtain the necessary conditions

$$2\mathbf{\Phi}_{nn}\mathbf{H} + q\mathbf{a} + 2q'\mathbf{H} = 0\,, \tag{12.82}$$

$$\mathbf{H}^H\mathbf{a} = 1\,, \quad \text{and} \tag{12.83}$$

$$\mathbf{H}^H\mathbf{H} = K_0\,. \tag{12.84}$$

The solution to these equations is given by

$$\mathbf{H} = -\frac{1}{2}q(\mathbf{\Phi}_{nn} + q'\mathbf{I})^{-1}\mathbf{a} \qquad \text{and} \tag{12.85}$$

$$q = \frac{-2}{\mathbf{a}^H(\mathbf{\Phi}_{nn} + q'\mathbf{I})^{-1}\mathbf{a}} \tag{12.86}$$

with q' as an implicit parameter. With these equations, the vector of optimal filters can be written as

$$\mathbf{H}_{\text{MVDR,K0}} = \frac{(\mathbf{\Phi}_{nn} + q'\mathbf{I})^{-1}\mathbf{a}}{\mathbf{a}^H(\mathbf{\Phi}_{nn} + q'\mathbf{I})^{-1}\mathbf{a}} . \tag{12.87}$$

The constraint on the susceptibility leads to a *diagonal loading* term and thus improves the condition of the noise power spectral density matrix. The relation between the Lagrange multiplier q' and the desired susceptibility K_0, however, is implicit

$$K_0 = \frac{\mathbf{a}^H(\mathbf{\Phi}_{nn} + q'\mathbf{I})^{-H}(\mathbf{\Phi}_{nn} + q'\mathbf{I})^{-1}\mathbf{a}}{[\mathbf{a}^H(\mathbf{\Phi}_{nn} + q'\mathbf{I})^{-1}\mathbf{a}]^2} \tag{12.88}$$

with no closed form solution for q'. For a given K_0 the corresponding Lagrange multiplier can be found by using an iterative procedure [Dörbecker 1997]. Using the optimal q', the coefficient vector $\mathbf{H}$ can then be computed.

The results of the beamformer optimization with the susceptibility limited to $K_0 = 1$ are shown in Fig. 12.13 for the same conditions as in the previous figures.

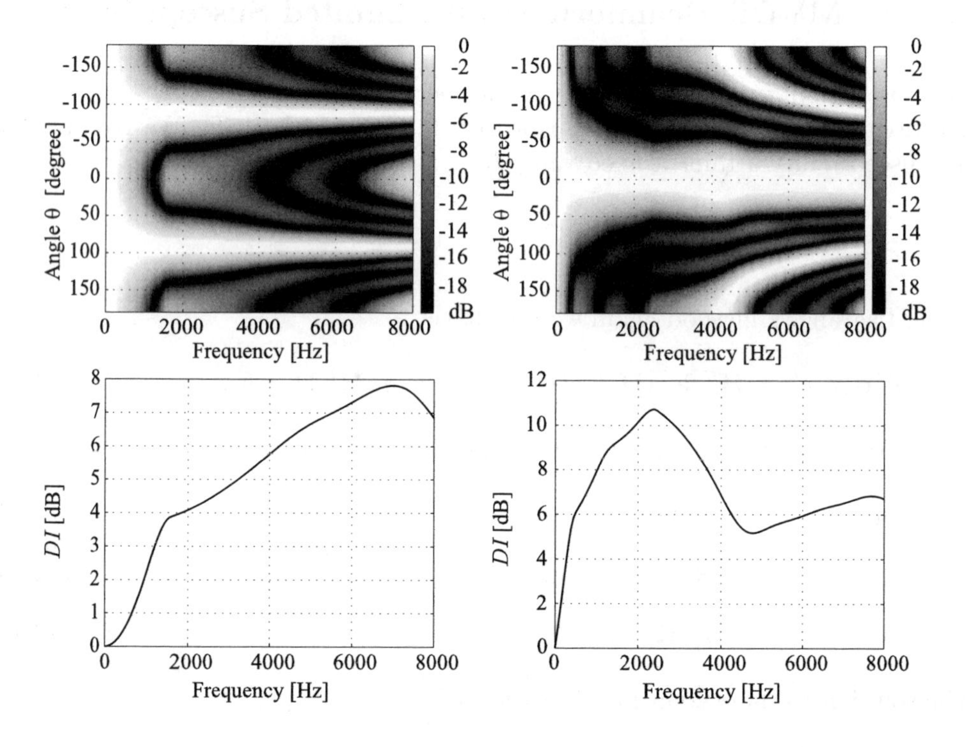

Figure 12.13: Directivity pattern and directivity index of the MVDR beamformer with susceptibility limited to $K_0 = 1$, $N_M = 4$ microphones, and $d = 4\,\text{cm}$ for the broadside (left) and endfire (right) orientation optimized for the ideal diffuse noise field

As a consequence of the limited susceptibility, the directivity at low frequencies is lower than the directivity without this constraint.

Besides the MVDR design, there are several other design methods for beamformers such as the design for nearfield conditions and constant beamwidth [Goodwin, Elko 1993], [Brandstein, Ward 2001]. Furthermore, the design of such arrays may be made robust against amplitude, phase, and microphone position errors by including suitable error distribution functions into a *least-square* design criterion [Doclo, Moonen 2003].

12.7 Multi-channel Wiener Filter and Postfilter

We now extend the well-known single channel Wiener filter to the multi-channel case. In the optimization of the minimum mean-square error, the source signal $s(k)$ is explicitly used and no constraint with respect to the look direction is imposed. However, as shall be seen below, the optimal filter may be decomposed into a distortionless response beamformer and a single channel Wiener-type *postfilter* [Edelblute et al. 1966], [Simmer et al. 2001].

The minimum mean square error of the filtered array output with respect to the reference signal $s(k)$ is given by

$$\begin{aligned} J &= \mathrm{E}\left\{(s(k) - \widehat{s}(k))^2\right\} \\ &= \mathrm{E}\left\{\left(s(k) - \sum_{\ell=1}^{N_M}\sum_{m=0}^{M} h_\ell(m) y_\ell(k-m)\right)^2\right\} . \end{aligned} \tag{12.89}$$

With $\mathrm{E}\{s(k)y_{\ell'}(k-i)\} = \varphi_{y_{\ell'}s}(i)$ and $\mathrm{E}\{y_\ell(k-m)y_{\ell'}(k-i)\} = \varphi_{y_{\ell'}y_\ell}(i-m)$, the differentiation with respect to the i-th coefficent of the ℓ'-th filter leads to

$$\sum_{\ell=1}^{N_M}\sum_{m=0}^{M} h_\ell(m)\varphi_{y_{\ell'}y_\ell}(i-m) = \varphi_{y_{\ell'}s}(i) . \tag{12.90}$$

The Fourier transform of (12.90) yields

$$\sum_{\ell=1}^{N_M} H_\ell(e^{j\Omega})\Phi_{y_{\ell'}y_\ell} = \Phi_{y_{\ell'}s} \quad \text{for} \quad \ell' = 1, \ldots N_M . \tag{12.91}$$

Therefore, using vector matrix notation, these equations may be stacked to yield

$$\mathbf{\Phi}_{yy}\mathbf{H} = \mathbf{\Phi}_{ys} , \tag{12.92}$$

and, for invertible $\mathbf{\Phi}_{yy}$, the optimal solution

$$\mathbf{H} = \mathbf{\Phi}_{yy}^{-1}\mathbf{\Phi}_{ys} \,. \tag{12.93}$$

In the case of additive noise which is not correlated with the source signal s, and identical noise and source power spectral densities at all microphones, we find, for farfield conditions,

$$\mathbf{\Phi}_{ys} = \Phi_{ss}\mathbf{a} \tag{12.94}$$

and

$$\mathbf{\Phi}_{yy} = \mathbf{\Phi}_{ss} + \mathbf{\Phi}_{nn} = \Phi_{ss}\mathbf{a}\mathbf{a}^H + \mathbf{\Phi}_{nn} \,. \tag{12.95}$$

Hence,

$$\mathbf{H} = \left(\Phi_{ss}\mathbf{a}\mathbf{a}^H + \mathbf{\Phi}_{nn}\right)^{-1}\Phi_{ss}\mathbf{a} \,. \tag{12.96}$$

Using the matrix inversion lemma [Haykin 1996],

$$\mathbf{A} = \mathbf{B}^{-1} + \mathbf{C}\mathbf{D}^{-1}\mathbf{C}^H \Leftrightarrow \mathbf{A}^{-1} = \mathbf{B} - \mathbf{B}\mathbf{C}(\mathbf{D} + \mathbf{C}^H\mathbf{B}\mathbf{C})^{-1}\mathbf{C}^H\mathbf{B} \tag{12.97}$$

with $\mathbf{B} = \mathbf{\Phi}_{nn}^{-1}$, $\mathbf{C} = \sqrt{\Phi_{ss}}\mathbf{a}$, and $\mathbf{D} = 1$ we obtain

$$\begin{aligned}\mathbf{H} &= \left[\mathbf{\Phi}_{nn}^{-1} - \frac{\Phi_{ss}\mathbf{\Phi}_{nn}^{-1}\mathbf{a}\mathbf{a}^H\mathbf{\Phi}_{nn}^{-1}}{1 + \Phi_{ss}\mathbf{a}^H\mathbf{\Phi}_{nn}^{-1}\mathbf{a}}\right]\Phi_{ss}\mathbf{a} \\ &= \left[\mathbf{\Phi}_{nn}^{-1}\mathbf{a} - \frac{\Phi_{ss}\mathbf{\Phi}_{nn}^{-1}\mathbf{a}}{(\mathbf{a}^H\mathbf{\Phi}_{nn}^{-1}\mathbf{a})^{-1} + \Phi_{ss}}\right]\Phi_{ss}\end{aligned} \tag{12.98}$$

and the final result

$$\begin{aligned}\mathbf{H} &= \frac{\mathbf{\Phi}_{nn}^{-1}\mathbf{a}}{\mathbf{a}^H\mathbf{\Phi}_{nn}^{-1}\mathbf{a}} \frac{\Phi_{ss}}{\left(\Phi_{ss} + \left(\mathbf{a}^H\mathbf{\Phi}_{nn}^{-1}\mathbf{a}\right)^{-1}\right)} \\ &= \mathbf{H}_{\mathrm{MVDR}}\frac{\Phi_{ss}}{\Phi_{ss} + \left(\mathbf{a}^H\mathbf{\Phi}_{nn}^{-1}\mathbf{a}\right)^{-1}} \,.\end{aligned} \tag{12.99}$$

The optimal Wiener solution thus comprises a distortionless MVDR beamformer and a single channel Wiener *postfilter*. Since

$$\mathbf{H}_{\mathrm{MVDR}}^H\mathbf{\Phi}_{nn}\mathbf{H}_{\mathrm{MVDR}} = \left(\mathbf{a}^H\mathbf{\Phi}_{nn}^{-1}\mathbf{a}\right)^{-1} \tag{12.100}$$

it can be seen that $\left(\mathbf{a}^H\mathbf{\Phi}_{nn}^{-1}\mathbf{a}\right)^{-1}$ represents the noise power at the output of the beamformer. Therefore, to compute the single channel postfilter, the speech power spectral density at the microphones and the noise power spectral density at the output of the beamformer are required. There are several proposals (for a survey, see [Simmer et al. 2001]) for computing these quantities using the available microphone signals. For example, [Zelinski 1988] uses cross-periodograms averaged over all microphone pairs to estimate the power spectral densities of the speech signal and of the noisy signal. This approach rests on the assumption that noise signals are mutally uncorrelated. Due to the large variance of the cross-periodograms and the residual correlation of noise components, additional post-processing of the estimates is necessary to suppress undesired fluctuations in the output of the postfilter. Zelinski [Zelinski 1988] then combines a delay-and-sum beamformer with this postfilter. This work is extended in [Marro et al. 1998] to account for the frequency response of a filter-and-sum beamformer. Proposals based on the coherence function are detailed in [Allen et al. 1977] and [Le Bouquin, Faucon 1990].

12.8 Adaptive Beamformers

When the spatial properties of the acoustic noise field are *a priori* unknown and possibly time varying, the beamformer coefficients cannot be precomputed but must be adapted on-line. While in [Frost 1972] the coefficients of a filter-and-sum beamformer are adapted using a constrained version of the LMS algorithm, most practical implementations are based on the *generalized side-lobe canceller* (GSC) or *Griffiths-Jim* beamformer [Griffiths, Jim 1982]. In this section, we discuss both the Frost and the GSC designs.

12.8.1 The Frost Beamformer

Figure 12.14 depicts the block diagram of an adaptive filter-and-sum beamformer as proposed in [Frost 1972], where the filter coefficients are now functions of time. The adaptation rule minimizes the noise power at the output while maintaining a constraint on the filter response in look direction. For the development of the Frost beamformer, we stack the coefficients of the N_M FIR filters into a single coefficient vector

$$\widetilde{\mathbf{h}} = (h_1(0), h_2(0), \ldots, h_{N_M}(0), h_1(1), \ldots, h_1(M), \ldots, h_{N_M}(M))^T \quad (12.101)$$

and the corresponding microphone signal samples into the vector

$$\widetilde{\mathbf{y}}(k) = \big(y_1(k), y_2(k), \ldots, y_{N_M}(k), y_1(k-1), \ldots, \\ \ldots, y_1(k-M), \ldots, y_{N_M}(k-M)\big)^T . \quad (12.102)$$

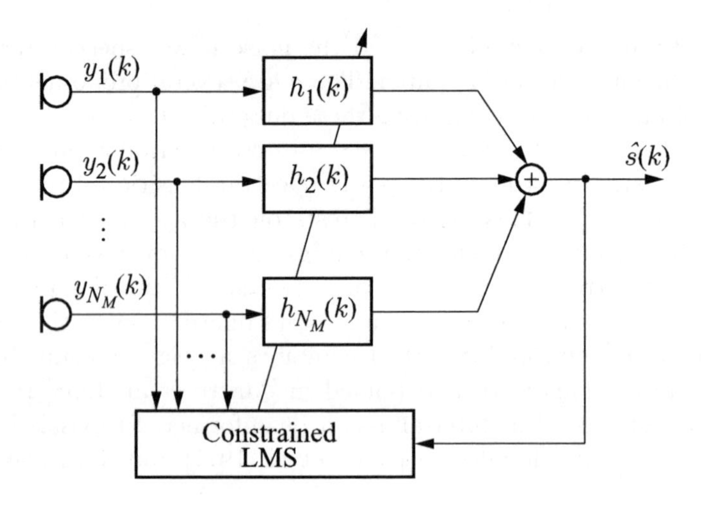

Figure 12.14: Frost beamformer

For each time lag, a constraint on the sum of the beamformer coefficients is applied. These constraints may be written as

$$\widetilde{\mathbf{C}}^T \widetilde{\mathbf{h}} = \widetilde{\mathbf{F}} \tag{12.103}$$

where the constraint matrix of dimensions $(M+1)N_M \times (M+1)$ is given by

$$\widetilde{\mathbf{C}} = \begin{pmatrix} \mathbf{1} & \mathbf{0} & \mathbf{0} & \cdots & \mathbf{0} \\ \mathbf{0} & \mathbf{1} & \mathbf{0} & \cdots & \mathbf{0} \\ \vdots & \vdots & \vdots & \ddots & \vdots \\ \mathbf{0} & \mathbf{0} & \mathbf{0} & \cdots & \mathbf{1} \end{pmatrix} \tag{12.104}$$

and

$$\mathbf{0} = (0, 0, \ldots, 0)^T \quad \text{and} \quad \mathbf{1} = (1, 1, \ldots, 1)^T \tag{12.105}$$

are vectors of dimension N_M.

For example, the undistorted and delay-free array response in the broadside look direction implies a constraint

$$\sum_{\ell=1}^{N_M} \sum_{i=0}^{M} h_\ell(i)\delta(k-i) = \delta(k-k') \quad \text{for} \quad k' = k-M, \ldots k \tag{12.106}$$

on the array response of the filter-and-sum beamformer, which can be also cast into the matrix equation (12.103) with (12.104) and the $M+1$-dimensional vector

$$\widetilde{\mathbf{F}} = (1, 0, \ldots, 0)^T . \tag{12.107}$$

An optimal stationary solution is again achieved by minimizing the output power $\widetilde{\mathbf{h}}^T \mathbf{R}_{\widetilde{y}\widetilde{y}} \widetilde{\mathbf{h}}$ while maintaining the constraint (12.103). Using the method of the Lagrange multiplier, we obtain

$$\mathfrak{L}(\widetilde{\mathbf{h}}, q) = \widetilde{\mathbf{h}}^T \mathbf{R}_{\widetilde{y}\widetilde{y}} \widetilde{\mathbf{h}} + q \left(\widetilde{\mathbf{C}}^T \widetilde{\mathbf{h}} - \widetilde{\mathbf{F}} \right) . \tag{12.108}$$

Differentiating with respect to $\widetilde{\mathbf{h}}$ and q results in the necessary conditions

$$2\mathbf{R}_{\widetilde{y}\widetilde{y}} \widetilde{\mathbf{h}} + q\widetilde{\mathbf{C}} = 0 \quad \text{and} \quad \widetilde{\mathbf{C}}^T \widetilde{\mathbf{h}} = \widetilde{\mathbf{F}} . \tag{12.109}$$

Solving the first condition for $\widetilde{\mathbf{h}}$, and, using the result in the second condition, we obtain

$$\widetilde{\mathbf{h}} = -\frac{1}{2} q \mathbf{R}_{\widetilde{y}\widetilde{y}}^{-1} \widetilde{\mathbf{C}} \quad \text{and} \quad q = \frac{-2}{\widetilde{\mathbf{C}}^T \mathbf{R}_{\widetilde{y}\widetilde{y}}^{-1} \widetilde{\mathbf{C}}} \widetilde{\mathbf{F}} . \tag{12.110}$$

Thus, the optimal linearly constrained solution is given by

$$\widetilde{\mathbf{h}}_{\mathrm{LC}} = \widetilde{\mathbf{R}}_{\widetilde{y}\widetilde{y}}^{-1} \widetilde{\mathbf{C}} \left(\widetilde{\mathbf{C}}^T \widetilde{\mathbf{R}}_{\widetilde{y}\widetilde{y}}^{-1} \widetilde{\mathbf{C}} \right)^{-1} \widetilde{\mathbf{F}} . \tag{12.111}$$

Based on the time domain solution in (12.111), deterministic and stochastic adaptive solutions may be developed [Frost 1972]. From the gradient of (12.108), a deterministic update rule for the coefficient vector may be derived as

$$\widetilde{\mathbf{h}}(k+1) = \widetilde{\mathbf{h}}(k) - \mu \left(\mathbf{R}_{\widetilde{y}\widetilde{y}} \widetilde{\mathbf{h}} + q\widetilde{\mathbf{C}} \right) \tag{12.112}$$

where μ is a positive stepsize parameter. Applying the constraint (12.103) to (12.112) leads to

$$\widetilde{\mathbf{F}} = \widetilde{\mathbf{C}}^T \widetilde{\mathbf{h}}(k+1) = \widetilde{\mathbf{C}}^T \widetilde{\mathbf{h}}(k) - \mu \left(\widetilde{\mathbf{C}}^T \mathbf{R}_{\widetilde{y}\widetilde{y}} \widetilde{\mathbf{h}} + q\widetilde{\mathbf{C}}^T \widetilde{\mathbf{C}} \right) . \tag{12.113}$$

Solving for q and using the result in (12.112) yields the deterministic update rule

$$\begin{aligned} \widetilde{\mathbf{h}}(k+1) = \widetilde{\mathbf{h}}(k) - \mu \left(\mathbf{I} - \widetilde{\mathbf{C}}(\widetilde{\mathbf{C}}^T \widetilde{\mathbf{C}})^{-1} \widetilde{\mathbf{C}}^T \right) \mathbf{R}_{\widetilde{y}\widetilde{y}} \widetilde{\mathbf{h}} \\ + \widetilde{\mathbf{C}}(\widetilde{\mathbf{C}}^T \widetilde{\mathbf{C}})^{-1} \left(\widetilde{\mathbf{F}} - \widetilde{\mathbf{C}}^T \widetilde{\mathbf{h}}(k) \right) . \end{aligned} \tag{12.114}$$

Using the sample covariance $\widetilde{\mathbf{y}}(k)\widetilde{\mathbf{y}}^T(k)$ as an approximation to $\mathbf{R}_{\widetilde{y}\widetilde{y}}$, and with

$$\mathbf{P} = \mathbf{I} - \widetilde{\mathbf{C}}(\widetilde{\mathbf{C}}^T \widetilde{\mathbf{C}})^{-1} \widetilde{\mathbf{C}}^T \tag{12.115}$$

$$\widetilde{\mathbf{f}} = \widetilde{\mathbf{C}}(\widetilde{\mathbf{C}}^T \widetilde{\mathbf{C}})^{-1} \widetilde{\mathbf{F}} \tag{12.116}$$

$$\widehat{s}(k) = \widetilde{\mathbf{h}}^T(k) \widetilde{\mathbf{y}}(k) , \tag{12.117}$$

the *stochastic constrained least-mean-square* (CLMS) algorithm can be written as

$$\widetilde{\mathbf{h}}(k+1) = \widetilde{\mathbf{h}}(k) + \mathbf{P}\left(\widetilde{\mathbf{h}}(k) - \mu\widehat{s}(k)\widetilde{\mathbf{y}}(k)\right) + \widetilde{\mathbf{f}} \tag{12.118}$$

with the initial condition $\widetilde{\mathbf{h}}(0) = \widetilde{\mathbf{f}}$. In each iteration, the algorithm updates the coefficient vector such that the constraint is met [Frost 1972].

12.8.2 Generalized Side-Lobe Canceller

An improved solution to the constrained adaptive beamforming problem decomposes the adaptive filter-and-sum beamformer into a fixed beamformer and an adaptive multi-channel noise canceller. The resulting system is termed the generalized side-lobe canceller [Griffiths, Jim 1982], a block diagram of which is shown in Fig. 12.15. Here, the constraint of a distortionless response in look direction is established by the fixed beamformer. The noise canceller can then be adapted without a constraint. The implementation of an adaptive, distortionless response beamformer is therefore much facilated.

The fixed beamformer can be implemented via one of the previously discussed methods, for example, as a delay-and-sum beamformer. To avoid distortions of the desired signal, the input to the adaptive noise canceller must not contain

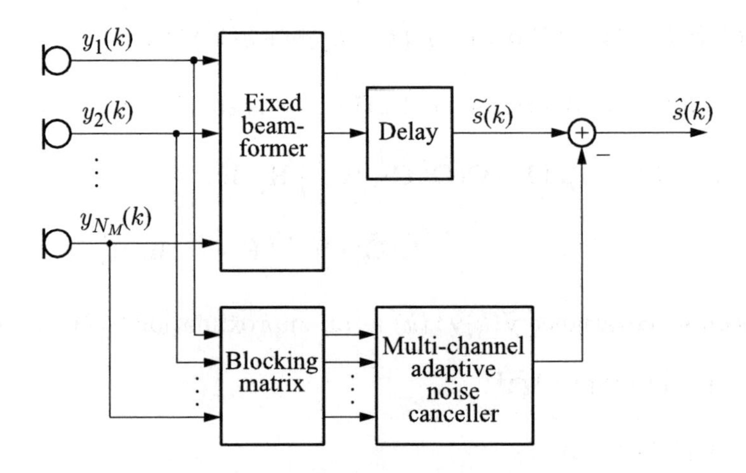

Figure 12.15: Generalized side-lobe canceller

the desired signal. Therefore, a *blocking matrix* $\mathbf{B}$ is employed such that the noise signals

$$\begin{pmatrix} \widehat{y}_1(k) \\ \widehat{y}_2(k) \\ \vdots \\ \widehat{y}_{N_M-1}(k) \end{pmatrix} = \mathbf{B} \begin{pmatrix} y_1(k) \\ y_2(k) \\ \vdots \\ y_{N_M}(k) \end{pmatrix} \tag{12.119}$$

are free of the desired signal. It is generally assumed that the colums of $\mathbf{B}$ are linearly independent. In our scenario, the blocking of a single desired signal by means of a linear operation on the microphone signals reduces the number of independent signal components by one. Therefore, $\mathbf{B}$ is a $N_M \times (N_M - 1)$ matrix.

When the desired signal originates from a point source under farfield conditions, and is received from the broadside look direction, blocking of the desired source can be achieved by a pairwise subtraction of the microphone signals. In this case, the blocking matrix may be written as

$$\mathbf{B} = \begin{pmatrix} 1 & -1 & 0 & 0 & \cdots & 0 & 0 \\ 0 & 1 & -1 & 0 & \cdots & 0 & 0 \\ \vdots & \vdots & \vdots & \vdots & \ddots & \vdots & \vdots \\ 0 & 0 & 0 & 0 & \cdots & 1 & -1 \end{pmatrix}. \tag{12.120}$$

The adaptive noise canceller then estimates the noise components at the output of the fixed beamformer and subtracts the estimate. Since both the fixed beamformer and the multi-channel noise canceller might delay their respective input signals, a delay in the signal path is required.

In reverberant environments, it is in general difficult to prevent the desired speech signal from leaking into the noise cancellation branch. A number of countermeasures have been proposed [Hoshuyama, Sugiyama 2001], such as

- the use of an adaptive blocking matrix
- improved target tracking
- adaptation-mode control
- coefficient and coefficient-norm constraint adaptive filters.

One of these designs is explained in the next section. Furthermore, for hands-free, full-duplex speech communication, it is desirable to combine beamforming microphone arrays with echo cancellation. The GSC and extensions towards combined beamforming and acoustic echo cancellation are discussed in detail in [Herbordt 2005].

12.8.3 Generalized Side-lobe Canceller with Adaptive Blocking Matrix

In reverberant environments the blocking matrix in (12.120) does not deliver sufficient attenuation of the desired signal since it does not account for multipath propagation effects. An improved implementation of the GSC principle is proposed in [Hoshuyama et al. 1999] where adaptive filters block the desired signal.

The resulting algorithm is shown in Fig. 12.16. The output of the fixed beamformer is fed into adaptive filters which minimize the error with respect to the microphone signals. Thus, any coherent signal originating from the look direction of the fixed beamformer can be reduced by the adaptive blocking method. For the adaptive blocking filters, a coefficient constrained adaptive filter (CCAF) is used. The coefficients of the adaptive filter are constrained to produce the dominant peak for those taps which correspond to the look direction with some tolerance interval (typically ±20 taps).

The blocked microphone signals are then fed into a multi-channel noise canceller which is implemented as a norm-constrained adaptive filter (NCAF). The norm constraint on the coefficient vector prevents excessive growth of the filter coeffi-

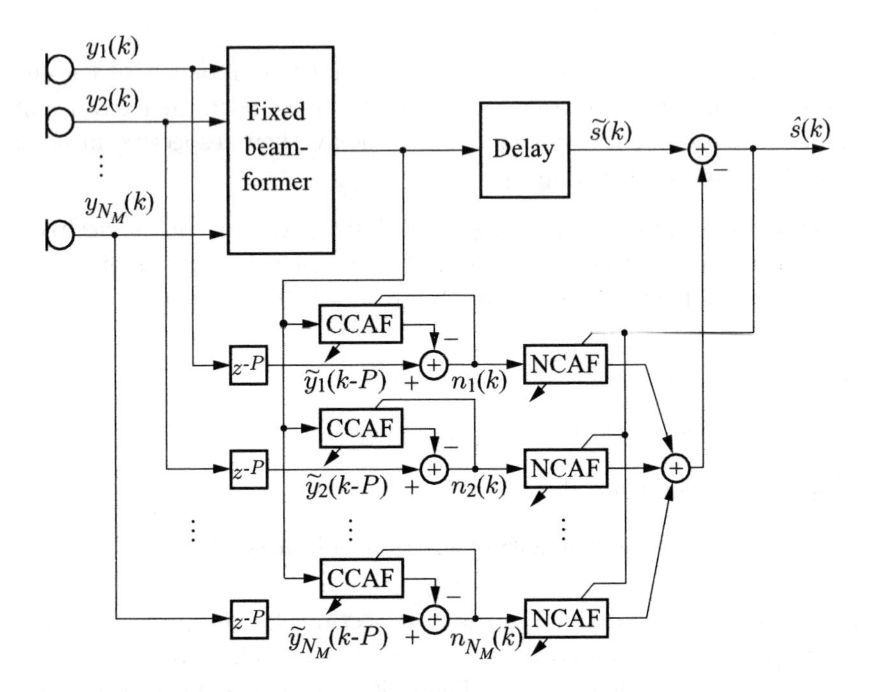

Figure 12.16: GSC with adaptive blocking matrix [Hoshuyama et al. 1999]; © 1999 IEEE

cients and thus leakage of the desired signal into the output of the multi-channel noise canceller. For adapting both the blocking and the noise cancellation filters, the NLMS algorithm with the above modifications is used.

12.9 Optimal Non-linear Multi-channel Noise Reduction

In analogy to the single channel case, non-linear optimal estimators may be derived which account for the probability distribution of the acoustic signals. For instance, in the short-term spectral domain the short-term spectral amplitude of the desired signal can be estimated using information about the spectral coefficients of all microphone signals.

Thus, as discussed in Chapter 11, MMSE and MAP estimators may be developed for the short-term spectral amplitude in the μ-th frequency bin of the desired signal. The MMSE solution is again given by the conditional expectation

$$\widehat{A}_\mu = \mathrm{E}\left\{A_\mu \mid Y_{\mu,1}, \ldots, Y_{\mu,N_M}\right\} \tag{12.121}$$

which is now conditioned on the complex spectral amplitudes $Y_{\mu,\ell}$ of the noisy signal. This expectation yields a closed form solution for Gaussian distributed signals [Balan, Rosca 2002].

MAP solutions are easier to compute and are known for Gaussian and super-Gaussian distributed spectral coefficients. Furthermore, since the condition on the complex spectral coefficients $Y_{\mu,1}, \ldots, Y_{\mu,N_M}$ introduces a phase dependency, it is advantageous to condition on the spectral amplitudes $R_{\mu,\ell} = |Y_{\mu,\ell}|$ only [Lotter et al. 2003a] and, hence, to solve

$$\widehat{A}_\mu = \arg\max_{A_\mu} p(A_\mu \mid R_{\mu,1}, \ldots, R_{\mu,N_M}) \tag{12.122}$$

$$= \arg\max_{A_\mu} p(R_{\mu,1}, \ldots, R_{\mu,N_M} | A_\mu)\, p(A_\mu)\,. \tag{12.123}$$

The solution to this optimization problem is detailed in [Lotter et al. 2003b], [Lotter 2004] and [Lotter 2005] for Gaussian and super-Gaussian distributed spectral coefficients.

Bibliography

Allen, J. B.; Berkley, D. A.; Blauert, J. (1977). Multimicrophone Signal-Processing Technique to Remove Room Reverberation from Speech Signals, *Journal of the Acoustical Society of America*, vol. 62, no. 4, October, pp. 912–915.

Balan, R.; Rosca, J. (2002). Microphone Array Speech Enhancement by Bayesian Estimation of Spectral Amplitude and Phase, *Proceedings of the IEEE Sensor Array and Multichannel Signal Processing Workshop*, Rosslyn, Virginia, pp. 209–213.

Brandstein, M.; Ward, D. (eds.) (2001). *Microphone Arrays*, Springer Verlag, Berlin, Heidelberg, New York.

Cox, H.; Zeskind, R. M.; Kooij, T. (1986). Practical Supergain, *IEEE Transactions on Acoustics, Speech and Signal Processing*, vol. 34, no. 3, June, pp. 393–398.

Crochiere, R. E.; Rabiner, L. R. (1981). Interpolation and Decimation of Digital Signals – A Tutorial Review, *Proceedings of the IEEE*, vol. 69, no. 3, March.

Crochiere, R. E.; Rabiner, L. R. (1983). *Multirate Digital Signal Processing*, Prentice Hall, Englewood Cliffs, New Jersey.

Doclo, S.; Moonen, M. (2003). Design of Broadband Beamformers Robust Against Microphone Position Errors, *Proceedings of the International Workshop on Acoustic Echo and Noise Control (IWAENC)*, Kyoto, Japan, pp. 267–270.

Dörbecker, M. (1997). Small Microphone Arrays with Optimized Directivity for Speech Enhancement, *Proceedings of the European Conference on Speech Communication and Technology (EUROSPEECH)*, Rhodes, Greece, pp. 327–330.

Dörbecker, M. (1998). *Mehrkanalige Signalverarbeitung zur Verbesserung akustisch gestörter Sprachsignale am Beispiel elektronischer Hörhilfen*, PhD thesis. *Aachener Beiträge zu digitalen Nachrichtensystemen*, vol. 10, P. Vary (ed.), RWTH Aachen University (in German).

Edelblute, D. J.; Fisk, J. M.; Kinnison, G. L. (1966). Criteria for Optimum-Signal-Detection Theory for Arrays, *Journal of the Acoustical Society of America*, vol. 41, no. 1, pp. 199–205.

Elko, G. W.; Nguyen Pong, A.-T. (1997). A Steerable and Variable First-Order Differential Microphone Array, *Proceedings of the IEEE International Conference on Acoustics, Speech, and Signal Processing (ICASSP)*, Munich, Germany, pp. 223–226.

Frost, O. L. (1972). An Algorithm for Linearly Constrained Adaptive Array Processing, *Proceedings of the IEEE*, vol. 60, no. 8, pp. 926–935.

Gabriel, W. F. (1992). Adaptive processing array systems, *Proceedings of the IEEE*, vol. 80, no. 1, pp. 152–162.

Goodwin, M. M.; Elko, G. W. (1993). Constant Beamwidth Beamforming, *Proceedings of the IEEE International Conference on Acoustics, Speech, and Signal Processing (ICASSP)*, Minneapolis, Minnesota, USA, vol. I, pp. 169–172.

Griffiths, L. J.; Jim, C. W. (1982). An Alternative Approach to Linearly Constrained Adaptive Beamforming, *IEEE Transactions on Antennas and Propagation*, vol. 30, no. 1, pp. 27–34.

Haykin, S. (1996). *Adaptive Filter Theory*, 3rd edn, Prentice Hall, Englewood Cliffs, New Jersey.

Haykin, S. (ed.) (1985). *Array Signal Processing*, Prentice Hall, Englewood Cliffs, New Jersey.

Herbordt, W. (2005). *Sound Capture for Human/machine Interfaces - Practical Aspects of Microphone Array Signal Processing*, vol. 315 of *Lecture Notes in Control and Information Sciences*, Springer Verlag, Berlin, Heidelberg, New York.

Hoshuyama, O.; Sugiyama, A. (2001). Robust Adaptive Beamforming, *in* M. Brandstein; D. Ward (eds.), *Microphone Arrays*, Springer Verlag, Berlin, Heidelberg, New York.

Hoshuyama, O.; Sugiyama, A.; Hirano, A. (1999). A Robust Adaptive Beamformer for Microphone Arrays with a Blocking Matrix Using Constrained Adaptive Filters, *IEEE Transactions on Signal Processing*, vol. 47, no. 10, October, pp. 2677–2683.

Kummer, W. H. (1992). Basic Array Theory, *Proceedings of the IEEE*, vol. 80, no. 1, pp. 127–140.

Kuttruff, H. (2004). *Akustik*, S. Hirzel Verlag, Stuttgart (in German).

Laakso, T. I.; Välimaki, V.; Karjalainen, M.; Laine, U. K. (1996). Splitting the Unit Delay - Tools for Fractional Filter Design, *IEEE Signal Processing Magazine*, vol. 13, no. 1, January, pp. 30–60.

Le Bouquin, R.; Faucon, G. (1990). On Using the Coherence Function for Noise Reduction, *in* L. Torres; E. Masgrau; M. A. Lagunas (eds.), *Signal Processing V: Theories and Applications*, pp. 1103–1106.

Lotter, T. (2004). *Single and Multimicrophone Speech Enhancement for Hearing Aids*, PhD thesis. *Aachener Beiträge zu digitalen Nachrichtensystemen*, vol. 18, P. Vary (ed.), RWTH Aachen University.

Lotter, T. (2005). Single and Multimicrophone Spectral Amplitude Estimation using a Super-Gaussian Speech Model, *in* J. Benesty et al. (eds.), *Speech Enhancement*, Springer Verlag, Berlin, Heidelberg, New York.

Lotter, T.; Benien, C.; Vary, P. (2003a). Multichannel Direction-Independent Speech Enhancement Using Spectral Amplitude Estimation, *Eurasip Journal on Applied Signal Processing, Special Issue on Signal Processing for Acoustic Communication Systems*, , September, pp. 1147–1156.

Lotter, T.; Benien, C.; Vary, P. (2003b). Multichannel Speech Enhancement using Bayesian Spectral Amplitude Estimation, *Proceedings of the IEEE International Conference on Acoustics, Speech, and Signal Processing (ICASSP)*, Hong Kong, China.

Lotter, T.; Vary, P. (2006). Dual Channel Speech Enhancement by Superdirective Beamforming, *EURASIP Journal on Applied Signal Processing. Special Issue: Advances in Multimicrophone Speech Enhancement*, to appear.

Marro, C.; Mahieux, Y.; Simmer, K. U. (1998). Analysis of noise reduction and dereverberation techniques based on microphone arrays with postfiltering, *IEEE Transactions on Speech and Audio Processing*, vol. 6, no. 3, pp. 240–259.

Monzingo, R. A.; Miller, T. W. (1980). *Introduction to Adaptive Arrays*, John Wiley & Sons, Inc., New York.

Simmer, K. U.; Bitzer, J.; Marro, C. (2001). Post-Filtering Techniques, *in* M. Brandstein; D. Ward (eds.), *Microphone Arrays*, Springer Verlag, Berlin, Heidelberg, New York.

Zelinski, R. (1988). A Microphone Array with Adaptive Post-Filtering for Noise Reduction in Reverberant Rooms, *Proceedings of the IEEE International Conference on Acoustics, Speech, and Signal Processing (ICASSP)*, New York City, USA, pp. 2578–2581.

13

Acoustic Echo Control

In this chapter we introduce algorithms for feedback control in handsfree voice communication systems. At the center of our discussion are adaptive algorithms for acoustic echo cancellation. In principle, the acoustic echo canceller can remove the annoying echo without distorting the near-end speech. Acoustic echo cancellation is therefore an essential component of high-quality full-duplex handsfree communication devices. As it turns out, it is a quite challenging application of adaptive filters and other signal processing techniques. We present and discuss several algorithms for the adaptation of acoustic echo cancellers in the time and frequency domain as well as additional measures for echo control.

13.1 The Echo Control Problem

With the advent of mobile communications many speech communication systems are equipped with so-called handsfree devices. For this, a loudspeaker and a microphone are used instead of a hand-held telephone receiver, in order to increase user comfort or for safety reasons, e.g., in a car. Applications of handsfree systems comprise not only car phones and comfort telephones, but also multimedia systems with speech input, human–machine interfaces, or teleconferencing facilities.

The basic set-up of a handsfree communication system is illustrated in Fig. 13.1. The echo control problem arises as a consequence of the acoustic coupling between

Digital Speech Transmission: Enhancement, Coding and Error Concealment
Peter Vary and Rainer Martin

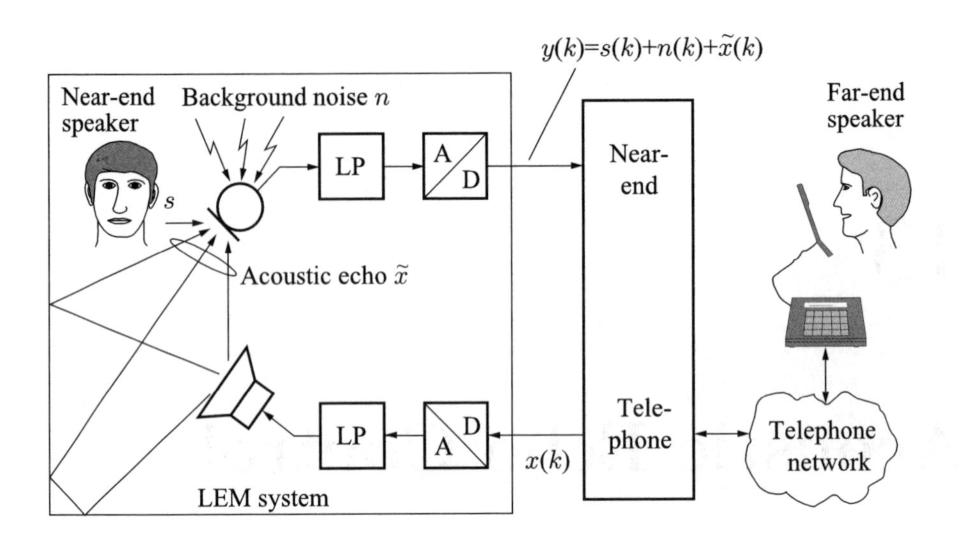

Figure 13.1: Loudspeaker–enclosure–microphone (LEM) system of a handsfree telephone with digital input and output signals

the loudspeaker and the microphone. The microphone picks up not only the desired signal s of the near-end speaker but also undesirable background noise n, and in particular the signal of the far-end speaker, denoted $\tilde{x}$, which is received via the electro-acoustic transmission path from the loudspeaker to the microphone. The signal $\tilde{x}$ results from multiple acoustic reflections. It is commonly called the acoustic echo signal (in distinction from the electric line echoes of the telephone network).

To simplify the discussion, we will assume that the far-end speaker's signal $x(k)$ and the transmission signal $y(k)$ are digitized with sampling rate f_s. These digital signals are available within ISDN or digital mobile radio systems (GSM or UMTS). In analog transmission systems they are generated by using directional filters (4-wire/2-wire hybrids) and analog-to-digital converters. In the following we do not differentiate between the acoustic or analog signals and their digital counterparts. Only the discrete time signals $x(k)$, $s(k)$, $n(k)$, etc., i.e., digital versions of band-limited analog signals will be used.

The microphone signal $y(k)$ of the handsfree device in Fig. 13.1 is thus given by

$$y(k) = s(k) + n(k) + \tilde{x}(k) . \tag{13.1}$$

The task of acoustic echo cancellation is to prevent the echo signal $\tilde{x}(k)$ from being fed back to the far-end user. Thus, the stability of the electro-acoustic loop is ensured, even if the far-end user uses a handsfree telephone as well.

However, acoustic echo cancellation is also compulsory for another reason, even if the far-end participant uses a telephone handset or a headset. Most of the wire-

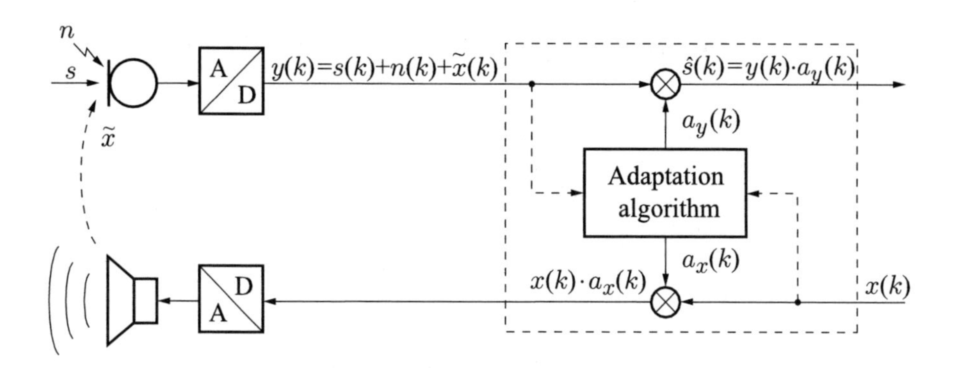

Figure 13.2: Loudspeaking telephone with voice-controlled echo suppressor

less digital telephone networks cause a relatively long signal delay, e.g., approx. 90 ms in GSM mobile radio networks or approx. 200 – 240 ms for transmission via geostationary satellites. Consequently, without echo cancellation, the far-end user receives an echo of his or her own voice. As the round-trip echo delay (which equals twice the transmission delay) might be large, the echo is not masked by the far-end user's own speech and thus interferes with the user's speech.

Simple solutions to the echo control problem employ voice-controlled *echo suppressors* which consist of one variable attenuator in the transmit branch and in the receive branch, respectively. Depending on the speech activity of the two speakers the transmit and the receive branch are differently attenuated such that the total attenuation in the echo path is not below a minimum of, for instance, 40 dB. This principle can be easily realized in analog or digital technology. A digital solution is shown in Fig. 13.2. When only one of the two speakers is active (*single talk*) the echo suppressor can achieve a high level of echo suppression. However, as the variable attenuation factors should fulfill the condition

$$-\big(20 \lg a_x(k) + 20 \lg a_y(k)\big) = 40\,\text{dB}\,, \tag{13.2}$$

simultaneous communication of the two users (*double talk*) is possible only in a very limited way. This limitation can be circumvented by using a handsfree telephone with *echo cancellation* as depicted in Fig. 13.3.

The basic idea of acoustic echo cancellation is to model the electro-acoustic echo path (loudspeaker–enclosure–microphone system) as a linear system and to identify the impulse response of this system by means of an adaptive digital filter. Due to the band limitation of the microphone signal, the electro-acoustic transmission of the far-end speaker's signal $x(k)$ via the loudspeaker–enclosure–microphone system (LEM system) can be described by a discrete time, linear system with the causal impulse response $\mathbf{g}$. As a result of the physical characteristics of the LEM system,

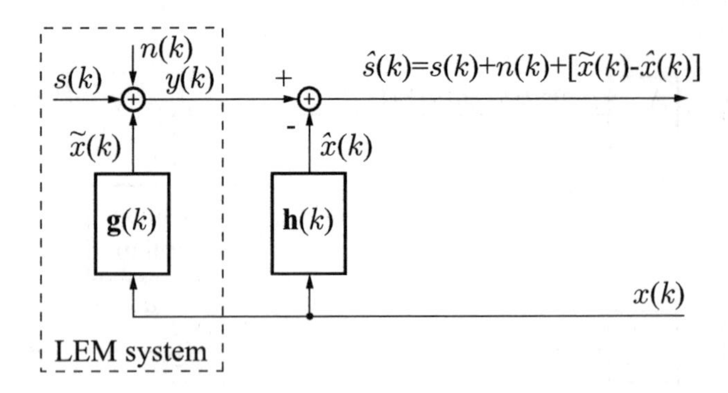

Figure 13.3: Discrete time model of a handsfree telephone with echo cancellation

this impulse response is in principle infinite (IIR). Two typical impulse responses of LEM systems are shown in Fig. 13.4. While the initial part of the LEM impulse response is dominated by the direct path from the loudspeaker to the microphone and by distinct peaks of early reflections, the reverberation manifests itself in a large number of decaying impulses which are best described by statistical models. The exponential decay of the late reverberation is usually characterized by a time constant, the *reverberation time* T_H. It describes the time span in which, according to an exponential law, the sound pressure drops to 10^{-6} of its initial value after turning off the sound source. Using *Sabine's reverberation formula* (e.g., [Kuttruff 1990]), the reverberation time as a function of the room volume V, the wall areas A_i with absorption factors α_i, and the sound velocity c can be approximately determined as

$$T_H = \frac{24 \ln(10)\, V}{c \sum\limits_i A_i\, \alpha_i} \ .$$

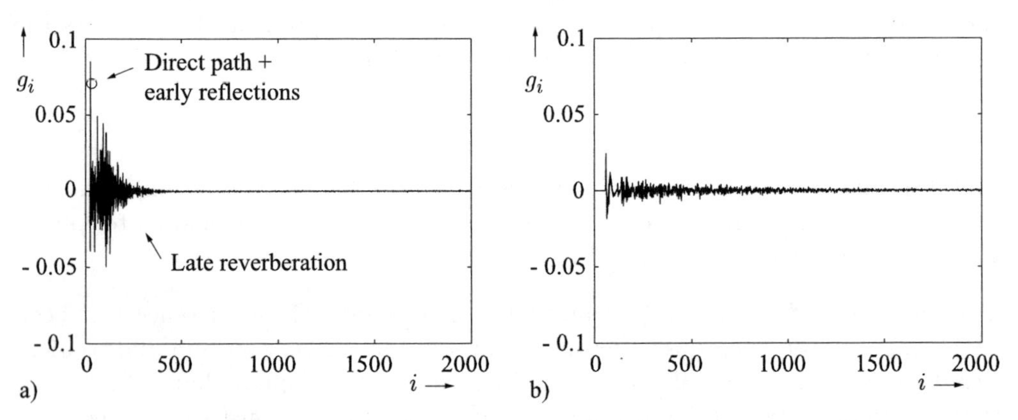

Figure 13.4: Measured impulse responses of LEM systems at $f_s = 8\,\text{kHz}$
a) Car
b) Office room

The reverberation time in a car for an estimated acoustically relevant volume of, say, $V = 1.3\,\mathrm{m}^3$ amounts to $T_H = 0.065\,\mathrm{s}$; in an office room with $V = 100\,\mathrm{m}^3$ it can be as large as $T_H = 0.7\,\mathrm{s}$ [Martin 1995]. The large number of quasi-random reflections in a reverberant enclosure also implies that any accurate model of the LEM system requires many degrees of freedom. Since IIR filters of low order cannot provide this flexibility, the LEM is usually modeled by high-order FIR filters.

Because of the exponential decay of a typical LEM response and because of amplifier and quantization noise in the microphone signal, the impulse response can be limited for all practical purposes to a finite number of coefficients m', provided that m' is chosen sufficiently large. The LEM system is thus modeled as a transversal (FIR) filter of order $m' - 1$ with impulse response

$$\mathbf{g}(k) = \big(g_0(k), g_1(k), \ldots, g_{m'-1}(k)\big)^T \tag{13.3}$$

and with excitation signal $x(k)$. The impulse response is in general time variant as it is influenced by movements of the near-end speaker and by other changes of the acoustic environment.

In order to compensate the undesirable echo signal $\tilde{x}(k)$, it may therefore be reproduced by using a transversal filter (FIR) with the time-variant coefficient vector of length m

$$\mathbf{h}(k) = \big(h_0(k), h_1(k), \ldots, h_{m-1}(k)\big)^T . \tag{13.4}$$

The estimated echo, $\hat{x}(k)$, is then subtracted from the microphone signal $y(k)$. If the impulse responses $\mathbf{g}$ and $\mathbf{h}$ match exactly and $m = m'$ is sufficiently large, the echo signal will be eliminated from the transmit branch.

The underlying signal processing problem therefore consists in identifying the time-varying system impulse response $\mathbf{g}(k)$. In practice, the impulse response $\mathbf{h}(k)$ of the cancellation filter is adapted with an iterative algorithm. One of the most prominent adaptation algorithms is the *normalized least-mean-square* (NLMS) algorithm, which has gained widespread acceptance due to its simple adaptation rule and stability. Tracking of fast-changing impulse responses is, however, not easily accomplished [van de Kerkhof, Kitzen 1992]. Fast-converging algorithms, such as the affine projection (AP) algorithm and its fast variants, as well as frequency domain implementations of adaptive filtering concepts, are therefore also of significant interest. A vast amount of literature deals with these adaptive algorithms and their application to echo cancellation. Excellent surveys are given in, for example, [Hänsler 1992], [Hänsler 1994], [Breining et al. 1999], [Gay, Benesty 2000], [Benesty et al. 2001], [Hänsler, Schmidt 2004].

To fully appreciate the difficulties of the echo cancellation problem a first estimate of the required order of the cancellation filter will be derived. In Fig. 13.5 an LEM system with two sound propagation paths is depicted: one direct path leads from

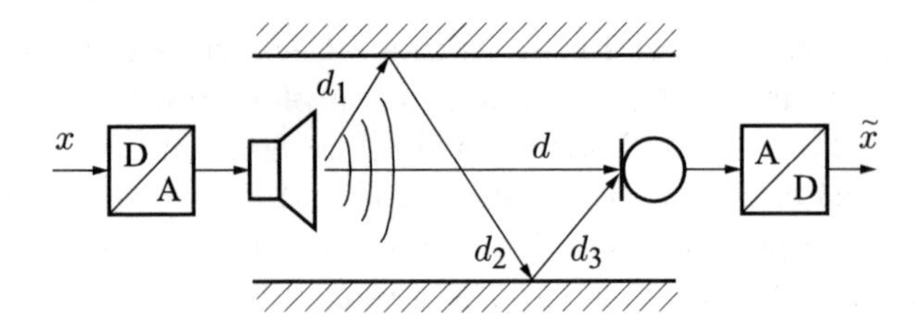

Figure 13.5: Sound propagation paths in the LEM system

the loudspeaker to the microphone via the distance d, and one indirect path with two reflections via the total distance of $d_1 + d_2 + d_3$. The length of the impulse response $\mathbf{h}$ of the cancellation filter should cover the corresponding propagation delay τ of the acoustic signal.

With a sampling rate of $f_s = 8\,\text{kHz}$ (corresponding to a sampling interval of $T = 125\,\mu\text{s}$) and the sound velocity $c \approx 343\,\text{m/s}$, the minimal required filter order amounts to

$$m - 1 \geq \frac{\sum_i d_i}{c} \cdot f_s \approx T_H \cdot f_s \,. \tag{13.5}$$

For a distance of, for example, $d = 20\,\text{cm}$ between loudspeaker and microphone, only $m = 6$ coefficients are needed to cancel the direct sound path. For the indirect sound, however, due to multiple reflections, total distances of some tens to hundreds of meters, depending on the size and the acoustic characteristics of the room (reverberation time), must be assumed. The average total distance can be obtained from the mean free paths [Kuttruff 1990] between reflections.

For a cumulative mean free path of $\bar{d} = c \cdot T_H = 343\,\text{m/s} \cdot 0.065\,\text{s} \approx 22.3\,\text{m}$, for example, which might be obtained in a car, a filter order of $m \approx 500$ is needed, while the application in reverberating office rooms ($0.2\,\text{s} \leq T_H \leq 0.5\,\text{s}$) requires a canceller with 2000–4000 coefficients. At a sampling rate of f_s the computational complexity of this filter is about

$$m \cdot f_s = 16\text{–}32\,\text{MOPS (Mega Operations Per Second)}. \tag{13.6}$$

To this, the computational complexity for the adaptation algorithm, when the convolution is implemented in the time domain, must be added, which increases (13.6) by a factor of at least 2–3. The high complexity of a full-size canceller combined with the difficulty of adapting a canceller in the presence of acoustic noise and double talk usually leads to implementations with significantly fewer canceller coefficients. Thus, the echo cannot be fully cancelled and additional measures as discussed in Section 13.10 become necessary.

13.2 Evaluation Criteria

Since the echo problem has a profound effect on the communication quality, numerous international recommendations and standards have been established to guide the development of handsfree devices. The required echo suppression depends on the signal delay of the transmission path and on the double talk condition. The actual requirements are frequently specified in terms of the *talker echo loudness rating* (TELR), i.e., the level difference between the original far-end voice and the resulting echo. The TELR which is deemed acceptable during single talk is given by 35 dB and 52 dB for transmission delays of 30 ms and 200 ms, respectively. For transmission delays of more than 200 ms, even more echo attenuation is necessary. During double talk some of the echo is masked by the near-end signal [Gilloire 1994]. In this case good listening quality is achieved when the single talk TELR requirement is lowered by no more than 4 dB [ITU-T G.131 2003].

The required high level of echo suppression, particularly needed for long transmission delays, can in practice only be achieved with an echo canceller in conjunction with additional measures such as an echo suppressor or echo reduction postfilter. However, the contribution of an echo suppressor to the required TELR has a profound effect on the double talk capability of the handsfree terminal. Thus, handsfree terminals are further categorized based on the attenuation range which is applied to the near-end signal by the echo suppressor [ITU-T P.340 2000]. Full duplex capability, for instance, is achieved when the contribution of the echo suppressor to the required TELR entails less than 3 dB of near-end signal attenuation.

Since the TELR measure comprises the entire electro-acoustic echo loop – including the telephone network and the handsfree terminal – it is often not practical for the fair comparison of echo control devices. For an instrumental evaluation of echo cancellation and echo suppression algorithms, two instrumental criteria, which characterize the system identification error and the level of echo reduction, are commonly used.

a) System Distance

For $m = m'$ the distance vector

$$\mathbf{d}(k) = \mathbf{g}(k) - \mathbf{h}(k) \tag{13.7}$$

of the impulse responses allows us to define the relative *system distance*

$$D(k) = \frac{||\mathbf{d}(k)||^2}{||\mathbf{g}(k)||^2} \tag{13.8}$$

as a performance measure for system identification. $||\mathbf{d}(k)||^2 = \mathbf{d}^T(k)\,\mathbf{d}(k)$ denotes the squared vector norm. Generally, the logarithmic distance $10\ \lg\,(D(k))$ (in dB) is used. If the coefficient vectors $\mathbf{g}(k)$ and $\mathbf{h}(k)$ differ in length, the shorter vector is padded with zero values. According to definition (13.8), the system distance $D(k)$ does not depend directly on the signal $x(k)$. However, since the cancellation filter is adapted using the signal $x(k)$, the system distance obtained at time instant k depends in fact on the particular excitation $x(k)$. Since the impulse responses of real LEM systems are generally not known, the system distance as defined in (13.8) is primarily an important evaluation criterion for off-line simulations with a given impulse response $\mathbf{g}(k)$.

b) Echo Return Loss Enhancement

One criterion which is more closely related to the subjectively perceived performance is the achievable reduction of the power of the echo signal $\tilde{x}(k)$. The corresponding measure is called *echo return loss enhancement* (ERLE) and is defined as

$$\frac{\text{ERLE}(k)}{\text{dB}} = 10 \lg \left(\frac{\text{E}\{\tilde{x}^2(k)\}}{\text{E}\{\left(\tilde{x}(k) - \hat{x}(k)\right)^2\}} \right). \tag{13.9}$$

This criterion depends on the echo signal $\tilde{x}(k)$ and its estimate $\hat{x}(k)$. The ensemble average $\text{E}\{\cdot\}$ provides a measure of echo reduction for every time instant k. In practice, the expected values in (13.9) are replaced by short-term smoothed estimates. Due to the length of the smoothing window, the short-term estimate of the echo return loss enhancement then has a limited time resolution.

A small system distance $D(k)$ implies a high level of echo return loss enhancement $\text{ERLE}(k)$. The inverse conclusion does not hold, which is easily demonstrated with a narrowband excitation signal $x(k)$. To achieve a high level of echo return loss enhancement, the frequency response of the cancellation filter must match the frequency response of the LEM system only for the frequencies which are actually excited by $x(k)$. Deviations at other frequencies do not have an impact on the measured $\text{ERLE}(k)$.

In a real system, the residual echo

$$e(k) = \tilde{x}(k) - \hat{x}(k) \tag{13.10}$$

is only accessible for $s(k) = 0$ (cf. Fig. 13.3), i.e., during speech pauses of the near-end speaker (*single talk*), and only if there is no background noise, i.e., for $n(k) = 0$. In simulation experiments, however, $e(k)$ can be calculated before the signals $s(k)$ and $n(k)$ are added so that an observation of the echo return loss enhancement over time is possible for double talk and additive background noise as well.

Besides $\text{ERLE}(k)$ and the system distance, measures for near-end speech quality and for double talk capability are used to characterize a handsfree device. Since modern handsfree devices are neither time invariant nor linear, special measurement procedures are required. An example based on *composite source signals* is given in [Gierlich 1992]. The most relevant standard procedures are summarized in [Carini 2001], [Gierlich, Kettler 2005].

13.3 The Wiener Solution

To lay the ground for the development of adaptive algorithms we briefly review the minimum mean square error solution (Wiener filter) to the echo cancellation problem.

The starting point is the minimization of the mean square error of the residual echo

$$\mathrm{E}\{e^2(k)\} = \mathrm{E}\{\big(\tilde{x}(k) - \mathbf{h}^T(k)\ \mathbf{x}(k)\big)^2\} \tag{13.11}$$

with the excitation vector

$$\mathbf{x}(k) = \big(x(k), x(k-1), \ldots, x(k-m+1)\big)^T \tag{13.12}$$

and the coefficient vector (tap weight vector)

$$\mathbf{h}(k) = \big(h_0(k), h_1(k), \ldots, h_{m-1}(k)\big)^T . \tag{13.13}$$

For wide sense stationary input signals and a time invariant LEM system response $\mathbf{g}$, the general solution to the FIR Wiener filter is given by (cf. Section 5.12.1)

$$\mathbf{h}_0 = \mathrm{E}\left\{\mathbf{x}(k)\mathbf{x}^T(k)\right\}^{-1} \cdot \mathrm{E}\left\{\tilde{x}(k)\mathbf{x}(k)\right\} = \mathbf{R}_{xx}^{-1}(k) \cdot \mathrm{E}\left\{\tilde{x}(k)\mathbf{x}(k)\right\} \tag{13.14}$$

provided that the $m \times m$ auto-correlation matrix $\mathbf{R}_{xx}(k) = \mathrm{E}\left\{\mathbf{x}(k)\mathbf{x}^T(k)\right\}$ is invertible.

In the echo cancellation context with $m \leq m'$ and $\tilde{x}(k) = \sum_{i=0}^{m'-1} g_i\, x(k-i)$, the cross-correlation vector $\mathrm{E}\left\{\tilde{x}(k)\mathbf{x}(k)\right\}$ may be written as

$$\mathrm{E}\left\{\tilde{x}(k)\mathbf{x}(k)\right\} = \begin{pmatrix} \sum\limits_{i=0}^{m'-1} g_i\, \varphi_{xx}(i) \\ \sum\limits_{i=0}^{m'-1} g_i\, \varphi_{xx}(i-1) \\ \vdots \\ \sum\limits_{i=0}^{m'-1} g_i\, \varphi_{xx}(i-m+1) \end{pmatrix} \tag{13.15}$$

$$= \begin{pmatrix} \varphi_{xx}(0) & \varphi_{xx}(1) & \cdots & \varphi_{xx}(m'-1) \\ \varphi_{xx}(1) & \varphi_{xx}(0) & \cdots & \varphi_{xx}(m'-2) \\ \vdots & \vdots & & \vdots \\ \varphi_{xx}(m-1) & \varphi_{xx}(m-2) & \cdots & \varphi_{xx}(m'-m) \end{pmatrix} \begin{pmatrix} g_0 \\ g_1 \\ \vdots \\ g_{m'-1} \end{pmatrix}$$

with $\varphi_{xx}(i) = \mathrm{E}\left\{x(k)x(k+i)\right\}$.

Combining (13.14) and (13.15) we conclude that for $m = m'$ perfect identification $\mathbf{h}_0 = \mathbf{g}$ is in principle possible. For $m < m'$ the FIR Wiener filter identifies the first m coefficients without error if the auto-correlation $\varphi_{xx}(i)$ is equal to zero for $1 \leq i \leq m'$, i.e., for a white noise excitation signal $x(k)$. If the auto-correlation does not vanish for $1 \leq i \leq m'$ and $m < m'$ the estimate $\mathbf{h}_0$ is biased.

The Wiener filter provides a solution for stationary signals. In general, the statistics of the excitation signal as well as the target impulse response $\mathbf{g}(k)$ are time varying. Therefore, adaptive algorithms are of great importance and will be studied in greater detail below.

13.4 The LMS and NLMS Algorithms

13.4.1 Derivation and Basic Properties

In analogy to the iterative adaptation of a linear predictor by means of the *least-mean-square* (LMS) algorithm (see Section 6.3.2) we derive a rule for adjusting the cancellation filter. The gradient of the mean square error (13.11) results in the vector

$$\boldsymbol{\nabla}(k) = \frac{\partial \mathrm{E}\{e^2(k)\}}{\partial \mathbf{h}(k)} \tag{13.16-a}$$

$$= 2\,\mathrm{E}\left\{e(k)\,\frac{\partial e(k)}{\partial \mathbf{h}(k)}\right\} \tag{13.16-b}$$

$$= -2\,\mathrm{E}\{e(k)\,\mathbf{x}(k)\}\,. \tag{13.16-c}$$

In order to decrease the error, the tap weights $h_i(k)$ must be adapted in the direction of the negative gradient. The fundamental idea of the LMS algorithm is to replace the gradient by the instantaneous gradient

$$\hat{\boldsymbol{\nabla}}(k) = -2\,e(k)\,\mathbf{x}(k). \tag{13.17}$$

The adaptation algorithm may be written as

$$\mathbf{h}(k+1) = \mathbf{h}(k) + \beta(k)\,e(k)\,\mathbf{x}(k) \tag{13.18}$$

with the effective (and in general time-variant) stepsize parameter $\beta(k)$ (cf. (6.81-b), $\beta = 2\,\vartheta$).

It can be shown that the LMS algorithm converges in the mean square sense if and only if the stepsize parameter $\beta(k)$ satisfies

$$0 < \beta(k) < \frac{2}{\lambda_{\max}} \tag{13.19}$$

where $\lambda_{\max}$ is the largest eigenvalue of the input correlation matrix $\mathbf{R}_{xx}$ [Haykin 1996].

Viewed as a dynamic system, the speed of convergence of the LMS depends on the stepsize as well as on the characteristic modes as described by the eigenvalues of the correlation matrix. For a small stepsize parameter β we might describe the mean trajectory of the squared error $e^2(k)$ ("learning curve") by a single exponential with time constant [Haykin 1996]

$$\tau_{av} \approx \frac{1}{2\,\beta\,\frac{1}{m}\sum\limits_{i=1}^{m}\lambda_i}\,. \tag{13.20}$$

Since the stepsize must take the largest eigenvalue into account it follows that a large eigenvalue spread leads to a slow convergence. It is also obvious that for non-stationary signals the stepsize must be adapted to the time-varying eigenstructure of the input signal.

Since $\sum\limits_{i=1}^{m}\lambda_i = \text{trace}(\mathbf{R}_{xx})$, a conservative upper bound for the stepsize is given by

$$0 < \beta(k) < \frac{2}{\text{trace}(\mathbf{R}_{xx})} = \frac{2}{m\sigma_x^2}\,. \tag{13.21}$$

If $\beta(k)$ is chosen proportional to $\text{trace}(\mathbf{R}_{xx})^{-1}$, $\beta(k) = \alpha \cdot \text{trace}(\mathbf{R}_{xx})^{-1}$, the mean time constant is given by

$$\tau_{av} \approx \frac{m}{2\alpha} \tag{13.22}$$

which is a first indication that long adaptive filters converge less rapidly than short filters.

Since in a real system the residual echo signal $e(k)$ cannot be isolated, the microphone signal

$$\hat{s}(k) = s(k) + n(k) + e(k) \tag{13.23}$$

is used instead of $e(k)$, leading to the practical LMS adaptation

$$\begin{aligned}\mathbf{h}(k+1) &= \mathbf{h}(k) + \beta(k)\,\hat{s}(k)\,\mathbf{x}(k) \\ &= \mathbf{h}(k) + \beta(k)\,e(k)\,\mathbf{x}(k) + \beta(k)\,(s(k)+n(k))\,\mathbf{x}(k)\,.\end{aligned} \tag{13.24}$$

The near-end speaker's signal $s(k)$ and the background noise $n(k)$ must be considered as interfering signals for the adaptation process. Because of short-time correlations between $s(k)+n(k)$ and $x(k)$ the minimization of the power of $e(k)$

might be severely disturbed by the near signals. Consequently, to avoid misalignment of the canceller the adaptation must be frozen (or at least slowed down) as soon as the near-end speaker becomes active. This can be achieved by an adaptive stepsize control mechanism which will be discussed in more detail in Section 13.5.4.

The *normalized least-mean-square* (NLMS) algorithm may be developed as a modification of the LMS algorithm using the normalized time-varying stepsize

$$\beta(k) = \frac{\alpha(k)}{||\mathbf{x}(k)||^2} = \frac{\alpha(k)}{\mathbf{x}^T(k)\mathbf{x}(k)}. \tag{13.25}$$

It can be shown that the NLMS converges in the mean square for [Haykin 1996]

$$0 < \alpha(k) < 2. \tag{13.26}$$

Moreover, the specific properties of the normalized coefficient update

$$\mathbf{h}(k+1) - \mathbf{h}(k) = \frac{\alpha(k)}{\mathbf{x}^T(k)\mathbf{x}(k)}\, e(k)\, \mathbf{x}(k) \tag{13.27}$$

allow an interpretation of the NLMS algorithm and its adaptation in terms of geometric projections. In contrast to the LMS algorithm, the NLMS algorithm with (13.26) is stable not only in the mean but also deterministically in each iteration [Rupp 1993], [Slock 1993]. This will be analyzed in more detail in Section 13.6. In the next section we will first consider the performance of the LMS and NLMS algorithms in the context of the echo cancellation application.

13.5 Convergence Analysis and Control of the LMS Algorithm

We now return to the LMS algorithm for an in-depth analysis of its convergence behavior. To guarantee stability for stationary signals we will use a normalized stepsize

$$\beta(k) = \frac{\alpha}{m\,\sigma_x^2} \tag{13.28}$$

with $0 < \alpha < 2$ and

$$\sigma_x^2 = \mathrm{E}\{x^2(k)\}. \tag{13.29}$$

For wide sense stationary signals and large m the results of this analysis will approximately hold also for the NLMS algorithm since

$$||\mathbf{x}(k)||^2 \approx m\,\sigma_x^2. \tag{13.30}$$

In the simulation examples below, we will therefore use the NLMS algorithm. We will also show how the convergence analysis leads to the design of optimal stepsize parameters.

13.5.1 Convergence in the Absence of Interference

In this section we analyze the convergence behavior for stationary noise-like excitation signals $x(k)$ in the absence of interference, i.e., for $s(k) = 0$ and $n(k) = 0$. The impulse response of the LEM system is assumed to be time invariant, i.e., $\mathbf{g}(k) = \mathbf{g}$, and of same length as the response of the echo canceller, i.e., $m = m'$.

When the adaptive filter is excited with zero mean, white noise of variance σ_x^2 and when the vectors $\mathbf{d}(k)$ and $\mathbf{x}(k)$ are assumed to be statistically independent we have

$$\mathrm{E}\{e^2(k)\} = \mathrm{E}\{\left(\mathbf{d}^T(k)\,\mathbf{x}(k)\right)^2\} \tag{13.31-a}$$

$$= \sigma_x^2\ \mathrm{E}\{||\mathbf{d}(k)||^2\} \tag{13.31-b}$$

where

$$\mathbf{d}(k) = \mathbf{g} - \mathbf{h}(k) \tag{13.31-c}$$

denotes the distance vector as before. With these assumptions, the adaptation rule for the LMS algorithm (13.18) with (13.28) and (13.31-b) yields

$$\mathrm{E}\{||\mathbf{d}(k+1)||^2\} \approx \mathrm{E}\{||\mathbf{d}(k)||^2\} - \mathrm{E}\{e^2(k)\}\ \frac{\alpha}{m\,\sigma_x^2}\ (2-\alpha) \tag{13.32-a}$$

$$= \mathrm{E}\{||\mathbf{d}(k)||^2\}\ \left(1 - \frac{\alpha}{m}\ (2-\alpha)\right). \tag{13.32-b}$$

For $0 < \alpha < 2$ the ensemble average of the system distance decreases in each iteration. Choosing the initial vector $\mathbf{h}(k=0) = \mathbf{0}$, the average system distance at the beginning of the recursion equals

$$\mathrm{E}\{||\mathbf{d}(0)||^2\} = ||\mathbf{g}||^2\,,$$

and (13.32-b) can also be expressed as

$$\mathrm{E}\{||\mathbf{d}(k)||^2\} = ||\mathbf{g}||^2\ \left(1 - \frac{\alpha}{m}\ (2-\alpha)\right)^k. \tag{13.33}$$

This relation is illustrated in Fig. 13.6 for different values of α. With the definition of the relative system distance according to (13.8), we obtain

$$\frac{\mathrm{E}\{||\mathbf{d}(k)||^2\}}{||\mathbf{g}||^2} = \mathrm{E}\{D(k)\}\,.$$

As shown below, the best *mean* convergence is attained for $\alpha = 1$ (cf. Section 13.5.4).

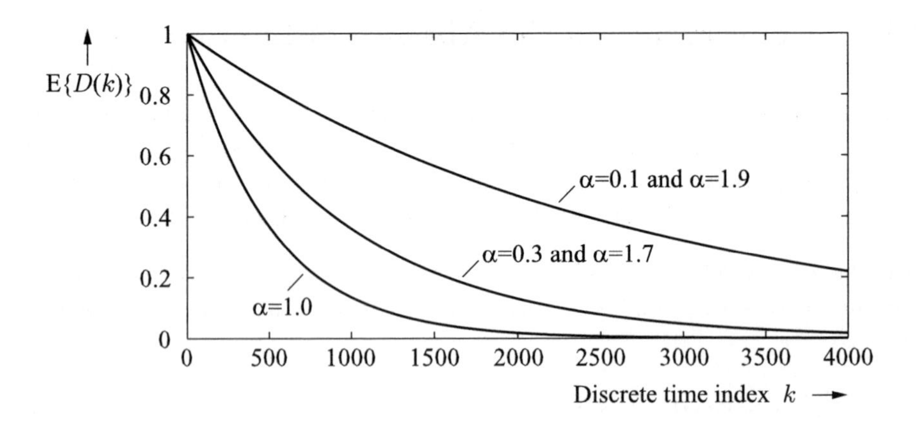

Figure 13.6: Convergence behavior of the LMS algorithm excited with white noise ($m = m' = 500$)

Next, the system excited with a colored noise signal $x(k)$ will be examined. We determine to what extent the convergence is influenced by a correlation of adjacent excitation samples. This investigation produces first qualitative statements on the efficiency of an NLMS-driven echo cancellation filter excited with speech signals.

The excitation signal will be derived from a white noise process $u(k)$ of power σ_u^2, by filtering with a first-order, recursive filter (first-order Markov process):

$$x(k) = b \cdot x(k-1) + u(k)\,; \qquad 0 \leq b < 1\,. \tag{13.34}$$

In analogy to (13.33) we find for $\alpha = 1$

$$\mathrm{E}\{||\mathbf{d}(k)||^2\} = ||\mathbf{g}||^2 \left(1 - \frac{1-b^2}{m}\right)^k . \tag{13.35}$$

Figure 13.7 shows the corresponding convergence behavior for different values of the parameter b. The curve for $b = 0$ is identical to the one in Fig. 13.6 for $\alpha = 1$.

With increasing correlation, the convergence speed of the algorithm obviously decreases. This leads to the conclusion that LMS- or NLMS-based echo cancellers will show poor adaptation performance when excited with speech signals. As outlined in Section 13.4 the eigenvalue spread of the correlation matrix of the excitation vector $\mathbf{x}(k)$ has a profound effect on the speed of adaptation [Haykin 1996].

The above analysis yields the mean convergence as a function of the discrete time index k in the sense of an ensemble average. The result of a simulation using the NLMS algorithm with colored noise is depicted in Fig. 13.8. It shows the logarithmic system distance $D(k)$ versus time for different values of the stepsize α. In general, the time evolution is in line with (13.35). However, deviations from the

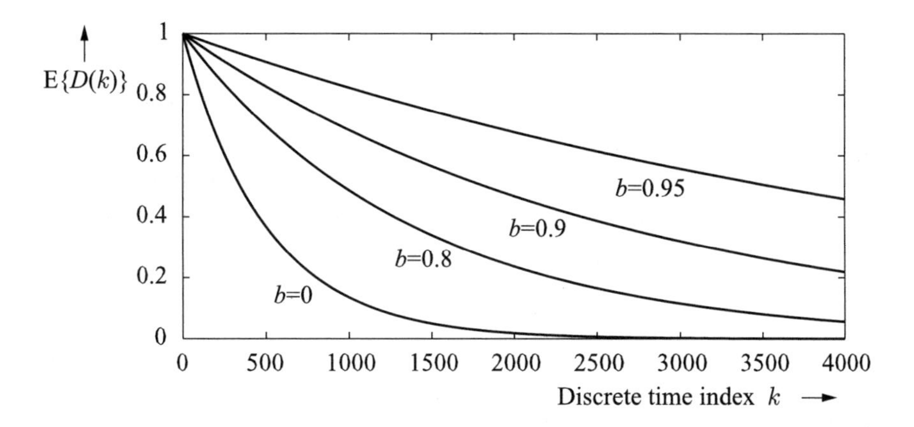

Figure 13.7: Convergence behavior of the LMS algorithm excited with colored noise ($m = m' = 500,\ \alpha = 1$)

analytical result (13.35) can be observed for the range $k < m$. This is primarily caused by the approximations of the equations (13.28) and (13.31-b), which are affected by the initialization of the coefficient and the excitation vectors.

The result of simulations with speech, colored, and white noise is presented in Fig. 13.9. The general conclusions from (13.35) for colored noise can be confirmed. The relatively poor convergence behavior for correlated noise and for speech signals can be clearly observed.

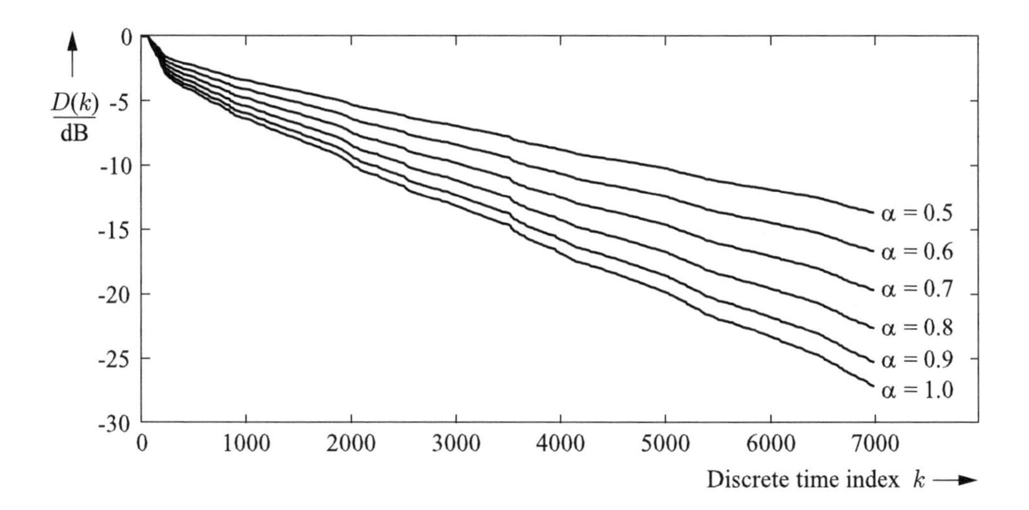

Figure 13.8: System distance with different stepsizes α in absence of interference (i.e., $s(k) = 0$, $n(k) = 0$); $m = m' = 500$, $x(k)$ = colored noise ($b = 0.8$)

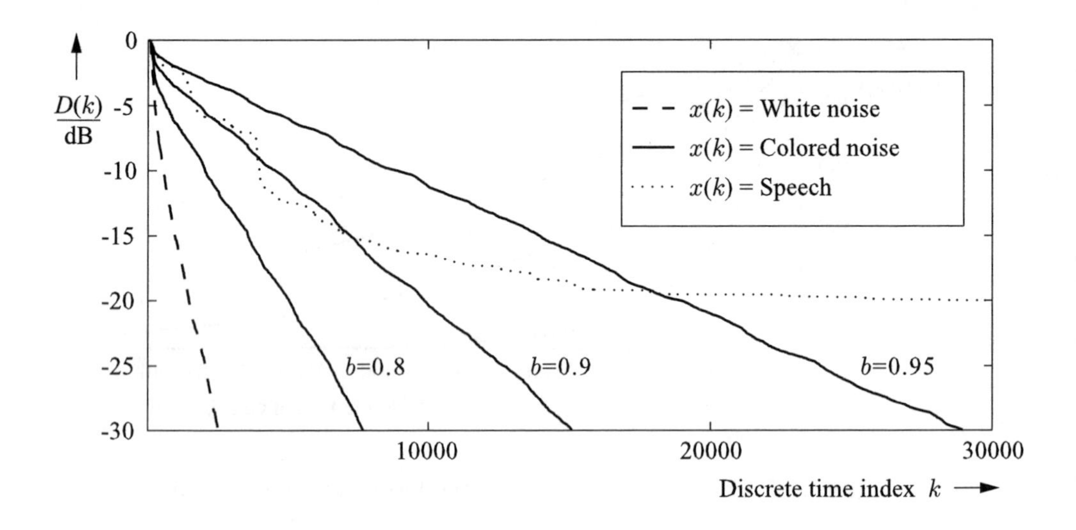

Figure 13.9: System distance with different input signals $x(k)$ in absence of interference (i.e., $s(k) = 0$, $n(k) = 0$); $m = m' = 500$, $\alpha = 1$

13.5.2 Convergence in the Presence of Interference

It was stated earlier that in a real system, the true instantaneous residual echo $e(k) = \tilde{x}(k) - \hat{x}(k)$ is not accessible and therefore the NLMS algorithm must use the signal

$$\hat{s}(k) = s(k) + n(k) + e(k)\,, \tag{13.36}$$

which comprises the speech signal of the near-end speaker and the background noise.

As a result, the steady-state echo reduction

$$\mathrm{ERLE}_\infty = \lim_{k\to\infty} \mathrm{E}\{\mathrm{ERLE}(k)\} \tag{13.37}$$

turns out to be lower, while the attainable system distance

$$D_\infty = \lim_{k\to\infty} \mathrm{E}\{D(k)\} \tag{13.38}$$

proves to be greater.

A universally valid analysis of the interfering factors is difficult to perform and cannot be done in an analytically closed form. The principal effects, however, can be demonstrated for the special case of a system excited with a stationary white noise signal $x(k)$ of power σ_x^2.

For the analysis below the signal of the near-end speaker is set to zero ($s(k) = 0$, i.e., speech pause). Furthermore, the existence of an interference $n(k)$ which is statistically independent of the signal $x(k)$ is assumed, leading to

$$\mathrm{E}\{x(k)\, n(\ell)\} = 0 \quad \text{and} \quad \mathrm{E}\{e(k)\, n(\ell)\} = 0\,. \tag{13.39}$$

With these assumptions the LMS algorithm with normalized stepsize can be written as

$$\mathbf{h}(k+1) = \mathbf{h}(k) + \frac{\alpha}{m\,\sigma_x^2}\Big(e(k) + n(k)\Big)\,\mathbf{x}(k)\,. \tag{13.40}$$

In analogy to (13.32-a), the mean system distance is given by

$$\mathrm{E}\{||\mathbf{d}(k+1)||^2\} = \mathrm{E}\{||\mathbf{d}(k)||^2\} - \mathrm{E}\{e^2(k)\}\frac{\alpha}{m\,\sigma_x^2}(2-\alpha) + \frac{\alpha^2}{m\,\sigma_x^2}\mathrm{E}\{n^2(k)\}\,. \tag{13.41-a}$$

Using (13.31-b) this equation can be modified to

$$\mathrm{E}\{||\mathbf{d}(k+1)||^2\} = \mathrm{E}\{||\mathbf{d}(k)||^2\}\left(1 - \frac{\alpha}{m}(2-\alpha)\right) + \frac{\alpha^2}{m\,\sigma_x^2}\mathrm{E}\{n^2(k)\}\,. \tag{13.41-b}$$

In comparison to the case without interference (13.32-b), an additional constant term has been added. This implies that with increasing k and for constant α the system distance cannot become arbitrarily small.

The steady-state solution of (13.41-b) is obtained for

$$\lim_{k\to\infty} \mathrm{E}\{||\mathbf{d}(k+1)||^2\} = \lim_{k\to\infty} \mathrm{E}\{||\mathbf{d}(k)||^2\} = ||\mathbf{d}_\infty||^2 \tag{13.42}$$

and is given by

$$||\mathbf{d}_\infty||^2 = \frac{\alpha}{2-\alpha}\,\frac{\mathrm{E}\{n^2(k)\}}{\sigma_x^2}\,. \tag{13.43}$$

With the stated assumptions the power of the echo signal $\tilde{x}(k)$ can be expressed as

$$\sigma_{\tilde{x}}^2 = ||\mathbf{g}||^2\,\sigma_x^2\,. \tag{13.44}$$

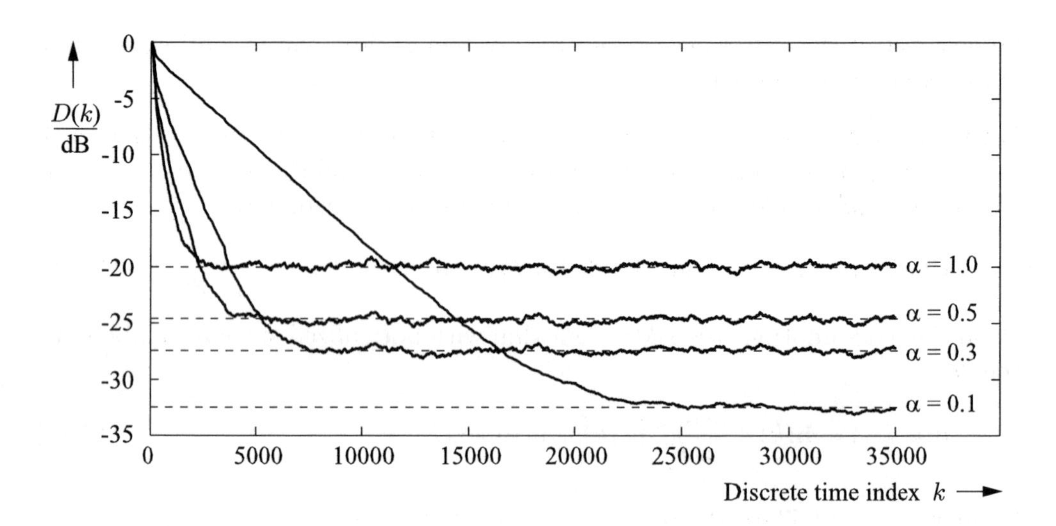

Figure 13.10: System distance for different stepsizes α in the presence of background noise with $10 \lg \left(\mathrm{E}\{n^2(k)\}/\mathrm{E}\{\tilde{x}^2(k)\}\right) = -20\,\mathrm{dB}$; $x(k), n(k) =$ white noise, $s(k) = 0$, $m = m' = 500$; [Antweiler 1995]

Therefore, the mean steady-state system distance results in

$$\lim_{k\to\infty} \mathrm{E}\{D(k)\} = D_\infty = \frac{||\mathbf{d}_\infty||^2}{||\mathbf{g}||^2} \tag{13.45}$$

$$= \frac{\alpha}{2-\alpha} \frac{\mathrm{E}\{n^2(k)\}}{\sigma_{\tilde{x}}^2}\,. \tag{13.46}$$

In the special case $\alpha = 1$, the achievable system distance is equivalent to the power ratio between the noise signal $n(k)$ and the echo signal $\tilde{x}(k)$. For $\alpha < 1$, the system distance can be improved at the expense of a slower convergence. Figure 13.10 shows the result of a simulation of the NLMS algorithm for different stepsize parameters α. The theoretical limit according to (13.46) is indicated by dashed lines.

Note that in the above derivation of the steady-state performance only the statistical independence of the echo signal $\tilde{x}(k)$ with respect to the noise $n(k)$ was assumed. In the case of a white excitation signal $x(k)$, the logarithmic echo return loss enhancement (ERLE) matches the negative logarithmic relative system distance in dB. If a speech signal is used for the adaptation process, the attainable system distance becomes even worse. The results of simulation examples are depicted in Fig. 13.11 and Fig. 13.12. Figure 13.12 shows that for a colored, stationary excitation signal $x(k)$ similar system distances as for white noise $x(k)$ in combination with $\alpha < 1$ (cf. Fig. 13.10) are obtained.

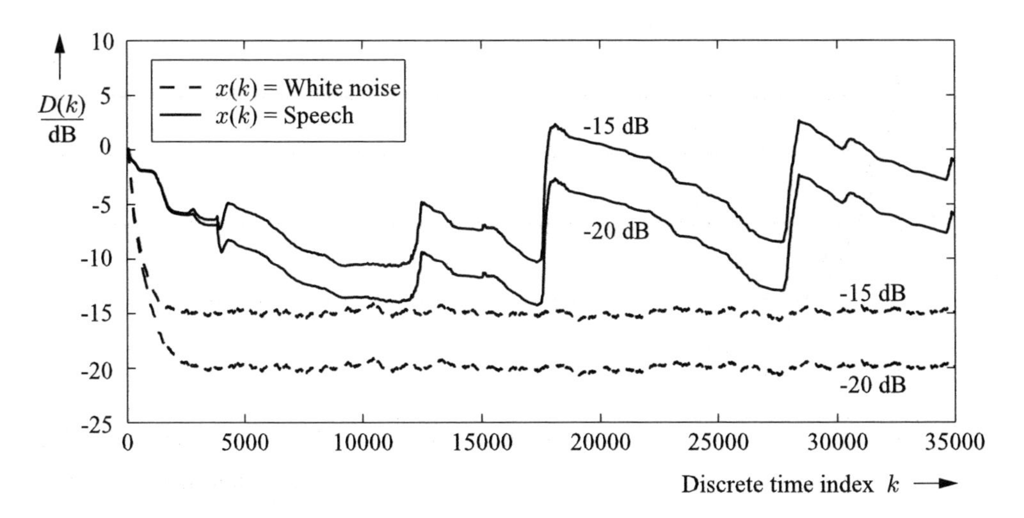

Figure 13.11: System distance for different excitation signals $x(k)$ in the presence of background noise with $10\ \lg\left(\mathrm{E}\{n^2(k)\}/\mathrm{E}\{\tilde{x}^2(k)\}\right) = -15\,\mathrm{dB}$ or $-20\,\mathrm{dB}$, respectively; $n(k)$ = white noise, $s(k) = 0$, $m = m' = 500$, $\alpha = 1.0$

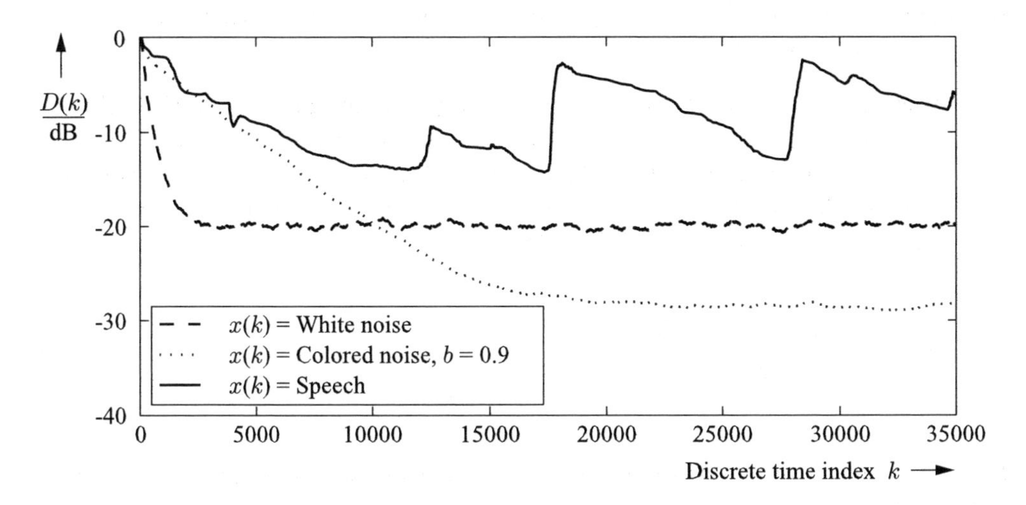

Figure 13.12: System distance for different excitation signals $x(k)$ in the presence of background noise with $10\ \lg\left(\mathrm{E}\{n^2(k)\}/\mathrm{E}\{\tilde{x}^2(k)\}\right) = -20\,\mathrm{dB}$; $n(k)$ = white noise, $s(k) = 0$, $m = m' = 500$, $\alpha = 1.0$

13.5.3 Filter Order of the Echo Canceller

Neglecting the computation of the stepsize parameter, the LMS algorithm requires $2m$ multiply–accumulate operations per sample, m operations for the adaptation of the coefficients, and m for the filtering, respectively. The order of the cancellation filter should therefore be kept as small as possible. This requirement is also

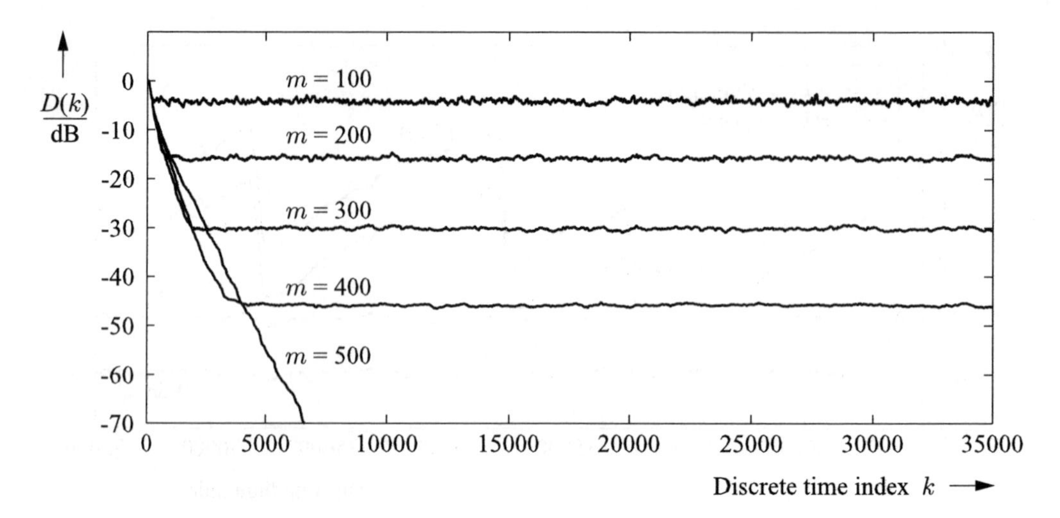

Figure 13.13: System distance for different filter lengths m in absence of interference (i.e., $s(k) = 0$, $n(k) = 0$); $x(k)$ = white noise, $m' = 500$, $\alpha = 1.0$

sustained by the speed of convergence, which slows down with increasing filter order as seen in (13.33).

A restriction in the length of the impulse response of the cancellation filter inevitably leads to a limitation of the attainable system distance. This limitation can easily be estimated as follows. With an ideal match of the impulse response of the compensation filter and the first m values of the LEM impulse response, the best possible system distance is given by

$$D_{\text{opt}} = \frac{||\mathbf{h} - \mathbf{g}||^2}{||\mathbf{g}||^2} = \frac{\sum_{i=m}^{\infty} g_i^2}{\sum_{i=0}^{\infty} g_i^2} . \tag{13.47}$$

By measuring an impulse response typical of the acoustic environment, the required minimum order of the cancellation filter can thus be determined in advance. For the LEM impulse response measured in a car (see Fig. 13.4-a), Fig. 13.13 clearly shows the limitation of the system distance caused by a limited filter order.

13.5.4 Stepsize Parameter

So far, the normalized stepsize parameter α was assumed to be constant. Below, an optimal and possibly adaptive adjustment of this parameter will be investigated. The optimization is based on the maximum improvement of the mean system distance from one time instant to the next.

First, a system excited with a white noise signal $x(k)$ is considered in the absence of interference. The insertion of a now time-variant (but deterministic) stepsize parameter $\alpha(k)$ into (13.32-a) results for every arbitrary but fixed time index k in

$$\mathrm{E}\{||\mathbf{d}(k+1)||^2\} \approx \mathrm{E}\{||\mathbf{d}(k)||^2\} - \mathrm{E}\{e^2(k)\} \, \frac{\alpha(k)}{m\,\sigma_x^2} \, \Big(2 - \alpha(k)\Big) \,. \tag{13.48}$$

Thus, the difference of the mean system distances at time instant k and time instant $k+1$ is given by

$$\begin{aligned}\Delta_{\mathrm{E}}^2(k) &= \mathrm{E}\{||\mathbf{d}(k)||^2\} - \mathrm{E}\{||\mathbf{d}(k+1)||^2\} \\ &\approx \mathrm{E}\{e^2(k)\} \, \frac{\alpha(k)}{m\,\sigma_x^2} \, \Big(2 - \alpha(k)\Big) \,. \end{aligned} \tag{13.49}$$

In the admissible range of the stepsize parameter

$$0 < \alpha(k) < 2 \,, \tag{13.50}$$

the quantity $\Delta_{\mathrm{E}}^2(k)$ only attains positive values. It is a quadratic function of $\alpha(k)$.

From the condition

$$\frac{\partial \Delta_{\mathrm{E}}^2(k)}{\partial \alpha(k)} \overset{!}{=} 0 \,, \tag{13.51}$$

the optimal stepsize for each time index k results in a constant stepsize parameter

$$\alpha(k) = \alpha = 1 \,, \tag{13.52}$$

which confirms the previous results. In the presence of interference (double talk, background noise), the residual echo $e(k)$ is not directly accessible. When $e(k)$ is replaced by $\hat{s}(k) = e(k) + s(k) + n(k)$ in the adaptation rule for the LMS algorithm the difference of the mean system distances at time instants k and $k+1$ follows in analogy to (13.41-a)

$$\Delta_{\mathrm{E}}^2(k) = \mathrm{E}\{e^2(k)\} \frac{\alpha(k)}{m\,\sigma_x^2} \Big(2 - \alpha(k)\Big) - \frac{\alpha^2(k)}{m\,\sigma_x^2} \Big(\mathrm{E}\{s^2(k)\} + \mathrm{E}\{n^2(k)\}\Big). \tag{13.53}$$

In order to obtain the best possible stepsize, the condition

$$\frac{\partial \Delta_{\mathrm{E}}^2(k)}{\partial \alpha(k)} = 2 \, \frac{\mathrm{E}\{e^2(k)\}}{m\,\sigma_x^2} - 2\,\alpha(k) \, \frac{\mathrm{E}\{\hat{s}^2(k)\}}{m\,\sigma_x^2} \overset{!}{=} 0 \tag{13.54}$$

with

$$\mathrm{E}\{\hat{s}^2(k)\} = \mathrm{E}\{e^2(k)\} + \mathrm{E}\{s^2(k)\} + \mathrm{E}\{n^2(k)\}$$

must be fulfilled. The general solution is [Schultheiß 1988], [Mader et al. 2000]

$$\alpha_{\text{opt}}(k) = \frac{\mathrm{E}\{e^2(k)\}}{\mathrm{E}\{\hat{s}^2(k)\}} \,. \tag{13.55}$$

In this, the interference-free case with $e(k) = \hat{s}(k)$ is included as the special case $\alpha(k) = 1$.

The optimal stepsize can also be written as

$$\alpha_{\text{opt}}(k) = \frac{\mathrm{E}\{e^2(k)\}}{\mathrm{E}\{\hat{s}^2(k)\}} = \frac{\mathrm{E}\{e^2(k)\}}{\mathrm{E}\{s^2(k)\} + \mathrm{E}\{n^2(k)\} + \mathrm{E}\{e^2(k)\}} \tag{13.56-a}$$

$$= \frac{1}{1 + \dfrac{\mathrm{E}\{s^2(k)\} + \mathrm{E}\{n^2(k)\}}{\mathrm{E}\{e^2(k)\}}} \leq 1 \,. \tag{13.56-b}$$

We see from (13.56-b) that during double talk phases the stepsize is reduced. For example, $\mathrm{E}\{s^2(k)\} = \mathrm{E}\{e^2(k)\}$ and $n(k) = 0$ yield

$$\alpha = 0.5 \tag{13.57}$$

and for $\mathrm{E}\{s^2(k)\} = 4\ \mathrm{E}\{e^2(k)\}$ we obtain

$$\alpha = 0.2 \,. \tag{13.58}$$

By reducing the stepsize, the signal-dependent bias of the echo canceller is minimized and thus distortions of the transmitted signal are avoided. During double talk, the adaptation need not be explicitly stopped so that slow improvements of the state of convergence are still possible. When the echo canceller has converged, the power of the residual echo $e(k)$ is in general significantly smaller than the power of the near-end speech signal $s(k)$ or the near-end noise signal $n(k)$. In this case the stepsize parameter is much smaller than unity, which in turn is a prerequisite to maintain a small system distance (see (13.46)).

Since the residual echo $e(k)$ does not exist as an isolated signal, the stepsize parameter $\alpha_{\text{opt}}(k)$ can only be approximated. In [Yamamoto, Kitayama 1982], [Schultheiß 1988] a solution which is based on an estimation of the instantaneous system distance is proposed. A delay of m_0 samples is applied to the microphone signal. This delays the effective LEM impulse response by m_0 samples as well, i.e., it introduces m_0 leading zero values. The m_0 leading taps of the echo canceller can now be used to estimate the system distance and hence the residual echo power.

This procedure is outlined in [Schultheiß 1988] where $m_0 = 20$ delay coefficients are used. Additional measures, however, are necessary to detect a change in the LEM response, otherwise the adaptation will freeze [Mader et al. 2000], [Hänsler, Schmidt 2004].

13.6 Geometric Projection Interpretation of the NLMS Algorithm

The operation of the NLMS algorithm can be interpreted in terms of a vector space representation of the individual adaptation step [Claasen, Mecklenbräuker 1981], [Sommen, van Valburg 1989]. However, before we develop this "geometric" approach to the echo cancellation problem we introduce the concept of orthogonal vectors and spaces in $\mathbb{R}^m$ (see, e.g., [Debnath, Mikusinski 1999]).

Two vectors $\mathbf{d}_1(k)$ and $\mathbf{d}_2(k)$ are said to be orthogonal if their inner product equals zero, i.e., $\mathbf{d}_1^T(k)\mathbf{d}_2(k) = 0$. By the same token, a vector $\mathbf{d}(k)$ is said to be orthogonal to a subspace $\mathbf{S}_x$ if $\mathbf{d}^T(k)\mathbf{x}(k) = 0$ for every $\mathbf{x}(k) \in \mathbf{S}_x$. The set of all vectors orthogonal to $\mathbf{S}_x$ are called the orthogonal complement of $\mathbf{S}_x$ and is denoted by $\mathbf{S}^\perp$.

Every element $\mathbf{d}(k) \in \mathbb{R}^m$ thus has a unique decomposition in the form

$$\mathbf{d}(k) = \mathbf{d}^{||}(k) + \mathbf{d}^{\perp}(k) \tag{13.59}$$

where $\mathbf{d}^{||}(k)$ is an element of a subspace $\mathbf{S}_x$ of $\mathbb{R}^m$ and $\mathbf{d}^{\perp}(k)$ an element of the orthogonal complement $\mathbf{S}^\perp$ of $\mathbf{S}_x$. $\mathbf{d}^{||}(k)$ is then called the projection of $\mathbf{d}(k)$ onto $\mathbf{S}_x$ and $\mathbf{P}_x$, with $\mathbf{P}_x\,\mathbf{d}(k) = \mathbf{d}^{||}(k)$ being the associated projection operator.

Projection operators satisfy two necessary conditions, namely

$$\mathbf{P}_x\,\mathbf{P}_x = \mathbf{P}_x \qquad \text{(idempotent)} \tag{13.60}$$

and

$$\left(\mathbf{P}_x\,\mathbf{d}_1(k)\right)^T\,\mathbf{d}_2(k) = \mathbf{d}_1^T\,\mathbf{P}_x\,\mathbf{d}_2(k) \qquad \text{(self-adjoint)} \tag{13.61}$$

for any $\mathbf{d}_1(k), \mathbf{d}_2(k) \in \mathbb{R}^m$.

If $\mathbf{S}_x$ is a subspace of dimension 1 and $\mathbf{x}(k) \in \mathbf{S}_x$, then the projection of $\mathbf{d}(k)$ onto $\mathbf{S}_x$ is given by

$$\mathbf{d}^{||}(k) = \mathbf{P}_x(k)\mathbf{d}(k) = \frac{\mathbf{x}(k)\,\mathbf{x}^T(k)}{||\mathbf{x}(k)||\,||\mathbf{x}(k)||}\,\mathbf{d}(k) \tag{13.62}$$

since $\mathbf{x}^T(k)\,\mathbf{d}(k)/||\mathbf{x}(k)||$ is the component of $\mathbf{d}(k)$ in the direction of unit vector $\mathbf{x}(k)/||\mathbf{x}(k)||$. In this case, the projection operator evaluates to an $m \times m$ matrix.

If we return to the NLMS-adapted echo canceller and consider the distance vector

$$\mathbf{d}(k) = \mathbf{g} - \mathbf{h}(k) \tag{13.63}$$

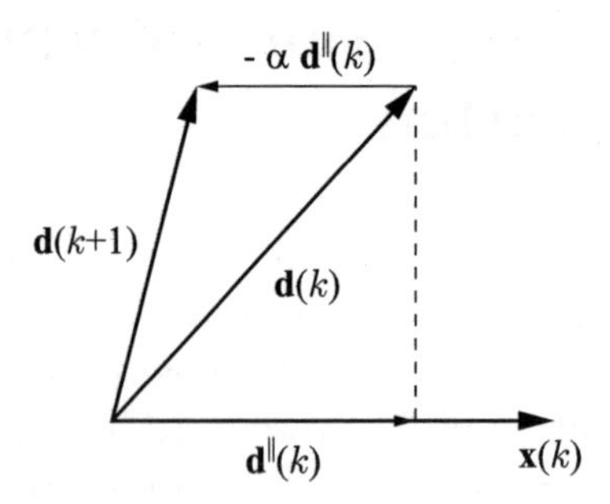

Figure 13.14: Geometric interpretation of the NLMS algorithm [Antweiler 1995]

the following equation holds for the NLMS update in an environment with no interference, i.e., $s(k) + n(k) = 0$,

$$\mathbf{d}(k+1) = \mathbf{d}(k) - \alpha \, \frac{\mathbf{d}^T(k)\,\mathbf{x}(k)}{||\mathbf{x}(k)||^2} \, \mathbf{x}(k) \tag{13.64-a}$$

$$= \mathbf{d}(k) - \alpha \frac{\mathbf{x}(k)\,\mathbf{x}^T(k)}{||\mathbf{x}(k)||\,||\mathbf{x}(k)||} \, \mathbf{d}(k) \tag{13.64-b}$$

$$= \mathbf{d}(k) - \alpha \, \mathbf{d}^{||}(k) \tag{13.64-c}$$

$$= (\mathbf{I} - \alpha \mathbf{P}_x(k))\, \mathbf{d}(k) \,. \tag{13.64-d}$$

The correction vector $\mathbf{d}^{||}(k)$ can therefore be interpreted as an orthogonal projection of the distance vector $\mathbf{d}(k)$ onto the signal vector $\mathbf{x}(k)$ and

$$\mathbf{P}_x(k) = \frac{\mathbf{x}(k)\,\mathbf{x}^T(k)}{||\mathbf{x}(k)||^2} \tag{13.65}$$

is the corresponding projection operator. It is easily verified that both necessary conditions (13.60) and (13.61) are satisfied. This concept is illustrated in Fig. 13.14 for $m = 2$.

The length of the distance vector $\mathbf{d}(k)$ is reduced by subtracting the component $\alpha\,\mathbf{d}^{||}(k)$, which is parallel to the signal vector $\mathbf{x}(k)$. Provided that the stepsize parameter is within the range

$$0 < \alpha < 2 \,, \tag{13.66}$$

the length of the distance vector $\mathbf{d}(k)$ always decreases. The convergence condition (13.26) is thus confirmed. For $\alpha < 1$ the projection does not fully eliminate the components in the space spanned by $\mathbf{x}(k)$ but amounts to a *relaxed* projection.

For $\alpha = 1$, we find $\mathbf{d}(k+1) = (\mathbf{I} - \mathbf{P}_x(k))\,\mathbf{d}(k)$ where $\mathbf{I} - \mathbf{P}_x(k)$ is the operator for projecting $\mathbf{d}(k)$ onto the orthogonal complement of $\mathbf{x}(k)$, since

$$\mathbf{P}_x(k)(\mathbf{I} - \mathbf{P}_x(k)) = \mathbf{P}_x(k) - \mathbf{P}_x(k)\mathbf{P}_x(k) = 0\,. \tag{13.67}$$

A system distance $D(k) = 0$ can only be obtained if, during the adaptation phase, the excitation signal vector points towards all directions in the m-dimensional vector space. This interpretation explains the fact that the best convergence is achieved for a *perfect sequence* excitation signal ([Antweiler, Dörbecker 1994], [Antweiler, Antweiler 1995]; see Section 13.10.5).

13.7 The Affine Projection Algorithm

The geometric interpretation leads to an interesting generalization of the NLMS algorithm, namely the *affine projection* (AP) algorithm [Ozeki, Umeda 1984].

For $s(k) = n(k) = 0$ and $m = m'$ the NLMS filter coefficient update can be written in terms of the projection operator $\mathbf{P}_x(k)$ as

$$\mathbf{h}(k+1) = \mathbf{h}(k) + \alpha\,\mathbf{P}_x(k)\,\mathbf{d}(k) \tag{13.68}$$

$$= [\mathbf{I} - \alpha\,\mathbf{P}_x(k)]\;\mathbf{h}(k) + \alpha\,\mathbf{P}_x(k)\,\mathbf{g}\,. \tag{13.69}$$

For $\alpha = 1$ we now have an affine projection of the form

$$\mathbf{h}(k+1) = \mathbf{P}_x^{\perp}(k)\,\mathbf{h}(k) + \mathbf{g}^{||}(k)\,. \tag{13.70}$$

The updated filter coefficient vector equals the current vector projected onto the orthogonal complement of $\mathbf{x}(k)$ plus the projection of the true impulse response $\mathbf{g}$ onto the vector $\mathbf{x}(k)$.

The AP algorithm generalizes this idea. In each iteration, the AP algorithm reduces the system distance not just in one but in several dimensions. Hence, we now use the projection onto the space spanned by $p < m$ vectors of the input signal

$$\mathbf{X}_p(k) = (\mathbf{x}(k), \mathbf{x}(k-1), \ldots, \mathbf{x}(k-p+1))\,. \tag{13.71}$$

If we consider $\mathbf{X}_p(k)$ to be a set of p non-orthogonal basis vectors for a subspace of $\mathbb{R}^m$, the inner product vector $\mathbf{X}_p^T(k)\,\mathbf{d}(k)$, in conjunction with an appropriate normalization, yields the components of $\mathbf{d}(k)$ with respect to this subspace basis. Therefore, the projection operator onto this subspace, $\mathbf{P}_{X_p}(k)$, may be written as $\mathbf{P}_{X_p}(k) = \mathbf{X}_p(k)\,\mathbf{A}\,\mathbf{X}_p^T(k)$ where the normalization matrix $\mathbf{A}$ is to be determined. Also, we note that the projection operator has to satisfy

$$\mathbf{P}_{X_p}(k)\,\mathbf{X}_p(k) = \mathbf{X}_p(k)\,. \tag{13.72}$$

With the above observations the operator can be constructed such that

$$\mathbf{X}_p(k)\,\mathbf{A}\,\mathbf{X}_p^T(k)\,\mathbf{X}_p(k) = \mathbf{X}_p(k) \tag{13.73}$$

where $\mathbf{A}$ is now identified as

$$\mathbf{A} = [\mathbf{X}_p^T(k)\,\mathbf{X}_p(k)]^{-1}\,. \tag{13.74}$$

We obtain

$$\mathbf{P}_{X_p}(k) = \mathbf{X}_p(k)\,[\mathbf{X}_p^T(k)\,\mathbf{X}_p(k)]^{-1}\,\mathbf{X}_p^T(k)\,. \tag{13.75}$$

It is easily verified that $\mathbf{P}_{X_p}(k)$ is indeed a projection operator, i.e.,

$$\mathbf{P}_{X_p}(k)\,\mathbf{P}_{X_p}(k) = \mathbf{P}_{X_p}(k) \text{ and} \tag{13.76}$$

$$(\mathbf{P}_{X_p}(k)\,\mathbf{d}_1(k))^T\,\mathbf{d}_2(k) = \mathbf{d}_1^T(k)\,\mathbf{P}_{X_p}(k)\,\mathbf{d}_2(k) \tag{13.77}$$

since $\mathbf{X}_p^T(k)\,\mathbf{X}_p(k)$ and its inverse are symmetric matrices.

In the noise-free case the AP algorithm is given by

$$\mathbf{h}(k+1) = \mathbf{h}(k) + \alpha\,\mathbf{X}_p(k)\,[\mathbf{X}_p^T(k)\mathbf{X}_p(k)]^{-1}\,\mathbf{X}_p^T(k)\,\mathbf{d}(k) \tag{13.78}$$

and in its general form with $\hat{\mathbf{s}}_p(k) = (\hat{s}(k), \hat{s}(k-1), \ldots, \hat{s}(k-p+1))^T$ by

$$\mathbf{h}(k+1) = \mathbf{h}(k) + \alpha(k)\,\mathbf{X}_p(k)\,[\mathbf{X}_p^T(k)\mathbf{X}_p(k) + \delta\,\mathbf{I}]^{-1}\,\hat{\mathbf{s}}_p(k) \tag{13.79}$$

with

$$\hat{\mathbf{s}}_p(k) = \mathbf{y}_p(k) - \mathbf{X}_p^T(k)\,\mathbf{h}(k) \tag{13.80}$$

$$= \mathbf{s}_p(k) + \mathbf{n}_p(k) + \tilde{\mathbf{x}}_p(k) - \mathbf{X}_p^T(k)\,\mathbf{h}(k)\,. \tag{13.81}$$

In (13.79) a regularization parameter δ has been added to decrease the condition number of the matrix before inversion. This is necessary since $\mathbf{X}_p^T(k)\,\mathbf{X}_p(k)$ is a rank deficient matrix for $p \to m$. Optimal regularization parameters are considered in [Myllylä, Schmidt 2002]. In the noise-free case fastest convergence is achieved for $\alpha = 1$. However, in the presence of near-end speech and noise $\alpha \ll 1$ results in better convergence. Figure 13.15 depicts the system distance for various far-end signals and $p = 10$. Also, in the case of a first-order Markov process (AR 1 in Fig. 13.15) convergence does not depend on the signal correlation. In the case of speech signals the AP algorithm leads to much faster convergence than the NLMS algorithm.

Figure 13.16 shows the impact of the projection order p on the convergence for speech signals and $\alpha = 1$. It is seen that a larger projection order is beneficial. Most of the gains are achieved, however, for a projection order of $p = 10$–20.

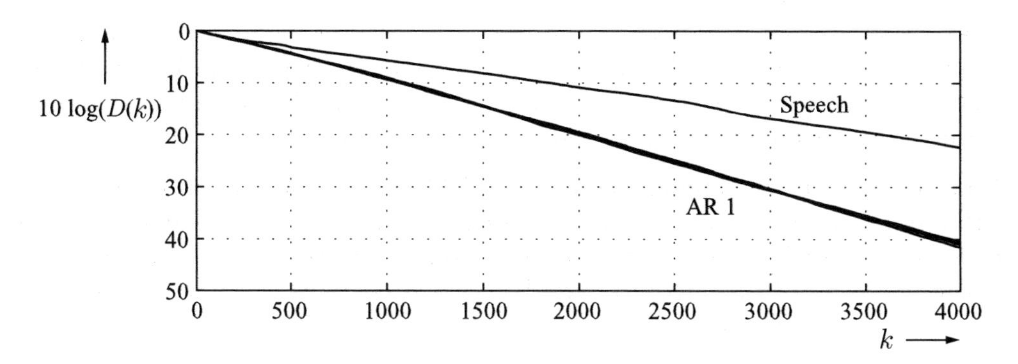

Figure 13.15: Convergence behavior of the AP algorithm with $p = 10$ and $\alpha = 1$ excited with colored noise (AR 1; $b = 0, 0.8, 0.9, 0.95$) and speech ($m = m' = 500$)

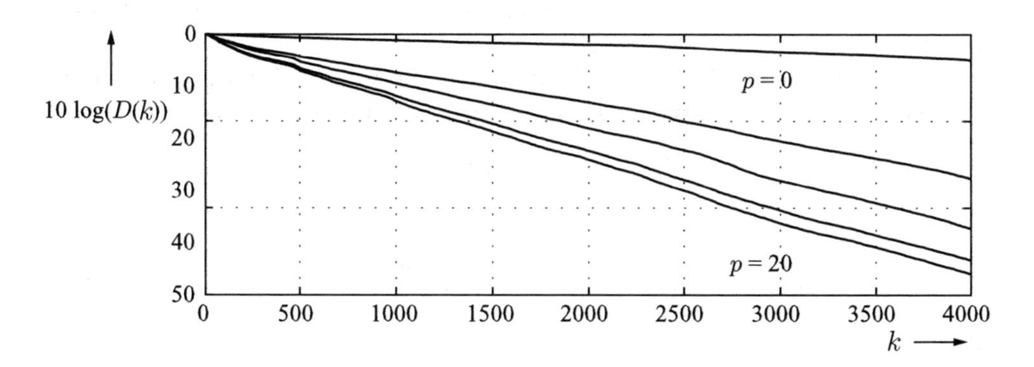

Figure 13.16: Convergence behavior of the AP algorithm with $\alpha = 1$ and $p = 1, 5, 10, 15, 20$ excited with speech ($m = m' = 500$)

Furthermore, fast implementations of the AP algorithm exist [Gay, Tavathia 1995], [Tanaka et al. 1995] which make the AP algorithm attractive for real-time applications. For a robust implementations of this algorithms see, for example, [Myllylä 2001]. A variation of the AP algorithm with exponentially weighted stepsize parameters is outlined in [Makino, Kaneda 1992]. In this algorithm a different stepsize parameter is assigned to each canceller coefficient.

13.8 Least-Squares and Recursive Least-Squares Algorithms

The Wiener solution requires knowledge about the first- and second-order statistics such as auto- and cross-correlation functions. These quantities must be estimated from the available data. In this section we will discuss a family of algorithms which relies directly on the minimization of the observed error on a block of data. We begin our presentation with the *weighted least-squares algorithm* which is, in the context of acoustic echo cancellation, more of theoretical than of practical interest since it requires the inversion of a possibly large matrix.

However, a recursive approximation, the RLS|seeRecursive least-squares algorithm*recursive least-squares (RLS) algorithm*, avoids this inversion and is in principle better suited for practical implementations. In contrast to the AP algorithm where $p < m$ the RLS solves the overdetermined case where the number of data vectors is larger than m.

13.8.1 The Weighted Least-Squares Algorithm

The *weighted least-squares approach* combines the actually measured adaptation error (including errors due to near-end speech and noise) on a block of data into a vector of dimensions $M \geq m$ and minimizes the weighted norm of this vector. With the $m \times M$ data matrix constructed from m-dimensional vectors $\mathbf{x}(k-i)$

$$\mathbf{X}_M(k) = (\mathbf{x}(k), \mathbf{x}(k-1), \ldots, \mathbf{x}(k-M+1)) \tag{13.82}$$

the error vector

$$\widehat{\mathbf{s}}_M(k) = (\widehat{s}(k), \widehat{s}(k-1), \ldots, \widehat{s}(k-M+1))^T \tag{13.83}$$

is given by

$$\widehat{\mathbf{s}}_M(k) = \mathbf{y}_M(k) - \mathbf{X}_M^T\, \mathbf{h}(k) \tag{13.84}$$

where $\mathbf{y}_M(k)$ denotes a vector of M samples of the microphone signal

$$\mathbf{y}_M(k) = (y(k), y(k-1), \ldots, y(k-M+1))^T. \tag{13.85}$$

The weighted least-squares algorithm minimizes the weighted error norm

$$\mathbf{J}_{\mathrm{LS}} = \widehat{\mathbf{s}}_M^T(k)\, \mathbf{W}\, \widehat{\mathbf{s}}_M(k) \tag{13.86}$$

where $\mathbf{W} = \mathrm{diag}\{(w_{11}, w_{22}, \ldots, w_{MM})^T\}$ is a positive-definite diagonal weighting matrix. The minimization of the error norm $\mathbf{J}_{\mathrm{LS}}$ with respect to the unknown coefficient vector $\mathbf{h}(k)$ requires the solution of

$$\frac{\partial \mathbf{J}_{\mathrm{LS}}}{\partial \mathbf{h}(k)} = -2\, \mathbf{X}_M(k) \mathbf{W}\, \mathbf{y}_M(k) + 2\, (\mathbf{X}_M(k)\, \mathbf{W}\, \mathbf{X}_M^T)\, \mathbf{h}(k) \stackrel{!}{=} \mathbf{0}\,. \tag{13.87}$$

If $\mathbf{X}_M(k)\, \mathbf{W}\, \mathbf{X}_M^T(k)$ is invertible (for which $M \geq m$ is a necessary and $rank(\mathbf{X}_M(k)) = m$ a sufficient condition), the solution is given by

$$\mathbf{h}_{\mathrm{LS}}(k) = \left(\mathbf{X}_M(k)\, \mathbf{W}\, \mathbf{X}_M^T(k)\right)^{-1} \mathbf{X}_M(k)\, \mathbf{W}\, \mathbf{y}_M(k)\,. \tag{13.88}$$

$\mathbf{X}_M(k)\mathbf{W}\,\mathbf{X}_M^T(k)$ and $\mathbf{X}_M(k)\,\mathbf{W}\,\mathbf{y}_M(k)$ are recognized as estimates of a weighted data auto-correlation matrix and weighted cross-correlation vector, respectively. For stationary signals and $M \to \infty$ the least-squares solution $\mathbf{h}_{\mathrm{LS}}$ approaches the FIR Wiener solution.

It can be shown that if the near-end signal $s(k)+n(k)$ is zero mean and if $s(k)+n(k)$ and $x(k)$ are statistically independent $\mathbf{h}_{\mathrm{LS}}$ is an unbiased estimator of $\mathbf{g}$, provided $m = m'$ holds.

Also, for $\mathbf{W} = \mathbf{I}$ the estimated echo vector

$$\tilde{\mathbf{x}}_M(k) = \mathbf{X}_M^T(k)\,\mathbf{h}_{\mathrm{LS}}(k) = \mathbf{X}_M^T(k)\,[\mathbf{X}_M(k)\mathbf{X}_M^T(k)]^{-1}\mathbf{X}_M(k)\,\mathbf{y}_M(k) \quad (13.89)$$

is the orthogonal projection of $\mathbf{y}_M(k)$ onto the space spanned by $\mathbf{X}_M(k)$.

With $\mathbf{W} = \mathbf{I}$ the least-squares solution is suited for stationary signals. However, in the context of acoustic echo cancellation where the excitation is speech an exponential weighting $w_{ii} = \lambda^{1-i}$ is much better suited. For large m or M the computational complexity of $O(M^3)$ associated with the matrix inversion prohibits the direct implementation of the weighted least-squares algorithm.

13.8.2 The RLS Algorithm

The computational effort for the least-squares algorithm can be significantly reduced when the inverse auto-correlation matrix is estimated in a recursive way. This leads to the *recursive least-squares* algorithm [Haykin 1996]. We note that

$$\mathbf{R}_{xx}(k) = \mathbf{X}_M(k)\,\mathbf{W}\,\mathbf{X}_M^T(k) \quad (13.90)$$

can also be written as

$$\begin{aligned}\mathbf{R}_{xx}(k) &= \Big(w_{11}\mathbf{x}(k), \dots, w_{MM}\mathbf{x}(k-M+1)\Big) \cdot \begin{pmatrix} \mathbf{x}^T(k) \\ \mathbf{x}^T(k-1) \\ \vdots \\ \mathbf{x}^T(k-M+1) \end{pmatrix} \\ &= \sum_{i=0}^{M-1} w_{i+1,i+1}\,\mathbf{x}(k-i)\,\mathbf{x}^T(k-i)\,. \end{aligned} \quad (13.91)$$

The estimated auto-correlation matrix is a sum over M rank 1 matrices. With $w_{i+1,i+1} = \lambda^{-i}$ we obtain

$$\begin{aligned}
\mathbf{R}_{xx}(k) &= \sum_{i=0}^{M-1} \lambda^{-i}\,\mathbf{x}(k-i)\,\mathbf{x}^T(k-i) \\
&= \mathbf{x}(k)\,\mathbf{x}^T(k) + \sum_{i=1}^{M-1} \lambda^{-i}\,\mathbf{x}(k-i)\,\mathbf{x}^T(k-i) \qquad (13.92)\\
&= \mathbf{x}(k)\,\mathbf{x}^T(k) + \lambda \sum_{i=0}^{M-2} \lambda^{-i}\,\mathbf{x}(k-i-1)\,\mathbf{x}^T(k-i-1)\,.
\end{aligned}$$

If $\lambda < 1$ and M is sufficiently large we may therefore write

$$\mathbf{R}_{xx}(k) \approx \lambda\,\mathbf{R}_{xx}(k-1) + \mathbf{x}(k)\,\mathbf{x}^T(k) \qquad (13.93)$$

and in the same way for the weighted cross-correlation vector

$$\begin{aligned}
\boldsymbol{\varphi}_{xy}(k) &= \mathbf{X}_M(k)\,\mathbf{W}\,\mathbf{y}_M(k) \\
&= \sum_{i=0}^{M-1} \lambda^{-i}\,\mathbf{x}(k-i)\,y(k-i) \qquad (13.94)\\
&\approx \lambda\,\boldsymbol{\varphi}_{xy}(k-1) + \mathbf{x}(k)\,y(k)\,.
\end{aligned}$$

The inversion of $\mathbf{R}_{xx}(k)$ can be avoided if we make use of the matrix inversion lemma [Haykin 1996]

$$\mathbf{A} = \mathbf{B}^{-1} + \mathbf{C}\mathbf{D}^{-1}\mathbf{C}^T \Leftrightarrow \mathbf{A}^{-1} = \mathbf{B} - \mathbf{B}\mathbf{C}(\mathbf{D} + \mathbf{C}^T\mathbf{B}\mathbf{C})^{-1}\mathbf{C}^T\mathbf{B} \qquad (13.95)$$

with

$$\mathbf{A} = \mathbf{R}_{xx}(k), \mathbf{B}^{-1} = \lambda\,\mathbf{R}_{xx}(k-1), \mathbf{C} = \mathbf{x}(k), \mathbf{D} = 1\,. \qquad (13.96)$$

We obtain

$$\begin{aligned}
\mathbf{R}_{xx}^{-1}(k) &= \lambda^{-1}\mathbf{R}_{xx}^{-1}(k-1) - \lambda^{-1}{\mathbf{R}_{\mathbf{xx}}}^{-1}(k-1)\mathbf{x}(k) \\
&\qquad \cdot (1 + \mathbf{x}^T(k)\,\lambda^{-1}\mathbf{R}_{xx}^{-1}(k-1)\,\mathbf{x}(k))^{-1} \cdot \mathbf{x}^T(k)\,\lambda^{-1}\mathbf{R}_{xx}^{-1}(k-1) \\
&= \lambda^{-1}\mathbf{R}_{xx}^{-1}(k-1) - \frac{\lambda^{-2}\mathbf{R}_{xx}^{-1}(k-1)\,\mathbf{x}(k)\,\mathbf{x}^T(k)\,\mathbf{R}_{xx}^{-1}(k-1)}{1 + \lambda^{-1}\mathbf{x}^T(k)\,\mathbf{R}_{xx}^{-1}(k-1)\,\mathbf{x}(k)}
\end{aligned}$$

or

$$\mathbf{R}_{xx}^{-1}(k) = \lambda^{-1}\,\mathbf{R}_{xx}^{-1}(k-1) - \lambda^{-1}\,\mathbf{k}(k)\,\mathbf{x}^T(k)\,\mathbf{R}_{xx}^{-1}(k-1) \qquad (13.97)$$

where

$$\mathbf{k}(k) = \frac{\lambda^{-1}\,\mathbf{R}_{xx}^{-1}(k-1)\,\mathbf{x}(k)}{1+\lambda^{-1}\,\mathbf{x}^T(k)\,\mathbf{R}_{xx}^{-1}(k-1)\,\mathbf{x}(k)} = \mathbf{R}_{xx}^{-1}(k)\,\mathbf{x}(k) \tag{13.98}$$

is also known as the *Kalman gain* vector. The last equality can be verified by rearranging the first equality in (13.98)

$$\mathbf{k}(k) = \lambda^{-1}\,\mathbf{R}_{xx}^{-1}(k-1)\,\mathbf{x}(k) - \lambda^{-1}\,\mathbf{k}(k)\mathbf{x}^T(k)\,\mathbf{R}_{xx}^{-1}(k-1)\,\mathbf{x}(k) \tag{13.99}$$

and by substituting the right hand side of (13.97).

These equations can now be used to develop a recursive update equation for the filter coefficients. Using the least-squares solution (13.88) and (13.94) we find

$$\mathbf{h}_{\mathrm{RLS}}(k) = \lambda\,\mathbf{R}_{xx}^{-1}(k)\,\boldsymbol{\varphi}_{xy}(k-1) + \mathbf{R}_{xx}^{-1}(k)\,\mathbf{x}(k)\,y(k) \tag{13.100}$$

and with (13.97)

$$\begin{aligned}\mathbf{h}_{\mathrm{RLS}}(k) &= \mathbf{R}_{xx}^{-1}(k-1)\,\boldsymbol{\varphi}_{xy}(k-1) - \mathbf{k}(k)\,\mathbf{x}^T(k)\,\mathbf{R}_{xx}^{-1}(k-1)\,\boldsymbol{\varphi}_{xy}(k-1)\\ &\quad + \mathbf{R}_{xx}^{-1}(k)\mathbf{x}(k)\,y(k)\\ &= \mathbf{h}_{\mathrm{RLS}}(k-1) - \mathbf{k}(k)\,\mathbf{x}^T(k)\,\mathbf{h}_{\mathrm{RLS}}(k-1) + \mathbf{k}(k)\,y(k)\\ &= \mathbf{h}_{\mathrm{RLS}}(k-1) + \mathbf{k}(k)\left(y(k) - \mathbf{x}^T(k)\,\mathbf{h}_{\mathrm{RLS}}(k-1)\right).\end{aligned} \tag{13.101}$$

As an example, Fig. 13.17 depicts the system distance for several values of the forgetting parameter λ, for a speech excitation, and no noise. For $\lambda \approx 1$ the RLS converges much faster than the NLMS algorithm.

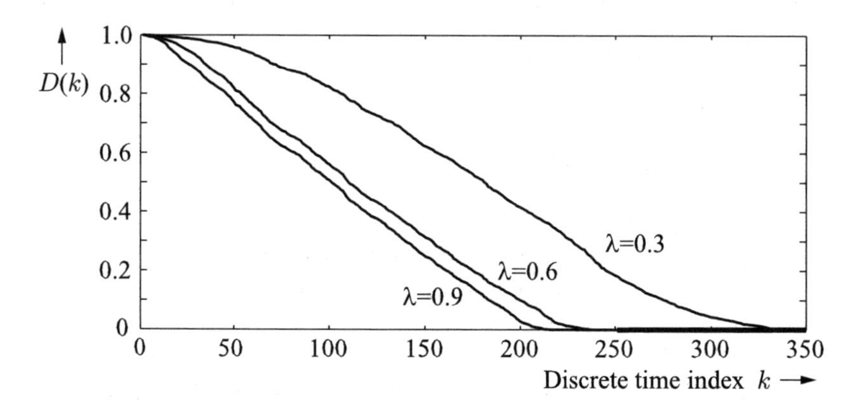

Figure 13.17: Convergence behavior of the RLS algorithm excited with speech ($m = m' = 200$, $\alpha = 1$)

Numerical stability is a critical issue for the RLS, especially for the fast $O(M)$ versions of the algorithm. A numerically stable version of the RLS is proposed in [Slock, Kailath 1991]. Some guidelines and initialization procedures are outlined in [Breining et al. 1999].

To conclude, we note that the convergence of the RLS algorithm is independent of the correlation properties of the input signal. Also, a number of other algorithms with convergence properties similar to the RLS algorithm have been developed (see, e.g., [Petillon et al. 1994]).

13.9 Block Processing and Frequency Domain Adaptive Filters

Depending on the acoustic environment, filters with some hundreds to thousands of coefficients are required for acoustic echo cancellation. When using the NLMS algorithm, a large filter order leads to a slow convergence (see Fig. 13.13). Furthermore, the computational effort increases linearly with the filter length m. Both problems can be alleviated to a certain degree by frequency domain processing.

The frequency domain approach reduces the computational complexity of the adaptive filter by using the FFT for fast convolution and correlation (see Section 3.6) or by using filter banks and multi-rate signal processing techniques. The improved convergence results from the inherent decorrelation of the signal and the associated reduction of eigenvalue spread. Among the many frequency domain approaches we will discuss the FFT-based *frequency domain adaptive filter* (FDAF) in greater detail as it allows for a reduction in complexity and a speeding up of convergence.

As a starting point for the FDAF, we consider the coefficient vector update

$$\mathbf{h}(k+1) = \mathbf{h}(k) + \beta\, e(k)\, \mathbf{x}(k) \qquad (13.102)$$

of the LMS algorithm with

$$e(k) = \tilde{x}(k) - \hat{x}(k) = \tilde{x}(k) - \mathbf{h}^T(k)\,\mathbf{x}(k) \qquad (13.103)$$

and a constant stepsize parameter β. In each time step, m MAC (multiply–accumulate) operations are required for the filtering $\mathbf{h}^T(k)\mathbf{x}(k)$ and m MAC operations for the coefficient vector update. The product $\beta \cdot e(k)$ and the residual error $e(k) = \tilde{x}(k) - \hat{x}(k)$ only have to be computed once per sample. Since $m \gg 1$, this part of the total computational effort can be neglected in further complexity estimations. Thus, the computational effort is split evenly between adaptation and filtering. Obviously, using of the fast convolution the computational effort for the filtering can be markedly reduced. As a first step towards frequency domain methods we consider the *block LMS* and the *exact block NLMS* algorithms.

13.9.1 Block LMS Algorithm

The *block LMS* algorithm [Clark et al. 1981] adjusts the filter coefficients not in each iteration, but only every L samples. Hence, the signal $x(k)$ is convolved with an impulse response that is constant over a time interval of $L \cdot T$. Also, L residual error samples $e(k), e(k+1), \ldots, e(k+L-1)$ are used to compute the new coefficient vector which is valid from time instant $k+L$ on,

$$\mathbf{h}(k+L) = \mathbf{h}(k) + \beta \sum_{\lambda=0}^{L-1} e(k+\lambda)\, \mathbf{x}(k+\lambda)\,. \tag{13.104-a}$$

In analogy to equation (13.102) the coefficient vector update is now composed of the sum of the increments $\beta\, e(k+\lambda)\, \mathbf{x}(k+\lambda)$. The output signal therefore differs from the original LMS algorithm, where each residual error sample $e(k)$ is generated with a different set of filter coefficients.

On the other hand (13.104-a) can also be written as

$$\mathbf{h}(k+L) = \mathbf{h}(k) + L\,\beta\, \frac{1}{L} \sum_{\lambda=0}^{L-1} e(k+\lambda)\, \mathbf{x}(k+\lambda) \tag{13.104-b}$$

$$= \mathbf{h}(k) + \widetilde{\beta}\, \hat{\nabla}_L(k+L-1)\,, \tag{13.104-c}$$

which indicates that the original instantaneous gradient is replaced by a temporally smoothed gradient

$$\hat{\nabla}_L(k+L-1) = \frac{1}{L} \sum_{\lambda=0}^{L-1} e(k+\lambda)\, \mathbf{x}(k+\lambda) \tag{13.105}$$

and $\widetilde{\beta} = L\beta$. Thus, for a stationary excitation the gradient is estimated more accurately. However, a slower convergence results, because, compared to the conventional LMS algorithm, the maximal stepsize parameter must be reduced by the factor $1/L$, in order to guarantee convergence (e.g., [Clark et al. 1981]). For a white noise signal $x(k)$, the conventional LMS algorithm and the block LMS algorithm converge towards the Wiener solution. When using a frequency domain implementation (see below), the speed of convergence can be significantly improved by the use of frequency- and time-dependent stepsize parameters.

13.9.2 The Exact Block NLMS Algorithm

As was pointed out earlier, the block LMS is not exactly equivalent to the LMS algorithm. It is, however, possible to develop mathematically exact block realizations of the adaptive algorithms presented so far. We demonstrate this for the NLMS algorithm [Nitsch 1997], thus also taking the time-varying stepsize normalization of the NLMS into account.

In analogy to (13.104-a) a block version of the NLMS algorithm may be written as

$$\mathbf{h}(k+L) = \mathbf{h}(k) + \sum_{\lambda=0}^{L-1} \alpha(k+\lambda)\, \frac{e(k+\lambda)\,\mathbf{x}(k+\lambda)}{||\mathbf{x}(k+\lambda)||^2} \tag{13.106}$$

where $e(k+\lambda), \lambda = 0, \ldots, L-1$, is computed with a fixed coefficient vector $\mathbf{h}(k)$,

$$e(k+\lambda) = y(k+\lambda) - \mathbf{x}^T(k+\lambda)\,\mathbf{h}(k)\,, \quad \lambda = 0, \ldots, L-1\,. \tag{13.107}$$

The difference between (13.106) and L iterations of the NLMS algorithm lies in the computation of the error signal.

We may write the exact NLMS error $e^{(\mathrm{NLMS})}(k+\lambda) = e(k+\lambda) + e_c(k+\lambda)$ as a sum of the block NLMS error $e(k+\lambda)$ and a correction term

$$e_c(k+\lambda) = \begin{cases} 0 & \lambda = 0 \\ -\displaystyle\sum_{i=1}^{\lambda} \frac{\alpha(k+i-1)}{||\mathbf{x}(k+i-1)||^2}\, e^{(\mathrm{NLMS})}(k+i-1) & \\ \cdot\, \mathbf{x}^T(k+\lambda)\,\mathbf{x}(k+i-1) & \lambda = 1, \ldots, L-1 \end{cases}$$

which accounts for the intermediate coefficient vector updates. Thus, for $\lambda = 1, \ldots, L-1$ the exact normalized stepsize–error product is given by

$$\begin{aligned} \breve{e}^{(\mathrm{NLMS})}(k+\lambda) &= \frac{\alpha(k+\lambda)}{||\mathbf{x}(k+\lambda)||^2}\, e^{(\mathrm{NLMS})}(k+\lambda) \\ &= \frac{\alpha(k+\lambda)}{||\mathbf{x}(k+\lambda)||^2} \Big(e(k+\lambda) + e_c(k+\lambda) \Big) \end{aligned} \tag{13.108}$$

with

$$e_c(k+\lambda) = \begin{cases} 0 & \lambda = 0 \\ -\displaystyle\sum_{i=1}^{\lambda} \breve{e}^{\mathrm{NLMS}}(k+i-1)\mathbf{x}^T(k+\lambda)\mathbf{x}(k+i-1) & \text{else}\,. \end{cases}$$

For $\lambda = 0, \ldots, L-1$, (13.108) represents a set of simultaneous linear equations for the exact normalized stepsize–error product $\check{e}^{(\mathrm{NLMS})}$. These equations are stacked into a set of implicit equations for $\check{e}^{(\mathrm{NLMS})}(k)$,

$$\check{\mathbf{e}}^{(\mathrm{NLMS})}(k) + \mathrm{diag}\{\boldsymbol{\beta}(k)\}\, \mathbf{R}(k)\, \check{\mathbf{e}}^{(\mathrm{NLMS})}(k) = \mathrm{diag}\{\boldsymbol{\beta}(k)\}\mathbf{e}(k) \tag{13.109}$$

where

$$\mathbf{e}(k) = (e(k), e(k+1), \ldots, e(k+L-1))^T$$

$$\mathbf{R}(k) = \begin{pmatrix} 0 & 0 & \cdots & 0 & 0 \\ \mathbf{x}^T(k+1)\mathbf{x}(k) & 0 & \cdots & 0 & 0 \\ \mathbf{x}^T(k+2)\mathbf{x}(k) & \mathbf{x}^T(k+2)\mathbf{x}(k+1) & & \vdots & \vdots \\ \vdots & \vdots & \ddots & 0 & 0 \\ \mathbf{x}^T(k+L-1)\mathbf{x}(k) & \mathbf{x}^T(k+L-1)\mathbf{x}(k+1) & \cdots & \mathbf{x}^T(k+L-1)\mathbf{x}(k+L-1) & 0 \end{pmatrix}$$

$$\check{\mathbf{e}}^{(\mathrm{NLMS})}(k) = (\check{e}^{(\mathrm{NLMS})}(k), \check{e}^{(\mathrm{NLMS})}(k+1), \ldots, \check{e}^{(\mathrm{NLMS})}(k+L-1))^T$$

and

$$\mathrm{diag}\{\boldsymbol{\beta}(k)\} = \begin{pmatrix} \frac{\alpha(k)}{||\mathbf{X}(k)||^2} & 0 & \cdots & 0 \\ 0 & \frac{\alpha(k+1)}{||\mathbf{X}(k+1)||^2} & \cdots & 0 \\ \vdots & \vdots & \ddots & \vdots \\ 0 & 0 & \cdots & \frac{\alpha(k+L-1)}{||\mathbf{X}(k+L-1)||^2} \end{pmatrix},$$

the solution of which is given by

$$\check{\mathbf{e}}^{(\mathrm{NLMS})}(k) = \left[\mathrm{diag}\{\boldsymbol{\beta}(k)\}^{-1} + \mathbf{R}(k)\right]^{-1} \mathbf{e}(k). \tag{13.110}$$

The correct normalized stepsize–error product can now be used in (13.106) to yield a mathematically exact block version of the NLMS algorithm. Due to the diagonal structure of $\mathbf{R}(k)$, the set of equations can be solved efficiently. Also, the elements of $\mathbf{R}(k)$ can be computed recursively. Furthermore, the coefficient vector might be partitioned to reduce the block delay [Nitsch 1997] and the algorithm can be performed in a transform domain to reduce the complexity as outlined below.

13.9.3 Frequency Domain Adaptive Filter (FDAF)

The frequency domain implementation of block adaptive algorithms comprises two steps. First, the convolution of the far-end signal with the filter impulse response is implemented as a fast convolution using the overlap-save scheme (see Section 3.6.3). Secondly, the coefficient vector is adapted in the frequency domain. For large m, both steps result in significant computational savings.

13.9.3.1 Fast Convolution and Overlap-Save

To develop the *frequency domain adaptive filter* (FDAF) the time index k is replaced by

$$k = \kappa\, L + \lambda; \qquad \lambda = 0, 1, \dots, L-1\,, \tag{13.111}$$

where κ is the block index and λ is the index of samples within a block. The convolution is performed in the frequency domain using the FFT and the *overlap-save* algorithm. In each block, L new output samples $\hat{x}(k)$ are computed, while the coefficients of the filter are constant for at least L time steps.

Figure 13.18 illustrates the basic structure. For an impulse response of order m, the length M of the FFT must be chosen to accommodate L valid output samples, therefore

$$M = m - 1 + L\,. \tag{13.112}$$

Hence, the coefficient vector is padded with zeros to a vector $\mathbf{h}'(\kappa)$ of length M with elements

$$h'_i(\kappa) = \begin{cases} h_i(\kappa \cdot L) & i = 0, 1, \dots, m-1 \\ 0 & i = m, m{+}1, \dots, M{-}1\,. \end{cases} \tag{13.113}$$

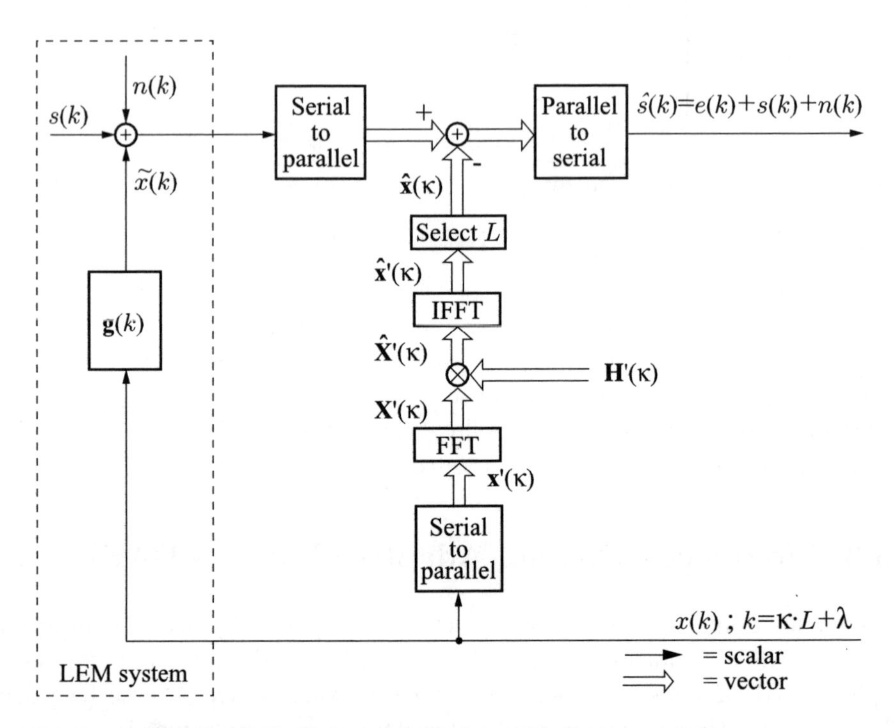

Figure 13.18: Echo cancellation with fast convolution

The input vector $\mathbf{x}'(\kappa)$ consists of the M elements

$$\begin{aligned} \mathbf{x}'(\kappa) &= (x'_0(\kappa), x'_1(\kappa), \dots, x'_{M-1}(\kappa))^T \\ &= \big(x(\kappa L - m + 1), \dots, x(\kappa L), x(\kappa L + 1), \dots, x(\kappa L + L - 1)\big)^T \end{aligned} \tag{13.114}$$

and is also fed into an FFT of length M. In contrast to the definition (13.12) of the state vector of the time domain echo canceller, the input vector $\mathbf{x}'(\kappa)$ contains the samples in the order of increasing sampling indices.

The transformation of the coefficient vector

$$\mathbf{h}'(\kappa) = \big(h'_0(\kappa), h'_1(\kappa), \dots, h'_{M-1}(\kappa)\big)^T$$

is written as

$$\mathbf{H}'(\kappa) = \text{FFT}\{\mathbf{h}'(\kappa)\},$$

with

$$\mathbf{H}'(\kappa) = \big(H'_0(\kappa), H'_1(\kappa), \dots, H'_{M-1}(\kappa)\big)^T.$$

The transformed vectors $\mathbf{H}'(\kappa)$ and $\mathbf{X}'(\kappa)$ are multiplied element by element, i.e.,

$$\widehat{X}'_\mu(\kappa) = H'_\mu(\kappa) X'_\mu(\kappa). \tag{13.115}$$

The resulting vector $\widehat{\mathbf{X}}'(\kappa)$, after inverse transformation, yields the vector $\widehat{\mathbf{x}}'(\kappa)$ of length M,

$$\widehat{\mathbf{x}}'(\kappa) = (\widehat{x}'_0(\kappa), \widehat{x}'_1(\kappa), \dots, \widehat{x}'_{M-1}(\kappa))^T \tag{13.116}$$

which is equivalent to a cyclic convolution (see Fig. 13.19) of the vectors $\mathbf{x}'(k)$ and $\mathbf{h}'(k)$. Consequently, the first $m-1$ values $\widehat{x}'_i(\kappa)$, $i = 0, 1, \dots, m-2$, are affected by cyclic effects. The selection of L valid output samples may be described by an elementwise multiplication with window $\mathbf{w}$, where

$$w_i = \begin{cases} 0 & i = 0, 1, \dots, m-2 \\ 1 & i = m-1, m, \dots, M-1 \end{cases} \tag{13.117}$$

are the components of $\mathbf{w}$. The L valid output values,

$$\widehat{x}(\kappa \cdot L + \lambda) = \widehat{x}'_{m-1+\lambda}(\kappa), \qquad \lambda = 0, 1, \dots, L-1, \tag{13.118}$$

correspond to the result of the linear convolution. Finally, the undisturbed compensation error is obtained from

$$e(\kappa \cdot L + \lambda) = \tilde{x}(\kappa \cdot L + \lambda) - \widehat{x}(\kappa \cdot L + \lambda) \tag{13.119}$$

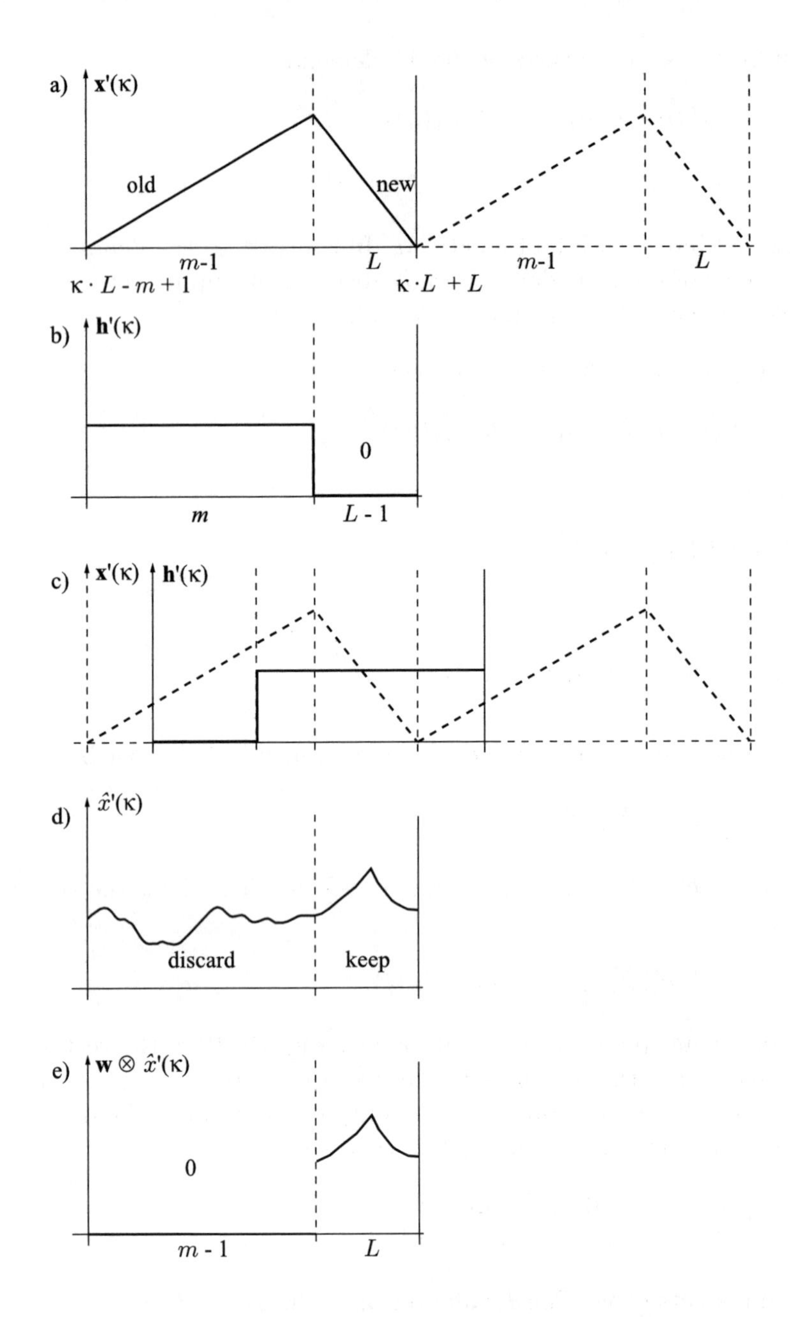

Figure 13.19: Illustration of the cyclic convolution
a) Input vector $\mathbf{x}'(\kappa)$ with periodic extension
b) Coefficient vector $\mathbf{h}'(\kappa)$
c) Cyclic convolution
d) Result of cyclic convolution
e) Selection of L valid samples

whereas in practical applications only the disturbed error

$$\hat{s}(\kappa \cdot L + \lambda) = y(\kappa \cdot L + \lambda) - \hat{x}(\kappa \cdot L + \lambda) \tag{13.120}$$

is available. The serial-to-parallel operation comprises the buffering of L successive signal samples. L samples must be collected before the next block of the transmit signal can be computed. The block processing therefore introduces a delay of $L-1$ samples. However, at the expense of increased computational complexity, L can be chosen to be much smaller than the transform length M. The FDAF therefore provides a flexible framework for balancing algorithmic delay and computational complexity.

However, when m is large and when a small algorithmic delay L is required, the computational complexity might not be acceptable. In [Sommen 1989], [Soo, Pang 1990] a partitioning of the coefficient vector, i.e., a distribution of the long impulse response onto several partial filters with shorter impulse responses, is proposed to deal with this problem. For Q partitions the *partitioned block frequency domain adaptive filter* (PBFDAF) is illustrated in Fig. 13.20. Instead of FFTs of length $m+L-1$ this scheme uses FFTs of length $m/Q+L-1$.

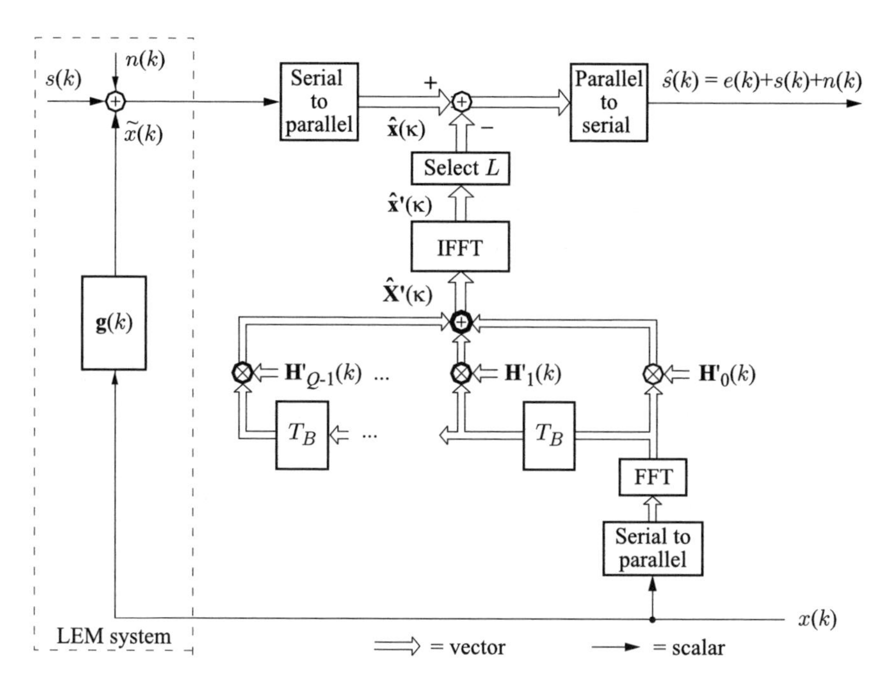

Figure 13.20: Partitioned block frequency domain adaptive filter. T_B denotes a unit delay of frequency domain vectors

13.9.3.2 FLMS Algorithm

In the next step, the adaptation of the coefficients is transformed in the frequency domain, as well. This leads to the *fast LMS* (FLMS) algorithm [Ferrara 1980], [Clark et al. 1983]. The benefits of an adaptation in the frequency domain are an additional reduction of the computational effort and an improvement of the convergence behavior using a time- *and* frequency-dependent stepsize control.

For this purpose, the adaptation rule of the block LMS algorithm according to (13.104-c) is inspected again. As before, we develop this algorithm on the basis of the undisturbed error signal. The coefficient update equation (13.104-b) is now written in terms of the block index κ,

$$\mathbf{h}'(\kappa+1) = \mathbf{h}'(\kappa) + \beta\, \hat{\boldsymbol{\nabla}}'(\kappa)\,, \tag{13.121}$$

or more generally as

$$\mathbf{h}'(\kappa+1) = \mathbf{h}'(\kappa) + \Delta\mathbf{h}'(\kappa) \tag{13.122}$$

where in the case of a constant stepsize β the coefficient update vector $\Delta\mathbf{h}'(\kappa)$ is computed as $\Delta\mathbf{h}'(\kappa) = \beta\, \hat{\boldsymbol{\nabla}}'(\kappa)$. The components $\hat{\nabla}_i'(\kappa)$ of the gradient vector $\hat{\boldsymbol{\nabla}}'(\kappa)$ are given by

$$\hat{\nabla}_i'(\kappa) = \begin{cases} \sum\limits_{\lambda=0}^{L-1} e(\kappa\cdot L+\lambda)\; x(\kappa\cdot L+\lambda-i) & i = 0,1,\ldots,m-1 \\ 0 & i = m, m{+}1,\ldots,M{-}1\,. \end{cases} \tag{13.123}$$

When the discrete Fourier transform is applied to both sides of (13.122), we obtain

$$\mathbf{H}'(\kappa+1) = \mathbf{H}'(\kappa) + \Delta\mathbf{H}'(\kappa)\,. \tag{13.124}$$

Thus, the update loop can be implemented in the frequency domain as shown in Fig. 13.21-a. In the simple case of a constant stepsize β we have $\Delta\mathbf{H}'(\kappa) = \beta\, \hat{\boldsymbol{\nabla}}'(\kappa)$ where in principle the frequency domain gradient vector $\hat{\boldsymbol{\nabla}}'(\kappa)$ could be obtained via the FFT of $\hat{\boldsymbol{\nabla}}'(\kappa)$.

However, the computational complexity can be further reduced when the update vector $\Delta\mathbf{H}'(\kappa)$ is computed in the frequency domain as well. As a function of i, the individual component $\hat{\nabla}_i'(\kappa)$ for fixed κ can be interpreted as a correlation between a segment of the residual error signal $e(k)$ of length L and the signal $x(k)$. In analogy to the fast convolution this correlation can be implemented as a fast correlation using the FFT. This is illustrated in Fig. 13.22.

In order to employ the FFT, the signal vectors must be constructed such that the cyclic convolution yields the components of the gradient vector $\hat{\nabla}_i'(\kappa)$ as in

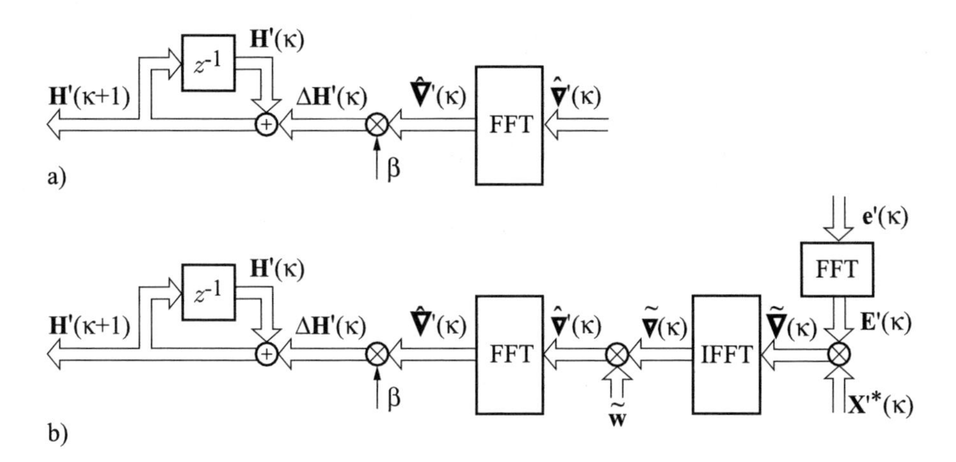

Figure 13.21: Derivation of the adaptation rule of the FLMS algorithm
a) Adaptation of the coefficient vector in the frequency domain
b) Fast computation of the gradient vector

(13.123). This can be achieved by padding $\mathbf{e}(\kappa) = (e(\kappa L), \ldots, e(\kappa L + L - 1))^T$ with $m-1$ leading zeros

$$e'_i(\kappa) = \begin{cases} 0 & i = 0, 1, \ldots, m-2 \\ e(\kappa \cdot L + i - m + 1) & i = m-1, m, \ldots, M-1, \end{cases} \tag{13.125}$$

and by using the vectors $\mathbf{x}'(\kappa)$ and $\mathbf{e}'(\kappa)$ as inputs to the fast correlation. If we compute the DFT $\mathbf{E}'(\kappa)$ and $\mathbf{X}'(\kappa)$ of vectors $\mathbf{e}'(\kappa)$ and $\mathbf{x}'(\kappa)$, respectively, the componentwise multiplication of the vectors $\mathbf{E}'(\kappa)$ and $\mathbf{X}'^*(\kappa)$,

$$\hat{\nabla}'_\mu(\kappa) = E'_\mu(\kappa) X'^*_\mu(\kappa), \quad \mu = 0, \ldots, M-1, \tag{13.126}$$

and a subsequent inverse transform produce the desired values $\hat{\nabla}'_i(\kappa)$ in the first m elements of the inverse of $\hat{\nabla}'_\mu(\kappa)$.

The desired gradient vector $\hat{\boldsymbol{\nabla}}'$ is thus generated by applying a (gradient) constraint $\tilde{\mathbf{w}}$ to IDFT$\{\hat{\boldsymbol{\nabla}}'(\kappa)\}$ which sets the last $L-1$ samples to zero. The constraint window $\tilde{\mathbf{w}}$ is given by

$$\tilde{w}_i = \begin{cases} 1 & i = 0, 1, \ldots, m-1 \\ 0 & i = m, m+1, \ldots, M-1 \end{cases} \tag{13.127}$$

and the resulting FLMS algorithm is illustrated in Fig. 13.21-b. Including the two transformations for fast convolution (Fig. 13.18), a total of five transformations of length M are performed in order to determine L values of the estimated echo signal $\hat{x}(k)$. For $M = m + L - 1 \approx m + L$ the computational complexity of the constraint FLMS can then be estimated as follows.

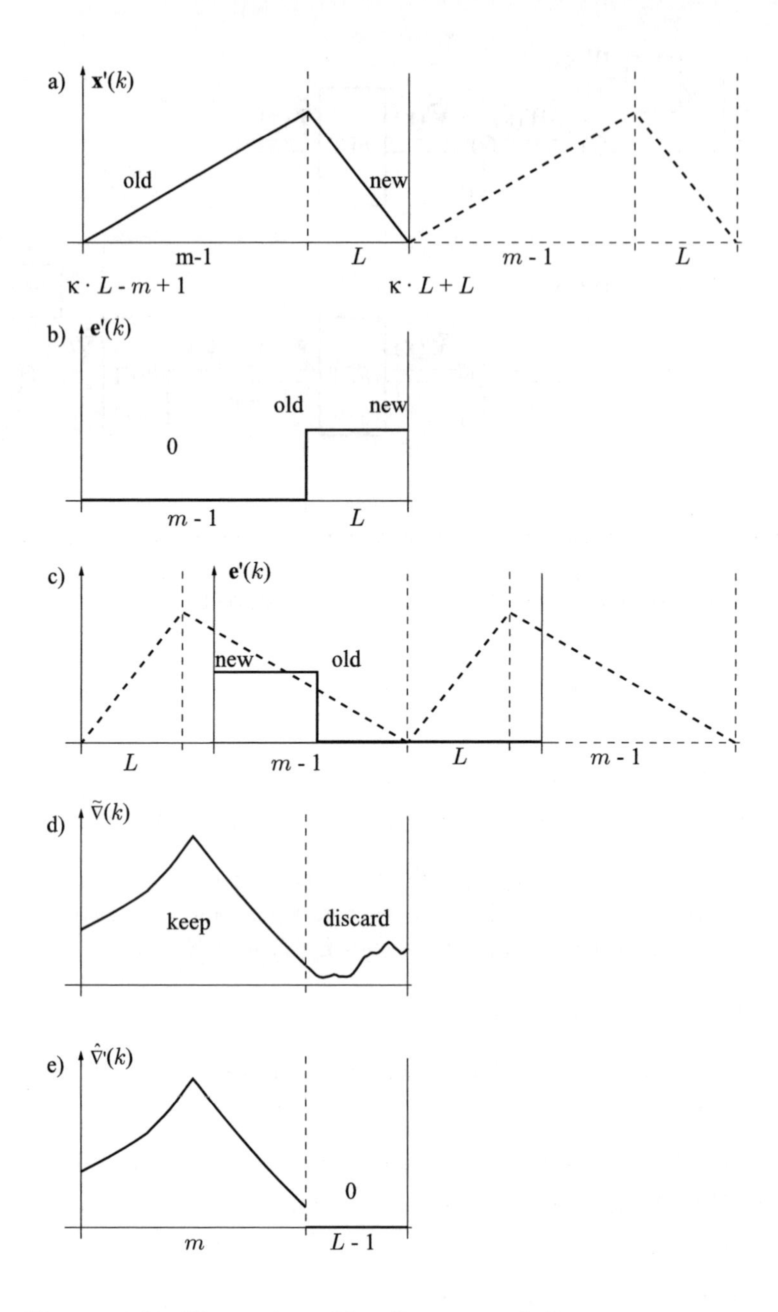

Figure 13.22: Illustration of fast linear correlation
a) Input vector $\mathbf{x}'(\kappa)$ with periodic extension
b) Error vector $\mathbf{e}'(\kappa)$
c) Cyclic correlation
d) Result of cyclic correlation
e) Selection of m valid samples

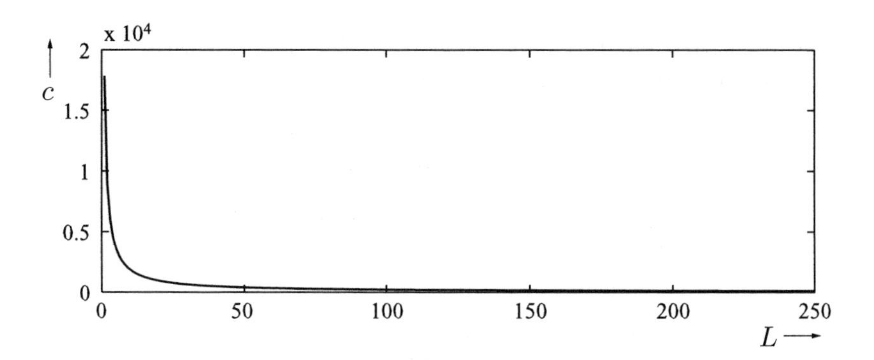

Figure 13.23: Computational complexity of the FLMS vs. block length L

The FFT of a real-valued sequence of length M requires about $(M/4)\ \mathrm{ld}(M/2)$ complex arithmetic operations. A total of five transformations are needed. For real-valued signals the symmetry of the DFT vectors can be exploited. Therefore, approximately $4 \cdot M/2 = 2M$ complex operations are performed to multiply the components of $\mathbf{E}'(\kappa)$ and $\mathbf{X}'^{*}(\kappa)$, to multiply the components of $\mathbf{X}'(\kappa)$ and $\mathbf{H}'(\kappa)$, and to adapt the coefficient vector. If we equate one complex operation with four real operations, the computational complexity for one block of L samples is given by $5(m+L)\,\mathrm{ld}((m+L)/2) + 8(m+L)$. Per output sample, we thus obtain a computational complexity of

$$c = \frac{5\,(m+L)\,\mathrm{ld}((m+L)/2) + 8\,(m+L)}{L}\,.$$

For $m = 500$, Fig. 13.23 plots c as a function of $0 \le L \le m/2$. Clearly, c attains its minimum for the maximum value of L. Fortunately, the computational complexity exhibits a sharp decline for increasing L. For example, for $m = 500$, we already achieve a significant complexity reduction for $L > 40$.

Furthermore, the complexity of the FLMS may be related to the complexity $2m$ of the time domain LMS algorithm. The relative complexity is given by

$$\rho = \frac{5\,((m+L)/2)\,\mathrm{ld}((m+L)/2) + 4\,(m+L)}{mL}\,. \tag{13.128}$$

For $m = L$, Table 13.1 gives typical values. For increasing L, and a corresponding increase of the block delay, ρ decreases significantly.

In comparison to the LMS algorithm, the computational complexity can be significantly reduced for large L. In contrast, the required memory capacity increases, because even with a skillful organization of the processes needed for an *in-place* FFT computation several vectors of the dimension M must be stored.

Table 13.1: Relative computational complexity of the FLMS algorithm in comparison to the LMS algorithm ($M = 2\,L$; $m = L$), according to [Ferrara 1985]

L	16	32	64	256	1024
ρ	1.75	1.03	0.59	0.19	0.057

The computational complexity can be reduced if the gradient constraint is neglected. Compared to the constrained FLMS, the unconstrained FLMS [Mansour, Gray 1982] saves two FFT/IFFT operations. However, in this case, the coefficient update must rely on a circular convolution. Therefore, the algorithm does not approach the Wiener solution in steady state. A biased filter coefficient vector results and convergence deteriorates for $L \approx m$ [Haykin 1996]. Furthermore, the *partitioned* FDAF does not converge well without the gradient constraint. The *soft-partitioned* FDAF [Enzner, Vary 2003a] provides a good compromise between computational complexity, speed of convergence, and roundoff noise in fix-point implementations.

13.9.3.3 Improved Stepsize Control

In addition to computational advantages, the frequency domain approach provides the possibility to control the stepsize parameter β as a function of frequency and time, in order to positively influence the convergence behavior during double talk phases ($s(k) \neq 0$), for additive interference ($n(k) \neq 0$), and for time variance of the LEM impulse response. It can be shown [Nitsch 2000] that by minimizing the average convergence state the optimal stepsize is given as a function of frequency by

$$\beta_\mu(\kappa) = \frac{\alpha_\mu(\kappa)}{\mathrm{E}\{|X_\mu(\kappa)|^2\}} = \frac{\mathrm{E}\left\{|E_\mu(\kappa)|^2\right\}}{\mathrm{E}\left\{|\widehat{S}_\mu(\kappa)|^2\right\}}\,\frac{1}{\mathrm{E}\{|X_\mu(\kappa)|^2\}}\,. \tag{13.129}$$

The PSDs $\mathrm{E}\left\{|\widehat{S}_\mu(\kappa)|^2\right\}$ and $\mathrm{E}\{|X_\mu(\kappa)|^2\}$ are determined from the signals $\hat{s}(k)$ and $x(k)$, respectively. The residual echo PSD $\mathrm{E}\left\{|E_\mu(\kappa)|^2\right\}$ or, with $\mathrm{E}\left\{|E_\mu(\kappa)|^2\right\} = |D_\mu(\kappa)|^2\mathrm{E}\left\{|X_\mu(\kappa)|^2\right\}$, the convergence state $|D_\mu(\kappa)|^2$ of the adaptive filter must be estimated as outlined, for example, in [Mader et al. 2000], [Enzner et al. 2002], [Hänsler, Schmidt 2004]. A simple and robust method for the continuous estimation of the convergence state, which does not need double talk detection mechanisms, is proposed in [Enzner, Vary 2003b].

13.9.4 Subband Acoustic Echo Cancellation

Besides the FDAF approach, subband acoustic echo cancellation is another widely used frequency domain method. In subband acoustic echo cancellation we use a digital filter bank (e.g., QMF or PPN, see Chapter 4) instead of the DFT or FFT [Kellermann 1985], [Kellermann 1988]. The far-end speaker's signal $x(k)$ and the microphone signal $y(k)$ are separated by a filter bank into M subband signals with a reduced sampling rate (Fig. 13.24). In each subband an individual echo canceller is utilized, which due to the downsampling has a correspondingly shortened impulse response in comparison to the full-band echo canceller. The individual cancellation filters are adapted with, for instance, the NLMS algorithm. By means of a synthesis filter bank, the compensated subband signals are interpolated and superimposed. As the spectrum of each individual subband signal is relatively flat within the respective frequency band, a favorable convergence behavior results.

In practical applications, *oversampled* filter banks are preferred [Kellermann 1985]. Due to the inevitable spectral overlaps of adjacent channels of the critically sampled filter bank, the sampling rate reduction must be selected to be smaller than M, e.g., $r_0 = M/2$. Compared to the critically sampled filter bank, this increases the effort for the subband cancellers, whose impulse responses then contain approximately $m' = m/r_0 = 2m/M$ coefficients. We refer to the literature [Kellermann 1985], [Kellermann 1989], [Shynk 1992] for further details. Critically sampled filter banks and IIR filter banks are possible as well but then additional cross-channel filters [Gilloire, Vetterli 1992] or notch filters [Naylor et al. 1998] must be employed to cancel aliasing components.

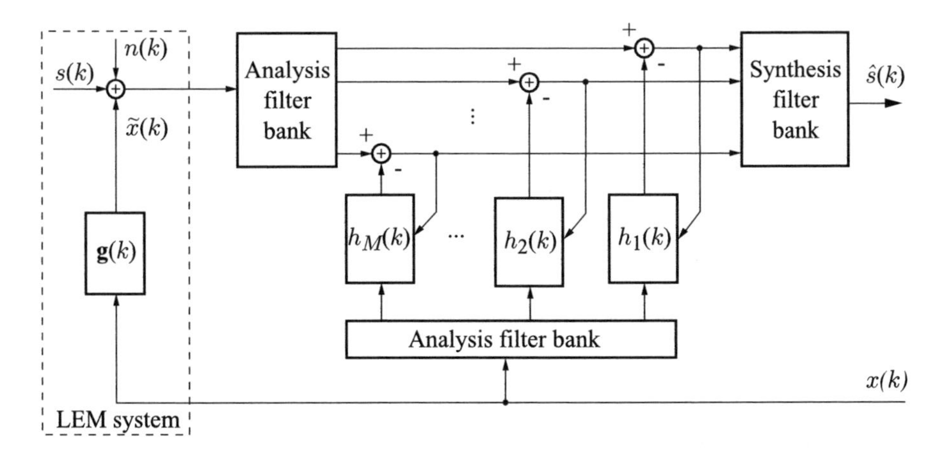

Figure 13.24: Subband acoustic echo cancellation

13.10 Additional Measures for Echo Control

It must be assumed that in practice the cancellation filter does not always deliver a sufficient amount of echo suppression. This is the case, for example, if a sudden change of the LEM impulse response occurs after the canceller has converged and a small system distance $D(k)$ and a small stepsize parameter has been achieved. For this reason, additional measures are needed to further reduce the residual echo. Such measures are also advisable if, due to the limited realization effort, the impulse response of the cancellation filter is much shorter than the actual impulse response of the LEM system.

The required echo suppression depends on the signal delay of the transmission path. Quality standards for handsfree devices have been established in international recommendations as discussed in Section 13.2.

13.10.1 Echo Canceller with Center Clipper

A low-level residual echo signal $e(k)$ might be audible during speech pauses of the near-end speaker, especially if a long transmission delay exists.

In such a situation the residual echo signal can be effectively suppressed with a non-linear *center clipping device*. The non-linear function of the center clipper, which, as depicted in Fig. 13.25, processes the output signal $\hat{s}(k)$ of the echo canceller, is

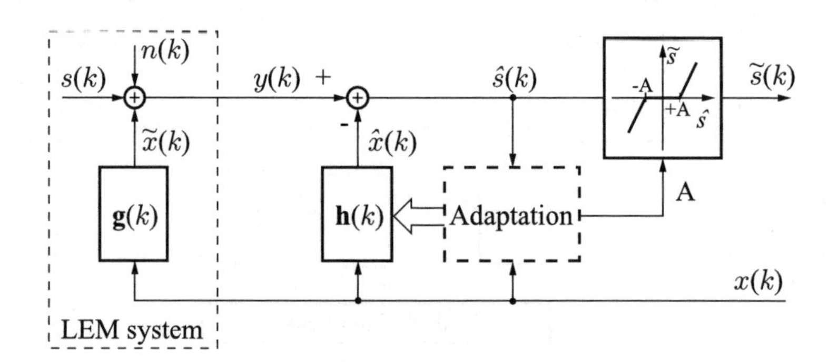

Figure 13.25: Echo canceller with center clipper according to (13.130) for the suppression of the residual echo

given by

$$\tilde{s}(k) = \begin{cases} \hat{s}(k) - A & \hat{s}(k) > +A \\ 0 & |\hat{s}(k)| \leq +A \\ \hat{s}(k) + A & \hat{s}(k) < -A \end{cases} \tag{13.130}$$

or by

$$\tilde{s}(k) = \begin{cases} \hat{s}(k) & \hat{s}(k) > +A \\ 0 & |\hat{s}(k)| \leq +A \\ \hat{s}(k) & \hat{s}(k) < -A\,. \end{cases} \tag{13.131}$$

Both variants provide similarly good results. The threshold value A, which can also be adapted, should be as small as possible.

13.10.2 Echo Canceller with Voice-Controlled Switching

With the center clipper, residual echoes can only be effectively suppressed without any perceptible distortion of the actual desired signal $s(k)$ if the echoes are already at a relatively low level. During the initialization phase of the cancellation filter, and after sudden changes of the LEM impulse response, this is not ensured. In [Armbrüster 1988] a system with an additional voice-activated *soft switching* as outlined in Fig. 13.26 was proposed in order to solve this problem.

The total attenuation of the far-end signal in the transmission loop with input $x(k)$ and output $\tilde{s}(k)$ consists of the two contributions of the voice-controlled switching (a_x, a_y) and the contribution of the echo canceller (a_c, see Fig. 13.27).

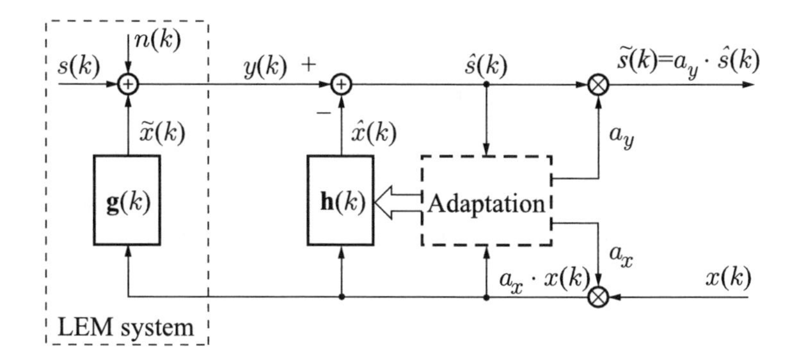

Figure 13.26: Echo canceller with voice-controlled soft switching supporting the initialization phase and suppressing the residual echo

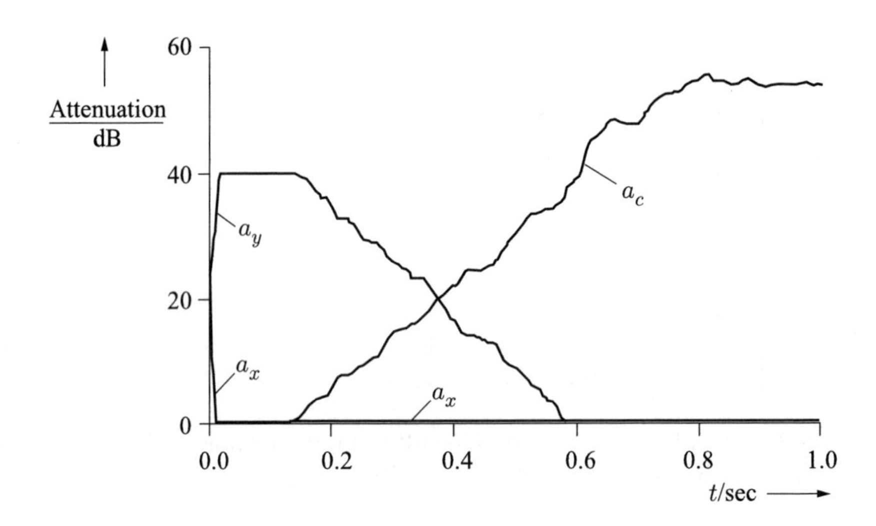

Figure 13.27: Example of a combination of echo cancellation and voice-controlled soft switching [Armbrüster 1988]

In the example of Fig. 13.27, a minimal suppression of 40 dB is required. At the beginning, as

$$20\ \lg(a_c) = 10\ \lg \frac{\hat{\mathrm{E}}\{\hat{s}^2(k)\}}{\hat{\mathrm{E}}\{y^2(k)\}} \approx 0\,,$$

the echo canceller does not contribute, and the two attenuation factors are set to $20\ \lg(a_x) = 20\ \lg(a_y) = 20\,\mathrm{dB}$. Then, the far-end speaker begins to talk. In this early stage of adaptation, the attenuation of the receive branch is switched to 0 dB ($a_x = 1$) by the voice-controlled switching and that of the transmit branch to 40 dB ($a_y = 0.01$). Full-duplex double talk is not yet possible. While the cancellation filter is adapted in each iteration using the signal $x(k)$ it can increasingly contribute to the total attenuation.

The attenuation a_y of the transmit branch can be reduced accordingly, until the cancellation filter finally reaches the desired attenuation of 40 dB, and the weighting factor assumes the value $a_y = 1$. Now the system is fully double talk capable.

After a change in the impulse response of the LEM, the echo canceller typically achieves a reduction of less than 40 dB. When this condition is detected, the voice-controlled switching becomes active again, with the required additional reduction of $(40 - 20\ \lg(a_c))$ dB half in the transmit and half in the receive branch.

13.10.3 Echo Canceller with Adaptive Postfilter in the Time Domain

An alternative to the additional echo suppressor was proposed in [Martin, Altenhöner 1995], [Martin, Gustafsson 1996], [Gustafsson et al. 1998]. An additional adaptive postfilter to reduce the residual echo is used in the transmit branch, as shown in Fig. 13.28. This filter is a transversal filter which, in contrast to the voice-activated soft switching, works in a frequency selective fashion. The spectrum of the signal $\hat{s}(k)$ is weighted as a function of frequency and time, depending on the instantaneous, spectral shape of the residual echo $e(k)$, the desired signal $s(k)$, and the background noise $n(k)$. In this, the psychoacoustic effect of masking is exploited. The echo is suppressed when it has significantly more power in a given frequency band than the near-end signals, i.e., when it is audible and not masked by the near-end signals. In the speech pauses of the far-end speaker, the signal $\hat{s}(k)$ is not influenced by the postfilter. The filter with the time-variant impulse response $\mathbf{c}(k)$ of the order $m_c = 20$ can be adjusted, for example, with the NLMS algorithm. It achieves a significant additional reduction of the residual echo. Therefore, the order of the echo canceller can be significantly reduced, thus leading to very efficient implementations. In fact, the combined echo cancellation and postfiltering approach opens up a twofold perspective on the design of echo control systems: a relatively short canceller can be used to guarantee *stability* of the transmission loop while the *audibility* of echoes can be tackled by the less expensive postfilter. During double talk the latter device can be operated without significant amounts of echo reduction since much of the echo is masked by the near-end speech and stability is provided by the echo canceller.

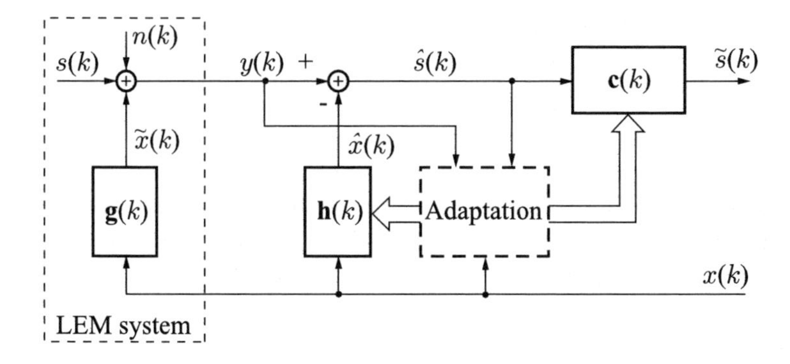

Figure 13.28: Combination of an echo canceller with an adaptive filter for a frequency selective attenuation of the echo

13.10.4 Echo Canceller with Adaptive Postfilter in the Frequency Domain

The adaptive postfilter can be realized in the frequency domain, as well, and can then be combined with measures for noise reduction [Faucon, Le Bouquin-Jeannès 1995], [Martin, Vary 1996], [Ayad et al. 1996], [Le Bouquin-Jeannès et al. 2001], [Hänsler, Schmidt 2004]. Also, the frequency domain implementation allows us to compute psychoacoustic masking thresholds and thus provides means for a more accurate computation of filter weights. A psychoacoustically motivated approach along these lines has been proposed in [Gustafsson et al. 2002].

In principle, any of the well-known noise reduction methods can also be employed for combined acoustic echo and noise reduction. For instance, the Wiener filter solution to the postfilter problem minimizes in our context the mean square error

$$\mathrm{E}\{(\tilde{s}(k) - s(k))^2\}$$

under a linear filtering constraint and also provides an interesting perspective on the relation of echo cancellation and residual echo suppression [Hänsler, Schmidt 2000], [Enzner et al. 2002], [Hänsler, Schmidt 2004].

In the frequency domain the Wiener filter results in a frequency response for the residual echo suppression filter

$$C_\mu(\kappa) = \frac{\mathrm{E}\left\{|S_\mu(\kappa)|^2\right\}}{\mathrm{E}\left\{|\widehat{S}_\mu(\kappa)|^2\right\}} = \frac{\mathrm{E}\left\{|S_\mu(\kappa)|^2\right\}}{\mathrm{E}\left\{|S_\mu(\kappa)|^2\right\} + \mathrm{E}\left\{|N_\mu(\kappa)|^2\right\} + \mathrm{E}\left\{|E_\mu(\kappa)|^2\right\}}$$

$$C_\mu(\kappa) = \frac{\dfrac{\mathrm{E}\left\{|S_\mu(\kappa)|^2\right\}}{\mathrm{E}\left\{|N_\mu(\kappa)|^2\right\} + \mathrm{E}\left\{|E_\mu(\kappa)|^2\right\}}}{1 + \dfrac{\mathrm{E}\left\{|S_\mu(\kappa)|^2\right\}}{\mathrm{E}\left\{|N_\mu(\kappa)|^2\right\} + \mathrm{E}\left\{|E_\mu(\kappa)|^2\right\}}} \tag{13.132}$$

provided that $s(k), n(k)$, and $e(k)$ are statistically independent. Compared to the standard noise reduction case we now have to account for the disturbing residual echo whose PSD $\mathrm{E}\left\{|E_\mu(\kappa)|^2\right\}$ must be estimated from the available signals. In analogy to the Wiener filter for noise reduction, the filter $C_\mu(\kappa)$ can be controlled by the *a priori* SNR

$$\frac{\mathrm{E}\left\{|S_\mu(\kappa)|^2\right\}}{\mathrm{E}\left\{|N_\mu(\kappa)|^2\right\} + \mathrm{E}\left\{|E_\mu(\kappa)|^2\right\}} .$$

In [Gustafsson et al. 2002] it is proposed to estimate the *a priori* SNR separately with respect to $\mathrm{E}\left\{|N_\mu(\kappa)|^2\right\}$ and $\mathrm{E}\left\{|E_\mu(\kappa)|^2\right\}$ and to combine it as

$$\frac{\mathrm{E}\left\{|S_\mu(\kappa)|^2\right\}}{\mathrm{E}\left\{|N_\mu(\kappa)|^2\right\}+\mathrm{E}\left\{|E_\mu(\kappa)|^2\right\}} = \frac{1}{\left[\frac{\mathrm{E}\{|S_\mu(\kappa)|^2\}}{\mathrm{E}\{|N_\mu(\kappa)|^2\}}\right]^{-1} + \left[\frac{\mathrm{E}\{|S_\mu(\kappa)|^2\}}{\mathrm{E}\{|E_\mu(\kappa)|^2\}}\right]^{-1}} \tag{13.133}$$

where for each term in the denominator on the right hand side of (13.133) a *decision-directed* [Ephraim, Malah 1984] estimator can be used.

When the ambient near-end noise is negligible, i.e., $\mathrm{E}\left\{|N_\mu(\kappa)|^2\right\} = 0$, we may write the Wiener filter for residual echo suppression as

$$C_\mu(\kappa) = \frac{\mathrm{E}\left\{|\widehat{S}_\mu(\kappa)|^2\right\} - \mathrm{E}\left\{|E_\mu(\kappa)|^2\right\}}{\mathrm{E}\left\{|\widehat{S}_\mu(\kappa)|^2\right\}}\,. \tag{13.134}$$

In conjunction with the optimal stepsize parameter of the FDAF (13.129), it is now straightforward to show that

$$\alpha_\mu(\kappa) + C_\mu(\kappa) = 1\,. \tag{13.135}$$

Thus, it turns out that the control of the FDAF-based echo canceller and of the Wiener postfilter are closely coupled. In both cases the PSD of the residual echo is the most critical control parameter. The residual echo power, however, depends on the frequency response $|D_\mu(\kappa)|^2$ of the distance vector $|\mathbf{d}(k)|^2$ which is not directly measurable. However, an efficient statistical approach for the estimation of this quantity is outlined in [Enzner, Vary 2003b], [Enzner, Vary 2005]. In this work, a synergy of FDAF, optimal stepsize control, Wiener postfilter, and convergence state estimation is established on the basis of Kalman filter theory.

13.10.5 Initialization with Perfect Sequences

The convergence behavior, in the sense of a fast system equalization, can be improved if a suitable auxiliary signal is applied in the initialization phase of the canceller.

In [Antweiler 1995] so-called *perfect sequences* [Lüke 1988], [Lüke, Schotten 1995], [Ipatov 1979] are proposed for this purpose. It is shown that, for an undisturbed adaptation, i.e., for $s(k) = 0$ and $n(k) = 0$, the NLMS algorithm converges in only m steps, and thus exactly identifies the impulse response of the LEM system.

Access to this solution is given by the geometric interpretation of the NLMS algorithm according to Fig. 13.14. The adaptation algorithm (13.64-d) shortens the system distance vector $\mathbf{d}(k)$ by subtracting the parallel component $\alpha\, \mathbf{d}^{||}(k)$. With an undisturbed adaptation, the normalized stepsize parameter can be set to $\alpha = 1$. As a result, the component of the distance vector $\mathbf{d}(k)$ which is parallel to the vector $\mathbf{x}(k)$ is completely eliminated.

With the assumption that all m successive state vectors $\mathbf{x}(k)$, $\mathbf{x}(k-1), \ldots,$ $\mathbf{x}(k+m-1)$ are orthogonal in the m-dimensional vector space, the complete identification of the unknown LEM system can be achieved in m steps. Periodically applied perfect sequences $p(\kappa)$ ($\kappa = 0, 1, \ldots, m-1$) fulfill the requirements of an optimal excitation signal with

$$x(\lambda \cdot m + \kappa) = p(\kappa)\,; \quad \lambda \in \mathbb{Z}\,, \tag{13.136}$$

since they are characterized by their periodic auto-correlation function $\tilde{\varphi}_{pp}(i)$, which vanishes for all out-of-phase values

$$\tilde{\varphi}_{pp}(i) = \varphi_{xp}(i) = \sum_{\kappa=0}^{m-1} p(\kappa)\; x(k+i) \tag{13.137-a}$$

$$= \sum_{\kappa=0}^{m-1} p(\kappa)\; x(\lambda \cdot m + \kappa + i) \tag{13.137-b}$$

$$= \begin{cases} \tilde{\varphi}_{pp}(0) & i \bmod m = 0 \\ 0 & \text{otherwise}\,. \end{cases} \tag{13.137-c}$$

All m phases of the perfect sequences are thus ideally orthogonal in the m-dimensional vector space.

The state vector $\mathbf{x}(k)$ meets the orthogonality requirement for $k \geq m$, as it only contains a complete period of the sequence $p(\kappa)$ from this time instant on ($\lambda \geq 1$ in (13.136)).

The convergence behavior for a perfect sequence excitation is illustrated as an example in Fig. 13.29, and compared to the adaptation using a speech or a white noise signal.

The simulation confirms the behavior to be expected from the geometric interpretation. In the initialization phase the complete identification of the LEM system (within computational precision) takes $2m$ iterations. After the initialization, during the runtime of the simulation, m iterations are already sufficient for a new equalization; for example, after a sudden change of the time-variant room (see Fig. 13.29 at $k = 4000$).

Since white noise only approximately fulfills the orthogonality requirement, the algorithm converges more slowly under excitation with a white noise signal than with a perfect sequence.

The most frequently used, so-called *odd-perfect* sequences [Lüke, Schotten 1995] are symmetrical, quasi-binary sequences, which, except for a (leading) zero, only

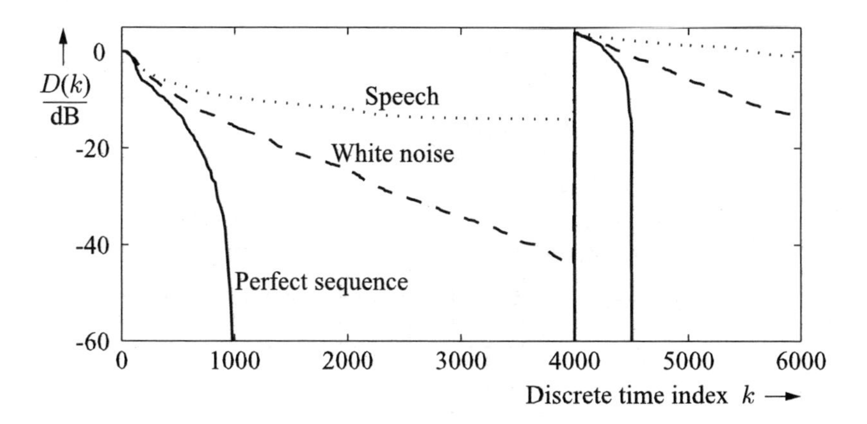

Figure 13.29: System distance for different excitation signals $x(k)$; undisturbed adaptation, change of the LEM system at $k = 4000$ ($\alpha = 1; s(k) = 0; n(k) = 0; x(k) = p(k_{|\text{mod } m})$); $m = m' = 500$, [Antweiler 1995], odd–perfect sequence $p(\kappa)$ [Lüke, Schotten 1995]

take two values $p(\kappa) \in \{+a, -a\}$, $\kappa = 1, 2, \ldots, m-1$. As the period length must be adapted to the length of the cancellation filter, it is a particular advantage that odd–perfect sequences can be generated for every length $m = p^K + 1$ with a prime number $p > 2$, $K \in \mathbb{N}$.

In practice, it is sufficient to apply a few periods of the perfect sequence to the system, only in the initialization phase or following strong changes of the room impulse response. In any case the power of the perfect sequence must be carefully controlled in order not to disturb the near-end listeners.

13.11 Stereophonic Acoustic Echo Control

Multi-channel sound transmission provides spatial realism and is of importance for many applications such as teleconferencing and multimedia systems. When many different talkers are involved, a realistic rendering of the acoustic scene provides valuable cues about the activity of the various talkers and thus contributes to the naturalness of the presentation. The simplest case of multi-channel reproduction, i.e., stereophonic reproduction, will be discussed in greater detail below. While early work focused on pseudo-stereo systems [Minami 1987] we will consider here sound rendering with two arbitrary loudspeaking signals. In particular we will discuss the relation between the cross-correlation of the two loudspeaker signals and the performance of the echo canceller. The more general case of a larger number of reproduction and recording channels is treated for example in [Benesty, Morgan 2000b], [Buchner, Kellermann 2001].

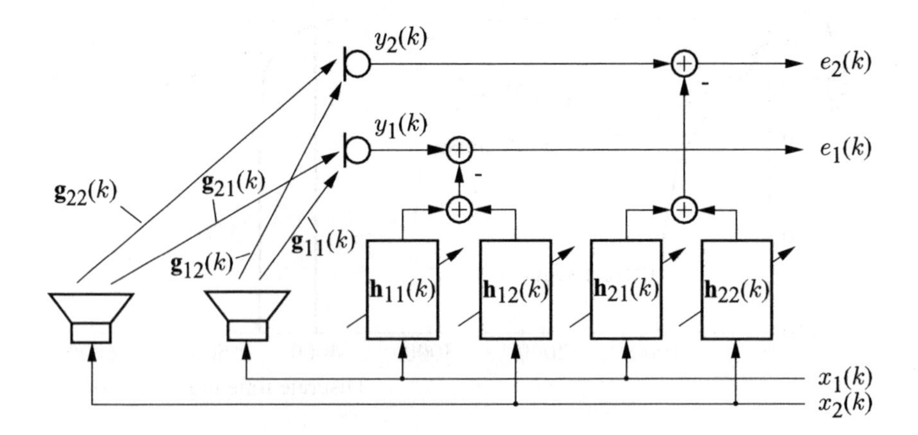

Figure 13.30: Basic block diagram of stereo acoustic echo cancellation (near-end side)

In the stereophonic transmission set-up the basic echo cancellation model in Fig. 13.3 must be extended since each microphone picks up sound from two loudspeakers. Hence, we have to identify two acoustic paths per microphone and will in general need two cancellers per microphone. This is shown in Fig. 13.30, where two acoustic paths $\mathbf{g}_{11}(k)$ and $\mathbf{g}_{12}(k)$ contribute to the microphone signal $y_1(k)$ and two cancellers with impulse responses $\mathbf{h}_{11}(k)$ and $\mathbf{h}_{12}(k)$ are used.

Typically, the two loudspeaker signals which are recorded at the far-end side and are transmitted to the near-end side originate from the same sources such as far-end speakers or far-end loudspeakers. In this case the received loudspeaker signals $x_1(k)$ and $x_2(k)$ in Fig. 13.30 can be written as a sum of the convolution of S far-end source signals $s_\ell(k)$, $\ell = 1, \ldots, S$, with the impulse responses $f_{1\ell}(k)$ and $f_{2\ell}(k)$ of the acoustic system at the far-end,

$$x_1(k) = \sum_{\ell=1}^{S} \sum_{i=0}^{m'_f-1} f_{1\ell}(k)\, s_\ell(k-i) = \sum_{\ell=1}^{S} f_{1\ell}(k) * s_\ell(k) \tag{13.138}$$

$$x_2(k) = \sum_{\ell=1}^{S} \sum_{i=0}^{m'_f-1} f_{2\ell,i}(k)\, s_\ell(k-i) = \sum_{\ell=1}^{S} f_{2\ell,i}(k) * s_\ell(k) \tag{13.139}$$

where $*$ denotes the linear convolution.

When only a single source $s_\ell(k)$ is active we may convolve (13.138) by $f_{2\ell}(k)$ and (13.139) by $f_{1\ell}(k)$ and obtain

$$f_{2\ell}(k) * x_1(k) = f_{2\ell}(k) * f_{1\ell}(k) * s_\ell(k) \tag{13.140}$$

$$f_{1\ell}(k) * x_2(k) = f_{1\ell}(k) * f_{2\ell}(k) * s_\ell(k)\,. \tag{13.141}$$

For time-invariant far-end responses it follows that

$$f_{2\ell}(k) * x_1(k) = f_{1\ell}(k) * x_2(k) . \tag{13.142}$$

Hence, $x_1(k)$ and $x_2(k)$ are linearly related. However, as shown below, a large amount of correlation of the loudspeaker signals is detrimental to the fast convergence of a stereophonic echo canceller [Sondhi et al. 1995], [Benesty et al. 1998], [Gänsler, Benesty 2000].

13.11.1 The Non-uniqueness Problem

As a consequence of the linear relation of the two loudspeaker signals $x_1(k)$ and $x_2(k)$ the minimization of the power of the cancellation error signals does not lead to an unambiguous identification of the near-end acoustic system. For the first microphone in Fig. 13.30 we obtain the error signal

$$e_1(k) = y_1(k) - x_1(k) * h_{11}(k) - x_2(k) * h_{12}(k) . \tag{13.143}$$

However, since $f_{21}(k) * x_1(k) - f_{11}(k) * x_2(k) = 0$ we also have for arbitrary $b \in \mathbb{R}$

$$e_1(k) = y_1(k) - x_1(k) * (h_{11}(k) - bf_{21}(k)) - x_2(k) * (h_{12}(k) + bf_{11}(k)) .$$

Therefore, the minimization of $E\{e_1^2(k)\}$ cannot result in a unique solution for $h_{11}(k)$ and $h_{12}(k)$. Moreover, the cancellation error depends on the impulse responses $f_{1\ell}(k)$ and $f_{2\ell}(k)$ of the far-end side. Any change of the far-end source position or alternating far-end speakers will have an immediate effect on the error signal and thus on the convergence of the coefficient vectors. Even for fairly stationary conditions and wideband excitation signals the error signal might be small without proper identification of the near-end acoustic paths. A small error signal, however, does not help in the adaptation of the coefficient vectors.

13.11.2 Solutions to the Non-uniqueness Problem

The non-uniqueness can be resolved if the linear relation between the loudspeaker signals is weakened. In the simplest case this might be achieved by adding independent white noise to the loudspeaker signals on the near-end side. Improved solutions use spectrally shaped noise to hide the noise below the masked threshold of the audio signal [Gilloire, Turbin 1998].

Another possibility to reduce the correlation of the loudspeaker signals is to use time-varying allpass filters [Ali 1998] or a non-linear processor [Benesty et al. 1998] such as

$$\begin{aligned} \widetilde{x}_1(k) &= x_1(k) + \frac{\alpha}{2}\left(x_1(k) + \mid x_1(k) \mid\right) \\ \widetilde{x}_2(k) &= x_2(k) + \frac{\alpha}{2}\left(x_2(k) - \mid x_2(k) \mid\right) . \end{aligned} \tag{13.144}$$

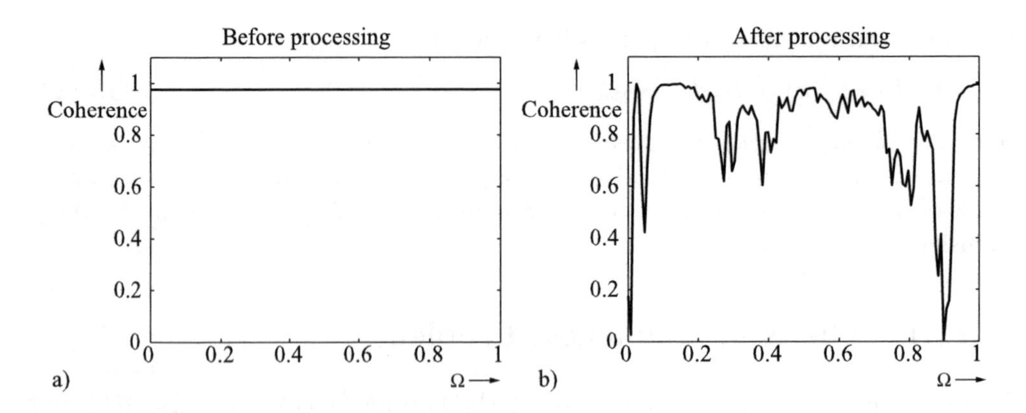

Figure 13.31: Magnitude squared coherence of two linearly related loudspeaker signals before and after the application of the nonlinear processor (13.144) with $\alpha = 0.5$

Adding a half-wave rectified version of the signal to itself will result in distortions of the signal. Because of the harmonic structure of voiced speech and simultaneous masking effects, these distortions are hardly noticeable for small values of $\alpha < 0.5$. This method has proven to be very effective without disturbing stereo perception [Gänsler, Benesty 2000].

To illustrate the effect of this non-linear processor on the correlation we depict the magnitude squared coherence function

$$C(\Omega) = \frac{|\Phi_{x1x2}(\Omega)|^2}{\Phi_{x1x1}(\Omega)\,\Phi_{x2x2}(\Omega)} \tag{13.145}$$

of the loudspeaker signals in Fig. 13.31 before and after applying the non-linear processor to a monophonic speech signal. For the processed signals we clearly observe a reduction of the coherence.

Algorithms for the adaptation of the stereophonic echo canceller may be designed by extending the single channel techniques to the multiple channel case [Benesty et al. 1995], [Shimauchi, Makino 1995]. When the loudspeaker signals are independent, any of the extended single channel echo cancellation algorithms can be successfully applied. As pointed out above, difficulties arise when the two loudspeaker signals are linearly related. Fast convergence is only achieved when the loudspeaker signals are not fully correlated and when the adaptive algorithms take the correlation of these signals into account. Therefore, multi-channel RLS-type algorithms converge much faster as they mutually decorrelate the input signals. Standard NLMS-type algorithms converge much slower in general since in most practical situations the signal covariance matrix is not well conditioned and the amount of additive independent noise or non-linear distortions that can be applied is limited. For the derivation of a multi-channel NLMS algorithm, the gradient computation can be modified to exploit the cross-channel correlation [Benesty et al. 1996]. The multi-channel RLS and FDAF algorithms are especially useful in this context [Benesty, Morgan 2000a]. Efficient implementations of such algorithms have been considered in [Buchner, Kellermann 2001].

Bibliography

Ali, M. (1998). Stereophonic Acoustic Echo Cancellation System Using Time-varying All-pass Filtering for Signal Decorrelation, *Proceedings of the IEEE International Conference on Acoustics, Speech, and Signal Processing (ICASSP)*, vol. 6, pp. 3689–3692.

Antweiler, C. (1995). *Orthogonalizing Algorithms for Digital Compensation of Acoustic Echoes*, PhD thesis. *Aachener Beiträge zu digitalen Nachrichtensystemen*, vol. 1, P. Vary (ed.), RWTH Aachen University (in German).

Antweiler, C.; Antweiler, M. (1995). System Identification with Perfect Sequences Based on the NLMS Algorithm, *Archiv für Elektronik und Übertragungstechnik*, pp. 129–134.

Antweiler, C.; Dörbecker, M. (1994). Perfect Sequence Excitation of the NLMS Algorithm and its Application to Acoustic Echo Control, *Annales des Télécommunication*, vol. 49, no. 7–8, July–August, pp. 386–397.

Armbrüster, W. (1988). High Quality Hands-Free Telephony Using Voice Switching Optimized with Echo Cancellation, *Proceedings of the European Signal Processing Conference (EUSIPCO)*, pp. 495–498.

Ayad, B.; Faucon, G.; Le Bouquin-Jeannès, R. (1996). Optimization of a Noise Reduction Preprocessing in an Acoustic Echo and Noise Controller, *Proceedings of the IEEE International Conference on Acoustics, Speech, and Signal Processing (ICASSP)*, pp. 953–956.

Benesty, J.; Amand, F.; Gilloire, A.; Grenier, Y. (1995). Adaptive Filtering Algorithms for Stereophonic Acoustic Echo Cancellation, *Proceedings of the IEEE International Conference on Acoustics, Speech, and Signal Processing (ICASSP)*, pp. 3099–3102.

Benesty, J.; Duhamel, P.; Grenier, Y. (1996). Multi-channel Adaptive filtering Applied to Multi-channel Acoustic Echo Cancellation, *Proceedings of the European Signal Processing Conference (EUSIPCO)*, pp. 1405–1408.

Benesty, J.; Gänsler, T.; Morgan, D. R.; Sondhi, M. M.; Gay, S. L. (2001). *Advances in Network and Acoustic Echo Cancellation*, Springer Verlag, Berlin, Heidelberg, New York.

Benesty, J.; Morgan, D. R. (2000a). Frequency-Domain Adaptive Filtering Revisited, Generalization to the Multi-Channel Case, and Application to Acoustic Echo Cancellation, *Proceedings of the IEEE International Conference on Acoustics, Speech, and Signal Processing (ICASSP)*, vol. 2, pp. 789–792.

Benesty, J.; Morgan, D. R. (2000b). Multi-channel Frequency-domain Adaptive Filtering, in S. L. Gay, J. Benesty (eds.), *Acoustic Signal Processing for Telecommunication,* Kluwer Academic, Dordrecht, pp. 121–133.

Benesty, J.; Morgan, D. R.; Sondhi, M. M. (1998). A Better Understanding and an Improved Solution to the Specific Problems of Stereophonic Acoustic Echo Cancellation, *IEEE Transactions on Speech and Audio Processing*, vol. 6, pp. 156–165.

Breining, C.; Dreiseitel, P.; Hänsler, E.; Mader, A.; Nitsch, B.; Puder, H.; Schertler, T.; Schmidt, G.; Tilp, J. (1999). Acoustic Echo Control, *IEEE Signal Processing Magazine*, vol. 16, no. 4, pp. 42–69.

Buchner, H.; Kellermann, W. (2001). Acoustic Echo Cancellation for Two and More Reproduction Channels, *Proceedings of the International Workshop on Acoustic Echo and Noise Control (IWAENC)*, pp. 99–102.

Carini, A. (2001). The Road of an Acoustic Echo Controller for Mobile Telephony from Product Definition till Production, *Proceedings of the International Workshop on Acoustic Echo and Noise Control (IWAENC)*, pp. 5–9.

Claasen, T. A. C. M.; Mecklenbräuker, W. F. G. (1981). Comparison of the Convergence of Two Algorithms for Adaptive FIR Digital Filters, *IEEE Transactions on Acoustics, Speech and Signal Processing*, vol. 29, no. 3, pp. 670–678.

Clark, G. A.; Mitra, S. K.; Parker, S. R. (1981). Block Implementation of Adaptive Digital Filters, *IEEE Transactions on Acoustics, Speech and Signal Processing*, vol. 29, no. 3, June, pp. 744–752.

Clark, G. A.; Parker, S. R.; Mitra, S. K. (1983). A Unified Approach to Time- and Frequency-Domain Realization of FIR Adaptive Digital Filters, *IEEE Transactions on Acoustics, Speech and Signal Processing*, vol. 31, no. 5, pp. 1073–1083.

Debnath, L.; Mikusinski, P. (1999). *Introduction to Hilbert Spaces with Applications*, 2nd edn, Academic Press, San Diego.

Enzner, G.; Martin, R.; Vary, P. (2002). Partitioned Residual Echo Power Estimation for Frequency-Domain Acoustic Echo Cancellation and Postfiltering, *European Transactions on Telecommunications*, vol. 13, no. 2, pp. 103–114.

Enzner, G.; Vary, P. (2003a). A Soft-Partitioned Frequency-Domain Adaptive Filter for Acoustic Echo Cancellation, *Proceedings of the IEEE International Conference on Acoustics, Speech, and Signal Processing (ICASSP)*, vol. V, pp. 393–395.

Enzner, G.; Vary, P. (2003b). Robust and Elegant, Purely Statistical Adaptation of Acoustic Echo Canceler and Postfilter, *Proceedings of the International Workshop on Acoustic Echo and Noise Control (IWAENC)*, pp. 43–46.

Enzner, G.; Vary, P. (2005). Frequency-Domain Adaptive Kalman Filter for Acoustic Echo Control in Handsfree Telephones, *Signal Processing.* Special Issue on Speech and Audio Processing (to appear).

Ephraim, Y.; Malah, D. (1984). Speech Enhancement Using a Minimum Mean-Square Error Short-Time Spectral Amplitude Estimator, *IEEE Transactions on Acoustics, Speech and Signal Processing*, vol. 32, no. 6, December, pp. 1109–1121.

Faucon, G.; Le Bouquin-Jeannès, R. (1995). Joint System for Acoustic Echo Cancellation and Noise Reduction, *Proceedings of the European Conference on Speech Communication and Technology (EUROSPEECH)*, pp. 1525–1528.

Ferrara, E. R. (1980). Fast Implementation of LMS Adaptive Filters, *IEEE Transactions on Acoustics, Speech and Signal Processing*, vol. 28, no. 4, pp. 474–475.

Ferrara, E. R. (1985). *Frequency-Domain Adaptive Filtering,* in Adaptive Filters, C. F. N Cowan, P. M. Grant (eds.), Englewood Cliffs, New Jersey, pp. 145–179.

Gänsler, T.; Benesty, J. (2000). Stereophonic Acoustic Echo Cancellation and Two-Channel Adaptive Filtering: An Overview, *International J. of Adaptive Control and Signal Processing*, vol. 14, pp. 565–586.

Gay, S. L.; Benesty, J. (eds.) (2000). *Acoustic Signal Processing for Telecommunication*, Kluwer Academic, Dordrecht.

Gay, S.; Tavathia, S. (1995). The Fast Affine Projection Algorithm, *Proceedings of the IEEE International Conference on Acoustics, Speech, and Signal Processing (ICASSP)*, vol. 5, pp. 3023–3026.

Gierlich, H. W. (1992). A Measurement Technique to Determine the Transfer Characteristics of Hands-Free Telephones, *Signal Processing*, vol. 27, pp. 281–300.

Gierlich, H. W.; Kettler, F. (2005). Advanced Speech Quality Testing of Modern Telecommunication Equipment: An Overview, *Signal Processing.* Special Issue on Speech and Audio Processing (to appear).

Gilloire, A. (1994). Performance Evaluation of Acoustic Echo Control: Required Values and Measurement Procedures, *Annales des Télécommunications*, vol. 49, no. 7–8, pp. 368–372.

Gilloire, A.; Turbin, V. (1998). Using Auditory Properties to Improve the Behavior of Stereophonic Acoustic Echo Cancellers, *Proceedings of the IEEE International Conference on Acoustics, Speech, and Signal Processing (ICASSP)*, pp. 3681–3684.

Gilloire, A.; Vetterli, M. (1992). Adaptive Filtering in Subbands with Critical Sampling: Analysis, Experiments, and Application to Acoustic Echo Cancellation, *IEEE Transactions on Signal Processing*, vol. 40, no. 8, pp. 1862–1875.

Gustafsson, S.; Martin, R.; Jax, P.; Vary, P. (2002). A Psychoacoustic Approach to Combined Acoustic Echo Cancellation and Noise Reduction, *IEEE Transactions on Speech and Audio Processing*, vol. 10, no. 5, pp. 245–256.

Gustafsson, S.; Martin, R.; Vary, P. (1998). Combined Acoustic Echo Control and Noise Reduction for Hands-free Telephony, *Signal Processing*, vol. 64, pp. 21–32.

Hänsler, E. (1992). The Hands-Free Telephone Problem – An Annotated Bibliography, *Signal Processing*, vol. 27, pp. 259–271.

Hänsler, E. (1994). The Hands-Free Telephone Problem - An Annotated Bibliography Update, *Annales des Télécommunication*, vol. 49, no. 7–8, pp. 360–367.

Hänsler, E.; Schmidt, G. (2000). Hands-Free Telephones – Joint Control of Echo Cancellation and Postfiltering, *Signal Processing*, vol. 80, no. 11, November, pp. 2295–2305.

Hänsler, E.; Schmidt, G. (2004). *Acoustic Echo and Noise Control – A Practical Approach*, John Wiley & Sons, Ltd., Chichester.

Haykin, S. (1996). *Adaptive Filter Theory*, 3rd edn, Prentice Hall, Upper Saddle River, New Jersey.

Ipatov, V. P. (1979). Ternary Sequences with Ideal Perfect Autocorrelation Properties, *Radio Engineering Electronics and Physics*, vol. 24, pp. 75–79.

ITU-T G.131 (2003). *Talker Echo and its Control*, International Telecommunication Union (ITU).

ITU-T P.340 (2000). *Transmission Characteristics and Speech Quality Parameters of Hands-free Terminals*, International Telecommunication Union (ITU).

Kellermann, W. (1985). Kompensation akustischer Echos in Frequenzteilbändern, *Frequenz*, vol. 39, no. 7/8, pp. 209–215, (in German).

Kellermann, W. (1988). Analysis and Design of Multirate Systems for Cancellation of Acoustical Echoes, *Proceedings of the IEEE International Conference on Acoustics, Speech, and Signal Processing (ICASSP)*, pp. 2570–2573.

Kellermann, W. (1989). *Zur Nachbildung physikalischer Systeme durch parallelisierte digitale Ersatzsysteme im Hinblick auf die Kompensation akustischer Echos*, PhD thesis, *Fortschrittsberichte VDI*, Series 10: Informatik/Kommunikationstechnik, Nr. 102, TU Darmstadt, (in German).

Kuttruff, H. (1990). *Room Acoustics*, 3rd edn, Applied Science Publishers, Barking.

Le Bouquin-Jeannès, R.; Scalart, P.; Faucon, G.; Beaugeant, C. (2001). Combined Noise and Echo Reduction in Hands-Free Systems: A Survey, *IEEE Transactions on Speech and Audio Processing*, vol. 9, no. 8, pp. 808–820.

Lüke, H. D. (1988). Sequences and Arrays with Perfect Periodic Correlation, *IEEE Transactions on Aerospace and Electronic Systems*, vol. 24, no. 3, pp. 287–294.

Lüke, H. D.; Schotten, H. (1995). Odd-perfect, Almost Binary Correlation Sequences, *IEEE Transactions on Aerospace and Electronic Systems*, vol. 31, pp. 495–498.

Mader, A.; Puder, H.; Schmidt, G. (2000). Step-size Control for Acoustic Echo Cancellation Filters – An Overview, *Signal Processing*, vol. 80, no. 9, September, pp. 1697–1719.

Makino, S.; Kaneda, Y. (1992). Exponentially Weighted Step-Size Projection Algorithm for Acoustic Echo Cancellers, *IEICE Transactions on Fundamentals of Electronics, Communications and Computer Sciences*, vol. E75-A, no. 11, pp. 1500–1508.

Mansour, D.; Gray, A. H. (1982). Unconstrained Frequency-Domain Adaptive Filters, *IEEE Transactions on Acoustics, Speech and Signal Processing*, vol. 30, October, pp. 726–734.

Martin, R. (1995). *Hands-free Systems with Multi-channel Echo Cancellation and Noise Reduction*, PhD thesis. *Aachener Beiträge zu digitalen Nachrichtensystemen*, vol. 3, P. Vary (ed.), RWTH Aachen University (in German).

Martin, R.; Altenhöner, J. (1995). Coupled Adaptive Filters for Acoustic Echo Control and Noise Reduction, *Proceedings of the IEEE International Conference on Acoustics, Speech, and Signal Processing (ICASSP)*, pp. 3043–3046.

Martin, R.; Gustafsson, S. (1996). The Echo Shaping Approach to Acoustic Echo Control, *Speech Communication*, vol. 20, pp. 181–190.

Martin, R.; Vary, P. (1996). Combined Acoustic Echo Control and Noise Reduction for Hands-Free Telephony - State of the Art and Perspectives, *Proceedings of the European Signal Processing Conference (EUSIPCO)*, pp. 1107–1110.

Minami, S. (1987). An Acoustic Echo Canceller for Pseudo Stereophonic Voice, *Globecom*, pp. 1355–1360.

Myllylä, V. (2001). Robust Fast Affine Projection Algorithm for Acoustic Echo Cancellation, *Proceedings of the International Workshop on Acoustic Echo and Noise Control (IWAENC)*.

Myllylä, V.; Schmidt, G. (2002). Pseudo-optimal Regularization for Affine Projection Algorithms, *Proceedings of the IEEE International Conference on Acoustics, Speech, and Signal Processing (ICASSP)*, vol. II, pp. 1917–1920.

Naylor, P. A.; Tanrikulu, O.; Constantinides, A. G. (1998). Subband Adaptive Filtering for Acoustic Echo Control Using Allpass Polyphase IIR Filterbanks, *IEEE Transactions on Speech and Audio Processing*, vol. 6, no. 2, pp. 143–155.

Nitsch, B. (1997). The Partitioned Exact Frequency Domain Block NLMS Algorithm, *Proceedings of the International Workshop on Acoustic Echo and Noise Control (IWAENC)*, pp. 45–48.

Nitsch, B. H. (2000). A Frequency-selective Stepfactor Control for an Adaptive Filter Algorithm Working in the Frequency Domain, *Signal Processing*, vol. 80, pp. 1733–1745.

Ozeki, K.; Umeda, T. (1984). An Adaptive Filtering Algorithm Using an Orthogonal Projection to an Affine Subspace and its Properties, *Electronics and Communications in Japan*, vol. 67-A, no. 5, pp. 19–27.

Petillon, T.; Gilloire, A.; Theodoridis, S. (1994). The Fast Newton Transversal Filter: An Efficient Scheme for Acoustic Echo Cancellation in Mobile Radio, *IEEE Transactions on Signal Processing*, vol. 42, no. 3, pp. 509–518.

Rupp, M. (1993). The Behavior of LMS and NLMS Algorithms in the Presence of Spherically Invariant Processes, *IEEE Transactions on Signal Processing*, vol. 41, no. 3, pp. 1149–1160.

Schultheiß, U. (1988). *Über die Adaption eines Kompensators für akustische Echos*, PhD thesis, *Fortschrittsberichte VDI*, Series 10: Informatik/Kommunikationstechnik, Nr. 90, TU Darmstadt, (in German).

Shimauchi, S.; Makino, S. (1995). Stereo Projection Echo Canceller with True Echo Path Estimation, *Proceedings of the IEEE International Conference on Acoustics, Speech, and Signal Processing (ICASSP)*, pp. 3059–3062.

Shynk, J. (1992). Frequency-Domain and Multirate Adaptive Filtering, *IEEE Signal Processing Magazine*, vol. 9, no. 1, pp. 14–37.

Slock, D. T. M. (1993). On the Convergence Behavior of the LMS and the Normalized LMS Algorithms, *IEEE Transactions on Signal Processing*, vol. 41, no. 1, pp. 2811–2825.

Slock, D. T. M.; Kailath, T. (1991). Numerically Stable Fast Transversal Filters for Recursive Least Squares Adaptive Filtering, *IEEE Transactions on Signal Processing*, vol. 39, no. 1, pp. 92–113.

Sommen, P. C. W. (1989). Partitioned Frequency Domain Adaptive Filters, *Proceedings of the 23rd Asilomar Conference on Signals, Systems, and Computers*, pp. 677–681.

Sommen, P. C. W.; van Valburg, C. J. (1989). Efficient Realisation of Adaptive Filter Using an Orthogonal Projection Method, *Proceedings of the IEEE International Conference on Acoustics, Speech, and Signal Processing (ICASSP)*, pp. 940–943.

Sondhi, M. M.; Morgan, D. R.; Hall, J. L. (1995). Stereophonic Acoustic Echo Cancellation – An Overview of the Fundamental Problem, *Signal Processing Letters*, vol. 2, pp. 148–151.

Soo, J. S.; Pang, K. K. (1990). Multidelay Block Frequency Domain Adaptive Filter, *IEEE Transactions on Acoustics, Speech and Signal Processing*, vol. 38, no. 2, pp. 373–376.

Tanaka, M.; Kaneda, Y.; Makino, S.; Kojima, J. (1995). Fast Projection Algorithm and Its Step Size Control, *Proceedings of the IEEE International Conference on Acoustics, Speech, and Signal Processing (ICASSP)*, pp. 945–948.

van de Kerkhof, L. M.; Kitzen, W. J. W. (1992). Tracking of a time-varying acoustic impulse response by an adaptive filter, *IEEE Transactions on Signal Processing*, vol. 40, no. 7, pp. 1285–1294.

Yamamoto, S.; Kitayama, S. (1982). An Adaptive Echo Canceller with Variable Step Gain Method, *Transactions of the IECE of Japan*, vol. E65, pp. 1–8.

Appendix A

Codec Standards

For various application areas a great number of codecs exist, which differ with respect to speech quality, bit rate B, complexity, and signal delay.

In this appendix, some of the most common codecs are presented with their distinctive features. Table A.1 provides an overview.

Table A.1: Overview of the most common codec standards
ITU : International Telecommunication Union
ETSI: European Telecommunications Standards Institute
TIA : Telecommunications Industry Association

Standard	Name	$B/(\text{kbit/s})$
ITU-T/G.726	Adaptive Differential Pulse Code Modulation (ADPCM)	32 (16, 24, 40)
ITU-T/G.728	Low-Delay CELP Speech Coder (LD-CELP)	16
ITU-T/G.729	Conjugate-Structure Algebraic CELP Codec (CS-ACELP)	8
ITU-T/G.722	7 kHz Audio Coding within 64 kbit/s	64 (48, 56)
ETSI-GSM 06.10	Full Rate Speech Transcoding	$12.2 + 0.8$
ETSI-GSM 06.20	Half Rate Speech Transcoding	5.6
ETSI-GSM 06.60	Enhanced Full Rate Speech Transcoding	13
ETSI-GSM 06.90	Adaptive Multi-Rate Speech Transcoding	4.75–12.2
ITU-T/G.722.2	Adaptive Multi-Rate Wideband Speech Transcoding	6.6–23.85
ETSI/3GPP 26.290	Extended Adaptive Multi-Rate Wideband Codec (AMR-WB$^+$)	12.8–38.4
TIA IS-96	Speech Service Option Standard for Wideband Spread-Spectrum Systems	1–8
INMARSAT/IMBE	Improved Multi-Band Excitation Codec (IMBE)	4.15

Digital Speech Transmission: Enhancement, Coding and Error Concealment
Peter Vary and Rainer Martin

A.1 Evaluation Criteria

This section will deal with the four fundamental codec evaluation criteria of *speech quality*, *bit rate*, *complexity*, and *signal delay*.

a) Speech Quality

Speech codecs for telephone systems are specified by international standardization groups. Usually a decision has to be taken between different competing codec candidates. In the evaluation and selection process, the speech quality at the given bit rates is the most important criterion. The main problem in speech quality assessment of different codecs is that so far no universally valid, "objective" instrumental measures exist (see also Appendix B.2). As already noted, the signal-to-noise ratio is in general not suitable for a comparative evaluation of hybrid coding algorithms.

Therefore, all speech codec standards have been selected on the basis of very expensive large-scale formal listening tests (see Appendix B.1).

However, this does not solve the problem of quality and conformity tests within the type approval of devices produced by the manufacturers. For this reason, the codec algorithms are generally specified bit-precisely and test sequences, i.e., speech sample sequences and their corresponding coded bit sequences, are established as part of each codec standard for instrumental verification. This allows separate testing of the encoder and the decoder.

b) Bit Rate

The allowed bit rate of a speech codec is determined by the application. The specification of the target bit rate may depend on several factors. Besides the interrelation between bit rate on the one hand and speech quality, signal delay, and computational complexity on the other hand, the allowed bit rate depends on other system constraints. In a cellular phone system, the bit rate needed for speech coding has a strong impact on the economy of frequency and on the design of the error protection scheme, i.e., ultimately the error robustness.

The bit rate may be constant (e.g., GSM full-rate codec, see Appendix A.6), it may be adjusted dynamically in discrete steps by *network control* (e.g., GSM adaptive multi-rate codec, see Appendix A.9), or it may be *source controlled* (e.g., IS-95/QCELP, see Appendix A.12).

c) Complexity

The complexity is generally given in terms of the required computational power and memory capacity. The computing power is often specified by the number of arithmetic operations per time unit (MOPS: *Mega Operations Per Second*) which are necessary for real-time operation. For codec realization with programmable signal processors this number can only be a coarse measure of the actual resources needed due to the architectural differences of specific programmable digital signal processors.

Since speech codecs are mainly implemented on programmable signal processors, the required computational complexity depends within certain limits on how well the architecture and instruction set of the processor satisfy the demands of the bit-accurate specification of the algorithm.

For this reason the European Telecommunications Standards Institute (ETSI) has specified the instruction set of a hypothetical 16 bit fixed point arithmetic DSP, which has been used for prototype implementation and for complexity evaluation of coding algorithms. The instructions are available as a set of macros in the programming language C (see also ITU Recommendation G.191, Chapter 12, ITU-T Basic Operators). Thus, the computational complexity of competing codec proposals can be compared easily if these macros are used for the codec simulation. Overheads for data moves, address calculations, initialization of program loops, subroutine groups, etc., are not counted. Therefore, the final implementation on any specific DSP usually has a somewhat higher computational load. However, this approach allows quite a realistic and fair comparison in terms of the *weighted mega operations per second* (wMOPS) as different instructions of the hypothetical DSP have different complexity weights between 1 and 30. For example, the addition of two fixed point variables has a complexity weight of 1, whereas the 16 bit division of two variables has a complexity weight of 18. All instructions are specified bit-precisely. In the case of a division both variables must be positive, and the second variable must be larger than or equal to the first.

d) Signal Delay

The last criterion to be discussed is the signal delay, which is primarily caused by block processing in the transmitter. A delay by two to three blocks of length 20 ms each is permissible for fixed telephone networks, whereas in radio-based systems, the allowed total delay should be utilized preferably for error protection as far as possible, especially to break up burst errors by time interleaving.

A.2 ITU-T/G.726: Adaptive Differential Pulse Code Modulation (ADPCM)

Block diagram: coding and decoding

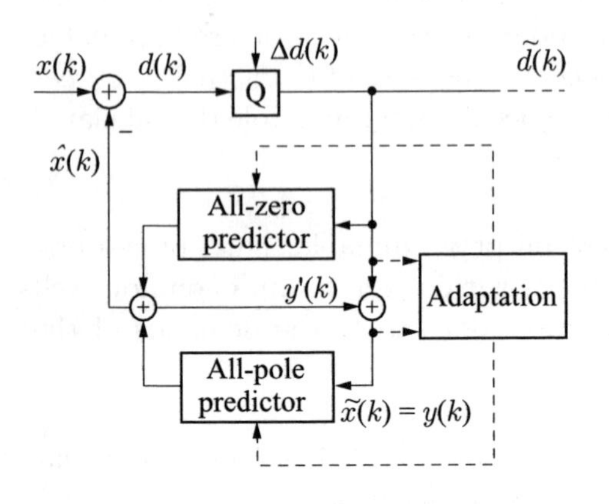

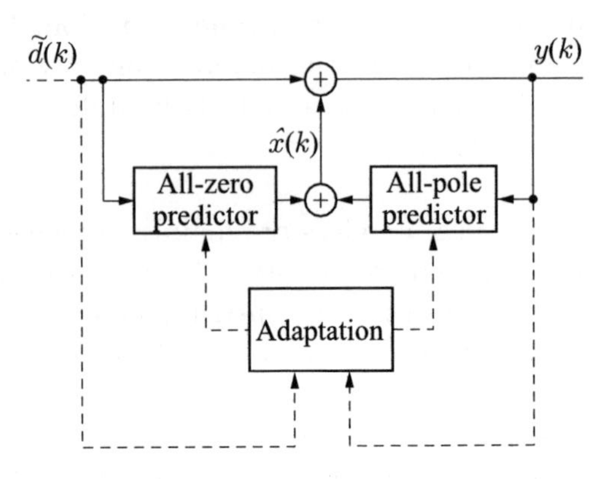

Sampling: $f_s = 8\,\text{kHz}$

Bit rates: $B = 16, 24, 32, 40$ kbit/s

Quality:
- signal form coding, nearly the same quality as PCM with 64 kbit/s
- modem signals (fax) up to 4.8 kbit/s (with $B = 40$ kbit/s)

Applications:
- cordless digital phones according to the DECT (*Digital Enhanced Cordless Telecommunications*) standard
- "digital circuit multiplier" (transatlantic cables)

Algorithm:
- backward-adaptive predictor (see Section 8.3.4) with poles and zeros
- adaptation of the predictor by means of the sign-LMS-algorithm (see Section 8.3.4)
- adaptive quantization of the prediction error signal according to the AQB method with $w = 2, 3, 4,$ or 5 (see Section 7.5)
- quantization with fixed step size Δd for coding of modem signals

Special feature: sign-LMS-algorithm without multiplication

$$a_i(k+1) = (1-2^{-8})\ a_i(k) + 2^{-7}\operatorname{sign}\{\tilde{d}(k)\}\ \cdot\ \operatorname{sign}\{\tilde{d}(k-i)\}$$

$$b_1(k+1) = \frac{255}{256}\ b_1(k) + \frac{3}{256}\operatorname{sign}\{y'(k)\}\ \cdot\ \operatorname{sign}\{y'(k-1)\}$$

$$y'(k) = \tilde{d}(k) + \sum_{i=1}^{6} a_i(k)\ \cdot\ \tilde{d}(k-i)$$

$$b_2(k+1) = \frac{127}{128}\ b_2(k) + \frac{1}{128}\operatorname{sign}\{y'(k)\}\ \cdot\ \operatorname{sign}\{y'(k-2)\}$$

$$-\ \frac{1}{128}\ f[b_1(k)]\operatorname{sign}\{y'(k)\}\ \cdot\ \operatorname{sign}\{y'(k-1)\}$$

$$f[b_1(k)] = \begin{cases} 4\cdot b_1(k) & |\,b_1(k)\,| \ \le\ 0.5 \\ 2\operatorname{sign}\{b_1(k)\} & |\,b_1(k)\,| \ >\ 0.5 \end{cases}$$

References:

Bonnet, M.; Macchi, O.; Jaidane-Saidane, M. (1990). Theoretical Analysis of the ADPCM CCITT Algorithm, *IEEE Transactions on Communications*, 38(6), pp. 847–858.

ITU-T Rec. G.726 (1992). *40, 32, 24, 16 kbit/s Adaptive Differential Pulse Code Modulation (ADPCM).*

Jayant, N. S.; Noll, P. (1984). *Digital Coding of Waveforms*, Section 6.5.3, Prentice Hall, Englewood Cliffs, New Jersey.

A.3 ITU-T/G.728: Low-Delay CELP Speech Coder

Block diagram: coding

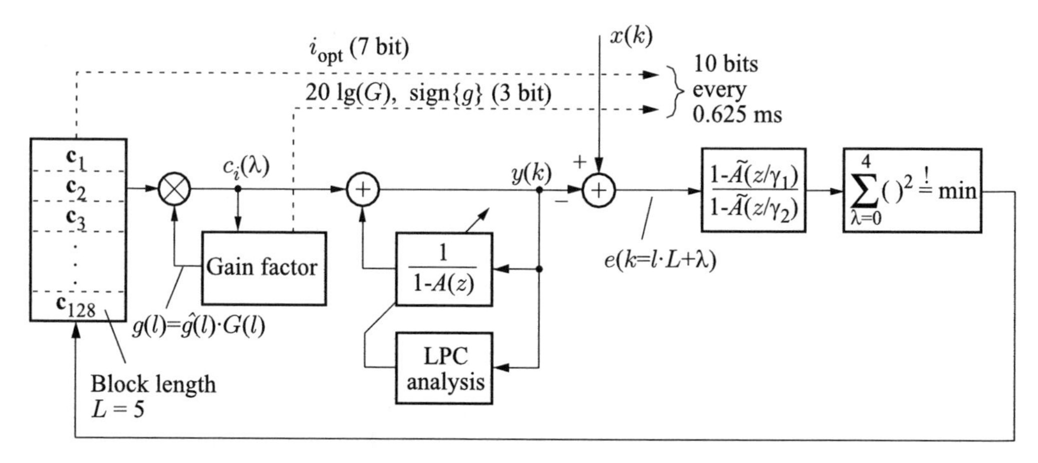

Sampling: $f_s = 8\,\text{kHz}$

Bit rate: $B = 16$ kbit/s

Quality:
- for speech comparable to (or partly better than) 32 kbit/s ADPCM (ITU-T/G.726)
- modem signals (fax) up to 4.8 kbit/s

Applications:
- video telephony, voice over IP
- "digital circuit multiplication"

Algorithm:
- backwards adaptation of an LPC predictor of order $n = 50$ (blockwise, Levinson–Durbin algorithm, see Section 6.3.1)
- no long-term predictor
- logarithmic–differential quantization of the gain factor g with backwards-adaptive prediction of order 10
- block length $L = 5$ samples or $\tau = 0.625\,\text{ms}$
- code book with $K = 128$ entries $\mathbf{c}_i$
- modified weighting filter of order 10
- adaptive postfilter (see Section 8.6)
- signal delay $< 2\,\text{ms}$
- complexity approx. 20 wMOPS (weighted Mega Operations Per Second)

Bit allocation: 10 bits every $5 \cdot 0.125\,\text{ms} = 0.625\,\text{ms}$

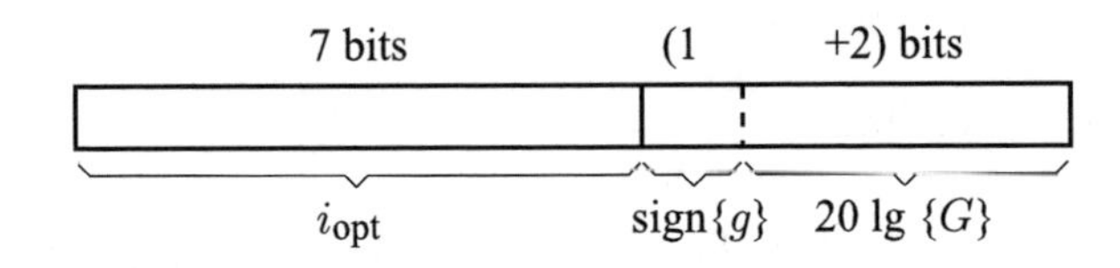

Special features: – logarithmic–differential quantization of the gain factor:

$$|\,g(l)\,| = |\,\hat{g}(l)\,| \cdot G(l) \quad ; \quad l = \text{block index}$$

$$v(l) = 20 \text{ lg } |\,g(l)\,| = 20 \text{ lg } |\,\hat{g}(l)\,| + 20 \text{ lg } G(l)$$

$$= \hat{v}(l) + \Delta v(l)$$

- quantization of $\Delta v(l) = v(l) - \hat{v}(l)$ with only four levels (2 bits) according to the analysis-by-synthesis criterion

- adaptive prediction of $\hat{v}(l)$

$$\hat{v}(l) = \sum_{j=1}^{10} b_j \cdot \overline{v}(l-j)$$

$$\text{with} \quad \overline{v}(l-j) = 20 \lg \sqrt{\sum_{\lambda=0}^{4} c_{i\text{opt}}^2(\lambda) \cdot g^2(l-j)}$$

$$\mathbf{c}_{i\text{opt}} = \text{optimal vector for block index } l-j$$

– calculation of weighting filter

$$W(z) = \frac{1 - \tilde{A}(z/\gamma_1)}{1 - \tilde{A}(z/\gamma_2)} \quad ; \quad \gamma_1 = 0.9\,, \quad \gamma_2 = 0.6$$

of order 10 from the unquantized input signal

– LPC analysis for synthesis filter ($A(z), n = 50$), weighting filter ($\tilde{A}(z)$, $\tilde{n} = 10$), and predictor for gain factor ($B(z), n' = 10$) with "hybrid" window $w_m(k)$ with recursive exponential and non-recursive sinusoidal part

$$w_m(k) = \begin{cases} b \cdot \alpha^{-\left(k-(m-N-1)\right)} & k \leq m-N-1 \\ -\sin\left(c \cdot (k-m)\right) & m-N \leq k \leq m-1 \\ 0 & k \geq m \end{cases}$$

"hybrid window" $w_m(k)$ for $b = 0.9889$, $\alpha = 0.9928$, and $c = 0.0478$:

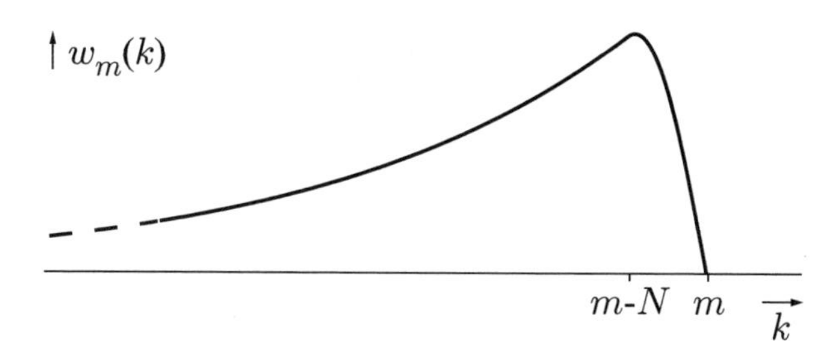

– adaptive postfilter with LTP and LPC part (see Section 8.6)

References:

Chen, J. H. (1995). Low-delay Coding of Speech, *Speech Coding and Synthesis*, W. B. Kleijn, K. K. Paliwal (eds.), Chapter 6, Elsevier, Amsterdam.

Chen, J. H.; Cox, R. V.; Lin, Y.-C.; Jayant, N. S.; Melchner, M. J. (1992). A Low-delay CELP-coder for the CCITT 16 kbit/s Speech Coding Standard, *IEEE Journal of Selected Areas in Communications*, pp. 830–849.

ITU-T Rec. G.728 (1992). *Coding of Speech at 16 kbit/s Using Low-delay Code Excited Linear Prediction.*

Murphy, M. T.; Cox, C. E. M. (1994). A Real Time Implementation of the ITU-T/G.728 LD-CELP Fixed Point Algorithm on the Motorola DSP56156, *Signal Processing VII: Theories and Applications*, M. J. J. Holt, C. F. N. Cowan, P. M. Grant, W. A. Sandham (eds.), Elsevier, Amsterdam, pp. 1617–1620.

A.4 ITU-T/G.729: Conjugate-Structure Algebraic CELP Codec (CS-ACELP)

Block diagram: decoding

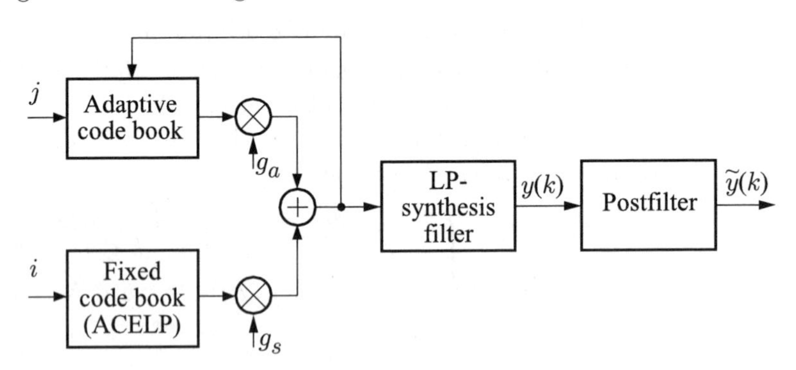

Sampling:	$f_s = 8\,\text{kHz}$
Bit rate:	$B = 8$ kbit/s
Quality:	– for speech comparable to (or for some instances slightly inferior to) 32 kbit/s ADPCM – modem signals (fax) up to 4.8 kbit/s – not suitable for music
Applications:	– video telephony, multimedia
Algorithm:	– *Code Excited Linear Prediction* (CELP) – LP synthesis filter of order $n = 10$ – frame length $T_N = 10\,\text{ms}$, subframe length $\frac{1}{2}T_N = 5\,\text{ms}$ – vector quantization of the filter parameter in form of LSF coefficients with 18 bits (see also Section 8.4.3-f) – long-term prediction (adaptive code book, Section 8.5.3) with so-called *open-loop* pitch analysis and *closed-loop* pitch search

of the delay N_0 (or j_{opt}) applying high-resolution pitch detection with interpolation by a factor of 3

- coding of the delay parameter N_0 of the first subframe with 8 bits, differential coding of the corresponding parameter of the second subframe with 5 bits
- fixed *algebraic code book* (ACELP) with effectively 2^{17} vectors of length 40; each with only four non-zero values (see below)
- adaptive weighting filter for the error signal

$$W(z) = \frac{1 - A(z/\gamma_1)}{1 - A(z/\gamma_2)} ,$$

  with γ_1 and γ_2 depending on the instantaneous spectral envelope
- two-step vector quantization of the gain factors g_a, g_s with 7 bits
- adaptive postfilter (see Section 8.6)
- algorithmic signal delay 15 ms
- complexity approx. 18 wMOPS (weighted Mega Operations Per Second)

Bit allocation:

Parameter	Quantization	Subframe 1 (5 ms)	Subframe 2 (5 ms)	Bits per 10 ms
LP filter 10 coefficients (*Line Spectral Frequencies*)	4 vector quantizers			18
Adaptive code book	scalar			
• N_0 (or j_1, j_2)	j_1	8		
	$j_1 - j_2$		5	13
• parity bit for j_1		1		1
Fixed code book	pulse positions			
• ACELP indices	$3 \times 3 + 4$	13	13	26
• sign	4×1	4	4	8
Gain factors g_a, g_s	2 vector quantizers $3 + 4$	7	7	14
Bits per 10 ms				80

Special feature: – advantageous algebraic ternary code book (see also Section 8.5.3.2) with

$$c_i(\lambda) \in \{+1, 0, -1\} \quad ; \quad \lambda = 0, 1, \ldots, 39$$

and only four non-zero elements per code book vector $\mathbf{c}_i$

$$\begin{aligned} c_i(\lambda) &= s_0\,\delta(\lambda - \mu_0) + s_1\,\delta(\lambda - \mu_1) \\ &\quad + s_2\,\delta(\lambda - \mu_2) + s_3\,\delta(\lambda - \mu_3) \end{aligned}$$

– separate coding of the pulse positions $\mu_0, \mu_1, \mu_2, \mu_3$ and the signs $s_0, s_1, s_2, s_3 \in \{+1, -1\}$

Pulse position	Sign	Positions	Bits
μ_0	s_0	0, 5, 10, 15, 20, 25, 30, 35	3 + 1
μ_1	s_1	1, 6, 11, 16, 21, 26, 31, 36	3 + 1
μ_2	s_2	2, 7, 12, 17, 22, 27, 32, 37	3 + 1
μ_3	s_3	3, 8, 13, 18, 23, 28, 33, 38 4, 9, 14, 19, 24, 29, 34, 39	4 + 1

– efficient code book search exploiting the code vectors' structure
– compatible variant G.729A with lower complexity (approx. 9 wMOPS, slightly reduced speech quality in case of codec tandem and background noise)

References:

ITU-T Rec. G.729 (1995). *Coding of Speech at 8 kbit/s Using Conjugate-structure Algebraic Code-excited Linear Prediction (CS-ACELP).*

Salami, R.; Laflamme, C.; Kataoka, A.; Lamblin, C.; Kroon, P. (1995). Description of the Proposed ITU-T 8 kbit/s Speech Coding Standard, *IEEE Workshop on Speech Coding for Telecommunications*, Annapolis, USA pp. 3–4. (Additional contributions regarding details of the standard appear in the same conference proceedings).

Salami, R.; Laflamme, C.; Bessette, B.; Adoul, J.-P. (1997). Description of ITU-T Recommendation G.729 Annex A: Reduced Complexity 8 kbit/s CS-ACELP Codec, *Proceedings of the IEEE International Conference on Acoustics, Speech, and Signal Processing (ICASSP)*, Munich, Germany, pp. 775–778.

A.5 ITU-T/G.722: 7 kHz Audio Coding within 64 kbit/s

Block diagram: coding and decoding

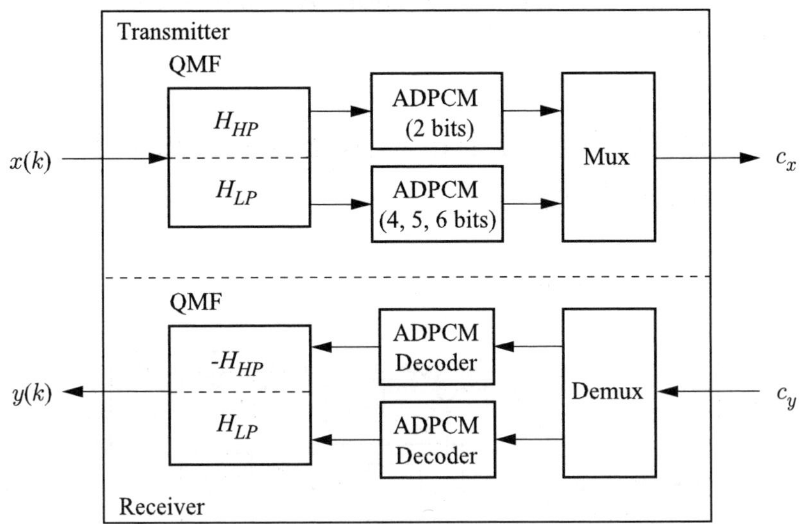

Sampling: $f_s = 16$ kHz

Bit rates: $B = 48, 56, 64$ kbit/s

Quality:
- high quality for speech and music
- increased bandwidth $0.05\,\text{kHz} \leq f \leq 7.0\,\text{kHz}$

Applications:
- video and audio conferences
- ISDN premium phones

Algorithm:
- subband coding with two subbands (see Section 4.3)
- band division with QMF–lowpass–highpass (see Section 4.3, $n = 24$)
- ADPCM (similar to G.726) in both subbands with $w = 4, 5$, or 6 (or 32, 40, 48 kbit/s) in the lower subband and $w = 2$ (or 16 kbit/s) in the upper subband
- algorithmic signal delay 1.5 ms
- complexity approx. 10 wMOPS (weighted Mega Operations Per Second)

Special features:
- possibility of simultaneous speech and data transmission with bit rate $B_D = 8$ or 16 kbit/s, with $B + B_D = 64$ kbit/s in total
- dimensioning of the quantizers for $w = 4, 5$ in such a way that for the simultaneous data transmission the last bit or the last two bits of the 6 bit quantizer are overwritten (*embedded coding*).

References:

ITU-T Rec. G.722 (1988). *7 kHz Audio Coding within 64 kbit/s*, vol. Fascicle III.4, Blue Book, pp. 269–341.

Taka, M.; Maitre, X. (1986). CCITT Standardization Activities on Speech Coding, *Proceedings of the IEEE International Conference on Acoustics, Speech, and Signal Processing (ICASSP)*, Tokyo, pp. 817–820.

A.6 ETSI-GSM 06.10: Full Rate Speech Transcoding

Block diagram: coding and decoding

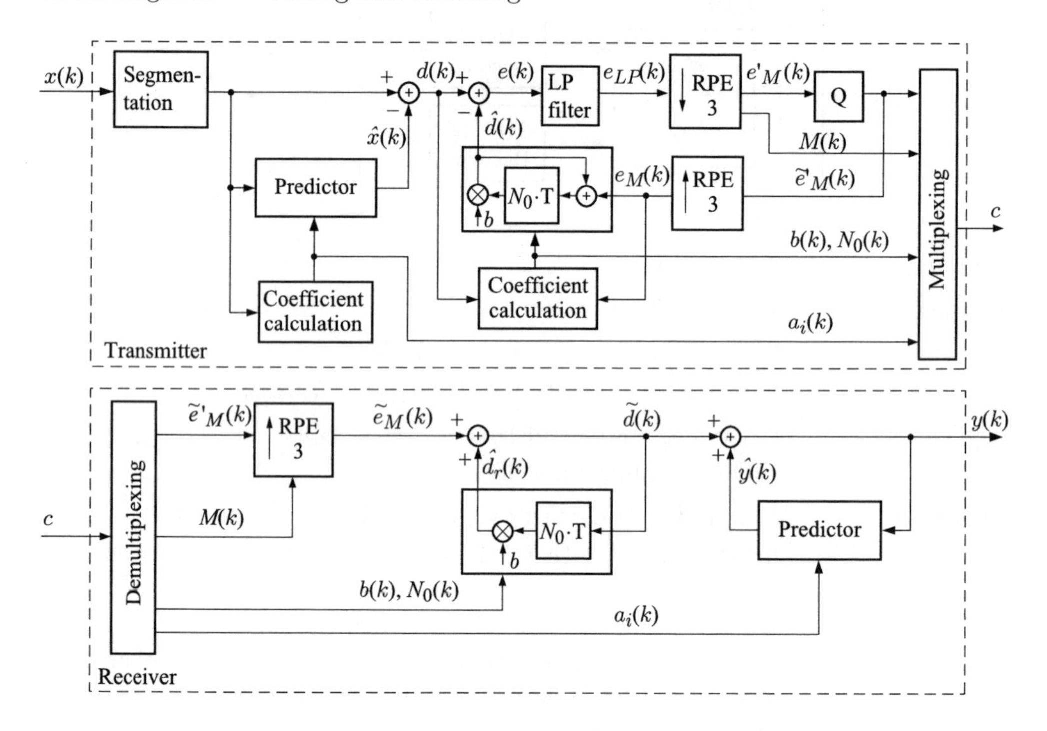

Sampling: $f_s = 8\,\text{kHz}$

Bit rate: $B = 13.0$ kbit/s

Quality:
- limited telephone quality (speech)
- not suitable for modem and music signals

Applications:
- mobile radio systems according to the GSM standard
- Internet telephony

Algorithm:
- predictive residual signal coding (see Section 8.5.2)

– algorithmic signal delay 20 ms
– complexity about 3 wMOPS (weighted Mega Operations Per Second)

Bit allocation (see also Section 8.5.2):

Parameter		Number of bits
8 reflection coefficients (LPC) $(2 \times 6, 2 \times 5, 2 \times 4, 2 \times 3$ bits)		36
4 LTP gain factors b	$(4 \times 2$ bits)	8
4 LTP delay values N_0	$(4 \times 7$ bits)	28
4 RPE grid positions	$(4 \times 2$ bits)	8
4 RPE block maxima	$(4 \times 6$ bits)	24
4×13 normalized RPE values	$(52 \times 3$ bits)	156
Bits every 20 ms		260

Special features: – interpolation (extrapolation) of corrupted frames by frame repetition

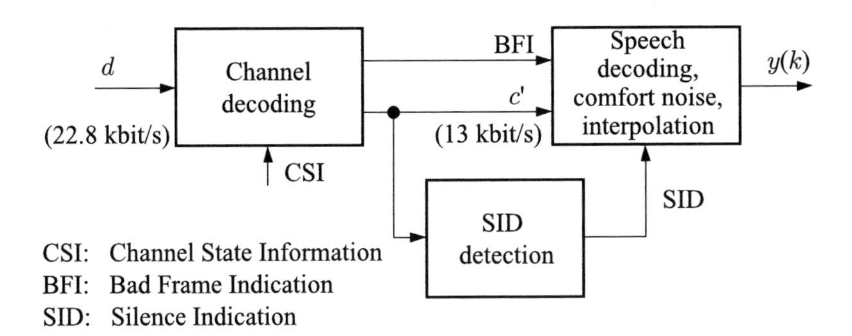

– marking of "bad" frames (260 bits) with residual errors in the group of the most important bits with BFI = 1 and repetition of the last "good" frame; muting after multiple repetition
– detection of speech pauses (SID: *Silence Indication*)
– when switching off the transmitter during the speech pauses (option of the network provider), an artificial background noise (*Comfort Noise*) is produced at the receiver by exciting the LPC-synthesis filter with white noise; the level is adapted according to the current background noise.

References:

ETSI Rec. GSM 06.10 (1988). *GSM Full Rate Speech Transcoding.*

Sluijter, R. J. (2005). *The Development of Speech Coding and the First Standard Coder for Public Mobile Telephony*, PhD thesis, Technical University Eindhoven.

Vary, P.; Hellwig, K.; Hofmann, R.; Sluijter, R. J.; Galand, C.; Rosso, M. (1988). Speech Codec for the European Mobile Radio System, *Proceedings of the IEEE International Conference on Acoustics, Speech, and Signal Processing (ICASSP)*, New York, USA, Contribution S6.1, April, pp. 227–230.

A.7 ETSI-GSM 06.20: Half Rate Speech Transcoding

Block diagram: decoding

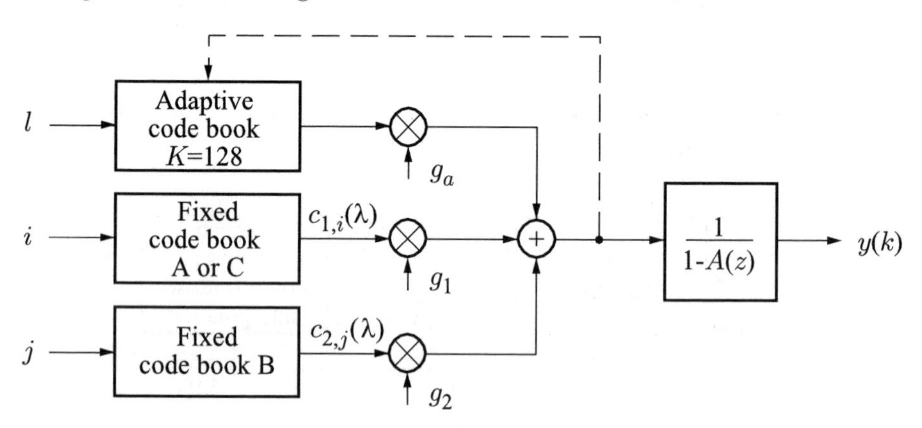

Sampling: $f_s = 8\,\text{kHz}$

Bit rate: $B = 5.6$ kbit/s

Quality:
- slightly lower speech quality than the full-rate codec (GSM 06.10)
- not suitable for modem and music signals

Applications:
- mobile radio systems according to the GSM standard (half-rate channel with $B' = 11.4$ kbit/s for speech and channel coding)
- modified versions with $B = 6.9$ kbit/s in the Japanese mobile radio system (JDC) and with $B = 7.95$ kbit/s in the American mobile radio system (*digital* AMPS)

Algorithm:
- *code excited linear prediction* (CELP) with special code books which are constructed from a set of orthogonal basis vectors (VSELP: *Vector Sum Excited Linear Prediction*, see below)

- frame length $T_N = 20\,\text{ms}$; subframe length $\frac{1}{4}T_N = 5\,\text{ms}$
- LPC filter of order $n = 10$; quantization of the reflection coefficients $k_1, k_2, \ldots, k_{10}$ with three vector quantizers of dimensions $2^{11} \times 3$ (k_1, k_2, k_3), $2^9 \times 3$ (k_4, k_5, k_6), and $2^8 \times 4$ (k_7, k_8, k_9, k_{10}); reduction of complexity using prequantization
- long-term prediction with combined *open-loop/closed-loop* search
- vector quantization of the gain factors
- adaptive error weighting filter (see Section A.4)
- adaptive postfilter (see Section 8.6)
- algorithmic signal delay 20 ms
- complexity about 18.5 wMOPS (weighted Mega Operations Per Second)

Bit allocation:

	Unvoiced frames: – no adaptive code book – 2 fixed code books A & B	Voiced frames: – adaptive code book – 1 fixed code book C
Parameters	Bits per frame (= 4 subframes)	Bits per frame (= 4 subframes)
Mode	1×2	1×2
Energy	1×5	1×5
Soft interpolation	1×1	1×1
10 reflection coeff.	1×28	1×28
Gain factors	4×5	4×5
Code book index A	4×7	
Code book index B	4×7	
Code book index C		4×9
LTP delay N_0		$1 \times 8 + 3 \times 4$
Bits per 20 ms	112	112

Special features:
- different coding of voiced and unvoiced segments based on a signal classification
- construction of the code vectors by weighted superposition of $M = 7$ (code book A, B) or $M = 9$ (code book C) orthogonal basis vectors; example: code book A with the basis vectors $\mathbf{a}_\mu = \big(a_\mu(0), \ldots, a_\mu(\lambda), \ldots, a_\mu(L-1)\big)^T$;

$$c_{1,i}(\lambda) = \sum_{\mu=1}^{7} \alpha_{\mu,i} \cdot a_\mu(\lambda)\ ; \quad \alpha_{\mu,i} \in \{-1, +1\}\ ;$$

only the $M = 7$ basis vectors of dimension $L = 40$ must be stored; instead of the code book indices, the weighting factors $\alpha_{\mu,i}$ are transmitted.

References:

ETSI Rec. GSM 06.20 (1994). *European Digital Cellular Telecommunications System; Speech Codec for the Half Rate Speech Traffic Channel.*

Gerson, J. A.; Jasiuk, M. A. (1990). Vector Sum Excited Linear Prediction (VSELP) Speech Coding at 8 kbit/s, *Proceedings of the IEEE International Conference on Acoustics, Speech, and Signal Processing (ICASSP)*, Albuquerque, USA pp. 461–464.

Gerson, J. A.; Jasiuk, M. A. (1991). Vector Sum Excited Linear Prediction (VSELP), *Advances in speech coding*, B. S. Atal, V. Cuperman, A. Gersho (eds.), Kluwer Academic, Dordrecht, pp. 69–79.

A.8 ETSI-GSM 06.60: Enhanced Full Rate Speech Transcoding

Block Diagram: decoding

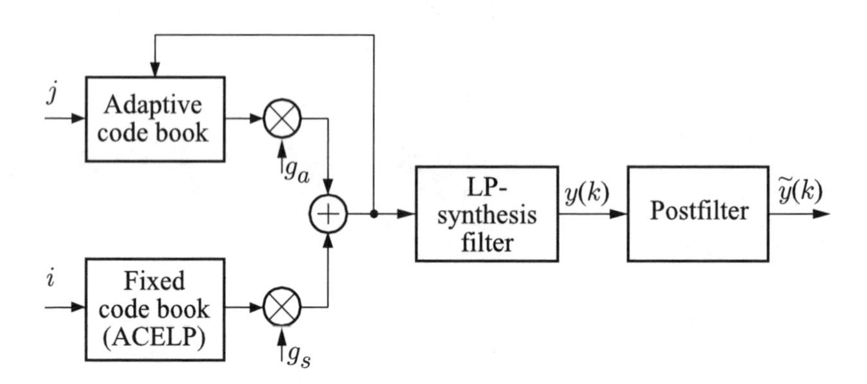

Sampling: $f_s = 8\,\text{kHz}$

Bit rate: $B = 13.0$ kbit/s (including 0.78 kbit/s for parity bits)

Quality:
- significantly higher speech quality than the full-rate codec according to GSM 06.10, comparable to 32 kbit/s ADPCM
- not suitable for modem and music signals

Applications:
- mobile radio systems according to the GSM standard

Algorithm:
- CELP codec with algebraic code book similar to the codec standard G.729 (see Section A.4)

- frame length $T_N = 20\,\text{ms}$; four subframes
- determination of 10 LP coefficients twice per frame; differential vector matrix quantization of 20 LSF coefficients
- adaptive code book with combined *open-loop/closed-loop* search and applying fractional pitch detection with interpolation by a factor of 6
- fixed algebraic code book (ACELP) with effectively 2^{35} vectors of dimension 40 with 10 non-zero values each
- predictive quantization of the gain factors
- adaptive error weighting filter
- adaptive postfilter (see Section 8.6)
- algorithmic signal delay 20 ms
- complexity about 15.2 wMOPS (weighted Mega Operations Per Second)

Bit allocation:

Parameter	Subframes 1 & 2 (5 ms each)	Subframes 3 & 4 (5 ms each)	Bits per frame 20 ms
2×10 LP coefficients (*line spectral frequencies*)			38
Adaptive code book			
• delay N_0	9	6	30
• gain g_a	4	4	16
Fixed code book (ACELP)			
• pulse positions and signs	35	35	140
• gain g_s	5	5	20
Bits per 20 ms			244

Special features:
- advantageous algebraic ternary code book according to the ACELP approach

$$c_i(\lambda) = \sum_{\mu=0}^{9} s_\mu \cdot \delta(\lambda - i_\mu) \quad ; \qquad s_\mu \in \{+1, -1\}$$

- for five tracks, selection of two pulses out of eight potential pulse positions i_μ (see Section 8.5.3)
- sign of first pulse encoded by 1 bit, sign of second pulse depends on the relative position of both pulses

Pulse position	Sign	Positions	Bits
i_0, i_5	s_0, s_5	$0, 5, 10, 15, 20, 25, 30, 35$	$2 \times 3 + 1$
i_1, i_6	s_1, s_6	$1, 6, 11, 16, 21, 26, 31, 36$	$2 \times 3 + 1$
i_2, i_7	s_2, s_7	$2, 7, 12, 17, 22, 27, 32, 37$	$2 \times 3 + 1$
i_3, i_8	s_3, s_8	$3, 8, 13, 18, 23, 28, 33, 38$	$2 \times 3 + 1$
i_4, i_9	s_4, s_9	$4, 9, 14, 19, 24, 29, 34, 39$	$2 \times 3 + 1$

- efficient, incomplete code book search exploiting the structure of the code vectors

References:

ETSI Rec. GSM 06.60 (1996). *Digital Cellular Telecommunications System; Enhanced Full Rate (EFR) Speech Transcoding.*

Järvinen, K.; Vainio, J.; Kapanen, P.; Salami, R.; Laflamme, C.; Adoul, J.-P. (1997). GSM Enhanced Full Rate Speech Codec, *Proceedings of the IEEE International Conference on Acoustics, Speech, and Signal Processing (ICASSP)*, Munich, Germany, pp. 771–774.

A.9 ETSI-GSM 06.90: Adaptive Multi-Rate (AMR) Speech Transcoding (AMR narrowband codec, AMR-NB)

Block diagram: decoding

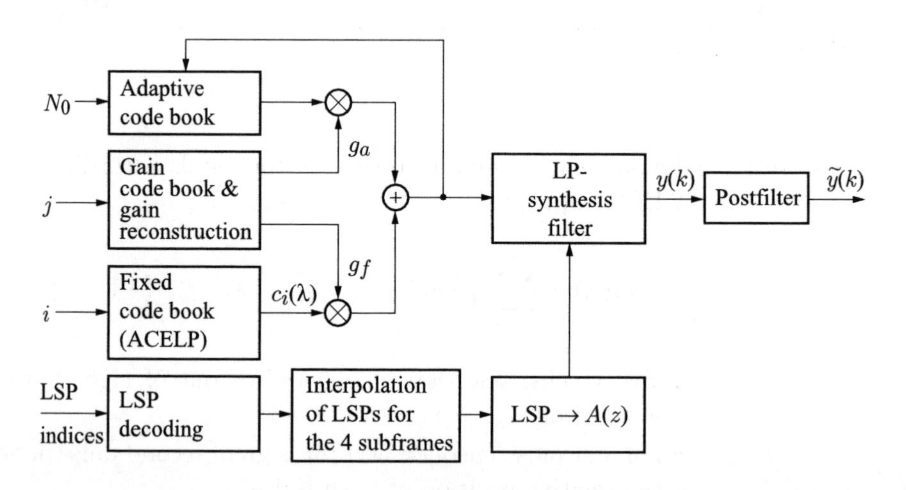

Sampling: $f_s = 8\,\text{kHz}$

Bit rates: (kbit/s) $B = 4.75$ / 5.15 / 5.90 / 6.70 (PDC-Japan) / 7.40 (DAMPS-IS136) / 7.95 / 10.2 / 12.2 (GSM-EFR)

network-controlled dynamic split of bit allocation between speech codec and channel codec depending on the instantaneous quality of the channel, codec mode can be changed every 20 ms

Quality:

- same as GSM-EFR at 12.2 kbit/s
- higher error robustness than GSM-EFR codec and better quality in adverse GSM channel conditions, extended quality operation region down to $C/I \geq 4$–7 dB (3–6 dB advantage compared to GSM-EFR codec)

Applications:

- mobile radio systems, according to GSM and UMTS standards
- packet-based *voice over Internet Protocol* (VoIP)

Algorithm:

- CELP codec with different algebraic code books similar to the standard ETSI-GSM 06.60 (EFR codec), which is part of the AMR codec ($B = 12.2$ kbit/s)
- one or two sets of 10 LSP coefficients depending on the bit rate (mode), interpolation of LSPs between subframes
- predictive split-matrix quantization of LSPs
- frame length $T_N = 20\,\text{ms}$, four subframes of 22.8 kbit/s
- constant gross bit rate including mode-specific channel coding
- inband signaling for rate switching
- closed-loop pitch search confined to a small number of lags around the open-loop pre-search
- fractional pitch lags by interpolation
- prediction of the fixed code book gains
- joint vector quantization of fixed and adaptive gains
- complexity around 16.8 wMOPS (weighted Mega Operations Per Second)

Special feature:

- ternary ACELP code books

$$c_i(\lambda) = \sum_{\mu=0}^{p-1} s_\mu \cdot \delta(\lambda - i_\mu)\,; \quad v_\mu \in \{+1, -1\}$$

Mode kbit/s	Number p of pulses	Bits per frame
12.2	10	35
10.2	8	31
7.95/7.4	4	17
6.7	3	14
5.9	2	11
5.15/4.75	2	9

Bit allocation:

Mode	Parameter	1st subframe	2nd subframe	3rd subframe	4th subframe	Total per frame
	2 LSP sets					38
12.2 kbit/s	Pitch delay	9	6	9	6	30
(GSM-EFR)	Pitch gain	4	4	4	4	16
	Algebraic code	35	35	35	35	140
	Code book gain	5	5	5	5	20
	Total					244
	LSP set					26
10.2 kbit/s	Pitch delay	8	5	8	5	26
	Algebraic code	31	31	31	31	124
	Gain	7	7	7	7	28
	Total					204
	LSP set					27
7.95 kbit/s	Pitch delay	8	6	8	6	28
	Pitch gain	4	4	4	4	16
	Algebraic code	17	17	17	17	68
	Code book gain	5	5	5	5	20
	Total					159
	LSP set					26
7.40 kbit/s	Pitch delay	8	5	8	5	26
(DAMPS-EFR)	Algebraic code	17	17	17	17	68
	Gain	7	7	7	7	28
	Total					148
	LSP set					26
6.70 kbit/s	Pitch delay	8	4	8	4	24
	Algebraic code	14	14	14	14	56
	Gain	7	7	7	7	28
	Total					134
	LSP set					26
5.90 kbit/s	Pitch delay	8	4	8	4	24
	Algebraic code	11	11	11	11	44
	Gain	6	6	6	6	24
	Total					118

<table>
<tr><th>Mode</th><th>Parameter</th><th>1st subframe</th><th>2nd subframe</th><th>3rd subframe</th><th>4th subframe</th><th>Total per frame</th></tr>
<tr><td></td><td>LSP set</td><td></td><td></td><td></td><td></td><td>23</td></tr>
<tr><td>5.15 kbit/s</td><td>Pitch delay</td><td>8</td><td>4</td><td>4</td><td>4</td><td>20</td></tr>
<tr><td></td><td>Algebraic code</td><td>9</td><td>9</td><td>9</td><td>9</td><td>36</td></tr>
<tr><td></td><td>Gain</td><td>6</td><td>6</td><td>6</td><td>6</td><td>24</td></tr>
<tr><td></td><td>Total</td><td></td><td></td><td></td><td></td><td>103</td></tr>
<tr><td></td><td>LSP set</td><td></td><td></td><td></td><td></td><td>23</td></tr>
<tr><td>4.75 kbit/s</td><td>Pitch delay</td><td>8</td><td>4</td><td>4</td><td>4</td><td>20</td></tr>
<tr><td></td><td>Algebraic code</td><td>9</td><td>9</td><td>9</td><td>9</td><td>36</td></tr>
<tr><td></td><td>Gain</td><td colspan="2">8</td><td colspan="2">8</td><td>16</td></tr>
<tr><td></td><td>Total</td><td></td><td></td><td></td><td></td><td>95</td></tr>
</table>

References:

Bruhn, S.; Blöcher, P.; Hellwig, K.; Sjöberg, J. (1999). Concepts and Solutions for Link Adaptation and Inband Signaling for the GSM AMR Speech Coding Standard, *IEEE Vehicular Technology Conference*, pp. 2451–2455.

Ekudden, E.; Hagen, R.; Johansson, I.; Svedberg, J. (1999). Relaxing Model-imposed Constraints Based on Decoder Analysis, *Proceedings of the IEEE Workshop on Speech Coding*, Porvoo, Finland, pp. 117–119.

ETSI Rec. GSM 06.90 (1998). *Digital Cellular Telecommunications System; Adaptive Multi-Rate (AMR) Speech Transcoding.*

Download GSM-AMR-NB codec from the 3GPP website/FTP server

- `26090-500.zip`: description of the standard
- `26073-530.zip`: ANSI C Software

A.10 ETSI/3GPP 26.190/ITU G.722.2: Adaptive Multi-Rate Wideband Speech Transcoding (AMR wideband codec, AMR-WB)

Block diagram: encoding

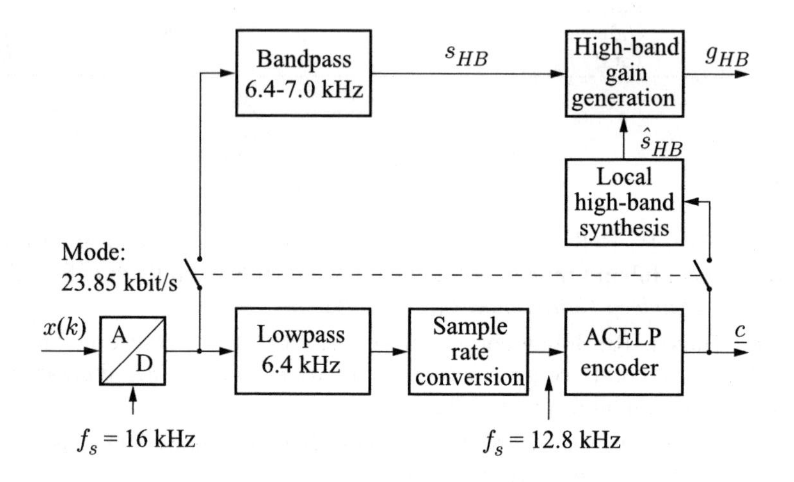

Block diagram: decoding

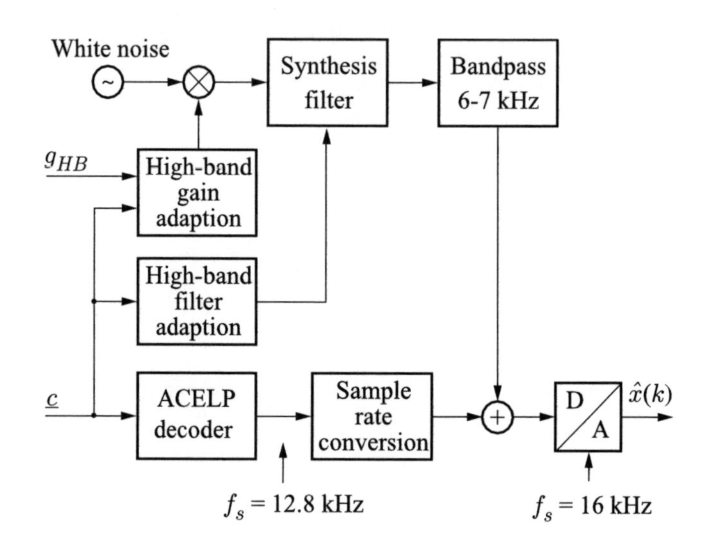

Sampling: $f_s = 16\,\text{kHz}$, audio bandwidth from 50 Hz to 7 kHz

Bit rates: (kbit/s) $B = 23.85$ / 23.05 / 19.58 / 18.25 / 15.85 / 14.25 / 12.65 / 8.85 / 6.60

network-controlled dynamic bit rate allocation between speech and channel coding similar to AMR-NB codec

Quality:

- clean speech quality for the six highest AMR-WB modes (23.85–14.25 kbit/s) equal to or better than ITU-T wideband codec G.722 at 64 kbit/s
- the 12.65 kbit/s mode is at least equal to G.722 at 56 kbit/s
- the 8.85 kbit/s mode gives equal quality as G.722 at 48 kbit/s
- the two lowest modes are used only during adverse radio channel conditions or during UMTS network congestion

Applications:

- GSM full-rate channel
- GSM EDGE Radio Access Network (GERAN, 8-PSK)
- 3G UMTS Terrestrial Radio Access Network (UTRAN, CDMA)
- packet-based *voice over Internet Protocol* (VoIP)

Algorithm:

- ACELP encoding of the frequency range 50 Hz to 6.4 kHz similar to AMR-NB codec
- 16 LP filter coefficients per 20 ms frame
- artificial wideband extension with or without side information in the range 6.4–7.0 kHz
- asymmetric analysis window of length 30 ms
- conversion of LP coefficients to ISP representation for quantization and interpolation (immittance spectral frequencies, instead of LSP)
- subframe size 64 samples
- ACELP code book with four tracks with 16 positions
- special weighting filter
- joint vector quantization of adaptive and fixed vector code book gains
- complexity 38.9 wMOPS (weighted Mega Operations Per Second)

Bit allocation:

	Codec mode (kbit/s)								
Parameter	6.60	8.85	12.65	14.25	15.85	18.25	19.85	23.05	23.85
VAD flag	1	1	1	1	1	1	1	1	1
LTP filtering flag	0	0	4	4	4	4	4	4	4
ISP	36	46	46	46	46	46	46	46	46
Pitch delay	23	36	30	30	30	30	30	30	30
Algebraic code	48	80	144	176	208	256	288	352	352
Gains	24	24	28	28	28	28	28	28	28
High-band energy	0	0	0	0	0	0	0	0	16
Total per frame	132	177	253	285	317	365	397	461	477

References:

3GPP TS 26.190 (2001). *AMR Wideband Speech Codec; Transcoding Functions.*

Bessette, B.; Salami, R.; Lefebvre, R.; Jelinek, M.; Rotola-Pukkila, J.; Vainio, J.; Mikkola, H.; Järvinen, K. (2002). The Adaptive Multirate Wideband Speech Codec (AMR-WB). *IEEE Transactions on Speech and Audio Processing*, vol. 10, no. 8, Nov., pp. 620–636.

Download GSM-AMR-WB codec from the 3GPP website/FTP server

- `26190-510.zip`: description of the standard
- `26173-580.zip`: ANSI C Software

A.11 ETSI/3GPP 26.290: Extended Adaptive Multi-Rate Wideband Codec (AMR-WB$^+$)

Block diagram: encoding

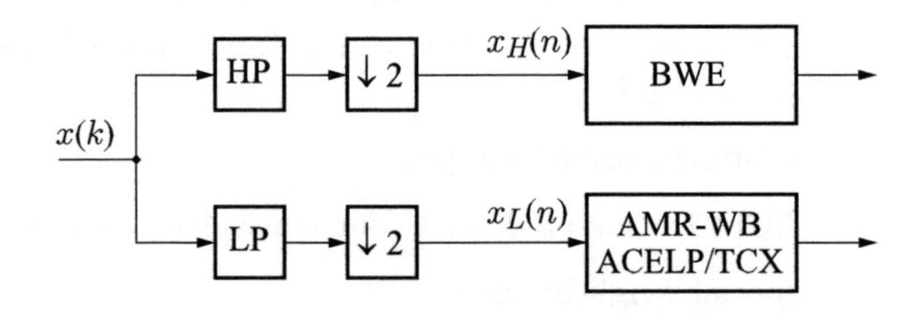

Sampling: $f_s = 12.8$–$38.4\,$kHz, audio bandwidth up to 16 kHz

Bit rates: a) all modes of the AMR-WB codec (ETSI/3GPP 26.190): 6.6–23.85 kbit/s

b) mono AMR-WB+: 10.4–24.0 kbit/s
c) stereo AMR-WB+: 18.0–32 kbit/s

Quality:
- the AMR-WB+ speech codec includes all functions of the AMR-WB coding schemes as well as extended functionality for encoding general audio signals such as music, speech, mixed, and other signals
- for music performance better than or equal to AAC Plus in most tests, and at low bit rate (less than 18 kbps) audio performance better than AAC Plus

Applications:
- GSM full-rate channel
- GSM EDGE Radio Access Network (GERAN, 8-PSK)
- 3G UMTS Terrestrial Radio Access Network (UTRAN, CDMA)

Algorithm:
- separation of the input signal into two bands, the low-frequency (LF) and the high-frequency (HF) bands
- critical downsampling of LF and HF signals at $f_s/2$
- encoding of LF signal using the core encoder/decoder which switches between ACELP and transform-codec excitation (TCX) mode
- in ACELP mode, the standard AMR-WB codec is used
- the HF signal is encoded with relatively few bits using a bandwidth extension (BWE method)
- complexity 38.9 wMOPS (AMR-WB) ... 72 wMOPS (stereo streaming)

References:

3GPP TS 26.290 (2005). *Extended Adaptive Multi-Rate Wideband (AMR-WB+) Codec; Transcoding Functions*, Release 6, March.

3GPP TS 26.304. *ANSI-C Code for the Floating Point Extended AMR Wideband Codec.*

3GPP TS 26.273. *ANSI-C Code for the Fixed Point Extended AMR Wideband Codec.*

http://www.voiceage.com: *AMR-WB+: Hi-Fi Audio Compression.*

A.12 TIA IS-96: Speech Service Option Standard for Wideband Spread-Spectrum Systems (QCELP: Variable Rate Speech Coder for CDMA Digital Cellular)

Block diagram: decoding

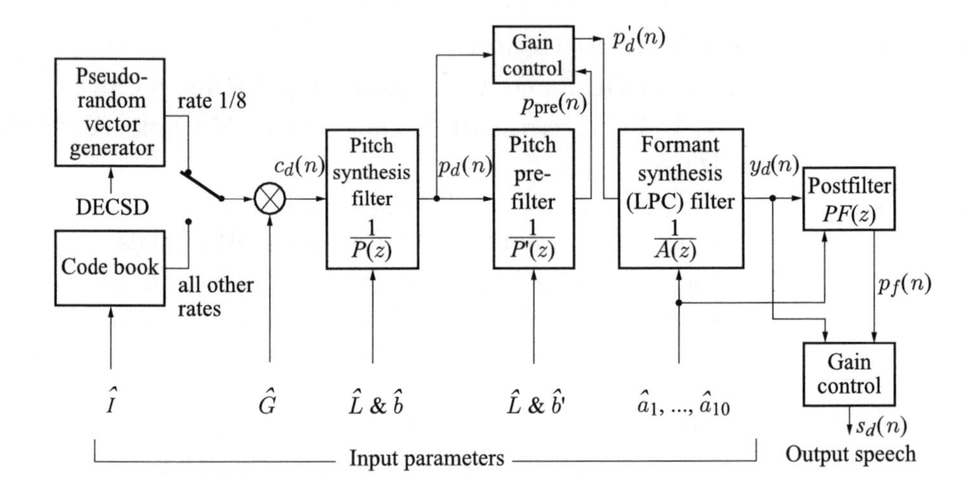

Sampling: $f_s = 8\,\text{kHz}$

Bit rates: dynamic selection of one out of four data rates every 20 ms depending on speech activity: $B = 8.0, 4.0, 2.0$, approx. 1 kbit/s

Quality: – near toll quality speech at an average data rate of under 4 kbit/s

Applications: – CDMA Digital Cellular Standard IS-54

Algorithm:
- conventional CELP structure
- division of 20 ms frame into subframes of different lengths, depending on the selected bit rate
  $B = 8$ kbit/s: 8 CB subframes of length 20, 4 pitch subframes
  $B = 4$ kbit/s: 4 CB subframes of length 40, 2 pitch subframes
  $B = 2$ kbit/s: 2 CB subframes of length 80, 1 pitch subframe
  $B \approx 1$ kbit/s: 1 CB subframe of length 160, no pitch subframe
- predictive quantization of 10 LSF coefficients
- integer pitch lags $N_0 = 17$–143 (7 bits)
- adaptive postfilter

References:

Gardner, W.; Jacobs, P.; Lee, C. (1993). QCELP: A Variable Rate Speech Coder for CDMA Digital Cellular, *Speech and Audio Coding for Wireless and Network Applications*, Atal, B. S.; Cuperman, V.; Gersho, A. (eds.), Kluwer Academic, Dordrecht, pp. 85–92.

TIA IS-96 (1998). *Speech Service Option Standard for Wideband Spread-Spectrum Systems - TIA/EIA-96C*, August.

A.13 INMARSAT: Improved Multi-Band Excitation Codec (IMBE)

Block diagram: decoding

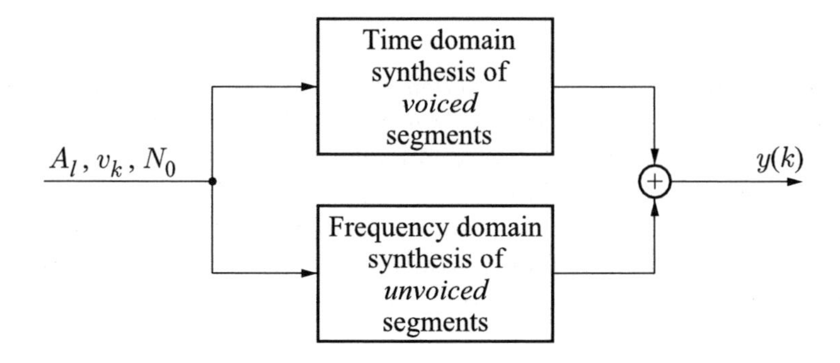

A_l: spectral amplitudes
v_k: voiced/unvoiced classification in frequency bands
N_0: pitch period

Sampling: $f_s = 8\,\text{kHz}$

Bit rate: $B = 4.15$ kbit/s

Quality:
- limited telephone quality (lower than with the GSM full-rate codec)
- further quality loss in case of background noise or background music
- not suitable for modem and music signals

Applications:
- mobile telephony via the INMARSAT satellite system
- radio systems of police and official authorities (USA)

Algorithm:
- classification of frequency bands as voiced or unvoiced based on a spectral analysis (FFT)

- block length 20 ms
- determination and quantization of the spectral envelope
- analysis of the instantaneous pitch period
- synthesis of *voiced* segments at the receiver in the time domain by summing the outputs of a band of sinusoidal oscillators running at the harmonics of the fundamental frequency
- synthesis of *unvoiced* segments at the receiver in the frequency domain by summing weighted bandpass filtered white noise and transformation into the time domain via inverse FFT
- final synthesized signal by summing the voiced and the unvoiced synthesized segments
- algorithmic signal delay 20 ms
- complexity approx. 7 wMOPS (weighted Mega Operations Per Second)

References:

DVS (1991). *Methods for Speech Quantization and Error Correction*, incl. INMARSAT-M Codec, Digital Voice Systems Inc., Patent PCT/US91/09135, June.

Griffin, D. W.; Lim, J. S. (1988). Multiband Excitation Vocoder, *IEEE Transactions on Acoustics, Speech and Signal Processing*, 36, pp. 1223–1235.

Hardwick, J. C.; Lim, J. S. (1991). The Application of the IMBE Speech Coder to Mobile Communications, *Proceedings of the IEEE International Conference on Acoustics, Speech, and Signal Processing (ICASSP)*, Toronto, pp. 249–252.

Appendix B

Speech Quality Assessment

The assessment of the *speech quality* and more generally the analysis of the *quality of service* (QoS) is a fundamental issue in the process of designing, standardizing, implementing, and operating components, application protocols, or even complete speech transmission networks.

Human perception of sound is a very complex process involving auditory and cognitive aspects. So far no universal model or theoretical framework exists which reflects all aspects of auditive perception. Speech quality assessment is a multidisciplinary research area of its own, which is based on psychoacoustics, e.g., [Zwicker 1982], [Zwicker, Fastl 1999], [Blauert 1997], [Blauert 2005].

There is a vast amount of literature on speech quality assessment and there are several ITU and ETSI standards dealing with specific quality aspects, such as evaluating the impairments introduced by speech coding, noise suppression, echo cancellation, bit error, and packet-loss concealment. An extensive literature review is beyond the scope of this book. In this appendix, only a few key definitions and procedures are described, which have been established in the area of telecommunications for the evaluation of speech quality under specific constraints.

B.1 Auditive Speech Quality Measures

The perceived subjective quality of speech signals can to a certain extent be explained by psychoacoustic effects such as masking, non-linear loudness transfor-

Digital Speech Transmission: Enhancement, Coding and Error Concealment
Peter Vary and Rainer Martin

mation, and non-uniform frequency grouping. The influence of these effects on the subjectively perceived quality was discussed, for instance, in the context of open- and closed-loop prediction and noise shaping (see Chapter 8). Although the signal-to-quantization-noise ratio cannot be improved by open-loop prediction, the speech quality is much better than plain PCM quantization. This is due to the fact that the spectrally shaped quantization noise is (partly) masked by the speech signal. This example illustrates that the *signal-to-noise ratio* (SNR) is not generally appropriate as a measure of speech quality.

Speech coding standards are specified on the bit level, which allows type approval of speech transmission devices by using specified test sequences and by comparing the resulting bit patterns and samples.

Especially in the context of standardization, formal listening tests are the basis for selecting the best candidate proposal. The rules for these tests have been standardized by the ITU [ITU-T Rec. P.800 1996], [ITU-T Rec. P.830 1996]. Many test persons judge the quality of speech samples which have been processed by the device under test, either in an absolute category (*absolute category rating*, ACR) or with reference to some other speech samples (*comparison category rating*, CCR, or *degradation category rating*, DCR).

In ACR listening tests, a five-point scale is used, ranging from "excellent" (score = 5) to "bad" (score = 1). The resulting quality of each speech sample is expressed as the average of the scores given by many test listeners, the so-called *mean opinion score* (MOS).

The MOS results of a listening test may still be biased by the particular laboratory conditions. Therefore, speech samples with a reference disturbance are included in the listening test. The reference disturbance is imposed by the *Modulated Noise Reference Unit* (MNRU) [ITU-T Rec. P.810 1996] (see Fig. B.1, Fig. B.2), which produces additive proportional disturbance according to

$$\tilde{x}(k) = x(k) \cdot [1 \; + \; g \cdot w(k)]$$

where $x(k)$ denotes the clean speech signal, $w(k)$ white noise and g the gain factor which is usually expressed as

$$g = 10^{-Q/20} .$$

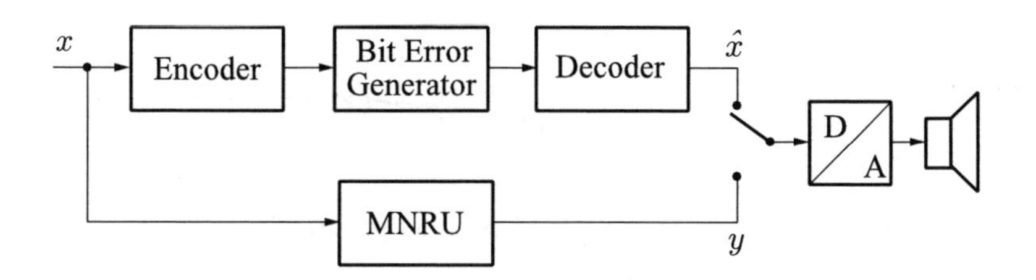

Figure B.1: Calibration of ACR listening tests using MNRU (Modulated Noise Reference Unit) references

Thus, the signal-to-noise ratio is

$$SNR = 10 \log \left(\frac{\mathrm{E}\{x^2\}}{\mathrm{E}\{[g \cdot w(k) \cdot x(k)]^2\}} \right)$$
$$= Q \,,$$

with $\mathrm{E}\{w^2(k)\} = 1$ and

$$\mathrm{E}\{[g \cdot w(k) \cdot x(k)]^2\} = g^2 \cdot \mathrm{E}\{w^2(k)\} \cdot \mathrm{E}\{x^2(k)\}$$
$$= 10^{-Q/10} \cdot \mathrm{E}\{x^2(k)\} \,.$$

The additive MNRU noise has the same characteristics as the noise produced by A-law and μ-law quantizers. The MOS of the MNRU samples allows for a calibration of all MOS results and facilitates a comparison between MOS scores of different listening tests, as indicated in Fig. B.3.

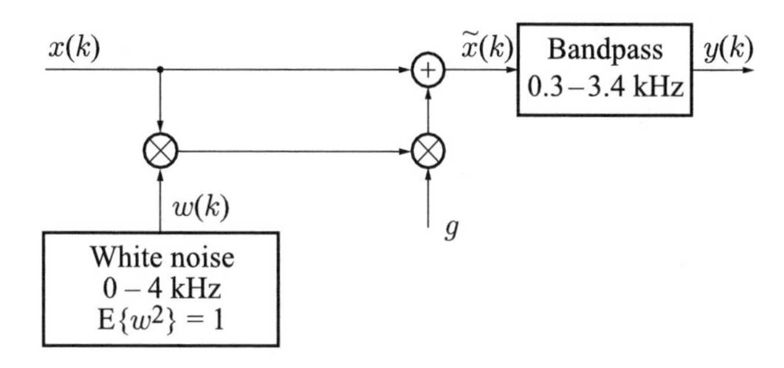

Figure B.2: Digital Modulated Noise Reference Unit (MNRU) for narrowband speech signals (ITU-T P.810)

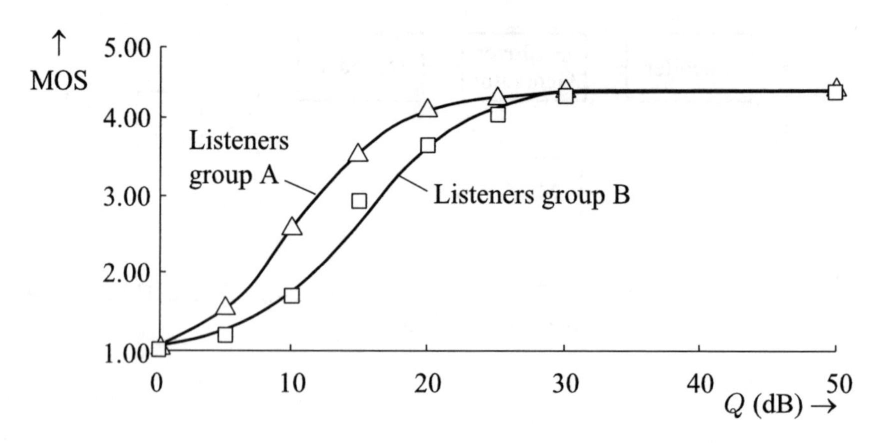

Figure B.3: MOS and MNRU Q-values

ABSOLUTE CATEGORY RATING (ACR) TEST
INSTRUCTIONS TO SUBJECTS:

In this experiment you will be listening to short groups of sentences via the telephone handset, and giving your opinion of the speech you hear. On the table in front of you is a box with five illuminated press buttons. When all the lamps go on, you will hear ... sentences. Listen to these, and when the lamps go out, press the appropriate button to indicate your opinion on the following scale.

Effort required to understand the meanings of sentences

5: Complete relaxation possible, no effort required.

4: Attention necessary; no appreciable effort required.

3: Moderate effort required.

2: Considerable effort required.

1: No meaning understood with any feasible effort.

The button you have pressed will light up for a short time. Then the lamp will go out, and there will be a brief pause before all the lamps go on again for the next group of ... sentences.
There will be a longer pause after every ... groups (each calling for an opinion).
There will be a total of ... groups in this visit and a similar number in your subsequent visit(s).
Thank you for your help in this experiment.

COMPARISON CATEGORY RATING (CCR) TEST
INSTRUCTIONS TO SUBJECTS:
"Evaluation of the influence of various environmental noises on the quality of different telephone systems"

In this experiment you will hear pairs of speech samples that have been recorded through various experimental telephone equipment. You will listen to these samples through the telephone handset in front of you.

What you will hear is one pair of sentences, a short period of silence, and another pair of sentences. You will evaluate the quality of the second pair of sentences compared to the quality of the first pair of sentences.

You should listen carefully to each pair of samples. Then, when the green light is on, please record your opinion about the quality of the second sample relative to the quality of the first sample using the following scale:

The Quality of the Second Compared to the Quality of the First is:

3: Much Better

2: Better

1: Slightly Better

0: About the Same

-1: Slightly Worse

-2: Worse

-3: Much Worse

You will have five seconds to record your answer by pushing the button corresponding to your choice. There will be a short pause before the presentation of the next pair of sentences.

We will begin with a short practice session to familiarize you with the test procedure. The actual tests will take place during sessions of 10 to 15 minutes.

DEGRADATION CATEGORY RATING (DCR) TEST
INSTRUCTIONS TO SUBJECTS:
"Evaluation of the influence of various environmental noises on the quality of different telephone systems"

In this experiment you will hear pairs of speech samples that have been recorded in different noise environments. You will listen to these samples through the telephone handset in front of you.

You will hear one pair of sentences, a short period of silence, and another pair of sentences. The first pair is the reference sample. You will evaluate the quality of the second pair of sentences compared to the quality of the first pair of sentences.

Please listen carefully to each pair of samples. Then, when the green light is on, please record your opinion about the quality of the second sample compared to the quality of the first sample using the following scale:

5: Degradation is inaudible

4: Degradation is audible but not annoying

3: Degradation is slightly annoying

2: Degradation is annoying

1: Degradation is very annoying

You will have five seconds to record your answer by pushing the button corresponding to your choice. There will be a short pause before the presentation of the next pair of sentences.

We will begin with a short practice session to familiarize you with the test procedure. The actual tests will take place during sessions of 10 to 15 minutes.

B.2 Instrumental Speech Quality Measures

Listening tests are time consuming and expensive. During the development phase of a speech processing algorithm, it is necessary to obtain estimates of the speech quality without expensive auditive tests. Therefore, *objective speech quality measures* or *instrumental speech quality measures* have been developed, which try to predict the outcome of a subjective listening test as accurately as possible, e.g., [Quackenbush et al. 1988], [Vary et al. 1998].

These measures are often based on a comparison of the original and distorted speech signals in an auditory domain. The transformation to the auditory domain is based on models of psychoacoustic mechanisms. Objective speech quality measures are often limited to specific situations, e.g., speech codec assessment or transmission distortions in mobile radio networks. If both original and degraded speech signals are needed, an objective speech quality measure is called *intrusive*, as it does not allow an automated, purely receiver-based quality measurement.

Intrusive Speech Quality Measures

Within ITU-T several proposals for objective quality measures have been investigated in the last 20 years. In 1996, the PSQM algorithm (Perceptual Speech Quality Measure, [ITU-T Rec. P.861 1996]) was standardized. In the meantime, this recommendation has been withdrawn and replaced by the PESQ algorithm (Perceptual Evaluation of Speech Quality, [ITU-T Rec. P.862 2001]). An extension of PESQ has been proposed to cover wideband speech as well [ITU-T Int. Doc. 2004].[1]

A simplified diagram of the structure of the PESQ algorithm is given in Fig. B.4 [Rix et al. 2001], [Voran 1999a], [Voran 1999b]. PESQ contains an adjustment of the power levels of the original and the degraded speech, a frame-wise time alignment, and models for the description of the following psychoacoustic effects:

- non-linear transformation of sound pressure level to perceived loudness (intensity warping)
- non-linear transformation of frequency to perceived pitch (frequency warping)
- masking in the frequency and time domains
- asymmetrical weighting of additive and subtractive disturbance
- effects of unbalanced memorization of disturbances, depending on the position within the speech sample.

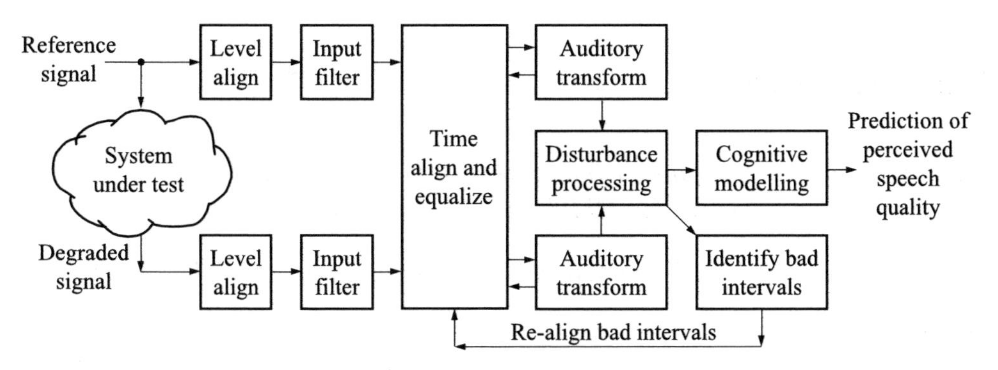

Figure B.4: Perceptual Evaluation of Speech Quality (ITU-T P.862) [Rix et al. 2001] © 2001 IEEE

[1]For wideband audio applications, other quality assessment methods have been proposed, which are based on the same principles as objective speech quality measurement algorithms [ITU-R Rec. BS.1387 1998].

When applied to correct measurement situations, PESQ offers a normalized correlation of $|\rho| > 0.9$ with MOS scores from subjective listening tests.

Non-intrusive Speech Quality Measures

In contrast to intrusive speech quality measures, *non-intrusive* speech quality measures allow an estimation of the perceived speech quality by analyzing only information that is available at the receiving end of a speech transmission system.

For example, the received speech material itself can be analyzed by using a database of statistical characteristics of typical speech segments. Alternatively, a speech production model may be used to identify natural speech components. These components can thereby be separated from artificial distortions [Kim, Tarraf 2004], [Picovici, Mahdi 2004]. This method is rather complex, but a first ITU standard has been developed [ITU-T Rec. P.563 2004].

Another possibility is the evaluation of link parameters of transmission networks which are normally used for other purposes (e.g., the frame error rate and bit error rate estimations in mobile radio systems such as GSM or UMTS). These parameters correlate well with the resulting speech quality if they are averaged in a suitable way (e.g., by L_P norms). Optimized functions of transmission parameter combinations have been shown to exhibit excellent correlations with reference speech quality scores [Karlsson et al. 1999], [Karlsson et al. 2002], [Werner et al. 2003], [Werner et al. 2004]. Although slightly less accurate than psychoacoustically motivated objective speech quality measures, these methods are well suited for an automated quality monitoring in speech transmission networks.

Bibliography

Blauert, J. (1997). *Spatial Hearing - The Psychoacoustics of Human Sound Localization*, revised edn, The MIT Press, Boston, Massachusetts.

Blauert, J. (ed.) (2005). *Communication Acoustics*, Springer Verlag, Berlin.

ITU-R Rec. BS.1387 (1998). Method for Objective Measurements of Perceived Audio Quality, *Recommendation BS.1387*, International Telecommunication Union (ITU).

ITU-T Int. Doc. (2004). Proposed Modification to Draft P.862 to Allow PESQ to be Used for Quality Assessment of Wideband Speech, ITU-T Internal Document ITU-T/COM-T/COM12/D/007E.

ITU-T Rec. P.563 (2004). Single-Ended Method for Objective Speech Quality Assessment in Narrow-Band Telephony Applications, *Recommendation P.563*, International Telecommunication Union (ITU).

ITU-T Rec. P.800 (1996). Methods for Subjective Determination of Transmission Quality, *Recommendation P.800*, International Telecommunication Union (ITU).

ITU-T Rec. P.810 (1996). Modulated Noise Reference Unit (MNRU), *Recommendation P.810*, International Telecommunication Union (ITU).

ITU-T Rec. P.830 (1996). Subjective Performance Assessment of Telephone-Band and Wideband Digital Codecs, , *Recommendation P.830*, International Telecommunication Union (ITU).

ITU-T Rec. P.861 (1996). Objective Quality Measurement of Telephone-band (300-3400 Hz) Speech Codecs, *Recommendation P.861*, International Telecommunication Union (ITU).

ITU-T Rec. P.862 (2001). Perceptual Evaluation of Speech Quality (PESQ), an Objective Method for End-to-End Speech Quality Assessment of Narrowband Telephone Networks and Speech Codecs, *Recommendation P.862*, International Telecommunication Union (ITU).

Karlsson, A.; Heikkila, G.; Minde, T. B.; Nordlund, M.; Timus, B.; Wiren, N. (1999). Radio Link Parameter Based Speech Quality Index - SQI, *Proceedings of the IEEE Workshop on Speech Coding*, Porvoo, Finland.

Karlsson, M.; Almgren, M.; Bruhn, S.; Larsson, K.; Sundelin, M. (2002). Joint Capacity and Quality Evaluation for AMR Telephony Speech in WCDMA Systems, *Proceedings of the IEEE Vehicular Technology Conference (VTC)*, Vancouver, Canada.

Kim, D. S.; Tarraf, A. (2004). Perceptual Model for Non-intrusive Speech Quality Assessment, *Proceedings of the IEEE International Conference on Acoustics, Speech, and Signal Processing (ICASSP)*, Montreal, Canada.

Picovici, D.; Mahdi, A. E. (2004). New Output-based Perceptual Measure for Predicting Subjective Quality of Speech, *Proceedings of the IEEE International Conference on Acoustics, Speech, and Signal Processing (ICASSP)*, Montreal, Canada.

Quackenbush, S. R.; Barnwell III, T. P.; Clements, M. A. (1988). *Objective Measures of Speech Quality*, Prentice Hall, Englewood Cliffs, New Jersey.

Rix, A. W.; Beerends, J. G.; Hollier, M. P.; Hekstra, A. P. (2001). Perceptual Evaluation of Speech Quality (PESQ) - A New Method for Speech Quality Assessment of Telephone Networks and Codecs, *Proceedings of the IEEE International Conference on Acoustics, Speech, and Signal Processing (ICASSP)*, Salt Lake City, Utah, USA, pp. 749–752.

Vary, P.; Heute, U.; Hess, W. (1998). *Digitale Sprachsignalverarbeitung*, B. G. Teubner, Stuttgart (in German).

Voran, S. (1999a). Objective Estimation of Perceived Speech Quality - Part I: Development of the Measuring Normalizing Block Technique, *IEEE Transactions on Acoustics, Speech and Signal Processing*, vol. 7, no. 4, July, pp. 371–382.

Voran, S. (1999b). Objective Estimation of Perceived Speech Quality - Part II: Evaluation of the Measuring Normalizing Block Technique, *IEEE Transactions on Acoustics, Speech and Signal Processing*, vol. 7, no. 4, July, pp. 383–390.

Werner, M.; Junge, T.; Vary, P. (2004). Quality Control for AMR Speech Channels in GSM Networks, *Proceedings of the IEEE International Conference on Acoustics, Speech, and Signal Processing (ICASSP)*, Montreal, Canada.

Werner, M.; Kamps, K.; Tuisel, U.; Beerends, J. G.; Vary, P. (2003). Parameter-based Speech Quality Measures for GSM, *IEEE International Symposium on Personal, Indoor, and Mobile Radio Communications (PIMRC)*, Beijing, China.

Zwicker, E. (1982). *Psychoakustik*, Springer Verlag, Berlin (in German).

Zwicker, E.; Fastl, H. (1999). *Psychoacoustics*, 2nd edn, Springer Verlag, Berlin.

Index

B

M

Reference List

Reproduced with permission of Elsevier:
Fig. 4.5, Fig. 11.17, Fig. 13.27

Reproduced with permission of IEEE:
Fig. 5.9, Fig. 5.10, Fig. 9.13, Fig. 9.16, Fig. 9.17, Fig. 11.12, Fig. 11.15, Fig. 11.16, Fig. 11.18, Fig. 12.16, Fig. B.4

Reproduced with permission of Schiele & Schön:
Fig. 8.29

Reproduced with permission of Springer Verlag:
Fig. 2.17, Fig. 2.18, Fig. 2.19, Fig. 2.20, Fig. 2.21, Fig. 2.22

Reproduced with permission of B. G. Teubner:
Fig. 5.2, Fig. 5.6, Fig 5.7, Fig. 5.8, Fig. 6.1, Fig. 6.6, Fig. 6.8, Fig. 6.12, Fig. 7.3, Fig. 7.4, Fig. 7.5, Fig 7.6, Fig. 7.8, Fig. 7.9, Fig. 7.11, Fig. 7.12, Fig. 7.16, Fig. 8.2, Fig. 8.4, Fig. 8.5, Fig 8.7, Fig. 8.8, Fig. 8.9, Fig. 8.10, Fig. 8.11, Fig. 8.17, Fig. 8.18, Fig. 8.19, Fig. 8.20, Fig 8.21, Fig. 8.22, Fig. 8.23, Fig. 8.25, Fig. 8.26, Fig. 8.30, Fig. 8.31, Fig. 8.36, Fig. 11.21, Fig 11.27, Fig. 13.2, Fig. 13.3, Fig. 13.5, Fig. 13.6, Fig. 13.7, Fig. 13.8, Fig. 13.9, Fig. 13.11, Fig 13.12, Fig. 13.21, Fig. 13.25, Fig. 13.26, Fig. 13.28, Fig. A.3-b, Fig. A.3-c, Fig. A.4, Fig. A.6-b, Fig. A.7, Fig. A.13